	IB	IIB	IIIA	IVA	VA	VIA	VIIA	VIIIA
								Helium 2 He 4.003
			Boron 5 B 10.81	Carbon 6 C 12.01	Nitrogen 7 N 14.01	Oxygen 8 O 16.00	Fluorine 9 F 19.00	Neon 10 Ne 20.18
			Aluminum 13 Al 26.98	Silicon 14 Si 28.09	Phosphorus 15 P 30.97	Sulfur 16 S 32.07	Chlorine 17 Cl 35.45	Argon 18 Ar 39.95
Nickel 28 Ni 58.69	Copper 29 Cu 63.55	Zinc 30 Zn 65.39	Gallium 31 Ga 69.72	Germanium 32 Ge 72.59	Arsenic 33 As 74.92	Selenium 34 Se 78.96	Bromine 35 Br 79.90	Krypton 36 Kr 83.80
Palladium 46 Pd 106.4	Silver 47 Ag 107.9	Cadmium 48 Cd 112.4	Indium 49 In 114.8	Tin 50 Sn 118.7	Antimony 51 Sb 121.8	Tellurium 52 Te 127.6	Iodine 53 I 126.9	Xenon 54 Xe 131.3
Platinum 78 Pt 195.1	Gold 79 Au 197.0	Mercury 80 Hg 200.6	Thallium 81 Tl 204.4	Lead 82 Pb 207.2	Bismuth 83 Bi 209.0	Polonium 84 Po (209)	Astatine 85 At (210)	Radon 86 Rn (222)

Gadolinium 64 Gd 157.3	Terbium 65 Tb 158.9	Dysprosium 66 Dy 162.5	Holmium 67 Ho 164.9	Erbium 68 Er 167.3	Thulium 69 Tm 168.9	Ytterbium 70 Yb 173.0	Lutetium 71 Lu 175.0
Curium 96 Cm (247)	Berkelium 97 Bk (247)	Californium 98 Cf (251)	Einsteinium 99 Es (252)	Fermium 100 Fm (257)	Mendelevium 101 Md (258)	Nobelium 102 No (259)	Lawrencium 103 Lr (260)

FUNDAMENTALS OF CHEMISTRY

FUNDAMENTALS OF CHEMISTRY

SECOND EDITION

Ralph A. Burns

St. Louis Community College–Meramec

With Special Contributions by
John W. Hill
University of Wisconsin–River Falls

PRENTICE HALL Englewood Cliffs, New Jersey 07632

Library of Congress Cataloging-in-Publication Data

Burns, Ralph A.
Fundamentals of chemistry/Ralph A. Burns.—2nd ed.
p. cm.
Includes index.
ISBN 0-02-317351-3
1. Chemistry. I. Title.
QD33.B8894 1995
540—dc20 94-39095
CIP

Editorial Director: Tim Bozik
Editor-in-Chief: Paul F. Corey
Director of Production and Manufacturing: David W. Riccardi
Managing Editor: Kathleen Schiaparelli
Production Supervisor: Elisabeth H. Belfer
Interior Designer: Laura Ierardi
Art Director: Heather Scott
Cover Designer: Anthony Gemmellaro
Cover Credit: Runk/Schoenberger/Grant Heilman Photography, Inc.
Photo Researcher: Barbara Scott
Manufacturing Manager: Trudy Pisciotti

Printed in the United States of America

10 9 8 7 6 5 4 3 2

ISBN 0-02-317351-3

Prentice-Hall International (UK) Limited, London
Prentice-Hall of Australia Pty. Limited, Sydney
Prentice-Hall Canada, Inc., Toronto
Prentice-Hall Hispanoamericana, S.A., Mexico
Prentice-Hall of India Private Limited, New Delhi
Prentice-Hall of Japan, Inc., Tokyo
Simon & Schuster Asia Pte. Ltd., Singapore
Editora Prentice-Hall do Brasil, Ltda., Rio de Janeiro

Brief Contents

Contents

3 Fundamental Measurements 43

4 Elements and Atoms 83

5 Electron Arrangements in Atoms 121

6 Periodic Properties of Elements 161

7 Chemical Bonds 191

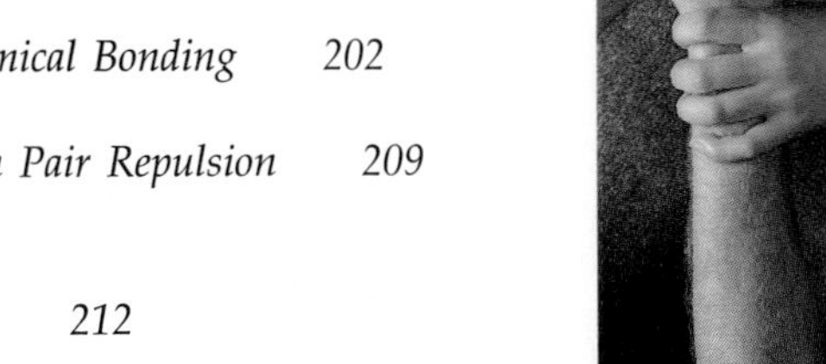

8 Names, Formulas, and Uses of Inorganic Compounds 225

9 Chemical Quantities 255

10 Chemical Reactions 283

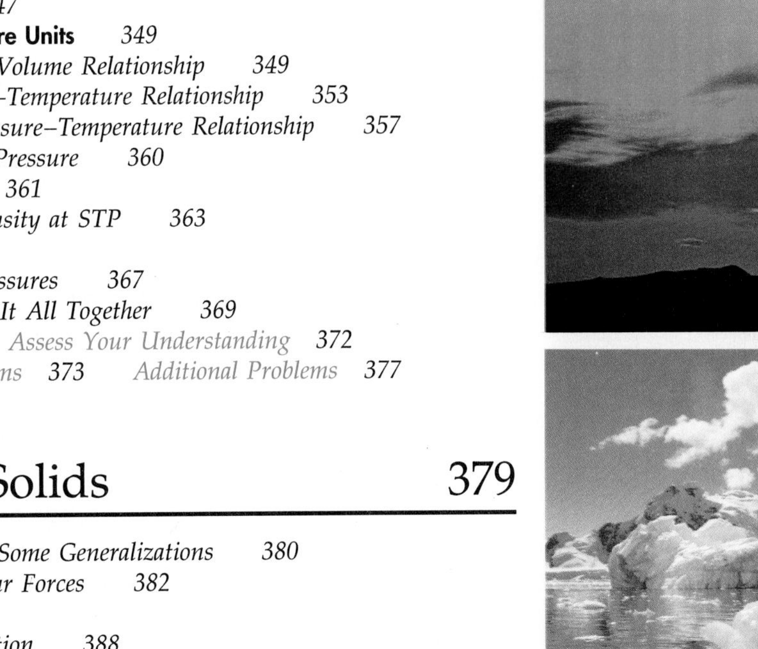

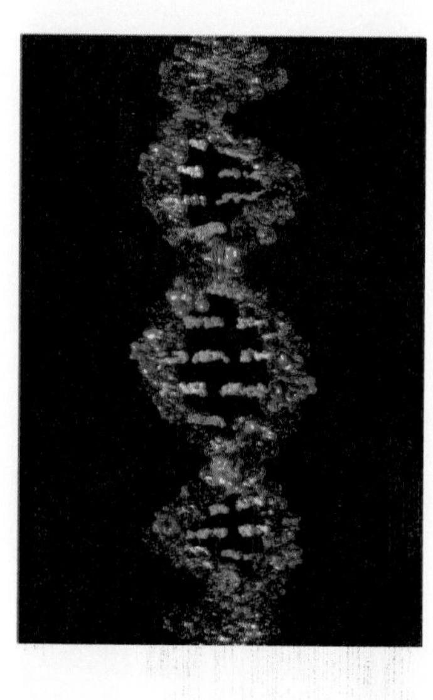

Preface

The wide acceptance and success of the first edition of this text have paved the way to refinements and modifications that are designed to make the text even more interesting and effective.

This text is written especially for students who have had no instruction—or limited instruction—in chemistry. It is for students who are in the process of preparing themselves for subsequent courses in chemistry and for students whose chemistry requirement will have been fulfilled upon completion of the course.

Although chemistry is everywhere—and without it life would not be possible—an excited anticipation of learning about chemicals and chemical reactions is generally not evident among students. Instead, students often approach the study of chemistry with considerable apprehension. Furthermore, chemistry educators at all levels express a genuine concern about the limited science background and mathematical skills of many students. What can be done to help students

- overcome their fears,
- approach chemistry with confidence,
- develop problem-solving skills,
- apply principles of chemistry when solving problems, and
- understand that chemistry is a part of everyday life?

Although no teacher or textbook can cause a student to learn chemistry, we have addressed these goals by providing a blend of approaches in this text that are directed at common problems students often have in understanding chemistry. Only when a student is competent will s/he be able to overcome fear and to have confidence.

To help students achieve thorough understanding, this text

addresses student attitudes from the start. Preceding Chapter 1 is a letter to students that addresses student apprehension. It lists six levels of thought processes and describes ten principles of productive study. Specific study techniques are recommended for use before lectures, during lectures, and before exams.

meets students where they are. Each chapter begins with an introduction that relates new topics to common applications and provides a rationale for learning. Throughout each chapter, care has been taken to provide a framework for learning that is built on thorough and simple explanations that interconnect new topics with terms, concepts, and problems of previous sections. Thus, major points are brought to the students' attention more than once. Repetition increases retention; applications increase understanding.

emphasizes learning in small steps. Beginning with Chapter 2, each chapter section contains one or more worked-out examples for a total of over 250 representative questions and problems. In-text exercises without answers have been added in this second edition; solutions are given in Appendix E. These examples and exercises provide students with a structured method of reviewing major points within a section before proceeding to the next section. Problem-solving strategies or "concept maps" and step-by-step explanations accompany the worked-out examples; they serve as models for working end-of-chapter problems.

provides matched pairs of problems. There are over 1300 end-of-chapter problems to provide ample opportunity for "learning by doing"—the *method* of science. Answers to odd-numbered problems appear in Appendix E. The Student Study Guide and Solutions Manual provides complete worked-out solutions to the odd-numbered problems. Because the odd-numbered and even-numbered problems make up two sets of equivalent difficulty, an instructor who does not want students to have access to answers can assign the even-numbered problems. The Instructor's Resource Manual contains complete solutions to all problems.

emphasizes the process of science. Problem solving cannot stand alone; it requires an understanding of terms and concepts. (Simple problems are often worked incorrectly when fundamental principles are confused.) Students also encounter the "process of science" through descriptions of experiments that led to major discoveries and through experimental data that support scientific laws. For example, the text describes experiments that led to the law of conservation of mass (Chapter 2), Dalton's atomic theory (Chapter 4), discoveries that gave us information about the atom (Chapter 5), and experiments that provide information about chemical bonding (Chapter 7) and gases (Chapter 12).

Where possible, results of experiments are presented before a law is stated to create a "need to know" learning environment. We intend to demonstrate that the search for answers—in the past, the present, and the future—can be both exciting and challenging.

emphasizes the language of chemistry. Chemical terms and chemical formulas are used to communicate chemistry; they must be understood before problems can be solved. Key terms appear in boldface print in the text where they are defined and are listed at the end of the chapter for review. One entire chapter provides a thorough treatment of chemical nomenclature; this "stand alone" chapter can be used earlier in the course if desired.

provides real-world applications. Special boxes, in-text descriptions, and margin notes present applications, word-picture illustrations, and points of historical emphasis for virtually every principle that is discussed. These materials help students understand that chemistry is a part of everyday life. Moreover, learning is made easier when new knowledge is related to the more familiar.

emphasizes the fundamentals. Special attention is given to providing clear and simple descriptions of topics that often confuse beginning students. Examples include (1) setting up and using conversion factors in solving problems, (2) developing a plan or "concept map" when solving problems, (3) using an appropriate number of significant figures in answers, (4) writing chemical formulas and chemical equations correctly, and (5) using the periodic table as a tool in predicting properties and reactions of chemicals.

helps student evaluate learning. A self-evaluation feature, Assess Your Understanding, precedes the end-of-chapter problems. This feature consists of a short list—keyed to chapter sections—of things a student should be able to do to demonstrate thorough understanding of the major points within the chapter.

is colorful and attractive. To get a student's attention the material must be inviting and interesting. Color is very much a part of chemistry and of our everyday lives. Four-color photographs and figures are used throughout this text to help students visualize chemistry.

Supplements

Student Study Guide and Solutions Manual

Instructor's Resource Manual

Fundamentals of Chemistry in the Laboratory, 2nd edition, and accompanying Instructor's Manual

Test Item File

3.5″ IBM Test Manager DOS

Mac Test Manager

Full-Color Transparencies

"How to Study Chemistry"

New York Times Contemporary View Program

Acknowledgments

I owe a special debt of gratitude to John W. Hill of the University of Wisconsin–River Falls for getting me started on the first edition of this text. His helpful suggestions and kind words of support were greatly appreciated. Besides being a good friend and a leader in chemical education, his writing style is an inspiration.

I am grateful to many people at Prentice Hall, especially to my editor, Paul F. Corey, who has been most helpful and supportive. He has always been available—almost around the clock—to answer any questions and to provide solid leadership. A special thanks goes to Elisabeth Belfer, production supervisor, who has an excellent eye for detail and organization.

I owe many thanks to the following individuals who consulted with Prentice Hall (and Macmillan) on the first and second editions, reviewed portions of the manuscript, and made contributions and suggestions. These most helpful instructors and their affiliations are:

Edward Alexander, San Diego Mesa College; Melvin Anderson, Lake Superior State University; Oren P. Anderson, Colorado State University; Joe Asire, Cuesta College; Caroline L. Ayers, East Carolina University; Jay Bardole, Vincennes University; Robert Batch, Canada College; O. H. Bezirjian, College of Marin; Rattan Bhatia, William Rainey Harper College; David Blackman, University of the District of Columbia; Doug Campbell, Eastern Oregon University; Katherine Craighead, University of Wisconsin–River Falls; David Dozark, Kirkwood Community College; Jerry Driscoll, University of Utah; Maureen Foley, Schoolcraft

College; Julie Frentrup, Eastern Michigan University; Verl G. Garrard, University of Idaho; Emerson Garver, University of Wisconsin–River Falls; William Givens, Grossmont College; James Hardcastle, Texas Woman's University; Victor Heasley, Point Loma Nazarene College; Paul Hoffman, Santa Ana College; Charles J. Horn, Mesa Community College; William Jensen, University of Cincinnati; James Johnson, Sinclair Community College; Joanna Kirvaitis, Moraine Valley Community College; Gerald Kokoszka, State University of New York–Plattsburgh; Dr. Doris K. Kolb, Bradley University; Robert Kowerski, San Mateo Community College; George F. Kraus, Charles County Community College; Kirklen J. Kupecz, Chaffey College; Irv Lillian, Miami Dade Community College; Roger Lloyd, Memphis State University; George G. Lowry, Western Michigan University; Katherine W. McLean, Phoenix College; Wendell H. Morgan, Hutchinson Community College; Susan Nurrenbern, University of Wisconsin–Stout; Ralph Petrucci, California State University–San Bernardino; George Potter, Schenectady County Community College; Fred Redmore, Highland Community College; George Salinas, Miami Dade Community College; Sue Sam, Lindenwood College; Doug Sawyer, Arizona State University; William D. Schulz, Eastern Kentucky University; Donald Showalter, University of Wisconsin–Stevens Point; Dr. Eileen Stitt, The University of Tulsa; Waltraut Sweeney, St. Petersburg Junior College; Danny V. White, American River College; Linda Wilkes, University of Southern Colorado.

Finally, I would like to thank my students at St. Louis Community College–Meramec who enthusiastically supported me and pinpointed mistakes in the first edition. Thanks also to my colleagues Michael Hauser, F. Axtell Kramer, Gee Krishnan, Lawrence S. Lynn, John Munch, and L. Gray Rueppel at St. Louis Community College–Meramec and Pauline Bellvance of Fontbonne College who provided feedback and suggestions on a number of ideas and features.

Teaching students and seeing them mature in chemistry during the course of a semester is a real joy. I wish the best for them and for all who use this text. Comments, corrections, and suggestions from users of this book are always welcome.

R. A. B.

From the desk of

RALPH A. BURNS

Dear Chemistry Student,

Welcome to an exciting adventure into the world of the atoms and molecules that make up everything you can touch or see or smell. Learning about the characteristics and interactions of these materials is at the heart of chemistry.

Many of you have varying degrees of apprehension as you approach the study of chemistry. Some of you are concerned about deficiencies in mathematics. Most of you have never taken a chemistry course and wonder what to expect. The author of this textbook and the person teaching this course know that chemistry can be fun and quite exciting, but it is also a very useful and practical science. We are convinced that you, too, will come to appreciate the part chemistry plays in our lives and learn to apply many of the fundamentals of chemistry, if you are willing to work with us as we guide you in this venture.

Did you ever *wish* you had a particular athletic skill, or that you could play a certain musical instrument? Unfortunately, *wishing* alone does not produce either an athlete or a musician. Similarly, wishing to succeed in chemistry is not enough! Only consistent hard work can produce the desired results for any of these endeavors. You must make a *conscious decision* to be successful by setting up a specific learning program with a definite time for study *each day*. Once this initial commitment is made, you can begin the ''training'' process. Begin by setting specific, small goals and mastering them one at a time. You move forward one step at a time, and one goal at a time, as you study *daily*.

The harder you work the luckier you get.
—Gary Player, Golfer

Memorization and Understanding

Sometimes students ask what they should memorize for the exam. In any field of study, a certain amount of memorization of key terms and definitions is necessary. This is also true in chemistry, but memorizing lecture notes or solutions to certain problems will not adequately prepare you for chemistry exams. *Memorization* may permit you to recall words, phrases, or equations, but this does not mean that you can apply the information or relationship to other knowledge. Questions in these chapters, and on exams, will also require a comparison or an *analysis* of data and facts. Combining information *(synthesis)* and arriving at conclusions *(evaluations)* are also necessary. This may be challenging, but with effective study, we expect you to be successful and proud of what you have learned. The approaches you will learn to use

People with goals succeed because they know where they're going.
—Earl Nightingale

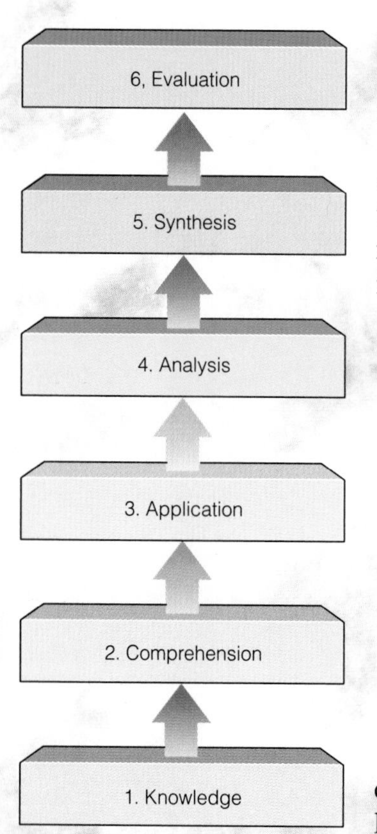

The six levels of thought

in solving chemistry problems are quite applicable to many disciplines. That is one reason chemistry is a required or recommended course for numerous disciplines.

As a student, you should be aware that studies on learning have shown that several levels of thought may be required to answer various types of questions or to solve problems. Most authorities on learning rank the levels of thought processes in the following levels of increasing complexity.

1. **Knowledge** involves *remembering* (memorizing) factual material.
2. **Comprehension** involves *interpreting* information, changing it from one form to another, and/or making predictions.
3. **Application** involves *using* facts and fundamental principles when *solving* problems or in laboratory investigations.
4. **Analysis** involves *identifying* and *sorting out* relevant and irrelevant facts to make comparisons.
5. **Synthesis** involves *combining* information and *developing* a plan or using original ideas.
6. **Evaluation** involves *judging* the value of observations and calculated results in order to reach a meaningful conclusion.

Although it is necessary to learn terminology and certain facts, you can see from this list that memorization only brings you to the lowest level of understanding. Higher levels of thought are often required when solving problems.

Keys to Productive Study

Active, creative thought is essential to meaningful study. Watching television, for example, does not require much active thought. Learning becomes active as you read and ask yourself questions, as you listen and ask questions, as you discuss what you have read and heard, and as you work problems.

The following "Ten Principles" of serious, meaningful, productive study can help you develop a formula for success, but there is no substitute for a positive attitude, whether you are pursuing chemistry or any other challenging course.

1. Learning occurs in *small steps*. Begin here and now—not tomorrow—to study and to solve problems.
2. Study *daily*. Don't expect to learn a lot the night before an exam.
3. First *scan*, then carefully read the material and ask yourself relevant questions. Write down questions you cannot answer.
4. *Read* the material again, take notes, and *list* key points. Learning is aided by repetition.
5. *Think* about interconnections with what you know, including applications.

6. *Visualize* structures, formulas, and key points until you can ''see'' them with your eyes shut.
7. *Write* down key points. You really don't know it if you can't write it.
8. *Think* about each key point. *Say* it! *Write* it! *Review* it! *Relate* key points to each other and compare their similarities and differences.
9. *Study sample problems* in the text. Consider the strategies used to solve these problems and how you would recognize and approach similar problems presented in the text and on a test.
10. *Solve problems* included at the end of each chapter before looking up the answers. Work problems daily. Become familiar with each type of problem and expect to solve similar problems on a test.

Successful students make mistakes, but they don't quit. They learn from them.

—Ralph Burns

Effective study takes a great deal of time with a commitment to active learning. More time spent does not necessarily guarantee more learning or a better grade. It's how you use your study time that counts.

Daily workouts and consistent *physical* conditioning are essential to success in any athletic event. Similarly, study on a daily basis and consistent *mental* conditioning are both essential to learning and to success on exams.

You cannot learn chemistry watching someone else ''do chemistry'' any more than you can learn to play tennis, golf, or baseball by watching someone else do it. Learning occurs as you raise questions, work problems, write down key points, talk about them, and apply them. Learning occurs as *you* ''do chemistry.''

Before Lectures

It is important that you skim over material in the text that relates to the next lecture; this will help you understand what your professor is presenting during the lecture. Read, review, and do the problems assigned each day. (Remember that learning occurs in small steps!) Quality study time involves daily, active study and does not include the time spent getting ready to study or time lost due to interruptions in concentration when your mind wanders. Falling behind is a major problem in chemistry. Make a commitment to yourself to apply the ''Ten Principles'' of learning outlined in this letter and to use the study helps provided in the text.

The greatest mistake a person can make is to be afraid of making one.

—Elbert Hubbard

During Lectures

Each lecture is like a step on a ladder. Missing lectures is like trying to climb a ladder that has missing steps. You are likely to slip and fall when you try to climb over missing steps. Actively think along with your professor. Don't try to write down every word spoken; *write down key points.* If you ask a few questions during lectures, you are likely to understand more, and to remember more. Never let your questions go unanswered. Doing so could cause you to miss questions on exams and to have difficulty understanding other topics. Major chemistry topics are interrelated and build on one another.

After Studying the Chapter

After studying a chapter, review the Chapter Summary, the points in Assess Your Understanding, and the list of Key Terms to be sure you know the major points and the terminology involved. Can you explain these terms and concepts to someone else? Try it! A person who understands a topic can use the vocabulary needed to discuss that topic. To obtain the greatest benefit for your efforts, use all study features in the text.

Before Exams

In the middle of difficulty lies opportunity.

—Albert Einstein

Taking exams will be less stressful if you are prepared! You will gain confidence daily as you work problems and answer questions much like those you expect to see on exams. Some students waste time trying to guess what topics might not be included on exams. Each topic assigned for study in your textbook or presented by your teacher is important, or it would not be included. Chemistry topics are presented in a definite, logical sequence; what you learn today will be the foundation for what you learn tomorrow. Dr. Henry A. Bent, a highly respected contemporary chemistry professor, points out that ''Nothing learned well is predictably irrelevant in the long run.'' Think about his profound words.

The large number of problems in this text are provided to help you develop confidence, accuracy, and speed. Review the notes you took while reading the text and during lectures. Review the topics listed at the end of each chapter and compare them with what your teacher presented in lectures. Also, review the exercises and problems you worked along with the examples shown in the text, the key terms, the figures and tables, and the chapter summary.

Studying for an exam can be compared to training for a marathon. It is better to train daily for six months, and take it easy the day before the event, than it is to procrastinate and put it off for six months, and then run 20 miles the day before the race. In other words, the night before the exam is too late to begin to study! A short review at that time should be enough. Get a good night's rest before taking an exam and you'll think better. Plan on arriving for an exam a little early, rested and confident.

If it's to be,
it's up to me.

—Unknown

Success is up to you! You must be committed. You can do it, but studying is real work! The thrill of victory does not come easily, in sports or in academic areas! Each success is a stepping-stone for future success.

Finally, let us compare success in learning to hiking up a majestic mountain. There are many pathways to the top. The novice needs a good map, appropriate equipment, and a good guide. You must climb one step at a time! Your teacher and this text and its supporting materials will guide you and point out the way. They will show you what to expect, what to watch out for, and how to handle problem situations as they guide you through the rough places. The climb to new heights may be difficult, but the "view from the top" is fantastic.

You must think of yourself as becoming the person you want to be.
—David Viscott

We truly hope you will enjoy your venture into the world of chemistry. There are also many chemistry courses beyond this one to interest and challenge you. Whether it is analyzing chemicals in the atmosphere, in the water, or in the soil, chemistry is there. Whether it is developing lightweight materials for sporting equipment or developing new semiconductor materials, chemistry is there. Whether it is developing aerosols that don't destroy the ozone layer or developing new chemicals to fight AIDS, cancer, and other medical mysteries, chemistry is there. The fundamentals of chemistry are central to the whole of science. As you will see in Chapter 1, chemistry is everywhere!

Sincerely yours,

Ralph A. Burns

Chemistry is a vital part of our daily lives.
From printing our books
to changing our looks,
From the clothes we wear
to doing our hair,
Chemistry is there.

1

Chemistry Is Everywhere

CONTENTS

Chemistry is everywhere! Everything you can touch or see or smell contains one or more chemicals. We live in a world of chemicals; many occur naturally, some are synthetic. Over 14 million chemicals are now known.

Chemicals are present in food, medicine, vitamins, clothing, paint, glue, household cleaning products, sporting equipment, and everything else you can buy. Chemicals are present in every kind of natural and synthetic fiber, but the chemicals that make up one kind of fiber are different from those in other fibers. Chemicals used in fertilizers differ greatly from those used in weed killers. Chemicals used in decongestants are different from those used in deodorants and detergents. The production of each of these products requires a variety of chemical processes.

Numerous types of chemicals are present in food. Some of these provide energy; others can cause allergic reactions in certain individuals. Each prescription medicine and over-the-counter drug contains chemicals that undergo specific chemical reactions within the body. Along with the beneficial effects there are side effects. There is often a trade-off between risk and benefit. For example, aspirin is a chemical that reduces both fever and pain, but it also thins the blood and may aggravate an ulcer condition.

Figure 1.1 *Chemicals are present in everything you can touch or see or smell. Chemicals are present in every natural and synthetic product. Chemistry is everywhere.*

Some chemicals can save lives; some can be lethal. Many chemicals are both useful and dangerous. Chemicals may be—and often are—both good and bad. It is how we handle and use them that makes the difference. An understanding of the fundamentals of chemistry is essential to the proper handling and use of chemicals, and to many careers. Furthermore, in our rapidly changing world, there is a new sense of urgency for all of us—regardless of career choices—to understand the fundamentals of chemistry. Chemicals are everywhere; without chemicals, life itself would not be possible.

1.1 Chemistry in Our World

Chemistry is the branch of science that deals with the characteristics and composition of all materials and with the changes they can undergo. Each chemical has specific characteristics. When a chemical change occurs, the chemicals produced are quite different from the starting materials. For example, iron metal slowly reacts chemically with oxygen in the air to form rust. Rust is quite different from the metal. Furthermore, iron metal can be obtained from iron ore through a series of chemical changes; the metal also is quite different from the ore. Complex chemical changes also occur when a plant produces carbohydrates. When you digest and utilize those carbohydrates and other food, a series of chemical reactions—collectively called metabolism—must take place as energy is released.

Not only does chemistry touch your life every moment, it also affects all of society. When a fuel is burned, chemical reactions occur that release energy that can provide power for transportation and electricity or heat for homes and businesses. However, some of the by-products of burning massive amounts of fuels are damaging our environment. Chemists are working on these problems. The natural chemicals in food can give you energy and help you stay healthy, but, occasionally, they can cause cancer. Chemists are working on these problems, too. In agriculture, chemists have helped reduce problems with insects, weeds, and disease and have increased crop yields. Chemists are also solving problems in health care. They have developed chemicals to assist in the diagnosis and treatment of many medical problems—chemicals to combat infection, relieve pain, control cancer, and detect diseases such as diabetes and AIDS. Chemistry is helping to improve the quality of life in many different areas.

From computer chips to liquid crystals to fiber optics, chemistry makes today's high technology possible. From space suits and swimwear to insulating materials and solar panels, or to tennis rackets and fishing rods, chemistry provides new materials for clothing, shelter, and recreation. Chemistry is fundamental to virtually everything produced or consumed by society.

Chemists are employed by companies producing all kinds of products from paint to plastics, from fertilizers to floor coverings, from bread to butter, from automobiles to airplanes. Some chemists analyze samples to check their quality. Other chemists are involved in the research and development of new products and work to insure their environmental safety. Still others are employed in government and university research and in teaching positions.

Figure 1.2 *From photographic and printing chemicals to pharmaceutical products and plastics, and from fibers to finishes to food production, chemists are working to improve the quality of life.*

Chemistry at Work

Chemistry and the Automobile Industry

Automobile production requires many materials that have special characteristics for special purposes. Advances in chemistry account for many of the developments and improvements in these materials. The paint and the protective finish on the paint contain a variety of chemicals. Chemists are employed to develop new colors and new finishes that are resistant to a harsh environment of dirt, grime, summer sun, and winter cold. The chemical reaction used to release gas in an air bag is a good example of chemistry being used to save lives.

The tire industry employs chemists who work to develop rubber materials with special characteristics for longer wear, for better grip in icy conditions, and for greater tolerance to the extreme heat of racing conditions. Materials used in tires, belts, and hoses must have special properties for their special functions.

The automobile industry relies heavily on chemistry. The metals, paints, plastics, fibers, finishes, and fuels all require research and development and quality control by chemists.

The preparation of steel, aluminum, chromium, and other metals used by the automobile industry requires a wide range of chemical processes. Plastics—widely used in automobile interiors—are becoming commonplace in bumpers, grilles, trim, and side panels. These light materials reduce the weight of the automobile and allow for greater fuel economy. Furthermore, plastics do not rust.

The production of glass for windshields and windows also involves chemical processes. Ceramic chemists work to improve properties of glass for each special function. Many automobile catalytic converters contain ceramic materials. Special ceramics are also being developed for use in small, lightweight engines.

The automobiles of the future are certain to contain a variety of new and better products of chemistry.

Of 42 basic industries in the United States, the chemical industry ranks first in worker safety. This is in sharp contrast to the popular belief that chemicals are extremely dangerous. Although some are, they can be used safely with proper precautions.

The chemical industry ranks as the fifth largest industry in the United States with sales of about $200 billion per year. Over 1 million people are employed by 10,000 chemical industries in the United States. Chemists employed by industries of all types are working to make manufacturing processes more efficient in terms of both materials and energy. The increased efficiency helps save our environment and saves money for the manufacturer and the consumer.

Chemical engineers are involved with the design and day-to-day operation of large chemical plants that produce the fertilizers, fibers, plastics, and other chemicals used by industries, businesses, and consumers. Many other professionals—in medicine, dentistry, building materials, construction, art, and numerous other areas—who are not usually thought of as being chemists, deal routinely with a wide variety of chemicals.

Virtually every industry or business that makes or sells a product is involved with chemicals and, therefore, with chemistry.

1.2 A Scientific Approach to Solving Problems

Each of us must solve problems daily. For example, suppose you need to run errands to several places such as the grocery store, the bank, the record shop, and the post office by 4:00 P.M. Here, the problem is to plan a route that will allow you to get the most errands completed by a certain time. This is the first step in solving any problem:

1. Identify and state the problem.

Past observations have given such data as the closing times for the bank and post office and the street construction project in front of the record shop. Also, the frozen food items from the grocery store need to be kept frozen. Previous experiences in driving to these locations—these can be thought of as previous experiments—have provided information about the approximate amount of time needed to go from one location to another. Here, we have carried out the second step in solving any problem:

2. Collect data pertaining to the problem by making observations and by carrying out experiments.

Based on the available data, including closing times and other facts, a possible solution—a tentative route—is proposed: The tentative order of stops on the route will be the bank, the post office, the record shop, and the grocery store. This is the third step in solving any problem:

3. Analyze the data and propose one or more possible solutions to the problem (or give an explanation for the observations).

Now we are ready to try something. Carry out the proposed plan. If this had been a scientific problem, we would now carry out experiments to see if our solution or explanation is reasonable. In the case of our proposed route to the bank and three other locations, we are ready to start our trip. This is the fourth step in problem solving:

4. Carry out the proposed plan or experiment.

Following the proposed route to four locations, we meet an unexpected situation. The street to the record shop is closed because of a fire, and there is a major traffic jam. Thus it will not be possible to go to all the locations by 4:00 P.M. With these new observations and data, it is time to start the problem-solving cycle all over again. It is often necessary to go through the cycle many times before arriving at a reasonable solution.

Although this was not a chemistry problem, the steps outlined here are essentially the same whether we are solving a chemistry problem, an engineering problem, or any other type of problem.

Solving Chemistry Problems

In the study of chemistry there are many opportunities to develop problem-solving skills. In fact, solving chemistry problems is one of the most important skills you will develop in this course and in any chemistry course. Some of the problems require mathematical calculations, others do not, but the steps to the solution follow the cycle described here.

Although each problem differs from all others in one or more ways, the general approach to solving problems involves five steps. Even simple problems—like that of running errands—can be broken into the same five steps. These "five steps in solving problems" can be used effectively as you solve problems that occur in chemistry or elsewhere. They are summarized here for easy reference. You may want to refer back to them as you work through problems in subsequent chapters.

Five Steps in Solving Problems

1. Identify the problem. Read the question carefully and write down precisely what is wanted.
2. Collect and write down the data and known facts related to the problem.
3. Analyze the data, identify the type of problem to be solved, and set up the problem by outlining a specific plan or pathway to the answer.
4. Carry out the proposed plan to a tentative solution.
5. Evaluate your answer to make sure it is a reasonable solution. If it is not, cycle through the steps again.

This logical stepwise approach to solving problems is applicable to solving problems in any field. That is one reason why a background in chemistry—with its emphasis on solving problems—is desirable for persons with careers in many areas. Companies of all types are looking for prospective employees who have good problem-solving skills.

1.3 Hypothesis to Theory: The Scientific Method

The results of research investigations are reported as quickly as possible in recognized scientific research publications called *journals*. If something is false, and the information cannot be verified by others, this is certain to be reported by other researchers.

Chemistry and other branches of science deal with much more than searching for answers to individual problems. Science is concerned with *explaining* nature, and the explanations must be tested by controlled research investigations that are frequently called **experiments**. We can learn certain things through personal experience, and we can learn about historical events, but the knowledge gained through science is different: It depends upon phenomena that can be verified through repeated testing.

Experimental observations are only a bare—but necessary—beginning to the intellectual process of science. Observations give rise to ideas that must be tested. Our understanding of nature is refined constantly by the interplay of ideas and observations. Some people assume that all of science is rigid, and unchanging, but this is not so. Our understanding of nature is often *tentative*, and must be modified to accept new findings. Chemistry—or any other body of scientific knowledge—is ever changing; it is constantly growing.

In seeking to explain nature, scientists seek explanations by following a set of procedures called the **scientific method**. The beginning steps of the scientific method are essentially those described as the "five steps in solving problems," but, as the next step, a researcher would look for common ways in which several investigations and other relevant facts are interrelated. Sometimes a meaningful generalization can be made about a particular pattern of facts. The statement is called a *natural law*.

A **natural law** is a statement that summarizes experimental facts about nature where behavior is consistent and has no known exceptions.

Researchers have no control over the natural (scientific) *laws* they discover. This is in contrast to laws in society that can be changed—and frequently are.

The law of conservation of mass—which will be described in Chapter 2—is one example of a scientific law. It states that mass is neither gained nor lost in a chemical reaction. The law may appear to be quite simple, but it took many years of investigation before this profound relationship was discovered.

A natural law summarizes the facts, but it does not attempt to explain the facts. The search for answers does not end with the discovery of a law. Scientists want to know *why* nature behaves in a particular way. A tentative but reasonable explanation of the phenomenon is called a *hypothesis.*

A **hypothesis** is a tentative, reasonable explanation of the facts or the law.

Once a reasonable explanation—a hypothesis—has been proposed, it is time to cycle through the steps of the scientific method again. This time experiments are designed to test the hypothesis, including experiments where the outcome is predicted on the basis of the hypothesis. When the hypothesis is thoroughly validated by extensive research, it eventually becomes accepted among scientists. Once the hypothesis becomes well established in the scientific community, it is called a *theory.*

A **theory** is a hypothesis that has withstood extensive testing.

There is no precisely defined point at which a hypothesis becomes a theory. There is no specific number of supporting researchers or research investigations required before a hypothesis reaches the status of a theory. Furthermore, a scientific theory does not represent absolute truth. It is not necessarily right or wrong; its value is dependent upon how useful it is in explaining phenomena and in making accurate predictions. Theories have limitations. They may provide explanations that are incomplete, or that are oversimplifications. In fact, theories may be—and often are—modified to account for new research findings. Although the facts that have been verified will not change, the theories that have been devised to explain the facts may need to be revised. The major steps of the scientific method are summarized in Figure 1.3.

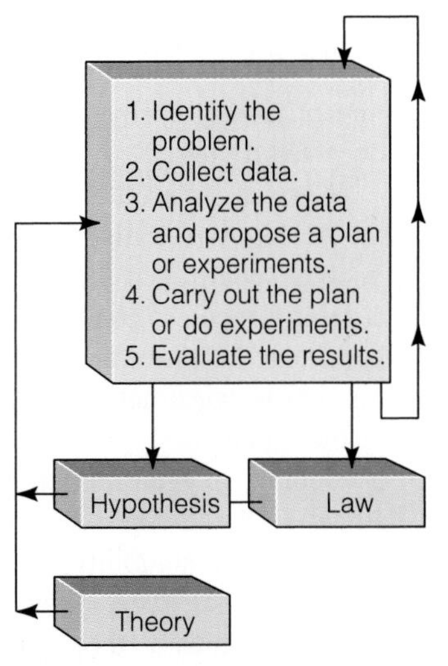

Figure 1.3 *Major steps of the scientific method.*

1.4 *Searching for Answers: Applied and Basic Research*

Research in chemistry is classified as either applied research or basic research. Chemists who are involved in *applied research* frequently work toward the development of a variety of new products to satisfy needs of business, industry, and the rest of society. These products include chemicals for use in medicine, agriculture, and other areas. Applied research also involves solving problems in industry and the environment, including analysis of food, fuels, and other consumer products or samples of air, water, and soil.

Other chemists are involved in *basic research*—the search for knowledge for its own sake. This research is of fundamental importance to society. It adds to the knowledge base used in both basic and applied research and provides information that helps us understand the secrets of nature.

Some chemists in basic research work with the structures and interactions of all sorts of chemicals; others synthesize new chemicals and determine their properties or measure energy changes that accompany complex chemical reactions. Basic research could involve a study of the attraction of certain materials for each other to help understand why some act like glue but others do not, or why two chemicals with totally different structures can both have a sweet taste.

> "There are two compelling reasons why society must support basic science. One is substantial: The theoretical physics of yesterday is the nuclear defense of today; the obscure synthetic chemistry of yesterday is curing disease today. The other reason is cultural. The essence of our civilization is to explore and analyze the nature of man and his surroundings. As proclaimed in the Book of Proverbs in the Bible, 'Where there is no vision, the people perish.'"
>
> Arthur Kornberg (1918–)
> American Biochemist
> Nobel prize in physiology and medicine, 1959

Chemistry in Our World

Chemistry and Deep-Sea Diving

A research paper on solubility was published in 1916 by Joel Hildebrand, a chemist and noted teacher at the University of California at Berkeley. In this theoretical paper, Hildebrand described basic research on gas solubility in liquids. He predicted that in any liquid, helium would be less soluble than any other gas. Hildebrand then proposed that a mixture of helium and oxygen—rather than air, a mixture of mostly nitrogen and oxygen—be used for deep-sea diving.

When air is used by a diver, nitrogen dissolves in the blood. As the diver rises toward the surface, bubbles of nitrogen separate from the blood, causing a painful medical condition called the *bends.* The bends can be prevented by using an oxygen–helium mixture because helium is less soluble than nitrogen in the blood.

Helium–oxygen mixtures are now used by professional deep-sea divers. Because Hildebrand was curious about the solubility of gases in liquids (a basic research topic), those divers need not worry as much about the bends as they once did (a practical application).

Because a chemist was curious about the solubility of gases in liquids—and because he followed through with basic research—deep-sea divers need not worry as much about the bends.

Information obtained through basic research is often *applied* at some point, but that is not the primary goal of the researcher. In fact, most of our modern technology is based on results obtained in basic research. Generally, considerable basic research in an area precedes a major useful product or technological development in a field. Most basic research is conducted at universities and in certain private and government research institutes. Most of the financial support comes from federal and state governments and foundations, but some large industries also provide funding for basic research.

1.5 Chemistry: A Central Science in Your Education

Chemistry is frequently called *the central science.* If you study biology, geology, or physics, you are sure to encounter topics that involve chemistry. When you study the makeup and changes in rocks, the soil, the atmosphere, semiconductor materials, metabolism in living systems, or numerous other topics, you are dealing with chemicals and chemical changes.

Solving problems in the environment, in industry, and in health-related areas often involves chemistry. If you search for an understanding of problems related to depletion of the ozone layer, or crystalline structures in rocks, or materials for superconductors, or metabolism and respiration, or the effects of medications on the body, chemistry is there. Chemists routinely do research in each of these areas.

Chemistry courses are required for numerous careers for valid reasons. Chemistry teaches us to deal with everyday problems, technical problems, and research problems, using a logical, planned approach, but that is not all. There is also a very human side to chemistry. Many of the personalities in chemistry have made significant contributions that impact the health and well-being of individuals and of society.

One example is Linus Pauling (1901–1994) who did basic research in chemical bonding that paved the way to understanding the gene, genetic coding, and heredity (Figure 1.4). He was awarded the 1954 Nobel prize in chemistry for his work in this area. As a humanitarian and a leader among scientists, Pauling was also awarded a second Nobel prize—the 1962 Nobel prize for peace—after he collected and submitted signatures of 11,000 scientists on a petition to stop nuclear weapons testing. Pauling carried out extensive work with vitamin C and was a strong advocate for its use in preventing the common cold. Some of his hypotheses about vitamin C have not been accepted by the medical community. Many questions about the action of the vitamin remain unanswered.

Figure 1.4 *Linus Pauling (1901–1994), winner of two Nobel prizes.*

The Balanced Focus of Chemistry

The focus of study in chemistry requires a balanced approach that includes at least six major areas: experiments, facts, terminology, laws, theories, and problem solving. Ultimately, chemistry teaches us to ask questions and to seek answers about properties and interactions of materials. However, chemistry does not require that a student immediately start solving sophisticated problems any more than a beginning medical student would be required to perform an organ transplant, or an engineering student would be required to oversee an industrial project.

Ultimately, chemistry teaches us to ask questions and to seek answers about properties and interactions of materials.

The chemistry student, the medical student, and the engineering student must begin by acquiring some fundamental knowledge. Today's science stands on the shoulders of persons who have gone before us who have collected, organized, and simplified great amounts of data. Some of these persons possessed brilliant minds, some had keen foresight, and some experienced lucky accidents, but their work makes it easier for us to understand nature. They have summarized the data in laws and have devised theories and models that help us explain the facts. We need to understand and apply these fundamentals as we learn to solve problems. The search for answers—in the past, the present, and the future—can be both exciting and challenging.

Chapter Summary

Everything that exists in the physical world is made up of one or more chemicals; many occur naturally, some are synthetic. An understanding of chemistry and its role in our society is beneficial to understanding our world. Along with beneficial effects, there are sometimes side effects from both natural and synthetic chemicals; there is often a trade-off between risk and benefit.

Chemistry deals with the characteristics and composition of materials, and with the changes they undergo. Virtually every industry or business that makes or sells a product is involved with chemicals and thus is dealing with chemistry.

Five steps to solving a problem are as follows: (1) identify the problem, (2) collect data, (3) develop a plan for the solution, (4) carry out the plan, and (5) evaluate the answer. The scientific method begins with these same steps. A statement that summarizes consistent information from many investigations is called a natural law. A tentative explanation for the law is called a hypothesis. Once the hypothesis becomes well established it is called a theory. Chemists who are involved in areas related to the development of commercial products work in applied research. Those who search for knowledge for its own sake work in basic research. Solving problems in chemistry often involves the use of experiments, facts, terminology, laws, and theories in the search for answers.

Assess Your Understanding

1. Define *chemistry* and give examples of chemicals and chemical changes. [1.1]
2. Identify and describe the steps of the scientific method. [1.2, 1.3]
3. Distinguish between applied and basic research. [1.4]
4. Give six areas of emphasis in a balanced study of chemistry. [1.5]

Key Terms

chemistry [1.1]
experiment [1.3]
hypothesis [1.3]
natural law [1.3]
scientific method [1.3]
theory [1.3]

Problems

INTRODUCTION

1.1 Which of the following are made up of chemicals?
a. vitamins b. fruit
c. cotton d. glass
e. carbohydrates

1.2 Which of the following are made up of chemicals?
a. proteins b. vegetables
c. nylon d. paint
e. brass

1.3 Name two chemicals that can help save lives and two that can be lethal.

1.4 Discuss the meaning of the statement, ''Without chemicals, life itself would not be possible.''

CHEMISTRY IN OUR WORLD

1.5 A definition of chemistry is given in the textbook. What are three key points in this definition?

1.6 What is *your* definition of chemicals and of chemistry?

1.7 Iron is a chemical. Give an example of a chemical change involving iron.

1.8 Carbohydrates are chemicals. Describe one or more chemical changes involving carbohydrates.

1.9 Fossil fuels contain chemicals. Describe some benefits of burning fossil fuels. Describe some problems. Is anyone doing anything about the problems?

1.10 Food naturally contains chemicals. Describe some benefits of foods. Describe some problems. Is anyone doing anything about the problems?

1.11 How are chemists and chemicals increasing crop yields in agriculture?

1.12 How are chemists and chemicals serving the medical professions?

1.13 Give three or more chemical products used in ''high technology.''

1.14 Give three or more chemical products used by the automobile industry.

1.15 Where does the chemical industry rank in terms of size and worker safety?

1.16 What are some of the jobs performed by chemists and chemical engineers?

A SCIENTIFIC APPROACH TO SOLVING PROBLEMS

1.17 What are the five steps used in solving problems scientifically?

1.18 From information given in ''Solving Chemistry Problems'' in Section 1.2, what do chemistry problems have in common with running errands and other everyday problems?

THE SCIENTIFIC METHOD

1.19 Discuss the terms ''experiment'' and ''natural law.''

1.20 Give one specific natural law. Why is it called a law?

1.21 In what ways is a hypothesis like a theory? How are they different?

1.22 Arrange the following terms so their order represents the usual steps of the scientific method: facts, law, theory, experiment, hypothesis.

APPLIED AND BASIC RESEARCH

1.23 What is the difference between basic and applied research?

1.24 In the box entitled ''Chemistry and Deep-Sea Diving'' in Section 1.4, was Hildebrand's initial work on solubility of gases an example of basic or applied research? How does his work illustrate the importance of basic research?

1.25 Classify the following as either basic research or applied research.
a. Research is conducted on interesting new materials that change colors at different temperatures.
b. The materials in a. were later called liquid crystals; further research was then conducted to see if the liquid crystals could be used in thin visual displays for TV sets and computer monitors.

1.26 Classify the following as either basic research or applied research.
a. Research is conducted to determine structural similarities in superconducting materials.
b. One of the materials is found to have superior reliability characteristics under extreme conditions and is selected for use in a telecommunications satellite.

CHEMISTRY: A CENTRAL SCIENCE

1.27 Criticize the statement, ''You don't need a chemistry course unless you are going to be a chemist or a chemical engineer.''

1.28 List six areas of emphasis in a balanced approach to studying chemistry.

Tremendous quantities of matter and energy are displayed in a powerful way by the Niagara falls. Each second, 6.0 million liters (1.6 million gallons) of water rush over these falls. The Niagara Falls, located on the border between Canada and the United States, separate Lake Erie and Lake Ontario.

2

Matter and Energy

CONTENTS

"What's that?" This is one of the first questions asked by nearly every one of us at the toddler stage, while pointing to various objects. Whether the item being pointed out is a ball, a spoon, ice cream, or something else, the toddler is generally satisfied after hearing the *name* of the object. The toddler may also want to examine the object. Basic questions about the chemical makeup—the composition—of materials are not asked at that age. Sooner or later the toddler learns that too many questions are not appreciated by others, so fewer questions are asked as time goes by. We want to reawaken your interest in asking basic questions about the composition and nature of everything in our physical world. No question is a "dumb" question. Don't hesitate to ask yourself and others the most fundamental questions, as you study these fundamentals of chemistry.

2.1 *Matter*

Matter can be described simply as the "stuff" that makes up all material things in the universe. Water, salt, sand, sugar, steel, the stars, and even the gases present in

Figure 2.1 *Astronaut John W. Young leaps from the lunar surface, where gravity pulls at him with only one-sixth the force on Earth.*

the air are all composed of matter. By definition, **matter** is anything that has mass (so, therefore, it must take up space). In fact, **chemistry** is a science that deals with matter and the changes that it undergoes.

Mass is a measure of the quantity of matter. Even air has mass, but you might not think about it unless you were walking against a strong wind. Mass is often confused with weight. **Weight** is the force of gravity acting on the mass of a particular object. (The strength of a planet's gravity depends on its mass and size.)

For most of its history, the human race was restricted to the surface of planet Earth—which exerts a relatively constant gravitational force on a given object—so the terms *mass* and *weight* were generally used interchangeably. If something has twice the mass of something else, it also weighs twice as much. When the exploration of space began, however, the distinct differences between mass and weight became more apparent, and easier to describe. The mass of an astronaut on the moon is the same as his or her mass on Earth. The amount of matter that makes up the astronaut does not change. The *weight* of the astronaut on the moon, however, is only one-sixth of his or her weight on Earth because the moon's pull is one-sixth as great as Earth's attraction (Figure 2.1). Weight varies with gravity; mass does not.

EXAMPLE 2.1

A certain astronaut has a mass of 65 kilograms (kg). Compare the astronaut's mass in the gravitational environments listed. Also, rearrange the list so the environment for greatest weight will be listed first.

(a) Earth's moon with a gravity 0.17 of Earth's gravity, (b) Earth, (c) space, (d) Mars, with a gravity that is 0.38 of Earth's gravity.

SOLUTION

The astronaut's mass does not change. The environment with greatest gravitational attraction is where the astronaut's weight is greatest. List this environment first.

See Problems 2.1–2.4.

(a) Earth's moon: third greatest weight for the astronaut; the moon has a smaller gravitational force than Earth or Mars.
(b) Earth: greatest weight for the astronaut (due to greatest gravity).
(c) Space: here, the astronaut is essentially weightless.
(d) Mars: second greatest weight for the astronaut; second greatest gravity.

Exercise 2.1
What would be your weight on Mars?

Figure 2.2 *Water in three states: solid (ice), liquid, and gas (vapor).*

2.2 *Matter Has States*

Depending on its temperature, a sample of matter can be a solid, a liquid, or a gas. These three forms of matter are called the **states of matter**, or simply, **physical states**. For water, different names are often used to designate its different physical states. Solid water is called **ice**. When heated sufficiently, the ice melts and gives liquid water. Further heating produces boiling water and a gas we call **steam**. Steam is actually invisible water vapor at a *high* temperature. A cloud that appears above the spout of a teakettle contains droplets of condensed liquid water (Figure 2.2).

$$\underset{\text{(ice)}}{\textbf{Solid}\text{ water}} \xrightarrow{\text{melting}} \textbf{Liquid}\text{ water} \xrightarrow{\text{boiling}} \underset{\text{(steam)}}{\textbf{Gaseous}\text{ water}}$$

Water that evaporates at room temperature also becomes water vapor.

Cooling the steam will make it **condense**; that is, it returns to a liquid. Lowering the temperature of liquid water sufficiently will **freeze** the water, making ice. Thus, the physical state of the water depends upon the temperature.

A **solid** has a definite shape and volume. Many solids are *crystalline*; they have a definite three-dimensional shape with surfaces at specific angles to each other. For example, sodium chloride (ordinary table salt) crystallizes in a cubic shape with surfaces (faces) at 90° angles (Figure 2.3). A crystal cleaves or splits when struck at certain angles so that crystal fragments retain the same characteristic angular shape.

The properties of solids can be explained at the atomic level in terms of a

Atomic level refers to the invisible submicroscopic level where the smallest individual particles of the material might be detected.

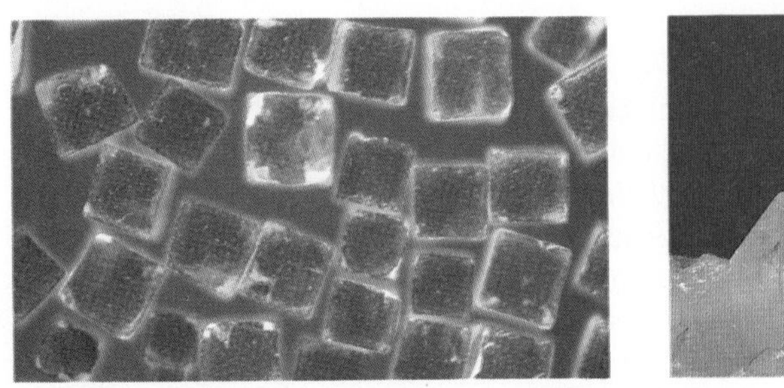

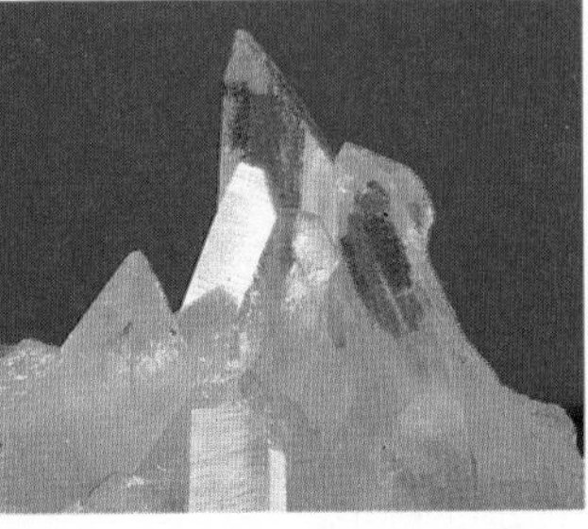

(a) (b) (c)

Figure 2.3 *Some pure crystalline solids at room temperature. (a) Sodium chloride, salt; (b) silicon dioxide, quartz; (c) pyrite, iron sulfide.*

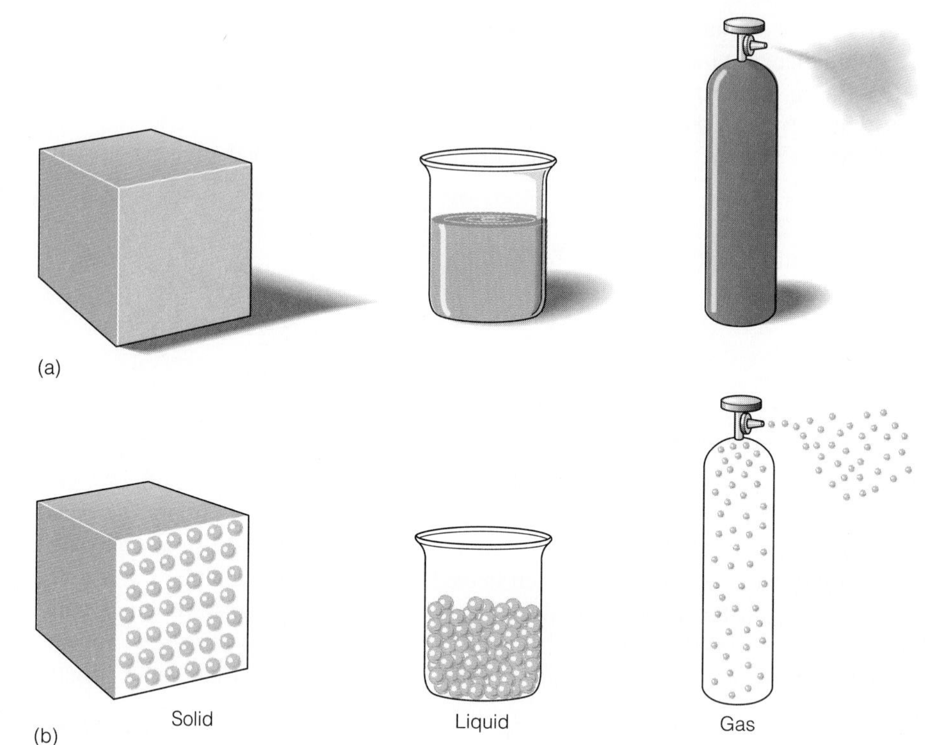

Figure 2.4 *Solids, liquids, and gases. (a) Bulk properties. (b) Interpretation of bulk properties at the atomic level.*

definite and regular arrangement of the tiny, invisible, individual particles that make up the solid (Figure 2.4). This arrangement is called an array, or a crystal lattice. The tightly packed particles usually are held together by strong attractive forces. There is little motion of the particles in a solid—only a slight vibration within the crystal lattice. Structures of several crystalline solids are discussed in greater detail in Chapter 13.

Unlike solids, **liquids** take the shape of their containers (except for a generally flat surface at the top). Like solids, however, liquids maintain a nearly constant volume. If you have a 12-ounce soft drink, you have 12 fluid ounces whether the soft drink is in a can, in a bottle, or spread out in a puddle on the floor—which demonstrates another property of liquids. Unlike solids, liquids flow readily, but some liquids flow more readily than others. The **viscosity** of a liquid is a measure of its resistance to flow, and is one of the special properties of each liquid. Viscous liquids, such as honey, flow at a slow rate; water and alcohol, with low viscosities, flow much faster.

Water and alcohol are two liquids that are **miscible**. This means that they will dissolve in one another. They can be mixed in any proportion—and remain mixed—without separating into layers. Vegetable oil and water are two liquids that are **immiscible.** When shaken together, two immiscible liquids form a cloudy mixture with tiny droplets of one liquid visibly suspended in the other. If allowed to stand, the immiscible liquids separate into two distinct layers (Figure 2.5).

When you watch a liquid flow, you can get an idea of what is happening at the atomic level if you imagine the smallest individual particles slipping, sliding, and gliding over one another. Individual particles in a liquid are close together, and their mutual attractions are fairly strong, yet they are free to move about. For example, tiny particles of oil or water come together to make visible droplets.

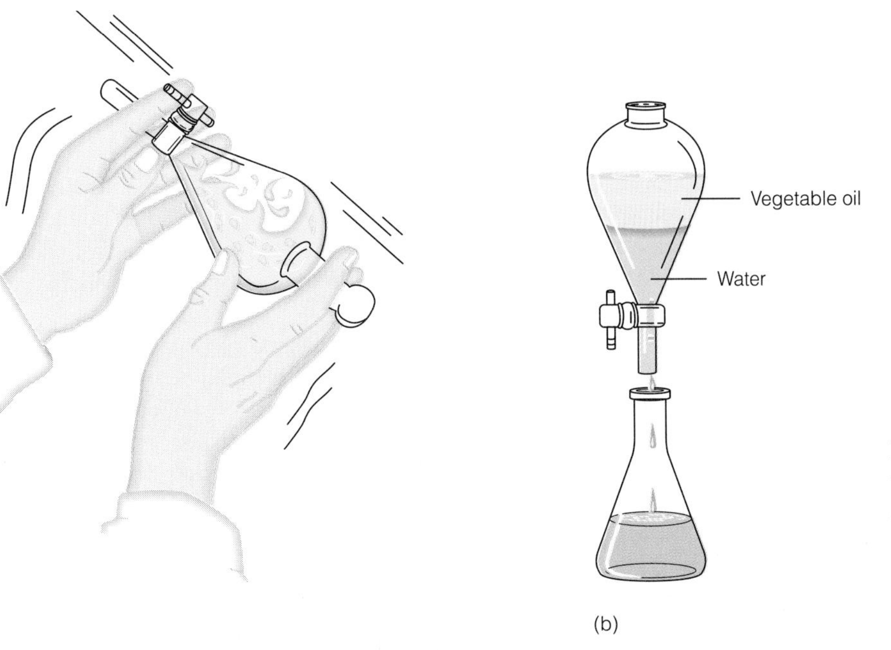

Figure 2.5 *Vegetable oil and water are immiscible. (a) The mixing of vegetable oil and water by vigorous shaking. (b) Separate layers form, with the oil on top and water on the bottom.*

Solids and liquids are virtually incompressible because their individual particles are closely spaced (see Figure 2.4).

Gases do not have a definite shape or volume; they take the shape and volume of the container they occupy. Partially blow up a balloon, and tie it shut. Squeeze it in one area and watch the gas flow into the less restricted area. Gases expand to fill completely whatever container they occupy, but they also can be compressed into smaller containers. For example, air can be compressed into a steel tank and used

Figure 2.6 *Helium gas is used to fill party balloons.*

by a diver, under water, for a period of time. Gases also **diffuse** rapidly; that is, they mix with other gases as they move to fill the available space. If skunk scent is released, the strong odor quickly permeates the area. Similarly, if you open a bottle of ammonia gas in a room, the irritating odor will soon spread throughout the entire room.

At the atomic level, visualize the particles of gas moving independently of one another, with much space between the particles. Visualize particles bouncing off the sides of the balloon, exerting sufficient force to keep the balloon from collapsing (Figure 2.6). At higher temperatures, the particles have more energy, thus causing the balloon to expand.

Air is a mixture of several gases including nitrogen (78%), oxygen (21%), a little argon (less than 1%), and traces of other gases including carbon dioxide and neon. Methane gas is the main component of natural gas, a commonly used fuel for heating homes in the United States and many other countries.

Tables 2.1 and 2.2 list examples of solids, liquids, and gases and give specific properties for each state.

Table 2.1 **Properties of Solids, Liquids, and Gases**

State	*Shape*	*Volume*	*Compressibility*	*Submicroscopic Properties*
Solid	Definite	Definite	Negligible	Particles touching and tightly packed in rigid arrays
Liquid	Indefinite	Definite	Very little	Particles touching but mobile
Gas	Indefinite	Indefinite	High	Particles far apart and independent of one another

Table 2.2 **Some Solids, Liquids, and Gases**

Solids	*Liquids*	*Gases*
Ice or snow*	Water*	Steam*
Aluminum	Mercury	Air
Copper	Gasoline	Helium
Salt	Vegetable oil	Carbon dioxide
Sugar	Alcohol	Acetylene
Sand	Vinegar	Argon
Lead	Motor oil	Krypton

*Ice, snow, and steam are forms of water—the same substance—in different states.

Acetylene is a gas that burns with a great deal of heat; it is often used in welding.

Krypton gas is now being used to fill certain superbright flashlight bulbs.

EXAMPLE 2.2

Identify the physical state of each of these items at room temperature.

(a) oxygen (b) water vapor (c) candle wax (d) alcohol

SOLUTION

(a) gas (b) gas (c) solid (d) liquid (See Table 2.2.)

See Problems 2.5–2.10.

2.3 Elements and Compounds

A **pure substance** is a single chemical composed of the same kind of matter—with the same kind of particles—throughout. It can be either an element or a compound.

Elements are the most fundamental substances from which all material things are constructed. The smallest particle that retains the property of the element is an **atom**. Atoms of a solid element are arranged in a regular pattern and are of the same type. All atoms of a piece of copper are copper atoms. All atoms of a piece of silver are silver atoms. Atoms of a particular element cannot be broken into simpler atoms. Gold has never been broken down into simpler kinds of atoms, thus demonstrating that it is an element.

Look at the inside front cover of this book and you will find what is called a periodic table. It contains symbols for more than 100 elements. Most of these elements are rather rare; only about 10 elements make up 99% of everything in the Earth's crust. In Chapter 4 we shall look more closely at the periodic table and the names and properties of the elements.

Compounds are pure substances that are made up of two or more kinds of elements, combined together in fixed proportions. Properties of compounds are different from those of the individual elements involved. Water was once thought to be an element, but we now know it to be a compound, composed of two elements, hydrogen and oxygen. The chemical formula for water, H_2O, indicates that two hydrogen atoms are combined with each atom of oxygen. In a laboratory, water can be decomposed into hydrogen and oxygen by passing an electric current through it (Figure 2.7). A particular compound always has a specific atom ratio and a specific percentage by mass of each element in the compound. This is a statement of the **law of definite composition**, which is also called the **law of definite proportions**.

Table salt can be broken down by first melting it and then passing an electric current through it, giving the elements sodium and chlorine. Table salt is a compound. It has a definite composition, 39.3% sodium and 60.7% chlorine by mass.

(a)

(b)

Figure 2.7 *Water (a) and table salt (b) are compounds.*

Table 2.3 **The Composition of Some Common Compounds**

Name of Compound	*Composition of Compound*	*Comparison of Properties*
Water	Hydrogen and oxygen	Hydrogen and oxygen are gases, but water is a liquid at room temperature.
Sugar	Carbon, hydrogen, and oxygen	Carbon can be a black solid, but hydrogen and oxygen are colorless gases. The compound sugar is a white solid that tastes sweet.
Table salt	Sodium and chlorine	Sodium is a silvery, reactive, solid metal, and chlorine is a pale green, poisonous gas. Salt is a white, crystalline solid.
Ammonia	Nitrogen and hydrogen	The elements are odorless, but ammonia has a strong odor.
Ethyl alcohol	Carbon, hydrogen, and oxygen	Carbon can be a black solid, but hydrogen and oxygen are colorless gases. The compound ethyl alcohol is a colorless, flammable liquid.
Hydrogen sulfide	Hydrogen and sulfur	Hydrogen is a colorless, odorless gas. Sulfur is a pale yellow solid. The compound hydrogen sulfide is a colorless gas that smells like rotten eggs.

Table 2.3 lists several common compounds and the elements that are combined in the compound. Notice that compounds should not be expected to look at all like the elements of which they are composed. Instead, compounds have unique, characteristic properties, different from the elements involved. The following example illustrates this point.

EXAMPLE 2.3

Explain how the compound sodium chloride (table salt) is easily distinguished from the elements that were combined to make the compound. (See Table 2.3 if you are not familiar with any of these substances.)

SOLUTION

By experience, sodium chloride (table salt) is a white crystalline compound that dissolves in water. The elements, sodium and chlorine (described in Table 2.3), look nothing at all like the compound. Sodium is a soft, silvery, reactive, solid metal; chlorine is a pale greenish yellow, poisonous gas.

See Problems 2.11–2.18.

2.4 *Pure Substances and Mixtures*

Any sample of matter can be classified as either a pure substance or a mixture. A pure substance can be either an element or a compound. The composition of a pure substance is definite and fixed. For example, pure water is a compound; it is always 11% hydrogen and 89% oxygen by mass. Pure (24 karat) gold is an element; it is 100% gold.

The composition of a **mixture** may vary. Orange juice is a mixture that contains juice, pulp, water, and a variety of natural chemicals and chemical additives, depending on the brand of juice purchased (Figure 2.8). A can of mixed nuts is also a mixture, where the proportions of cashews or pecans or peanuts are dependent on how much you are willing to pay per pound. Each cake, cookie, glass of iced tea, or bottle of cola is a mixture.

Figure 2.8 *Orange juice is a heterogeneous mixture containing pulp fibers and juice.*

As shown in Figure 2.9, pure substances are either elements or compounds, and mixtures are either *homogeneous mixtures* or *heterogeneous mixtures.* The mixed nuts and cake represent examples of heterogeneous mixtures. The prefix *hetero* means ''different.'' A **heterogeneous mixture** does not have uniform properties throughout; the composition of one part (or phase) differs from the composition of another part (or phase). An oil-and-water mixture is another example of a heterogeneous mixture.

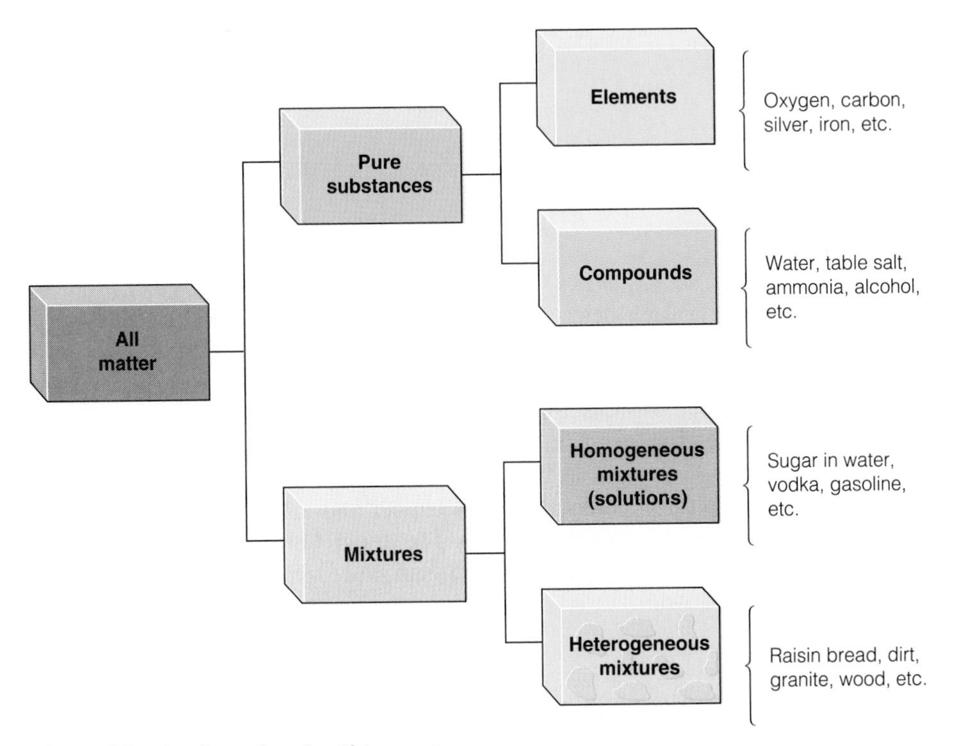

Figure 2.9 *A scheme for classifying matter.*

A **homogeneous mixture** is the same throughout. A **solution** is a homogeneous mixture; it has uniform composition and appearance throughout. You expect a cup of hot tea to taste the same from start to finish, and to have uniform composition throughout. Solids such as sugar and salt dissolve in water to make solutions. Mixtures of miscible liquids, such as alcohol and water, are solutions; they are the same throughout. Almost all metal alloys, such as bronze, brass, and steel, are really solutions of one solid dissolved in another solid; they are homogeneous. Several homogeneous mixtures (solutions) are listed in Table 2.4.

Table 2.4 **Some Common Solutions (Homogeneous Mixtures)**

Solution	*Composition*
Gaseous solutions	
Air	78.0% nitrogen, 20.9% oxygen, 0.9% argon, and traces of carbon dioxide and other gases
Natural gas	Methane and small amounts of several other gases
Liquid solutions	
Tincture of iodine	Iodine dissolved in alcohol
Rubbing alcohol	70% isopropyl alcohol and 30% water
Wine	10–12% ethyl alcohol
Beer	5% ethyl alcohol
Solid solutions (alloys)	
Brass	Copper (about 70%) with zinc (about 30%)
Bronze	Copper and tin in varying proportions
Carbon steel	1.0% manganese, 0.9% carbon, remainder is iron
Chromium steel	3.5% chromium, 0.9% carbon, remainder is iron
Sterling silver	92.5% silver with 7.5% copper
Yellow gold (14 K)	58% gold, 24% silver, 17% copper, 1% zinc
Yellow gold (10 K)	42% gold, 12% silver, 40% copper, 6% zinc

EXAMPLE 2.4

Classify the following as heterogeneous or homogeneous.

(a) scrambled eggs (b) gasoline (c) wood (d) brass (e) a pizza

SOLUTION

(a) Scrambled eggs are heterogeneous, with some portions having more egg white than other portions.
(b) Gasoline is a homogeneous mixture, having a blend of petroleum-based compounds that are miscible, and consistent throughout.
(c) Wood is a heterogeneous mixture of cellulose, sap, and other natural products.
(d) Brass is a homogeneous mixture of copper and zinc. (See Table 2.4.)
(e) A pizza is a heterogeneous mixture. The dough contains flour, oils, water, and yeast. The topping can contain tomato sauce, cheese, olives, sausage, or whatever you want.

See Problems 2.19–2.26.

2.5 *Characteristic Properties of Substances*

Sugar, water, and aluminum are different substances. Each substance has specific properties that do not depend on the *quantity* of the substance. Properties that can be used to identify or characterize a substance—and distinguish that substance from other substances—are called **characteristic properties**. They are subdivided into two categories: physical properties and chemical properties.

Ductility is the capability of a metal to be drawn out into a wire.

Malleability is the capability of a metal to be shaped when hammered or rolled into thin sheets.

The characteristic **physical properties** of a substance are those that identify the substance without causing a change in the composition of the substance. They do not depend on the *quantity* of substance. Color, odor, density, melting point, boiling point, hardness, metallic luster (shininess), ductility, malleability, and viscosity are all characteristic physical properties (Figure 2.10). For example, whether a

(a)

(b)

Figure 2.10 *Aluminum is a metal that is ductile (a) and malleable (b).*

A CLOSER LOOK

Properties of Sulfur

The intensive (characteristic) properties of sulfur include both physical and chemical properties. Each of the physical properties can be evaluated without changing the composition of the sulfur. Chemical properties are those exhibited when the substance undergoes a chemical change. Some characteristic physical and chemical properties of sulfur are listed here.

Physical Properties

It has a pale yellow color.
It is a brittle solid at room temperature.
It crumbles easily when crushed.
It does not conduct an electric current.
It does not dissolve in water.

Chemical Properties

It burns in oxygen to make a gas, sulfur dioxide, which has an irritating odor of burning matches.
It reacts with carbon to make carbon disulfide, a colorless liquid.
It reacts with iron to make iron sulfide, a solid.

Solid yellow sulfur burns with a blue flame to form sulfur dioxide, a gas.

Table 2.5 **Some Intensive Properties**

SOME PHYSICAL PROPERTIES			
Boiling point	Color	Odor	Electrical conductivity
Melting point	Taste	Hardness	Thermal (heat) conductivity
Shininess	Softness	Ductility	Viscosity (resistance to flow)
Volatility	Slipperiness	Malleability	Density (mass/volume ratio)

SOME CHEMICAL PROPERTIES		
Burns in air	Reacts with specific acids	Decomposes when heated
Explodes	Reacts with specific metals	Reacts with specific nonmetals
Tarnishes	Reacts with water	Is toxic

small pan of water is raised to its boiling point or a very large kettle of water is raised to its boiling point, the temperature at which the water boils is the same value, 100 °C or 212 °F. Similarly, the freezing point of water is 0 °C or 32 °F. These values are independent of quantity.

Characteristic properties that relate to how the substance changes in composition or how it interacts with other substances are called **chemical properties**. The following questions pertain to the chemical properties of a substance.

1. Does it burn in air?
2. Does it decompose (break up into simpler substance) when heated or left in the sun?
3. What happens when it is placed in an acid?
4. What other chemicals will it react with, and what substances are obtained from the reaction?

Chemical properties include the tendency to react with various substances, to tarnish, to corrode, to explode, or to act as a poison or carcinogen (cancer-causing agent).

Characteristic physical and chemical properties—those used to identify a substance—are also called **intensive properties**. Some intensive properties are listed

Table 2.6 **Characteristic Properties of Some Substances**

	Physical Properties				
Substance	*State**	*Melting Point*	*Color*	*Electrical Conductivity*	*Chemical Properties*
Sodium chloride (table salt)	Solid	801 °C	White	No as a solid, good when dissolved in water	Decomposed by electricity to sodium and chlorine
Sucrose (cane sugar)	Solid	185 °C	White	No for all conditions	Burns in oxygen to produce water and carbon dioxide
Ethyl alcohol	Liquid	−117 °C	Colorless	No	Flammable
Water	Liquid	0 °C	Colorless	No	Decomposed by electricity to hydrogen and oxygen
Helium	Gas	−272 °C	Colorless	No	Nonreactive
Aluminum	Solid	660 °C	Silvery	Good	Reacts with acids to produce hydrogen gas

*Physical state at room temperature.

in Table 2.5. See Table 2.6 for a list of characteristic properties of several common substances. **Extensive properties** of substances depend on the *quantity* of the sample, including measurements of mass, volume, and length. Intensive properties help identify or characterize a particular kind of matter; extensive properties relate to the amount present.

Measurements of mass, volume, and length will be discussed in the next chapter.

If a lump of candle wax is cut or broken into smaller pieces, or if it is melted (a change of state), the sample remaining is still candle wax. When cooled, the molten wax returns to a solid. In these examples, only a **physical change** has taken place; that is, the *composition* of the substance was not affected.

The burning of a candle (Figure 2.11) involves both physical and chemical changes. After the candle is lighted, the solid wax near the burning wick melts. This is a physical change; the composition of the wax does not change as it goes from solid to liquid. Some of the wax is drawn into the burning wick where a chemical change occurs. Here, wax in the candle flame reacts chemically with oxygen of the air to form carbon dioxide gas and water vapor. In any **chemical change**, one or more substances are used up while one or more new substances are formed. The new substances produced have their own unique physical and chemical properties.

Figure 2.11 *The melting of wax is a physical change. A chemical change occurs when the wax is burned. Candle wax burns in oxygen, producing carbon dioxide and water vapor.*

The apparent disappearance of something, like the candle wax, however, is not necessarily a sign that we are observing a chemical change. For example, when water evaporates from a glass and disappears, it has changed from a liquid to a gas (called water vapor), but in both forms it is water. This is a phase change (liquid to gas) and is a physical change. When attempting to determine whether a change is physical or chemical, one should ask the critical question: Has the fundamental *composition* of the substance been changed? In a chemical change (a reaction), it has, but in a physical change, it has not.

EXAMPLE 2.5

Classify each of the following as a physical property, chemical property, physical change, or chemical change.

(a) Alcohol is flammable.
(b) Alcohol is volatile; it evaporates readily.
(c) A sample of table salt dissolves in a glass of water.
(d) Over time, a flashlight battery loses its charge.

SOLUTION

(a) This is a chemical property; burning produces new substances.
(b) This is a physical property; the alcohol can change from a liquid to a gas.
(c) This is a physical change; the salt is still present even if the water is evaporated away.
(d) This is a chemical change; when electricity is generated, certain chemicals are used up as others are produced.

See Problems 2.27–2.32.

2.6 *The Law of Conservation of Mass*

During the burning of a candle, as was described in the previous section, mass is neither gained nor lost. The total mass of the wax and oxygen present before burning is equal to the total mass of carbon dioxide, water vapor, and unburned candle wax that remain after burning the candle.

Mass of wax + Mass of oxygen = Mass of carbon dioxide + Mass of water

Scientific laws are statements that summarize experimental facts regarding the behavior of matter for which no exceptions are known.

No change in mass occurs during the chemical reaction. Mass is *conserved.* This is what is meant by the **law of conservation of mass,** which states that mass is neither created nor destroyed during a chemical change.

The discovery of the law of conservation of mass was made in France by Antoine Lavoisier at about the same time as the American colonies were involved in the Revolutionary War. After carrying out many reactions in closed vessels—so no substances could get in or out—Lavoisier concluded that no change in total mass had occurred. This is the law of conservation of mass. It has been verified repeatedly, thus standing the test of time.

Because mass is conserved during reactions, matter must also be conserved; matter is neither created nor destroyed during a chemical reaction. In other words,

HISTORY MAKERS

Chemistry Becomes a *Quantitative* Science

Lavoisier was not the first to carry out this reaction, but he was the first to weigh all the substances present before and after the reaction. He was also the first to interpret the reaction correctly.

Antoine Lavoisier, a Frenchman (1743–1794), perhaps did more than anyone else to establish chemistry as a *quantitative* science. He convinced his contemporaries of the importance of accurate measurements in experimental investigations. One famous investigation that he performed was the heating of the red oxide of mercury. It decomposed and produced metallic mercury and a gas he named oxygen. There was no change in mass.

Lavoisier carried out many quantitative experiments. In one of his demonstrations, he decomposed water. He found that when coal was burned, it combined with oxygen to form carbon dioxide. Lavoisier was the first to understand the role of oxygen in combustion. Although Lavoisier did not discover oxygen, he named it. He also found that when a guinea pig breathed, oxygen was consumed and carbon dioxide was formed. Lavoisier then correctly concluded that respiration was related to combustion. In both cases oxygen is consumed while carbon dioxide and water vapor are formed.

In 1789 Lavoisier published his now famous *Elementary Treatise on Chemistry,* the first modern chemistry textbook. In that same year, the French Revolution broke out and the French government turned to Lavoisier to improve both the quantity and the quality of gunpowder.

Because of his work in establishing chemistry as a quantitative science, Lavoisier is often called the ''father of modern chemistry.''

Antoine Lavoisier lost his head (on the guillotine) during the French Revolution, but not because of his chemical research. In those days no one was a full-time chemist. Lavoisier, a member of the French nobility, was also a tax collector for Louis XVI. It was in that capacity that he incurred the wrath of French revolutionaries.

CHEMISTRY IN OUR WORLD

Getting Rid of Solid Waste

The law of conservation of mass applies to pollution and solid waste problems. The law states that matter is neither created nor destroyed during chemical reactions. No atoms are gained; none are lost. Matter does not go away. We must put the waste somewhere. But knowledge of chemistry does offer an alternative: we can change potentially harmful wastes to less dangerous substances. Indeed, such transformations of matter from one form to another are what chemistry is all about. Chemical transformations of this type do require energy. We must pay for the energy used in chemical processing; it is not free.

Matter is conserved; pollution problems don't just go away.

matter cannot be created from nothing; atoms cannot be created from nothing. Stated in still another way,

During chemical reactions, no atoms are gained; none are lost.

Thus, new materials can be made only by changing the way atoms are combined.

EXAMPLE 2.6

Considering the law of conservation of mass, explain how rusted iron—which is iron combined with oxygen—can weigh more than pure iron.

SOLUTION

The rusting of iron is much like the burning of candle wax that was described—the substance reacted with oxygen but there was no overall mass change. Similarly, as iron rusts, it combines with a specific mass of oxygen to produce a mass of iron oxide equal to the sum of the masses of iron and oxygen consumed in the process.

$$\text{Iron} + \text{Oxygen} \longrightarrow \text{Iron oxide (rust)}$$
$$\text{Mass before the reaction} = \text{Mass after the reaction}$$

The total masses of substances before and after the reaction must be equal.

See Problems 2.33–2.36.

EXERCISE 2.2

Ask a friend whether or not mass is actually lost when a log burns in a fireplace. Can you explain what happens?

2.7 Energy and Chemical Change

Energy can be defined as the capacity to do work or to transfer heat. **Work** is performed when a mass is moved through a distance. Common forms of energy include light, heat, electrical energy, mechanical energy, and chemical energy. Energy can be converted from one form to another form. For example, when you turn on a flashlight, chemical energy stored in batteries is converted into electrical energy and, finally, into light and some heat energy.

Each of the various forms of energy can be classified as either potential energy or kinetic energy. **Potential energy** is stored energy; it is the energy an object possesses because of its position or its chemical composition. Gasoline and table sugar possess potential energy due to their *chemical composition.* An automobile parked on a hill has potential energy due to its *position.*

Kinetic energy is energy of motion. As a parked automobile starts to roll down a hill, potential energy is transformed into kinetic energy. Mathematically, the kinetic energy (K.E.) of an object is equal to one-half its mass (m) times the square of its velocity (v).

$$\text{K.E.} = \tfrac{1}{2} mv^2$$

During *most* chemical reactions, the potential energy of the substances involved decreases. In other words, high-energy compounds are usually converted to low-energy compounds. When this occurs, energy is released to the surrounding, usually as heat. If a chemical reaction that releases energy is reversed (some can be), energy must be continually supplied to keep the reaction going.

$$\text{High-energy compounds} \underset{\text{energy is absorbed}}{\overset{\text{energy is released}}{\rightleftharpoons}} \text{Low-energy compounds} + \text{Energy}$$

In terms of potential energy, materials in a chemical reaction are somewhat like a car on a hill. Potential energy is released as the car rolls down a hill, but energy is required to get it up the hill. A chemical process that is ''downhill'' (in terms of energy) in one direction must be ''uphill'' in the other direction.

Potential energy, stored in sugar and other types of food, is released when the food is used by living cells in a process known as metabolism. The process is quite complex, but it can be summarized as follows: sugar combines with oxygen to produce carbon dioxide, water, and energy. Thus, some energy is released as the high-energy (less stable) molecules of sugar and oxygen are converted—through chemical reactions—into the low-energy (more stable) molecules of carbon dioxide and water. In this reaction, represented here, energy is released.

Melvin Calvin, of the University of California, Berkeley, received the Nobel prize in chemistry in 1961 for his work on chemical reactions that occur during photosynthesis in plants.

$$\text{Sugar} + \text{Oxygen} \xrightarrow{\text{metabolism}} \text{Carbon dioxide} + \text{Water} + \text{Energy}$$

A reaction that *releases* heat energy is called an **exothermic** reaction. When the term **exergonic** is used instead of exothermic, it indicates that energy in forms other than heat can also be released. When heat or other forms of energy are *taken in* or *absorbed* during reactions, the reactions are called **endothermic** and **endergonic** reactions, respectively.

CHEMISTRY IN OUR WORLD

Photosynthesis and Metabolism

Some of the solar energy striking the Earth's surface is absorbed by green plants. Using a complex series of chemical reactions, collectively called *photosynthesis*, plants take up solar energy and store it as chemical energy. During photosynthesis, the light energy is used to convert carbon dioxide and water into sugar (and starch). This is an *endergonic reaction.*

During metabolism, sugar combines with oxygen to produce carbon dioxide, water, and energy. During photosynthesis, the reverse reaction occurs. Metabolism and burning are exergonic processes. Photosynthesis is an endergonic process. The reversible processes can be represented as shown below.

The starting materials for metabolism are shown on the left; the starting materials for photosynthesis are shown on the right. The substances produced by one reaction are the starting materials for the other reaction. One reaction—metabolism—releases energy to the surroundings, while the reverse reaction—photosynthesis—requires energy in order to proceed.

Metabolism in humans and photosynthesis in plants involve reverse processes.

$$\text{Sugar} + \text{Oxygen} \underset{\text{photosynthesis}}{\overset{\text{metabolism or burning}}{\rightleftharpoons}} \text{Carbon dioxide} + \text{Water} + \text{Energy}$$

EXAMPLE 2.7

Does each process listed here represent a chemical change or a physical change, and is there an increase or decrease in the potential energy of the materials involved?

(a) A glass bowl falls to the floor.
(b) A bicycle is pushed to the top of a hill.
(c) Hydrogen and oxygen gases explode with a bang when ignited, producing water.
(d) An electric current decomposes water into hydrogen and oxygen.

SOLUTION

(a) A physical change. The potential energy of the bowl decreases. As the bowl falls, some of the potential energy is converted into kinetic energy.
(b) A physical change. The potential energy of the bicycle increases as it is pushed up the hill to greater heights.
(c) A chemical change. Potential energy of the chemicals decreases. Noise and other forms of energy are released as the chemicals react.

(d) A chemical change Potential energy of the chemicals increases. This is the reverse reaction for the previous example. Here, energy must be supplied.

See Problems 2.37–2.40.

EXERCISE 2.3
When butane gas burns, is energy released or absorbed? Is the reaction exothermic or endothermic? Explain.

2.8 *The Law of Conservation of Energy*

Any time a chemical reaction takes place, there is also a change in energy. Either energy is released by the reaction, or else it is required continually to keep the reaction going. As described for the reverse processes of metabolism and photosynthesis, if the forward reaction releases energy, the reverse reaction must absorb energy. There is an explanation for this phenomenon. Energy is either *released* or *absorbed* during a chemical reaction, but

Energy is *neither created nor destroyed* during chemical processes.

This is known as the **law of conservation of energy**; it is one way of stating what is also known as the **first law of thermodynamics**.

If you take more courses in chemistry or physics, you are certain to encounter more discussion of the first law of thermodynamics.

For an explanation of how energy can be gained or lost without being created or destroyed, let's say you have a certain amount of money. It might be available as cash, or "stored" in a checking account, but the total amount has not changed. The money in the bank, the "stored" money, is somewhat like "stored" potential energy. You cannot take one hundred dollars from your account (for spending) without having the account decrease in value by one hundred dollars, but none is lost. If energy is released from the stored condition and made available for use, then less energy remains stored, but none is lost.

If part of the stored potential energy of chemicals is converted into available heat energy, then the potential energy of the remaining substances must be less than in the beginning. Energy is neither created nor destroyed, but is, instead, transferred from stored, potential energy, into available heat energy or work.

EXAMPLE 2.8
For each item given in Example 2.7, explain how energy is conserved.

SOLUTION

(a) As the bowl falls to the floor, the potential energy of the bowl decreases—it went from a higher position to a lower position. The energy from the fall went into the breaking of the bowl. Total energy is unchanged.
(b) As the bicycle is pushed up the hill, the person pushing the bicycle expends energy (and gets tired) while the bicycle gains potential energy. Energy is gained by the bicycle; energy is lost by the person pushing it.
(c) During the explosion, the energy released by the chemicals is transferred into sound and motion as gases escape. Total energy is unchanged.
(d) The electrical energy is transferred to the chemicals and stored as chemical (potential) energy in the gases. The stored energy can later be released, as described in part (c). No energy is lost.

See Problems 2.41–2.44.

2.9 The Conversion of Matter to Energy

Figure 2.12 *Albert Einstein (1879–1955) was born in Germany. He was not an outstanding student in school, but by the age of 25 he was regarded as an outstanding physicist. He received the Nobel prize in physics in 1921. In 1933, when Jews were being persecuted in Germany, he fled to the United States and worked with a number of other noted scientists at Princeton University until his death.*

One of the most noted advances of this century was worked out with a pencil and notepad. These are not the usual tools associated with major discoveries in the sciences. Albert Einstein may well be the best known scientist of all time, yet his achievements were those of the mind, not the laboratory (Figure 2.12).

By 1905, Einstein had worked out his theory of relativity. In doing this, he derived a relationship between matter and energy. The now famous **Einstein equation** is often written

$$E = mc^2 \qquad \text{or, more precisely,} \qquad \Delta E = \Delta mc^2$$

where ΔE represents a change in energy, Δm represents a change in mass, and c is the speed of light. According to Einstein's equation, a definite mass is always converted into a definite amount of energy. The equation takes on greater significance as it is understood that 1.0 gram of matter—if totally converted to energy—contains enough energy to heat a house for about a thousand years.

Einstein's reasoning was not verified until 40 years later. That verification shook the world.

The conservation of mass was described in Section 2.6, and the conservation of energy was described in Section 2.8. Now, as a result of Einstein's work, it is clear that matter and energy should be treated together in a combined law of conservation of matter and energy. In simple terms, we can say that the sum total of matter and energy in the universe is constant.

For chemical reactions, the energy change is extremely small when compared with the energy change for nuclear reactions. Chemical reactions are quite different from nuclear reactions. When the relative energy change is quite small, as it is for chemical reactions, any change in mass is far too small to be detected. For all practical purposes, there is conservation of mass and conservation of energy during chemical reactions; there is no measurable conversion between the two.

In a nuclear explosion, less than 1% of the available matter is actually converted to energy, but even then, the energy liberated is tremendous.

Chapter Summary

Chemistry is a science that deals with matter and its changes. Matter is anything that takes up space and has mass. Mass depends upon the quantity of matter present, but weight depends upon the gravitational force. Matter can exist in any one of three physical states—solid, liquid, or gas—depending on the temperature. Properties of each state were summarized in Table 2.1.

Pure substances are either elements, having one kind of atom, or compounds, with two or more kinds of atoms bonded chemically. A heterogeneous mixture is not uniform throughout. All solutions are homogeneous mixtures, with uniform composition. Most alloys are solid solutions.

Characteristic properties or intensive properties of substances can include both chemical properties and physical properties. Examples of these were provided. Mass is neither created nor destroyed during physical or chemical changes, according to the law of conservation of mass.

Energy is the capacity to do work. An object in motion has kinetic energy.

Potential energy (stored energy) is due to an object's position or chemical composition. Energy is released during exothermic or exergonic processes, but energy must be continuously supplied for an endothermic or endergonic reaction. There are many forms of energy, including heat, light, sound, electrical, chemical, and mechanical energy. Energy can be converted from one form to another, but energy is neither created nor destroyed. This fact is summarized by the law of conservation of energy and the first law of thermodynamics.

Matter and energy are related. The Einstein equation describes the mathematical relationship between matter and energy. In nuclear reactions, extremely small quantities of matter are converted to energy, but the total of mass and energy in the universe remains constant.

Assess Your Understanding

1. Distinguish between mass and weight. [2.1]
2. Describe three or more macroscopic properties and a submicroscopic property for each of the three states of matter. [2.2]
3. Distinguish between elements and compounds. [2.3]
4. Distinguish between pure substances and mixtures. [2.4]
5. Classify a specific mixture as being homogeneous or heterogeneous. [2.4]
6. Distinguish between physical and chemical properties of substances. [2.5]
7. Give two examples to illustrate the law of conservation of mass. [2.6]
8. Give two examples that show an increase in potential energy and two examples that show a decrease in potential energy. [2.7]
9. Give two examples to illustrate the law of conservation of energy. [2.8]
10. Describe the implications of the relation between mass and energy. [2.9]

Key Terms

atom [2.3]
characteristic properties [2.5]
chemical change [2.5]
chemical properties [2.5]
chemistry [2.1]
compound [2.3]
condense [2.2]
diffuse [2.2]
Einstein equation [2.9]
element [2.3]
endergonic reaction [2.7]
endothermic reaction [2.7]
energy [2.7]
exergonic reaction [2.7]
exothermic reaction [2.7]
extensive properties [2.5]
first law of thermodynamics [2.8]
freeze [2.2]
gas [2.2]
heterogeneous mixture [2.4]
homogeneous mixture [2.4]
ice [2.2]
immiscible [2.2]
intensive properties [2.5]
kinetic energy [2.7]
law of conservation of energy [2.8]
law of conservation of mass [2.6]
law of definite proportions [2.3]
liquid [2.2]
mass [2.1]
matter [2.1]
miscible [2.2]
physical change [2.5]
physical properties [2.5]
physical states [2.2]
potential energy [2.7]
pure substance [2.3]

scientific laws [2.6]
solid [2.2]
solution [2.4]
states of matter [2.2]
steam [2.2]
viscosity [2.2]
water vapor [2.2]
weight [2.1]
work [2.7]

Problems

MATTER

2.1 Describe why any specific rock has more weight on Earth than it does on the moon. How does the mass of the rock compare in these two environments?

2.2 How does the mass of your body compare on Earth, in space, and on the moon? Mass and weight are sometimes used interchangeably. Why is this incorrect?

2.3 Which of the following contain matter, and which do not. Why?
a. air b. heat
c. paint d. sunshine

2.4 Which of the following contain matter, and which do not. Why?
a. light b. electricity
c. beef steak d. chocolate

STATES OF MATTER

2.5 Vegetable oil and water are (a) miscible (b) immiscible.

2.6 Vinegar and water are (a) miscible (b) immiscible.

2.7 Which state of matter involves particles packed tightly, sometimes in crystal lattices?

2.8 Which state of matter involves particles that touch but can slide or flow over one another?

2.9 Use Table 2.6 as you determine the physical state of
a. ethyl alcohol at −115 °C.
b. sodium chloride at 803 °C.

2.10 Use Table 2.6 as you determine the physical state of
a. aluminum at 642 °C.
b. helium at −270 °C.

2.11 Compare properties for the elements hydrogen and oxygen with the compound water, which is composed of these same elements. (See Table 2.3.)

2.12 Compare properties for the elements carbon, hydrogen, and oxygen, with sugar, which is a compound containing these same elements. (See Table 2.3.)

2.13 Compare properties of table salt with the elements that combine to make table salt.

2.14 Compare properties of ammonia with the elements that combine to make this compound.

2.15 Name the smallest unit of an element that has the properties of that element.

2.16 What is the name for a pure substance that contains two or more elements?

2.17 Two or more elements combine chemically to make (choose one answer).
a. compounds b. new elements

2.18 Which of the following are substances?
a. elements b. compounds
c. water d. light

PURE SUBSTANCES AND MIXTURES

2.19 Is ethyl alcohol an element, a compound, or a mixture? (See Table 2.3.)

2.20 Is hydrogen sulfide an element, a compound, or a mixture? (See Table 2.3.)

2.21 Is bronze heterogeneous or homogeneous? Why?

2.22 Is high-carbon steel heterogeneous or homogeneous? Why?

2.23 Mouthwash is (choose one answer)
a. an element
b. a compound
c. a heterogeneous mixture
d. a homogeneous mixture

2.24 Sterling silver is (choose one answer). (See Table 2.4.)
a. an element
b. a compound
c. a heterogeneous mixture
d. a homogeneous mixture

2.25 Iodine is (choose one answer)
a. an element
b. a compound
c. a heterogeneous mixture
d. a homogeneous mixture

2.26 A bowl of cereal with sugar and milk is (choose one answer)
a. an element
b. a compound
c. a heterogeneous mixture
d. a homogeneous mixture

CHARACTERISTIC PROPERTIES

2.27 Which of the following are physical properties, and which are chemical properties, of copper?
   a. It melts at 1284 °C.
   b. Its density is 8.96 g/cm^3.
   c. It is a good conductor of heat.
   d. It turns greenish when exposed to chlorine.
   e. It is malleable.

2.28 Which of the following are physical properties, and which are chemical properties, of vinegar?
   a. It is a colorless liquid.
   b. It tastes sour.
   c. It has a strong aroma.
   d. It reacts with lime deposits on faucets.
   e. It produces carbon dioxide gas when mixed with baking soda.

2.29 Classify each of the following as a chemical or a physical change.
   a. lighting a butane lighter
   b. expansion of water as it freezes
   c. evaporation of alcohol
   d. rusting of an iron nail

2.30 Classify each of the following as a chemical or a physical change.
   a. tarnishing of silver
   b. sharpening a pencil
   c. digestion of a candy bar
   d. melting of solder

2.31 Classify each of the following as a chemical or physical property.
   a. Charcoal lighter fluid is flammable.
   b. Alcohol evaporates quickly.
   c. Silver is a good conductor of both heat and electricity.
   d. Magnesium metal reacts with sulfuric acid producing hydrogen gas.

2.32 Classify each of the following as a chemical or physical property.
   a. Charcoal lighter fluid evaporates readily.
   b. Alcohol can be burned in an engine.
   c. Aluminum metal reacts with acid.
   d. Salt dissolves in water.

THE LAW OF CONSERVATION OF MASS

2.33 Explain how iron rust can have a mass greater than pure iron.

2.34 Sulfur burns in oxygen to produce sulfur dioxide. Explain how the mass of sulfur dioxide produced will compare to the original sulfur.

2.35 When limestone is heated in a kiln, the material left after the heating has less mass than the original limestone. Explain what has happened.

2.36 A full tank of gasoline in your automobile is completely used up. What happened to all this mass? Is the law of conservation of mass violated?

ENERGY AND CHEMICAL CHANGE

2.37 Which has greater potential energy, a diver on the 1-m board or the same diver on the 10-m platform?

2.38 Which has greater potential energy, a roller coaster as it starts to climb the first hill or as it reaches the top of the same hill?

2.39 Which of the following changes are exothermic, and which are endothermic?
   a. a firecracker when lit
   b. a burning candle
   c. melting the wax around the wick of a candle
   d. a plant making sugar by photosynthesis
   e. metabolism of the sugar in candy

2.40 Which of the following changes are exothermic, and which are endothermic?
   a. digesting food
   b. melting ice
   c. an electric current decomposing sodium chloride (table salt)
   d. burning natural gas
   e. explosion of an ordinary (not nuclear) bomb

THE LAW OF CONVERSION OF ENERGY

2.41 What is a scientific law?

2.42 Give a concise statement that summarizes both the law of conservation of energy and the first law of thermodynamics.

2.43 Explain how the following statement can be true. Energy is released by burning wood, but energy is neither created nor destroyed.

2.44 Explain how the following statement can be true. Energy is absorbed during evaporation of water, but energy is neither created nor destroyed.

THE CONVERSION OF MATTER TO ENERGY

2.45 Although the Einstein equation shows that changes in energy and mass are related, why does mass not appear to be lost during a chemical reaction where energy is released?

2.46 Rewrite the Einstein equation solved for m and describe how the magnitude of m relates to energy. (Recall that c represents the speed of light.)

Additional Problems

2.47 Does burning a piece of wood involve a chemical change or a physical change? Is energy created? Is energy destroyed? Explain.

2.48 A piece of chalk fell to the floor and broke into several pieces. Explain what happened in terms of potential energy, kinetic energy, and total energy.

2.49 During chemical reactions, which is *not* true?
a. For some reactions, energy is released.
b. For some reactions, energy is absorbed.
c. Energy is sometimes created, and sometimes destroyed.
d. Potential energy is sometimes converted into heat energy.

2.50 Classify each of the following as either a chemical change or a physical change.
a. A can of soda releases gas when opened.
b. Photographic film exposed to light shows an image when developed.
c. a fireworks aerial display
d. removing a stain from fabric with a bleach

2.51 Classify the following as heterogeneous or homogeneous.
a. a pure sample of gold
b. a fruit cake
c. glass
d. a concrete floor

2.52 Classify the following as heterogeneous or homogeneous.
a. table sugar
b. apple pie
c. gasoline
d. a cheeseburger

2.53 Which has more energy, ice or liquid water? Explain.

2.54 Which has more energy, liquid water or steam? Explain.

2.55 When you step on the brake pedal of a car, what happens to the kinetic energy of motion of the car? How does this relate to the law of conservation of energy?

2.56 Compare the energy changes in the freezing of water versus the melting of ice. Is energy created or destroyed in either of these two changes? Explain.

If you measure length in meters and volume in liters,
At what temperature in degrees will water freeze?

3

Fundamental Measurements

CONTENTS

Chemists, engineers, and persons in every science-related field, including the medical profession, must make decisions based on scientific data. This means making accurate measurements of length, volume, mass, and temperature.

Understanding the details of recording and working with these measurements is fundamental to success in all science-related fields. A measured value has three component parts—the **numerical quantity**, the **unit**, and the **name of the substance**—which should be included any time data is recorded. For example, consider the following quantity.

125 mg vitamin C

- numerical quantity (125)
- unit (mg)
- name of substance measured (vitamin C)

Any time one of these three parts of a measured quantity is missing or is in error, accurate calculations and interpretations of results are in jeopardy. Errors in medical analysis, engineering, industrial operations, scientific research, and space exploration can often be traced to errors in measurement or the interpretation of those measurements. Examples include instances such as the patient who died when 17.5 grams (g) of medication was administered instead of the prescribed 17.5 milligrams (mg) (Figure 3.1).

Figure 3.1 *For 17.5 milligrams of medication, just how many grams of the medication should be administered?*

Only the United States and three other countries, Myanmar (formerly Burma), Brunei, and South Yemen, have not fully adopted metric or SI units.

Many systems of measurement have been used around the world. The familiar English system of feet, quarts, and pounds is slowly but steadily being phased out in the United States, one of the last nations to do so. A February 1994 U.S. Federal Trade Commission regulation requires all consumer packaging to include metric measurements.

3.1 *Metric and SI Units*

Scientists around the world have long used the **metric system**. Now, an updated metric plan, called the **International System** or **SI** (from the French *Système International*), is being adopted worldwide. In this book we shall generally refer to metric or SI values interchangeably.

The metric system was first adopted in France in the 1790s and has grown in worldwide use by nearly all countries since that time. In 1866 the U.S. Congress endorsed the system, and in 1975 Congress passed the Metric Conversion Act, which set up a U.S. Metric Board to report on the progress of voluntary changeover to the system. That changeover has been slow, but should now be obvious to any consumer (Figure 3.2).

Many consumer products show both metric and English units (Figure 3.3). Some companies are now packaging products in standard metric sizes. For example, soft drinks are bottled in 500-milliliter (mL), 1.0-liter (L), and 2.0-L sizes. Alcoholic beverages are bottled in 750-mL, 1.0-L, and 1.5-L sizes.

The metric system (SI) is based on the decimal system, and that is its beauty. Unlike fractions, decimal quantities can be quickly added or subtracted in the same way money is added or subtracted in your checkbook. Converting between large and small SI units requires multiplying or dividing by factors of 10, 100, and 1000. This is simpler than multiplying or dividing by a fractional value, as is often the case for conversion within the English system. For example, in the English system, 1 barrel (bbl) is equivalent to 4.08 cubic feet (ft^3) of dry measure or 31.5 gallons (gal) for most liquids, but for petroleum products, 1 barrel is 42.0 gal. Also, an imperial quart equals 1.2009 U.S. quarts in liquid measure or 1.0320 U.S. quarts in dry measure. There are numerous other examples of unnecessary confusion caused by nonuniform units. The metric system helps eliminate this confusion.

In SI, the base unit of length is the **metre** (spelled **meter** in the United States). The meter (m) is about equal to one long step, a distance slightly longer than a yard. The SI base unit of mass is defined as the **kilogram** (kg), a quantity slightly

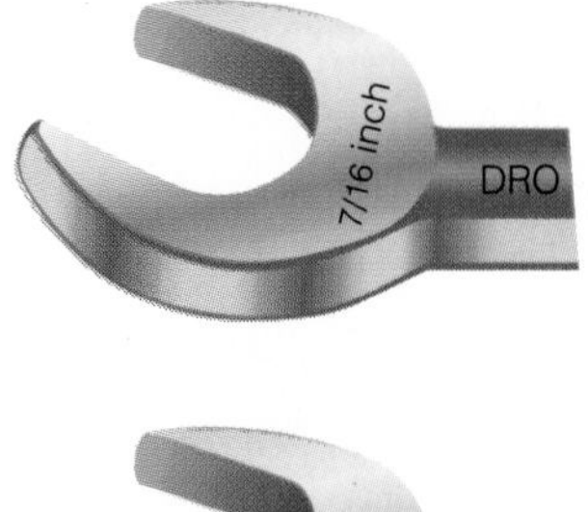

Figure 3.2 *If you want a wrench that is the next size smaller than an 11-mm wrench, it's easy—use a 10-mm wrench. SI units are easier to work with than English units. If you have a $\frac{7}{16}$-inch wrench, the next one smaller is marked $\frac{3}{8}$-inch, not $\frac{6}{16}$-inch (assuming the wrenches are in $\frac{1}{16}$-inch intervals).*

Figure 3.3 *Metric units are now used for all types of products for measurements of volume, length, and mass.*

Table 3.1 Some Fundamental SI Units

Quantity	*Name of SI Unit*	*Symbol*
Length	Meter	m
Mass	Kilogram	kg
Temperature	Kelvin	K
Time	Second	s
Amount of substance	Mole	mol

greater than 2 pounds. Table 3.1 lists some important SI base units. Other units of measurement can be derived from these base units.

Study Hint
You should learn the common prefixes identified in Table 3.2.

To express quantities that are larger or smaller than the base units, prefixes are used. Table 3.2 shows a list of prefixes with their decimal and exponential equivalents along with phonetic spelling and symbols. As shown in the table, a prefix changes the size of a unit by multiples of 10. For example, the prefix *milli-* means 1/1000 or 0.001 times the base unit. Thus, one miligram (mg) equals 1/1000 gram or 0.001 gram. As a mathematical equation, we can write

$$1 \text{ mg} = 0.001 \text{ g}$$

or, multiplying by 1000,

$$1000 \text{ mg} = 1 \text{ g}$$

The use of prefixes is illustrated by the following examples.

Table 3.2 Metric Prefixes and Equivalents*

Prefix	*Phonetic*	*Symbol*	*Decimal Equivalent*	*Exponential Equivalent*
Tera-	TER-uh	T	1,000,000,000,000	10^{12}
Giga-	GIG-uh	G	1,000,000,000	10^{9}
Mega-	MEG-uh	M	1,000,000	10^{6}
Kilo-	KILL-uh	k	1,000	10^{3}
Hecto-	HEK-tuh	h	100	10^{2}
Deka-	DEK-uh	da	10	10
Deci	DES-uh	d	0.1	10^{-1}
Centi-	SENT-uh	c	0.01	10^{-2}
Milli-	MILL-uh	m	0.001	10^{-3}
Micro-	MY-crow	μ†	0.000 001	10^{-6}
Nano-	NAN-uh	n	0.000 000 001	10^{-9}
Pico-	PEA-ko	p	0.000 000 000 001	10^{-12}
Femto-	FEM-toe	f	0.000 000 000 000 001	10^{-15}
Atto-	AT-toe	a	0.000 000 000 000 000 001	10^{-18}

*The most commonly used units are shown in blue.

†The micro prefix is sometimes symbolized as mc, so a 2-microgram sample could be written as 2 mcg.

EXAMPLE 3.1

Use Table 3.2 to assist you in answering the following.

(a) Kilo- is equivalent to the number ______, so 1.000 kg = ______ g.
(b) Centi- has a decimal equivalent of ______, so 1.000 cm = ______ m.
(c) Mega- is equivalent to ______, so 1.6 MW (megawatts) = ______ W.

SOLUTION

(a) Kilo- means 1000, so 1.000 kg = 1000 g, since 1000 can be substituted for k.
(b) Centi- means 0.01, so 1.000 cm = 0.01 m, since 0.01 can be substituted for c.
(c) Mega- means 1 million, so 1.6 MW = 1.6 million watts.

See Problems 3.1–3.4.

3.2 Metric Length Measurement and Approximations

The SI base unit of length is the meter. Originally, the meter was defined as one ten-millionth of the distance from the North Pole to the equator. It was redefined in 1875 as the distance between two lines on a certain platinum-iridium bar (resistant to corosion). Now, a meter is defined more precisely for technological reasons as 1,650,763.73 times a certain wavelength of orange-red light emitted by the element krypton under specified conditions. Table 3.3 lists common metric units of length. You need to learn to write lengths using the units shown in the left-hand column and also in meters shown in the middle column. The exponential equivalent values shown in the table will be discussed later in this chapter.

Table 3.3 **Common Metric Units of Length**

Unit	*Abbreviation*	*Meter Equivalent*	*Exponential Equivalent*
Kilometer	km	1000 m	10^3 m
Meter	m	1 m	10^0 m
Decimeter	dm	0.1 m	10^{-1} m
Centimeter	cm	0.01 m	10^{-2} m
Millimeter	mm	0.001 m	10^{-3} m
Micrometer	μm	0.000001 m	10^{-6} m
Nanometer	nm	0.000000001 m	10^{-9} m

Besides knowing the meaning of the various metric lengths and prefixes, it is also important to be able to make approximations of metric lengths. Figure 3.4 provides a guide for making those approximations.

As you work with metric units, learn to make an approximation (an educated guess) about the size of an object

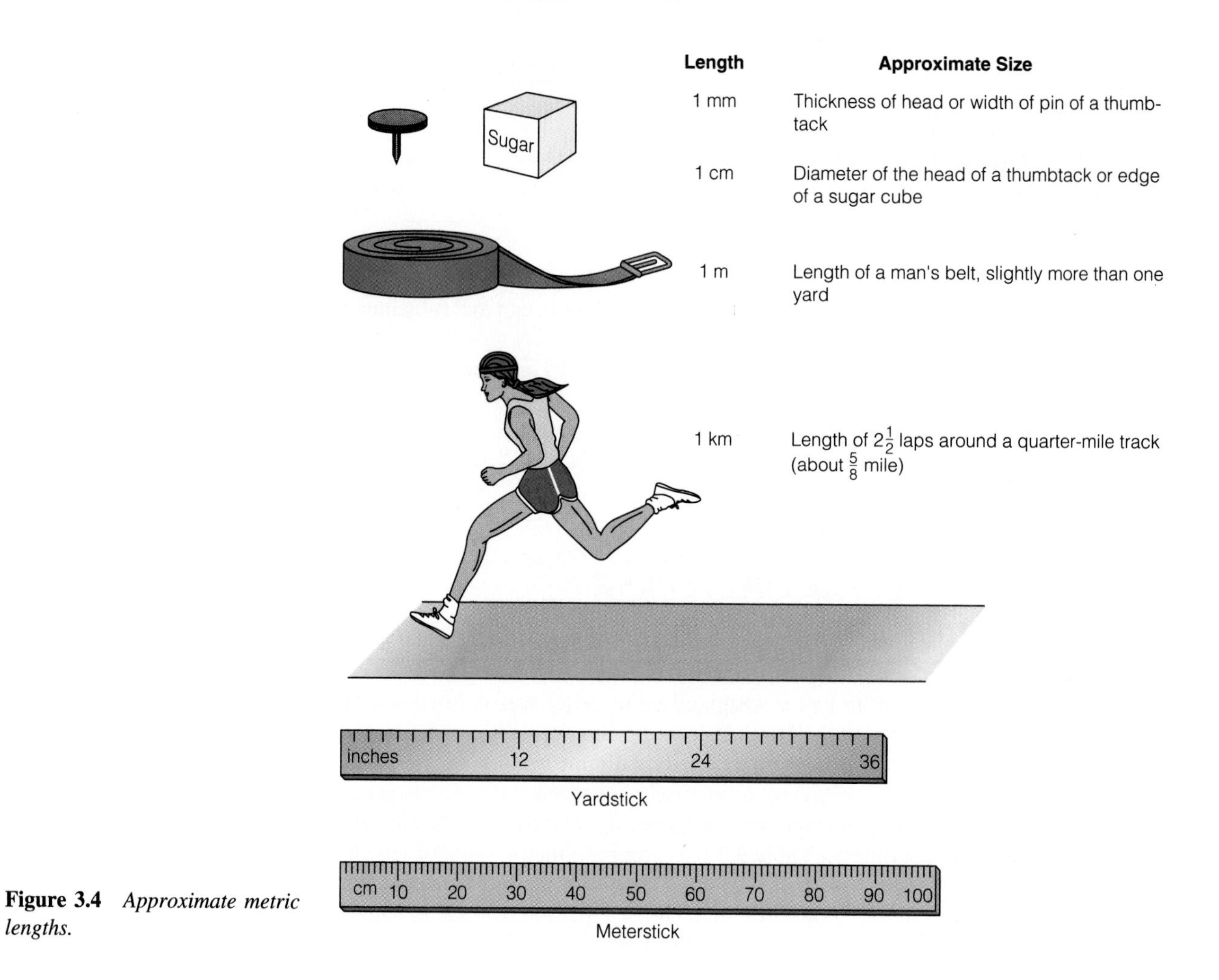

Figure 3.4 *Approximate metric lengths.*

EXAMPLE 3.2

From the approximations provided in Figure 3.4, select the best answer for each of the following.

(a) A penny coin has a diameter of about
(1) 2 mm (2) 0.2 cm (3) 2 cm (4) 20 dm

(b) The distance across a paper clip is about
(1) 80 mm (2) 8 mm (3) 8 cm (4) 0.08 cm

(c) An adult's height may be about
(1) 1.7 km (2) 1.7 mm (3) 1.7 cm (4) 1.7 m

(d) The thickness of a blunt pencil mark is about
(1) 1 mm (2) 0.01 cm (3) 10 mm (4) 0.01 m

SOLUTION

(a) A penny is about 2 cm, answer (3).
(b) The distance across a paperclip is about 8 mm, answer (2).
(c) A height of 5 feet (ft) 7 inches (in.) is about 1.7 m, answer (4).
(d) A pencil mark is about 1 mm, answer (1).

See Problems 3.5–3.6.

Besides making approximations with metric values, it is essential that you learn to convert any metric quantity into any other equivalent metric quantity, such as centimeters into millimeters or millimeters into centimeters or meters. The best way to learn to do this is to use **conversion factors**, as described in the next section. You may be tempted to take a shortcut, but until you are very good at this, shortcuts can lead to incorrect answers.

Shortcuts provide easy routes to wrong answers!

3.3 *Conversion Factors and Dimensional Analysis*

A widely used approach to problem solving is called **dimensional analysis** or the **factor-label method**. This approach involves the multiplication of the given or known quantity (and its units!) by one or more *conversion factors* to obtain the answer in the desired *units.*

Known quantity and unit(s) × Conversion factor(s) = Answer in desired units

Any mathematical equality can be written as a conversion factor. For an example, let us use a familiar equality.

$$1 \text{ ft} = 12 \text{ in.} \qquad (1)$$

We can divide both sides by 12 in. to give

$$\frac{1 \text{ ft}}{12 \text{ in.}} = 1 \qquad (2)$$

Any quantity divided by an equivalent quantity equals the number 1.

or we can invert (turn upside down) the fraction to give the reciprocal

$$\frac{12 \text{ in.}}{1 \text{ ft}} = 1 \qquad (3)$$

thus making two fractions that are equivalent and equal to the number 1. From equations (2) and (3) we can obtain the following two conversion factors.

$$\frac{1 \text{ ft}}{12 \text{ in.}} \quad (4) \qquad \text{and} \qquad \frac{12 \text{ in.}}{1 \text{ ft}} \quad (5)$$

If we wish to convert feet to inches, we can do so by choosing one of the fractions as a conversion factor. Which one do we choose? Choose the conversion factor that allows cancellation of the unwanted unit! When a distance in *feet* is to be converted to inches, conversion factor (5) should be used, so the same unit, *feet,* will be present in both the numerator and the denominator.

$$\text{Distance in feet} \times \frac{12 \text{ in.}}{1 \text{ ft}} = \text{Answer in inches}$$

For example, we can convert 12 ft to inches, as shown here.

$$12\ \cancel{\text{ft}} \times \frac{12\ \text{in.}}{1\ \cancel{\text{ft}}} = 144\ \text{in.}$$

Similarly, when a known distance in *inches* is to be converted to feet, conversion factor (4) should be used, so the answer will be in feet, as desired.

EXAMPLE 3.3

126 in. = _______ ft.

SOLUTION

Start with the known quantity, 126 in., and follow a plan using conversion factors to obtain feet. This can be done in one step.

Plan: Inches ⟶ feet

$$126\ \cancel{\text{in.}} \times \frac{1\ \text{ft}}{12\ \cancel{\text{in.}}} = 10.5\ \text{ft}$$

Notice: Some conversion factors can be obtained from information given in problems or in tables, but you should know the common conversion factors.

The dimensional analysis approach is even more useful when several conversion factors are needed to solve a problem, as shown in Example 3.4.

EXAMPLE 3.4

If your heart beats at a rate of 72 times per minute, how many times will your heart beat each year?

SOLUTION

Start with the known quantity that must be converted, and develop a planned series of conversions that leads to the desired units.

Plan: Beats/min × Series of conversion factors ⟶ Beats/yr

The other conversions that you will need should be recalled from memory.

60 min/hr
24 hr/day
365 days/yr

Since beats/min is given and we want beats/yr, the key to solving the problem is converting minutes to year. The plan becomes

Beats/min ⟶ Beats/hr ⟶ Beats/day ⟶ Beats/yr

Start with the known quantity, 72 beats/min, and use conversion factors that permit cancellation of unwanted units to obtain the desired answer in beats per year.

$$\frac{72\ \text{beats}}{1\ \cancel{\text{min}}} \times \frac{60\ \cancel{\text{min}}}{1\ \cancel{\text{hr}}} \times \frac{24\ \cancel{\text{hr}}}{1\ \cancel{\text{day}}} \times \frac{365\ \cancel{\text{days}}}{1\ \text{yr}} = \frac{37{,}843{,}200\ \text{beats}}{\text{yr}}$$

Exercise 3.1
4.5 days = ______ min (Use two conversion factors.)

Now, let us use the same approach with metric units. If you want to convert a length in centimeters to meters, first develop a plan of attack

$$\text{Centimeters} \longrightarrow \text{Meters}$$

and look for the appropriate metric equality that can be used to develop a conversion factor. For example, 1 m = 100 cm (exactly), so we can write two conversion factors.

$$\frac{1 \text{ m}}{100 \text{ cm}} \quad \text{and} \quad \frac{100 \text{ cm}}{1 \text{ m}}$$

See Table 3.4 for more examples of common conversion factors for metric lengths. Refer to this table as you work through the following examples. Then, do the related problems at the end of the chapter.

Table 3.4 **Some Conversion Factors for Metric Lengths**

Given: 1 m = 100 cm
Conversion factors:

$$\frac{1 \text{ m}}{100 \text{ cm}} \quad \text{or} \quad \frac{100 \text{ cm}}{1 \text{ m}}$$

Given: 1 m = 1000 mm
Conversion factors:

$$\frac{1 \text{ m}}{1000 \text{ mm}} \quad \text{or} \quad \frac{1000 \text{ mm}}{1 \text{ m}}$$

From above: 100 cm = 1000 mm
so 1 cm = 10 mm
Conversion factors:

$$\frac{1 \text{ cm}}{10 \text{ mm}} \quad \text{or} \quad \frac{10 \text{ mm}}{1 \text{ cm}}$$

EXAMPLE 3.5
A small bolt is 2.3 cm in length. What is its length in millimeters?

SOLUTION
Write down the quantity given.

$$2.3 \text{ cm}$$

Develop a plan using conversion factors to obtain the desired units.

$$\text{Plan: cm} \longrightarrow \text{mm}$$

Use the appropriate conversion factor(s) to eliminate centimeters and give millimeters.

$$\frac{1 \text{ cm}}{10 \text{ mm}} \quad \text{or} \quad \frac{10 \text{ mm}}{1 \text{ cm}}$$

Multiply the original quantity by the appropriate conversion factor to eliminate centimeters, the unit to be converted, and give the desired answer in millimeters.

$$2.3 \text{ cm} \times \frac{10 \text{ mm}}{1 \text{ cm}} = 23 \text{ mm}$$

EXAMPLE 3.6
A piece of tubing has an inside diameter of 7.0 mm. What is this diameter in centimeters?

SOLUTION
Write down the quantity given.

$$7.0 \text{ mm}$$

Develop a plan using conversion factors to obtain the desired units.

$$\text{Plan: mm} \longrightarrow \text{cm}$$

Select the appropriate conversion factor to eliminate (cancel) millimeters and give centimeters.

See Problems 3.7–3.10.

$$7.0\ \text{mm} \times \frac{1\ \text{cm}}{10\ \text{mm}} = 0.70\ \text{cm}$$

EXERCISE 3.2

0.00451 km = _______ cm

3.4 *Metric Volume Measurement and Conversions*

The volume of a box is obtained by multiplying the length *(l)* times the width *(w)* times the height *(h)* of the box.

$$\text{Volume of a rectangular solid} = l \times w \times h$$

If the box is a cube that is 10 cm on each edge (see Figure 3.5), the volume is 1000 cm^3, or 1 dm^3.

Notice that the *units* are also cubed: cm × cm × cm = cm^3 or dm × dm × dm = dm^3.

$$\text{Volume} = 10\ \text{cm} \times 10\ \text{cm} \times 10\ \text{cm} = 1000\ \text{cm}^3$$
$$1\ \text{dm} \times 1\ \text{dm} \times 1\ \text{dm} = 1\ \text{dm}^3$$

The volume 1000 cm^3 is read as 1000 cubic centimeters. It is sometimes abbreviated 1000 cc in medical applications. Because volume units are derived from linear measurements, they are said to be *derived* units.

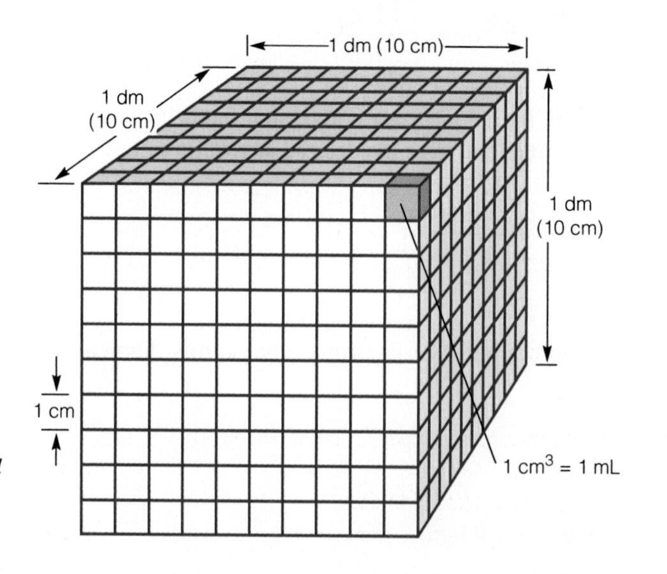

Figure 3.5 *One cubic decimeter. One liter has a volume equal to 1000 cm^3; 1.00 mL equals 1.00 cm^3.*

Volumes of solids or liquids can be measured in cubic centimeters, or cubic meters. A convenient volume unit for liquids is the **liter** (pronounded LEE-ter), which is identical in volume to 1000 cm^3. The liter is slightly larger than a quart. Two-liter plastic soft drink bottles—slightly larger than 2 qt—are common metric containers. To avoid confusion between the number 1 and a lowercase letter l, a capital L should be used as the symbol for liter.

The volume unit equal to one thousandth of a liter is the **milliliter**. As shown in Figure 3.5, a milliliter is equivalent in volume to a cubic centimeter. It is a small quantity that is about the size of a sugar cube or 15 to 20 drops of water. The symbol for milliliter is mL. Notice the capital L. Small volumes are generally measured in milliliters, while large volumes are generally measured in liters, but the convenience lies in the fact that one can convert a number of milliliters into liters simply by dividing the quantity in milliliters by 1000. This is much quicker than converting teaspoons into quarts! One **microliter** (μL) is much smaller; it is one millionth of a liter.

The following are fundamental equivalent metric volumes.

$$1 \text{ L} = 1000 \text{ mL} = 1000 \text{ cm}^3$$

thus

$$1 \text{ mL} = 1 \text{ cm}^3$$

also

$$1 \text{ mL} = 1000\ \mu\text{L}$$

It is customary to use the following volume units for solids and liquids.

Volumes of Solids	*Volumes of Liquids*
Meter^3 (m^3)	Liter (L)
Centimeter^3 (cm^3)	Milliliter (mL), also cm^3 or cc
	Microliter (μL)

Volume	Approximate Size
1 mL	20 drops from a medicine dropper
5 mL	1 teaspoon
250 mL	1 cup
1 L	slightly over 1 quart

Figure 3.6 *Approximate metric volumes.*

Convenient approximations of metric volumes are shown in Figure 3.6. Laboratory equipment for accurate measurement of liquid volumes is shown in Figure 3.7

EXAMPLE 3.7

From the approximations provided in this section, select the best answer for each of the following volumes.

(a) The volume in a typical medicine dropper is approximately
  (1) 0.01 cc (2) 1 mL (3) 100 cc (4) 0.08 L

(b) A small glass of orange juice is approximately
  (1) 2.0 L (2) 2.0 cc (3) 200 mL (4) 0.02 L

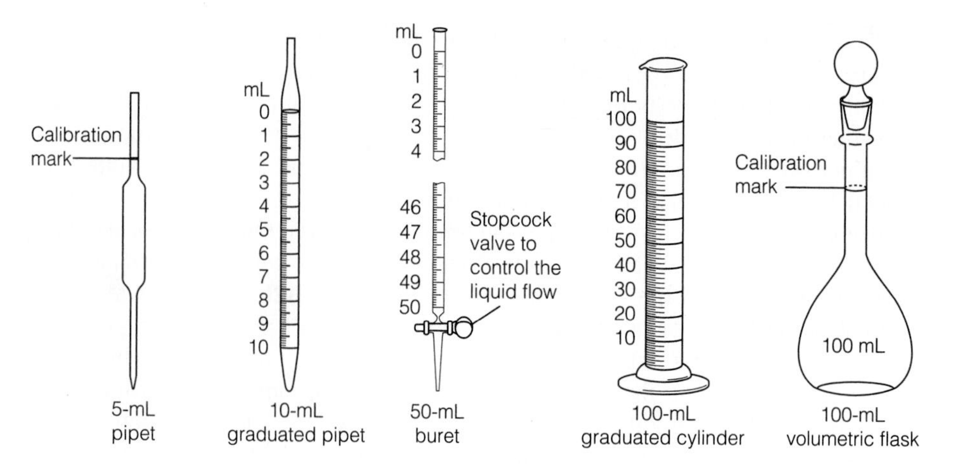

Figure 3.7 *Laboratory apparatus for measuring liquid volumes.*

SOLUTION

(a) A medicine dropper holds about 20 drops or 1 mL, answer (2).
(b) A small glass of orange juice is about 200 mL, answer (3). This is a little smaller in volume than a can of soda.

See Problems 3.11–3.20.

EXERCISE 3.3

A gallon of milk is about equal to

(1) 380 cc (2) 380 cm^3 (3) 3.8 dL (4) 3.8 L

EXAMPLE 3.8

Convert the following exact metric volumes to the units required.

(a) 150 mL = ______ cc (b) 2.4 mL = ______ L

SOLUTION

(a) 150 cc, since 1 mL is equivalent to 1 cm^3 or 1 cc

(b) 0.0024 L, since $2.4\ \text{mL} \times \frac{1\ \text{L}}{1000\ \text{mL}} = 0.0024\ \text{L}$

EXERCISE 3.4

What is the volume in milliliters of 0.075 L?

EXAMPLE 3.9

In the United States, the usual soft drink can holds 355 mL. How many such cans could be filled from one 2-L bottle?

SOLUTION

Write down the given quantity: 2.00 L. Then write down the equality provided by the problem: 1 can =355 mL. Also, since volumes are in both liters and milliliters, we use 1 L = 1000 mL.

Develop a planned series of conversions leading to the desired units.

Plan: L ⟶ mL ⟶ cans

$$2.00\ \text{L} \times \frac{1000\ \text{mL}}{1\ \text{L}} \times \frac{1\ \text{can}}{355\ \text{mL}} = 5.63\ \text{cans} \quad \text{(rounded to hundredths)}$$

3.5 Metric Mass Measurement and Conversions

The SI base unit of mass is the **kilogram** (kg), which is equal to 1000 **grams** (g). The gram is approximately equal to the mass of four thumbtacks and is a convenient unit for most laboratory measurements. Figure 3.8 shows several metric mass approximations. A standard kilogram mass made from a platinum-iridium alloy is housed in specially controlled conditions in France, but duplicates are stored by several other countries. The kilogram, gram, **milligram**, and **microgram** (μg) are common metric masses.

Mass	Approximate Size
1 mg	Pinch of salt
250 mg	Mass of 1 thumbtack
1 g	Mass of 4 thumbtacks
1 kg	Mass of quart of milk
100 kg	Mass of football player

Figure 3.8 *Approximate metric masses.*

$$1 \text{ kg} = 1000 \text{ g} \quad \text{or} \quad 0.001 \text{ kg} = 1 \text{ g}$$

$$1 \text{ g} = 1000 \text{ mg} \quad \text{or} \quad 0.001 \text{ g} = 1 \text{ mg}$$

$$1 \text{ mg} = 1000\ \mu\text{g} \quad \text{or} \quad 0.001 \text{ mg} = 1\ \mu\text{g}$$

All these quantities are exact.

The gram was originally defined as the mass of 1.000 cm^3 (1.000 mL) of water at 4 °C, the temperature at which a gram of water occupies the smallest volume. Thus, 1 L of water has a mass of 1 kg. Although the exact volume of water changes slightly at different temperatures, for practical purposes, 100 g of water has a volume of 100 mL.

Mass of Water	*Volume*
1 g of water	1 mL
100 g of water	100 mL
1 kg of water	1 L

EXAMPLE 3.10

Make the following metric mass approximations.

(a) An aspirin tablet has a mass of about
(1) 30 g (2) 3 g (3) 0.3 g (4) 0.03 g

(b) A full 8-oz cup of coffee has a mass of about
(1) 500 g (2) 250 g (3) 25 g (4) 0.50 kg

(c) Two liters of cola has a mass (weight) of about
(1) 2 kg (2) 1 kg (3) 200 g (4) 2 g

SOLUTION

(a) An aspirin has a mass of about $\frac{1}{3}$ g or 0.3 g, answer (3).

(b) The 8-oz cup of coffee or any liquid that is nearly all water is one-fourth of a quart. A quart is about equal to a liter (actually slightly less than a liter), so a cup (8 oz) is about equal to $\frac{1}{4}$ L (250 mL) and has a mass of 250 g, answer (2).

(c) Since the mass of 1 L of water is 1 kg (1000 g), 2 L of liquid that is mostly water has a mass of about 2000 g or 2 kg, answer (1).

EXERCISE 3.5

A petite cheerleader weighs about

(1) 500 kg (2) 100 kg (3) 50 kg (4) 5 kg

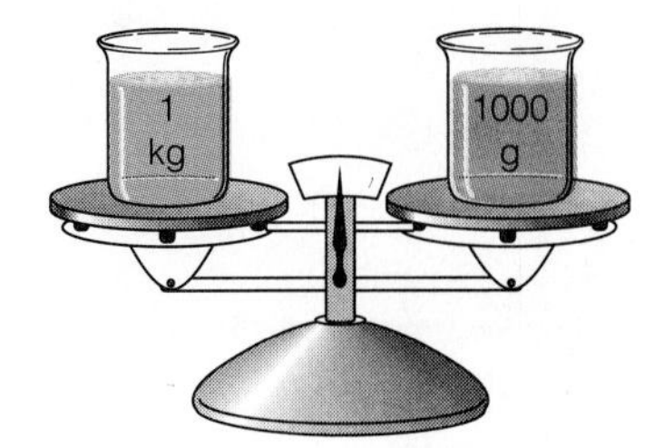

Figure 3.9 *A simple balance.*

If two masses are equal, the Earth's gravitationsl attraction for both masses will be equal (Figure 3.9). The oldest balances used for comparing equal masses consisted primarily of a bar balanced in the center with a basket hanging from each

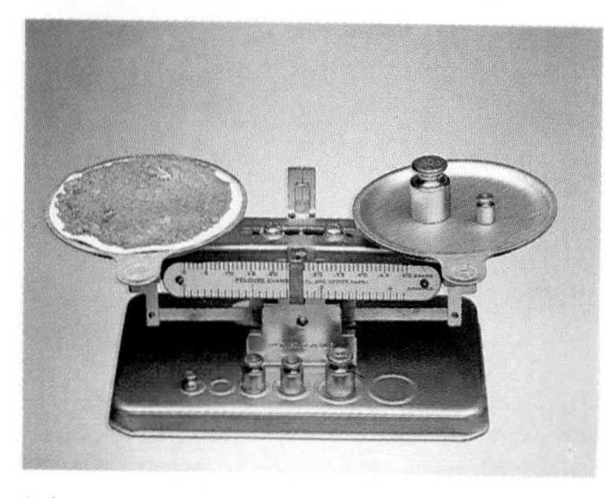

(a)

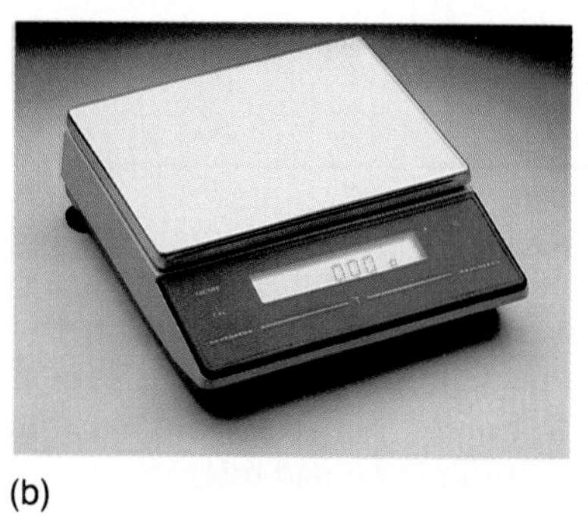

(b)

(c)

Figure 3.10 *Laboratory balances. (a) Classic double-pan balance. (b) Top-loading balance determines masses to the nearest 0.01 g. (c) Single-pan electronic analytic balance determines masses to the nearest 0.0001 g.*

end. The object to be weighed could be placed in one basket, while stones used as weights could be added to the other basket until the beam was balanced. Balances of this type were used by ancient Egyptians in about 5000 B.C. for weighing gold dust and commercial goods.

Sensitive double-pan **analytical balances** were developed in the 1860s. Although weighings took considerable time, this development allowed routine weighings to be made to the nearest 0.1 mg and thus paved the way to accurate analysis and to many discoveries in chemistry (Figure 3.10a).

0.1 mg = 0.0001 g. That's equal to about one small crystal of salt.

Approximately 100 years later, in the 1960s, single-pan analytical balances became common in chemistry laboratories. With these balances, the mass of a sample could be determined to the nearest 0.0001 g in less than a minute. These analytical balances are sometimes called mechanical balances because masses are mechanically added to, or removed from, a counterbalanced beam. By the mid-1980s, electronic balances became available. These balances, with digital readout, shortened the weighing procedure even further. Simply press a button or bar to "zero" the balance. Then place the item on the balance pan and read the mass. **Top-loading balances** with no enclosed cage around the pan were also developed for quickly determining masses to the nearest 0.01 g or 0.001 g (Figure 3.11). Modern technology has eliminated much of the tedium of science.

Electronic balances use an electromagnet with electronic circuitry that can vary the magnetic force needed to counterbalance the force of gravity on the item being weighed.

EXAMPLE 3.11

The top-loading balance pictured in Figure 3.11 shows a total mass measured to the nearest 0.01 g, that is, to the nearest

(a) decigram (tenths of a gram).
(b) centigram (hundredths of a gram).
(c) milligram (thousandths of a gram).
(d) tenths of a milligram (ten-thousandths of a gram).

SOLUTION

The mass is measured to the nearest 0.01 g, which is equivalent to the nearest centigram (b).

Figure 3.11 *An electronic top-loading balance.*

EXAMPLE 3.12

Make the following metric mass conversions.

(a) 0.600 kg = ______ g
(b) A 250-mg vitamin C tablet = ______ g

SOLUTION

(a) 600 g, since $0.600\ \text{kg} \times \dfrac{1000\ \text{g}}{1\ \text{kg}} = 600\ \text{g}$

(b) 0.250 g, since $250\ \text{mg} \times \dfrac{1\ \text{g}}{1000\ \text{mg}} = 0.250\ \text{g}$

See Problems 3.21–3.24.

EXERCISE 3.6

What is the mass in milligrams of 0.497 g?

3.6 *Conversion of Metric and English Units*

If all measurements were in metric (SI) units, like those just described, conversions between small and large units would be rather simple. Unfortunately, many measurements are not in metric units, and conversion factors are not always multiples of 10, but the same approach we have used in solving the simpler problems—dimensional analysis—is an excellent approach to solving more complex problems. As described for metric conversions, we start with the *known quantity* and multiply by one, two, or more conversion factors to convert the value to the desired units.

Known quantity and unit(s) × Conversion factor(s) = Answer in desired units

A list of some metric and English conversions is provided in Table 3.5. Additional conversions are provided in Appendix A. Use these when you work the following sample problems and similar problems at the end of the chapter.

Table 3.5 **Metric and English Conversions***

Length	*Volume*	*Mass*
1 in. = 2.54 cm (exact)	1 qt = 946 mL	1 lb = 454 g
1 m = 39.37 in.	1 L = 1.057 qt	1 kg = 2.20 lb
1 mi = 1.609 km	1 m^3 = 1057 qt	1 oz (avoir.) = 28.35 g
1 km = 0.6215 mi	1 $in.^3$ = 16.39 cm^3	1 oz (troy) = 31.10 g
	1 fl oz = 29.6 mL	

*The most frequently used conversions appear in blue.

EXAMPLE 3.13

Suppose your height is 5 ft 9 in., but the job application form asks for your height in meters. What is it?

SOLUTION

The given quantity, 5 ft 9 in, is 5.75 ft in decimal form (since 9 in. = $\frac{9}{12}$ ft or 0.75 ft). Write down this quantity and the unit needed in the answer: meters.

Develop a planned series of conversions leading to the desired units.

Plan: ft $\longrightarrow$ in. $\longrightarrow$ cm $\longrightarrow$ m

Use conversion factors from memory, or from tables, as needed.

$$5.75\ \text{ft} \times \frac{12\ \text{in.}}{1\ \text{ft}} \times \frac{2.54\ \text{cm}}{1\ \text{in.}} \times \frac{1\ \text{m}}{100\ \text{cm}} = 1.75\ \text{m} \quad \text{(rounded to three figures)}$$

EXERCISE 3.7

What is your height in meters?

EXAMPLE 3.14

A sprinter who runs the 100-m dash in 11.0 s is running at what rate in kilometers per hour?

SOLUTION

Write down the given quantity.

$$\frac{100\ \text{m}}{11.0\ \text{s}}$$

Then, write down the unit needed in the answer: kilometers per hour.

Develop a plan of conversions leading to the desired units.

Numerator conversion plan: Convert m $\longrightarrow$ km

Denominator conversion plan: Convert s $\longrightarrow$ min $\longrightarrow$ hr

Use conversion factors as needed.

$$\underbrace{\frac{100\ \text{m}}{11.0\ \text{s}} \times \frac{1\ \text{km}}{1000\ \text{m}}}_{\text{m} \longrightarrow \text{km}} \times \underbrace{\frac{60\ \text{s}}{1\ \text{min}} \times \frac{60\ \text{min}}{1\ \text{hr}}}_{\text{s} \longrightarrow \text{hr}} = 32.7\ \text{km/hr} \quad \text{(rounded to three figures)}$$

Meters were converted to kilometers and then two factors were used to convert seconds to hours.

See Problems 3.25–3.34.

3.7 Uncertainty in Measurement

No measurement is ever 100% exact. A manufactured machine part can be made to specifications measured to thousandths of an inch, or hundredths of a millimeter, but magnification of the measured item reveals that the measurement is still inexact! All measurements are **uncertain** at some point.

When several measurements are taken that have close agreement, we say the measurements have good **precision**. When the range of values is small, precision is increased, but just because the figures are in close agreement does not mean they are accurate. If a person steps on the bathroom scales three or four times, the weights obtained may have good precision, within one or two pounds, but if the scales are out of adjustment, the values are not accurate. **Accuracy** relates to how closely the measurements agree with the true value.

Like trying to hit the bull's-eye of the dart board, a chemist or analyst attempts to "hit" the correct or true value for the measurement. Both precision and accuracy are achieved when several darts are thrown in a cluster around the bull's-eye (Figure 3.12). Darts thrown all to one side of the dart board may have good precision, but poor accuracy. Similarly, repeated chemical analysis of a blood sample may show a high cholesterol level with good *precision,* but if the analysis was performed with an instrument that was not calibrated properly, the analysis will not be *accurate.*

As long as the equipment is calibrated and functioning correctly, and the method of analysis is appropriate for the sample involved, greater precision will usually lead to greater accuracy. Precision often depends on how finely the units are marked on the device. A thermometer that shows tenths of degrees allows more precise readings to be taken than does a thermometer that is marked off only in degrees. One could use an ordinary watch calibrated in seconds to time an event, but greater precision can be obtained by using a stopwatch calibrated in tenths of a second. To achieve greater precision and accuracy, more sophisticated—more expensive—equipment is usually required.

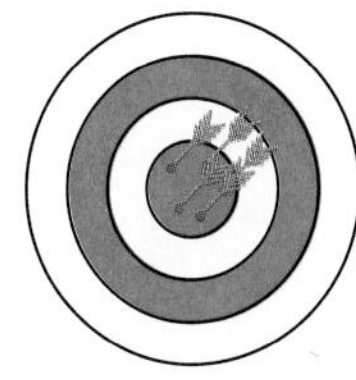

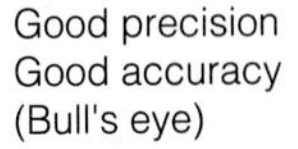

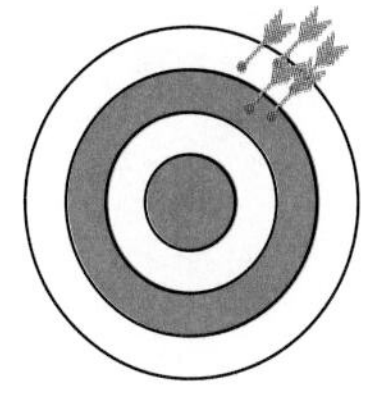

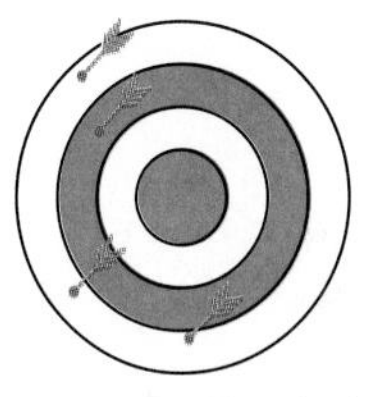

Figure 3.12 *A dart board analogy of precision and accuracy.*

EXAMPLE 3.15

The mass of a beaker with a solid sample was determined to the nearest 0.01 g on a top-loading balance. The following values were recorded in the sequence listed: 104.01 g, 104.02 g, 103.99 g, 104.01 g. The "true" or standard value was later found to be 103.03 g. Discuss precision and accuracy for the weighings.

SOLUTION

While the values recorded had good precision, they were poor in accuracy. Some possible sources of error include the following: (1) failure to initially adjust the balance to a "zero" position, (2) consistent misreading of the balance, (3) since the readings were all too high, a few crystals may have been spilled onto the balance pan, giving precise, but inaccurate results.

3.8 Significant Figures

When you read the automobile mileage on the odometer of many cars, you may notice that the number of tenths of a mile is part way between values. You can read each number in the mileage with certainty except the tenths-of-a-mile digit, which must be approximated. For the odometer reading of 45,206.3 all figures are known with certainty, except for the final digit, 3. The 3 in the tenths position is an estimate; it is uncertain. The number of **significant figures** in a measured value equals the number of digits that are **certain**, plus one additional rounded off (esti-

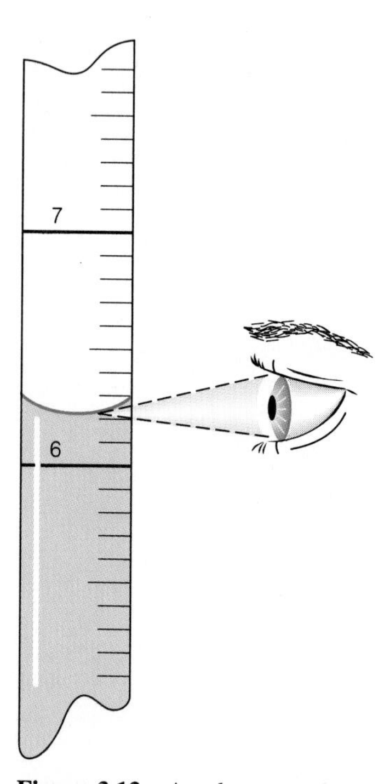

Figure 3.13 *A volume can be approximated to ±0.01 mL when the calibrations are shown in tenths, as they are in this figure. The volume 6.23 mL has three significant figures.*

mated) digit, which is an **uncertain digit**. In the odometer reading there are five certain digits and one uncertain digit, making six significant figures. To summarize,

Number of significant figures = All certain digits + One uncertain digit

The mass of a thumbtack was obtained with a top-loading balance and was reported as 0.24 g. When the thumbtack was placed on an analytical balance, the mass was found to be 0.2436 g. The first mass was recorded with two significant figures, but the second mass was recorded to four significant figures. The number of significant figures indicates the reliability of the measurement.

The volume pictured in Figure 3.13 should be read with the eye looking squarely at the bottom of the **meniscus**, that is, at the crescent-shaped liquid surface caused by the attraction of the liquid to the glass. Since the bottom of the meniscus is between the 6.2-mL and 6.3-mL marks, we can estimate the volume to hundredths, as 6.23 mL. The number 6.23 has two figures that are certain (6 and 2) and one estimated digit that is uncertain (the 3). The two certain figures with one uncertain figure give a total of three significant figures. Any attempt to approximate the volume to thousandths or beyond is misleading and unjustified when the instrument shows only tenths. Only one uncertain digit is included when recording a number or in counting significant figures.

For any measured amount, you should record the *number,* the *units,* and a *label* or name for the substance measured. The amount always has a specific number of significant figures that indicates the precision of the measurement. See the examples in Table 3.6.

Exact Numbers

For all measured values, the last digit to the right is estimated; it is an uncertain figure.

Some numbers are **exact numbers** by definition; they have no uncertain digit since there is no approximation involved. In the definitions 1 m = 1000 mm and 1 L = 1000 mL, the 1 m and 1 L are exact numbers. They can be thought of as having an infinite number of zeros (1.00000 . . .) when performing calculations. Pure numbers, like 2 or 10, and pure fractions, like $\frac{2}{3}$ or $\frac{1}{4}$, are also exact. They are not approximated quantities; they contain no uncertain figures.

Table 3.6 **Significant Figures**

Quantity	*Certain Digits*	*Uncertain Digit*	*Number of Significant Figures**
14.379 g	1 4 3 7	9 (thousandths)	5
6.02 mL	6 0	2 (hundredths)	3
120.586 m	1 2 0 5 8	6 (thousandths)	6
7.5 km	7	5 (tenths)	2
0.0037 g	3	7 (ten-thousandths)	2
1.940 g	1 9 4	0 (thousandths)	4

*The position of the decimal point has nothing to do with the number of significant figures.

Zeros in Numbers

When there are zeros in a measured value, the number of significant figures is not always the same as the total number of digits. The number 0.007, for example, has only one significant figure because the zeros are only ''place holders'' to identify where the 7 belongs. The following rules should be used to determine the number of significant figures when zeros are involved.

Rules for Determining Significant Figures

1. All nonzero integers are significant.

2. All zeros to the left of (or preceding) the first nonzero digit are not significant, since they are used to locate the decimal point, as explained in the previous paragraph. Examples:

When counting significant figures, begin at the first nonzero digit on the left, regardless of the decimal location.

0.00567 has three significant figures (5, 6, and 7).

significant figures

0.0089 has two significant figures (8 and 9).

significant figures

3. All zeros between nonzero digits are significant. Examples:

207.08 has five significant figures (2, 0, 7, 0, and 8).
0.0401 has three significant figures (4, 0, and 1).

4. All zeros at the end of a number that has a decimal point are significant. Examples:

34.070 has five significant figures (all are significant).
0.0670 has three significant figures (6, 7, and the final 0).
400. has three significant figures. (Notice the decimal point). Unfortunately not all scientific writing follows this example. Omitting the decimal point (see rule 5) leads to confusion.

5. Zeros at the end of a whole number that has no decimal point cause confusion since they may—or may not—be significant. For example, it is impossible to tell how many significant figures are represented by 300 mL. The number may have one, two, or three significant figures, depending on the precision of the measurement. It may mean that the 300-mL quantity was measured to the nearest whole milliliter, with a precision of three significant figures. It may also represent an approximation rounded to the nearest multiple of 10, as 300 ± 10 mL. Finally, the number may be accurate only to the nearest multiple of 100, as 300 ± 100 mL. The confusion described here can be avoided by writing the number in scientific notation. This approach will be described later in this chapter.

EXAMPLE 3.16

How many significant figures are in each of the following amounts?

(a) 60.1 g (b) 6.100 g (c) 0.061 g (d) 6100 g

SOLUTION

(a) three significant figures (rules 1 and 3)
(b) four significant figures (rules 1 and 4)
(c) two significant figures (rules 1 and 2)
(d) uncertain: could be two, three, or four significant figures (rule 5)

See Problems 3.35–3.38.

3.9 Calculations with Significant Figures

In the process of working with significant figures, it will be necessary to round off numbers. Follow these rules.

Rules for Rounding Off Numbers

1. If the digit you want to drop is less than 5, drop that digit and all others to the right of that digit. Digits to be dropped are in blue.

   Rounding 86.0234 g to three significant figures gives 86.0 g.
   Rounding 0.07893 m to three significant figures gives 0.0789 m.

2. When the digit you drop is 5 or greater, increase the value of the last digit retained by one.

   Rounding 0.06587 L to three significant figures gives 0.0659 L.
   Rounding 586.52 g to three significant figures gives 587 g.

No calculations involving measured quantities can give results that are more precise than the least precise measurement. Measured quantities with three or four significant figures cannot be added, subtracted, multiplied, or divided to give answers with six or seven significant figures! Rules for addition and subtraction differ from multiplication and division, as described below.

Addition or Subtraction

When measured quantities are either added or subtracted, the answer retains the same number of digits to the right of the decimal point as were present in the value with the fewest number of digits to the right of the decimal point. In the following example, uncertain digits are underlined.

Add: 46.1 g, 106.22 g, and 8.357 g.

46.1 g	The 46.1 is only measured to tenths of a gram
8.357 g	and is the least precise value, so the answer
106.22 g	should be rounded to tenths of a gram.
160.677 g	Rounding to tenths gives 160.7 g.

The calculator answer of 160.677 does not give the appropriate number of significant figures. Digits to be dropped are in blue.

Multiplication or Division

When measured quantities are either multiplied or divided, the answer must contain the same number of significant figures as were present in the measurement with the fewest number of significant figures. The following calculations were made with a calculator.

Digits to be dropped are in blue.

Multiplication:

$$80.2 \text{ cm} \times 3.407 \text{ cm} \times 0.0076 \text{ cm} = 2.0766346 \text{ cm}^3 \quad \text{(calculator answer)}$$

Report the answer as 2.1 cm^3. The answer should contain only two significant figures because one of the numbers (0.0076) has only two significant figures.

Division:

$$\frac{425.0 \text{ m}}{424.7 \text{ s}} = 9.5078299 \text{ m/s} \quad \text{(calculator answer)}$$

Report the answer as 9.51 m/s, showing only three significant figures.

EXAMPLE 3.17

Do the following calculations and round off the answer to the proper number of significant figures.

Be sure you actually *do* all example problems. The answers are given here to allow you to check yourself. The best way to gain confidence in your ability to solve problems is to *solve* problems.

(a) $913.1 \text{ m} \times 0.0165 \text{ m} \times 1.247 \text{ m} =$ ______
(b) 500. g is a standard metric size. Convert this to pounds. (See Table 3.5.)
(c) $3.0278 \text{ g} + 110.4 \text{ g} + 49.34 \text{ g} =$ ______

SOLUTION

(a) $913.1 \text{ m} \times 0.0165 \text{ m} \times 1.247 \text{ m} = 18.8 \text{ m}^3$ (Round calculator answer to three significant figures, since 0.0165 m only has three significant figures.)
(b) $500. \text{ g} \times 1 \text{ lb}/454 \text{ g} = 1.10 \text{ lb}$ (Round calculator answer to three significant figures, since 500. and 454 have three significant figures; 1 lb is exact.)
(c) $3.0278 \text{ g} + 110.4 \text{ g} + 49.34 \text{ g} = 162.8 \text{ g}$ (Round to tenths, since the least precise value is measured to tenths.)

3.10 *Scientific Notation*

Some numbers used in chemistry are so large—or so small—that they can boggle the mind. For example, light travels at 30,000,000,000 cm/s. There are 602,200,000,000,000,000,000,000 carbon atoms in 12.0 g of carbon. On the small side, the diameter of an atom is about 0.0000000001 m. The diameter of an atomic nucleus is about 0.000000000000001 m. It is obviously difficult to keep track of the zeros in quantities like these. Such numbers can be stated more precisely, and they are easier to work with when they are written in **scientific notation**, a form that uses **powers of 10**. Table 3.2 contains a list of some numbers as exponentials in powers of 10.

A number in scientific notation has two quantities multiplied, in the form

$$n \times 10^p$$

where n is a number between 1 and 10 that is multiplied by 10 raised to a power, p. To write a number in scientific notation, first move the decimal point of the number to the right or left so that only one nonzero digit is to the left of the decimal point. This gives a number that lies between 1 and 10. Then, show this number multiplied by 10 raised to a power equal to the number of places the decimal point was moved because each decimal place corresponds to a factor of 10.

For numbers that are larger than 10, the decimal must be moved to the left, so the exponent is a positive number. For example,

$$345.8 = 3.458 \times 10^2$$

Here, the decimal was moved two places to the left, which is equivalent to a factor of 100 or 10^2.

A CLOSER LOOK

How Big Is a Billion?

Most of us find it difficult to comprehend the meaning of very large and very small numbers like those often used in scientific calculations. How long do you think it would take to count to 1 **million**? Try counting a million of something. For example, there are about

- 1 million bricks in a typical two-story campus library building.
- 1 million letters in 15 pages of newspaper want ads.
- 1 million pages in 2000 textbooks like this one.
- 1 million cupfuls of gasoline in seven gasoline tanker trucks (assuming each truck hauls 9000 gallons).

If you personally count a million items—one at a time—how long would it take? If you count one item per second, it would obviously take 1 million seconds, but how many days would pass before you would finish? Let's find out by using conversion factors and the exponential notation described in this chapter.

$$\frac{1 \times 10^6 \text{ s}}{1} \times \frac{1 \text{ min}}{60 \text{ s}} \times \frac{1 \text{ hr}}{60 \text{ min}} \times \frac{1 \text{ day}}{24 \text{ hr}} = 11.6 \text{ days}$$

Let's take a larger number. Just how big is a **billion**? Consider these

- 1 billion drops of gasoline would be enough to fill one and one-half large (9000-gal) gasoline tanker trucks.

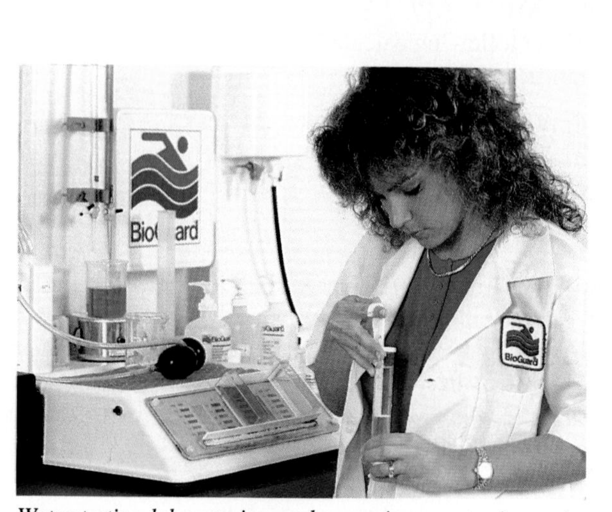

Water-testing laboratories analyze various contaminants in water samples in the parts per billion range.

- 1 billion drops of water would fill a typical residential swimming pool.

If you count one item every second, how long (in years) would it take you to count a billion of something?

$$\frac{1 \times 10^9 \text{ s}}{1} \times \frac{1 \text{ min}}{60 \text{ s}} \times \frac{1 \text{ hr}}{60 \text{ min}} \times \frac{1 \text{ day}}{24 \text{ hr}} \times \frac{1 \text{ yr}}{365 \text{ days}}$$

$$= 31.7 \text{ yr}$$

Could you determine this without using conversion factors?

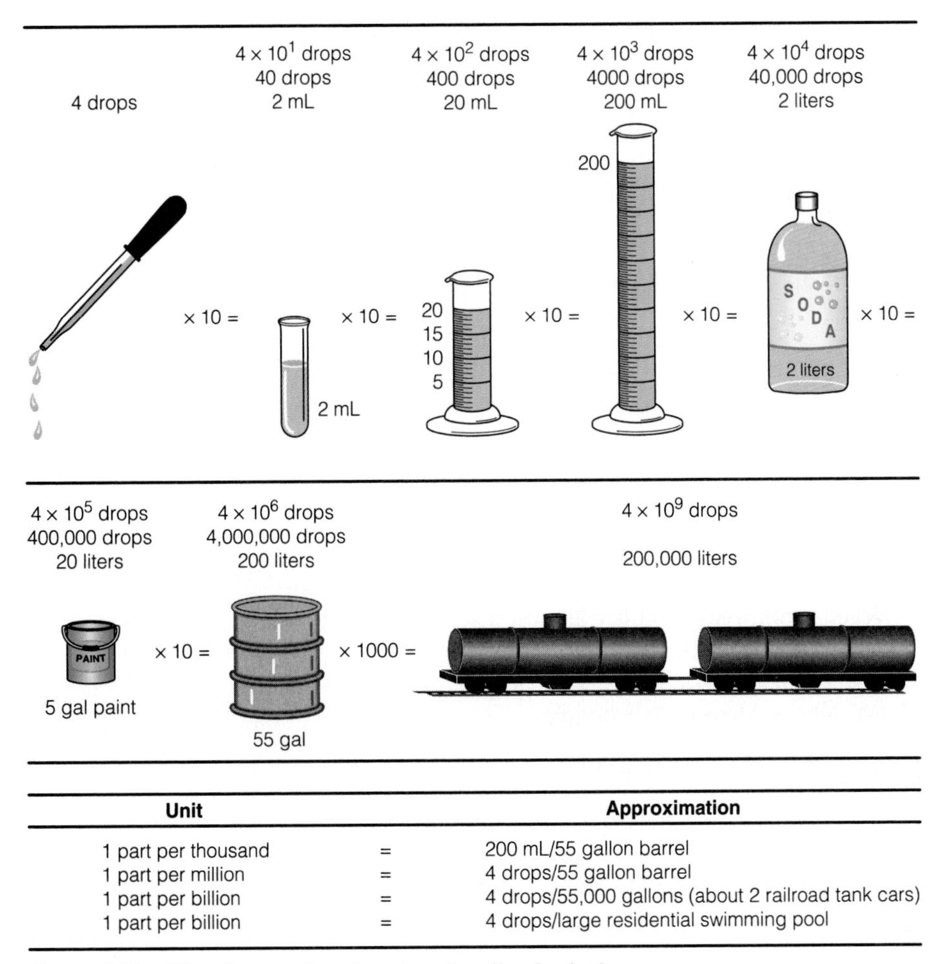

Unit		Approximation
1 part per thousand	=	200 mL/55 gallon barrel
1 part per million	=	4 drops/55 gallon barrel
1 part per billion	=	4 drops/55,000 gallons (about 2 railroad tank cars)
1 part per billion	=	4 drops/large residential swimming pool

Figure 3.14 *What fraction is a drop in a 5-gallon bucket?*

For numbers between 0 and 1, the decimal must be moved to the right, so the exponent is a negative number. For example,

$$0.00456 = 4.56 \times 10^{-3}$$

Here, the decimal was moved three places to the right, which is equivalent to a factor of 1/1000 or 10^{-3}.

Now, look back at those numbers with many zeros at the beginning of this section. The speed of light in scientific notation is 3.00×10^{10} cm/s to three significant figures. The number of carbon atoms in 12.0 g of carbon is 6.02×10^{23} atoms, and the nucleus of an atom has a diameter of about 1×10^{-15} m (the decimal was moved 15 places to the right).

The quantities represented in Figure 3.14 increase sequentially by factors of 10. These quantities are written in scientific notation so you can quickly see what happens when a volume is multiplied by 10, 100, 1000, and so on.

Exponential numbers are often used in calculations. For multiplication and division, two rules must be followed.

Multiplying and Dividing Exponential Numbers

1. To multiply numbers expressed in scientific notation, add the exponents.

Expressed algebraically: $(a^x)(a^y) = a^{x+y}$

Examples: $(10^6)(10^4) = 10^{6+4} = 10^{10}$

$(10^6)(10^{-4}) = 10^{6+(-4)} = 10^2$

2. To divide numbers expressed in scientific notation, subtract the exponents.

Expressed algebraically: $\dfrac{a^x}{a^y} = a^{x-y}$

Examples: $\dfrac{10^{14}}{10^6} = 10^{14-6} = 10^8$

$\dfrac{10^{-6}}{10^{-23}} = 10^{-6-(-23)} = 10^{17}$

When multiplying and dividing numbers with both nonexponential and exponential terms, the nonexponential terms can be grouped together and taken care of first, in the usual way. The exponential terms are then multiplied and divided as shown above. Here is an example.

$$\frac{(4.34 \times 10^4)(2.76 \times 10^{-2})}{1.67 \times 10^5} = \frac{4.34 \times 2.76}{1.67} \times \frac{10^4 \times 10^{-2}}{10^5} = 7.17 \times 10^{4-2-5} = 7.17 \times 10^{-3}$$

Now, work through the following example yourself.

EXAMPLE 3.18

Perform the indicated calculations, and give the answer to the proper number of significant figures.

$$\frac{(6.02 \times 10^{23})(2.2 \times 10^{-3})}{168} = \_\_\_\_\_\_$$

SOLUTION

First, multiply and divide the nonexponential terms.

$$\frac{6.02 \times 2.2}{168} = 0.0788333 \quad \text{(calculator display)}$$

$= 0.079$ to two significant figures or 7.9×10^{-2} in scientific notation

Then, combine this value with other exponential terms:

$$(7.9 \times 10^{-2}) \times 10^{23} \times 10^{-3} = \_\_\_\_\_\_$$

Combining exponential terms gives

See Problems 3.39–3.42.

7.9×10^{18} (answer)

A CLOSER LOOK

How Much Is One Part Per Billion?

Twenty years ago, we could measure chemical impurities in the parts per million (ppm) range, which means one unit per million of the same units. That would be like finding one person in a city with a population the size of San Diego, Denver, metropolitan Kansas City, or metropolitan New Orleans. Today it is possible to analyze impurities in the range of parts per billion. That is like finding five persons in the world (population = 5 billion).

How much is **1 part per billion** (1 ppb)? Let's start with a large 75.0 ft × 32.0 ft × 5.0 ft deep swimming pool and—using conversion factors—determine the number of drops of liquid that would be equivalent to 1 ppb.

First, find the pool volume.

$$75.0 \text{ ft} \times 32.0 \text{ ft} \times 5.0 \text{ ft deep} = 12{,}000 \text{ ft}^3$$

Then, set up the units to convert cubic feet to milliliters.

$$\text{Conversion plan: ft}^3 \longrightarrow \text{in.}^3 \longrightarrow \text{cm}^3 \longrightarrow \text{mL}$$

$$\frac{12{,}000 \text{ ft}^3}{1} \times \frac{(12 \text{ in.})^3}{1 \text{ ft}^3} \times \frac{(2.54 \text{ cm})^3}{1 \text{ in.}^3} \times \frac{1 \text{ mL}}{1 \text{ cm}^3}$$

$$= \underline{\qquad\qquad}$$

Be sure to cube numbers as well as units; for example, $(12 \text{ in.})^3 = 12^3 \times \text{in.}^3 = 1728 \text{ in.}^3$.

$$\frac{12{,}000 \text{ ft}^3}{1} \times \frac{1728 \text{ in.}^3}{1 \text{ ft}^3} \times \frac{16.4 \text{ cm}^3}{1 \text{ in.}^3} \times \frac{1 \text{ mL}}{1 \text{ cm}^3}$$

$$= 3.4 \times 10^8 \text{ mL (pool volume)}$$

Now multiply this volume by the factor of 1 part per billion, or $1 \text{ mL}/10^9 \text{ mL}$.

$$\frac{3.4 \times 10^8 \text{ mL}}{1} \times \frac{1 \text{ mL}}{10^9 \text{ mL}} = 0.34 \text{ mL}$$

Adding this volume (0.34 mL) to the filled pool would be adding 1 part per billion. To convert this volume to drops, use the factor 20. drops/mL.

$$\frac{0.34 \text{ mL}}{1} \times \frac{20. \text{ drops}}{1 \text{ mL}} = 6.8 \text{ drops or about 7 drops}$$

Thus, 7 drops of liquid in a large swimming pool of water is equivalent to about 1 part per billion (1 ppb).

3.11 *Density and Specific Gravity*

Density is an important characteristic property of matter. When one speaks of lead as being "heavy" or aluminum as "light," one is actually referring to the density of these metals. **Density** is defined as mass per unit volume.

$$\text{Density} = \frac{\text{Mass}}{\text{Volume}} \quad \text{or} \quad d = \frac{m}{V}$$

Densities for solids are reported in grams per cubic centimeter (g/cm^3), and densities for liquids are usually expressed in grams per milliliter (g/mL). Remember that 1 mL of liquid takes up the same space as 1 cm^3, so the density for a liquid in grams per milliliter could also be reported as grams per cubic centimeter. For gases, the densities are reported in grams per liter. Densities of several substances are listed in Table 3.7.

Table 3.7 **Densities of Several Materials at Room Temperature**

Solids	*g/cm³*	*Liquids*	*g/mL*	*Gases*	*g/L*
Balsa wood (approx.)	0.13	Gasoline (approx.)	0.67	Hydrogen	0.090
Pine wood (approx.)	0.42	Ethyl alcohol	0.79	Helium	0.177
Ice (−10 °C)	0.917	Cottonseed oil	0.926	Ammonia	0.771
Magnesium	1.74	Water (20 °C)	0.998	Neon	0.901
Aluminum	2.70	Water (4 °C)	1.000	Nitrogen	1.25
Iron	7.86	Methylene chloride	1.34	Air (dry)	1.29
Copper	8.96	Chloroform	1.49	Oxygen	1.42
Lead	11.4	Sulfuric acid	1.84	Carbon dioxide	1.96
Gold	19.3	Mercury	13.55	Chlorine	3.17

If the *volumes* of two different substances A and B are equal, but the mass of A is greater than the mass of B, the *density* of A is greater than the density of B (see Figure 3.15). Every substance has a unique density; this characteristic physical property can be used to help identify a substance.

EXAMPLE 3.19

A flask filled to the 25.0-mL mark contained 27.42 g of a salt–water solution. What is the density of this solution?

SOLUTION

$$d = \frac{m}{V} \quad \text{or} \quad \frac{27.42\text{ g}}{25.0\text{ mL}} = 1.0968\text{ g/mL} \quad \text{(calculator display)}$$

$$= 1.10\text{ g/mL} \quad \text{(answer to three significant figures)}$$

EXAMPLE 3.20

What would be the volume of 461 g of mercury? (*Hint:* The density of mercury is listed in Table 3.7 as 13.55 g/mL.)

SOLUTION

Plan: g ⟶ mL

Start with the *quantity given* in grams, and use density as the conversion factor

Here, the factor representing density is inverted so grams will cancel.

$$461\text{ g} \times \frac{1\text{ mL}}{13.55\text{ g}} = 34.0\text{ mL} \quad \text{(three significant figures)}$$

Exercise 3.8

What is the mass of 2.5 L of gasoline? (See Table 3.7 for densities.)

See Problems 3.43–3.50.

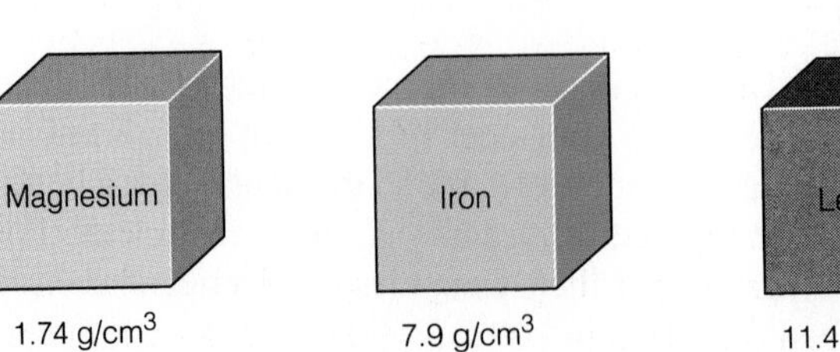
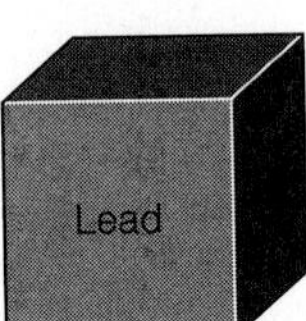
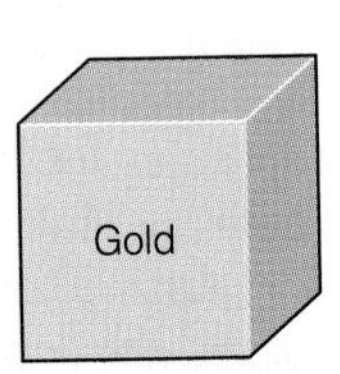

Figure 3.15 *The masses of 1.00 cm³ of several metals. For the metals shown, magnesium is the least dense and gold is the most dense.*

The **density of water** is 1.00 g/mL at 4.0 °C. This nice round number is no accident. The metric system originally defined the gram so this would be the case. If you measure out 249.00 g of pure water, it will have a volume of 249.00 mL at 4.0 °C, but even at normal room temperatures, the volume remains quite close to 249 mL. One can, therefore, quickly approximate the volume of water when mass is known, or approximate the mass of water when volume is known.

Oil floats on water because the liquids are immiscible and the density of oil is less than the density of water. Mercury and methylene chloride are liquids that have densities greater than the density of water. Relative densities of several liquids are pictured in Figure 3.16.

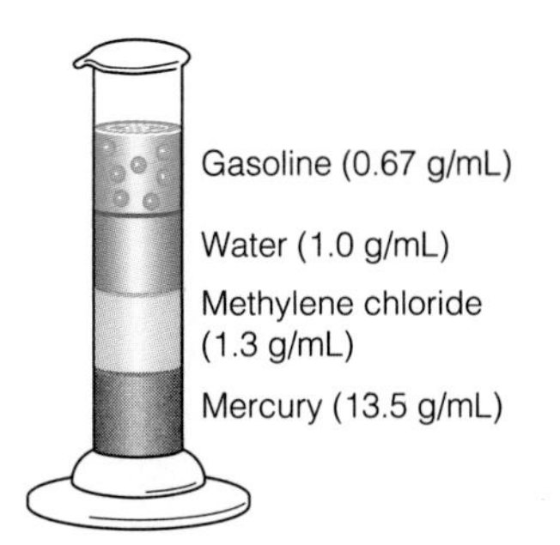

Figure 3.16 *The relative densities of some liquids. Immiscible liquids will separate into layers, with the most dense liquid going to the bottom and the least dense liquid rising to the top.*

A term similar to density is **specific gravity** (sp. gr.). Specific gravity is the ratio of the mass of any substance to the mass of an equal volume of water at the same conditions. This ratio is, therefore, also equivalent to the density of a substance divided by the density of water.

$$\text{Specific gravity of a substance} = \frac{\text{Density of the substance}}{\text{Density of water}}$$

The specific gravity of water itself, therefore, is exactly 1. Specific gravity is a number without units. This is because two values with the same units are divided to give a number with no units.

If we work in SI units, where the density of water is very close to 1 g/mL at usual temperatures, then the specific gravity of a substance is numerically the same as its density.

EXAMPLE 3.21

The density of a liquid is 1.5 g/mL. What is its specific gravity?

SOLUTION

$$\frac{1.5 \text{ g/mL}}{1.0 \text{ g/mL}} = 1.5 \quad \text{(Specific gravity has no units.)}$$

The specific gravity of a liquid is frequently measured by a device called a **hydrometer**. When the hydrometer is placed in the liquid, it sinks to a depth that depends upon the density of the liquid (Figure 3.17a). The hydrometer is calibrated so the specific gravity can be determined directly by reading the number on the stem of the hydrometer at the surface of the liquid.

For precise determinations of the density or specific gravity of a liquid, a small bottle called a **pycnometer** or specific gravity bottle (Figure 3.17b) is weighed empty, then filled with the ''unknown liquid,'' and reweighed to determine the mass of the liquid. The volume of the pycnometer can be obtained by determining the mass of water that the pycnometer will hold and then multiplying this mass by the density of water (which is inverted so units will cancel). To calculate density, the mass of the ''unknown liquid'' is divided by its volume. The specific gravity is obtained by simply dividing the mass of the ''unknown liquid'' by the mass of water that the pycnometer will hold.

EXAMPLE 3.22

An empty pycnometer with a mass of 25.0224 g was filled with pure water; the total mass was 34.9495 g. When the pycnometer was filled with an antifreeze

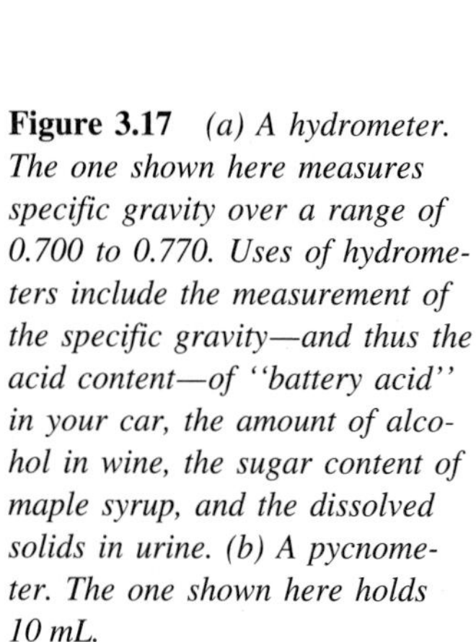

Figure 3.17 *(a) A hydrometer. The one shown here measures specific gravity over a range of 0.700 to 0.770. Uses of hydrometers include the measurement of the specific gravity—and thus the acid content—of "battery acid" in your car, the amount of alcohol in wine, the sugar content of maple syrup, and the dissolved solids in urine. (b) A pycnometer. The one shown here holds 10 mL.*

solution, the total mass was 35.9858 g. Determine the density and the specific gravity of the antifreeze solution.

SOLUTION

Mass of Antifreeze

Pycnometer + Antifreeze	= 35.9858 g
Empty pycnometer	= 25.0224 g
Mass of antifreeze	= 10.9634 g

Mass of Water in Pycnometer

Pycnometer + Water	= 34.9495 g
Empty pycnometer	= 25.0224 g
Mass of water	= 9.9271 g

$$\text{Volume} = 9.9271 \text{ g water} \times \frac{1 \text{ mL}}{1.0000 \text{ g water}} = 9.9271 \text{ mL}$$

$$\text{Density} = \frac{\text{Mass of antifreeze}}{\text{Volume}} = \frac{10.9634 \text{ g}}{9.9271 \text{ mL}} = 1.1044 \text{ g mL}$$

$$\text{Specific gravity} = \frac{\text{Mass of antifreeze}}{\text{Mass of water}} = \frac{10.9634 \text{ g}}{9.9271 \text{ g}} = 1.1044 \quad \text{(no units)}$$

See Problems 3.51 and 3.52.

3.12 Measurement of Temperature

Most people in the United States are most familiar with the **Fahrenheit** temperature scale. On this scale, the freezing point of water is 32 °F and the boiling point of water is 212 °F. Between these two temperatures, the scale has 212 − 32 = 180 units. Each unit is a Fahrenheit degree.

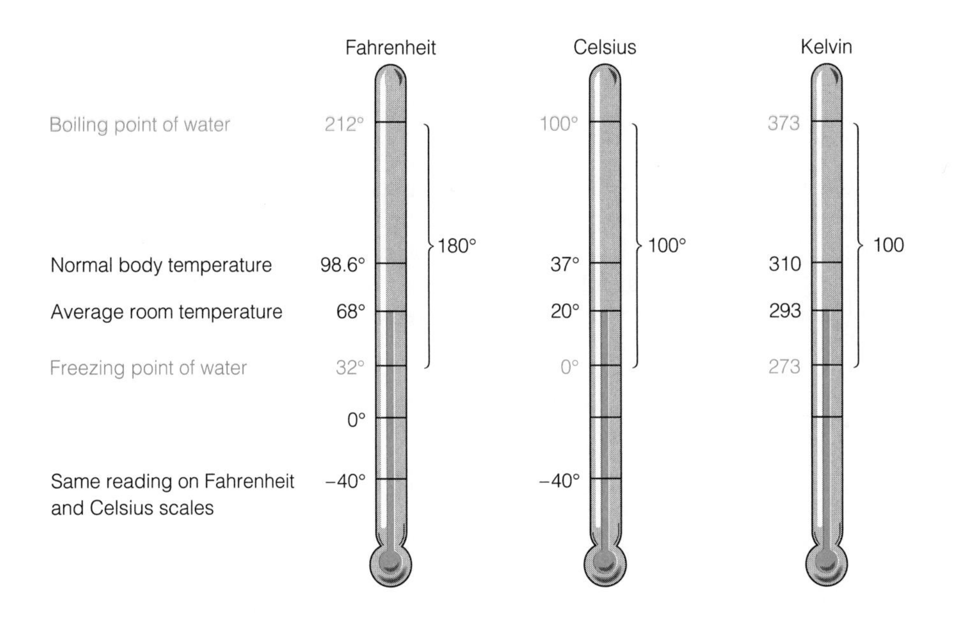

Figure 3.18 *A comparison of the Fahrenheit, Celsius, and Kelvin temperature scales.*

Most people in the world—and all who work with scientific information—use temperatures in degrees **Celsius** (°C). By definition, the freezing point of water is 0 °C, and the boiling point of water is 100 °C. Thus, between the freezing and boiling points of water, there are exactly 100 units on the Celsius scale and 180 units difference on the Fahrenheit scale (Figure 3.18). Therefore, a change of 180 °F is equivalent to a change of 100 °C, so it takes 1.80 °F to be equal to one Celsius degree.

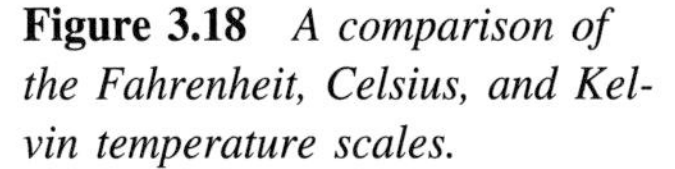

Originally, metric temperatures were in degrees centigrade, but the scale was renamed after its inventor, Anders Celsius, a Swedish astronomer.

A temperature change of 1.80 °F = A change of 1 °C (exactly)

EXAMPLE 3.23

For a temperature change of 2 °C, what would be the equivalent temperature change in degrees Fahrenheit?

SOLUTION

$$2\ \cancel{°C} \times \frac{1.8\ °F}{1\ \cancel{°C}} = 3.6\ °F$$

As shown in Figure 3.18, a temperature change of 180 °F is equal to a change of 100 °C. From this relationship, we can derive an equation where the temperature in °F divided by 180 units is proportional to the temperature in °C divided by 100 units. Because 0 °C is equal to 32 °F, we must subtract 32 degrees from the Fahrenheit temperature.

$$\frac{°F - 32}{180} = \frac{°C}{100}$$

We can multiply both sides of the equation by 180 and obtain

$$°F - 32 = \frac{180}{100} \times °C$$

A simplified form of this equation follows.

$$°F - 32 = 1.8 \times °C$$

When a Celsius temperature is known, use that value in the equation where degrees Celsius appears, and solve the equation for degrees Fahrenheit. When a Fahrenheit temperature is known, use that value where degrees Fahrenheit appears in the equation to solve for degrees Celsius. We can also rearrange the simplified equation to obtain two different forms, with one solved for degrees Fahrenheit and the other solved for degrees Celsius.

$$°F = (1.8 \times °C) + 32 \quad \text{and} \quad °C = \frac{°F - 32}{1.8}$$

EXAMPLE 3.24

If a person has a body temperature of 40. °C, what would this be in degrees Fahrenheit?

SOLUTION

$$°F = (1.8 \times °C) + 32$$
$$°F = (1.8 \times 40.) + 32$$
$$°F = 72 + 32 = 104\ °F$$

EXAMPLE 3.25

The temperature in Tucson reached 113 °F on a summer day. What temperature would that be on the Celsius scale?

SOLUTION

$$°C = \frac{°F - 32}{1.8}$$
$$°C = \frac{113 - 32}{1.8} = \frac{81}{1.8} = 45\ °C$$

EXERCISE 3.9

If it is −10. °C in St. Louis, what is the Fahrenheit temperature?

Notice that the kelvin unit is not capitalized but that the name Kelvin scale and symbol K are capitalized.

The SI unit for temperature is the **kelvin** (K), named after an English physicist, Lord Kelvin. Notice that the units are in kelvins, not degrees Kelvin. The kelvin unit has the same size temperature change as a degree Celsius, so a change of 50 °C is equivalent to a change of 50 kelvins. The lowest possible temperature reading has been determined to be −273.15 °C. This is defined as the zero point on the Kelvin scale and is known as **absolute zero.** Thus, the Kelvin scale has no negative temperatures. To convert from degrees Celsius to kelvins, add 273.15, or simply 273 (using three significant figures) to the Celsius temperature.

$$K = °C + 273$$

EXAMPLE 3.26

What is the boiling point of water in kelvins? The boiling point of water is 100. °C.

SOLUTION

$$K = °C + 273 = 100. + 273 = 373\ K$$

EXAMPLE 3.27

Voyager I determined the surface temperature of Saturn's largest moon, Titan, to be 94 K. What is the temperature in degrees Celsius?

SOLUTION

Since K = °C + 273, then °C = K − 273.

$$°C = 94 - 273 = -179\ °C$$

See Problems 3.53–3.58.

3.13 *Temperature and Heat Energy*

We have been working with temperatures, but have not yet defined the term. **Temperature** is a measure of the *hotness* or *coldness* of matter and is usually expressed in degrees Fahrenheit, degrees Celsius, or kelvins. Temperature measures the *intensity* of energy of the particles in a substance. For example, the particles of water in a cup of hot water at 50 °C are more energetic, on the average, than are particles in a glass of cold water at 10 °C. Temperature and heat are related, but they are often confused.

Heat is the form of energy that is transferred between samples of matter because of a difference in their temperatures. A cup of hot water at 40 °C can have the same temperature as a bathtub of water, but the bathtub of water will melt more ice than will the cup of water (Figure 3.19). More heat can flow out of the bathtub of water than can flow out of the cup of water at the same temperature. For another example, suppose a pan full of water and a pan half full of water are both heated for the same time period, and the same amount of heat energy is transferred to both samples of water. After heating, the temperature will be higher in the half-full pan than in the full pan of water. This can be explained in the following way: when a

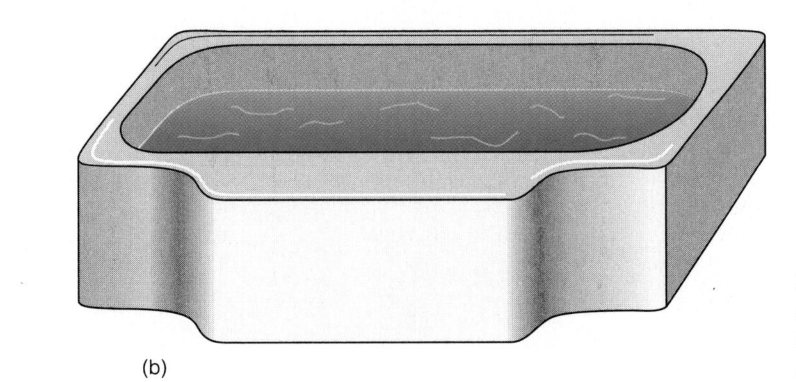

Figure 3.19 *A bathtub of water at a given temperature has more heat than does a glass of water at the same temperature.*

A *spontaneous* process is one that can occur when left alone.

A 75-watt light bulb uses 75 J of energy every second that it is lit.

g-°C means g × °C (grams times degrees Celsius).

Water, with its large specific heat, is one of the better materials for storing heat in solar-energy heating systems.

quantity of energy is distributed over fewer particles, each particle gets more of the energy, causing a greater temperature rise.

When heat energy flows spontaneously from one object to another, it always flows from the hot object to the cool object. When ice is placed in warm water, the temperature of the water drops as heat flows into the ice and melts the ice.

The SI unit of heat is the **joule** (J), but the familiar **calorie** (cal) is also a metric heat unit. Since both the joule and the calorie represent very small quantities of energy, the kilojoule (kJ) and kilocalorie (kcal) are often used.

$$1 \text{ cal} = 4.184 \text{ J}$$

$$1000 \text{ cal} = 1 \text{ kcal} = 4184 \text{ J}$$

$$1 \text{ kcal} = 4.184 \text{ kJ}$$

The **large Calorie** (note the capital C) is used for the measurement of the energy content of foods. The large Calorie is equivalent to a kilocalorie, so a 50-Calorie chocolate chip cookie actually has 50,000 calories. A dieter may know that a banana split contains 1500 Cal (kcal), but if the dieter realized that is 1,500,000 calories, giving up the banana split might be easier.

A **calorie** is the amount of heat required to raise the temperature of 1 g of water by 1 °C. Notice that heat was defined in terms of an energy change for water. Each kind of substance, such as iron, brass, or water, for example, takes a different amount of heat to raise the temperature of a 1-g sample by 1 °C. This value is called the **specific heat**.

$$\text{Specific heat is expressed in } \frac{\text{cal}}{\text{g-°C}} \quad \text{or} \quad \frac{\text{J}}{\text{g-°C}}$$

Substances with small specific heat values absorb little energy in warming, and give off little energy in cooling, as compared to substances like water, which has one of the highest specific heat values. Table 3.8 gives specific heats of several substances.

Table 3.8 **Specific Heats of Some Substances at 25 °C**

	Specific Heat	
*Substance**	*J/g-°C*	*cal/g-°C*
Aluminum (s)	0.900	0.215
Brass (s)	0.385	0.092
Copper (s)	0.385	0.0922
Ethyl alcohol (l)	2.45	0.586
Gold (s)	0.129	0.0308
Iron (s)	0.448	0.107
Lead (s)	0.129	0.0308
Magnesium (s)	1.02	0.244
Mercury (l)	0.139	0.0332
Silver (s)	0.236	0.0564
Steel (stainless) (s)	0.50	0.12
Water (l)	4.184	1.000
Zinc (s)	0.385	0.0922

*Solid (s), liquid (l), gas (g).

When heat is transferred to a certain quantity of a substance, the temperature changes. A temperature change is often symbolized as ΔT (called ''delta T'' because of the Greek letter used). The heat that a substance gains or loses as its temperature changes can be calculated by the following equation.

$$\text{Mass of substance} \times (\Delta T) \times \text{Specific heat} = \text{Heat gained or lost}$$

$$\text{Grams} \times {}^{\circ}\text{C} \times \text{J/g-}{}^{\circ}\text{C} = \text{Joules}$$
$$\text{Grams} \times {}^{\circ}\text{C} \times \text{cal/g-}{}^{\circ}\text{C} = \text{calories}$$

Any one of the four terms in the equation can be determined when the other three values are known. For these calculations, the mass must be in grams and the temperature change must be in either degrees Celsius or kelvins (since the units are the same size). When heat is expressed in calories, the specific heat must be in cal/g-°C. When heat is expressed in joules, the specific heat must be in J/g-°C.

EXAMPLE 3.28

How many joules are required to change the temperature of 225 g of lead from 5.0 °C to 25.0 °C?

SOLUTION

Simply substitute appropriate values into the equation given to calculate heat in joules. The mass of lead is 225 g, the temperature change is 25.0 − 5.0 = 20.0 °C, and the specific heat of lead from the table is 0.129 J/g-°C.

$$225\ \text{g} \times 20.0\ {}^{\circ}\text{C} \times \frac{0.129\ \text{J}}{\text{g-}{}^{\circ}\text{C}} = 581\ \text{J}$$

See Problems 3.59–3.62.

EXAMPLE 3.29

Suppose your diet is 2100 Calories (2100 kcal) per day and your body weight is 68 kg (assumed to be 100% water for this problem). Start with an initial body temperature of a normal 37 °C, and use a specific heat of 1.00 cal/g-°C or 1.00 kcal/kg-°C.

(a) Calculate the maximum temperature your body could reach by absorbing all 2100 kcal at once.
(b) Try to explain why your body does not reach temperatures in this range.

This problem is not as hard as it looks! It demonstrates a practical example involving heat.

SOLUTION

(a) Rearrange the equation given in this section to solve for the change in temperature, ΔT.

$$\Delta T = \frac{\text{Total heat}}{\text{Mass} \times \text{Sp heat}} = \frac{2100\ \text{kcal}}{68\ \text{kg} \times 1.00\ \text{kcal/kg-}{}^{\circ}\text{C}} = 31\ {}^{\circ}\text{C}$$

$$\text{Maximum temperature} = \text{Original temperature} + 31\ {}^{\circ}\text{C} = 37 + 31 = 68\ {}^{\circ}\text{C}$$

(b) Your body maintains the 37 °C temperature by metabolizing food at a relatively constant rate and by a cooling process involving the evaporation of perspiration.

Notice how these problems have involved several of the topics discussed in this chapter, including mass, significant figures, temperatures, specific heat, and heat energy in joules and calories.

Chapter Summary

Measurement in chemistry—and all other sciences—involves the use of metric or SI units. The essential base units in this system are the meter (length), the liter (volume), and the kilogram (mass). These base units are shown in Table 3.1. Larger and smaller units are obtained by using appropriate prefixes that represent multiples of 10. Working with decimals rather than simple fractions like fourths, eighths, and sixteenths makes calculations much easier. Common prefixes include kilo- (10^3), centi- (10^{-2}), milli- (10^{-3}), micro- (10^{-6}), and others shown in Table 3.2.

A volume of 1 liter is equal to 1000 mL, and 1 mL has the same volume as 1 cubic centimeter (cm^3 or cc). Mass is defined as the quantity of a substance and is measured in grams; weight varies in proportion to the gravitational attraction for the object. Density is the ratio of mass to volume for a given sample. Specific gravity is the ratio of the density of a certain substance to the density of water at the same conditions.

Dimensional analysis allows problem solving to become a logical, sequential process with the use of appropriate conversion factors. A conversion factor can be obtained from any two quantities that are equivalent, whether they are in metric units, English units, or a combination of the two systems. For example, the equality 2.54 cm = 1 in. is exact, by definition, and can be used to write two conversion factors: 1 in./2.54 cm and 2.54 cm/in. With dimensional analysis, a given quantity is multiplied by one or more conversion factors to obtain the answer in the units desired.

Pure numbers and defined quantities may be exact, but measured quantities always have some uncertainty. Precision relates to the closeness of several measured values to each other, but accuracy relates to the closeness of measured values to the accepted or true value. No calculated value can be more precise than the least precise number used in the calculation. The precision of a number is indicated by using an appropriate number of significant figures. For multiplication and division, the number of significant figures in the answer should be equal to the number of significant figures in the least precise value used. For addition and subtraction, the answer is rounded to the same decimal place as the least precise value.

The Fahrenheit temperature scale remains in wide use in the United States, but scientific measurements are based on Celsius or Kelvin temperatures. The zero point on the Kelvin scale is absolute zero, the lowest possible temperature. Mathematical equations are used to convert temperature readings.

Temperature is a measure of the hotness or coldness of matter. Heat is a measure of the quantity of energy transferred. Heat units include joules, kilojoules, calories, and kilocalories. The heat gained or lost, mass, temperature change, and specific heat are interrelated as shown by the equation

$$\text{Heat absorbed or released} = \text{Mass} \times \text{Temperature change} \times \text{Specific heat}$$

SI and metric units, conversion factors, and the dimensional analysis approach to problem solving will continue to be used throughout this text. This is the language of science.

Assess Your Understanding

1. Make approximations involving SI units of length, volume, mass, and temperature. [3.2, 3.4, 3.5, 3.12]
2. Convert metric lengths, volumes, and masses to other equivalent metric units. [3.3–3.5]
3. Use dimensional analysis and conversion factors to set up and solve problems involving both metric and English quantities. [3.6]
4. Use experimental data to discuss uncertainty in measurement. [3.7]
5. Determine the number of significant figures in data and in calculations. [3.8, 3.9]
6. Write numbers in scientific notation, and use these in calculations. [3.10]
7. Calculate density, specific gravity, volume, or mass from experimental data. [3.11]
8. Make conversions involving Fahrenheit, Celsius, and Kelvin temperatures. [3.12]
9. Make calculations involving heat (joules or calories), specific heat, mass, and change in temperature. Explain the meaning of these terms. [3.13]

Key Terms

absolute zero [3.12]
accuracy [3.7]
analytical balance [3.5]
calorie [3.13]
Calorie (the large) (kcal) [3.13]
Celsius [3.12]
conversion factors [3.2]
density [3.11]
dimensional analysis [3.3]
exact number [3.8]
factor-label method [3.3]
Fahrenheit [3.12]
gram [3.5]
heat [3.13]
hydrometer [3.11]
International System (SI) [3.1]
joule [3.13]
kelvin [3.12]
kilogram [3.1, 3.5]
liter [3.4]
meniscus [3.8]
meter (metre) [3.1]
metric system [3.1]
microgram [3.5]
microliter [3.4]
milligram [3.5]
milliliter [3.4]
parts per billion (ppb) [3.10]
powers of ten [3.10]
precision [3.7]
pycnometer [3.11]
rounding off numbers [3.9]
scientific notation [3.10]
significant figure(s) [3.8]
specific gravity [3.11]
specific heat [3.13]
temperature [3.13]
top-loading balance [3.5]
uncertain measurement [3.7]
uncertain digit [3.8]

Problems

METRIC AND SI UNITS (with prefixes)

3.1 For the following terms, identify the numerical quantity, unit, and name of the substance.

a. 10 lb sugar
b. 5 kg potatoes
c. 2 L cola

3.2 For the following items, identify the numerical quantity, unit, and name of the substance.
a. 1 gal milk
b. 500 mg vitamin C
c. 35 mm film (for camera)

3.3 Use Tables 3.1 and 3.2 to assist you with prefixes for the following.
a. Milli- is equivalent to the number ______, so 1.000 mg = ______ g.
b. Micro- is equivalent to the number ______, so 1.000 μL = ______ L.
c. Hecto- is equivalent to the number ______, so 1.000 hm = ______ m.

3.4 Use Tables 3.1 and 3.2 to assist you with prefixes for the following.
a. Pico- is equivalent to the number ______, so 1.000 ps = ______ s.
b. Kilo- is equivalent to the number ______, so 1.000 km = ______ m.
c. Deci- is equivalent to the number ______, so 1.000 dg = ______ g.

METRIC CONVERSIONS AND APPROXIMATIONS

3.5 Use Figure 3.4 to assist you with approximations of the following lengths.
a. The length of a dollar bill is about.
(1) 1.5 cm (2) 15 cm (3) 15 mm
(4) 1.5 m
b. The length of a football field is about
(1) 1 cm (2) 10 m (3) 100 m
(4) 1 km

3.6 Use Figure 3.4 to assist you with approximations of the following lengths.
a. The thickness of the wire in a paper clip is about
(1) 1 mm (2) 10 mm (3) 1 cm
(4) 10 cm
b. The thickness of this book is about
(1) 3 mm (2) 3 cm (3) 30 cm
(4) 0.3 cm
c. The width of a sheet of typing paper is about
(1) 22 m (2) 22 dm (3) 22 cm
(4) 22 mm

3.7 Show how you would set up each of the following problems, using the appropriate conversion factor(s). Then solve for the answer.
a. 1820. m to kilometers
b. 1400. cm to kilometers
c. 1700. mm to meters

3.8 Show how you would set up each of the following problems, using the appropriate conversion factor(s). Then solve for the answer.
a. 0.062 m to centimeters
b. 3000. m to kilometers
c. 875 μm to kilometers

3.9 Make the following conversions of metric lengths.
a. 12.5 cm = ______ mm
b. 345 cm = ______ m
c. 34.5 mm = ______ μm
d. 10.5 mm = ______ cm
e. 42.5 m = ______ cm
f. 0.092 m = ______ mm

3.10 Make the following conversions of metric lengths.
a. 200. m = ______ km
b. 0.829 cm = ______ μm
c. 52.8 nm = ______ mm
d. 4.5 km = ______ m
e. 6.5 μm = ______ mm
f. 105 mm = ______ nm

3.11 Use Figures 3.6 and 3.7 to assist you in making these approximations if necessary.
a. 20. mL is about how many teaspoons?
(1) 2 (2) 4 (3) 6
(4) 8 (5) 10
b. One aluminum can of soda is about how many milliliters?
(1) 3.5 mL (2) 35 mL (3) 350 mL
(4) 3500 mL
c. To measure out 86 mL of acid, use a
(1) 10.-mL graduated pipet
(2) 100.-mL cylinder
(3) 50.-mL buret

3.12 Use Figures 3.6 and 3.7 to assist you in making these approximations if necessary.
a. A pint is a little smaller than which of the following?
(1) 500. mL (2) 50. mL (3) 5 mL
(4) 5000. cm^3
b. 10 drops is about
(1) 5 mL (2) 0.5 mL (3) 0.05 mL
(4) 0.005 mL
c. To measure out 27.2 mL of a liquid accurately, use a
(1) 100.-mL graduated cylinder
(2) 10.0-mL pipet
(3) 50.0-mL buret

3.13 Make the following conversions of metric volumes.
a. 25 mL = ______ L

b. 0.005 mL = ______ μL
c. 50. μL = ______ mL
d. 750 mL = ______ cm^3

3.14 Make the following conversions of metric volumes.
a. 0.050 L = ______ mL
b. 0.8 μL = ______ mL
c. 8.9 cm^3 = ______ mL
d. 75 cc = ______ L

3.15 What is the volume in cubic centimeters of a 15 cm $\times$ 0.050 m $\times$ 8.0 mm rectangular solid?

3.16 What is the volume in cubic meters of a 6.0 m $\times$ 50. cm $\times$ 800. mm rectangular solid?

3.17 How many cubic decimeters are equal to 2 m^3?

3.18 One cubic decimeter is equal to how many liters?

3.19 If a 2-L bottle of cola costs $1.79 and 6 cans (354 mL each) cost $1.99,
a. What is the cost per liter for the cola in bottles?
b. What is the cost per liter for the cola in cans?
c. Which is the better bargain, for the prices quoted?

3.20 If a can of soft drink (354 mL) from a vending machine costs $0.50, and a 2-L bottle of the soft drink costs $1.37,
a. What is the cost per liter for the beverage from the vending machine?
b. What is the cost per liter for the beverage in bottles?
c. Which is the better bargain, for the prices quoted?

3.21 Use Figure 3.8 to assist you in making these approximations if necessary.
a. A penny weighs about
(1) 300 mg (2) 3 g (3) 30 g (4) 30 mg
b. A person with a mass of 130 kg
(1) is apparently rather slim
(2) has an average weight
(3) needs a weight control plan

3.22 Use Figure 3.8 to assist you in making these approximations if necessary.
a. To make 1 kilogram, it would take the mass of about
(1) 1.5 cans of soda (2) 3 cans of soda
(3) a 6-pack of soda
b. To have a mass of 1 g would take about
(1) 0.5 aspirin tablet (2) 9 aspirin tablets
(3) 3 aspirin tablets

3.23 Make the following conversions of metric masses.
a. 0.1 mg = ______ g
b. 250. g of water = ______ mL
c. 0.5 mg = ______ μg
d. 52.4 cg = ______ g

3.24 Make the following conversions of metric masses.
a. 5.4 g = ______ mg
b. 0.725 kg = ______ mg
c. 25 μg = ______ g
d. 50. mL of water = ______ g

CONVERSIONS INVOLVING METRIC AND ENGLISH UNITS

3.25 Show problem setups and answers to three significant figures for the following problems. (Conversion factors are provided in Table 3.5.)
a. 165 mm = ______ in.
b. 1200. mL = ______ qt
c. 145 lb = ______ kg
d. 1.50 ft to centimeters
e. 500. mL to fluid ounces
f. 275 g = ______ lb

3.26 Show problem setups and answers to three significant figures for the following problems. (Conversion factors are provided in Table 3.5.)
a. 55.0 mi = ______ km
b. 1.25 m = ______ in.
c. 150. mL = ______ fl oz
d. 55.0 mi/hr to kilometers per hour
e. 4.00 L to quarts
f. 150. g to ounces (avoir.)

3.27 Basketball player Shaquille O'Neal weighs 310. lb. What is his weight (actually mass) in kilograms?

3.28 Basketball player Manute Bol is 7.5 ft. tall. What is his height (a) in meters (b) in centimeters?

3.29 The speed of light is 186,000 mi/s. Convert this to meters per second.

3.30 An outstanding runner completed 1500. m in 3.00 min, 39.0 s. What is this rate in meters per second?

3.31 If you travel at the posted speed limit of 55.0 mi/hr, what is this rate in meters per second?

3.32 How many seconds will it take light to travel from the sun to Earth, a distance of 93 million mi or 1.5×10^8 km? The speed of light is 3.00×10^8 m/s.

3.33 How many days would it take you to count 200,000 items assuming you count one item every second, without stopping?

3.34 If water drips from a faucet at 1 drop per second

how many liters could be collected after 24.0 hr? Assume 20. drops make 1 milliliter.

SIGNIFICANT FIGURES AND SCIENTIFIC NOTATION

3.35 How many significant figures are in each of the following numbers?
a. 2.2000 b. 0.0350
c. 0.0006 d. 0.0089
e. 24,000 f. 4.360×10^4

3.36 How many significant figures are in each of the following numbers?
a. 0.0708 b. 1200
c. 0.6070 d. 21.0400
e. 0.007 f. 5.80×10^{-3}

3.37 Round off each of the following numbers to three significant figures.
a. 86.048 b. 29.974
c. 6.1275 d. 0.008230

3.38 Round off each of the following numbers to three significant figures.
a. 800.7 b. 0.07864
c. 0.06995 d. 7.096

3.39 Express each of the following numbers in scientific notation.
a. 0.0000070 b. 25.3×10^4
c. 825,000. d. 826.7×10^{-5}

3.40 Express each of the following numbers in scientific notation.
a. 43,500. b. 65.0×10^{-5}
c. 0.000320 d. 0.0432×10^4

3.41 Perform the following calculations, and report the answer with the appropriate number of significant figures.
a. 146.20 g beaker + 23.1 g water + 335 mg vitamin C = ______
b. 11.2 cm × 8.0 mm × 0.0093 cm = ______
c. $\dfrac{(86.0 \times 10^6)(0.00543 \times 10^{-2})}{0.03952} =$ ______

3.42 Perform the following calculations, and report the answer with the appropriate number of significant figures.
a. 124 g flask + 65 g water + 10.827 g salt = ______
b. 1.584 m × 62.0 cm × 345 mm = ______
c. $\dfrac{(0.0630 \times 10^{-9})(2.30 \times 10^2)}{6.28 \times 10^{-2}} =$ ______

DENSITY AND SPECIFIC GRAVITY

3.43 What is the mass (in kilograms) of a lead block measuring 20. cm × 20. cm × 10. cm? Densities are given in Table 3.7.

3.44 What is the mass (in kilograms) of a block of aluminum measuring 20. cm × 20. cm × 10. cm? Densities are given in Table 3.7.

3.45 What volume of ethyl alcohol, in milliliters, should be used for a procedure that requires 500. g of the alcohol? Densities are given in Table 3.7.

3.46 What volume of sulfuric acid (density 1.84 g/mL) is needed for a procedure requiring 54.0 g of the acid?

3.47 An irregular chunk of metal weighing 109.2 g was placed in a graduated cylinder containing 21.0 mL of water. The water plus the chunk of metal made a total volume of 33.2 mL.
(a) What is the density of the metal?
(b) From the list of densities, what could the metal be?
(c) Why can one not be totally sure of the identity of this metal based on this analysis?

3.48 An unknown liquid was placed in a pycnometer. The empty pycnometer had a mass of 15.2132 g. The pycnometer plus unknown had a mass of 23.4478 g. The pycnometer was cleaned and reweighed with distilled water, giving a total mass of 25.9263 g. What is the density of the unknown?

3.49 A procedure called for 45 g of concentrated hydrochloric acid (density 1.19 g/mL). What volume in milliliters should be used?

3.50 What is the mass, in grams, of 350.0 mL of chloroform (density 1.49 g/mL)?

3.51 What is the specific gravity of a sample of methylene chloride? (Densities are listed in Table 3.7.)

3.52 Determine the specific gravity of a liquid with a mass of 11.023 g, when the same volume of water had a mass of 11.997 g.

TEMPERATURE AND HEAT CALCULATIONS

3.53 Convert the following temperatures.
a. 68 °F = ______ °C
b. 39 °C = ______ °F
c. 39 °C = ______ K
d. −10. °F = ______ °C
e. −10. °C = ______ °F

3.54 Convert the following temperatures
a. 25 °F = ______ °C
b. 20. °C = ______ °F
c. 298 K = ______ °C
d. 0. °F = ______ °C
e. −40. °C = ______ °F

3.55 The temperature of liquid nitrogen is −196 °C. What is this in degrees Fahrenheit?

3.56 The temperature of Mars at night can drop as low as −120. °F. What is the Kelvin temperature?

3.57 Which is hotter, 100. °C or 100. °F?

3.58 Place these temperatures in order from coldest to hottest: 0 K, 0 °C, 0 °F.

3.59 How many calories of heat would be required to raise the temperature of 50.0 g of water from 20.0 °C to 50.0 °C?

3.60 How many calories of heat would be required to raise the temperature of 50.0 g of silver (specific heat 0.0564 cal/g-°C) from 20.0 °C to 50.0 °C?

3.61 How much energy in joules is released as 100. g of iron (specific heat 0.448 J/g-°C) cools from 100.0 °C to 30.0 °C?

3.62 How much energy in joules is released as 100. g of aluminum (specific heat 0.900 J/g-°C) cools from 100.0 °C to 30.0 °C?

Additional Problems

3.63 The energy released by consuming one quarter-pound (4-oz) hamburger is listed as 202 Cal (202 kcal).
- (a) Convert this energy to joules.
- (b) This energy would be enough to heat how many grams of water from 20. °C to 37 °C (body temperature)?

3.64 One baked potato provides 31 mg of vitamin C on the average. If 5.0 lb of potatoes has 15 potatoes, how many milligrams of vitamin C are provided per pound of potatoes?

3.65 Assume your diet provides 2700 Cal (2700 kcal) per day and your weight is 75 kg.
- (a) How many calories are used each hour, on the average, by each kilogram of your body?
- (b) How many joules is this?

3.66 A certain automobile gasoline tank holds 15.0 gal of gasoline (density 0.670 g/mL).
- (a) What is the mass of a full tank of gasoline in grams?
- (b) What is this mass in pounds?

3.67 An engine has a displacement (maximum volume change) of 3.80 L. Convert this to cubic inches. (See Table 3.5 for conversion factors.)

3.68 A motorcycle has a 750.-cc engine. Convert this to cubic inches.

3.69 How many grams of dry air are in a room 15.0 ft × 18.0 ft × 8.0 ft. Use an average density of dry air as 1.168 g/L.

3.70 Through experimental work, it was determined that 143 J of heat was released by 32.2 g of an unknown metal as it cooled 10.3 °C. What is the specific heat of the unknown metal?

The elements shown here are (clockwise from the top) bromine, sulfur, carbon, aluminum, copper, mercury, and cobalt.

4

Elements and Atoms

CONTENTS

On the inside cover of this book—and probably on the wall of your classroom—you will find what is called a periodic table of the elements. This table lists symbols of over 100 known elements that are arranged so that those with similar properties are positioned near one another. As far as we can tell, everything in the entire universe is made from these elements. There is a different kind of atom for each kind of element. When these atoms combine together chemically, many different compounds are possible. Chemists have now identified over 10 million compounds. Your own body has thousands of compounds made from atoms of these elements.

In this chapter we will concentrate our study on the elements. Names and symbols of the elements are essential to further understanding. We will identify the elements that are abundant and those that are rare. We will compare these elements and discuss their distinguishing properties. Moving then to the atomic level, we will examine fundamental properties of atoms and describe the major subatomic particles.

4.1 Elements: Ancient and Modern Theory

Boyle's work with gases is described in Chapter 12.

It took many centuries for our present understanding of chemical elements to be developed. Robert Boyle, a quiet-mannered English scientist, included a definition of elements in his book *The Sceptical Chymist*, published in 1661. He said that supposed **elements** must be tested to see if they really were simple. If a substance could be broken down into simpler substances, it was not an element. Boyle further stated that the simpler substances might be elements and would be so regarded until such time (if it ever came) as they in turn could be broken down into still simpler substances. When two or more elements combine, they form a distinctly different substance, called a **compound**.

Using Boyle's definition, Antoine Lavoisier (1743–1794), the noted Frenchman who discovered the law of conservation of mass, included a table of 33 elements in his chemistry textbook, *Elementary Treatise on Chemistry*, published in 1789 (see page 32). A few of the elements listed in his table were not really elements, but Lavoisier was the first to use modern, somewhat systematic, names for chemical elements.

We could say that modern chemistry began at about this time.

Since the time of Lavoisier, researchers have discovered many naturally occurring elements and have synthesized several others, bringing the total to more than

A CLOSER LOOK

Searching for the Elements

Identifying the elements present in each kind of substance was no easy task—it took centuries. Greek philosophers who lived in the fifth century B.C. believed that all matter was composed of only four fundamental elements: earth, air, fire, and water. It eventually became clear that none of the four is actually an element.

During the Middle Ages (about 500–1500 A.D.) alchemists, working in secrecy and surrounded by mysticism, searched for a universal solvent to change metals such as iron, copper, and zinc into gold. Because the alchemists did not really understand the nature of elements, they assumed that these common metals could be transformed into the valuable metal, gold. The alchemists failed to carry out this process, called transmutation, but their investigations in primitive laboratories led to many discoveries such as improved methods of distillation and extraction that are still used today.

The Alchemist, a painting done by the Dutch artist Cornelis Bega around 1860, depicts a laboratory of the seventeenth century.

100 elements. In the periodic table, the elements are numbered sequentially. Elements with numbers greater than 92 have not been found in nature, but 17 of these elements have been synthesized since 1940. The most recent discoveries occurred in 1984, when element 108 was first produced, and in 1981 and 1982, when elements 107 and 109, respectively, were synthesized. Three of the first 92 elements are believed to be absent from the Earth, but they have been synthesized.

Element	*Number*	*Date Synthesized*	*Location*
Technetium	43	1937	Italy
Astatine	85	1940	University of California
Promethium	61	1941	Ohio State University

Technetium does not occur naturally on the Earth, but astronomers believe the element exists in certain stars. This prediction is based on spectral analysis of stars. Synthetic technetium is used for medical diagnosis, especially in scans of brain, liver, and bone tissue.

EXAMPLE 4.1

List major contributions of each of the following.

(a) Robert Boyle (b) Lavoisier (c) modern researchers

See Problems 4.1–4.8.

SOLUTION

(a) Robert Boyle defined elements as substances that cannot be broken into simpler substances.
(b) Lavoisier's textbook included a list of elements using Boyle's criteria. He emphasized experimentation and collection of quantitative data.
(c) Since 1940, modern researchers have synthesized 17 elements with numbers greater than 92.

4.2 *Names and Symbols*

Symbols are often used in place of written statements. Many of the early symbols used to represent various chemical substances were derived from ancient mythology (Figure 4.1). But symbols were not standardized; different alchemists of the Middle Ages developed their own shorthandlike notations to keep their work hidden. For example, in one Italian manuscript written in the seventeenth century, the element mercury was represented by 20 different symbols and 35 different names.

J. J. Berzelius, a Swedish chemist (Figure 4.2), invented a simple system of chemical notation that he introduced in 1814. His symbols were letters taken from the name of the element. Today, these symbols are used worldwide.

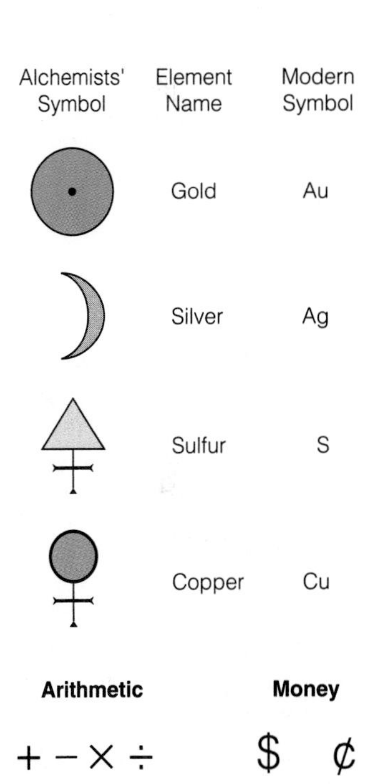

Figure 4.1 *The use of symbols is not unique to chemistry. Symbols can be quite helpful—when you know what they mean.*

The **symbol** for an element has no more than three letters. The first letter is always capitalized; the second and third letters, if used, are not capitalized. Only elements after number 103 in the periodic table have symbols with three letters. These symbols are derived from Latin names that represent the number of the element. For example, Uno is the symbol for element number 108, Unniloctium.

The names of elements, and therefore their symbols, are derived from many sources. Some names are from Latin, Greek, or German words that describe a characteristic property of the element. Others are named for the country or location of their discovery or to honor famous scientists. For example, the name for the element barium comes from the Greek word *barys,* meaning heavy. The name germanium was derived from *Germania,* the Latin name for Germany. Element number 99 was named einsteinium in honor of Albert Einstein. The names, sym-

The *symbol* Co represents cobalt, but CO (capital C and capital O) is the *formula* for carbon monoxide, which is made up of the elements carbon (C) and oxygen (O). Care must be taken to make sure each letter is clearly and properly written to avoid confusion.

Figure 4.2 *Jöns Jakob Berzelius (1799–1848) was the Swedish chemist who invented modern chemical symbols. He carried out over 2000 experiments in his simple laboratory and discovered the elements silicon, selenium, cerium, and thorium.*

bols, and origins of symbols, along with important uses of the most common elements are listed in Table 4.1.

EXAMPLE 4.2

Give the symbol and two important uses of the following elements.
(a) sodium (b) lead (c) mercury (d) zinc

SOLUTION

Element	*Symbol*	*Some Uses*
(a) Sodium	Na	Table salt (NaCl), lye (NaOH), and many other compounds
(b) Lead	Pb	Automobile batteries, lead shot in ammunition, and leaded ''crystal'' glassware (contains lead compounds)
(c) Mercury	Hg	Small batteries for cameras and hearing aids and thermometers for use in laboratories and hospitals
(d) Zinc	Zn	Cosmetics, pigments, and zinc-coated ''galvanized'' materials that prevent rusting (nails, buckets, garbage cans, etc.)

See Problems 4.9–4.16.

EXERCISE 4.1

Give the names of elements with the following symbols.
(a) K (b) Cu (c) Fe (d) Ag

Table 4.1 **Some Important Elements: Names, Symbols, and Uses**

Element	*Symbol*	*Origin of Name*	*Some Important Uses*
Aluminum	Al	Latin, *alumen* (alum)	Kitchen utensils, aircraft, containers, structural alloys
Antimony	Sb	Latin, *stibium* (mark)	Batteries, cable sheathing, flameproofing compounds
Argon	Ar	Greek, *argon* (inactive)	Electric light bulbs and fluorescent tubes
Arsenic	As	Latin, *arsenicum*	Semiconductors (It is poisonous.)
Barium	Ba	Greek, *barys* (heavy)	Paint, glassmaking, fireworks
Beryllium	Be	Greek, *beryl*	Alloy in springs, nonsparking tools, computer parts
Bismuth	Bi	German, *bisemutum* (white mass)	Low-melting alloys for castings and fire sprinklers
Boron	B	Arabic, *buraq*	Borax water softener, hardening steel, borosilicate glass
Bromine	Br	Greek, *bromos* (strong smell)	Fumigants, compounds used as medicines and dyes
Cadmium	Cd	Latin, *cadmia*	Low-friction bearings, batteries, electroplating
Calcium	Ca	Latin, *calx* (lime)	Cement, lime, alloys
Carbon	C	Latin, *carbo* (charcoal)	Diamonds, graphite lubricants, charcoal, tires, ink
Cesium	Cs	Latin, *caesius* (sky blue)	Photoelectric cells
Chlorine	Cl	Greek, *chloros* (greenish yellow)	Water purification, paper bleaching, dyes, solvents

Table 4.1 **Some Important Elements: Names, Symbols, and Uses (*continued*)**

Element	*Symbol*	*Origin of Name*	*Some Important Uses*
Chromium	Cr	Greek, *chroma* (color)	Chrome, stainless steel
Cobalt	Co	German, *Kobold* (goblin)	Magnets, tools, stainless steel
Copper	Cu	Latin, *cuprum*	Electrical wire, brass, coins, algae control
Fluorine	F	Latin, *fluere* (to flow)	Uranium production, etching glass, freon compounds
Gallium	Ga	Latin, *Gallia* (France)	Semiconductors, transistors
Germanium	Ge	Latin, *Germania* (Germany)	Semiconductors, transistors
Gold	Au	Latin, *aurum* (shining dawn)	Electrical conductors, jewelry, coinage
Helium	He	Greek, *helios* (sun)	Welding, weather balloons, gas mixture for divers
Hydrogen	H	Greek, *hydro* (water) and *genes* (forming)	Manufacture of ammonia, rockets, hydrogenation of vegetable oil
Iodine	I	Greek, *iodos* (violet)	Medicine, thyroid treatment, manufacture of chemicals
Iron	Fe	Latin, *ferrum*	Steel, iron alloys, magnets, machines, tools, automotive parts
Krypton	Kr	Greek, *kryptos* (hidden)	Bright light bulbs
Lead	Pb	Latin, *plumbum*	Automobile batteries, radiation shielding, ammunition, crystal
Lithium	Li	Greek, *lithos* (stone)	Lubricants, synthesis of organic chemicals, drying agent
Magnesium	Mg	Magnesia, Greece	Flashbulbs, flares, lightweight alloys
Manganese	Mn	Latin, *magnes* (magnet)	Steel alloys, dry cell batteries
Mercury	Hg	Ancient, *hydrargyrum*	Thermometers, electrical switches, batteries, explosives
Neon	Ne	Green, *neos* (new)	Neon advertising signs
Nickel	Ni	German for satan or Old Nick	Stainless steel, coinage, vaults, armor plate
Nitrogen	N	Greek, *nitron*	Ammonia synthesis, cryogenics (very cold temperatures, −196 °C)
Oxygen	O	Greek, *oxys* (acid) and *genes* (forming)	Respiration, combustion, thousands of organic compounds
Phosphorus	P	Greek for light producing	All plant and animal cells, fertilizers, detergents
Platinum	Pt	Spanish, *platina* (silver)	Jewelry, jet engines and missiles, corrosion resistance
Potassium	K	English, *potash* Latin, *kalium*	Fertilizers, present in thousands of compounds
Silicon	Si	Latin, *silex* (flint)	Semiconductors, computer chips, abrasives, tools, water repellents
Silver	Ag	Latin, *argentum*	Coinage, photographic chemicals, jewelry, silverware, electrical contacts, batteries
Sodium	Na	Latin, *natrium*	Present in many compounds, salt (NaCl), lye (NaOH)
Strontium	Sr	Strontian, Scotland	Flares and fireworks (for red color)
Sulfur	S	Sanskrit, *sulvere*	Gunpowder, automobile tires, sulfuric acid, paper, fumigants
Tin	Sn	Latin, *stannum*	Alloys (pewter, bronze, solder), coating of steel for tin cans
Tungsten	W	Swedish *tung sten* (heavy) German, *wolfram*	Highest melting point, light bulbs, dental drills, steel
Zinc	Zn	German, *zink*	Galvanized nails (zinc-coated), die castings, pigments, cosmetics

Study Hint
Begin now to learn the names and symbols for elements used in Table 4.1. You should be able to write the correct symbol when the name is given and to provide the correct spelling for the name of the element when the symbol is given. These symbols will be used extensively in writing chemical formulas of compounds and writing chemical equations.

Table 4.2 **The Most Abundant Elements in the Universe**

	Percentages of Total Atoms	
Element	*The Universe*	*Our Solar System*
Hydrogen	93%	85%
Helium	7%	15%
All others	0.1%	0.1%
Oxygen, Carbon, Nitrogen, Silicon	The percentages of these elements total less than 0.1%, but they are more abundant than other elements not listed.	

4.3 *Abundant and Rare Elements*

The abundance of different elements varies greatly throughout the universe. About 93% of all atoms in the entire universe are hydrogen atoms. Helium atoms account for about 7%, leaving a total of less than 0.1% for all other elements. The percentages for our own solar system are a little different, as shown in Table 4.2, but the first two elements in the periodic table—the simplest elements—account for over 99% of all atoms in the universe and in our solar system.

On our planet, 11 elements account for over 99% of the mass of the Earth's crust, the water of oceans and rivers, and the atmosphere. The percentages of these 11 most abundant elements are shown in Figure 4.3. Oxygen (almost 50%) and

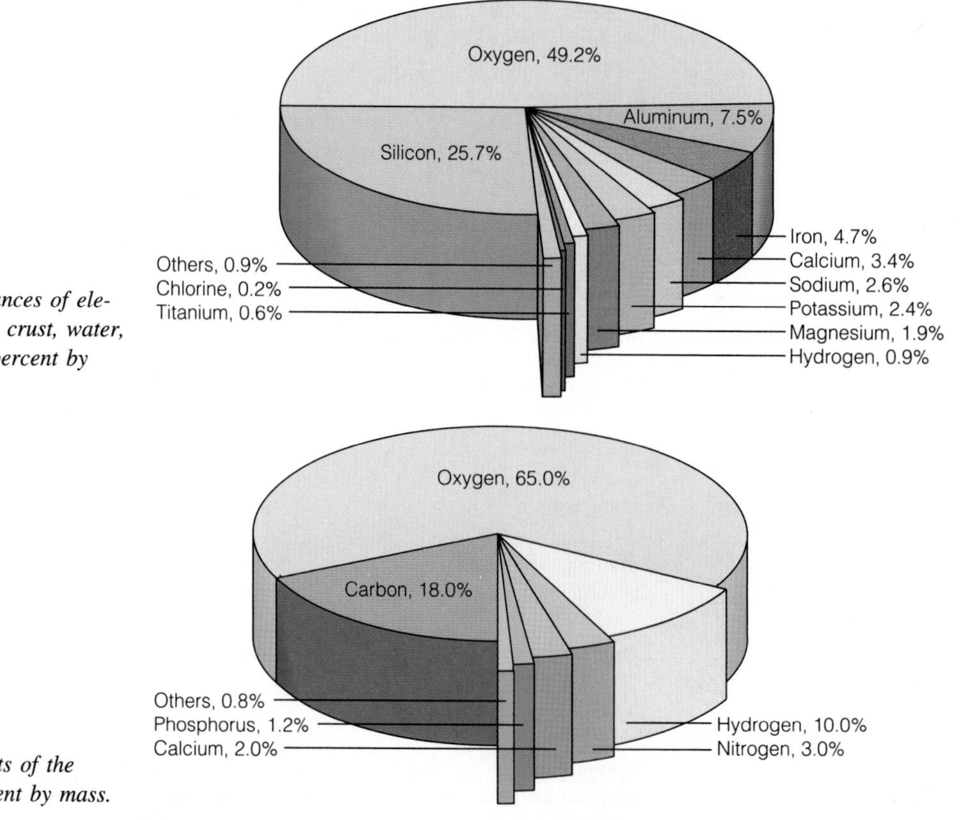

Figure 4.3 *Abundances of elements in the Earth's crust, water, and atmosphere in percent by mass.*

Figure 4.4 *Elements of the human body in percent by mass.*

silicon (25%) together make up 75% of this mass. The water that covers about 71% of the Earth's surface is about 89% oxygen, by mass. Sand and silicate compounds present in clay are high in both oxygen and silicon.

Only three elements—oxygen, carbon, and hydrogen—are responsible for 93% of the mass of the human body. These three elements plus nitrogen, calcium, and phosphorus make up 99% of the mass of the human body (Figure 4.4). Only trace amounts of other elements are in the human body, but several of the trace elements are of utmost importance to maintaining health.

EXAMPLE 4.3

List the two most abundant elements in the universe, the Earth's crust and atmosphere, and the human body.

SOLUTION

Rank (abundance)	*Universe*	*Earth's Crust*	*Human Body*
First	Hydrogen	Oxygen	Oxygen
Second	Helium	Silicon	Carbon

See Problems 4.17 and 4.18.

4.4 The Periodic Table of Elements

A periodic table, like the one pictured in Figure 4.5, provides a great deal of information about the elements. Each of the elements can be classified as a metal, a nonmetal, or a metalloid. The **metals**, shown at the left of the periodic table, are

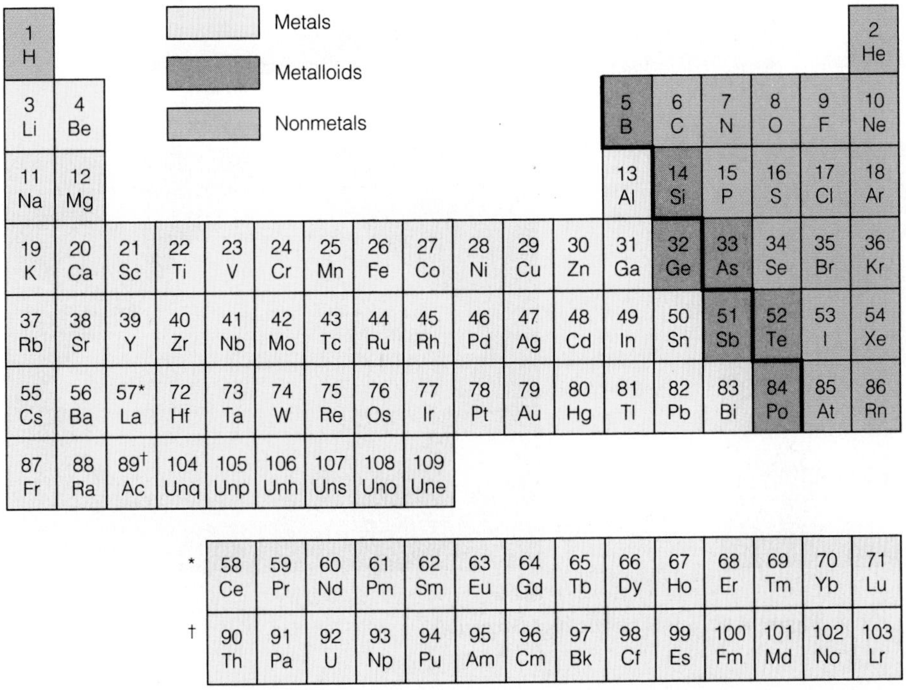

Metals
Metalloids
Nonmetals

1 H																	2 He
3 Li	4 Be											5 B	6 C	7 N	8 O	9 F	10 Ne
11 Na	12 Mg											13 Al	14 Si	15 P	16 S	17 Cl	18 Ar
19 K	20 Ca	21 Sc	22 Ti	23 V	24 Cr	25 Mn	26 Fe	27 Co	28 Ni	29 Cu	30 Zn	31 Ga	32 Ge	33 As	34 Se	35 Br	36 Kr
37 Rb	38 Sr	39 Y	40 Zr	41 Nb	42 Mo	43 Tc	44 Ru	45 Rh	46 Pd	47 Ag	48 Cd	49 In	50 Sn	51 Sb	52 Te	53 I	54 Xe
55 Cs	56 Ba	57* La	72 Hf	73 Ta	74 W	75 Re	76 Os	77 Ir	78 Pt	79 Au	80 Hg	81 Tl	82 Pb	83 Bi	84 Po	85 At	86 Rn
87 Fr	88 Ra	89† Ac	104 Unq	105 Unp	106 Unh	107 Uns	108 Uno	109 Une									

*	58 Ce	59 Pr	60 Nd	61 Pm	62 Sm	63 Eu	64 Gd	65 Tb	66 Dy	67 Ho	68 Er	69 Tm	70 Yb	71 Lu
†	90 Th	91 Pa	92 U	93 Np	94 Pu	95 Am	96 Cm	97 Bk	98 Cf	99 Es	100 Fm	101 Md	102 No	103 Lr

Figure 4.5 *Elements are classified as metals, metalloids, and nonmetals, as shown on this periodic table.*

Some Elements That Are Metals

Cobalt

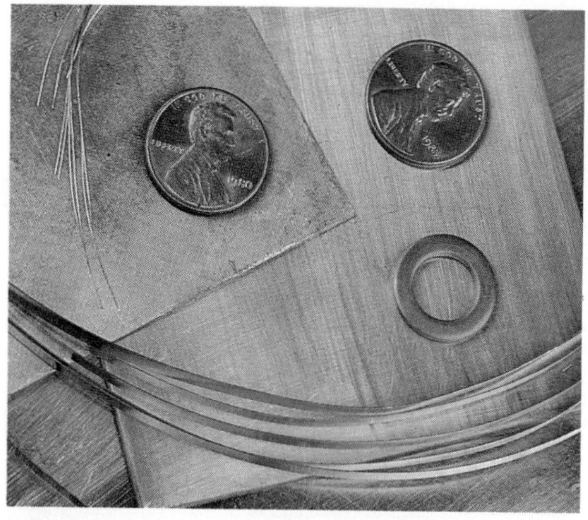
Copper

Magnesium

Some Elements That Are Nonmetals

Chlorine

Phosphorus

Sulfur

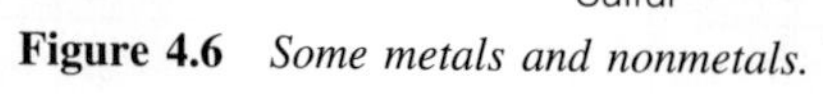
Figure 4.6 *Some metals and nonmetals.*

separated from the **nonmetals**, at the right of the periodic table, by a heavy, step-like, diagonal line. Notice that there are many more metals than there are nonmetals. Elements located adjacent to this diagonal line are called **metalloids**. They possess certain properties that are intermediate between those of typical metals and nonmetals. Horizontal rows of elements in the periodic table are called **periods** of elements. Vertical columns of elements in the table are called **groups** or, sometimes, **families** of elements. For example, Group IA metals, the family of **alkali metals**, are those in the first column at the far left of the periodic table.

Metals

Most metals, unlike the nonmetals, can be polished to give a shiny metallic luster (Figure 4.6). Metals do not tend to combine chemically with each other, but react with nonmetals to form many different compounds. Common ores of metals such as iron and aluminum contain the metal combined with oxygen. Metals in Group IA of the periodic table are the most reactive. These metals are never found in nature as "free," uncombined elements. Less reactive metals, such as copper, silver, and gold—located near the center of the periodic table—are much more likely to be found in nature as "free" elements.

Nonmetals

Nonmetals include two familiar gases—nitrogen and oxygen—that are present in the atmosphere. Carbon—present as diamond, graphite, and charcoal—and sulfur are nonmetals that can be found in nature as solids in the uncombined, elemental form. In minerals, metals are combined chemically with nonmetals such as oxygen, sulfur, nitrogen, and phosphorus. Nonmetals also combine with one another to form compounds such as carbon dioxide, CO_2, carbon monoxide, CO, sulfur dioxide, SO_2, methane, CH_4, and ammonia, NH_3. Fluorine is the most reactive nonmetal.

Diatomic Elements

Instead of existing as single atoms, seven of the nonmetallic elements—hydrogen, nitrogen, oxygen, fluorine, chlorine, bromine, and iodine—exist as atom pairs that are combined chemically to form **diatomic molecules** at typical room conditions (Figure 4.7). The subscript 2 is used in their chemical formulas—H_2, N_2, O_2, F_2, Cl_2, Br_2, and I_2—to indicate that each molecule has two atoms of the same element. In each diatomic molecule, the two atoms are held together by attractive forces called *chemical bonds.* The last four elements in this list belong to the same family of elements—the **halogen family**.

Notice the locations of the diatomic elements on the periodic table and learn their names.

To clarify a common point of confusion, once any one of the seven diatomic elements combines with another element to form a compound, the tendency to be in atom pairs no longer applies. In compounds, the numbers of atoms of each kind depend on which kinds of atoms are combined. For example, in ammonia gas, NH_3, each molecule is composed of *one* nitrogen atom and *three* hydrogen atoms; there are no atom pairs. In nitric acid, HNO_3, there is only one atom of hydrogen for each atom of nitrogen and three atoms of oxygen; there are no atom pairs.

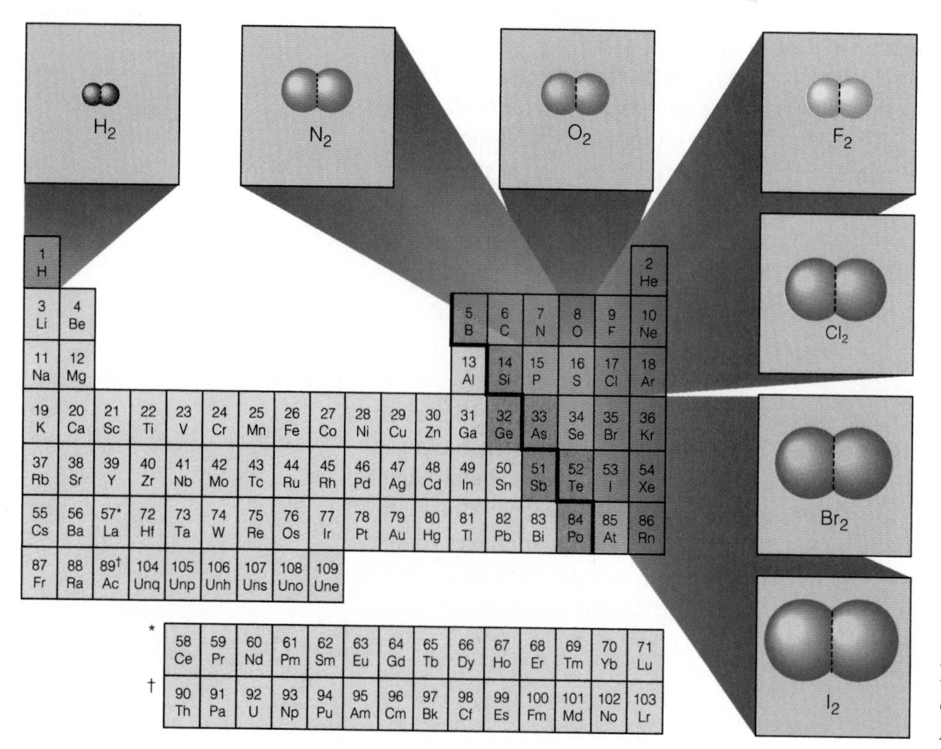

Figure 4.7 *The diatomic molecules: H_2, N_2, O_2, F_2, Cl_2, Br_2, I_2.*

Metalloids

Elements found in the intermediate region between metals and nonmetals in the periodic table are called metalloids. Their properties are also generally intermediate in character. For example, metals are good conductors of electricity, nonmetals are nonconductors, and metalloids are electrical semiconductors. This special property makes metalloids like silicon, germanium, arsenic, and boron especially useful in the electronics industry for the manufacture of transitors, computer chips, and solar electric cells. Silicon, the most abundant metalloid, is the fourth most abundant element. It is never found in nature in the elemental form, but silicates—complex compounds of silicon, oxygen, and various metals—are present in soils, clays, and sand. Quartz, sand, agate, amethyst, and flint all contain forms of impure silicon dioxide, SiO_2. Bricks, glass, cement, and ceramics all contain silicon compounds (Figure 4.8).

Computer chips and semiconductor devices require high-purity silicon; any impurities can be no greater than 1 part per million.

EXAMPLE 4.4

Using the periodic table, classify the following elements as metals, nonmetals, or metalloids.

(a) calcium (b) arsenic (c) iodine

SOLUTION

(a) Calcium is a metal.
(b) Arsenic is a metalloid.
(c) Iodine is a nonmetal.

See Problems 4.19–4.22.

EXERCISE 4.2
Are magnesium and phosphorus in the same family or the same period of elements?

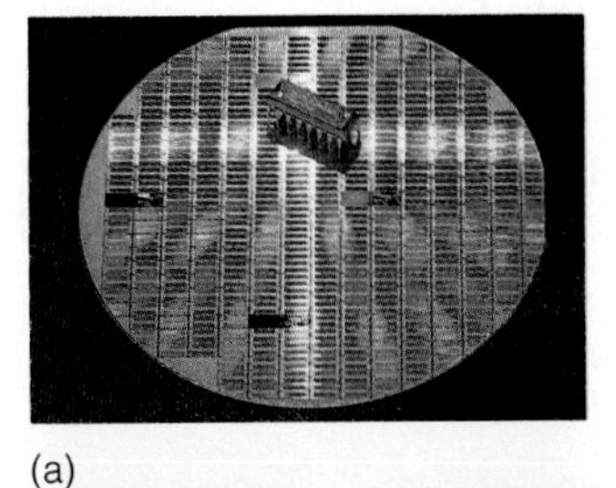

(a)

(b)

Figure 4.8 *(a) The element silicon is used extensively in computer chips. (b) Common sand (shown magnified) is composed of silicon dioxide, a compound of silicon and oxygen.*

4.5 Physical Properties of Elements

Special physical properties of different elements are responsible for nearly endless and varied applications. The semiconductor characteristics required for computer chips can be met by using silicon, but not silver. The lightweight structural material requirements for aircraft can be met by using aluminum and titanium, but not by using a reactive metal such as lithium or a dense metal such as lead. Thus, it is the unique properties of many elements that make them valuable. Some of these unique properties also help us classify elements as metals or nonmetals, as shown in Table 4.3.

Physical State

Whether an element is a solid, liquid, or gas at a particular temperature depends on its melting point and boiling point. Most elements are solids at room temperature (20 °C). Eleven elements are gases. Six of these gases are called **noble gases**; they do not ordinarily combine with other elements to form compounds. The noble gases—helium, neon, argon, krypton, xenon, and radon—are in the right-hand column of the periodic table. The other five elements that are gases are hydrogen, nitrogen, oxygen, fluorine, and chlorine.

Table 4.3 **Physical Properties of Metals and Nonmetals**

Property	*Metals*	*Nonmetals*
Physical state	Solids, except Hg	Solids, liquids, gases
Conductivity	Good conductors of heat and electricity Examples: Ag, Cu, Hg, Al	Poor conductors of heat and electricity Examples: S, Se, I_2
Luster	Shiny surface Examples: Ag, Au, Cr	Dull surface Examples: C (charcoal), sulfur, phosphorus
Malleability	Malleable; many can be hammered or rolled Examples: Fe, Au, Sn, Pb	Not malleable, brittle, crumble when struck Examples: S, C (charcoal)
Ductility	Ductile; many can be drawn into wire Examples: Al, Cu, Fe	Not ductile
Hardness	Some hard; some soft Examples: Hard metals: Cr, Fe, Mn Soft metals: Au, Pb, Na	Most—except diamond—not hard

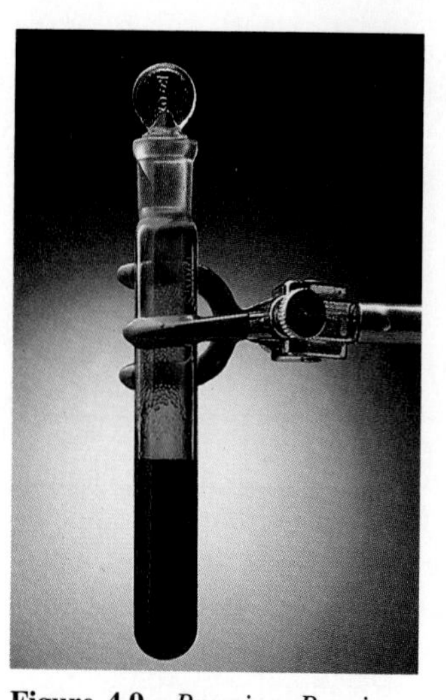

Figure 4.9 *Bromine, Br_2, is the only nonmetallic element that is a liquid at room temperature.*

Figure 4.10 *Mercury, Hg, is the only metallic element that is a liquid at room temperature.*

At room temperature (20 °C) only two elements are liquids: bromine and mercury. Bromine is a red liquid nonmetal (Figure 4.9). It is quite reactive and is never found in nature as an uncombined element. On an industrial scale, bromine can be extracted chemically from seawater or from brine wells. Bromine is used widely in the preparation of numerous important chemicals including prescription drugs.

Mercury metal is quite poisonous; it can be absorbed through the skin or the respiratory tract.

Mercury is a shiny, silvery liquid metal with a melting point of −39 °C and a boiling point of 357 °C (Figure 4.10). With this temperature range in the liquid state and good electrical conductivity, mercury is especially useful in mercury "quiet" switches (they allow electric lights to be turned on and off silently), thermostats, thermometers, and mercury vapor lamps used to light parking lots and other large open areas.

Tungsten has a higher melting point than any other element, (3407 °C or 6165 °F). It is used as a filament in incandescent electric light bulbs.

Except for mercury, all other metals are solid at room temperature (20 °C). Some have rather high melting points. These elements are clustered in the lower central region of the periodic table.

Conductivity

The **conductivity** of a substance is a measure of the relative ease with which a sample can transmit heat or electricity. Metals conduct both heat and electricity, while nonmetals are poor conductors. The best conductors are the metals copper, silver, gold (note their placement in the periodic table), and aluminum. Certain very reactive metals in Groups IA and IIA (at the left edge of the periodic table) are also good conductors. These include sodium, potassium, magnesium, and calcium. By contrast, good insulators can be made from various glass and ceramic materials containing complex compounds of nonmetals such as oxygen with metalloids such as silicon.

Luster, Malleability, Ductility, and Hardness

Polished surfaces of metals such as silver and chromium have a **luster** that results from their ability to reflect light. Nonmetals do not have a luster. Certain metals are

malleable; that is, they can be rolled or hammered into shape. The malleable metals include aluminum, copper, and steel—an alloy of iron with carbon and traces of several other elements. Metals like copper, aluminum, and iron are **ductile**; they can be drawn into wire. Nonmetals are not malleable or ductile. Although some metals like chromium, manganese, and nickel have a hard surface, other metals like gold, lead, copper, sodium, and potassium are soft. The unique physical properties of many elements make possible a wide variety of important uses for industrial and consumer products.

EXAMPLE 4.5

Which of the following characteristics are true for nonmetals?

See Problems 4.23–4.28.

(a) They are all gases at room temperature.
(b) They are not malleable or ductile.
(c) They have shiny surfaces.
(d) They are nonconductors.

SOLUTION

(a) False. One nonmetal, Br_2, is a liquid; the others are solids or gases.
(b) True.
(c) False. Their surfaces are dull.
(d) True.

EXERCISE 4.3

Name two elements that are liquids at room temperature.

Figure 4.11 *Democritus, an ancient greek philosopher, believed matter was composed of ultimate particles that he named* atoms.

4.6 Atoms: Democritus to Dalton

Most of the ancient Greek philosophers, including Aristotle (384–322 B.C.), believed matter was continuous and could be divided endlessly into smaller portions. They thought droplets of water could be divided into smaller droplets, indefinitely. But Leucippus, on the basis of intuition alone, concluded that there must be ultimate particles that could not be further subdivided. His pupil, Democritus (about 470–380 B.C.), gave these ultimate particles names (Figure 4.11). He called them *atomos* (Gr., *a*, "not" + *tomos*, "to cut") meaning "indivisible." It is from this name that we get the word "atom."

"To understand the very large, we must understand the very small."
—Democritus

Today we know that Democritus was right, although his was a minority view in his time. The popular "continuous" view of matter prevailed for 2000 years—until about 300 years ago. By that time, some scientists were making careful observations and accurate measurements.

Today, ideas about the nature of matter are based on what we learn from experimental research.

The Discovery of Oxygen

In 1774 Joseph Priestley, an English clergyman and scientist, prepared pure oxygen when he used a lens to focus the sun's rays on a compound containing mercury and oxygen. The gaseous product caused a candle to burn more brightly. He called the gas "perfect air," but he did not recognize it as a new element. Karl W.

Priestley (1733–1804) openly supported the French and American revolutions. He was harassed on several occasions, and his home, library, and laboratory were plundered. He fled England in 1794 and settled in Pennsylvania, where he lived until his death in 1804.

Scheele (1742–1786) was quite enthusiastic about scientific investigations. He said, "how glad is the enquirer when discovery rewards his diligence; then his heart rejoices."

Scheele, a Swedish apothecary (pharmacist), also discovered the element at about the same time—earlier, according to his dated notebook—but while his book was delayed in printing, Priestley announced his discovery.

The Explanation of Burning

Not long after oxygen was discovered, Antoine Lavoisier realized that this new element was the missing piece of the puzzle needed to explain combustion. He carried out quantitative experiments and formulated the correct theory of combustion: substances combine with oxygen of the air as they burn. He explained that respiration and combustion are similar, chemically. In both processes, a substance reacts with oxygen to produce carbon dioxide and water. Mass (matter) is conserved in both cases. This law of conservation of mass was described in Chapter 2.

The Law of Definite Proportions

Both Lavoisier and Proust were members of the French nobility, but Proust was working in Spain, temporarily safe from the ravages of the French Revolution. However, his laboratory was destroyed and he was reduced to poverty when Napoleon Bonaparte's troops occupied Madrid in 1808.

In 1799, Joseph Louis Proust showed that a substance called copper carbonate, whether prepared in the laboratory or obtained from natural sources, contained the same three elements—copper, carbon, and oxygen—and always in the same proportions by mass—5.3 parts of copper to 4.0 parts of oxygen to 1.0 part of carbon (Figure 4.12). Proust formulated a new law that summarized the results of this experiment and numerous others. A compound, he concluded, always contains elements in certain definite proportions, and in no other combinations. This generalization he called the **law of definite proportions**; it is sometimes called the *law of constant composition.*

The Electrolysis of Water

Henry Cavendish, a wealthy, eccentric Englishman, had observed in 1783 that water was produced when hydrogen was burned in oxygen. (It was Lavoisier, however, who correctly explained this experiment and used the names hydrogen and oxygen for these elements.) In 1800, the reverse reaction was carried out by two English chemists who passed an electric current through water and decomposed it into hydrogen and oxygen. The decomposition of a compound by an electric current is called **electrolysis**. The electrolysis of water always produces hydrogen and oxygen in a ratio of 2 to 1 by volume and further illustrates the law of definite proportions. This scientific breakthrough was a death blow to the ancient Greek idea that water is an element. It also paved the way for important developments that were about to occur in chemistry.

The first electrolysis of water was performed only six weeks after the electric battery was invented by an Italian, Alessandro Volta.

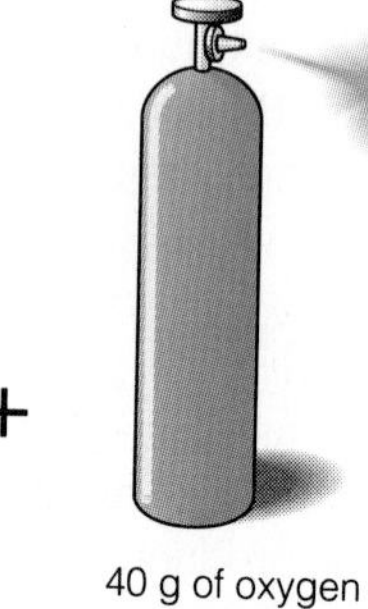

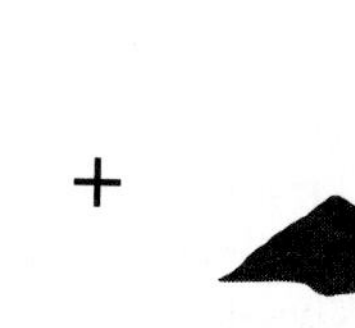

Figure 4.12 *Whether synthesized in the laboratory or obtained from various natural sources, copper carbonate always has the same composition. Analysis of this compound led Proust to formulate the law of definite proportions.*

A CLOSER LOOK

Experimental Verification

One of the earliest illustrations of the law of definite proportions is found in the work of a Swedish chemist, J. J. Berzelius (1779–1848). (See Figure 4.2.)

In a typical experiment, he heated 10.00 g of lead with varying amounts of sulfur to form lead sulfide. Since lead is a soft, grayish metal, sulfur is a pale yellow solid, and lead sulfide is a shiny black solid, it was easy to tell when all the lead had reacted. Excess sulfur was easily washed away by carbon disulfide, a liquid that dissolves sulfur but not lead sulfide. As long as he used at least 1.56 g of sulfur, he got exactly 11.56 g of lead sulfide. Any sulfur in excess of 1.56 g was left over, unreacted. If he used more than 10.00 g of lead with 1.56 g of sulfur, he got 11.56 g of lead sulfide with lead left over. These reactions are illustrated in Figure 4.13 and explained in Figure 4.15 according to Dalton's atomic theory.

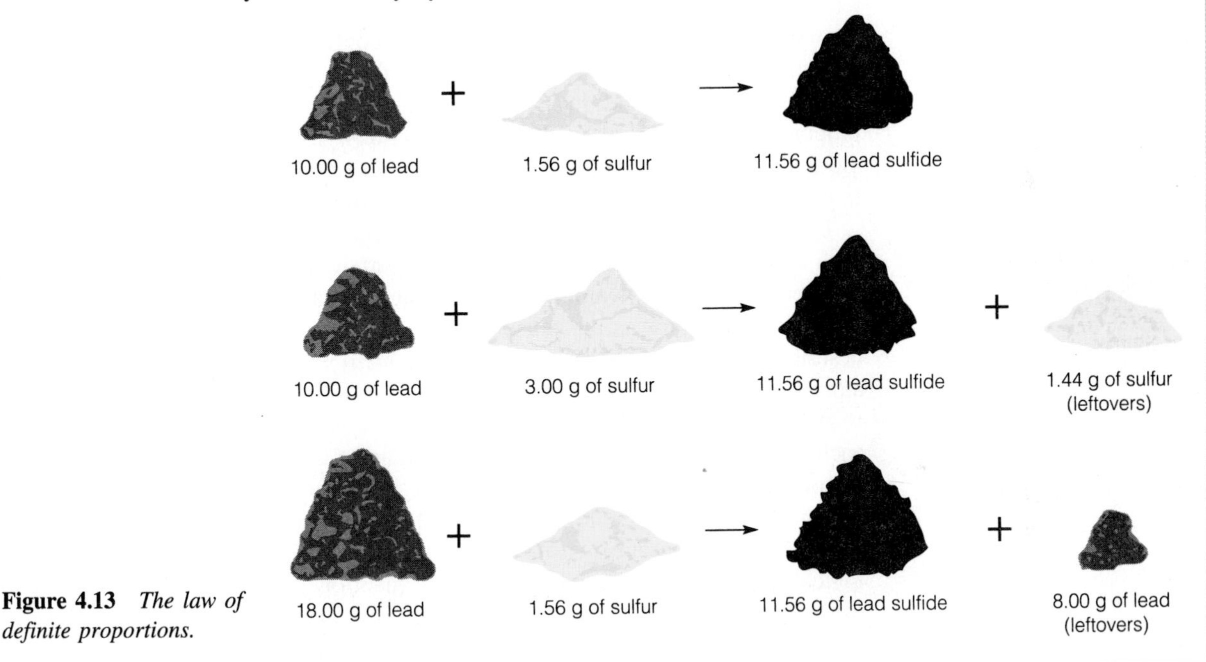

Figure 4.13 *The law of definite proportions.*

EXAMPLE 4.6

List key ideas or contributions for each of the following.

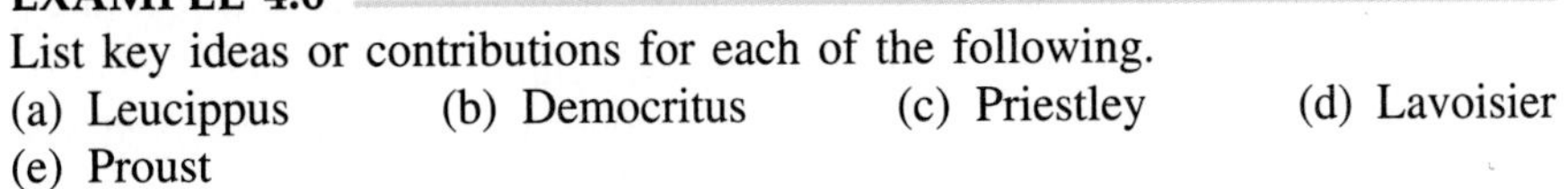
(a) Leucippus (b) Democritus (c) Priestley (d) Lavoisier
(e) Proust

SOLUTION

(a) Leucippus said matter is not continuous; it contains indivisible ultimate particles.
(b) Democritus named the ultimate particles atoms, meaning indivisible.
(c) Priestley discovered oxygen.
(d) Lavoisier explained burning and formulated the law of conservation of mass.
(e) Proust showed that compounds contain elements combined a definite proportions.

See Problems 4.29–4.36.

EXERCISE 4.4
What do Berzelius and Proust have in common? Explain.

Figure 4.14 *John Dalton (1766–1844) was able to explain experimental data gathered by several scientists when he proposed his now famous atomic theory. Although this Quaker schoolteacher was poor, colorblind, and not a good experimenter, his contributions had major impact on the development of modern chemistry.*

4.7 Dalton's Atomic Theory

Lavoisier's law of conservation of mass and Proust's law of definite proportions were repeatedly confirmed by laboratory experimentation. This led to attempts to formulate *theories* that would account for these laws. (In the sciences, a **theory** is a model that consistently explains observations.)

John Dalton (Figure 4.14), an English schoolteacher, proposed a comprehensive model to explain the accumulating experimental data regarding the nature of matter. As he developed the details of his model, he uncovered another ''law'' that his theory would also have to explain. Besides confirming Proust's conclusion that a compound contains elements in certain definite proportions, and only those proportions. Dalton also found that certain elements can combine in *more* than one set of proportions. His **law of multiple proportions** states that if two elements form more than one compound, the differing masses of one element that combine with a fixed mass of the second element are in a simple, whole-number ratio.

For example, Dalton found that three parts of carbon, by mass, can combine with either eight parts of oxygen or four parts of oxygen, by mass, to form two different compounds. Dalton explained that the first compound must have twice as many oxygen atoms as are present in the second compound. The first compound described was carbon dioxide, with the *chemical formula* CO_2 to represent a 1-to-2 atom ratio for carbon to oxygen.

As shown here, a **chemical formula** uses a numerical subscript to the right of each chemical symbol in the formula to indicate the atom ratios. (The subscript 1 is not written.) The second compound described was carbon monoxide, with the chemical formula, CO, to represent a 1-to-1 atom ratio for carbon to oxygen.

In 1803 Dalton set down the details of his now famous theory that provided a logical explanation for the laws we have mentioned. The principal ideas of **Dalton's atomic theory** are as follows.

Dalton's statements need some modification. Research has shown that atoms of an element do not all have precisely the same mass, that atoms do contain subatomic particles, and that—under certain conditions—atoms can be split.

1. All elements are made up of tiny, indivisible particles called atoms. Atoms can be neither created nor destroyed during chemical reactions.
2. All atoms of a given element are identical, but the atoms of one element differ from the atoms of every other element.
3. Atoms of different elements form compounds by combining in fixed, small, whole-number ratios such as 1 atom of A to 1 atom of B, 2 atoms of A to 1 atom of B, and 3 atoms of A to 2 atoms of B.
4. If the same elements form more than one compound, there is a different, but definite, small whole-number mass ratio and atom ratio for each compound.
5. A chemical reaction involves a change not in the atoms themselves, but in the way atoms are combined to form compounds.

Thinking It Through

Dalton's Theory Explains Experimental Results

The first point of Dalton's atomic theory says that you can't have fractional parts of atoms, and if the atoms are indestructible, then the very same atoms present before a chemical reaction must also be present after the reaction. Thus, the

A CLOSER LOOK

The Development of a Theory

Dalton's atomic theory is a good example of how useful theories are developed. A **theory** is a model that consistently explains observations and data. If the theory is a good one, it will accurately explain the existing data, and it will be useful in answering questions and making predictions about related situations. Also a good theory is one that can withstand some modifications, if the need arises, to take into account new research findings.

Dalton's atomic theory was such a theory. It helped explain how the "pieces of the puzzle"—the available data—fit together. It was not a perfect theory, but it was so simple and profund that minor modifications could not destroy the fundamental truths that it explained.

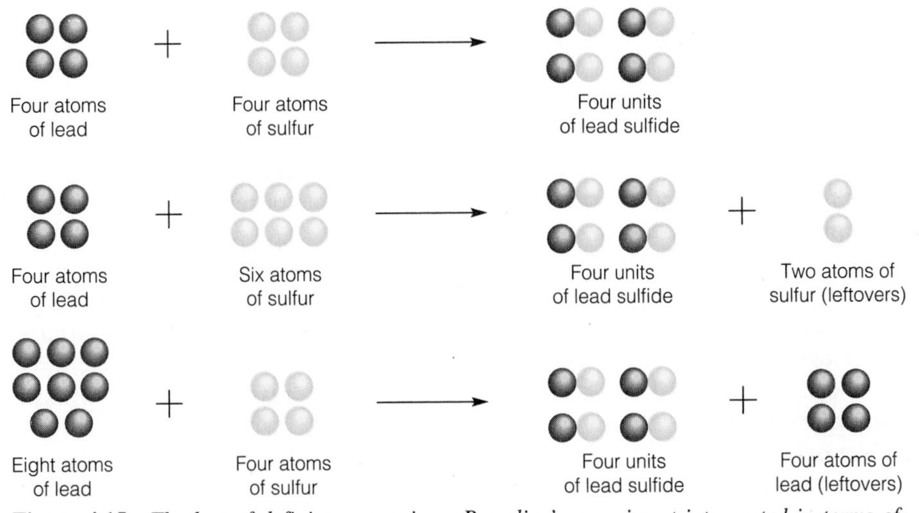

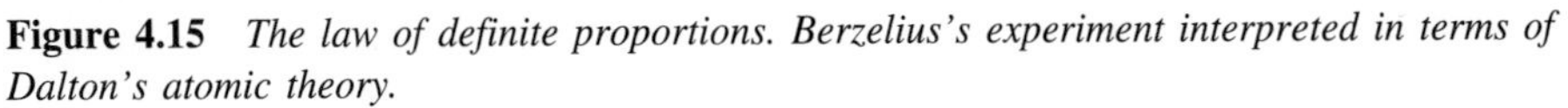

Figure 4.15 *The law of definite proportions. Berzelius's experiment interpreted in terms of Dalton's atomic theory.*

total mass before and after the reaction is unchanged. Dalton's theory, then, *explains* Lavoisier's law of conservation of mass.

The second point in Dalton's theory—that all atoms of a given element are identical—and the third—that atoms combine in fixed, whole-number ratios—together explain Proust's law of constant composition. Thus, each compound contains specific kinds of atoms that are combined in specific mass ratios, regardless of the size of the sample analyzed. The results of Berzelius's experiment (Figure 4.13) can be explained by Dalton's theory, as shown in Figure 4.15.

Dalton's fourth point summarizes his own law of multiple proportions. Since certain elements can combine in two or more *mass* ratios to form different compounds, these elements must, therefore, combine in two or more *atom* ratios.

Figure 4.16 illustrates the law of multiple proportions as well as the law of definite proportions. With two atoms of nitrogen, the *atom* ratios of oxygen for the

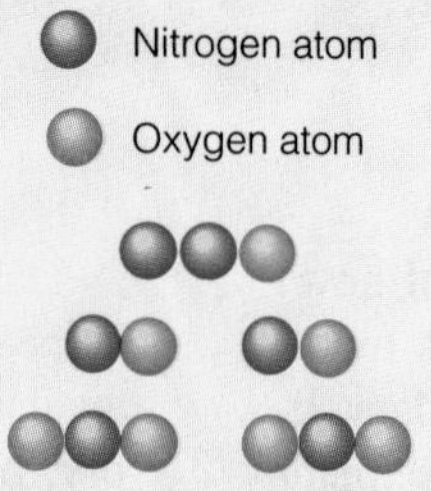

Number of Oxygen Atoms That Combine with 2 Nitrogen Atoms	Nitrogen atom / Oxygen atom	Compound	Mass of Oxygen That Combines with 28.0 g of Nitrogen	Simplest Mass Ratios for Oxygen
1		Nitrous oxide	16 g	1
2		Nitric oxide	32 g	2
4		Nitrogen dioxide	64 g	4

Figure 4.16 *The law of multiple proportions. With two atoms of nitrogen, the* atom *ratios of oxygen for the three compounds are 1:2:4. Also, with 28.0 g of nitrogen, the* mass *ratios of oxygen for the three compounds are 1:2:4. Thus both the mass ratios (last column) and the atom ratios (first column) are in the same simple proportions.*

three compounds are 1:2:4. Also, with 28 g of nitrogen, the *mass* ratios of oxygen for the three compounds are 1:2:4. Thus, both the mass ratios and the atom ratios are in the same simple proportions.

Not only did Dalton say that atoms of a given element are different from atoms of all other elements (the second point), but he proceeded to demonstrate how this is possible. He proposed that the mass of each kind of atom is different from the masses of all other kinds of atoms, and he set up a table of relative masses for the elements based on hydrogen, the lightest element.

Many of Dalton's relative atomic masses were wrong, mainly because he assumed water molecules have one hydrogen atom and one oxygen atom instead of *two* hydrogen atoms and one oxygen atom. While it was impossible to determine actual masses of any atoms until rather recently, Dalton was able to determine relative masses. The relative masses of elements are expressed in terms of **atomic mass units** (amu).

Not only did Dalton make a mistake in calculating the relative masses of some elements such as oxygen, he was also incorrect about atoms being indestructible. Present-day "atom-smashing" equipment can develop enough energy to split atoms into many fragments called subatomic particles.

Dalton's atomic theory was a great success, even with its inaccuracies. Why? Because it explained a large body of experimental data. It was useful then, and it is useful today because it enables us to predict how matter will behave under a wide variety of conditions. Dalton arrived at his theory on the basis of experimental data and reasoning. With certain modifications it has stood the test of time. Other scientists soon adopted his ideas and made some corrections and modifications, and a new era in chemistry began. The following examples are applications of Dalton's atomic theory. Be sure you understand the reasoning.

EXAMPLE 4.7

The formula of ammonia gas is written NH_3. This means that ammonia always has three hydrogen atoms combined with each nitrogen atom.

(a) The statement given illustrates one of the laws. Which one? Why?
(b) Three dozen nitrogen atoms would combine with ______ hydrogen atoms to make ______ ammonia molecules.

(c) 2×10^9 nitrogen atoms would combine with ______ hydrogen atoms to make ______ ammonia molecules.

(d) 6.02×10^{23} nitrogen atoms would combine with ______ hydrogen atoms to make ______ ammonia molecules.

1×10^9 is a billion.

You will soon learn more about this large number, 6.02×10^{23}; it is often used in chemistry.

SOLUTION

(a) The statement demonstrates the law of definite proportions. The hydrogen-to-nitrogen atom ratio is always 3 to 1, also written 3:1.

(b) Three dozen nitrogen atoms would combine with 9 dozen hydrogen atoms to make 3 dozen ammonia molecules.

(c) 2×10^9 nitrogen atoms would combine with 6×10^9 hydrogen atoms to make 2×10^9 ammonia molecules.

(d) 6.02×10^{23} nitrogen atoms would combine with $3(6.02 \times 10^{23})$ hydrogen atoms to make 6.02×10^{23} ammonia molecules.

EXAMPLE 4.8

The compositions of two samples, A and B, containing only copper and bromine were analyzed. Results of the analysis are as follows.

	Sample A	*Sample B*
Bromine	160. g	64.0 g
Copper	127 g	25.4 g

(a) Were these samples the same compound or different compounds?

(b) Do these data support the law of definite proportions or the law of multiple proportions or both laws?

(c) How much bromine is required to completely combine with 2.50 g of copper to produce a sample of compound A?

SOLUTION

Determine the mass of bromine for each gram of copper (g Br/g Cu).

Sample A	*Sample B*
$\dfrac{160.\text{ g Br}}{127\text{ g Cu}} = 1.26\text{ g Br/g Cu}$	$\dfrac{64.0\text{ g Br}}{25.4\text{ g Cu}} = 2.52\text{ g Br/g Cu}$

For one gram of copper, the mass ratio of bromine in sample B to sample A is

$$\frac{\text{Sample B}}{\text{Sample A}} = \frac{2.52\text{ g Br}}{1.26\text{ g Br}} = \frac{2}{1}$$

(a) Samples A and B are from two different compounds. The mass of bromine in sample B is twice as great as in sample A, for a fixed mass of copper.

(b) The data support the law of multiple proportions. There are different masses of bromine per gram of copper in each sample. These masses are in a simple 2-to-1 ratio, consistent with two different compounds.

(c) $2.50\text{ g Cu} \times \dfrac{1.26\text{ g Br}}{1\text{ g Cu}} = 3.15\text{ g Br}$ needed.

See Problems 4.37–4.46.

4.8 Atoms and Subatomic Particles

Dalton's atomic theory helped explain some experimental data, but as scientists looked for better ways to measure relative masses of the atoms, they encountered more questions than answers. It wasn't long before that simple picture of atoms underwent significant modification. Actually, evidence that suggested a more complicated structure of the atom was beginning to surface before Dalton's theory was published. The electrolysis of water in 1800 by two English chemists, William Nicholson and Anthony Carlisle, not only supported the law of definite proportions, but also showed that matter could somehow interact with electricity. Dalton's model of the atom failed to show how. Further evidence for the electrical nature of matter was soon obtained.

When Dalton explained that atoms cannot be broken, he was actually describing that they are not broken in *chemical reactions.* By the 1930s there was considerable evidence that atoms contain small, subatomic particles. Well over 100 **subatomic particles** have been discovered, but many of these have durations of less than a second. Only three main subatomic particles are needed to account for the masses and special chemical properties of atoms. These three particles are the **electron**, the **proton**, and the **neutron**.

Charged Particles

Both the electron and the proton have electrical charges. A particle with a "charge" can exert a force; that is, it can push or pull another particle that also has a "charge." There are two opposite kinds of charges; they are designated positive (+) and negative (−). The proton has a single positive charge (1+). The electron has a single negative charge (1−). Every electrically neutral atom has an equal number of protons and electrons. Neutrons have no charge.

Every atom is neutral in electrical charge because the number of electrons equals the number of protons.

Commercials call it "static cling" when items of clothing stick to one another.

What we often call "static electricity" is due to electrical charges. Touching an object after walking across a carpet sometimes causes a spark—a discharge of electrical energy. Opposite charges can cause a thin piece of plastic to cling to your fingers. If you are carrying one kind of charge and the plastic has picked up an

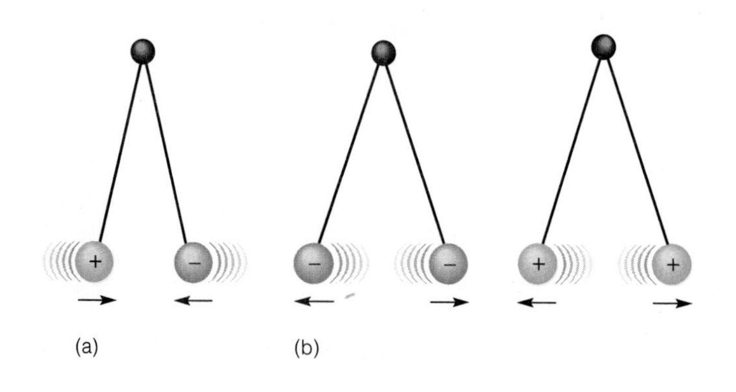

Figure 4.17 *Particles with unlike charges attract one another (a). Those with like charges repel one another (b).*

opposite charge, there will be an attraction that causes the plastic to cling to you. Opposite charges attract (Figure 4.17a). Your hair may have stood on end while blowing it dry or while combing vigorously on a cold day. When the individual hairs repel, they are carrying the same charge. Like charges repel (Figure 4.17b).

Electrons, Protons, and Neutrons

Protons and neutrons have virtually the same mass, 1.007276 amu and 1.008665 amu, respectively. That is equivalent to saying one person weighs 100.7 lb and another person weighs 100.9 lb; the difference is so small it is negligible. In most cases we use 1 amu for the masses of both the proton and the neutron.

Table 4.4 **Subatomic Particles**

Particle	*Symbol*	*Electrical Charge*	*Relative Mass (amu)*	*Mass (g)*
Electron	e^-	1−	$\frac{1}{1837}$	9.10953×10^{-28} g
Proton	p^+ or p	1+	1	1.67265×10^{-24} g
Neutron	n	0	1	1.67495×10^{-24} g

It would take 1837 electrons to have a total mass equal to the mass of just one proton, and no element has been discovered that has atoms with more than 109 electrons. In other words, electrons make up an exceedingly small fraction of the mass of an atom. For all practical purposes we can use a mass of 0 amu for an electron. The charges and relative masses (in atomic mass units) of these three particles are listed in Table 4.4. The table also lists the masses of these particles in grams to show just how small they really are (1 amu = 1.6606×10^{-24} g). To work with exact masses in grams would make calculations much more bothersome, so the relative atomic mass units are used. Since the mass of the electron is virtually zero, the mass of an atom is essentially that of its protons and neutrons. The sum of the protons and neutrons for an atom is called the **mass number**.

$$\text{Mass number} = \text{Number of protons} + \text{Number of neutrons}$$

The Size of an Atom

Atoms are far too tiny to see with the most powerful optical microscope. However, by 1970, Albert Crewe of the University of Chicago announced that images of single atoms of uranium and thorium had been photographed (Figure 4.18). In 1976, a group of scientists led by George W. Stroke of the State University of New York at Stony Brook photographed images showing the locations and relative sizes of tiny carbon, magnesium, and oxygen atoms in a section of crystal. By the late 1980s, images of atoms on surfaces of materials were being obtained by using the scanning tunneling microscope (STM) invented in 1981 and similar instruments such as the atomic force microscope (AFM) invented in 1985. Probes of these instruments can sense and map the atom's ''bumps'' on surfaces of materials (Figure 4.19).

The diameter of an atom is only a few tenths of a nanometer. It would take about 10 million atoms lined up, touching each other, to make a line 1 mm long.

At the center of each atom, the more massive subatomic particles—the protons

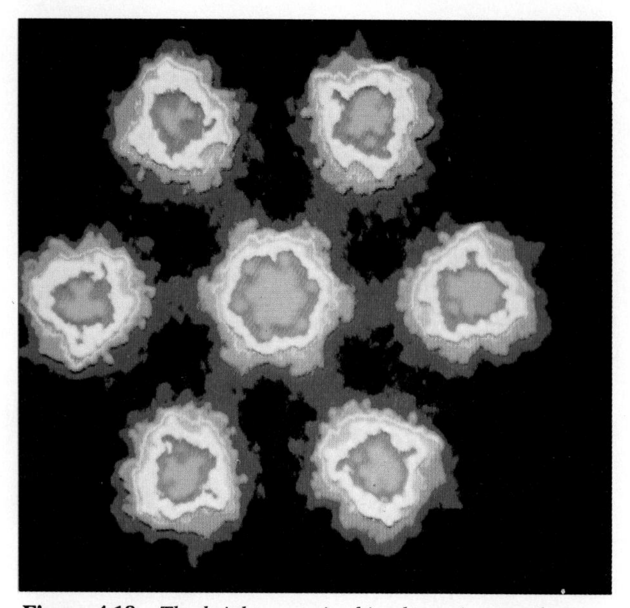

Figure 4.18 *The bright spots in this photomicrograph are images of seven uranium atoms that are 0.34 nm apart. The images were made with an electron microscope.*

Figure 4.19 *Image of the surface of silicon produced by a scanning tunneling microscope. The blue spots are individual silicon atoms, which are arranged in a regular pattern that repeats across the surface.*

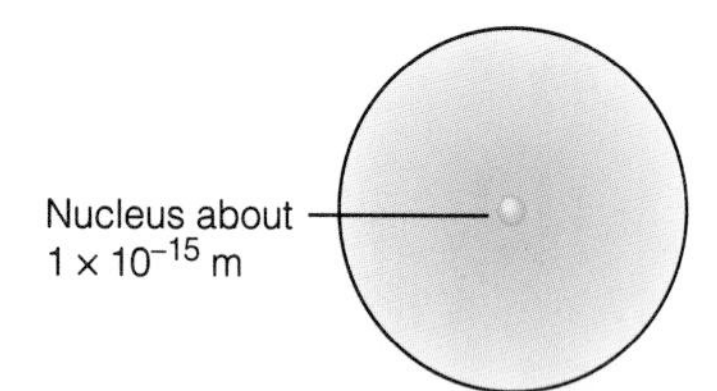

Figure 4.20 *Protons and neutrons are packed into the tiny nucleus, which has a diameter that is about one hundred-thousandth of the diameter of an atom.*

and neutrons—are packed into a tiny **nucleus** (Figure 4.20). The nucleus of an atom has a diameter of about 1×10^{-15} m. That is about one hundred-thousandth of the diameter of an atom. To picture just how small the nucleus is, imagine a balloon that stands 10 stories high. If the balloon were an atom, then its nucleus would be the size of a BB; the rest of the space in the balloon atom would be the domain of the electrons. For another example, if the nucleus could be enlarged to be the size of a period on this page, the diameter of the atom would be about 5 m, which is about the height of a two-story house. The electrons in an atom move about the nucleus in definite regions called orbitals. We shall discuss the arrangements of electrons in atoms in Chapter 5.

Atomic Number

All atoms of a particular kind of element have the same number of protons. This number of protons in the nucleus of an atom is defined as the **atomic number**.

- All atoms of hydrogen have one proton; the atomic number of hydrogen is 1.
- All atoms of oxygen have eight protons; the atomic number of oxygen is 8.
- All atoms of gold have 79 protons; the atomic number of gold is 79.

The number of protons determines the identity of each kind of element. Gold is gold because of its 79 protons, not 78, 80, or any other number.

> The atomic number for an element is the same as the number of protons in the nucleus of each atom of that element.

Look at the periodic table on the inside cover of this book and note that the elements are arranged by atomic number, beginning with hydrogen, atomic number 1. Each successive element in the periodic table has atoms with exactly one more proton than the previous element. For example, nitrogen (atomic number 7) is placed just before oxygen (atomic number 8). Atomic numbers are always whole, exact numbers since protons do not exist in fractional amounts.

EXAMPLE 4.9

Let's examine an atom of sodium that has 11 protons, 11 electrons, and a mass number of 23 amu.

(a) What is the total electrical charge of the atom?
(b) How many neutrons does this atom have?
(c) What is the atomic number of sodium?

SOLUTION

(a) The total charge is zero. The number of electrons (each with a charge of 1−) equals the number of protons (each with a charge of 1+).
(b) This atom has 12 neutrons.

$$\text{Number of neutrons} = \text{Mass number} - \text{Protons}$$
$$\text{Number of neutrons} = 23 - 11 = 12$$

(c) The atomic number of sodium is 11; it has 11 protons.

EXAMPLE 4.10

A particular kind of atom has 61 neutrons and a mass number of 108.

(a) How many protons does this atom have?
(b) How many electrons does this atom have?
(c) What is the atomic number of this element?
(d) What is the name of the element?

SOLUTION

(a) The atom has 47 protons

$$\text{Number of protons} = \text{Mass number} - \text{Number of neutrons}$$
$$\text{Number of protons} = 108 - 61 = 47$$

(b) The atom has 47 electrons, the same as the number of protons.
(c) The atomic number is 47. The atomic number is the number of protons.
(d) Silver. It is the only element with 47 protons. (See the periodic table.)

See Problems 4.47–4.52.

Exercise 4.5

Answer the same four questions for an atom that has 18 neutrons and a mass number of 35.

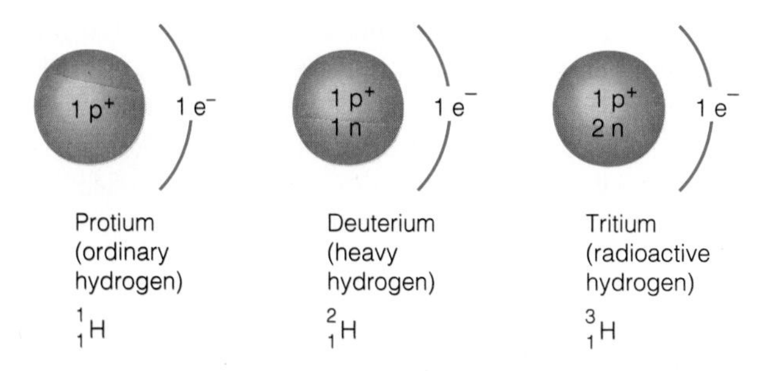

Figure 4.21 *The isotopes of hydrogen.*

4.9 Isotopes

All naturally occurring aluminum atoms have 13 protons, 13 electrons, and 14 neutrons. For many elements, however, all atoms do not have the same number of neutrons. For example, some atoms of chlorine have 18 neutrons and others have 20 neutrons. Atoms of chlorine that have different numbers of neutrons must also have different masses. Atoms of a particular kind of element that have different masses are called **isotopes**. Most elements have several isotopes. The element with the most stable isotopes is tin; it has 10 stable isotopes. All isotopes of an element have virtually the same chemical properties.

In addition to about 270 isotopes that occur naturally, over 1000 more have been synthesized using nuclear reactors. Many of these isotopes are used in chemical and biological research as well as in medicine.

Most hydrogen atoms have a nucleus consisting of a single proton and no neutrons, but about 1 hydrogen atom in 5000 has a nucleus containing a neutron along with its proton. Thus, some hydrogen atoms have a mass number of 1 and others have a mass number of 2. These atoms are isotopes. There is a third, very rare isotope of hydrogen, called tritium, that is radioactive; it has a mass number of 3 (1 proton + 2 neutrons). Figure 4.21 summarizes the subatomic particles present in these three isotopes of hydrogen.

Regardless of the particular isotope involved, hydrogen enters into the same chemical reactions. For example, either deuterium (hydrogen atoms with a mass number of 2) or protium (ordinary hydrogen) will react with oxygen to form water, H_2O. Molecules of water with one atom of oxygen (16 amu) and two atoms of deuterium are called *heavy water.* Ordinary water, with atoms of protium, has a mass of 18 amu (16 + 1 + 1), but heavy water has a mass of 20 amu (16 + 2 + 2). The heavier molecules can be expected to move around more slowly, but enter into the same chemical reactions.

Hydrogen is the only element with names for each of its isotopes. There are two other popular methods of showing which isotope is being discussed. In one method, the mass number is separated from the name of the element by a hyphen. For example, cobalt-60 identifies the isotope of cobalt with a mass number of 60 (27 protons + 33 neutrons). A second method takes the generalized form

Cobalt-60 is one radioactive isotope used in the treatment of cancer.

$$^A_ZX$$

where the subscript Z represents the atomic number (the number of protons), the superscript A represents the mass number, and the X is the symbol for the element. Thus, cobalt-60 can also be written in the form $^{60}_{27}Co$. Notice that the number of neutrons in an atom of the isotope can quickly be determined by subtracting the number of protons (27) from the mass number (60). Follow through with these techniques in the next examples.

EXAMPLE 4.11

The radioactive isotope iodine-131 is used in treatment of thyroid cancer and the measurement of liver activity and fat metabolism.

(a) What is the atomic number of this isotope? (Use the periodic table.)
(b) How many neutrons do atoms of this isotope contain?

SOLUTION

(a) The atomic number for this isotope, and all isotopes of iodine, is 53.
(b) The number of neutrons = mass number − protons = 131 − 53 = 78.

EXERCISE 4.6

You may have some radioactive americium at home! The isotope $^{241}_{95}Am$ is used in home smoke detectors of the ionization type.

(a) How many protons are present in each atom of this isotope?
(b) How many neutrons do atoms of this isotope contain?

EXAMPLE 4.12

(a) Which of the following are isotopes of the same element? We are using the letter X as a symbol for the element in each case

$$^{16}_{8}X \quad ^{16}_{7}X \quad ^{14}_{7}X \quad ^{14}_{6}X \quad ^{12}_{6}X$$

(b) Which of the five kinds of atoms have the same number of neutrons?

SOLUTION

(a) Both $^{16}_{7}X$ and $^{14}_{7}X$ are isotopes of the element nitrogen (N). Both $^{14}_{6}X$ and $^{12}_{6}X$ are isotopes of the element carbon (C).
(b) Both $^{16}_{8}X$ (16 − 8 = 8 neutrons) and $^{14}_{6}X$ (14 − 6 = 8 neutrons) have the same number of neutrons.

See Problems 4.53–4.58.

4.10 *Atomic Masses of the Elements*

Look at the periodic table and notice that each of the elements has an **average atomic mass**—often called the **atomic weight**—which is generally a decimal value rather than a whole number. The atomic mass shown in the periodic table for an element is actually a weighted average of the masses of all naturally occurring isotopes of that element. Most elements have several naturally occurring isotopes, but these are in varying proportions, depending on the element. Atoms of the carbon-12 isotope are defined as having a mass of exactly 12 amu. Relative masses of all other atoms are determined by comparing them to this standard. Although synthetic isotopes of virtually every element have also been produced in laboratories, these are not considered in determining average atomic masses.

Notice that the atomic masses of synthetic elements in the periodic table are listed as whole numbers in parentheses. Numerous isotopes can be prepared; only the mass of the most stable isotope is listed.

We'll use two examples to demonstrate what is meant by average atomic masses. In a sample of the element bromine, about half of the atoms have an

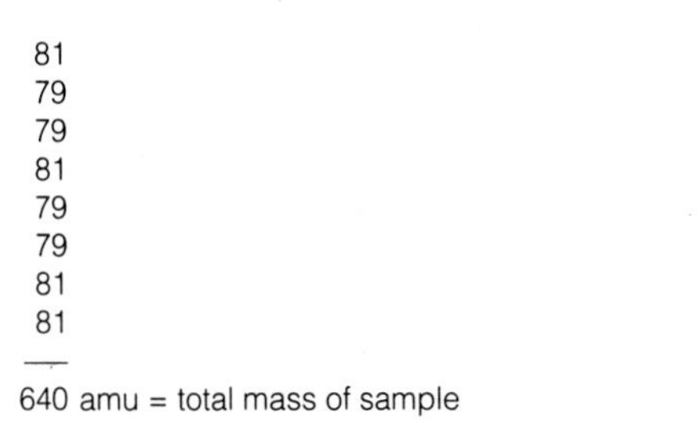

Average mass = $\frac{640 \text{ amu}}{8 \text{ atoms}}$ = 80 amu = atomic mass

Atomic masses (in amu) of atoms in sample

35
37
35
35
35
37
35
35

284 amu = total mass of sample

Sample
of chlorine

Average mass = $\frac{284 \text{ amu}}{8 \text{ atoms}}$ = 35.5 amu = atomic mass

Figure 4.22 *Atomic masses of elements are averages of isotopic masses.*

atomic mass of 79 amu and half have an atomic mass of 81 amu. With 50% bromine-79 and 50% bromine-81, the mass would be exactly 80 amu. This is very close to the mass of bromine, which is listed in the periodic table as 79.9 amu. As another example, about 75% of the atoms in a sample of chlorine gas have an atomic mass of 35 amu and about 25% have an atomic mass of 37 amu. The

Table 4.5 **Natural Isotopes of Selected Elements**

Isotope	*Mass (amu)*	*Natural Abundance (%)*	*Isotope*	*Mass (amu)*	*Natural Abundance (%)*
^{1}H	1.0078	99.985	^{35}Cl	34.9688	75.77
^{2}H	2.0140	0.015	^{37}Cl	36.9659	24.23
^{10}B	10.0129	20.0	^{63}Cu	62.9296	69.20
^{11}B	11.0093	80.0	^{65}Cu	64.9278	30.80
^{12}C	12.0000	98.89	^{79}Br	78.9183	50.69
^{13}C	13.0033	1.11	^{81}Br	80.9163	49.31
^{23}Na	22.9898	100.00	^{84}Sr	83.9134	0.50
^{24}Mg	23.9850	78.99	^{86}Sr	85.9094	9.90
^{25}Mg	24.9858	10.00	^{87}Sr	86.9089	7.00
^{26}Mg	25.9826	11.01	^{88}Sr	87.9056	82.60
^{27}Al	26.9815	100.00			

average atomic mass of chlorine is 35.5 amu. This average is much closer to the mass of the chlorine-35 isotope because there is much more of that isotope in the sample. These two examples are illustrated in Figure 4.22.

When the natural abundance of isotopes (as percentages) is known, we can calculate the average atomic mass for an element. Table 4.5 lists masses of isotopes for some elements and the percentages of each. To calculate the average atomic mass, first multiply each isotopic mass by the percentage of that isotope (written as a decimal). Each of these values represents the ''mass contribution'' of the isotope. The sum of all mass contributions gives the average atomic mass for the element. This explanation will be much clearer if you follow through with the next two examples.

EXAMPLE 4.13

Use values provided in Table 4.5 to calculate the average atomic mass for the element chlorine to four significant figures.

SOLUTION

For each isotope, list the accurate mass and multiply by the percentage written as a decimal. Add the resultant ''mass contributions'' to obtain the average mass of chlorine in atomic mass units.

Isotope	*Mass (amu)*	*Abundance (decimal)*	*Mass Contribution*
Chlorine-35	34.9688 amu ×	0.7577	= 26.49 amu
Chlorine-37	36.9659 amu ×	0.2423	= 8.96 amu
		Average mass	= 35.45 amu

Compare this value with the atomic mass for chlorine shown in the periodic table.

EXAMPLE 4.14

Use values provided in Table 4.5 to calculate the atomic mass for the element strontium to four significant figures.

SOLUTION

For each isotope, list the accurate mass and multiply by the percentage written as a decimal. Add the resultant ''mass contributions'' to obtain the average mass of strontium in atomic mass units, as shown.

Isotope	*Mass (amu)*	*Abundance (decimal)*	*Mass Contribution*
Strontium-84	83.9134 amu ×	0.0050	= 0.42 amu
Strontium-86	85.9094 amu ×	0.0990	= 8.51 amu
Strontium-87	86.9089 amu ×	0.0700	= 6.08 amu
Strontium-88	87.9056 amu ×	0.8260	= 72.61 amu
		Average mass	= 87.62 amu

Compare this value with the atomic mass for strontium shown in the periodic table.

See Problems 4.59 and 4.60.

Figure 4.23 *How could you determine the number of sugar crystals in the bowl of sugar?*

4.11 *Counting with Moles*

How would you go about counting the number of sugar crystals in a full bowl of sugar (Figure 4.23)? Could you count the number of sugar molecules in the bowl? If we know the mass of just one crystal of sugar, we can calculate the number of crystals of sugar in the bowl. Weighing the sugar in the sugar bowl and dividing this total mass by the mass of one crystal of sugar will give us the number of crystals of sugar in the bowl. A similar process can be used to calculate the number of sugar molecules present. Thus, the mass of the sample can be used to determine the number of particles in the sample if the mass of a representative particle is known; *mass and number of particles are proportional.*

Thinking It Through

Using Relative Atomic Masses

Chemists in a laboratory can't weigh individual atoms or molecules, but the average atomic masses shown on the periodic table provide a convenient way of obtaining equal numbers of different kinds of atoms. Since the relative atomic masses (shown on the periodic table) for hydrogen, carbon, and oxygen are 1.01 amu to 12.01 amu to 16.00 amu, respectively, we know that a dozen of each kind of these atoms will also have the same relative masses of 1.01 to 12.01 to 16.00. Similarly, a hundred dozen of each kind of these atoms will have the same relative masses of 1.01 to 12.01 to 16.00. But even a hundred dozen or a billion dozen is too small a number to use when counting out enough atoms to be visible.

The number of atoms, *N*, present in exactly 12.0000 g of the carbon-12 isotope has been chosen as a standard. This number of atoms, *N*, of hydrogen has an average mass of 1.008 g, while *N* atoms of oxygen has an average mass of 16.00 g. This number, *N*, is extremely large; it is called **Avogadro's number** in honor of the Italian chemist Amadeo Avogadro (1776–1856). Numerous research investigations with gases, crystals, and electroplating processes have been used to determine the value of Avogadro's number.

$$N = 602{,}200{,}000{,}000{,}000{,}000{,}000{,}000$$

$$N = 6.022 \times 10^{23} \text{ in scientific notation}$$

Avogadro did not determine the value of N himself, but researchers have calculated the value as 6.022045×10^{23}. This value has more significant figures than we normally need.

An amount of substance containing Avogadro's number of unit particles is called a **mole** (abbreviated mol). Just as a dozen contains 12 items, and a gross contains 144 items, a mole of the substance always contains 6.022×10^{23} unit particles—Avogadro's number of particles.

- A mole of carbon *atoms* contains 6.022×10^{23} *atoms* of carbon.
- A mole of water *molecules* contains 6.022×10^{23} *molecules* of water.
- A mole of any substance contains Avogadro's number of the unit particles of that substance.

Instead of counting individual atoms or other unit particles, which are too small to see, we count *moles* of a substance by weighing out a specific mass of the substance. The mass of a dozen dimes is different from the mass of a dozen nickels, but the number of coins is the same. Similarly, a mole of carbon atoms and a mole of copper atoms—or of any other substance—have different masses, but the number of atoms represented is the same; it's Avogadro's number.

- One mole of carbon atoms has a mass of 12.0 g and contains 6.02×10^{23} atoms.
- One mole of oxygen atoms has a mass of 16.0 g and contains 6.02×10^{23} atoms.
- One mole of sodium atoms has a mass of 23.0 g and contains 6.02×10^{23} atoms.
- One mole of chlorine atoms has a mass of 35.5 g and contains 6.02×10^{23} atoms.

It is important that you understand the following relationships involving atomic mass units, grams, and moles.

> If the atomic mass of a substance is x amu, then
>
> 1 mole of the substance has a mass of x grams, and
> 1 mole of the substance contains 6.02×10^{23} unit particles.

Officially, a **mole** is defined as the *amount of substance* containing as many formula units as there are atoms in exactly 12 g of carbon-12. The mass of a mole of any substance expressed in *grams* is called the **molar mass** (MM) of that substance. Essentially, it means that the mass of a mole of any monatomic element is simply the atomic mass expressed in grams (Figure 4.24). The molar mass of a substance contains Avogadro's number of **formula units** of that substance.

For a compound, the sum of the atomic masses in amu is called the **formula weight** (F.W.) or if the substance is composed of molecules, this sum can also be called the **molecular weight** (M.W.) for the compound. To determine the mass of 1 mole—the molar mass—of a compound, simply add up the atomic masses of all of the atoms represented by the formula and express this quantity in *grams* instead of atomic mass units. For example, the formula of carbon dioxide, CO_2, represents one atom of carbon and two atoms of oxygen or 1 mole of carbon and 2 moles of oxygen. The formula weight and molar mass of CO_2 can be determined as follows.

Figure 4.24 *A mole of each element has a different mass but contains the same number of atoms, Avogadro's number, 6.02×10^{23}. Shown here* (clockwise from the top) *are bromine, sulfur, carbon, aluminum, copper, mercury, and cobalt.*

Carbon dioxide is one by-product of burning substances containing carbon.

$$
\begin{aligned}
1 \times \text{atomic mass of C} = 1 \times 12 &= 12 \text{ amu} \\
2 \times \text{atomic mass of O} = 2 \times 16 &= \underline{32 \text{ amu}} \\
\text{Formula weight (molecular weight)} &= 44 \text{ amu} \\
1 \text{ mole of } CO_2 &= 44 \text{ g}
\end{aligned}
$$

The formula for ammonium phosphate, $(NH_4)_3PO_4$, represents 3 atoms of N, 12 atoms of H, 1 atom of P, and 4 atoms of O.

Each quantity within a *set of parentheses in a chemical formula* must be multiplied by the subscript that immediately follows the set of parentheses.

Thus, for $(NH_4)_3PO_4$ the $(NH_4)_3$ part of the formula means that everything within the parentheses is multiplied by 3, as shown here.

Ammonium phosphate is used in some fertilizers to supply both N and P to soils and plants.

$$
\begin{aligned}
3 \times \text{atomic mass of N} = 3 \times 14 &= 42 \text{ amu} \\
12 \times \text{atomic mass of H} = 12 \times 1 &= 12 \text{ amu} \\
1 \times \text{atomic mass of P} = 1 \times 31 &= 31 \text{ amu} \\
4 \times \text{atomic mass of O} = 4 \times 16 &= \underline{64 \text{ amu}} \\
\text{Formula weight} &= 149 \text{ amu} \\
1 \text{ mole of } (NH_4)_3PO_4 &= 149 \text{ g}
\end{aligned}
$$

With moles, one can work at both the atomic level and the macroscopic level (visible with the naked eye). For example, a molecule of water, H_2O, consists of two atoms of hydrogen and one atom of oxygen. On the macroscopic level, we work with 1 mole of H_2O molecules (18.0 g) made up of 2 moles of hydrogen atoms (2.0 g) and 1 mole of oxygen atoms (16.0 g). Study the following comparisons.

Chemical formula:	H_2O	consists of	2 H	+	O
Atomic level:	1 molecule 1 dozen molecules		2 H atoms 2 dozen H atoms		1 O atom 1 dozen O atoms
Laboratory size sample:	6.02×10^{23} H_2O molecules 1 mol of H_2O molecules		$2(6.02 \times 10^{23})$ H atoms 2 mol of H atoms		6.02×10^{23} O atoms 1 mol of O atoms
In grams:	18.0 g of H_2O		2.0 g of hydrogen		16.0 g of oxygen

Work through the following examples that involve calculations with moles. Especially examine the examples that deal with fractional parts of a mole.

EXAMPLE 4.15

Use atomic masses shown in the periodic table whenever necessary.

(a) How many atoms are in 1 mole of helium, iron, and gold?
(b) Determine the masses of 1 mole of helium, iron, and gold.
(c) Determine the masses of 0.600 mole of helium, iron, and gold. (In actual practice we are not likely to have exactly 1 mole in a laboratory sample.)

SOLUTION

(a) 1 mole of He, Fe, Au—or any other monatomic element—contains 6.02×10^{23} atoms.
(b) 1 mole He = 4.00 g   1 mole Fe = 55.8 g   1 mole Au = 197 g
(c) $0.600 \text{ mol He} \times \dfrac{4.00 \text{ g He}}{\text{mol He}} = 2.40 \text{ g He}$

$$0.600 \text{ mol Fe} \times \frac{55.8 \text{ g Fe}}{\text{mol Fe}} = 33.5 \text{ g Fe}$$

$$0.600 \text{ mol Au} \times \frac{197 \text{ g Au}}{\text{mol Au}} = 118 \text{ g Au}$$

EXAMPLE 4.16

Use the formula of calcium phosphate, $Ca_3(PO_4)_2$, when performing the following calculations.

Calcium phosphate is present in bone tissue

(a) What is the mass of 1 mole of calcium phosphate?
(b) What is the mass of 1.464 moles of calcium phosphate?
(c) How many phosphorus atoms are in 1 mole of calcium phosphate?

SOLUTION

(a)

$$
\begin{aligned}
3 \times \text{atomic mass of Ca} &= 3 \times 40.1 = 120 \text{ amu}\\
2 \times \text{atomic mass of P} &= 2 \times 31.0 = 62 \text{ amu}\\
8 \times \text{atomic mass of O} &= 8 \times 16.0 = \underline{128 \text{ amu}}\\
\text{Formula weight} &= 310 \text{ amu}\\
1 \text{ mole } Ca_3(PO_4)_2 &= 310 \text{ g}
\end{aligned}
$$

(b) $1.464 \text{ mol } Ca_3(PO_4)_2 \times \frac{310 \text{ g } Ca_3(PO_4)_2}{\text{mol } Ca_3(PO_4)_2} = 454 \text{ g } Ca_3(PO_4)_2$ (1 pound)

(c) $1 \text{ mol } Ca_3(PO_4)_2 \times \frac{6.02 \times 10^{23} \text{ formula units}}{\text{mol } Ca_3(PO_4)_2} \times \frac{2 \text{ P atoms}}{\text{formula unit}} =$ 1.20×10^{24} P atoms

See Problems 4.61–4.65.

EXERCISE 4.7
What is the mass (in grams) of 2.47 moles of calcium hydroxide, $Ca(OH)_2$?

These examples illustrate how we can work with both large and small quantities of chemicals in moles. The quantity can be expressed either as a mass or as a number of particles. It is important that you understand these calculations now; we will perform many more calculations with moles in Chapter 9. In the next chapter we will be investigating the electronic structure of atoms—the key to why atoms combine in specific proportions.

Chapter Summary

Current periodic tables list 109 elements. Those with atomic numbers greater than 92 do not occur in nature, but they have been synthesized. The first working definition of an element—a substance that cannot be broken down to simpler substances—was provided by Robert Boyle. Lavoisier was the first to use modern names of the elements, and Berzelius introduced the use of one- and two-letter symbols that were later adopted for worldwide use.

The same elements are present in all matter throughout the entire universe, but abundances of these elements differ for our solar system (Table 4.2), our planet (Figure 4.3), and the human body (Figure 4.4).

The periodic table summarizes a great deal of information, including a classification of elements as metals (on the left), nonmetals (on the right), and the metalloids, with intermediate properties. Certain nonmetallic elements exist as diatomic molecules (H_2, N_2, O_2, and the halogens). Different physical properties of metals and nonmetals are listed in Table 4.3.

Ancient Greek philosophers who had the best understanding of the atomic nature of matter were Leucippus and Democritus, but the first major experimental data concerning the atomic nature of matter was provided by eighteenth-century scientists like Priestley (discovery of oxygen), Lavoisier (explanation of combustion), Proust (law of definite proportions), and Cavendish (electrolysis of water). John Dalton's atomic theory provided a useful model to explain these experimental data. He also observed that certain elements combine in more than one set of proportions—the law of multiple proportions.

Research has shown that an atom is not indivisible; it can be split up into more than 100 subatomic particles. Many of these are unstable, short-lived fragments. The three principal subatomic particles are the electron, the proton, and the neutron. Each atom is neutral, with an equal number of protons (positive charge) and electrons (negative charge). The mass of the electron is essentially 0 amu. Protons and neutrons have a mass of 1 amu. The mass number for an atom is the sum of the number of protons and neutrons. Elements are arranged in the periodic table by atomic number—the number of protons. Atoms of an element having different numbers of neutrons are called isotopes; they are named by the methods described.

One mole of an element is the amount of that element that has Avogadro's number of atoms and a mass, in grams, equal to its atomic mass given in the periodic table. One mole of a compound is the amount of that compound that has Avogadro's number of formula units and a molar mass, in grams, equal to its formula weight or molecular weight.

Assess Your Understanding

1. Use correct spelling for the names and symbols of elements listed in Table 4.1. [4.1]
2. List the two most abundant elements in our solar system and the four most abundant elements on Earth and in the human body. [4.3]
3. Give formulas of the elements that exist as diatomic molecules. [4.4]
4. Use the periodic table to identify metals, nonmetals, and metalloids, and list general physical properties for each category. [4.4, 4.5]
5. Identify examples that illustrate the law of definite proportions and the law of multiple proportions. [4.6, 4.7]
6. List and explain the five key points of Dalton's atomic theory. [4.7]
7. Give names, symbols, charges, and masses (in atomic mass units) for the three major subatomic particles. [4.8]
8. Determine the number of protons, neutrons, atomic number, and mass number for isotopes. [4.9]
9. Express symbols of isotopes by two methods. [4.9]
10. Determine the average atomic mass when isotopic abundances are given. [4.10]
11. Perform calculations involving moles, number of atoms, and grams. [4.11]
12. Use atomic masses to determine formula weights of compounds. [4.11]

Key Terms

alkali metals [4.4]
atomic mass units [4.7]
atomic number [4.8]
average atomic mass [4.10]
Avogadro's number [4.11]
chemical formula [4.7]
compound [4.1]
conductivity [4.5]
Dalton's atomic theory [4.7]
deuterium [4.9]
diatomic elements [4.4]
ductility [4.5]
electrolysis [4.6]
electron [4.8]
element [4.1]
family of elements [4.4]
formula unit [4.11]
formula weight (F.W.) [4.11]

group of elements [4.4]
halogen family [4.4]
isotopes [4.9]
law of definite proportions [4.6]
law of multiple proportions [4.7]
luster [4.5]
malleability [4.5]
mass number [4.8]
metalloids [4.4]
metals [4.4]
molar mass [4.11]
mole (mol) [4.11]
molecular weight (M.W.) [4.11]
neutron [4.8]
noble gases [4.5]
nonmetals [4.4]
nucleus [4.8]
parentheses in formulas [4.11]
period of elements [4.4]
protium [4.9]
proton [4.8]
subatomic particles [4.8]
symbols of elements [4.2]
theory [4.7]
tritium [4.9]

Problems

ELEMENTS: DISCOVERIES AND NAMES

4.1 Who were the alchemists?

4.2 In what ways did the alchemists contribute to science?

4.3 What was Robert Boyle's definition of an element?

4.4 What was Robert Boyle's definition of a compound?

4.5 Who was the first scientist (the "father of chemistry") who used modern, systematic names for chemical elements?

4.6 Who recognized the importance of recording quantitative data and formulated the law of conservation of mass?

4.7 Which of the following elements has been synthesized but does not occur naturally? (It has been found in stars.)
a. arsenic b. boron c. cobalt
d. rubidium e. technetium

4.8 Which of the following elements does not occur naturally?
a. astatine b. boron c. cesium
d. titanium e. uranium

4.9 Who first used a two-letter system of symbols for elements?

4.10 Which letter in a symbol for an element is always capitalized when there is more than one letter in the symbol?

4.11 Give correct symbols for the following elements.
a. sodium b. magnesium c. chromium
d. iron e. mercury f. silver

4.12 Give correct symbols for the following elements.
a. potassium b. manganese c. copper
d. gold e. phosphorus f. fluorine

4.13 Give the names of the elements represented by these symbols.
a. Sr b. Br c. Cl
d. Sn e. W f. Pb

4.14 Give the names of the elements represented by these symbols.
a. As b. Ba c. Sb
d. Si e. Pt f. N

4.15 Consult Table 4.1 and list two or three important uses for each of the following elements.
a. argon b. bromine c. calcium
d. magnesium e. phosphorus

4.16 Consult Table 4.1 and list three important uses for each of the following elements.
a. antimony b. boron c. chlorine
d. manganese e. tin

4.17 Which of the following elements are *not* among the four most abundant elements in the Earth's crust?
a. aluminum b. hydrogen
c. silicon d. iron

4.18 Which of the following elements are *not* among the four most abundant elements in the human body?
a. nitrogen b. hydrogen
c. carbon d. iron

METALS, NONMETALS, AND METALLOIDS

4.19 Classify the following elements as metals, nonmetals, or metalloids.
a. potassium b. bromine c. calcium
d. silicon e. phosphorus f. germanium

4.20 Classify the following elements as metals, nonmetals, or metalloids.

a. boron b. beryllium c. chlorine
d. sodium e. sulfur f. fluorine

4.21 Which of the following are incorrect formulas for elements?
a. Ar_2 b. Fe c. Cu_2
d. I_2 e. O_2

4.22 Which of the following are incorrect formulas for elements?
a. H_2 b. Br_2 c. He_2
d. Cr_2 e. F_2

4.23 List all elements that are gases at room temperature (other than the noble gases). Classify them as metals, nonmetals, or metalloids.

4.24 List all elements that are liquids at room temperature. Classify them as metals, nonmetals, or metalloids.

4.25 Which one of these elements has the highest melting point?
a. mercury b. sodium c. tungsten
d. iron e. bromine f. nitrogen

4.26 Which of the following are *not* good conductors?
a. aluminum b. sulfur c. copper
d. gold e. phosphorus f. sodium

4.27 Which one of the following elements exhibits the *least* amount of luster?
a. aluminum b. phosphorus c. chromium
d. silver e. platinum

4.28 Which of the following elements are *not* ductile?
a. carbon b. silver c. copper
d. sulfur e. iron

ATOMS: DEMOCRITUS TO DALTON

4.29 Is matter continuous or atomistic? Explain.

4.30 Which important Greek philosophers thought of matter as being atomistic, and which philosophers believed matter was continuous?

4.31 Name two persons who discovered oxygen independently. Which one is credited with the discovery? Why?

4.32 Who carried out quantitative experiments and correctly explained burning?

4.33 What happens during combustion?

4.34 Who first stated that a chemical compound always contains elements in certain definite proportions?

4.35 In the laboratory, electrolysis of a sample of water produced 20 mL of hydrogen and 10 mL of oxygen. For another sample, 28 mL of hydrogen and 14 mL of oxygen were produced. Do these data support the law of definite proportions or the law of multiple proportions, or both?

4.36 How did the electrolysis of water disprove the Greek philosopher's belief that water was an element?

DALTON'S ATOMIC THEORY

4.37 What is a scientific model? Give an example.

4.38 On the atomic level, Dalton explained that atoms combine in whole-number ratios. How does this explain Berzelius's experiment showing that 10.00 g of lead would always produce no more than 11.56 g of the compound, lead sulfide, even when more sulfur was used?

4.39 Three samples were analyzed and found to contain only copper and chlorine.

	Sample A	*Sample B*	*Sample C*
Chlorine	5.50 g	20.0 g	12.0 g
Copper	10.0 g	18.0 g	21.8 g

a. Which laws are demonstrated by compounds A and C?
b. Which laws are demonstrated by compounds A and B?
c. What is the ratio of grams of chlorine in sample B to sample A when 1.00 g of copper is used?

4.40 For compound A described in problem 4.39, how many grams of chlorine are needed to completely combine with 26.0 g of copper?

4.41 In one of the compounds of nitrogen and oxygen, 14.0 g of nitrogen combines with 32.0 g of oxygen. Use the law of definite proportions as you determine how much oxygen is required to combine with a sample of 10.5 g of nitrogen in producing this same compound.

4.42 One gaseous compound containing nitrogen and oxygen atoms is used for anesthesia in dentistry. Analysis of a sample of the gas showed 2.80 g of nitrogen and 1.60 g of oxygen. How much oxygen is combined with 10.5 g of nitrogen?

4.43 The formula of methane (present in natural gas) is CH_4. If a sample of methane contains 6×10^{10} atoms of carbon, how many atoms of hydrogen must also be present? To answer this question, were we applying the law of multiple proportions or the law of definite proportions?

4.44 In molecules of propane gas, C_3H_8, how many atoms of hydrogen are combined with 6×10^{24} atoms of carbon? If you had 6×10^{24} atoms of carbon, would this be enough to be visible?

4.45 Explain what is meant by a scientific law. How does it differ from a governmental law? Give examples.

4.46 Explain what is meant by a scientific theory. Give an example.

ATOMS AND SUBATOMIC PARTICLES

4.47 What are the names, electrical charges, and mass numbers of the three main subatomic particles?

4.48 Why are atoms neutral? Give locations of the particles within an atom that make it neutral.

4.49 The atomic number is always the same as the number of which kind of subatomic particle?

4.50 When the number of protons, neutrons, and electrons is known for an atom, how is the mass number determined?

4.51 Use the periodic table to help you determine the number of protons and electrons in atoms of the following elements.
a. sodium b. radium c. nitrogen d. fluorine

4.52 Use the periodic table to help you determine the number of protons and electrons in atoms of the following elements.
a. calcium b. lead c. phosphorus d. neon

ISOTOPES

4.53 The following table describes four atoms.

	Atom A	*Atom B*	*Atom C*	*Atom D*
No. of protons	10	11	11	10
No. of neutrons	11	10	11	10
No. of electrons	10	11	11	10

a. Are atoms A and B isotopes of the same element?
b. Are atoms A and D isotopes of the same element?
c. What is the mass number for atom A?
d. What is the mass number for atom D?

4.54 The following questions pertain to the four atoms described in problem 4.53.
a. Are atoms A and C isotopes of the same element?
b. Are atoms B and C isotopes of the same element?
c. What is the mass number for atom B?
d. What is the mass number for atom C?

4.55 For an atom of deuterium, write the symbol for this isotope in the form $^A_Z X$ and list the atomic number, mass number, and numbers of protons, neutrons, and electrons.

4.56 For an atom of radioactive tritium, write the symbol for this isotope in the form $^A_Z X$ and list the atomic number, mass number, and numbers of protons, neutrons, and electrons.

4.57 For an atom of radioactive strontium-90, give the number of a. protons b. neutrons c. electrons.

4.58 For an atom of naturally occurring, radioactive radon-222, give the number of a. protons b. neutrons c. electrons.

4.59 Calculate the average atomic mass for the element copper, which has two isotopes. Use the data provided in Table 4.5.

4.60 Calculate the average atomic mass for the element boron, which has two isotopes. Use the data provided in Table 4.5.

4.61 What is the mass of 1 mole of carbon? How many atoms are represented by this mass?

4.62 What is the mass of 1 mole of calcium? How many atoms are represented by this mass?

4.63 Determine the mass of a mole of each substance.
a. $CaCO_3$ (calcium carbonate, in limestone and marble and used in antacids)
b. NH_4NO_3 (ammonium nitrate, used in fertilizers and explosives)
c. Na_3PO_4 (sodium phosphate, used to clean surfaces before painting)

4.64 Determine the mass of a mole of each substance.
a. SO_2 (sulfur dioxide, a gas that can cause pollution problems)
b. Na_2CO_3 (sodium carbonate, used as a water-softening agent)
c. H_2SO_4 (sulfuric acid, an important industrial acid)

4.65 Calculate the mass of 1.22 moles of each substance listed in Problem 4.63.

4.66 Calculate the mass of 1.22 moles of each substance listed in Problem 4.64.

Additional Problems

4.67 Explain what is represented by the following: (a) Co and CO, (b) Pb and PB.

4.68 When we burn a 10-kg piece of wood, only 0.50 kg of ash is left. Explain this apparent contradiction of the law of conservation of mass.

4.69 The complete burning of 3.0 g of carbon uses

8.0 g of oxygen to produce 11 g of carbon dioxide. What weight of carbon dioxide would be formed from burning 3.0 g of carbon in 50 g of oxygen?

4.70 Give the mass in grams and number of atoms present in 1.50 mol of tungsten. Do the same for 1.50 mol of silver.

4.71 The diameter of an atom is about 1 Å (1 angstrom = 1×10^{-10} m). How many atoms would have to be lined up touching each other to make a line 1.00 mm long?

4.72 The natural abundances of isotopes of magnesium are listed in Table 4.5. Use these data to calculate the average atomic weight of magnesium.

4.73 The radioactive isotope technetium-99 has a number of medical diagnostic uses. Symbolize this isotope in the form $^{A}_{Z}X$ and list the atomic number, mass number, and numbers of protons, neutrons, and electrons.

4.74 Determine the numbers of protons, electrons, and neutrons for the following isotopes.
a. chromium-52 b. argon-40
c. iron-59 d. gold-197

4.75 How many molecules are represented by a mole of sulfur trioxide, SO_3?

4.76 How many atoms of oxygen are present in a mole of SO_2?

We see the rainbow colors of visible light,
But many frequencies are far beyond our sight.

5

Electron Arrangements in Atoms

CONTENTS

What causes the different bright colors of fireworks? Why do certain minerals glow under ''black light''? What is the difference between ultraviolet and infrared radiation? What are ions? We can answer questions like these when we understand some of the properties of excited electrons in atoms. In this chapter we will consider some of the unique properties of excited atoms and the arrangement of electrons within atoms. The electron arrangements in atoms are responsible for many of the properties of different elements. But, first, we will examine some discoveries that revealed important information about atoms. Reviewing these investigations in chronological order allows us to follow the reasoning that led to our present understanding of atoms and subatomic particles.

5.1 Probing the Atom with New Tools

Dalton's atomic theory and his work with relative atomic masses paved the way for many experimental investigations in the nineteenth century, but it was the discoveries of electricity and radioactivity that provided the best new tools for probing the atom. We will briefly look at some of these discoveries, beginning where we left off in the previous chapter, with Dalton.

Specific Heat and Atomic Mass

J. J. Berzelius (in 1814) and two Frenchmen, Pierre Dulong and Alexis Petit (in 1819), observed that the specific heats of metals decreased as the atomic mass increased. They proceeded to use these data to determine accurately the atomic masses of many metals.

Although Dalton had provided a list of relative atomic masses of some elements, the list was not complete, and some of his values were not accurate. The confusion over relative atomic masses attracted considerable attention in the early 1800s, but determining these atomic masses was no easy task at that time.

By 1869, relative atomic masses of many elements had been determined rather accurately, and Mendeleev, a Russian chemist, published a chemistry textbook with his "periodic table" that showed elements arranged in order of increasing atomic mass so that elements having similar chemical properties were grouped together into families. Mendeleev's periodic table was quite similar to modern periodic tables. Several periodic properties of elements will be discussed in Chapter 6.

Using Electricity to Study Atoms

Of major significance was the experimental work of two British scientists, William Crookes, a chemist (Figure 5.1), and Joseph J. Thomson, a physicist (Figure 5.2). In 1879, Crookes carried out studies in an evacuated glass tube. Into the tube were inserted two metal disks, called **electrodes**—one at each end of the tube. When the

Figure 5.1 *Sir William Crookes (1832–1919) was the British scientist who invented the cathode ray tube. His work paved the way to the discovery of the electron.*

Figure 5.2 *J. J. Thomson (1856–1940) was the British scientist who determined the charge-to-mass ratio of the electron. He received the Nobel prize in physics in 1906.*

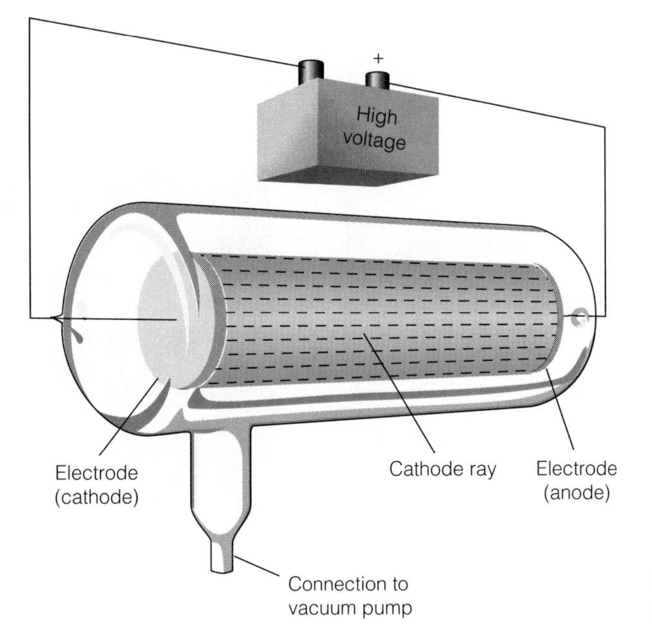

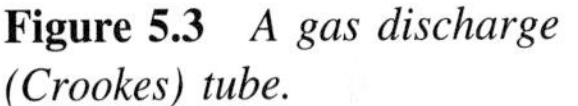

Figure 5.3 *A gas discharge (Crookes) tube.*

electrodes were connected to the voltage source by separate wires, one disk became positively charged and the other disk became negatively charged. Crookes also attached the tube to a vacuum pump to remove most of the air inside.

When a high voltage was applied to the **cathode** (the negative electrode) and the **anode** (the positive electrode) the evacuated tube began to glow (Figure 5.3). The tube is called a *gas discharge tube.* Crookes also observed that the beam was deflected when a magnet was placed near the tube. He was convinced that this glowing beam, now called a **cathode ray**, consisted of charged particles.

The Crookes tube can be thought of as the forerunner to the modern fluorescent tube—a partially evacuated tube fitted with electrodes and filled with mercury vapor and a little argon gas.

But just what were these cathode rays? J. J. Thomson obtained the answer in 1897. He showed that the cathode rays were deflected in an electric field. To one side of the Crookes gas discharge tube was placed a metal plate that carried a positive charge, and to the opposite side was placed a plate with a negative charge. Rays traveling from the cathode to the anode were attracted by the positive plate and repelled by the negative plate (Figure 5.4). Thus, the rays must be composed of negatively charged particles. Thomson called these particles *electrons.*

Although Thomson was unable to measure either the charge or the mass of electrons, he was able to measure the ratio of the charge, e, to the mass, m, of the electrons by determining the amount of deflection in a magnetic field of known strength. This value, known as the e/m ratio, is -1.76×10^8 coulombs per gram. Knowing this ratio but not being able to calculate the mass of the electron is similar to knowing that an apple has a mass 1000 times as great as the mass of an apple seed, but not being able to determine the mass of either the apple or the seed.

The e/m ratio for an electron has a negative value because electrons carry a negative charge. The coulomb is the SI unit of electrical charge. It represents quite a large charge—about the amount of charge that passes through a 100-watt bulb in a second.

The properties of the cathode ray were the same regardless of the material from which the cathode was made. Cathodes made from *different* elements emitted streams of particles with identical mass and charge. Thus, electrons came to be regarded as constituents of all matter. Dalton's model of the atom could not account for these experimental results. (If atoms were indivisible, it would not be possible for them to give off particles.)

In 1886, a German scientist named Eugen Goldstein conducted some experiments with a modified Crookes tube that had a cathode made from a metal disk

The modern television picture tube is a cathode ray tube (CRT) that is based on the same principles used by Thomson. A changing electric field is used to direct the path of a beam of charged particles across a phosphorescent surface.

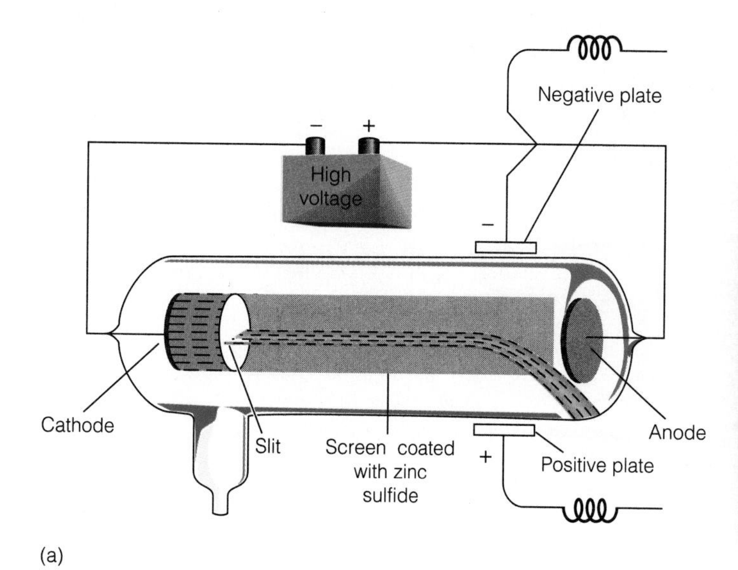

(a)

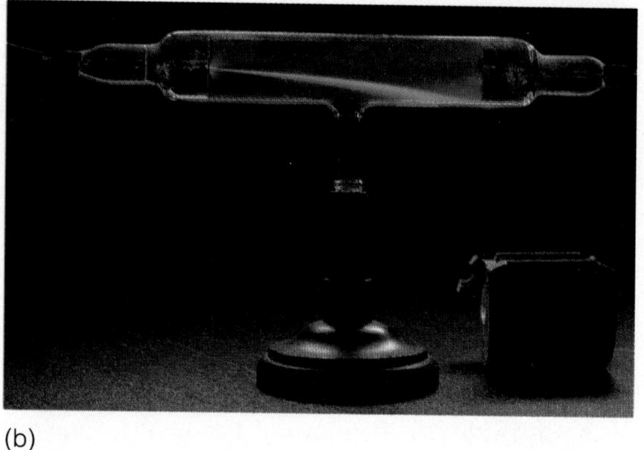

(b)

Figure 5.4 *(a) Thomson's apparatus showing deflection of an electron beam in an electric field. The screen is coated with zinc sulfide, a substance that glows when struck by electrons. (b) Deflection of the beam in a magnetic field.*

Figure 5.5 *Goldstein's apparatus for study of positive particles.*

"In science, a wrong theory can be valuable and better than no theory at all."

—Sir William L. Bragg

Sir William and his father, Sir W. H. Bragg, shared the 1915 Nobel prize in physics for studies of crystals with X-rays.

filled with holes (Figure 5.5). With this apparatus, he not only observed the stream of electrons emitted by the cathode, but he also observed positive rays (called canal rays) in the region behind the cathode. We now know that these positive charges are formed when cathode rays knock electrons from neutral gaseous atoms. Different positive particles are produced when different gases are used, but all the positive charges are multiples of the value obtained when hydrogen gas is used.

By 1904 enough evidence had been collected to suggest that atoms were definitely made up of smaller particles. J. J. Thomson had verified the existence of electrons, and Goldstein had shown that positive particles could also be formed. Now, Thomson offered an explanation; it is called the "plum pudding" model of the atom (Figure 5.6). He visualized an atom as having negative charges (electrons) spread out among an equal number of positive charges (protons). Thomson's model was later found to be incorrect, but it provided an explanation of the facts known at that time.

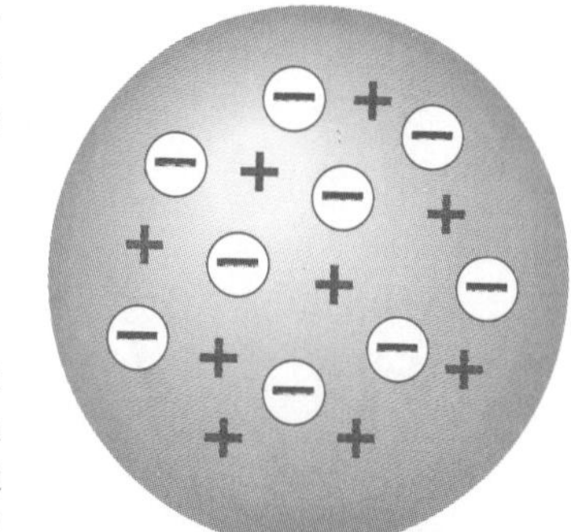

Figure 5.6 *Thomson's "plum pudding" model of the atom.*

The Charge of the Electron

It was an American scientist, Robert A. Millikan, who performed experiments in 1909 at the University of Chicago that allowed him to determine the charge of an electron. The essential parts of Millikan's apparatus for his "oil drop" experiment are pictured in Figure 5.7. An atomizer spray bottle was used to spray a mist of oil into the upper chamber of the apparatus. As an oil droplet fell through a hole into the lower chamber, it picked up electrons produced by irradiating the air with X-rays. By adjusting the electrical force on the plates, the movement of the falling, negatively charged drop could be slowed or even stopped.

Millikan could calculate the amount of charge on a drop because he knew the amount of electrical force on the plates, and he could determine the mass of the drop. After many experiments, Millikan found that the electrical charge on an oil drop was not always the same, but was a multiple of -1.60×10^{-19} coulomb. He reasoned that this value is the charge on the electron since a drop could pick up one or more electrons. Once the charge of the electron was known, the mass of the electron could be calculated because Thomson had already determined a value for the *e*/*m* ratio. Each electron has a mass of 9.110×10^{-28} g.

The mass of an oil drop could be determined by observing its rate of fall in the absence of an electric field. This experiment was not totally original with Millikan; other scientists had used water drops, but evaporation caused the mass to change continuously.

Robert Millikan was awarded the 1923 Nobel prize in physics.

X-Rays and Radioactivity: Two New Tools to Probe Atoms

After Crookes's and Goldstein's discoveries, but before Millikan's oil drop experiment was performed, other important discoveries concerning atoms were rapidly taking place.

Two important discoveries in the 1890s can be described as lucky accidents. A "lucky accident" is the term used to describe something unexpected that happens when, as a result, whole new areas of study are opened. Such "lucky" accidents would be overlooked unless they took place in the presence of a trained observed who can grasp their significance.

One must do some type of investigation to discover something, but what is discovered may be the unexpected—a lucky accident!

One lucky accident occurred in 1895 when Wilhelm Roentgen, a German scientist, was working in a darkroom, studying certain substances that glowed while exposed to cathode rays. To his surprise, he observed this glow in a chemically

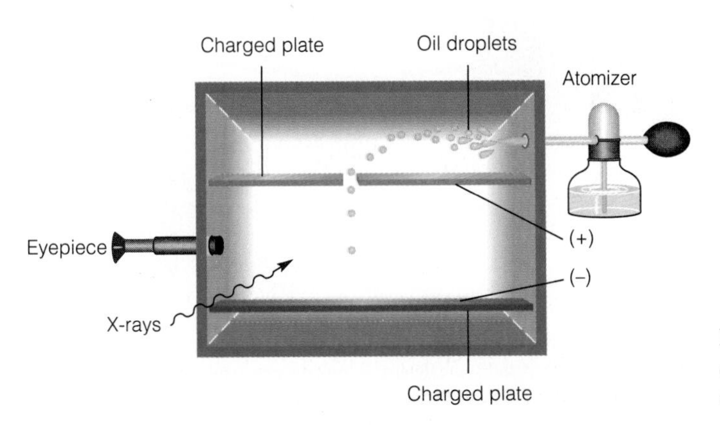

Figure 5.7 *Millikan's oil drop experiment determined the charge of the electron.*

treated piece of paper held some distance from the cathode ray tube. The paper even glowed if the tube was located in the next room. Roentgen had discovered a new type of ray, one that could travel through walls! These penetrating rays, named **X-rays** by Roentgen (see Figure 5.8), were given off from the anode whenever the cathode ray tube was operating. Unlike cathode rays, which are streams of charged particles, the X-rays are related to light energy—both are forms of electromagnetic radiation.

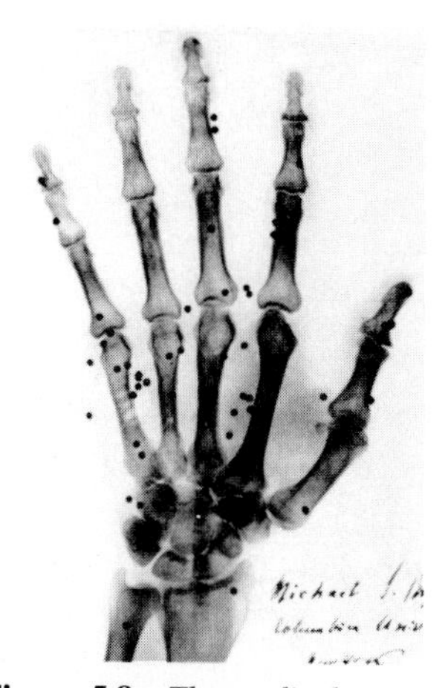

Figure 5.8 *The medical community immediately recognized the significance of penetrating X-rays. This picture was taken in February 1896, within two months of the publication of Roentgen's discovery, by Professor Michael Pupin of Columbia University. The round black dots are shotgun pellets, identified for removal by surgery. (Roentgen would not patent his discovery and thus refused to reap its commercial rewards.)*

As is often the case with an exciting new discovery, many other scientists began to study X-rays.* Antoine Becquerel, a Frenchman, had been studying fluorescence, a phenomenon that occurs as certain chemicals glow when they are exposed to strong sunlight. Was this phenomenon, he wondered, related to X-rays?

Before Becquerel's studies of fluorescence had progressed very far, another lucky accident occurred. After only slight exposure of one sample to sunlight, his experiment was interrupted by several cloudy days, so he placed the covered photographic plate and the uranium compound in a drawer. Later, when the experiment was completed, and the photographic plate was developed, he found—to his great surprise—that it showed images of the uranium sample. Further experiments showed that this radiation had no connection with fluorescence; instead, it was a characteristic property of the element uranium.

Becquerel's discovery attracted the attention of many other scientists who also began to study this new type of radiation. Marie Curie was one of Becquerel's colleagues who studied this new phenomenon, which she named **radioactivity.**

Rutherford's Nuclear Model of the Atom

The radiation from uranium, radium, and other radioactive elements was investigated by Rutherford, a New Zealander working at McGill University in Montreal. When he passed this radiation through a strong magnetic field, he observed that the rays were separated—they were deflected in different directions (Figure 5.9). The **alpha**, α, rays that Rutherford discovered have a doubly positive charge, 2+. They were deflected toward the negatively charged plate, and were found to have a mass four times that of the hydrogen atom. The **beta**, β, rays were found to be identical

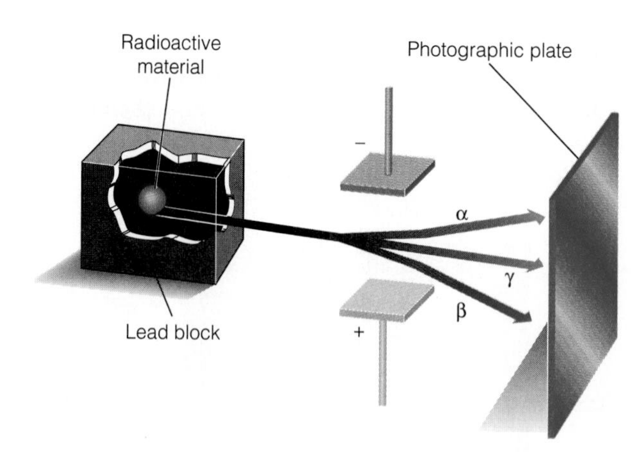

Figure 5.9 *Behavior of radioactive rays in an electric field.*

*X-rays are important in basic science and medicine, but their indiscriminate use resulted in a number of cases of severe burns and some deaths. It took people some time to realize that rays passing through the body could also inflict biological damage.

HISTORY MAKERS

The Curies

Marie Curie in her laboratory.

Marie Sklodowska was born in Poland in 1867. She went to Paris to work for her doctorate in mathematics and physics. There she met and married Pierre Curie, a respected French physicist. In 1898–1900 Marie Curie, with assistance from Pierre, isolated samples of two new elements from the uranium ore called pitchblende. She named one of the elements polonium, for her native country of Poland. She named the other element radium, for its highly radioactive properties. Radium soon became widely used for treating cancer.

Marie Curie was the first person to win two Nobel prizes in the sciences. The Curies along with Becquerel shared the 1903 Nobel prize in physics for their discovery of natural radioactivity. Pierre Curie was killed in a traffic accident three years later, but Marie continued to work with radioactivity, and in 1911 she won a second Nobel prize, in chemistry, for the discovery of radium and polonium. The Curies' daughter, Irène, and Irène's husband, Frédéric Joliot, later won the 1935 Nobel prize in chemistry for work with radioactive materials. This is the only family to win three Nobel prizes.

Marie Curie died in 1934 of pernicious anemia, perhaps brought on by exhausting work and long exposure to radiation from the materials she had studied. The scientific contributions of the Curies were quite remarkable.

to cathode rays, which are streams of electrons with a negative charge, 1−. The **gamma**, γ, rays were not deflected by the magnetic field. They were found to be very much like X-rays but even more penetrating. Gamma rays have no mass or charge. These properties are summarized in Table 5.1.

Rutherford's discovery of alpha particles soon led to another important discovery. When Rutherford placed a highly radioactive material in a lead-lined box with a tiny hole, some alpha particles could escape through the hole in the box, forming a narrow stream of very-high-energy particles. This apparatus could then be aimed like a gun at a "target."

Ernest Rutherford received the 1908 Nobel prize in chemistry for his work at McGill University with radioactive substances.

When Hans Geiger—inventor of the Geiger counter—asked his colleague, Rutherford, to suggest a research project for Earnest Marsden, a 20-year-old undergraduate student, Rutherford recommended that Marsden might investigate the effect of alpha particles on various thicknesses of metal foil (Figure 5.10). A few days later, Geiger, in great excitement, went to Rutherford and explained what they had observed while using a gold foil. Most of the alpha particles went through the foil, but some were sharply deflected. A few of the alpha particles even came

The search for new information can be exciting, compelling, and rewarding, but it is a venture into the unknown. Answers are rarely obtained on a time schedule.

Table 5.1 **Types of Radioactivity**

Name	*Symbol*	*Mass (amu)*	Charge
Alpha	α	4	2+
Beta	β	$\frac{1}{1837}$	1−
Gamma	γ	0	0

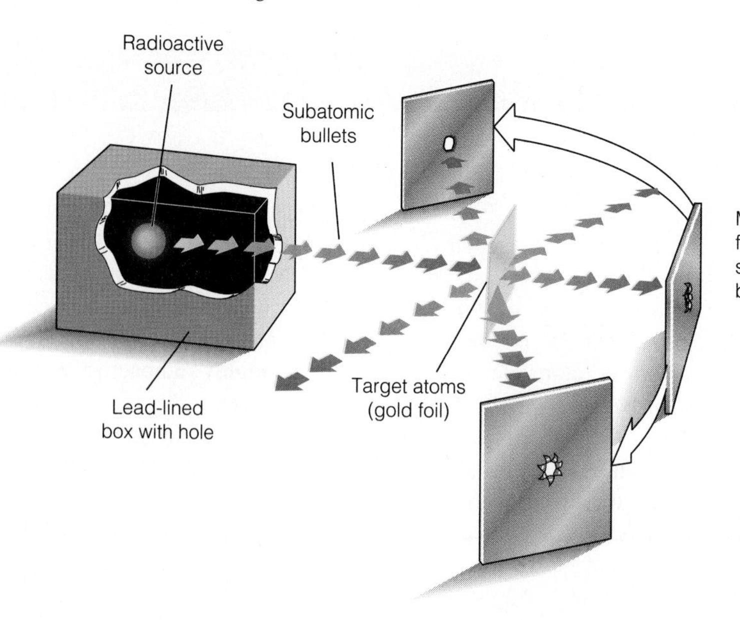

Figure 5.10 *Rutherford's gold foil experiment.*

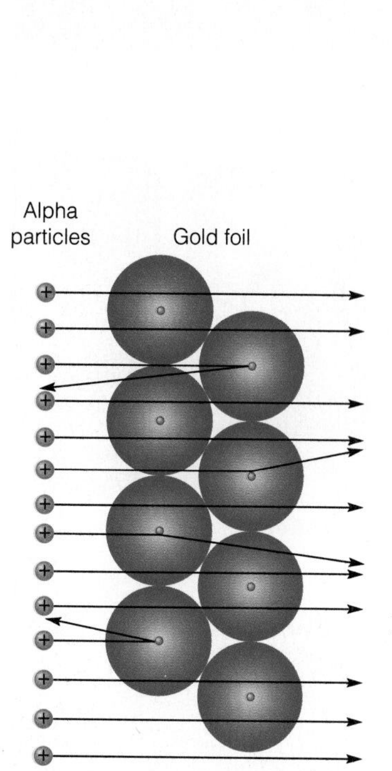

Figure 5.11 *Model explaining the results of Rutherford's experiment.*

directly backward. Rutherford thought this was incredible and had no immediate explanation. He had not expected any of the particles to be so sharply deflected, and he did not immediately recognize the significance of these investigations.

It was in 1911, almost two years later, that Rutherford told Geiger that he now knew what the atom looked like. He had concluded that all of the positive charge and virtually all of the mass of an atom are concentrated in its extremely tiny nucleus. That is what caused the alpha particles to be so sharply deflected (Figure 5.11). Rutherford's nuclear theory of the atom represents a major step in understanding the structure of the atom.

If an atom could be enlarged to the size of the New Orleans Superdome, the nucleus would be proportional to the size of a BB in the center of this volume. The electrons would be like tiny mosquitoes scattered throughout the stadium-sized atom.

Among other unanswered questions was the relationship of mass and charge. While the 2+ charge of the helium nucleus is twice as great as the 1+ charge of the hydrogen nucleus, the mass of the helium nucleus is four times the mass of the hydrogen nucleus. This apparent "excess mass" of helium puzzled scientists until 1932, when the British physicist James Chadwick discovered the neutron—a particle with nearly the same mass as a proton, but with *no* electrical charge. The excess mass of helium could now be explained; the helium nucleus contains two neutrons and two protons.

The alpha particle, with a mass number of 4 and a charge of 2+, is identical to the nucleus of a helium atom. An alpha particle can be represented as ${}^{4}_{2}He^{2+}$.

Before we discuss additional discoveries that revealed important facts about the arrangement of electrons in atoms, you will need to have a general understanding of the electromagnetic spectrum. We will present these fundamentals in the next section. At that point, we can then explain the connection between excited electrons and certain chemical properties of atoms.

EXAMPLE 5.1

Consider the Rutherford gold foil experiment.

(a) List three observations that relate to the gold foil experiment.
(b) List three conclusions that can be drawn from this experiment.

See Problems 5.1–5.18

SOLUTION

(a) Observations: (1) Most of the alpha particles went through the gold foil. (2) Some alpha particles were deflected as they went through the foil. (3) A small number of the alpha particles bounced backward.
(b) Conclusions: (1) The atom is mostly empty space, so most of the alpha particles could pass through the foil. (2) The nucleus must occupy a very small fraction of the total volume, since only a small fraction of the alpha particles were bounced backward. (3) The nucleus must contain nearly all of the mass of the atom in order to cause the alpha particles to be bounced backward.

Figure 5.12 *The rainbow is an example of a continuous spectrum.*

5.2 The Electromagnetic Spectrum

Light, such as that emitted by the sun or by an incandescent bulb, is a form of radiant energy. When white light from an incandescent lamp (an ordinary light bulb) is passed through a prism, it separates into a **continuous spectrum** or rainbow of colors (Figures 5.12 and 5.13). When sunlight passes through a raindrop, the same phenomenon occurs. The different colors of light represent different amounts of radiant energy. Blue light, for example, packs more energy than does red light of the same intensity.

Figure 5.13 *A glass prism disperses a beam of white light into a continuous spectrum or rainbow of colors.*

Besides visible light, there are several other forms of radiant energy, such as gamma rays, ultraviolet radiation, and infrared radiation. All these forms of radiant energy, or **electromagnetic radiation**, travel through space at a speed of 3.00×10^8 m/s—the speed of light. Electromagnetic radiation travels in waves, much like the waves formed on a lake on a windy day or formed when a pebble is tossed into a pond. As shown in Figure 5.14, the distance between peaks (or any other equivalent points) of consecutive waves is called the **wavelength**; it is represented by the Greek letter λ (lambda). The number of peaks that pass a particular point in 1 second is called the **frequency**; it is represented by the Greek letter ν (nu, pronounced ''new''). The speed of a wave is obtained when the wavelength is multiplied by the frequency. For electromagnetic radiation, the speed of the wave is the speed of light, represented by the symbol c.

$$\lambda \nu = c$$

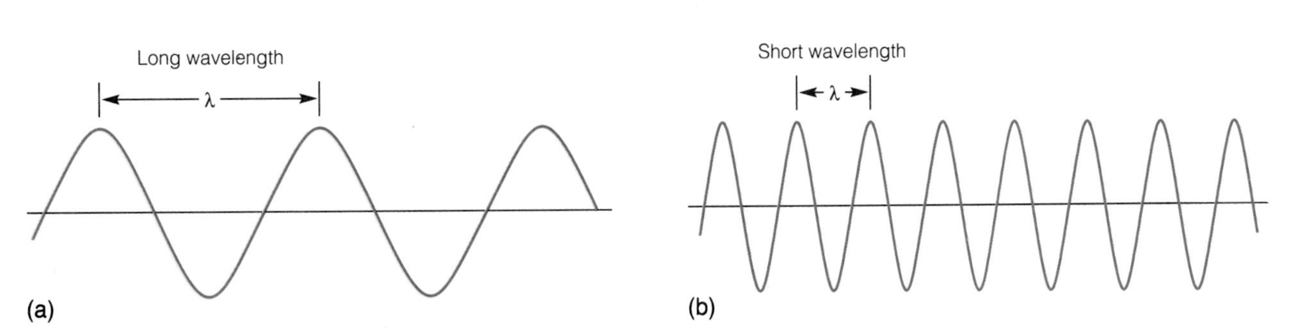

Figure 5.14 *As the wavelength (λ) of a wave decreases, the frequency (ν) increases. Electromagnetic waves with a low frequency (a) have a long wavelength. Electromagnetic waves with a high frequency (b) have a short wavelength.*

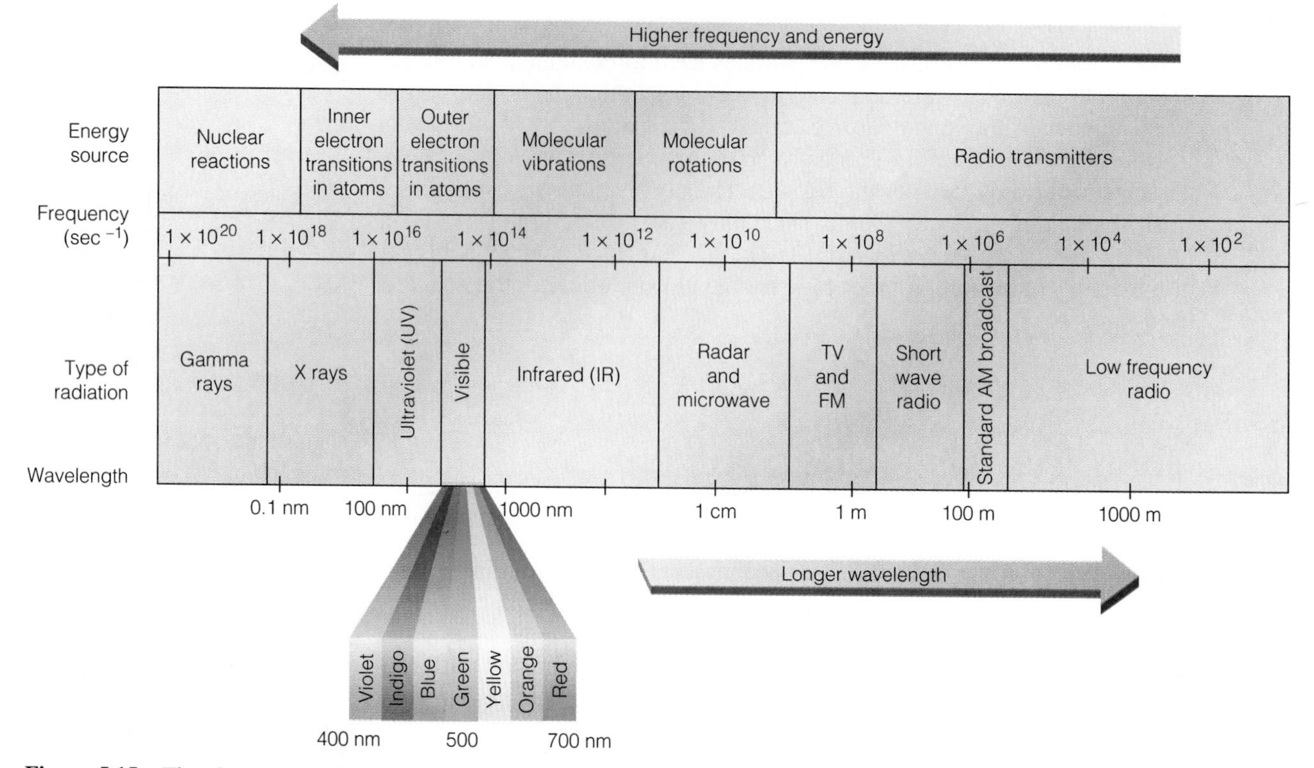

Figure 5.15 *The electromagnetic spectrum.*

For electromagnetic radiation,

$$\lambda\nu = 3 \times 10^8 \text{ m/s}$$

The frequency increases as wavelength decreases, and vice versa.

Visible Light

White light is a mixture of all wavelengths in the visible spectrum.

Our eyes can sense wavelengths as short as about 400 nm (violet light) and as long as about 750 nm (red light). All wavelengths within this range of values are in the **visible spectrum**. Each specific color of visible light—red, orange, yellow, green, blue, indigo, and violet—has a different wavelength and frequency. Red light has a long wavelength and a low frequency; blue light has a short wavelength and a high frequency.

Visible light covers only a small fraction of the entire electromagnetic spectrum, as shown in Figure 5.15. The total electromagnetic spectrum covers a broad span from the high-energy gamma rays with high frequency and short wavelengths to the low-energy radio waves with low frequency and long wavelengths.

One should not look directly at a UV light source, since the high-energy rays can severely damage the eye. Most UV rays are absorbed by glass, so window glass can provide some protection from these rays.

Ultraviolet Radiation: "Black Light"

Figure 5.15 shows that ultraviolet (UV) radiation has wavelengths that are shorter than visible light. The longest UV wavelengths—those near visible light—are called ''near UV.'' Shorter UV wavelengths are called ''far UV.'' Three categories, UV-A, UV-B, and UV-C are also used to identify long, middle, and short UV wavelengths, respectively. The UV-C wavelengths are filtered out by our atmosphere.

What is sometimes called "black light" is really UV—primarily near UV. When UV radiation strikes certain rocks or certain types of paint, the objects **fluoresce**; they appear to give off light of their own while being bombarded by the UV rays. This occurs when the electrons in atoms of a material absorb UV rays and then give up this energy by emitting lower energy visible light. The light emitted makes the material appear to glow.

Sunscreens that offer both UV-A and UV-B protection filter both near-UV and far-UV radiation. Common sunscreens block UV-B rays that can tan or sunburn you. The UV-A rays penetrate deeply into the dermis, breaking down the tissue and causing premature aging of the skin. Both UV-B and UV-A wavelengths can induce cancer.

Infrared Radiation

What we usually think of as radiant heat is actually **infrared** (IR) radiation. Infrared rays are too long to be visible by the human eye, but they have the proper frequencies to interact with molecules and cause molecular vibrations.

Microwaves and Radar

Microwave and **radar** wavelengths are similar (about 1 cm). They are longer than IR wavelengths. Microwaves have the appropriate energy to cause molecules to *rotate.* This property makes it possible for microwave ovens to warm a hot dog quickly. As the microwaves pass through the hot dog, the energy of the microwaves causes molecules of water at the center of the hot dog as well as at the surface of the hot dog to rotate. The heat produced by the rotating molecules quickly warms the entire hot dog, or whatever is being heated.

It should now be obvious that visible light frequencies make up only a small part of the entire electromagnetic spectrum. In the next section we will discuss important connections between frequencies of light and energies of electrons.

Because of microwaves, it is possible for thousands of telephone conversations to be transmitted simultaneously over great distances without wires. When a radar signal is transmitted, the elapsed time for its return to the source can be used to measure the distance to an object or the speed of the object.

CHEMISTRY IN OUR WORLD

Depletion of Our Ozone Layer

In the outer part of our atmosphere, a "blanket" of ozone, O_3, filters out much of the solar *ultraviolet radiation* that otherwise would strike the surface of the earth. More ultraviolet radiation would cause an increase in skin cancers and perhaps upset the delicate balance in marine food chains.

Scientists are concerned that various Freon gas compounds (chlorofluorocarbons)—used for air conditioning, refrigeration, and aerosols—may be causing a breakdown of this ozone layer, especially over the North and South poles. New refrigerants are being developed that do not contribute to these environmental problems.

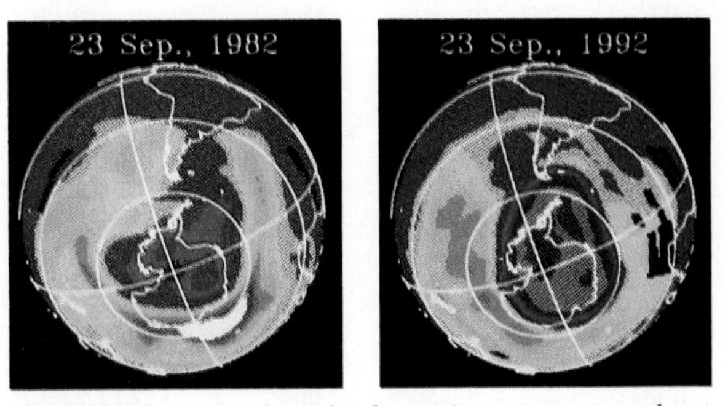

These NASA photos show the decreasing upper atmosphere ozone values near the South Pole. Ozone levels are monitored by a satellite-based Total Ozone Mapping Spectrometer (TOMS).

CHEMISTRY AT WORK

IR Spectrometers

Infrared spectrometers are electronic instruments used by chemists to help identify many types of molecules. The substance being tested is bombarded with different IR frequencies, and the interactions are recorded. The recorded IR spectrum for a compound can be used to help identify a compound, somewhat like the way a fingerprint can be used to help identify a particular person. Molecules of ethyl alcohol, cholesterol, and various drugs, for example, all have different IR spectra. Each substance—hero or villain—has its telltale spectrum.

Each chemical substance has a unique infrared spectrum. The IR spectrum displayed here is for acetaminophen. A portion is enlarged on the computer screen at the left.

EXAMPLE 5.2

Refer to the electromagnetic spectrum as you answer the following.

(a) Which color of visible light has the highest frequency?
(b) Which color of visible light has the longest wavelength?
(c) Arrange in order of decreasing frequencies IR, UV, visible light.

SOLUTION

See Problems 5.19–5.30.

(a) violet (b) red (c) UV > visible light > IR

EXERCISE 5.1

What form of electromagnetic radiation causes (a) molecules to vibrate, (b) molecules to rotate, and (c) outer electrons in atoms to be excited? (See Figure 5.15.)

5.3 Excited Electrons and Spectra

Light from a flame in which a certain chemical is being heated can be passed through a prism, but only narrow colored lines can be observed (Figure 5.16) instead of the continuous spectrum observed when white light is passed through a

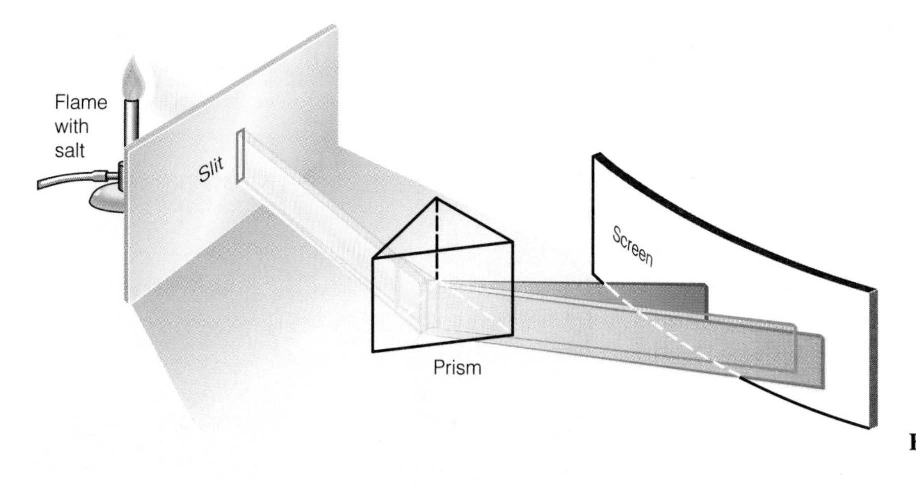

Figure 5.16 *A line spectrum.*

prism. Each line corresponds to light having a definite energy and a definite frequency. The specific pattern of colored lines (or frequencies) emitted by each element—its **line spectrum**—is a characteristic property of that element. A line spectrum can be used to identify the element (Figure 5.17). An instrument called a **spectroscope** can be used to view this spectrum. Robert Bunsen and Gustav Kirchoff, two German scientists, reported in 1859 that each element has a characteristic spectrum. In 1885, J. J. Balmer was able to calculate wavelengths for lines in the visible spectrum of hydrogen.

Using a spectroscope, Bunsen and Kirchoff discovered the elements cesium and rubidium, which were present in mineral water samples and mineral deposits.

Scientists have used line spectra to determine the chemical makeup of the stars and the atmosphere of the planets. Until recently, everything we knew about these heavenly bodies had to be deduced from our examination of this light. During the solar eclipse of 1868, a new line in the solar spectrum was identified by Pierre Janssen, a French astronomer. This line was caused by the presence of the element helium, which had not yet been discovered on the Earth.

The name helium is derived from the Greek word *helios,* the sun.

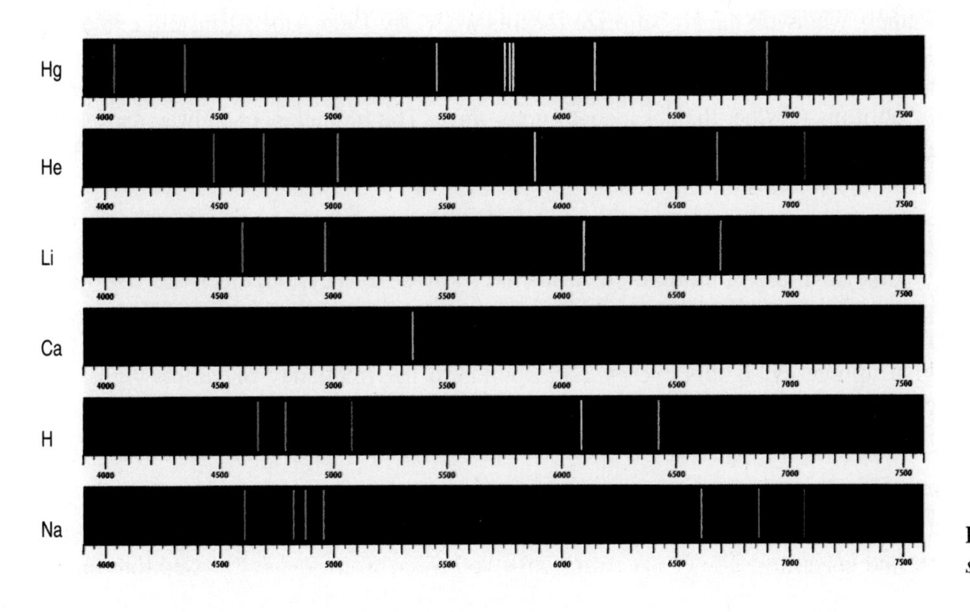

Figure 5.17 *Characteristic line spectra of some elements.*

CHEMISTRY IN OUR WORLD

Fireworks and Flame Colors

Distinctive colors can be observed when certain chemicals are heated in a flame. These colors are determined by the particular metallic element present in the compound. Fireworks with colors of brilliant red, gold, and green emit frequencies that are characteristic of the metal present. Typical flame colors include

Type of Compound	*Flame Color*
Lithium	Brilliant red
Sodium	Golden yellow
Potassium	Lavender
Calcium	Orange
Strontium	Red
Barium	Yellow-green
Copper	Blue and green

The bright red, blue, and yellow colors of fireworks are produced by using compounds of strontium, copper, and sodium, respectively.

Perhaps you have observed the yellow flame produced when table salt (sodium chloride) is sprinkled into the flame of a barbecue. When this yellow flame is viewed through a spectroscope, only a yellow line is apparent. If the yellowish light from a sodium vapor street lamp is also viewed through the spectroscope, the same line spectrum can be observed—sodium is present in both cases. The yellow color in a flame—even without the aid of a spectroscope—can be used to identify the presence of sodium in a sample.

By the year 1900, an explanation for the frequencies of light emitted by very hot solids—called the **quantum theory**—was proposed by the German physicist Max Planck, who won the Nobel prize for this work. By 1905 Albert Einstein extended this theory to include all forms of light. According to this quantum theory, light is emitted in a discontinuous fashion—in discrete or definite packets called **quanta** or **photons**—rather than as a continuous wave. The frequency of light, ν, increases proportionally with an increase in energy, E, where h, Planck's constant, is 6.63×10^{-34} joule-second.

$$E = h\nu$$

Like anything in motion, electrons have kinetic energy, but electrons also have potential energy. Electrons, in this respect, are like rocks on a cliff. When rocks fall, they give up potential energy. If electrons fall toward the nucleus, they too give up energy. If this energy corresponds to the frequency of yellow light, for example, then that is the color that can be observed. Thus, a specific energy drop for excited electrons results in a specific frequency or color.

EXAMPLE 5.3

Provide examples of sources that produce (a) a continuous spectrum and (b) a line spectrum.

SOLUTION

(a) Continuous spectrum: the sun, the bright electronic flash of a camera.
(b) Line spectrum: a compound containing any of the metals listed with visible flame colors.

See Problems 5.31–5.36.

EXERCISE 5.2
What is a photon? How is the frequency of light related to energy?

5.4 Electrons in Atoms

How does this information about excited electrons and spectra relate to the electronic structure of atoms? The first satisfactory answer was supplied in 1913 by Niels Bohr, a Danish physicist, who included the work of Planck and Einstein in his formula that allowed him to calculate the frequencies of light in the hydrogen spectrum. Bohr made the revolutionary suggestion that electrons of atoms exist in specific **energy levels**. Electrons cannot have just any amount of energy, but must have certain specified values. This is something like saying you can stand on specific rungs of a ladder, but not between the rungs. You can move up a specified number of rungs, to a higher level, or back down to a ''ground state.'' In moving from one rung to another, the potential energy (energy due to position) changes by definite amounts, or quanta.

HISTORY MAKERS

Niels Bohr

Niels Bohr (1885–1962), Danish, is known for his work on the hydrogen atom and for his planetary model of electrons in atoms. He theorized that electrons in atoms have specific energy values, that they exist in specific energy levels—he called them shells—and that electrons absorb or emit discrete quantities of energy during transitions between specific allowable energy states.

In 1913 Bohr developed a mathematical equation that he used to calculate the amount of energy and the wavelength of light emitted or absorbed when the electron of a hydrogen atom jumps from one level to another. The values Bohr calculated for hydrogen matched values obtained by experiment, but when the equation was applied to other atoms, it did not work at all. Bohr's model was incorrect, but it paved the way for other theories, especially the more sophisticated quantum mechanical model of the atom that is used today.

In 1922 Niels Bohr, at the age of 37, was awarded the Nobel prize in physics for his work with atomic structure and spectra. When Bohr came to the United States in 1939, he brought with him the news about the tremendous amount of energy that might be released during the nuclear fission of uranium.

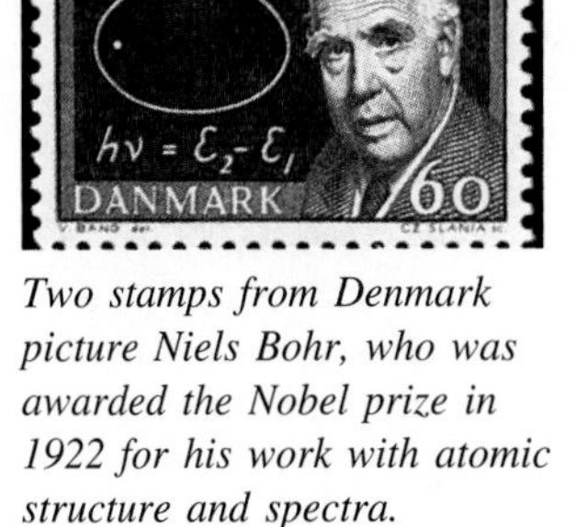

Two stamps from Denmark picture Niels Bohr, who was awarded the Nobel prize in 1922 for his work with atomic structure and spectra.

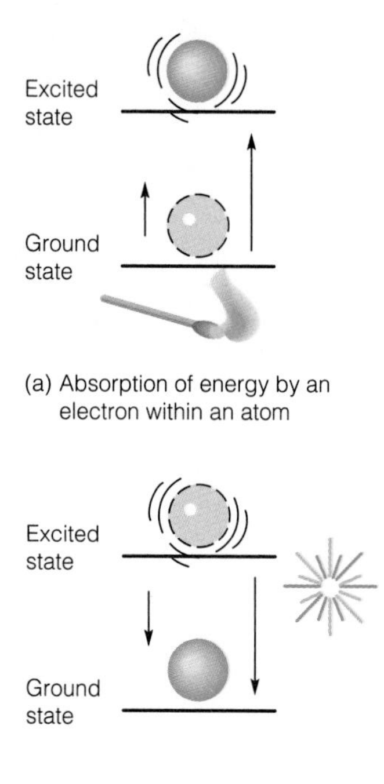

Figure 5.18 *The processes of absorption (a) and emission (b) of energy by electrons within an atom.*

For an electron, its total energy (both potential and kinetic) is changed as it moves from one energy level to another within an atom. By absorbing a photon or quantum of energy (an extremely small energy packet with a definite size), an electron is elevated to a higher energy level—an **excited state**. When the electron then falls to lower energy levels, energy is given up in specific quanta (Figure 5.18). The energy released has specific frequencies and shows up as a line spectrum—rather than as a continuous spectrum (Figure 5.19).

According to Bohr's model of the atom, electrons orbit about the nucleus much like the way planets orbit the sun (Figure 5.20). Different energy levels can be pictured as different orbits. An electron in a hydrogen atom, the simplest atom, is usually in the first energy level (the lowest level nearest the nucleus). Atoms with all electrons in their lowest energy levels are said to be in their **ground state**. When energy in the form of heat or light, for example, causes electrons to jump to higher energy levels, the atom is said to be in an excited state. An atom in an excited state then emits energy (light) as electrons jump back down to one of the lower levels.

Ionization

If an atom receives a sufficient amount of energy, one or more of its electrons can be stripped from the atom. This phenomenon is called **ionization. Ionization energy** is defined as the *energy required to remove an electron from a gaseous atom in its ground state.* When an electron is lost from an atom, the remaining charged particle is called an **ion**. Ions are formed as atoms, or groups of atoms, either gain or lose electrons. The formation of ions is essential to many chemical reactions.

Bohr Atoms and Valence Electrons

Bohr was able to deduce that each energy level of an atom could only handle a certain number of electrons at one time (see Figure 5.21). We shall simply state

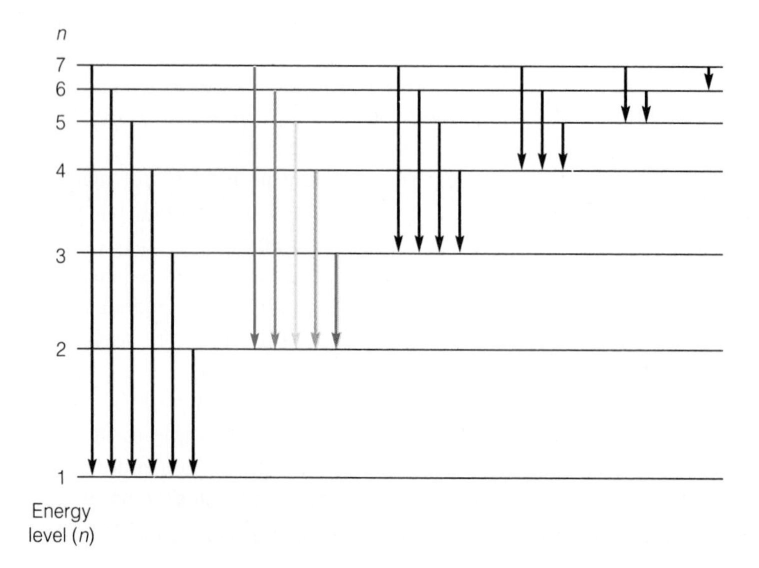

Figure 5.19 *Possible electron shifts between energy levels in atoms to produce the lines found in spectra. Not all the lines are in the visible portion of the spectrum. Electron transitions from higher levels to* n = *1 produce ultraviolet light, transitions to* n = *2 give visible light, and transitions to* n = *3 give infrared radiation.*

Bohr's finding in this regard. The maximum number of electrons—or population—that can be in a certain energy level is given by the formula $2n^2$, where n is equal to the energy level number being filled.

- For the first energy level ($n = 1$), the maximum population is $2(1)^2$, or 2.
- For the second energy level ($n = 2$), the maximum population is $2(2)^2$, or 8.
- For the third energy level, the maximum number of electrons is $2(3)^2$, or 18.

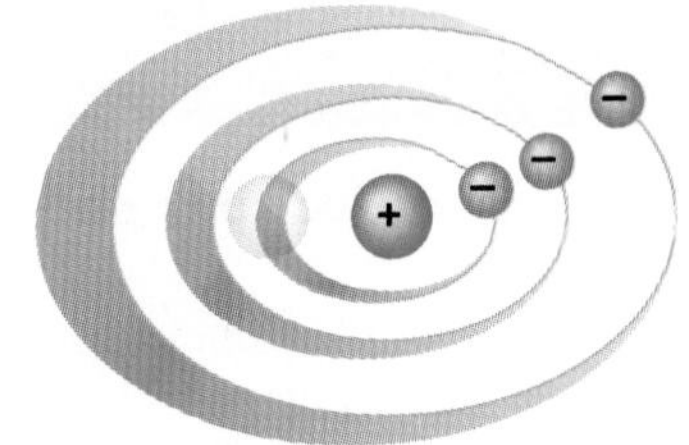

Figure 5.20 *Bohr visualized the atom as planetary electrons circling a nuclear sun.*

Imagine building up atoms by adding one electron to the proper energy level as protons are added to the nucleus, keeping in mind that electrons will go to the lowest energy level available. For hydrogen, H, with a nucleus of only one proton, its one electron goes into the first energy level. For helium, He, with a nucleus of two protons (and two neutrons), two electrons go into the first energy level. Since the first energy level can hold only two electrons, that energy level is filled in each helium atom.

Bohr actually used the term "shells" labeled K, L, M, N, . . . , but today we refer to "energy levels" numbered 1, 2, 3, 4, ...

Lithium, Li, has three electrons. Two of these electrons are in the first energy level, but the third electron is in the second energy level. The electrons in the outer energy level are called **valence electrons**. Lithium has one valence electron—the single electron in the second energy level. This process of adding electrons is continued until the second energy level is filled with eight electrons, giving the same electronic structure as the neon atom, Ne, which has two of its ten electrons in its first energy level and the remaining eight in its second energy level. Carbon, for example, has a total of six electrons—the four electrons in its second energy level are valence electrons. Similarly, nitrogen has five valence electrons, oxygen has six valence electrons, and fluorine has seven valence electrons.

For each element being described, notice its placement in the periodic table.

The sodium atom, Na, has 11 electrons. Of these, 2 are in the first energy level, another 8 electrons fill the second energy level, and the remaining electron is in the third energy level. We can use a Bohr diagram to indicate this arrangement.

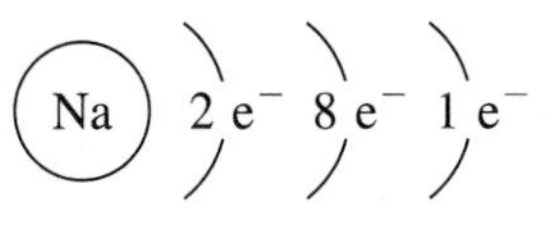

The circle with the symbol indicates the sodium nucleus. The arcs represent the energy levels—the one closest to the nucleus represents the first energy level, the next arc represents the second level, and so on. A sodium atoms has one valence electron. If this electron is removed, a positively charged sodium ion, Na^+, is formed.

As shown in the Bohr diagrams pictured in Figure 5.22, we could continue to add electrons to the third energy level until we get to argon, Ar, which has 18 electrons.

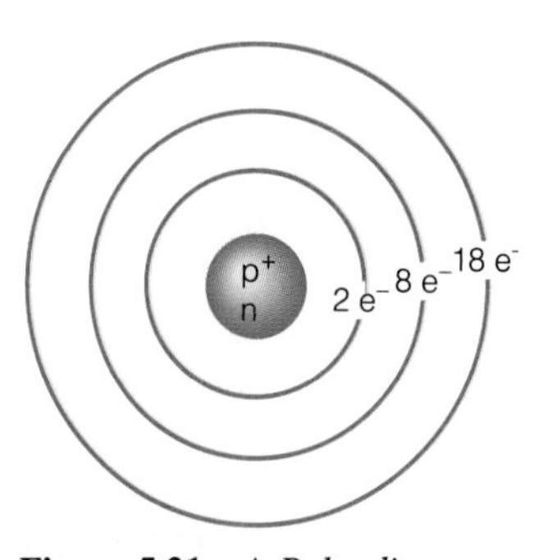

Figure 5.21 *A Bohr diagram for the atom of an element pictures specified numbers of electrons in distinct energy levels.*

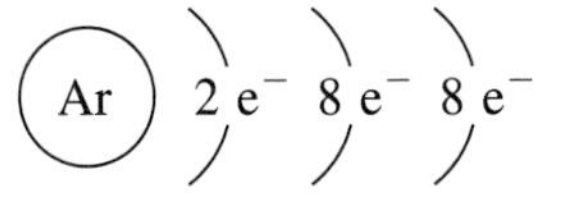

The placement of an element in the periodic table can be used to determine the number of energy levels and the number of valence electrons of an atom. With this

CHEMISTRY IN OUR WORLD

Excited Electrons and Spectra

Energy in the form of heat, electron bombardment, light (ultraviolet, visible, lasers, etc.), and chemical reactions can be used to excite electrons within atoms and push them to higher energy states (levels).

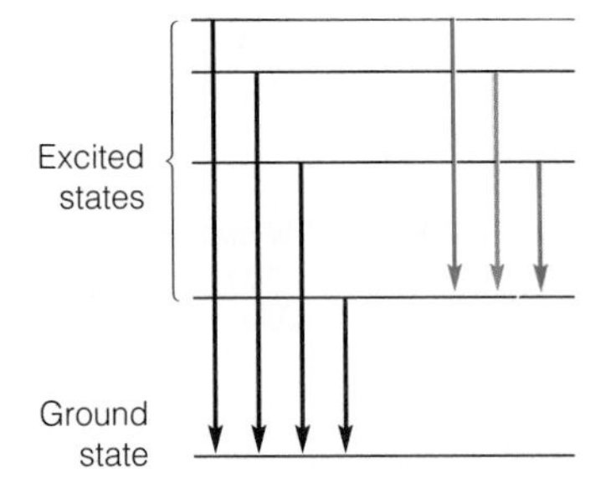

Electrons in excited states of atoms or ions fall back to lower energy levels and, ultimately, to their ground states.

When a sodium-containing compound is placed in a flame, a yellow color can be observed as electrons of sodium are excited by heat. The yellow glow of a sodium vapor street lamp is due to electrons being excited by electron bombardment. In both cases, the excited electrons of sodium fall back to the most stable energy state while producing a unique line spectrum. This characteristic line spectrum is the same regardless of how the electrons of sodium are excited.

In producing the bluish white light of mercury vapor street lamps and fluorescent lights, electrons in mercury atoms are excited by electron bombardment. The orange glow of neon signs and the red glow of helium–neon lasers also result from electrons becoming excited and falling back to lower energy states within atoms. Each line spectrum is unique.

Light can also excite electrons within atoms. When fluorescent paint glows in the presence of ultraviolet light, the UV light is the energy source that excites electrons in molecules of the paint.

Certain chemical reactions can also excite electrons and produce visible light. Examples include the yellow-green glow of the firefly and the chemical lightsticks that glow for several hours. Once again, energy—in the form of visible light—is released as excited electrons in atoms fall back to lower energy states.

As shown by the examples given here, excited electrons are very much a part of our daily lives.

Brightly colored visible light is emitted by a neon sign as excited electrons of the gas fall back to their most stable ground states.

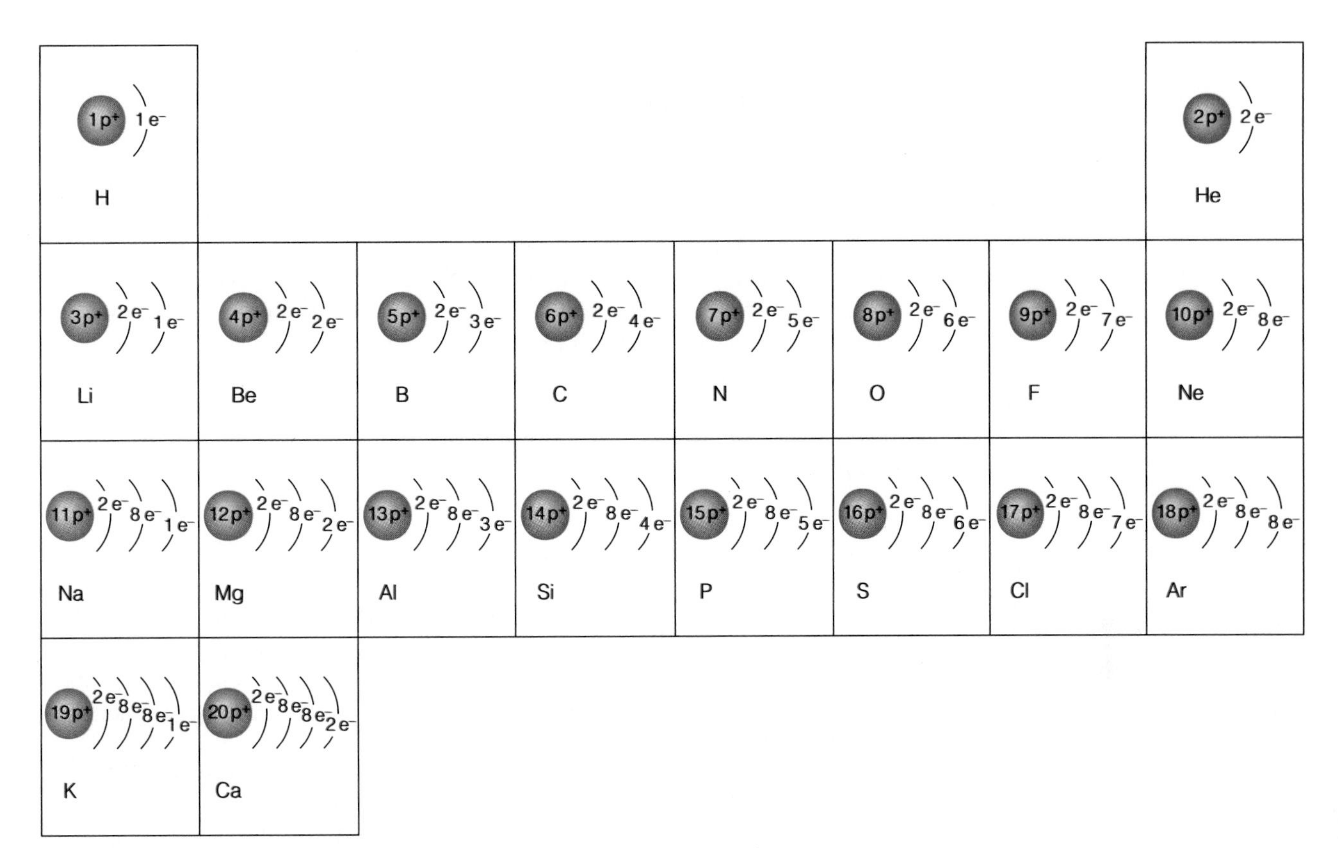

Figure 5.22 *Bohr diagrams of the first 20 elements.*

information, we can construct Bohr diagrams for many elements to show electron distributions within an atom. Practice drawing Bohr diagrams to show electron arrangements of the first 18 elements. The number of valence electrons has much to do with the chemistry of each element.

EXAMPLE 5.4

Draw a Bohr diagram for fluorine.

SOLUTION

The atomic number of fluorine is 9; it has nine electrons. Two of these go into the first energy level; the remaining seven go into the second energy level.

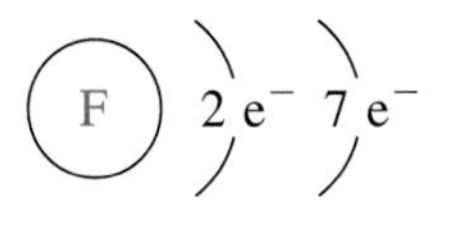

EXERCISE 5.3

Draw a Bohr diagram for phosphorus. How many valence electrons are in an atom of phosphorus?

See Problems 5.37–5.44.

5.5 The Quantum Mechanical Model of the Atom

The Bohr model of the atom clearly established the concept of definite electron energy levels within atoms, but certain questions remained unanswered. For example, Bohr's equation could be used to predict the values of frequencies for lines in the hydrogen spectrum, but it was unsuccessful in correctly predicting spectral lines of more complex atoms. It was not long before the Bohr model was replaced by more sophisticated models. There is a trade-off: in order to present a more accurate model of the atom, more sophisticated mathematical interpretations are required, and some of the simplicity of the Bohr model is sacrificed. Fortunately, however, we can make use of the results of these calculations without carrying them out ourselves.

De Broglie's Proposal

Louis de Broglie was awarded the 1929 Nobel prize in physics for his explanation of the wave nature of electrons.

Louis de Broglie, a French physics graduate student, wrote a doctoral thesis in 1924 that presented the idea that if light waves exhibit some characteristics of particles, then perhaps particles of matter can show wave characteristics. In other words, de Broglie said that a beam of electrons should exhibit wave characteristics and behave like a beam of light.

Electron microscopes that make use of the wave nature of electrons are used to obtain close-up images of the surfaces of many substances such as metals, plastics, and fibers.

At that time there was no experimental evidence to support de Broglie's idea that particles of matter have wave characteristics, but in 1927 two Americans (Clinton Davisson and Lester Germer) at the Bell Telephone Laboratories in New York and George Thomson (son of J. J. Thomson) in England independently reported that a beam of electrons is deflected as it passes through a crystal. Since that time, the wave nature of matter has been verified many times; de Broglie was right.

Schrödinger's Wave Equation

In 1926 Erwin Schrödinger (Figure 5.23), who was one of Bohr's graduate students, developed elaborate mathematical equations based on de Broglie's work. Schrödinger's equations combined the wave properties and particle nature of an electron with quantum restrictions in complex probability equations.

With Schrödinger's equations, values that correspond to regions of high electron probability around a nucleus can be obtained. Like an electron cloud, the regions of high electron probability are not the definite planetarylike orbits of the Bohr model but, instead, represent less defined energy levels as well as regions called **sublevels** or subshells. Each of these sublevels contains one or more orbitals. Each **orbital** is a region for a maximum of two electrons with opposite spin. Charge cloud representations of two types of orbitals—the *s* and *p* orbitals—are shown in Figure 5.24.

Figure 5.23 *Erwin Schrödinger (1887–1961) was the Austrian physicist who developed sophisticated mathematical equations that can be used to calculate and to graph the probability distribution of electrons in atoms. The equations took into account both the particle and the wavelike aspects of electron behavior. As a result of his work in quantum mechanics, Schrödinger was one of the Nobel prize winners in physics in 1933.*

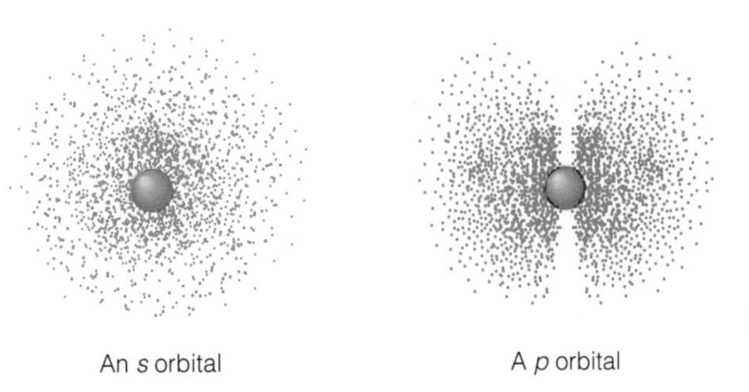

Figure 5.24 *Charge cloud representations of atomic orbitals.*

Heisenberg's Uncertainty Principle

Another blow to the Bohr planetary model of atoms occurred in 1927, when Werner Heisenberg, a German physicist and student of Niels Bohr, came to the conclusion that it is impossible to determine precisely both the position and the energy of an electron. If the electron acts like a particle, we should be able to determine precisely its location, but if it is a wave, as de Broglie had proposed, then we cannot know its precise location. Thus, according to this **uncertainty principle**, the specific path of an electron cannot be determined; it is uncertain. With the

Werner Heisenberg received the 1932 Nobel prize in physics for his "uncertainty principle" and work in quantum mechanics.

A CLOSER LOOK

The Uncertainty Principle

To help get an idea of the meaning of Heisenberg's uncertainty principle, picture in your mind two photographs of Formula One racing cars traveling at 220 miles per hour. A photograph taken with a fast shutter speed of 1/1000 second can "freeze" the action to give a sharp picture. This photograph could be used to locate the position of the vehicle, but the path of movement cannot be determined.

A second photograph taken with a slow shutter speed of about a half second will appear blurred. In this example, the general path of the racing cars can be determined, but there is greater doubt about the position of a particular vehicle. As was concluded by Heisenberg, one cannot precisely determine both the position and the momentum of a particle moving at the speed of electrons.

Thinking of electrons in atoms as freeze-action (left) *and blurred* (right) *photographs of race cars is one way to help visualize the Heisenberg uncertainty principle: The more precisely we know a particle's position, the less precisely we know its path. In other words, we must live with fuzzy pictures of electrons.*

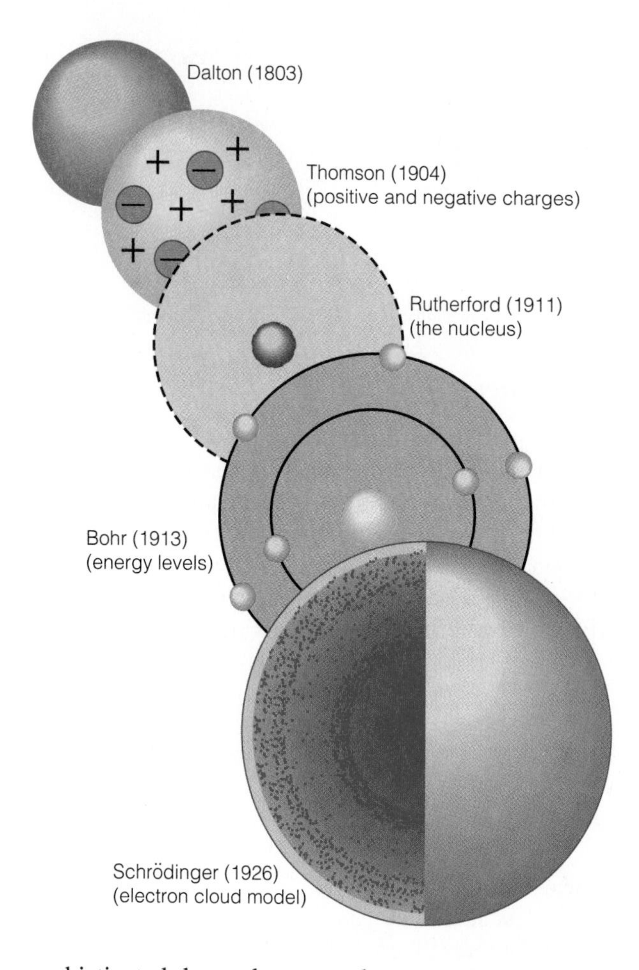

Figure 5.25 *From the time of Dalton to Schrödinger, our model of the atom has undergone many modifications.*

sophisticated theory known today as **quantum mechanics**, it is possible to calculate the probability of finding an electron at specific locations within an atom or molecule. The development of this theory during the 1920s is the result of contributions by noted scientists, including Einstein, Planck, de Broglie, Bohr, Schrödinger, and Heisenberg.

Along with the many discoveries that took place from the time of Dalton to Schrödinger, our understanding of the atom changed greatly, and a more accurate but sophisticated model of the atom emerged (see Figure 5.25).

EXAMPLE 5.5

Name the scientist who is credited with each of the following contributions to our understanding of electrons in atoms.

(a) He showed that electrons exhibit wave characteristics, as well as mass properties.
(b) He said that it is impossible to determine precisely an electron's position and energy.
(c) He developed a probability equation that included the wave properties, particle nature, and quantum restrictions of electrons.

SOLUTION

(a) de Broglie (b) Heisenberg (c) Schrödinger

See Problems 5.45 and 5.46.

Table 5.2 **The Maximum Number of Electrons Allowed per Principal Energy Level**

Principal Energy Level, n	*Maximum Number of Electrons Allowed per Energy Level* = $2n^2$	*Principal Energy Level, n*	*Maximum Number of Electrons Allowed per Energy Level* = $2n^2$
1	$2 \times (1)^2 = 2$	4	$2 \times (4)^2 = 32$
2	$2 \times (2)^2 = 8$	5	$2 \times (5)^2 = 50$*
3	$2 \times (3)^2 = 18$	6	$2 \times (6)^2 = 72$*

*No atom for any known element actually has enough electrons to completely fill these energy levels.

5.6 Energy Levels of Electrons

According to modern theory of quantum mechanics, each **principal energy level**, designated by the letter n, is assigned a positive whole number 1, 2, 3, . . . , beginning with $n = 1$ for the first energy level nearest the nucleus. Electrons in energy levels with higher numbers are farther from the nucleus. The total possible numbers of electrons in the first four energy levels are 2, 8, 18, and 32 (Table 5.2). On the periodic table, elements in the second period—the second horizontal row—have their first energy level filled and also have from one to eight outer electrons in the second energy level. Similarly, elements in the third horizontal row have two completely filled energy levels along with one to eight outer electrons in the third energy level.

Although it is not actually possible, suppose you had a camera that could photograph electrons, and you left the shutter open while the electron zipped around the nucleus. When you develop the picture, you would have a record of where the electron had been. Doing the same thing with an electric fan that was turned on would give you a picture where the blades of the fan look like a solid disk. The blades move so rapidly that their photographic image is blurred (Figure 5.26). If we had been looking at electrons, this camera model of electrons in the first energy level would look like a fuzzy ball that is **spherically symmetrical** (Figure 5.24). Higher energy levels can be represented by more complex charge cloud models.

(a)

(b)

Figure 5.26 *The individual blades, visible in (a), of an operating electric fan (b) give the appearance of a fuzzy blurred disk. If we could see a picture of electrons in the first energy level of an atom moving about the nucleus, we would see a fuzzy spherically shaped cloud of negative charge.*

EXAMPLE 5.6

Indicate the number of electrons in each principal energy level for the following elements.

(a) Al (b) Cl (c) N

SOLUTION

(a) Al (13 total electrons) = 2, 8, 3 (b) Cl (17 total electrons) = 2, 8, 7
(c) N (7 total electrons) = 2, 5

5.7 Valence Electrons and Lewis Symbols

The valence electrons—those in the outer energy level of an atom—have special importance; they become involved in chemical reactions. Gilbert N. Lewis, an American chemist, is widely known for his use of simple symbolic representations

Discoveries That Led to Our Model of the Atom

It is not intended that you attempt to memorize this list of discoveries; the list is provided to help you understand the sequence of interrelated events that has led to our present model of the atom.

1803 **John Dalton** (English) used his atomic theory to explain why atoms react in simple, whole-number proportions. He added his law of multiple proportions.

1814 **J. J. Berzelius** (Swedish) determined relative atomic masses from specific heats and proposed two-letter symbols for elements. He also discovered Ce, Se, Si, and Th.

1819 **P. Dulong** and **A. Petit** (French) determined atomic masses from specific heats.

1859 **Robert Bunsen** and **Gustav Kirchoff** (Germans) showed that each element has a characteristic spectrum. They also discovered Cs and Rb.

1868 **Pierre Janssen** (French astronomer) saw a new line in the sun's spectrum that he claimed was from a new element. Most scientists didn't believe him. The element was named helium by Sir Joseph Lockyer (English).

1869 **Dmitri Mendeleev** (Russian) published a textbook that included his periodic table. He correctly predicted properties of several undiscovered elements.

1879 **William Crookes** (British) concluded that cathode rays in an evacuated tube consist of a stream of charged particles produced at the cathode.

1885 **J. J. Balmer** (Swiss) developed a simple mathematical equation to calculate wavelengths of lines in the hydrogen spectrum.

1886 **E. Goldstein** (German) observed "canal rays" (protons) in a modified Crookes tube.

1895 **Wilhelm Roentgen** (German) discovered X-rays produced in a Crookes tube.

1896 **Antoine Becquerel** (French) discovered natural radioactivity.

1897 **J. J. Thomson** (British) showed that cathode rays are beams of negatively charged particles (electrons) and calculated their charge-to-mass ratio.

1898 **Marie Curie** (Polish) and her husband, Pierre (French), discovered polonium and radium, which they isolated from pitchblende (a lead ore) by chemical processes.

1900 **Max Planck** (German) proposed a quantum theory that explains that an excited atom emits light in *discrete* units called quanta or photons.

1904 **J. J. Thomson** proposed a "plum pudding" model of the atom with electrons embedded in a sea of positive charges. He verified positive charges in 1907.

1905 **Albert Einstein** (German) published a paper relating mass and energy.

1909 **Robert Millikan** (American) determined the charge of the electron (1.60×10^{-19} coulomb) with his oil drop experiment.

1911 **Lord Rutherford** (English) bombarded gold foil with alpha particles from a radium source and found that the atom's mass is nearly all in its tiny, positively charged nucleus.

1912 **J. J. Thomson** separated nonradioactive isotopes neon-20 and -22.

1913 **Niels Bohr** (Danish) used the Balmer formula to show that electrons of hydrogen atoms exist only in specific spherical orbits (energy levels).

1913 **Henry Moseley** (English) observed that each element emits characteristic X-rays that depend on nuclear charge (protons). The number of protons in the nucleus is the atomic number.

1924 **Louis de Broglie** (French) combined equations from Einstein and Planck to show that electrons have specific wavelengths and wavelike properties.

1925 **Wolfgang Pauli** (German) stated—in his exclusion principle—that only two electrons with opposite spin are allowed per orbital.

1926 **Erwin Schrödinger** (Austrian) described the motion of electrons in atoms by a mathematical equation that combined the particle nature of an electron, wave properties, and quantum restrictions, in a probability relationship.

1927 **Werner Heisenberg** (German) explained in his uncertainty principle that we cannot know the path of an electron, that is, its exact position and momentum.

1927 **Frederick Hund** (German) is known for Hund's rule that electrons in sublevels have maximum unpairing and unpaired electrons have the same spin.

1932 **James Chadwick** (British) discovered neutrons. In an investigation involving radioactivity in 1930, a scientist named Bothe produced neutrons, but he was not aware he had.

1934 **Harold Urey** (American) received the Nobel prize in chemistry for his discovery to deuterium, the isotope called heavy hydrogen.

1970 **Albert Crewe** (University of Chicago) used a modified electron microscope to photograph single atoms of thorium.

1979 **Samuel Hurst** (Oak Ridge, Tennessee) used a laser-based technique to detect and identify single atoms of all but four elements.

1985 **Gerd Binning** (IBM's Zurich Laboratory) mapped the surface of graphite with an atomic force microscope (AFM) made from a scanning tunneling microscope (STM).

RESEARCH CONTINUES, OUR MODEL OF THE ATOM IS NOT YET COMPLETE.

of elements that show valence electrons as dots. These representations are called **Lewis electron-dot symbols**.

It is more convenient to represent a sodium atom by its Lewis electron-dot symbol, which is Na with one dot, Na· (representing one valence electron), than it is to use a Bohr diagram. The Lewis symbol for a chlorine atom is simply Cl surrounded by seven dots.

	Bohr Diagram		*Lewis Symbol*
Sodium	(11 p, 12 n) $2\ e^-$ $8\ e^-$ $1\ e^-$	is represented as	Na·
Chlorine	(17 p, 18 n) $2\ e^-$ $8\ e^-$ $7\ e^-$	is represented as	:C̈l̤·

It is easy to write Lewis electron-dot symbols for elements in the first three periods (horizontal rows) of the periodic table. The number of valence electrons is the same for all elements in the same group (or family) in the periodic table. All elements in Group IA—including H, Li, Na, and K—have one valence electron. Lewis symbols of these *alkali metals* all have one dot near the symbol, as was shown for the sodium atom.

Symbolism, like shorthand, is a convenient way of conveying a lot of information in compact form. It is the chemist's most efficient form of communication. Learning this symbolism is much like learning a foreign language. Once you master a certain basic "vocabulary," the rest is easier.

Elements in Group IIA—including Be, Mg, and Ca—are called the **alkaline earth metals**; they have two valence electrons. Their Lewis symbols would show two dots. Table 5.3 gives the Lewis electron-dot symbols for the first 20 elements. Notice that there is a pattern to the way in which the dots are drawn. For elements with no more than four valence electrons, the electron dots are placed separately around the symbol of the atom—isolated and unpaired. For example, Lewis symbols for carbon and silicon (with four valence electrons) use four unpaired dots. Elements with more than four valence electrons are symbolized with both paired and unpaired electrons as shown in Table 5.3. Noble gases with eight valence electrons are said to have a complete **octet of electrons** and are symbolized by four sets of electron pairs. Helium is the only noble gas with only two valence electrons; the first energy level can hold only two electrons.

A thorough familiarity with energy levels, valence electrons, and Lewis symbols for the valence electrons is essential to further study involving chemical bonding of atoms and the formation of compounds.

Table 5.3 **Electron-Dot Symbols for Selected Elements**

IA	*IIA*	*IIIA*	*IVA*	*VA*	*VIA*	*VIIAA*	*VIIIA*
H·							He:
Li·	·Be·	·Ḃ·	·Ċ̣·	:Ṇ̇·	:Ö̤·	:F̤̈:	:N̈e̤:
Na·	·Mg·	·Ȧl·	·Ṩi·	:Ṗ̣·	:S̤̈·	:C̈l̤:	:Ä̤r:
K·	·Ca·					:B̈r̤:	:K̈r̤:
Rb·	·Sr·					:Ï̤:	:Ẍe̤:
Cs·	·Ba·						

When there are only two dots—for two valence electrons—they are sometimes shown as a pair, but we will place the dots on opposite sides of the symbol because we will later see how these two electrons can be shared with two other atoms in certain compounds.

HISTORY MAKERS

Gilbert N. Lewis

Gilbert N. Lewis (1875–1946) made numerous outstanding contributions to chemistry, including electron-dot structures, chemical bonding, acid–base theory, and the application of thermodynamics to chemistry. He was on the faculty of the Massachusetts Institute of Technology until 1912, when he became professor of chemistry at the University of California, Berkeley.

Describing his experience as Lewis's research assistant (from 1937 to 1939), Glenn T. Seaborg has said, "We'd make simple observations, and then he would draw conclusions from them—correct conclusions. . . . I was awestruck as I worked with him."

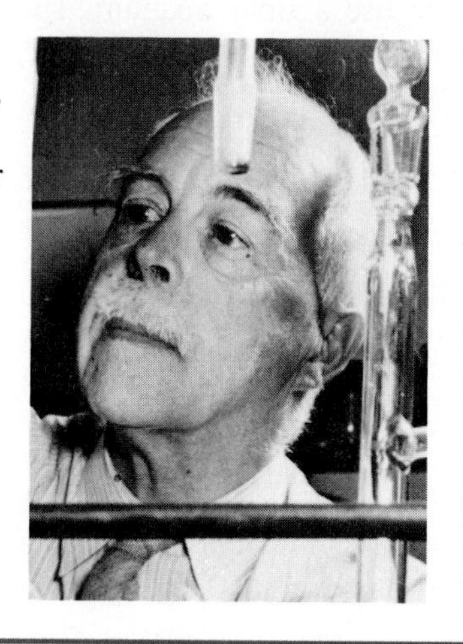

Gilbert N. Lewis (1875–1946) helped to make the chemistry of elements easier for students to understand when he introduced his electron-dot symbols for valence electrons. Variations of his electron-dot symbols are used by virtually every chemist when predicting and explaining chemical bonding in compounds.

EXAMPLE 5.7

Write Lewis electron-dot symbols for the following atoms.

(a) Cl (b) Br (c) O (d) S (e) C (f) Al

SOLUTION

(a) :C̈l̤· (b) :B̈r̤· (c) :Ö· (d) :S̈· (e) ·Ċ· (f) ·Ȧl·

Both Cl and Br are halogens with seven valence electrons. Oxygen and sulfur have six valence electrons each. Carbon is symbolized with four unpaired valence electrons. Aluminum is symbolized with three unpaired valence electrons

See Problems 5.47 and 5.48.

EXERCISE 5.4

Write Lewis electron-dot symbols for atoms of phosphorus, potassium, and barium.

Electronic Structure: Optional Additional Insights

Instructor's Note
Depending on the nature and needs of a particular class, some or all of the following topics may be optional.

5.8 Energy Sublevels and Orbitals

The use of quantum mechanics and Schrödinger's equation has provided additional insight into the electron structure of atoms. According to calculations in quantum mechanics, each energy level of an atom consists of one or more **sublevels** (also sometimes called **subshells**). The first energy level has only one sublevel. The second energy level has two sublevels. The third energy level has three sublevels, and so on. In other words, energy level n has n sublevels.

Table 5.4 **Electron Energy Levels, Sublevels, and Orbitals**

Principal Energy Level, n	*Number of Sublevels*	*Type of Orbital*	*Number of Orbitals*	*Maximum Number of Electrons per Sublevel*	*Maximum Total Number of Electrons*
1	1	1*s*	1	2	2
2	2	2*s*	1	2	
		2*p*	3	6	8
3	3	3*s*	1	2	
		3*p*	3	6	
		3*d*	5	10	18
4	4	4*s*	1	2	
		4*p*	3	6	
		4*d*	5	10	
		4*f*	7	14	32

Each sublevel has one or more atomic orbitals with a specific three-dimensional shape. The orbitals are designated by the lowercase letters *s, p, d,* and *f.* Furthermore, each **orbital** can contain a maximum of two electrons—a pair—but electrons in this pair must have opposite spin. The idea that if two electrons occupy the same orbital they must have opposite spin was an important contribution made in 1925 by Wolfgang Pauli. This is known as the **Pauli exclusion principle**. Table 5.4 summarizes key information about energy levels, sublevels, atomic orbitals, and the distribution of electrons within sublevels.

Wolfgang Pauli was an Austrian physicist who won the 1945 Nobel prize in physics for his discovery of the exclusion principle.

Electrons in s Orbitals

The first two electrons in each energy level are in a region where the electron probability is represented by a spherically symmetrical ''*s*'' orbital. These orbitals are designated 1*s*, 2*s*, 3*s*, and so forth. The 3*s* orbital, for example, is larger than the 1*s* orbital, as shown in Figure 5.27. Thus, each and every energy level has an *s* sublevel with a single, spherically shaped *s* orbital, and that orbital can contain a pair of electrons of opposite spin. Electrons in orbitals farther from the nucleus in an atom have more energy than do electrons near the nucleus.

Electrons in p Orbitals

Beginning with the second energy level, and for all subsequent energy levels, there are an *s* sublevel and also a ''*p*'' sublevel. Each *p* sublevel consists of three *p* orbitals that have equal energy but have different orientations in space. The charge cloud representation for each *p* orbital is dumbbell shaped, with two *lobes,* or regions, oriented along the axis where electron density is the greatest. Figure 5.28 shows that one *p* orbital with two lobes is oriented in space along the *x* axis, a second orbital with two lobes is oriented along the *y* axis, and the third orbital is oriented along the *z* axis. Each orbital can accommodate only two electrons with opposite spin. As represented in Figure 5.29, a *p* sublevel, with its three orbitals, can accommodate a maximum of six electrons (three pairs).

1s

2s

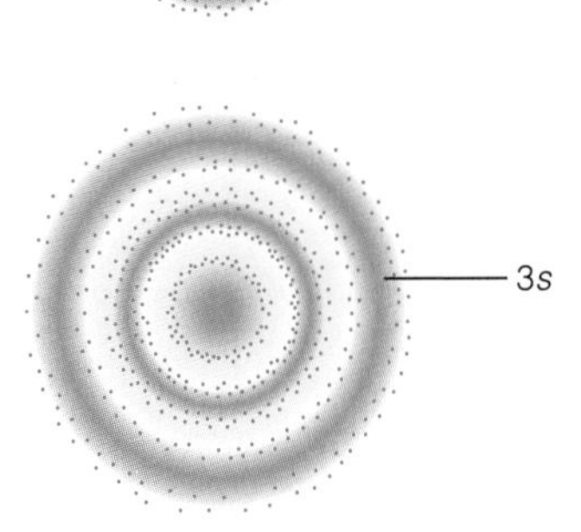

Figure 5.27 *Charge cloud representations of 1*s*, 2*s*, and 3*s* orbitals.*

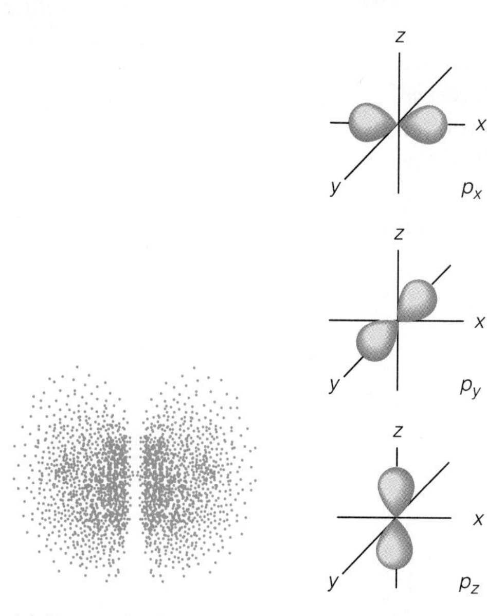

(a) Charge cloud representation of a *p* orbital.

(b) Outline representations of *p* orbitals. The three *p* orbitals differ only in their orientation in space.

Figure 5.28 *Representations of* p *orbitals.*

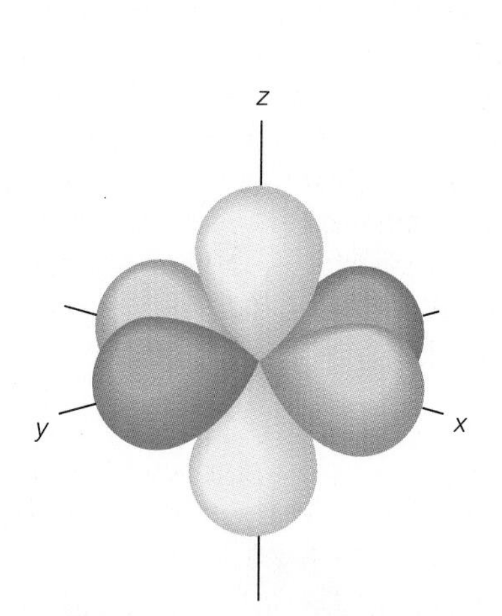

Figure 5.29 *Representation of the three* p *orbitals on the same axes to give a* p *sublevel that can accommodate a maximum of three pairs of, or six, electrons.*

Sublevel	Orbital Shapes	Orbitals per Sublevel	Lobes per Orbital
s	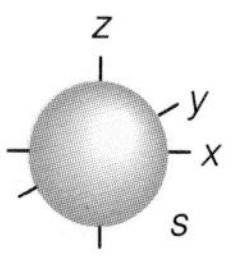 *s*	1	1
p	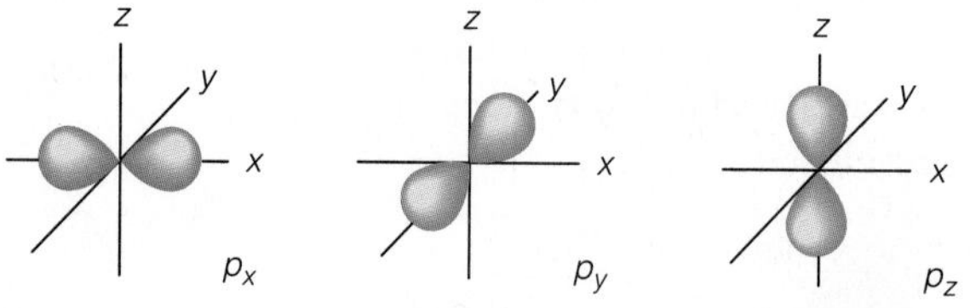 p_x p_y p_z	3	2
d	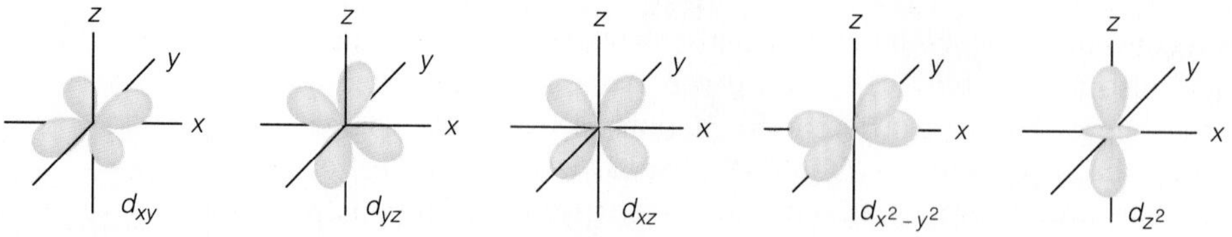 d_{xy} d_{yz} d_{xz} $d_{x^2-y^2}$ d_{z^2}	5	4*
f	Seven complex shapes	7	8*

*The number of lobes for one *d* orbital and three of seven *f* orbitals is less than the number specified.

Figure 5.30 *Atomic orbitals.*

Electrons in d and f Orbitals

Beginning with the third energy level, and for all subsequent energy levels, there is also a "*d*" sublevel with a set of five orbitals, capable of holding a total of five pairs of electrons, to give a maximum of 10 *d* electrons in a *d* sublevel. The shapes corresponding to *d* orbitals are more complex than those of *s* and *p* orbitals. Figure 5.30 shows the regular increase in the number of orbitals per sublevel and the corresponding number of lobes per orbital.

Beginning with the fourth energy level, and for all subsequent energy levels, there is an "*f*" sublevel with a set of seven orbitals, capable of holding a total of seven pairs of electrons, to give a maximum of 14 *f* electrons in an *f* sublevel. The shapes of *f* orbitals are even more complex than the *d* orbitals—most have eight lobes—but we do not actually need to concern ourselves with the specific shapes of the *f* orbitals.

EXAMPLE 5.8

For the first three energy levels, designate the kind of sublevels available, the number of orbitals per sublevel, the maximum number of electrons per sublevel, and the maximum number of electrons for the entire energy level.

SOLUTION

Energy Level	*Designation for Sublevel*	*Orbitals per Sublevel*	*Maximum Electrons per Sublevel*	*Maximum Electrons in Each Energy Level*
1	1*s*	1	2	2
2	2*s*	1	2	
	2*p*	3	6	8
3	3*s*	1	2	
	3*p*	3	6	
	3*d*	5	10	18

5.9 *Energy Sublevels and the Periodic Table*

The electron arrangement for atoms of an element can be determined quickly by locating the symbol of the element in the periodic table. For example, as you study Figure 5.31, you will notice that elements in the first two columns at the left of the periodic table have one or two outer (valence) *s* electrons, identified as s^1 and s^2. Elements in this region of the periodic table are sometimes referred to as the *s* block of elements.

Elements shown in the six columns at the far right of the periodic table (except helium) all have two electrons in an *s* sublevel, plus one to six electrons in a *p* sublevel. These elements are sometimes referred to as the *p* block of elements. All elements located in the first two columns and the last six columns of the periodic table are often called the **representative elements**. We can easily draw Lewis electron-dot symbols for any of the representative elements; they all have from one to eight valence electrons.

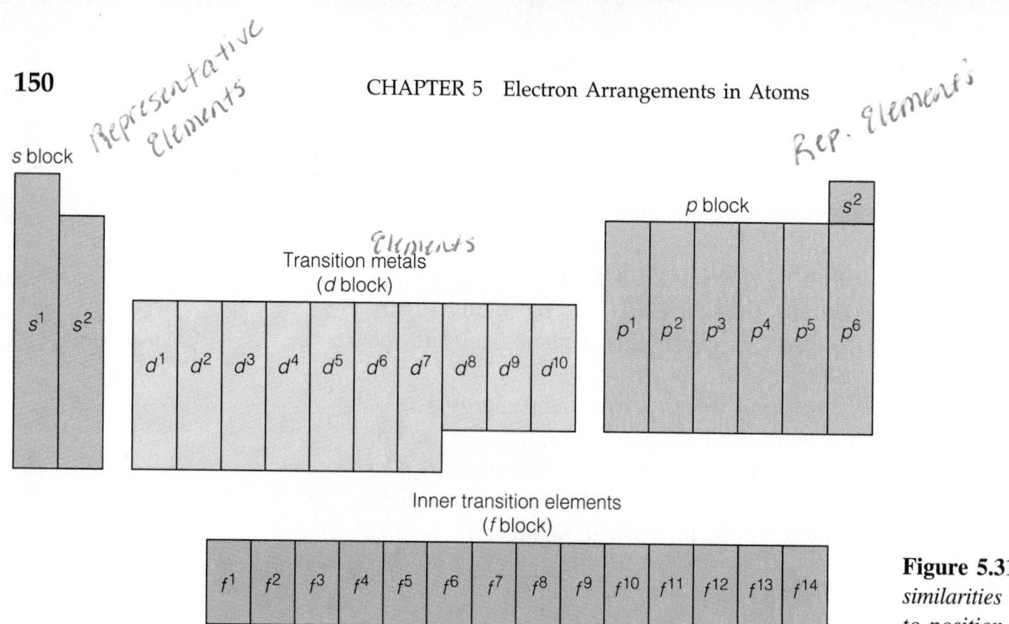

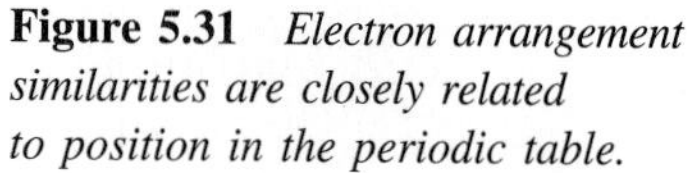

Figure 5.31 *Electron arrangement similarities are closely related to position in the periodic table.*

The Transition Elements

Elements with outer electrons in *d* orbitals are known as the **transition elements**. The transition elements—the *d* block of elements—are located in the central region of the periodic table, as shown in Figure 5.31, where there are 10 columns, d^1 through d^{10}, corresponding to the 1 to 10 electrons possible in a *d* sublevel.

Many compounds that contain transition metals are brightly colored. These colors are related to the frequencies of light absorbed and emitted when electrons in partially filled *d* sublevels become excited and then fall back to their ground states. Crystalline solids such as sodium chloride (table salt) that contain only representative elements are white. They dissolve in water to give colorless solutions.

The Order of Filling Sublevels

Study Hint
Locate these sublevels on the periodic table while repeating this order of filling.

Electrons are filled in atoms in the order shown for the periodic table in Figure 5.32, with the lower energy levels and sublevels filled first. For the first three periods of elements in the periodic table, electrons fill each *s* and *p* sublevel available in the following sequence: 1*s*, then 2*s* and 2*p*, then 3*s* and 3*p*. Argon (atomic number 18 with 18 electrons) is at the end of the third period of elements. It has exactly the right number of electrons to fill all sublevels completely up through the 3*p* sublevel.

Beginning with the fourth period of elements, the order of filling sublevels becomes more complex. Potassium and calcium, the first two elements in the fourth period, have 1 or 2 valence electrons, respectively, in the 4*s* sublevel. Figure 5.32 shows that the next sublevel to be filled after the 4*s* sublevel is the 3*d* sublevel, which can accommodate a maximum of 10 3*d* electrons—corresponding to the first 10 transition elements.

The 4*d* transition elements and 5*d* transition elements are positioned right below the 3*d* transition elements (Figure 5.32). In each case, notice that the *d* sublevel has an energy level number that is consistently one less than the energy level number for the *s* block and *p* block of elements in the same period. Thus, the 4*d* sublevel is filled after the 5*s* sublevel, but before the 5*p* sublevel.

The inner transition elements with partially filled 4*f* and 5*f* sublevels are grouped together at the bottom of the periodic table. For the most part, the 4*f*

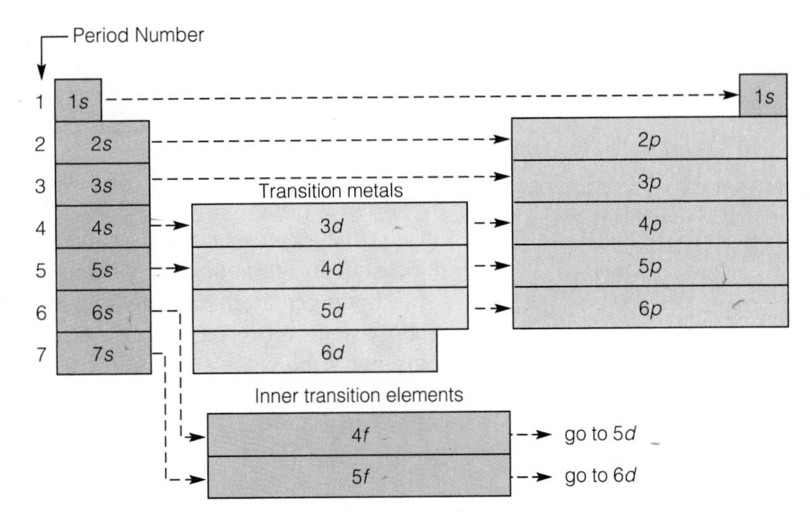

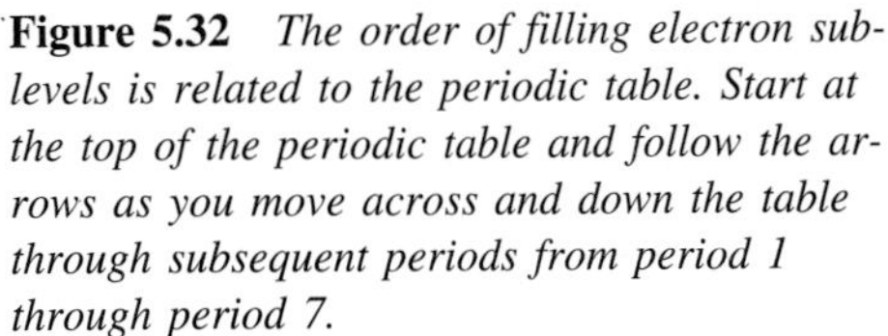
Figure 5.32 *The order of filling electron sublevels is related to the periodic table. Start at the top of the periodic table and follow the arrows as you move across and down the table through subsequent periods from period 1 through period 7.*

sublevel is filled after the 6*s* sublevel and before the 5*d* sublevel. Similarly, the 5*f* sublevel is filled after the 7*s* but before the 6*d* sublevel. Thus, inner transition elements have electrons in *f* sublevels that are two energy levels behind the outermost *s* electrons, but the order of filling is actually more irregular than is described here. The periodic table itself, then, can be used as an excellent tool to predict the order of filling electrons in sublevels of atoms, but other methods of showing the regularity of this process are provided in Figure 5.33.

EXAMPLE 5.9

For each of the sublevels listed, indicate the next sublevel to be filled using the order described in this section.

(a) 3*s* (b) 4*s* (c) 5*s* (d) 6*s* (e) 4*d* (f) 4*f*

SOLUTION

Refer to Figure 5.33 for assistance in determining order of filling.

(a) 3*p* (b) 3*d* (c) 4*d* (d) 4*f* (e) 5*p* (f) 5*d*

See Problems 5.49–5.52.

Flevel

Filled

1s 2s 3s 4s 5s 6s 7s
2p 3p 4p 5p 6p 7p
3d 4d 5d 6d
4f 5f

sorsit

(a)

1s*			
2s			2p
3s			3p
4s		3d	4p
5s		4d	5p
6s	4f	5d	6p
7s	5f	6d	7p

* Read this diagram in the normal left-to-right sequence from top to bottom as you do for the periodic table: 1s, 2s, 2p, 3s, 3p, 4s, 3d, 4p, ...

(b)

Figure 5.33 *The regular order of filling electrons in sublevels of atoms is shown in both diagrams.*

5.10 Electron Configurations and Orbital Diagrams

Electrons of atoms in their ground states fill the lower energy sublevels first, as described in Section 5.9, but we need a way to represent this arrangement concisely; that representation is called an **electron configuration**. To illustrate how electron configurations are written, let's begin with the simplest atom, hydrogen.

A hydrogen atom has only one electron. As long as that electron is in its lowest energy state—its ground state—that electron is in the first energy level, which has only one sublevel, the 1*s* sublevel. We write the electron configuration of hydrogen, then, as $1s^1$.

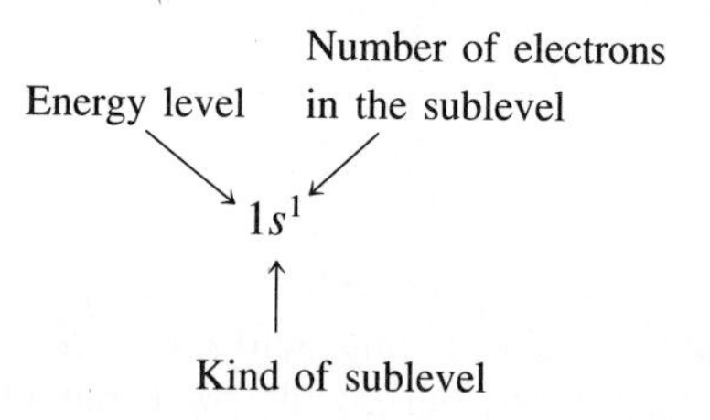

The superscript (1 in this case) placed after the sublevel designation gives the number of electrons in the sublevel. Similarly, the two electrons of a helium atom can both be held in its 1*s* sublevel; helium's electronic structure is $1s^2$. As we present the electron configurations of elements, be careful to consider their placement in the periodic table.

Lithium has three electrons: two in the first energy level, the 1*s* sublevel, and a third electron that must go in the *s* sublevel of the second energy level. The electron configuration for Li is written $1s^22s^1$. For a lithium ion, Li^+, simply remove the single 2*s* valence electron, and write $1s^2$. Beryllium has four electrons; its electron configuration is $1s^22s^2$. Boron has five electrons: two in the 1*s* sublevel, two in the 2*s* sublevel, and one in the 2*p* sublevel. The electron configuration of boron is $1s^22s^22p^1$. All three electrons in boron's second energy level—the 2*s* and 2*p* sublevels—are valence electrons.

The 2*p* sublevel, with three *p* orbitals, can hold a maximum of six electrons—two per orbital. The electron configurations of boron, carbon, nitrogen, oxygen, fluorine, and neon require the filling of the 2*p* sublevel with one through six electrons, respectively. With neon, the 2*p* sublevel, and hence the second energy level, is completely filled. The electron configuration of neon is $1s^22s^22p^6$. Electron configurations for all elements in a particular column of the periodic table fit a pattern, as summarized in Figure 5.34. Electron configurations for the first 20 elements are listed in Table 5.5.

An electron configuration allows us to show concisely the number of electrons within each sublevel of an atom, but an **orbital diagram** can be used to represent the *distribution* of electrons within orbitals. The orbital diagrams of the first 20 elements are shown in Table 5.5 alongside the electron configurations. In this table, circles are used to represent orbitals: an *s* orbital is represented by a single circle, and the three *p* orbitals are represented by three circles. A single arrow within the circle represents a single electron; a pair of arrows pointed in opposite directions represents an electron pair with opposite spins. The following two methods of showing an orbital with an electron pair can be used interchangeably.

(⇅) or ⇅

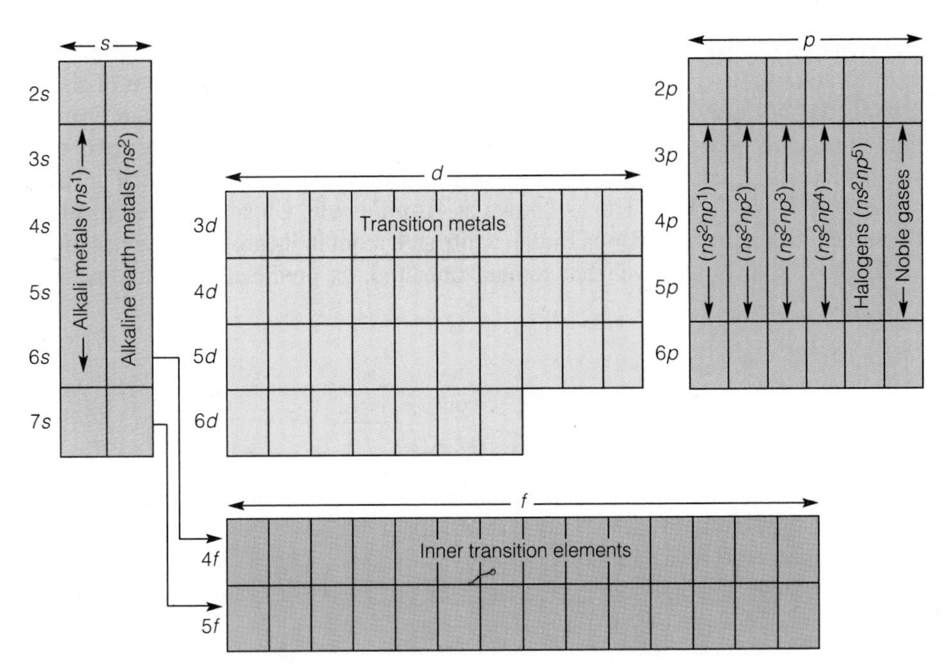

Figure 5.34 *Electron configurations and the periodic table.*

Table 5.5 **Electron Configurations and Orbital Diagrams for Atoms of the First 20 Elements**

Name	*Atomic Number*	*Electron Configuration*	*Orbital Diagram* s	p
PERIOD 1				
Hydrogen	1	$1s^1$	(↑)	
Helium	2	$1s^2$	(↑↓)	
PERIOD 2				
Lithium	3	$1s^22s^1$	[He] (↑)	
Beryllium	4	$1s^22s^2$	[He] (↑↓)	
Boron	5	$1s^22s^22p^1$	[He] (↑↓)	(↑)( )( )
Carbon	6	$1s^22s^22p^2$	[He] (↑↓)	(↑)(↑)( )
Nitrogen	7	$1s^22s^22p^3$	[He] (↑↓)	(↑)(↑)(↑)
Oxygen	8	$1s^22s^22p^4$	[He] (↑↓)	(↑↓)(↑)(↑)
Fluorine	9	$1s^22s^22p^5$	[He] (↑↓)	(↑↓)(↑↓)(↑)
Neon	10	$1s^22s^22p^6$	[He] (↑↓)	(↑↓)(↑↓)(↑↓)
PERIOD 3				
Sodium	11	$1s^22s^22p^63s^1$	[Ne] (↑)	
Magnesium	12	$1s^22s^22p^63s^2$	[Ne] (↑↓)	
Aluminum	13	$1s^22s^22p^63s^23p^1$	[Ne] (↑↓)	(↑)( )( )
Silicon	14	$1s^22s^22p^63s^23p^2$	[Ne] (↑↓)	(↑)(↑)( )
Phosphorus	15	$1s^22s^22p^63s^23p^3$	[Ne] (↑↓)	(↑)(↑)(↑)
Sulfur	16	$1s^22s^22p^63s^23p^4$	[Ne] (↑↓)	(↑↓)(↑)(↑)
Chlorine	17	$1s^22s^22p^63s^23p^5$	[Ne] (↑↓)	(↑↓)(↑↓)(↑)
Argon	18	$1s^22s^22p^63s^23p^6$	[Ne] (↑↓)	(↑↓)(↑↓)(↑↓)
PERIOD 4				
Potassium	19	$1s^22s^22p^63s^23p^64s^1$	[Ar] (↑)	
Calcium	20	$1s^22s^22p^63s^23p^64s^2$	[Ar] (↑↓)	

Whenever an atom has *p* electrons, we need to represent a complete *p* sublevel—with three orbitals (represented by three circles or lines)—even if some of the *p* orbitals are not occupied. In order for two electrons to occupy the same orbital, they must have opposite spins—represented by a pair of arrows pointing in opposite directions. Electrons do not pair up in an orbital until each orbital in that sublevel has one electron. This is known as **Hund's rule**. Unpaired electrons have the same (parallel) spins. For example, a nitrogen atom in its ground state has three electrons in the $2p$ sublevel that remain unpaired, as predicted by Hund's rule.

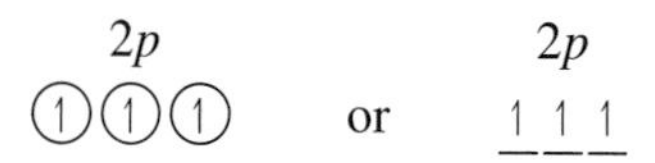

Figure 5.35 shows energy level diagrams for the order of filling electrons in atomic orbitals. Orbitals at the bottom of the diagrams have the lowest energy and are filled before orbitals of higher energy as shown. The difference in energy between sublevels progressively decreases as electrons fill orbitals of higher energy. The order of predicted filling obeys Hund's rule.

To reduce the space required to write electron configurations and orbital diagrams of elements with many electrons, a shortened notation is often used. To use this shortened notation, locate the specified element in the periodic table and write the symbol of the immediately preceding noble gas in brackets. Then show only the electron configuration for the remaining outer electrons. For example, the electron configuration for sodium can be shortened to [Ne]$3s^1$.

EXAMPLE 5.10

Write the electron configuration and an orbital diagram for nitrogen, N, assuming the electrons are in their ground state.

SOLUTION

A nitrogen atom has seven electrons. Two electrons go in the $1s$ sublevel, two electrons go in the $2s$ sublevel, and the remaining three electrons are in the $2p$ sublevel. In the orbital diagram, the electrons in the $2p$ sublevel are shown in separate orbitals. This is in agreement with Hund's rule.

	Electron Configuration	*Orbital Diagram*
		$2s$ $2p$
Nitrogen	$1s^2 2s^2 2p^3$	[He] (⇅) (↑)(↑)(↑)

EXERCISE 5.5

Write out the electron configuration and an orbital diagram for oxygen, assuming that the electrons are in their ground state.

We can continue to build up electron configurations by following a simple procedure. First, find the atomic number for the element, using the periodic table; then, place that number of electrons in the lowest possible energy sublevels. Keep in mind that the maximum number of electrons in an *s* sublevel is two and the maximum number of electrons in a *p* sublevel is six.

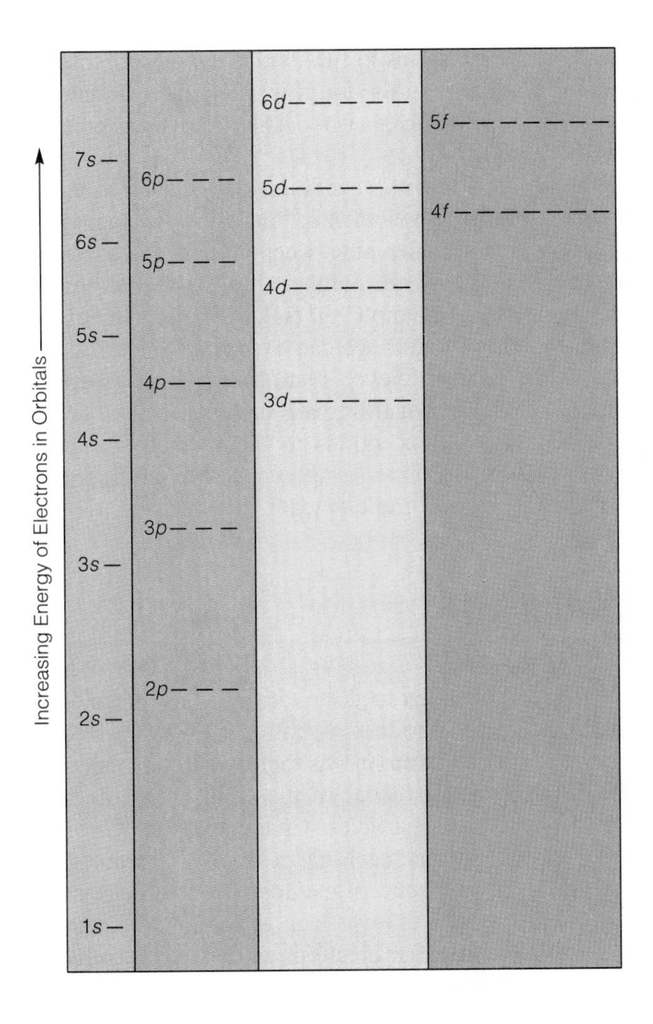

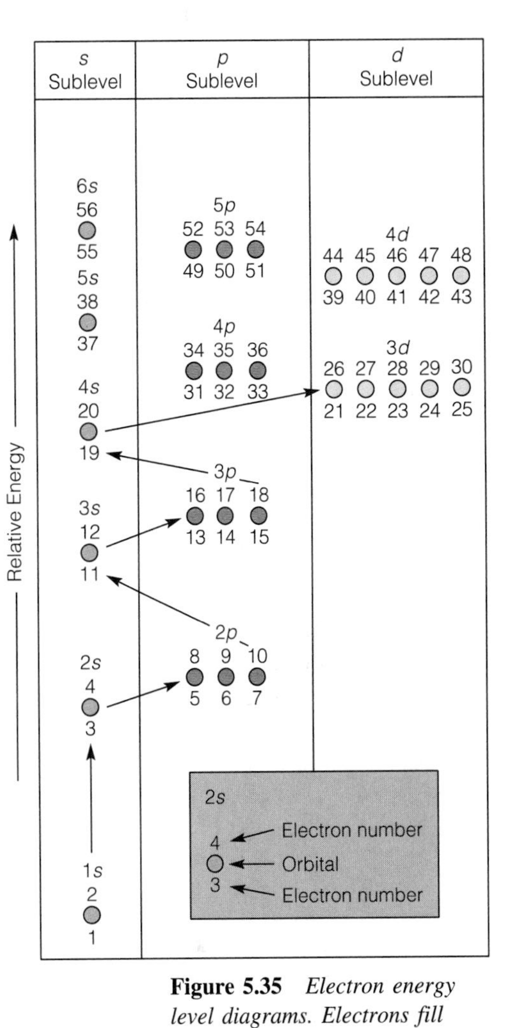

Figure 5.35 *Electron energy level diagrams. Electrons fill lower energy orbitals first, in the order specified. Notice that higher sublevels are closer together in energy than are lower sublevels.*

EXAMPLE 5.11

Write out the electron configuration and an orbital diagram for a chlorine atom, assuming the electrons are in their ground state.

SOLUTION

Chlorine atoms have 17 electrons, as indicated by the periodic table. The electron configuration and orbital diagram for chlorine are

	Electron Configuration	*Orbital Diagram*
		3s 3p
Chlorine	$1s^2 2s^2 2p^6 3s^2 3p^5$	[Ne] (⇅) (⇅)(⇅)(↑)

Notice that the *total* of the superscripts in the electron configuration is 17.

See Problems 5.53 and 5.54

The third period of the periodic table ends with the noble gas argon, which has eight electrons in its second energy level, and eight electrons in its third energy level, to give the following electron configuration: $1s^2 2s^2 2p^6 3s^2 3p^6$.

We might expect that potassium, K, with 19 electrons, would have the argon configuration plus the 19th electron in the 3*d* sublevel. It doesn't. Instead, the potassium atom has the electron configuration $1s^2 2s^2 2p^6 3s^2 3p^6 4s^1$. The 4*s* sublevel fills before the $3d$ sublevel begins to fill. Notice the position of potassium on the periodic table, at the beginning of the fourth period, in a column with an outer $4s$ electron. Calcium, with 20 electrons, has two $4s$ electrons; its shortened electron structure is $[Ar]4s^2$.

The $3d$ sublevel begins to fill with scandium, the 21st element, $[Ar]4s^23d^1$. Scandium is the first of the transition elements. Notice its placement in the *d* block of elements in the center of the periodic table. Keep in mind that a *d* sublevel that is being filled always has an energy-level number that is one less than that of the preceding *s* sublevel number. For example, $3d$ follows $4s$ and $4d$ is filled after the $5s$ sublevel. According to Hund's rule, we can expect $3d$ electrons for transition metals to remain unpaired until the sublevel is half full. Actually, there are a couple of exceptions to the predicted order of filling a *d* sublevel, but we will not present these here. There are even more exceptions to the predicted order of filling *f* sublevels. We will not present the more complex electron configurations and orbital diagrams for these inner transition elements.

Chapter Summary

Investigations with Crookes tubes led to discoveries of both the electron and the proton. Thomson measured the charge-to-mass ratio for electrons, and when Millikan measured the charge of an electron, it was then possible to calculate the electron's mass. Rutherford proposed a model of the atom with a dense nucleus to account for the scattering of alpha particles. Chadwick later discovered that an atom can have one or more neutrons; each has a mass of 1 amu.

Niels Bohr accounted for line spectra of excited atoms by reasoning that electrons are in specific energy levels. The work in quantum mechanics by Planck, Einstein, and de Broglie indicated that electrons must have wavelike properties. Pauli explained that electrons appear to have spin, but the Heisenberg uncertainty principle pointed out that we cannot determine the precise path of an electron. Schrödinger's wave equation made it possible to calculate regions of high electron density that correspond to the *s, p, d,* and *f* orbitals.

White light, which can be split into many frequencies, is only one form of electromagnetic radiation. Other forms of electromagnetic radiation include X-rays, UV, IR, microwaves, and various radio frequencies. The energy and frequency of electromagnetic radiation increase as wavelength decreases. The color (frequencies) of light emitted by excited atoms is a characteristic property. Spectra can be used to identify the kinds of atoms present. Ionization occurs when one or more electrons are ripped from an atom, leaving a charged particle, called an ion.

Bohr diagrams can be used to show the number of electrons in each energy level, but the most important electrons—the valence electrons—can clearly be shown with Lewis electron-dot symbols.

The *n*th energy level has *n* sublevels. Energy sublevels, designated as *s, p, d,* and *f,* have 1, 3, 5, and 7 orbitals, respectively. Each orbital can hold a maximum of two electrons, which must have opposite spins. Electrons fill orbitals with the lowest energy first, as shown in Figures 5.32 and 5.33. Electron configurations and orbital diagrams can be used to specify the electron arrangements in sublevels of atoms.

The placement of elements in the periodic table can be used to predict the electron configurations of atoms. The representative elements have typical *s* and *p* valence electrons, but for transition elements, the highest energy electrons go into orbitals in *d* sublevels. Inner transition elements are involved with the filling of orbitals in *f* sublevels as shown in Figure 5.34.

Once you understand the general electron structure of atoms, we can pursue other properties related to the periodic table and to chemical bonding and reactions. These topics will be presented in subsequent chapters.

Assess Your Understanding

1. Describe major discoveries that revealed information about atoms. [5.1]
2. Compare the Rutherford and the Bohr models of the atom. [5.1, 5.4]
3. Explain major developments that led to our model of the atom. [5.5]
4. Explain how X-rays and the discovery of radioactivity led to important information about atoms. [5.1]
5. Compare frequency, wavelength, and energy for electromagnetic radiation. [5.2]
6. Give the types of electromagnetic radiation and their relative energies. [5.2]
7. Define and give an example of ionization. [5.4]
8. Draw Bohr diagrams for the first 20 elements. [5.4]
9. Draw Lewis symbols for the representative elements. [5.7]
10. Write electron configurations and orbital diagrams for the representative elements and some transition metals [5.8, 5.10]
11. Relate sublevels and orbitals to position in the periodic table. [5.9]

Key Terms

alkali metals [5.7]
alkaline earth metals [5.7]
alpha rays [5.1]
anode [5.1]
beta rays [5.1]
cathode [5.1]
cathode ray [5.1]
continuous spectrum [5.2]
electrodes [5.1]
electromagnetic radiation [5.2]
electron configuration [5.10]
energy level [5.4]
excited state [5.4]
fluorescence [5.2]
frequency [5.2]
gamma rays [5.1]
ground state [5.4]
Hund's rule [5.10]
infrared radiation [5.2]
ion [5.4]
ionization [5.4]
ionization energy [5.4]
Lewis electron-dot symbol [5.7]
line spectrum [5.3]
microwaves [5.2]
octet of electrons [5.7]
orbital [5.5, 5.8]
orbital diagram [5.10]
Pauli exclusion principle [5.8]
photons [5.3]
principal energy level [5.6]
quanta [5.3]
quantum mechanics [5.5]
quantum theory [5.3]
radar [5.2]
radioactivity [5.1]
representative elements [5.9]
spectroscope [5.3]
spherically symmetrical [5.6]
sublevel [5.8]
subshell [5.8]
transition elements [5.9]
ultraviolet radiation [5.2]
uncertainty principle [5.5]
valence electrons [5.4]
visible spectrum [5.2]
wavelength [5.2]
X-rays [5.1]

Problems

PROBING THE ATOM WITH NEW TOOLS

5.1 What did Berzelius have in common with Dulong and Petit?

5.2 What did Mendeleev do to apply the information obtained by Dulong and Petit?

5.3 What was observed when an evacuated Crookes tube was connected to a source of direct current?

5.4 What was observed when a magnet was placed perpendicular to Crookes's tube?

5.5 What modern device is closely related in shape and operation to a Crookes tube?

5.6 What appearance did cathode rays exhibit in a Crookes tube, and what electrical charge do cathode rays carry?

5.7 Who determined the *e/m* ratio for an electron? What does the term ''*e/m* ratio'' mean? Why was this determination important?

5.8 How did Thomson's experiments relate to a modern TV picture tube?

5.9 How did Goldstein modify the Crookes tube? What new facts did he learn?

5.10 Describe Thomson's ''plum pudding'' model of the atom. How was it incorrect?

5.11 What new information was obtained by Robert Millikan?

5.12 Describe Millikan's apparatus and experiment.

5.13 What new rays did Roentgen discover when using a Crookes tube?

5.14 What did Becquerel discover? Why were the Curies awarded Nobel prizes?

5.15 Describe differences in alpha, beta, and gamma rays.

5.16 Why do most alpha rays go through a thin gold foil?

5.17 List three observations made by Marsden in Rutherford's gold foil experiment. Describe three conclusions to this experiment and the important model of the atom that was then developed.

5.18 Describe the makeup of an atom if it were the size of the Superdome.

THE ELECTROMAGNETIC SPECTRUM

5.19 List three forms of electromagnetic radiation (in order) with higher frequencies than visible light.

5.20 List three or more forms of electromagnetic radiation (in order) with lower frequencies than visible light.

5.21 Describe the mathematical relationship of wavelength and frequency, and define the symbols used.

5.22 List the colors of visible light, beginning with the color having the highest frequency.

5.23 What form of radiation is often called ''black light''?

5.24 What makes some objects fluoresce?

5.25 What form of electromagnetic radiation can have the proper frequency to cause molecular vibrations? *Hint*: See Figure 5.15.

5.26 Besides microwaves, what other form of radiation can have a wavelength near 1 cm?

5.27 Does yellow light or red light have a longer wavelength?

5.28 Arrange these in order from high to low frequencies: visible light, UV, shortwave radio.

5.29 What form of radiation could have a wavelength of 3000 nm?

5.30 What form of radiation could have a wavelength of 90 cm?

EXCITED ELECTRONS AND SPECTRA

5.31 Explain the statement: ''Each element has a different line spectrum.''

5.32 What types of spectra are produced by the sun and by the bluish color of mercury vapor lamps often found in large parking lots?

5.33 Fireworks with orange-red to red can be produced by using compounds containing which element?

5.34 The golden yellow flame of burning logs in a fireplace results from the presence of which element? Can you explain why this element is present?

5.35 How are energy and frequency related—mathematically? Describe what happens to the frequency as energy increases.

5.36 What is another name for a small packet of light energy?

ELECTRONS IN ATOMS

5.37 How did Niels Bohr explain the line spectrum of the hydrogen atom?

5.38 Describe the difference between an atom in its ''ground state'' and in an ''excited state.''

5.39 Define the terms (a) ion, (b) ionization, (c) ionization energy.

5.40 List three ways that can be used to excite electrons of atoms.

5.41 When light is emitted by an atom, what change has occurred within the atom?

5.42 What is the *maximum* number of electrons possible in the first, second, third, and fourth energy levels of atoms?

5.43 Draw Bohr diagrams and give the number of valence electrons for atoms of the following elements.
a. magnesium b. calcium c. nitrogen
d. sulfur e. fluorine

5.44 Draw Bohr diagrams and give the number of valence electrons for atoms of the following elements.
a. potassium b. aluminum c. carbon
d. oxygen e. chlorine

5.45 How did contributions of Schrödinger and Heisenberg change the model of the atom?

5.46 How does the charge cloud model of the atom differ from Bohr's model?

VALENCE ELECTRONS AND LEWIS SYMBOLS

5.47 Using the periodic table, write Lewis electron-dot symbols for the following elements.
a. lithium b. aluminum c. phosphorus
d. oxygen e. bromine

5.48 Using the periodic table, write Lewis electron-dot symbols for the following elements.
a. potassium b. boron c. nitrogen
d. sulfur e. chlorine

5.49 According to the Pauli exclusion principle, how many electrons can occupy an orbital, and what are the restrictions on those electrons?

5.50 For the second energy level, describe the kinds of sublevels available and the number of orbitals within each sublevel.

5.51 What sublevel is being filled for transition elements, and what is unique about the order of filling?

5.52 What sublevel is being filled for inner transition elements, and what is unique about the order of filling?

5.53 Write out the electron configuration and shortened notation form of the orbital diagram for each of the following elements.
a. beryllium b. boron c. nitrogen
d. sulfur e. chlorine

5.54 Write out the electron configuration and shortened notation form of the orbital diagram for each of the following elements.
a. lithium b. aluminum c. phosphorus
d. oxygen e. bromine

Additional Problems

5.55 Since atoms of hydrogen have only one electron, why is it possible for the spectrum of hydrogen to have several visible lines?

5.56 Use Hund's rule and other information provided in this chapter to write out electron configurations and orbital diagrams for the following transition elements.
a. manganese b. iron

5.57 What do atoms of fluorine, chlorine, bromine, and iodine have in common in terms of electrons? What is the name of this group of elements?

5.58 What do atoms of lithium, sodium, potassium, rubidium, and cesium have in common in terms of electrons? What is the name of this group of elements?

5.59 In what way was each of these important scientists wrong in his understanding of atoms?
a. Dalton b. Thomson c. Bohr

5.60 If the scientists listed in Problem 5.59 were wrong, why was each one of them so important to our present-day understanding of atoms?

5.61 How many unpaired electrons are in an atom of nitrogen in its ground state?

5.62 How many unpaired electrons are in an atom of oxygen in its ground state?

5.63 In terms of valence electrons, what do lithium, sodium, and potassium have in common? Where are these elements placed in the periodic table?

5.64 In terms of valence electrons, what do beryllium, magnesium, and calcium have in common? Where are these elements placed in the periodic table?

5.65 Using the electron configuration and the periodic table, give the name of the element and the number of valence electrons.
a. $1s^2 2s^2 2p^4$
b. $1s^2 2s^2 2p^6 3s^2 3p^3$

5.66 Using the electron configuration and the periodic table, give the name of the element and the number of valence electrons.
a. $1s^2 2s^2 2p^6 3s^2 3p^1$
b. $1s^2 2s^2 2p^5$

There is beauty in a ROSE *that arises from the sequential buildup of petals—layer upon layer—to make a beautiful bud or blossom.*

There is beauty in MUSIC *that arises from a sequence of tones that build—octave after octave—from a fundamental set of pitches to form a total spectrum of sound.*

There is beauty in CHEMISTRY *that arises from a sequence of properties of elements that are repeated over and over—period after period—in the periodic table.*

6

Periodic Properties of Elements

CONTENTS

As chemists studied the various properties of elements and their reactions, they learned that these properties relate directly to electron structure—presented in Chapter 5—and that these properties occur periodically among the elements. In this chapter we will explain how the periodic table can be a tremendous asset in predicting certain properties of the elements. We will also include some practical uses of several elements to demonstrate how the uses of elements depend on their properties.

IA	IIA											IIIA	IVA	VA	VIA	VIIA	VIIIA
1 **H** 1.01																	2 **He** 4.00
3 **Li** 6.94	4 **Be** 9.01											5 **B** 10.8	6 **C** 12.0	7 **N** 14.0	8 **O** 16.0	9 **F** 19.0	10 **Ne** 20.2
11 **Na** 23.0	12 **Mg** 24.3											13 **Al** 27.0	14 **Si** 28.1	15 **P** 31.0	16 **S** 32.1	17 **Cl** 35.5	18 **Ar** 40.0
19 **K** 39.1	20 **Ca** 40.1	21 **Sc** 45.0	22 **Ti** 47.9	23 **V** 50.9	24 **Cr** 52.0	25 **Mn** 54.9	26 **Fe** 55.9	27 **Co** 58.9	28 **Ni** 58.7	29 **Cu** 63.5	30 **Zn** 65.4	31 **Ga** 69.7	32 **Ge** 72.6	33 **As** 74.9	34 **Se** 79.0	35 **Br** 79.9	36 **Kr** 83.8
37 **Rb** 85.5	38 **Sr** 87.6	39 **Y** 88.9	40 **Zr** 91.2	41 **Nb** 92.9	42 **Mo** 95.9	43 **Tc** (99)	44 **Ru** 101	45 **Rh** 103	46 **Pd** 106	47 **Ag** 108	48 **Cd** 112	49 **In** 115	50 **Sn** 119	51 **Sb** 122	52 **Te** 128	53 **I** 127	54 **Xe** 131
55 **Cs** 133	56 **Ba** 137	57 **La** 139	72 **Mf** 179	73 **Ta** 181	74 **W** 184	75 **Re** 186	76 **Os** 190	77 **Ir** 192	78 **Pt** 195	79 **Au** 197	80 **Hg** 201	81 **Tl** 204	82 **Pb** 207	83 **Bi** 209	84 **Po** (210)	85 **At** (210)	86 **Rn** (222)
87 **Fr** (233)	88 **Ra** (226)	89 **Ac** (227)	104	105	106	107	108	109									

Figure 6.1 *The central element in each triad of elements has properties with numerical values midway between those of the other elements in the triad.*

6.1 *Making Order Out of Chaos: Discoveries of Periodicity*

The Law of Octaves
In 1864, John A. R. Newlands, an English chemist, proposed his "law of octaves" after he noticed that when the elements are arranged in order of increasing atomic mass, every eighth element has similar properties. It should be mentioned that the noble gases had not been discovered at that time. The work of Newlands was ridiculed by other scientists of the Royal Chemical Society who refused to publish his work, but many years later he was honored by the society for his important contributions.

Let's go back to the 1800s. New elements were discovered with surprising frequency. By 1830 there were 55 known elements, all with seemingly different properties and with no apparent order. People—not just scientists—tend to look for regular patterns in nature and attempt to find order in apparent chaos. Several attempts were made to arrange the elements in some sort of systematic fashion. By 1817 J. W. Dobereiner, a professor of chemistry in Germany, showed that the atomic mass of strontium is very close to an average of the atomic masses of two similar metals, calcium and barium. He later found that there are other *triads* of similar elements, such as lithium, sodium, and potassium or chlorine, bromine, and iodine (Figure 6.1). Dobereiner recommended that elements be classified by triads, but he was unable to find enough triads to make the system useful. Although his proposal did not gain acceptance, he is recognized for these pioneering attempts to classify the elements. Other classification schemes, including one by Newlands, were also proposed.

The most successful arrangement of the elements was developed by Dmitri Ivanovich Mendeleev (1834–1907), a Russian chemistry professor (Figure 6.2). He pointed out that *both the physical and chemical properties of the elements vary periodically with increasing atomic mass.* This is known as the **periodic law**. In 1869, at the age of 35, Mendeleev published a **periodic table** of the elements that resembles our modern periodic table. His table had the elements arranged in order of increasing atomic mass, and in periods, so that elements with similar chemical properties were grouped together, although there were a few instances where he had to place an element with a slightly greater atomic mass before an element with a slightly lower mass. For example, he placed tellurium (atomic mass of 127.6)

To Mendeleev, periodicity was related to atomic mass, but today we use atomic number. See Moseley's work in Section 6.2.

ahead of iodine (atomic mass of 126.9) because tellurium resembled sulfur and selenium in its properties, whereas iodine was similar to chlorine and bromine.

Notice the placement of tellurium and iodine in the periodic table.

Mendeleev left a number of gaps in his table. Instead of looking upon those blank spaces as defects, he boldly predicted the existence of elements that had not been discovered. Furthermore, he even predicted the properties of some of those missing elements. In Mendeleev's lifetime several new elements, including scandium (Sc), gallium (Ga), and germanium (Ge), were discovered; their chemical properties closely matched the properties predicted by Mendeleev.

Although Mendeleev is credited with the discovery of the periodic table, Lothar Meyer, a German chemist, had independently developed his own periodic table in 1868, but his work was not published until 1870—one year after Mendeleev's publication. Mendeleev, however, had already gained considerable notoriety by boldly predicting chemical properties of certain undiscovered elements.

Mendeleev missed receiving the Nobel prize in chemistry by just one vote in 1906, and died before the next year's election. Element number 101 (discovered in 1955) was named Mendelevium in his honor.

EXAMPLE 6.1

For the Li, Na, and K triad, determine the atomic mass of Na to the nearest 0.1 amu by averaging the masses of Li and K. (Atomic masses are shown in the periodic table on the inside front cover of this book.) Compare your answer with the atomic mass of sodium shown in the table.

SOLUTION

$$\frac{\text{Li amu} + \text{K amu}}{2} = \frac{6.9 \text{ amu} + 39.1 \text{ amu}}{2} = 23.0 \text{ amu}$$

This atomic mass for sodium, Na, is the value shown in the periodic table.

See Problems 6.1–6.6.

6.2 *The Periodic Table Today*

Since the time of Mendeleev, the periodic table has undergone numerous changes to include new elements, more accurate values, and different ways of labeling columns of elements in the table. A modern periodic table appears on the inside front cover of this book. You will want to refer to it often. The sequence of elements in the modern periodic table coincides with the increase in atomic number, that is, in the number of protons in the nucleus of each element.

About 45 years after Mendeleev's development of the periodic table, but only a few years after Rutherford's investigations of the nucleus, Henry Moseley developed a technique of determining the size of the positive charge for a nucleus. He concluded that each element differs from all other elements by having a different number of protons (or atomic number). For most elements, the increase in atomic number coincides with an increase in atomic mass, but there are a few exceptions. When elements in the periodic table are arranged by atomic number, tellurium and iodine fall in appropriate families of elements; this was not the case when Mendeleev used atomic mass. The same is true for the metals cobalt and nickel, and for argon and potassium. Today, elements in periodic tables are arranged in order of increasing *atomic number* rather than atomic mass. Notice that there are no missing atomic numbers for the first 109 elements listed in the table. We can, therefore, be

Figure 6.2 *Dmitri Mendeleev (1834–1907) was the Russian chemist who invented the periodic table of the elements. He was born in Siberia and was the youngest of 17 children.*

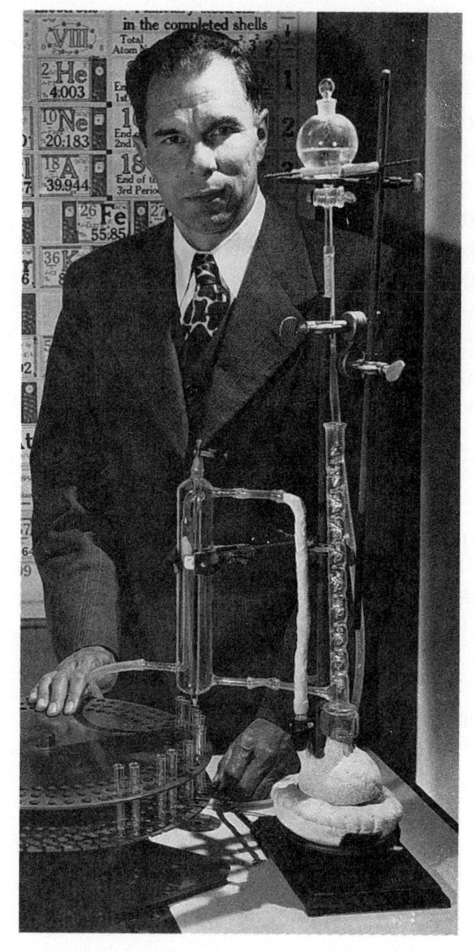

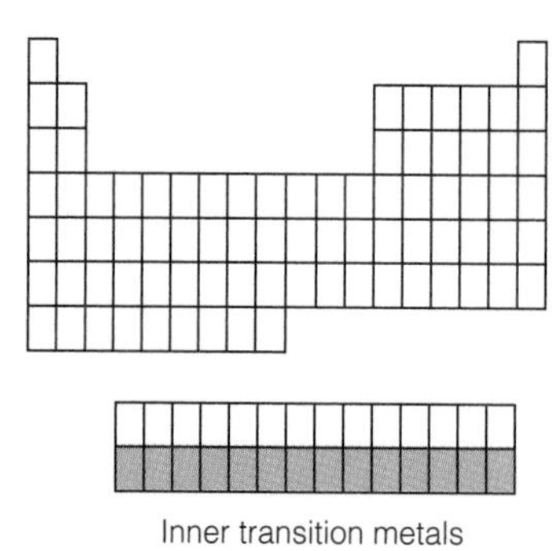

Figure 6.3 *Glenn T. Seaborg (1912–), who had been involved with the discovery and preparation of several transuranium elements, also made history when he changed the periodic table by placing the actinides beneath the main body of the table along with the other inner transition elements, as shown below. He was one of the recipients of the 1951 Nobel prize in chemistry. In 1994, it was proposed that element 106 be named seaborgium, Sg, in his honor.*

assured that no element will be discovered that falls between any two of these 109 elements.

The two rows of inner transition metals—placed below the body of most periodic tables today—were not a part of Mendeleev's periodic table. Until Glenn T. Seaborg (1912–) rejected the advice of colleagues and published his own version of the periodic table, elements having atomic numbers 90, 91, and 92 (thorium, Th, protactinium, Pa, and uranium, U) followed actinium in the main body of the periodic table. Seaborg was convinced that Th, Pa, U, and the transuranium elements—precisely the elements that Seaborg and associates had synthesized at the University of California, Berkeley—had properties like the other inner transition elements positioned below the main body of the periodic table. His modification of the periodic table made it possible to accurately predict properties of undiscovered transuranium elements. Synthesis of several of these elements proved that Seaborg was correct; his insight changed the shape of periodic tables in use today.

Periods of Elements: General Trends

A *period*—or horizontal row—of elements in the periodic table has a variation in physical and chemical properties that closely parallels the variation in properties of other periods of elements. For example, the second and third periods of elements begin with shiny, reactive metals at the left, followed by dull solids, and reactive nonmetals. Each period ends with a colorless, nonreactive noble gas. This trend in appearance accompanies the trend from metallic to nonmetallic character within a period of elements.

The increase in valence electrons of elements in the third period parallels the increase in valence electrons for elements in the second period. The first element in each period has one valence electron in its highest energy level. For example, lithium has one valence electron in its second energy level, sodium has one valence electron in its third energy level, and potassium has one valence electron in its fourth energy level. In summary, the periodic changes in properties of elements coincide with their placement in the periodic table.

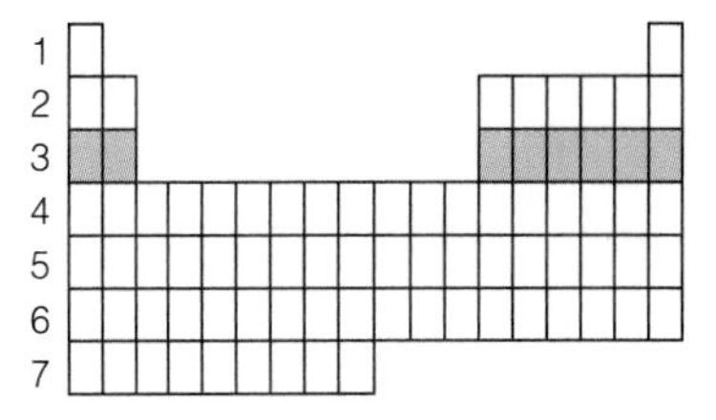

Figure 6.4 *There are seven periods of elements in the periodic table. Period 3 is emphasized here.*

Groups and Families of Elements

Vertical columns of elements in the periodic table are called *groups* when the number of the column is included, or *chemical families* when the family name is included. For example, metals in Group IA are in the family of alkali metals (Figure 6.5). Elements in the same family tend to have similar chemical properties.

In the United States, group headings for the **representative elements** (those in the first two columns and last six columns of elements) have traditionally been designated by a Roman numeral followed by the letter A, for Groups IA through VIIIA (the noble gases). Transition metals are easy to identify in the B groups near the center of the table. In Europe, different headings are used: all groups to the left of Group VIII in the transition metals are designated as A groups, and groups to the right of Group VIII in the transition metals are designated as the B groups. Numerous proposals have been submitted to resolve this dilemma. An American Chemical Society proposal that is gaining acceptance gets around the A and B notation by simply numbering the groups sequentially 1 through 18. Since the issue of labeling groups has not been resolved, chemistry students will need to accept the use of periodic tables with different systems of notation. In this book, we will use the popular A and B group labeling system preferred by many chemistry educators.

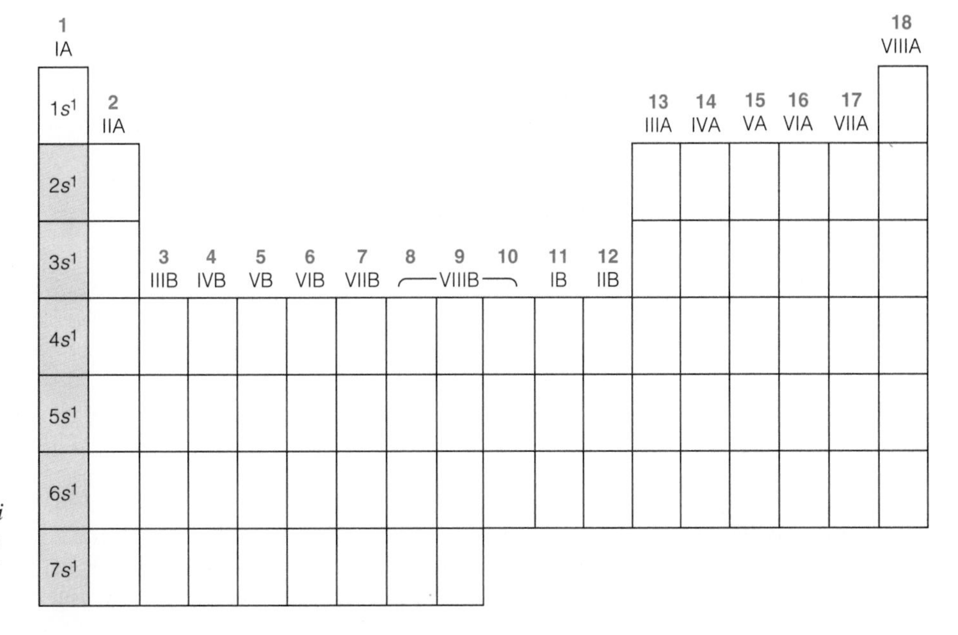

Figure 6.5 *Group IA, the alkali metal family, is the group that is positioned at the far left edge of the periodic table.*

As described in Chapter 5, all elements in the same group have the same number of valence electrons. All elements in Group IA have one valence electron, which is an *s* electron. All elements in Group IIA have two *s* valence electrons. These two groups of elements make up the *s* block of elements. Elements in Groups IIIA, IVA, VA, VIA, VIIA, and VIIIA have one to six electrons, respectively, in an outer *p* sublevel. These elements make up the *p* block of elements. The groups of transition metals, located near the center of the periodic table, have one to ten electrons in a *d* sublevel.

EXAMPLE 6.2

Give the name of the family of elements with atomic numbers 9, 17, 35, 53, and 85, and describe their similarities in terms of metallic or nonmetallic properties, number of valence electrons, and type of electrons.

SOLUTION

This is the halogen family of elements (identified in Chapter 4). Halogens are all nonmetals with seven valence electrons (two valence electrons are in an *s* sublevel and five valence electrons are in a *p* sublevel).

See Problems 6.7–6.20.

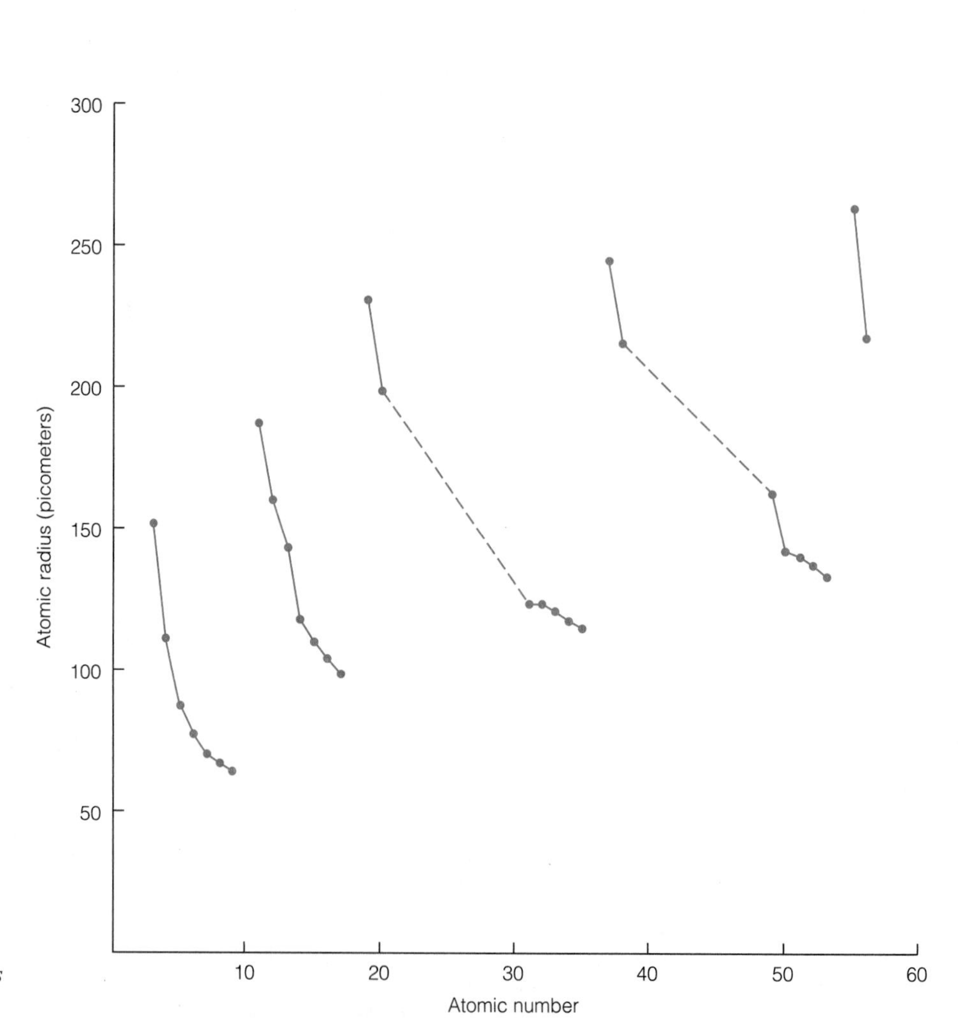

Figure 6.6 *Atomic size is a periodic property, as indicated by the graph of atomic radius versus atomic number. (Sizes of transition metals and noble gases are not included.)*

Exercise 6.1
Which periods of elements contain transition metals? Which periods do not?

6.3 *Atomic and Ionic Size*

It is not possible to determine an exact radius or volume for an atom since it is not a hard sphere with a definite boundary. The probability (or chance) of finding an electron decreases at greater distances from the nucleus in a manner that might be compared to the decrease in atmospheric oxygen at greater distances from the Earth. Thus, atom size is based on the average distance between the outer electrons and the nucleus. An atom's radius can be reported in angstroms (1 Å$=10^{-10}$ m), but SI units are in nanometers (1 nm = 10^{-9} m) or picometers (1 pm = 10^{-12} m). A sodium atom, for example, has a radius of 1.86 Å, 0.186 nm, or 186 pm.

Trends in Atomic Size

The variation in size of atoms is a periodic property, as shown in Figure 6.6. Notice that each peak on the graph represents the radius of an alkali metal with a large radius. For comparative purposes the atomic radii for the representative elements are shown in periodic table format in Figure 6.7. We can summarize trends in atomic size as follows.

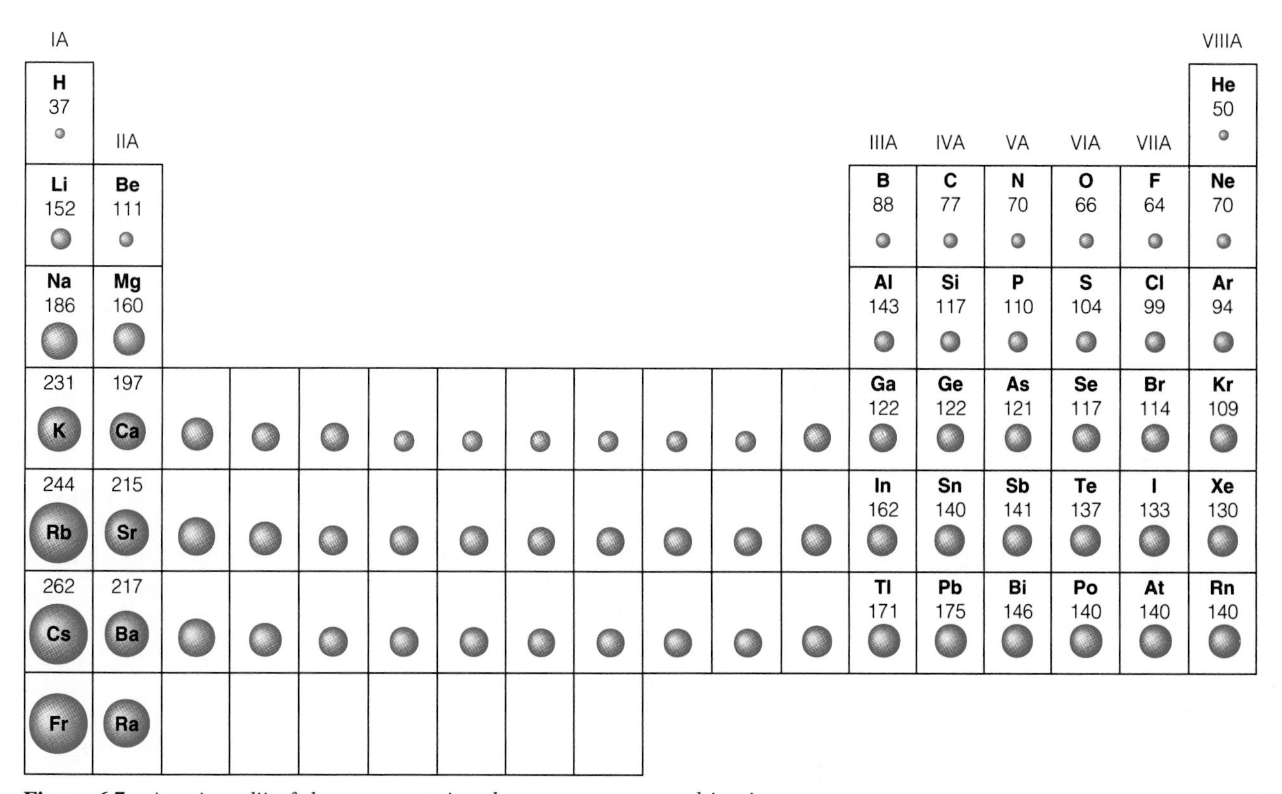

Figure 6.7 *Atomic radii of the representative elements are expressed in picometers.*

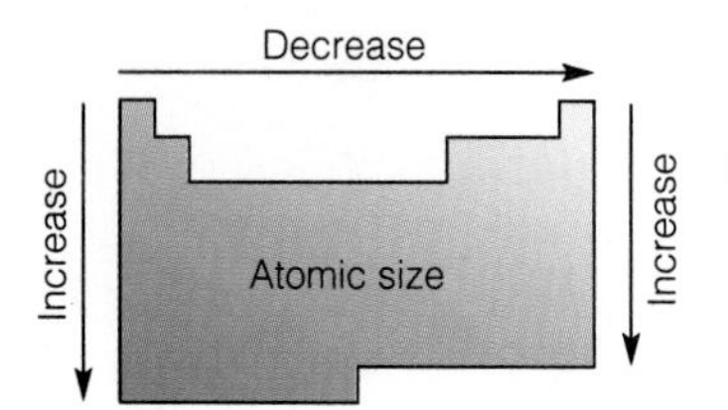

Figure 6.8 *General trends in atomic size.*

Atomic size within a PERIOD tends to decrease as atomic number increases. Atomic size within a GROUP increases as atomic number increases.

Within each group (or family) of elements, the size of atoms increases each time a new higher energy level is occupied by more electrons. However, within each period of elements, the size of atoms tends to decrease as electrons are added to a particular energy level. This is because each element in a period has one more proton than the previous element, and the increase in positive nuclear charge draws the electron cloud closer to the nucleus (Figure 6.8).

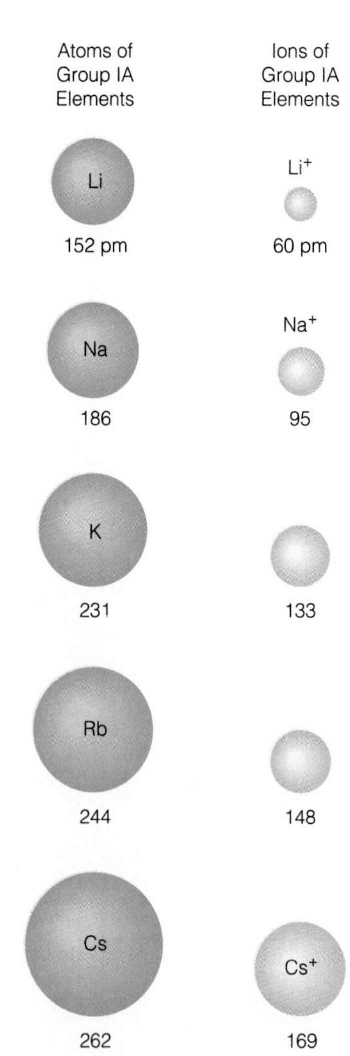

Figure 6.10 *A comparison of radii for Group IA atoms and ions expressed in picometers (pm).*

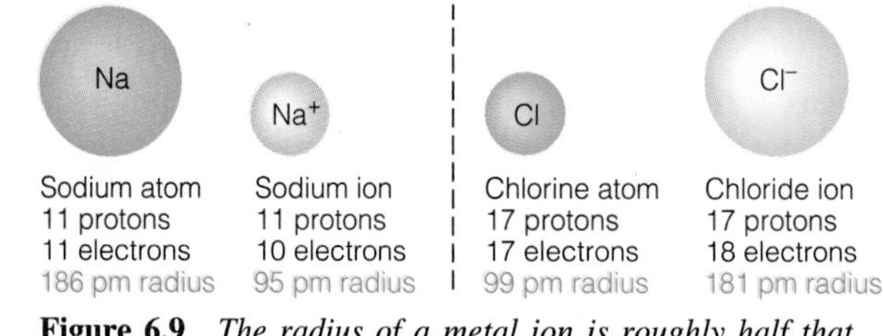

Figure 6.9 *The radius of a metal ion is roughly half that of the corresponding metal atom. The radius of a nonmetal ion is roughly twice that of the corresponding nonmetal atom.*

Ions and Trends in Ionic Size

An **ion** is a charged particle that is produced when any atom or group of atoms gains or loses one or more electrons. Metal atoms—nearly all of which have fewer than four valence electrons—tend to lose their valence electrons to form positive ions called **cations** (pronounced CAT-ions). A metal ion has a radius that is roughly half that of the corresponding metal atom (Figure 6.9).

Nonmetal atoms are those that tend to gain electrons to form negative **anions** (pronounced AN-ions). For example, the halogens, with seven valence electrons, readily gain one electron to fill an outer energy level completely with eight electrons—an octet of electrons. Because a halide ion—an ion from a halogen element—has one more electron than it has protons, its ion charge is 1−. A negative ion is considerably larger than the corresponding nonmetal atom. For example, the radius of a chloride ion (181 pm) is roughly twice the radius of a chlorine atom (99 pm). Within a family of elements, ion sizes increase with atomic number (Figure 6.10).

Let us compare the size of a sodium ion (represented as Na^+) with the size of a fluoride ion, F^-. These ions are **isoelectronic**; that is, they have the same total number of electrons. Both ions have 10 electrons, but the sodium ion has 11 protons (a nuclear charge of +11) and a greater attraction for its electrons than does a fluoride ion, which has 9 protons (a nuclear charge of +9). We can expect the sodium ion—with its greater nuclear charge—to have a smaller radius than a fluoride ion. We can make the following generalization.

For isoelectronic ions, the radius decreases as the positive nuclear charge increases.

EXAMPLE 6.3

Compare the size (radius) of a chloride ion, Cl^-, with the size of a potassium ion, K^+.

SOLUTION

The ions are isoelectronic with a total of 18 electrons each. Since the nucleus with a greater positive charge has greater attraction for its electrons, we can expect the potassium ion to be smaller than a chloride ion.

See Problems 6.21–6.28.

EXERCISE 6.2

Which is larger, a potassium atom or a potassium ion? Which is larger, a chlorine atom or a chloride ion?

6.4 *Ionization Energy*

It takes a specific amount of energy to remove an electron from a neutral atom since electrons of atoms are in definite energy levels. The amount of energy required to remove an electron from a gaseous atom in its ground state is the **ionization energy**. This is another periodic property of the elements; it is a measure of how tightly electrons are bound to atoms.

We can represent the ionization of a sodium atom, for example, by the equation

$$Na + Energy \longrightarrow Na^+ + 1\ e^-$$

For review, recall that a sodium atom has 11 electrons around a nucleus with 11 protons (and 12 neutrons). When enough energy is supplied to "rip off" a valence electron, as represented by the equation, a sodium ion is produced along with a free electron. The energy required to remove the most loosely held electron from an atom is called the **first ionization energy**. More energy is required to remove each additional electron—for the second and third ionization energies, etc.—because the positive charge increases by one as each subsequent electron is removed. Ionization energy can be expressed in several energy units, including kilojoules per mole, kilocalories per mole, and electronvolts per atom.

Ionization energy does *not* apply to the formation of negatively charged ions.

First ionization energies for elements in the first three periods are listed in Table 6.1. Notice that noble gases (atomic numbers 2, 10, 18, . . .), which are the most

Table 6.1 **First Ionization Energies (I.E.) for the First 18 Elements in kJ/mole**

Period 1	H							He
I.E.	1312							2371
Period 2	Li	Be	B	C	N	O	F	Ne
I.E.	520	900	800	1086	1402	1314	1681	2080
Period 3	Na	Mg	Al	Si	P	S	Cl	Ar
I.E.	496	738	577	786	1012	1000	1255	1520

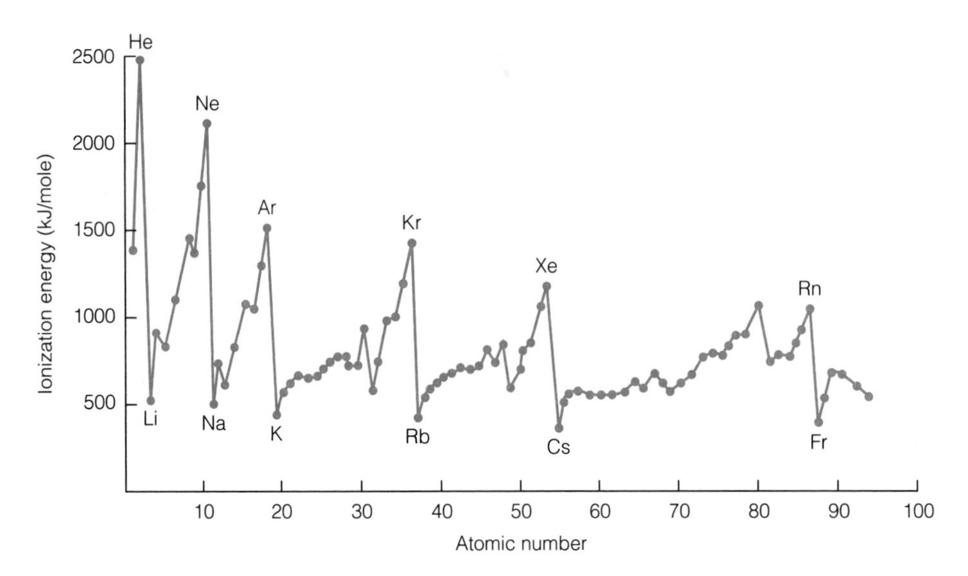

Figure 6.11 *The variation in ionization energy is a periodic property. First ionization energies are given here in kilojoules per mole.*

stable elements, chemically, have very high ionization energies, while the reactive metals of Group IA (atomic numbers 1, 3, 11, . . .) have low ionization energies. We can make the following generalizations.

Ionization energy for elements within each PERIOD increases as atomic number increases.

Ionization energy for elements within a GROUP decreases as atomic number increases.

The most metallic elements (Group IA) have the lowest ionization energies.

The periodic variation in first ionization energies is clearly evident in the graphical plot of ionization energy versus atomic number as illustrated in Figure 6.11. These trends closely parallel the variations in atomic size. In general, it takes more energy to remove an electron from a smaller atom; its outermost electron is closer to its nucleus. Notice that the ionization energies of the noble gases—the maximum points on the graph—consistently decrease as the atomic number increases. In other words, ionization energy decreases within a family or group of elements as the atomic size increases. Notice, also, that the ionization energies of the alkali metals—the minimum points on the graph—decrease as atomic number and size increase. A summary of these trends in ionization energies as related to the periodic table is given in Figure 6.12. Metals at the lower left corner of the periodic table have the lowest ionization energies; they readily form positive metal ions by losing electrons.

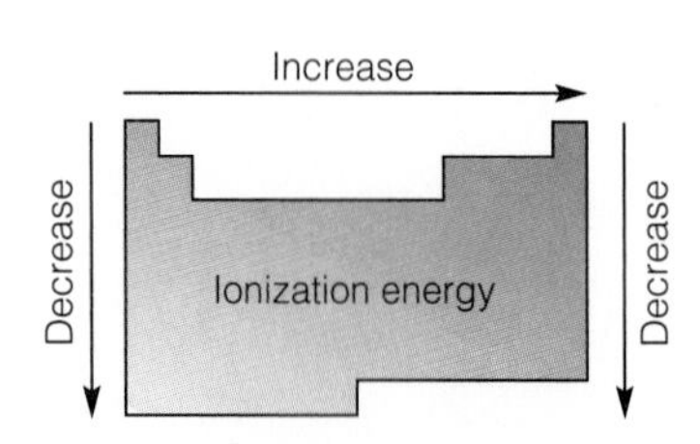

Figure 6.12 *Ionization energy for the elements within a period increases with increasing atomic number. Within a group ionization energy decreases as atomic number increases.*

EXAMPLE 6.4

Select the element from each pair that would be expected to have a lower ionization energy. Explain why. (Refer to the periodic table.)

(a) B and O

(b) Li and Cs

SOLUTION

(a) Boron would be expected to have the lower ionization energy since ionization energy increases within a period (left to right).
(b) Cesium would be expected to have a lower ionization energy since ionization energy decreases within a group as atomic number increases.

See Problems 6.29–6.38.

EXERCISE 6.3
As you look at the graph showing ionization energies, what can you conclude?

6.5 Melting Point and Boiling Point Trends

Trends in melting points and boiling points can be used as a convenient measure of the attractive forces between atoms or molecules. For example, the first two halogens, fluorine and chlorine, are both gases at room temperature. They are followed by bromine, a reddish liquid, and iodine, a steel gray solid. This trend from gas to liquid to solid is a dramatic example of the increases in melting points as the halogen atoms get larger and atomic number increases. There is also a parallel trend of increasing boiling points. On the other hand, Group IA metals show the opposite trend, with a decrease in melting points and boiling points due to the weakening of metallic bonds between atoms as size increases. Specific values for these two groups of elements are summarized in Table 6.2.

Table 6.2 **Melting Points and Boiling Points for Selected Elements**

Alkali Metals			*Halogens*		
Element	*Melting Point (K)*	*Boiling Point (K)*	*Element*	*Melting Point (K)*	*Boiling Point (K)*
Li	454	1615	F	53	85
Na	371	1156	Cl	172	239
K	336	1032	Br	266	332
Rb	312	961	I	387	458
Cs	301	944	At	575	610

Melting points in the second period of elements increase from left to right for the first four elements, which are solids, and then drop off sharply to low values for the last four elements, which are gases, as shown in Table 6.3. Notice that carbon

Table 6.3 **Melting Points and Densities for Period 2 Elements**

	Li	*Be*	*B*	*C*	*N*	*O*	*F*	*Ne*
Melting point (K)	454	1560	2300	4100	63	50	53	25
Density*	0.53	1.85	2.34	2.62	1.2†	1.4†	1.7†	0.90†

*Densities are in grams per cubic centimeter except for gases (†) with densities in grams per liter.

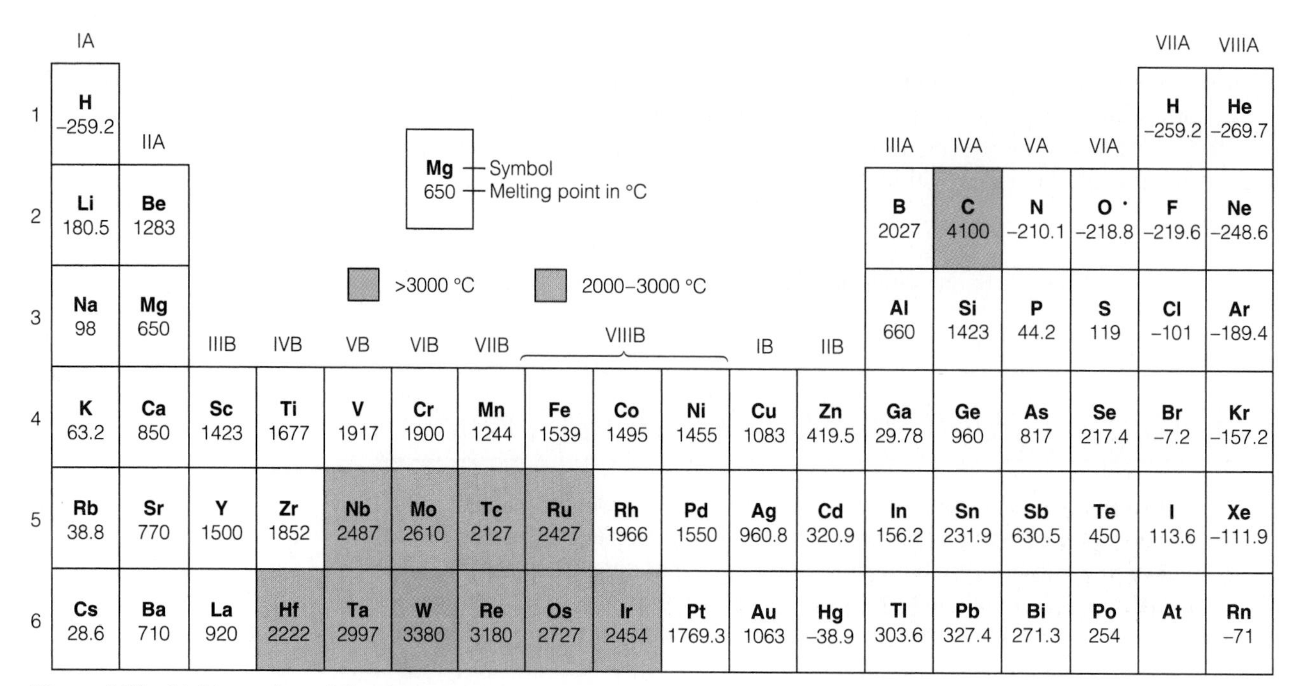

Figure 6.13 *Melting points of the elements.*

has the highest melting point (4100 K) for period 2 elements. The diamond form of carbon is the hardest and has a higher melting point than does any other element. Silicon, which is directly below carbon in the periodic table, has the highest melting point for elements in the third period. Within a period, then, melting points first increase, then decrease. Moving from left to right within a row, melting points increase sharply as the attractive forces change from strong metallic bonding with loose electrons to solids like carbon where outer electrons are tied up in a complex network. Melting points then drop off sharply for the nonmetals, which have very weak forces of attraction. Besides diamond, mentioned previously, tungsten (W) and other transition metals clustered near tungsten in periods 5 and 6 have very high melting points (Figure 6.13).

You should be able to point out regions in the periodic table where melting points of elements are the highest.

EXAMPLE 6.5

Which element in each pair has a higher melting point? Consider their positions in the periodic table.

(a) Li and K (b) F_2 and Br_2 (c) Fe and Se

SOLUTION

(a) Li has a higher melting point than K; melting points *decrease* top to bottom for Group IA.
(b) Br_2 has a higher melting point than F_2; melting points *increase* top to bottom for the halogens.
(c) Fe has a higher melting point than Se; within a period melting points first increase, then decrease.

See Problems 6.39–6.42.

6.6 *Density and Conductivity Trends*

With few exceptions, densities of elements within each group or family increase as atomic number increases. In a period of elements, densities first increase, then decrease (Figure 6.14). This trend is more dramatic for elements in periods 4, 5, and 6. For example, cobalt, nickel, and copper near the center of period 4 are the most dense elements in that period. The densities of elements in period 5 are nearly 50% greater than the densities of corresponding period 4 elements.

Elements with the greatest densities are metals in the center of period 6, including osmium (Os), iridium (Ir), and platinum (Pt), all with densities of about 22 g/cm^3. These densities are nearly twice as great as those for the most dense metals of period 5.

To summarize,

> Densities of elements within a GROUP increase as atomic number increases.
>
> Densities of elements within a PERIOD first increase, then decrease.
>
> Elements with the greatest densities are at the center of period 6.

Because sodium is a good thermal conductor and has a low melting point, it is used as a coolant in some nuclear power plants in much the same way as antifreeze and water are used as coolants in automobile engines. Also, certain spark plugs used in race cars are sodium filled to improve heat transfer.

All metals conduct both electricity and heat, but a few metals have especially high conductivities. These include, in order of decreasing conductivity, silver, copper, gold, aluminum, calcium, sodium, and magnesium. The conductivities of other metals are considerably less; nonmetals are nonconductors.

Mg — Symbol
1.74 — Density in g/cm^3 or, for gases, in g/L

8.0-11.9 g/cm^3 | 12.0-17.9 g/cm^3 | >18.0 g/cm^3

	IA	IIA	IIIB	IVB	VB	VIB	VIIB	VIIIB	VIIIB	VIIIB	IB	IIB	IIIA	IVA	VA	VIA	VIIA	VIIIA
1	H 0.071																H 0.071	He 0.126
2	Li 0.53	Be 1.8											B 2.5	C 2.26	N 0.81	O 1.14	F 1.11	Ne 1.204
3	Na 0.97	Mg 1.74											Al 2.70	Si 2.4	P 1.82w	S 2.07	Cl 1.557	Ar 1.402
4	K 0.86	Ca 1.55	Sc (2.5)	Ti 4.5	V 5.96	Cr 7.1	Mn 7.4	Fe 7.86	Co 8.9	Ni 8.90	Cu 8.92	Zn 7.14	Ga 5.91	Ge 5.36	As 5.7	Se 4.7	Br 3.119	Kr 2.6
5	Rb 1.53	Sr 2.6	Y 5.51	Zr 6.4	Nb 8.4	Mo 10.2	Tc 11.5	Ru 12.2	Rh 12.5	Pd 12.0	Ag 10.5	Cd 8.6	In 7.3	Sn 7.3	Sb 6.7	Te 6.1	I 4.93	Xe 3.06
6	Cs 1.90	Ba 3.5	La 6. 7	Hf 13.1	Ta 16.6	W 19.3	Re 21.4	Os 22.48	Ir 22.4	Pt 21.45	Au 19.3	Hg 13.55	Tl 11.85	Pb 11.34	Bi 9.8	Po 9.4	At —	Rn 4.4

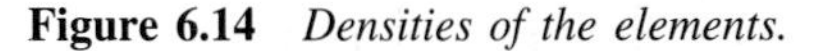

Figure 6.14 *Densities of the elements.*

EXAMPLE 6.6

Compare the trend in densities for Cu, Ag, and Au. Explain.

SOLUTION

The order of increasing densities is Cu < Ag < Au, with gold being the most dense of the three metals. These three are in the same group; densities increase from top to bottom as atomic number increases.

EXAMPLE 6.7

Which of the following elements are very good conductors of electricity: Al, Au, S, P, Cu?

SOLUTION

See Problems 6.43–6.46.

The metals Al, Au, and Cu are good conductors; nonmetals are nonconductors.

6.7 A Survey of Elements by Groups

A brief survey of properties and uses of selected elements by groups is presented here. It is not possible for us to provide a great deal of information about each element.

Hydrogen, a Unique Element

Hydrogen, the first element, is unique. It is, by far, the most abundant element in the universe. It combines with oxygen to give water, H_2O, which is the most abundant compound on the Earth—it covers three-fourths of the Earth's surface. Hydrogen is becoming an important fuel (Figure 6.15).

Hydrogen is usually placed in Group IA of the periodic table since it has one valence electron like all other elements in Group IA. However, hydrogen is a diatomic gas, H_2, with chemistry that is quite different from typical alkali metals. For this reason, it is positioned in different locations on different periodic tables.

Figure 6.15 *The Mazda Miata shown here is engineered to run on hydrogen. Mercedes-Benz and several other automobile manufacturers are also testing hydrogen-powered prototypes. There are trade-offs. The fuel burns cleanly but producing it is quite expensive.*

Since hydrogen, like fluorine and chlorine, needs one more electron to fill an energy level, some periodic tables show it placed next to helium—above fluorine—but the chemistry of hydrogen is not like the halogens. To emphasize the unique characteristics of hydrogen, some periodic tables show the element by itself at the top center of the periodic table.

Group IA: The Alkali Metals

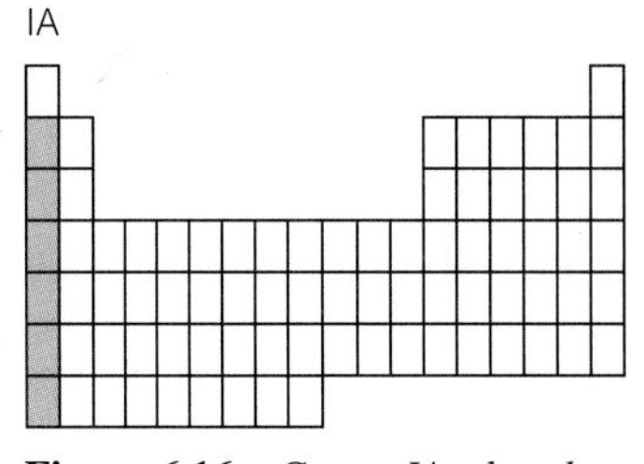

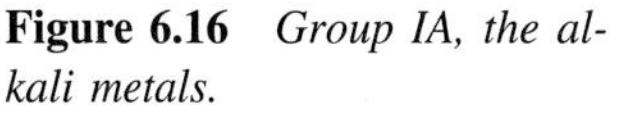
Figure 6.16 *Group IA, the alkali metals.*

The alkali metals—lithium (Li), sodium (Na), potassium (K), rubidium (Rb), cesium (Cs), and francium (Fr)—are silver gray, soft metals that can be cut with a knife (Figure 6.16). They have very low densities and are good conductors of heat and electricity. Group IA metals are quick to react with water, oxygen, and other chemicals and are never found as free (uncombined) elements in nature. Typical alkali metal compounds are water soluble and are present in seawater and salt deposits. Since these metals quickly react with oxygen, they are sold in evacuated containers but they usually are stored under mineral oil or kerosene. In this group, sodium and potassium are the most common, ranking sixth and seventh in abundance in the Earth's crust. Common table salt, the most common compound of sodium, is sodium chloride, NaCl (Figure 6.17). Potassium (usually as KCl) is an important ingredient in fertilizers. Major deposits of potassium compounds are present in New Mexico and California.

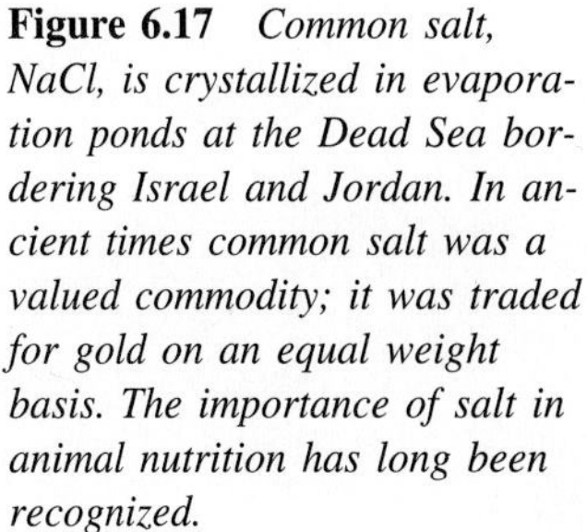
Figure 6.17 *Common salt, NaCl, is crystallized in evaporation ponds at the Dead Sea bordering Israel and Jordan. In ancient times common salt was a valued commodity; it was traded for gold on an equal weight basis. The importance of salt in animal nutrition has long been recognized.*

Group IIA: The Alkaline Earth Metals

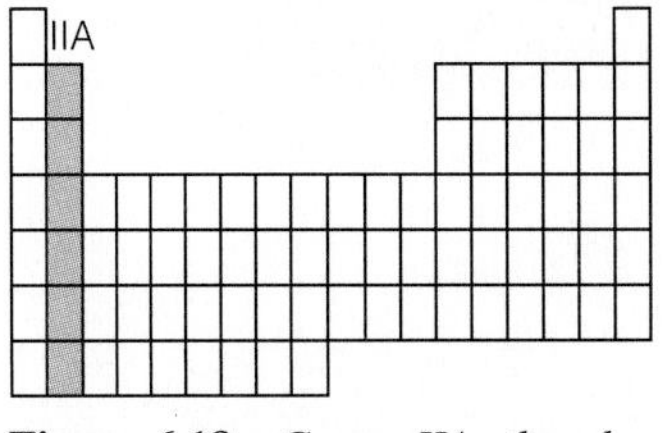

Figure 6.18 *Group IIA, the alkaline earths.*

Group IIA metals, the alkaline earth metals (Figure 6.18), include beryllium (Be), magnesium (Mg), calcium (Ca), strontium (Sr), barium (Ba), and radium (Ra). These metals have melting points that are higher than the Group IA metals. Their densities are all low but are still somewhat higher than densities of comparable alkali metals. They are less reactive than the alkali metals. All alkaline earth metals have two valence electrons and form ions with a positive two (2+) charge.

Calcium ranks fifth in abundance; about 4% of the Earth's crust is calcium or magnesium. Calcium carbonate, $CaCO_3$, is the compound in chalk, limestone, marble, and calcite. Lime, cement, bones, and seashell deposits are rich in calcium. Magnesium metal is used in flash powder, photographic flashbulbs, and in aluminum alloys, especially for airplanes and missiles. Most ''hard water'' contains both calcium and magnesium ions. Beryllium is expensive, but alloys of the metal are used in nonsparking tools, springs, and electrodes for spot welding. Beryllium and its compounds are toxic. Barium compounds are widely used in white pigments. Radium is radioactive.

Two alkaline earths, Mg and Ca, are important for human health, while two others, Be and Ba, are poisonous.

Group IIIA

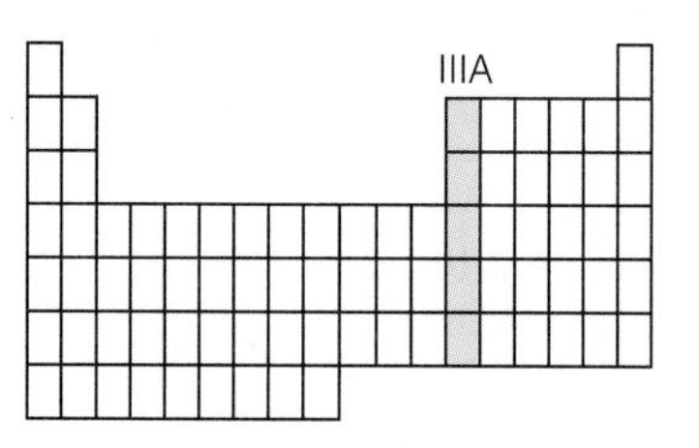

Figure 6.19 *Group IIIA.*

The first element in Group IIIA is boron (B), a metalloid with a very high melting point and predominantly nonmetallic properties. The other elements in this group include aluminum (Al), gallium (Ga), indium (In), and thallium (Tl), which form ions with a positive three (3+) charge (Figure 6.19). Density and metallic character increase as atomic number increases within this group.

Boron is not found free in nature, but is the key element in borax. There are extensive borax deposits near Boron, California (Figure 6.20). Borax is used as a water softener and in cleaning agents. Boric acid is the common mild antiseptic used as an eye wash. Boron compounds are widely used in Pyrex glassware, fiberglass, abrasives, cutting tools, and porcelain enamels and as a fire retardant. Chemically, boron behaves more like the metalloid silicon than like metallic aluminum.

Aluminum is adjacent to two metalloids in the periodic table, but its properties are predominantly metallic. Aluminum is a good conductor of heat and electricity and is a ductile metal used in lightweight wire. Aluminum is the most abundant metal in the Earth's crust (8%), but the metal is too reactive to be found free in

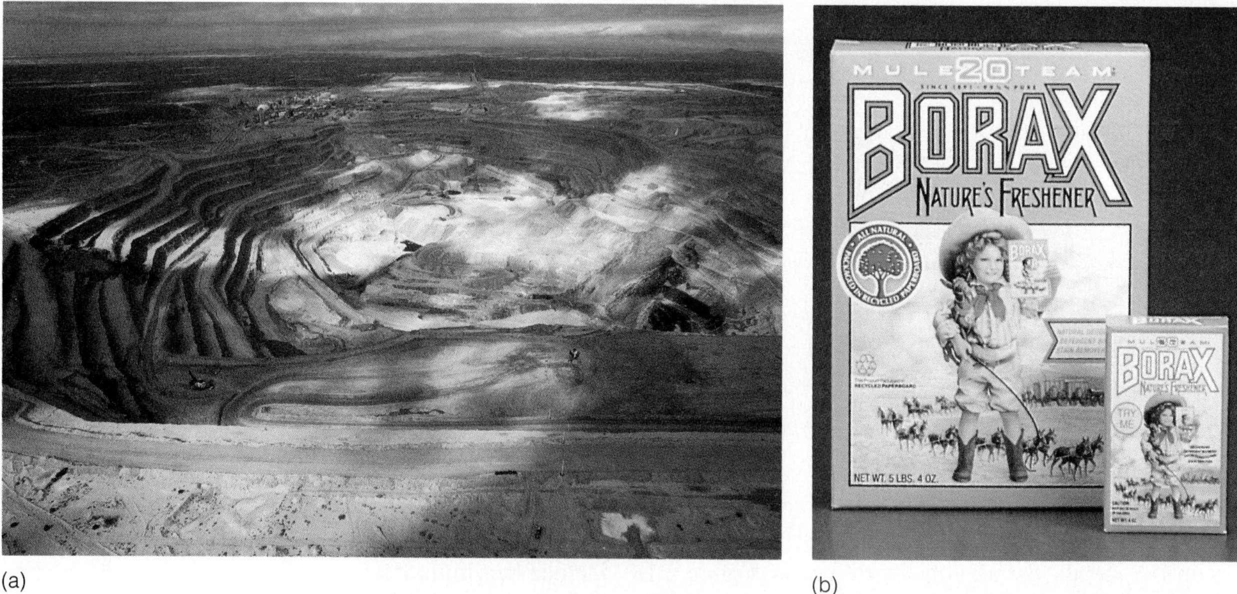

(a) (b)

Figure 6.20 *The most common boron-containing compounds are boric acid and borax. Borax had been scarce and expensive until major deposits of the ore (a) were discovered in California and Nevada in the mid-nineteenth century. During the 1880s, 20-mule teams moved thousands of tons of borax ore over a 165-mile trail from Death Valley, across the mountains, to the railroad. ''20 Mule Team'' is still associated with the product (b). Borax is often used in water softeners and washing compounds.*

(a)

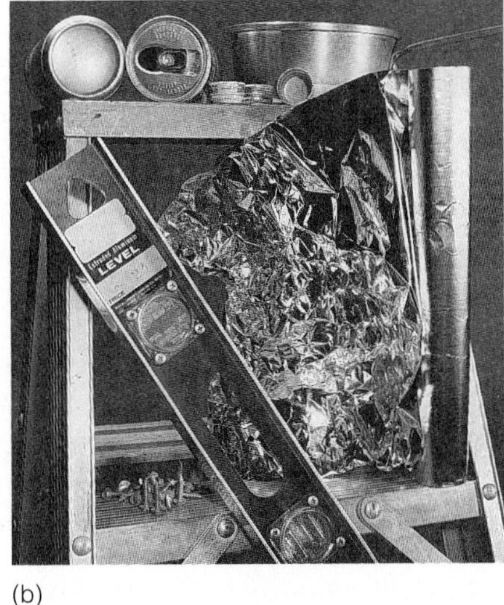

(b)

Figure 6.21 *Bauxite (a) is the main ore from which alumina (aluminum oxide) is obtained. Once the aluminum metal is freed from alumina—by an electrolytic process—the metal (b) is used in a variety of applications including aircraft, aluminum foil, and lightweight containers.*

nature. It is widely distributed in various minerals such as clay and feldspar where it is tightly bound in compounds with oxygen and silicon. Bauxite, mined in Jamaica, Australia, Arkansas, and elsewhere, is the main ore from which aluminum oxide (alumina) is extracted. The ore undergoes a series of chemical processes and energy from an electric current is used to liberate the aluminum metal (Figure 6.21). This electrolysis process was discovered in 1886 by Charles M. Hall, who was a student at Oberlin College in Ohio. Pure aluminum and aluminum alloys have a wide variety of uses including aircraft, electrical transmission wire, engines, automobiles (an increase of 48% in 10 years), cooking utensils, paint pigment, and aluminum foil.

By remarkable coincidence, the same electrolysis process was also discovered in France by Paul Héroult in the same year. Both Hall and Héroult were born in the same year, 1863, died in the same year, 1914, and made the same discovery in the same year.

Gallium melts at 29.8 °C, which is only slightly above room temperature. Demand for the metal is increasing; it has new applications in solid-state semiconductors for computers and solar cells (Figure 6.22). Indium is very soft; its uses include transistors and mirror coatings. Thallium and its compounds are toxic.

Figure 6.22 *A gallium arsenide electrical circuit chip is shown here. Gallium is used in manufacturing semiconductors for use in computers, solar cells, and other electrical applications.*

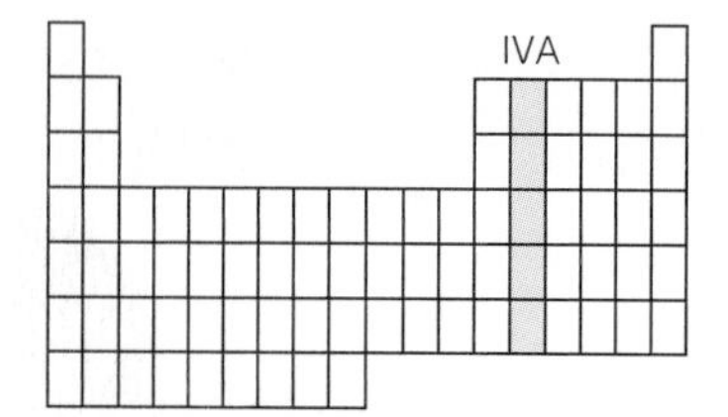

Figure 6.23 *Group IVA, the carbon family.*

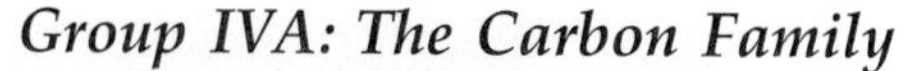

Group IVA: The Carbon Family

Metallic character increases from top to bottom for the Group IVA elements (Figure 6.23) carbon (C), silicon (Si), germanium (Ge), tin (Sn), and lead (Pb). Differences in the crystalline arrangements of carbon atoms account for the hardness of diamond and the slipperiness of black graphite (Figure 6.24). Different forms of the same element, such as these, are called **allotropes**. A new allotropic form of carbon with 60 atoms in a pattern like the surface of a soccer ball was discovered in the mid-1980s. These C_{60} spheres are often called buckyballs. Charcoal is a noncrystalline—or perhaps microcrystalline—allotropic form of carbon; it has no definite atomic pattern. Besides the two oxides of carbon, carbon dioxide (CO_2) and carbon monoxide (CO), carbon is present in over 8 million compounds. Among the organic (carbon-containing) compounds are the natural substances present in all living things. All petroleum products and synthetics from plastics to fibers to medicines are also organic compounds.

Figure 6.24 *Diamond (lower left) and graphite (upper right) are two allotropic forms of carbon.*

Silicon, the second member of this group, is a metalloid with predominantly nonmetallic properties. It is the second most abundant element in the Earth's crust (26%), but is not found as a free element. Quartz sand, silicon dioxide, is used in the production of glass and cement. Silicon has a gray, metallic luster. This metalloid has had an enormous impact on modern technology—extremely pure silicon is used in the manufacture of semiconductors and computer chips (Figure 6.25). Carborundum is silicon carbide, a compound of silicon and carbon used in cutting and grinding tools (Figure 6.26). Germanium is also a metalloid semiconductor and is involved in thousands of electronic applications.

Figure 6.25 *Silicon is used on the photosensitive drum of photocopiers, in solar cells, and for a wide variety of semiconductor applications. The Xerox copier shown here produces up to 135 prints per minute from electronically stored originals.*

Figure 6.26 *Common grinding wheels and grinding stones are made of Carborundum, the trade name for a product containing a compound of silicon and carbon called silicon carbide, SiC.*

Tin and lead, the last two elements in Group IVA, are typical metals. A major use of tin is in the manufacture of "tin" cans. These are actually made of steel covered with a thin protective coating of tin, which is much less reactive than iron. Some tin is also used in such alloys as bronze and in solder for joining metal parts. Lead is used extensively for electrodes in typical automobile lead storage batteries, for plumbing, and as a radiation shield. Certain white and yellow compounds of lead are used as pigments in some exterior house paints. Environmental concern with lead poisoning has led to restrictions regarding its use.

CHEMISTRY IN OUR WORLD

"Buckyballs"

Diamond and graphite have been recognized as two allotropes of carbon for more than 200 years, but in 1985 a new allotrope of carbon was discovered by a team of scientists at Rice University in Houston, Texas, who were trying to simulate the conditions found near giant red stars. Using a mass spectrometer, they frequently detected molecules composed of 60 carbon atoms. Proposing a geometric structure for a molecule with a C_{60} structure was a challenge.

The most stable C_{60} model that works has the 60 carbon atoms placed in a spherically shaped geodesic dome arrangement, like the surface of a soccer ball, which has 60 vertices that come together to form faces made up of 12 pentagons and 20 hexagons. Because the C_{60} arrangements of carbon atoms resemble the geodesic dome structures designed by the American architect, R. Buckminster Fuller, they were first named buckminsterfullerenes, then fullerenes, or "buckyballs" for short. The buckyball structure for C_{60} was confirmed in 1991. In 1994 researchers reported using vaporized buckyballs to deposit super-tough diamond films on machine tools. Electronics applications are now being investigated.

Researchers have produced a variety of fullerenes by attaching atoms or groups of atoms to the fullerene surface or with an atom caged inside the fullerene. For example, attaching fluorine atoms to each carbon produces a structure called a "fuzzball." Unlike diamond and graphite, fullerenes can be dissolved in various solvents. In the years ahead, we can anticipate more amazing developments in this new area of chemistry.

With 60 carbon atoms in the soccer-ball type of arrangement, the buckminsterfullerene structure shown here is responsible for the allotropic form of carbon that was discovered in 1985. In 1994, new research indicated that buckyballs have been present in nature all along.

Group VA

Elements in Group VA (Figure 6.27) include the nonmetals nitrogen (N) and phosphorus (P), the metalloids arsenic (As) and antimony (Sb), and the heavy metal bismuth (Bi). Thus, there is a dramatic change in appearance and properties from top to bottom in this group.

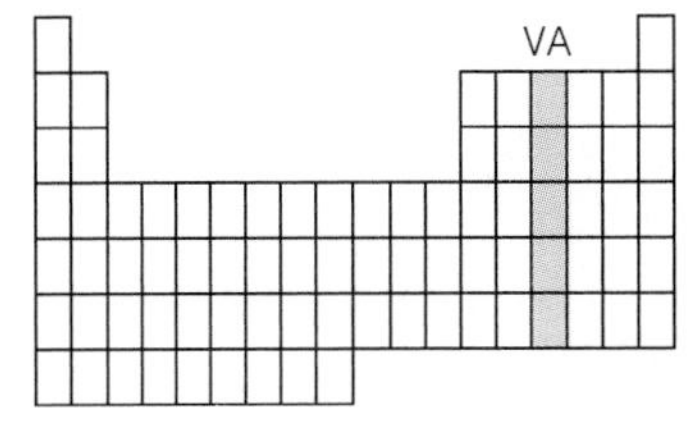

Figure 6.27 *Group VA.*

Diatomic nitrogen gas (N_2) makes up 78% of the air by volume. Both nitrogen and phosphorus are essential to life. Nitrogen is an essential element in all amino acids that make up all proteins. Nitrogen molecules in the air are not very reactive, but certain bacteria in soil can "fix" nitrogen by converting the element to ammonia that can then be taken up by the roots of plants. Industrially, nitrogen gas and hydrogen gas are combined to produce ammonia gas, NH_3, which is used as a fertilizer and also in the manufacture of nitric acid and various explosives.

Nitrogen Oxides and Pollution
At the high temperatures in automobile engines, small amounts of nitrogen and oxygen—from air taken into the engine—combine to produce various oxides of nitrogen. These compounds contribute to air pollution, especially in metropolitan areas.

Phosphorus is a reactive solid that is not found free in nature. One allotropic form of phosphorus is a purplish red, noncrystalline form that was once used in matches. Another allotropic form, with the formula P_4, has a yellowish white, crystalline, waxy appearance (Figure 6.28). It must be kept under water to prevent spontaneous combustion with the oxygen in the air. Phosphorus is used in the manufacture of matches, smoke bombs, tracer bullets, pesticides, and many other products. Phosphoric acid, H_3PO_4, is used in several soft drinks and to prepare

Figure 6.28 *There are two allotropic forms of phosphorus. Red phosphorus is a noncrystalline powder. White phosphorus is a crystalline, waxy solid that is stored under water because it spontaneously ignites when exposed to the air.*

other chemicals. Certain minerals containing phosphorus are very important as fertilizers. This element is essential to all plant and animal cells.

Arsenic is a metalloid with predominantly nonmetallic properties. Both the element and its compounds are toxic, partly because it can almost mimic the chemical behavior of phosphorus, but arsenic is unable to function as phosphorus in living tissue, with lethal results. Some agricultural insecticides and fungicides contain arsenic. The element is also used for semiconductor applications and in lasers.

Antimony is a metalloid that has predominantly metallic properties. The element is brittle and flaky with a metallic luster. It is used to increase the hardness of lead for automobile batteries, in coverings for cables, and in tracer bullets. Certain compounds of antimony are used in paint pigments, in ceramic enamels, and as flameproofing agents.

Bismuth is the only true metal in this group. It is used in making alloys such as pewter and low-melting alloys used in electric fuses and fire sprinkler systems. Certain bismuth compounds are used in face powders and cosmetics.

Group VIA

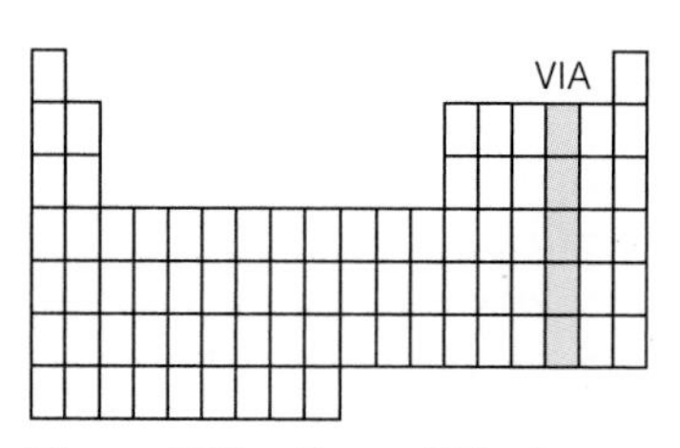

Figure 6.29 *Group VIA, the oxygen family.*

Group VIA elements, called the oxygen family (Figure 6.29), include oxygen (O), sulfur (S), selenium (Se), tellurium (Te), and polonium (Po). Although they all have six valence electrons, properties vary from nonmetallic to somewhat metallic as atomic number increases.

Oxygen gas, O_2, is essential to life; it is required to burn fossil fuels for energy and is required during human metabolism to burn carbohydrates. In both processes, carbon dioxide and water are by-products. Oxygen accounts for 21% by volume of the air and 49.5% by weight of the Earth's crust.

The other allotropic form of oxygen is ozone, with the formula O_3. It is more reactive than ordinary oxygen and can be formed from oxygen in an electric arc, such as the spark gap of an electric motor. Ozone can also be produced by the action of ultraviolet light on oxygen; it accounts for the "fresh" aroma of the air during thunderstorms.

Sulfur is the second nonmetal in Group VIA. At room temperature it is a pale yellow solid that occurs free in nature. It was known to the ancients and is referred to in the book of Genesis as brimstone. Sulfur molecules contain eight sulfur atoms connected in a ring; the formula is S_8. Sulfur is mined in Texas and Louisiana along the Gulf Coast. It is especially important in the manufacture of rubber tires and sulfuric acid, H_2SO_4. Other sulfur compounds are important in bleaching fruit and grain.

Selenium is a nonmetal that has interesting properties and uses. This element's conductivity of electricity increases with light intensity. Because of this photoconductivity, selenium has been used in light meters for cameras and in photocopiers, but concern about toxicity has led to diminished usage. Selenium can also convert alternating electric current to direct current; it has been used in rectifiers, such as the converters used with portable radios, tape recorders, and electric tools that can be recharged. The red color that selenium imparts to glass makes it useful in lenses for signal lights.

Tellurium looks metallic, but it is a metalloid with predominantly nonmetallic properties. It is used in semiconductors and to harden lead battery plates and cast iron. It occurs naturally in several compounds, but is not abundant. Polonium is a rare, radioactive element that emits alpha and gamma radiation; it is very dangerous to handle. Uses of the element are related to its radioactivity. Polonium was discovered by Marie Curie, who named it after her native Poland.

Group VIIA: The Halogens

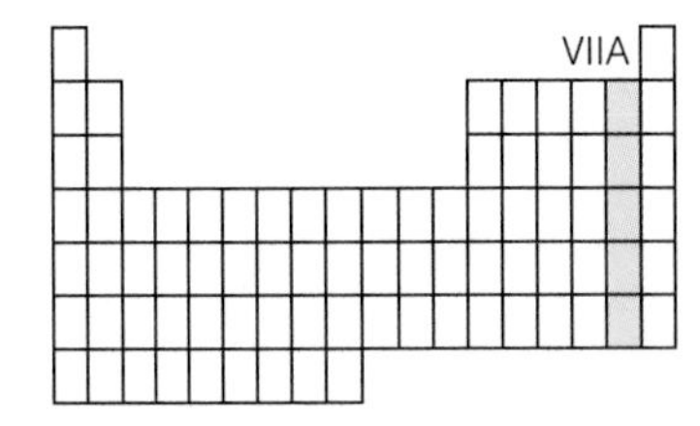

Figure 6.30 *Group VIIA, the halogens.*

Group VIIA, the halogen family (Figure 6.30), includes fluorine (F), chlorine (Cl), bromine (Br), iodine (I), and astatine (At). The family name, halogen, comes from Greek words meaning ''salt former.'' Each halogen atom has seven valence electrons. As elements, the halogens are all diatomic; they have two atoms per molecule. Halogens are too reactive to be found free in nature.

The first halogen, fluorine, is a pale yellow gas that is the most nonmetallic of all elements. It has a strong tendency to gain an electron to form fluoride ions, F^-. Both wood and rubber ignite spontaneously in fluorine gas. Fluorine is used to produce compounds with carbon called **fluorocarbons**, such as Freon-12, CCl_2F_2, which is used as a refrigerant in air conditioners. Teflon is a fluorocarbon that is a polymer; it has molecular units of two carbon atoms and four fluorine atoms that are repeated thousands of times in long chains. Fluorine compounds are also used in preventing tooth decay and in some lubricants.

Chlorine is a greenish yellow gas with an irritating odor that reacts with nearly all elements. In high concentrations it is poisonous; in low concentrations it can save lives—it is used to purify drinking water. Chlorine is used in the production of paper, textiles, bleaches, medicines, insecticides, paints, plastics, and many other consumer products.

Bromine is the only nonmetallic element that is a liquid at room temperature. This reactive blood-red liquid with a red vapor is both pungent and poisonous; it must be handled with great care. The element is obtained primarily by processing brine from wells in Arkansas and Michigan. Bromine also can be obtained from seawater, but this is no longer a major source of the element. Bromine is used in the production of photographic chemicals, dyes, and fire retardants and in the manufacture of a wide variety of other chemicals including pharmaceutical products (Figure 6.31).

Figure 6.31 *Iodine is a gray crystalline solid that sublimes to give a violet vapor. Bromine is a brownish red liquid at room temperature; its vapor is also red. Chlorine is a greenish yellow gas.*

Iodine is a steel-gray crystalline solid at room temperature. When heated, the iodine solid **sublimes**—it is transformed directly from the solid state to a vapor without passing through the liquid state. The iodine vapor has a beautiful bright violet color. Iodine, which is less abundant than the other halogens, is obtained from brine wells in oil fields in California and Louisiana. The element is also present in sea plants, such as kelp. Iodine compounds are used in photographic chemicals as well as in certain medicines. The human body requires traces of iodine to make the hormone thyroxine.

All isotopes of astatine are radioactive. The total amount of the element believed to exist in the Earth's crust is less than 30 g (1 oz). Traces of this unstable element were first synthesized at the University of California, Berkeley, in 1940.

Group VIIIA: The Noble Gases

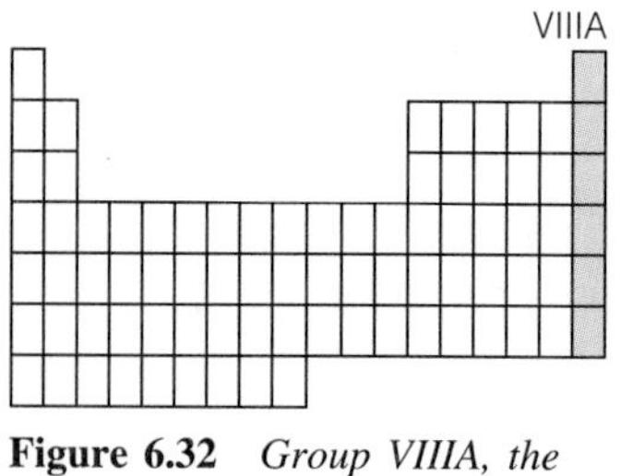

Figure 6.32 *Group VIIIA, the noble gases.*

Elements in Group VIIIA at the far right of the periodic table are known as the noble gases (Figure 6.32). This family includes helium (He), neon (Ne), argon (Ar), krypton (Kr), xenon (Xe), and radon (Rn). Noble gases all exist as monatomic (single) gaseous atoms that do not tend to enter into reactions with other elements.

All noble gases have a completely filled outer energy level of electrons (two for He and eight for the others). This stable electron arrangement accounts for the nonreactive nature of these elements. About 1% of Earth's atmosphere is argon; the other noble gases are present in only trace amounts. Except for helium, which is extracted from natural gas wells, these elements are separated from liquefied air.

During the 1890s, the Scottish chemist Sir William Ramsay and his associates discovered the existence of each of these elements except helium and radon. When Janssen, an astronomer, was using a spectroscope to study an eclipse of the sun in 1868, he observed a new line in the spectrum. It was concluded that the sun contained an undiscovered element that was later named helium, from the Greek *helios*, meaning ''the sun.'' The first discovery of helium's presence on Earth took place in 1895, when Ramsay found that helium gas was produced by a sample of uranium ore. Radon is a radioactive gas that was discovered in 1900 by Friedrich Dorn, a physicist, who found that this element was produced during the radioactive decay of the element radium.

Radioactive Radon
Where radon levels in the soil are high, the gas can enter homes through cracks in basement floors to cause high indoor radon levels. This increases the risk of lung cancer, especially for smokers.

Because of its low density and nonflammable nature, helium is used to fill balloons and dirigibles (blimps) and to pressurize liquid fuel in Saturn booster rockets. The distinctive property of the noble gases as a group is their ''inertness.'' For example, helium and argon are used in arc welding and metallurgical processes to prevent reactions of materials with oxygen and nitrogen in the air. Light bulbs and fluorescent tubes are filled with a mixture of argon and nitrogen as an inert atmosphere to prolong the filament life. Krypton is more expensive, but it is used to increase the efficiency and brightness in some flashlight bulbs and electronic flash attachments used in photography (Figure 6.33). The bright orange-red light of neon signs is produced when an electric current is passed through a tube of neon gas at low pressure. The nonreactive nature of noble gases makes them valuable.

Noble Gas Compounds
The first noble gas compound was prepared in 1962. Since then, researchers have prepared certain compounds of fluorine with xenon, krypton, and radon, but most of these are easily decomposed.

EXAMPLE 6.8

Use the periodic table as you pick two elements from each set with properties that are the most similar.

(a) oxygen, argon, silicon, magnesium, potassium, nitrogen, and carbon
(b) helium, chlorine, sodium, aluminum, beryllium, fluorine, and sulfur

SOLUTION

(a) Silicon and carbon have the most similar properties. Both are in Group IVA; they have four valence electrons and similar compounds.
(b) Chlorine and fluorine have the most similar properties. They are both nonmetallic, diatomic halogens with seven valence electrons and similar reactivities.

Figure 6.33 *A krypton-filled superbright flashlight bulb is about 60% brighter than a regular flashlight bulb.*

6.8 Transition Metals

The transition metals are located in the central region of the periodic table (Figure 6.34) and are readily identified by a Roman numeral followed by the letter ''B'' on many tables. Be aware, however, that some periodic tables use a different labeling system where the first transition metal groups are labeled as ''A'' groups and the last two transition metal groups are labeled as ''B'' groups. Still other tables use no ''A'' and ''B'' designations at all.

In general, the properties of the transition metals are rather similar. These metals are more brittle, they are harder, and they have higher melting points and boiling points than do other metals. The densities, melting points, and boiling points of transition metals first increase and then decrease within each period as atomic number increases. This trend is the most dramatic for transition metals in the sixth period. Transition metals are much less reactive than the alkali metals and alkaline earths. Although alkali metals like sodium and potassium are never found free in nature, relatively pure samples of several transition metals such as gold, silver, iron, and manganese have been recovered.

Transition metals can lose two valence electrons from an outermost *s* sublevel plus *d* electrons held loosely in the next lower energy level. Thus, a particular transition metal can lose a variable number of electrons to form positive ions with different charges. For example, iron can form the Fe^{2+} ion or the Fe^{3+} ion. We say that iron has *oxidation numbers* of +2 and +3. Many transition metal compounds are brightly colored due to varying numbers of unpaired electrons.

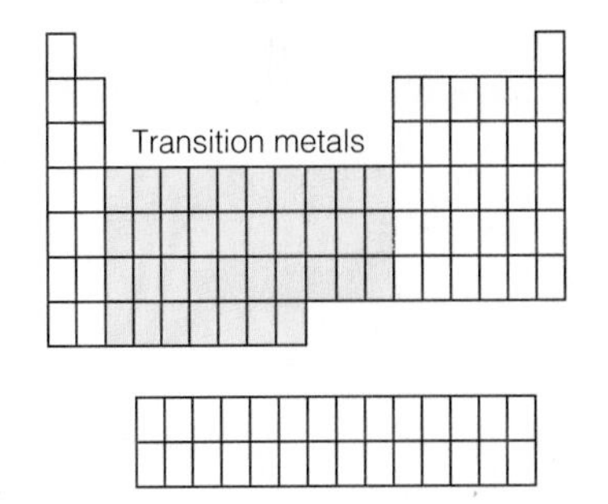

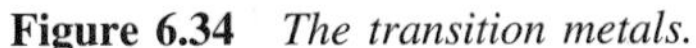

Figure 6.34 *The transition metals.*

Notice the placement of copper (Cu), silver (Ag), and gold (Au) in the periodic table. These metals are often called the coinage metals. All three are good conductors of heat and electricity. Copper has a distinctive reddish color that slowly darkens as it reacts with oxygen and sulfur compounds in the air. Copper is widely used in electrical applications, coinage, water pipes, and familiar alloys such as brass, bronze, and sterling silver.

Silver, with a brilliant metallic luster, is the best conductor of both heat and electricity. It is used in coinage, jewelry, electrical contacts, printed circuits, mirrors, batteries, and photographic chemicals. Gold is the most malleable and ductile metal. It is soft, but usually contains small amounts of other metals to make alloys that have greater strength. Gold is resistant to reaction with the air and most chemicals.

Other familiar transition metals include chromium, iron, cobalt, nickel, and zinc

Figure 6.35 *Structural steel shown here is made from iron that contains 0.3 to 0.7% carbon. Steel alloys for tools also contain small amounts of other transition metals such as tungsten, molybdenum, manganese, cobalt, and chromium.*

from the fourth period of the periodic table. These metals are used extensively in various tools and related applications. Iron is the fourth most abundant element and is the least expensive metal. Iron alloys, called *steel,* contain small amounts of metals such as chromium, manganese, and nickel for strength, hardness, and durability (Figure 6.35). Iron that is coated with a thin protective layer of zinc is said to be *galvanized.* About one-third of all the zinc produced is used in galvanizing wire, nails, and sheet metal. Zinc is also important in the production of brass, dry cell batteries, and die castings for automotive and hardware items.

EXAMPLE 6.9

Describe the trends in densities, melting points, and boiling points for transition metals in periods 4 and 5.

SOLUTION

Densities, melting points, and boiling points of transition metals all increase and then decrease within a period.

See Problems 6.47–6.62.

6.9 Inner Transition Metals

The two rows of elements at the bottom of the periodic table are known as **inner transition metals** (Figure 6.36). Locate lanthanum with atomic number 57 in the periodic table. The series of elements following lanthanum—the elements with atomic numbers 58 through 71—are known as the **lanthanides**. These elements have two outer electrons in the 6*s* sublevel plus additional electrons in the 4*f* sublevel. Similarly, the series of elements following actinium—the elements with atomic numbers 90 through 103—are known as the **actinides**. The actinides have two outer electrons in the 7*s* sublevel plus additional electrons in the 5*f* sublevel. In the past, inner transition metals were often called the ''rare earth elements,'' but this was not a good classification, since most are not as rare as some other elements; they are, however, quite difficult to separate.

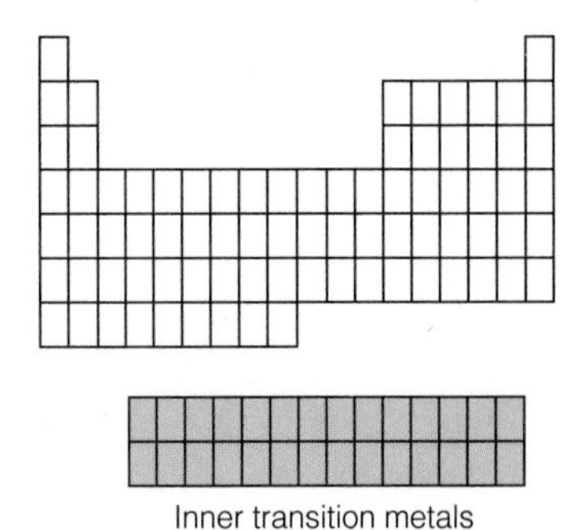

Figure 6.36 *The inner transition metals.*

The lanthanides and actinides have partially filled *f* sublevels. They have properties that are so similar to each other that it is difficult to separate them chemically, but newer methods have cut purification costs. These metals, unlike the transition metals, are soft and malleable. They are used in cigarette lighter flints, carbon arc lamps, lasers, glass coloring agents, and compounds that produce the intense red color needed for television picture tubes.

Transuranium Elements

Uranium, atomic number 92, is in the actinide series and has more protons than any other naturally occurring element. In 1940 a new element—with 93 protons—was synthesized at the University of California, Berkeley. This element, named neptunium, is the first member of the synthetic elements with atomic numbers greater than 92. These elements are known as the **transuranium elements**, all of which are radioactive. Plutonium was also synthesized in 1940; it is now produced as a by-product in nuclear reactors. Sixteen transuranium elements have now been produced; some are quite stable, while others readily undergo radioactive decay. Names for elements 95–103 were derived from places and important scientists. Elements with numbers 95, 97, and 98 were named for America (Figure 6.37), Berkeley, and California, respectively. Elements with numbers 96, 99, 100, 101, 102, and 103 were named, respectively, for the Curies, Albert Einstein, Enrico Fermi, Mendeleev, Alfred Nobel, and Ernest Lawrence (inventor of the cyclotron). In 1994 it was formally proposed that element 106 be named seaborgium (Sg) to honor Glenn T. Seaborg for his work with transuranium elements.

The synthesis of transuranium elements has added to our understanding of the nucleus and of nuclear processes. Americium is one transuranium element with commercial applications; very small amounts of it are used in smoke detectors.

Equipment required to produce new transuranium elements has become more complex, but there is no reason to doubt that additional elements will be synthesized or that new uses will be found for natural and synthetic elements.

Figure 6.37 *Many smoke detectors use small amounts of americium, a synthetic transuranium element, to alert people to the presence of smoke particles in the air.*

EXAMPLE 6.10

In what way are the electronic structures for the lanthanides and the actinides similar?

SOLUTION

Both series of elements have two valence electrons in an *s* sublevel along with additional electrons in an *f* sublevel that is ''buried'' two energy levels below the outer energy level. As a result their properties are similar?

See Problems 6.63 and 6.64.

Every material that exists anywhere in the universe is made of one or more of the 109 elements.

Chapter Summary

Early attempts to organize the elements include triads of elements described by Dobereiner and the law of octaves described by Newlands, but the most successful arrangement of elements was published by Mendeleev in 1869. His periodic law stated that the physical and chemical properties of the elements vary periodically with increasing atomic mass, but atomic number is the more appropriate term.

There are seven horizontal rows of elements in the periodic table; they are called periods. The variation in chemical and physical properties in one period roughly parallels the variation in properties for other periods, beginning with shiny, reactive metals at the left of the periodic table, followed by dull solids, reactive nonmetals, and finally a noble gas. Within vertical columns, called groups or families of elements, all atoms have the same number of valence electrons and enter into similar chemical reactions.

Atomic size within a period of elements tends to decrease as atomic number increases, but within a group, atomic size increases as atomic number increases. Metal ions are smaller than metal atoms, but ions of nonmetals are larger than atoms of nonmetallic elements. For two isoelectronic ions, the one with the greater number of protons is smaller. Ionization energy—the energy required to remove an electron—increases across each period as atomic number increases and is the lowest for metals in the lower left corner of the periodic table.

For the alkali metals, melting points and boiling points decrease as atomic number increases. For the halogens, melting points and boiling points increase as atomic number increases. Except for carbon (diamond), which has a higher melting point than any other element, tungsten and other metals clustered in the lower region of the transition metal group have the highest melting and boiling points.

Densities of elements in groups increase, but within a period, densities first increase, then decrease. The most dense elements are the transition metals in period 6. Three metals—Ag, Cu, and Au—in the same transition metal group are the best conductors of both heat and electricity.

While hydrogen has a single valence electron like the alkali metals, it also lacks one electron of having a completely filled outer energy level like the halogens. The various unique properties of hydrogen suggest that this element actually belongs in a group by itself. The alkali metals are very reactive metals, but the halogens are very reactive nonmetals.

The inner transition elements at the bottom of the periodic table include the lanthanides and the actinides. Elements with atomic numbers greater than 92 are called transuranium elements; they are all synthetic.

Although the unique characteristics of each element are important, the periodic table allows us to identify numerous periodic trends in both physical and chemical properties of the elements.

Assess Your Understanding

1. Describe the contributions of Newlands, Dobereiner, and Mendeleev. [6.1]
2. Describe trends in appearance within a period of elements. [6.2]
3. Identify all periods and groups shown on the periodic table. [6.2]
4. Compare sizes of atoms and ions within families of elements. [6.3]
5. Describe trends in ionization energies within groups and periods. [6.4]
6. Compare melting point and boiling point trends for the alkali metals and the halogens. [6.5]
7. Compare trends in densities within groups and periods. [6.6]
8. List general properties and some specific uses of common elements within each group. [6.7]

Key Terms

actinides [6.9]
allotropes [6.7]
anion [6.3]
cation [6.3]
chemical families [6.2]
first ionization energy [6.4]
fluorocarbons [6.7]
groups of elements [6.2]
inner transition elements [6.9]
ion [6.3]
ionization energy [6.4]
isoelectronic [6.3]
lanthanides [6.9]
periodic law [6.1]
periodic table [6.1]
periods of elements [6.2]
sublime [6.7]
transuranium elements [6.9]

Problems

DISCOVERIES OF PERIODICITY

6.1 What did Dobereiner notice about the atomic masses of Ca, Sr, and Ba?

6.2 Use the atomic masses of calcium (40.1) and barium (137.3) to approximate the atomic mass of strontium.

6.3 Why did Mendeleev leave gaps in his periodic table?

6.4 Why were the discoveries of gallium and germanium important to Mendeleev?

6.5 Although Meyer and Mendeleev independently developed periodic tables, give two reasons why Mendeleev is credited with this discovery.

6.6 State the periodic law as described by Mendeleev. What exceptions to this periodic law did Mendeleev observe?

THE PERIODIC TABLE TODAY

6.7 Why do we now state the periodic law in terms of atomic number rather than atomic mass, as described by Mendeleev?

6.8 What does atomic number mean? Who was first able to compare the nuclear charge for different elements? Approximately when was this discovery made?

6.9 What is a period of elements? How many periods of elements are there?

6.10 Compare the terms ''family of elements'' and ''groups of elements.''

6.11 Describe the variation in appearance of elements from left to right for period 2 and period 3.

6.12 Compare the variation in valence electrons for elements in periods 2 and 3.

6.13 Compare the number of valence electrons for each element in Group IIA.

6.14 Compare the number of valence electrons for each element in Group VIA.

6.15 Give the group number and the number of valence electrons for each of the following pairs of elements.
a. Li and K b. Cl and I
c. N and P d. Al and B

6.16 Give the group number and the number of valence electrons for each of the following pairs of elements.
a. F and Br b. Mg and Ca
c. C and Si d. He and Ar

6.17 Give group numbers for the alkali metals, the alkaline earths, and the halogens.

6.18 What is consistent about "B" group labeling used in this text and preferred by many chemistry educators? Are all periodic tables labeled this way? Explain.

6.19 How many elements are in period 1 and period 2?

6.20 How many elements are in period 3 and period 4?

ATOMIC AND IONIC SIZE

6.21 What group of elements on the periodic table has the greatest atomic size?

6.22 What group of elements on the periodic table has the smallest atomic size?

6.23 What is the trend in size of atoms from left to right for period 2? Explain.

6.24 What is the trend in size of atoms in Group II (with increasing atomic number)? Explain.

6.25 Compare the sizes of
a. a sodium atom and a sodium ion, Na^+.
b. a chlorine atom and a chloride ion, Cl^-.

6.26 Compare the sizes of
a. a potassium atom and a potassium ion, K^+.
b. a bromine atom and a bromide ion, Br^-.

6.27 Compare sizes of a fluoride ion, F^-, and a sodium ion, Na^+. Are these ions isoelectronic?

6.28 Compare sizes of a potassium ion, K^+, and a chloride ion, Cl^-. Are these isoelectronic?

IONIZATION ENERGY

6.29 What does "ionization energy" mean? Is this a periodic property?

6.30 What specifically does "first ionization energy" mean?

6.31 What is the general trend in first ionization energies for elements in period 2 and period 3?

6.32 Which group of elements has the lowest ionization energies? Do elements in this group normally tend to lose or to gain electrons?

6.33 Discuss the trend in first ionization energies for the alkali metals.

6.34 If we exclude the noble gases, which group of elements has the highest ionization energies? Do these elements tend to lose or to gain electrons?

6.35 Predict which element in each pair would be expected to have a smaller first ionization energy. Which element in each pair has the greater tendency to form a positive ion? Explain your reasoning.
a. Na or Cs b. Na or Si c. Si or Cl

6.36 Predict which element in each pair would be expected to have a smaller first ionization energy. Which element in each pair has the greater tendency to form a positive ion? Explain your reasoning.
a. Mg or S b. F or Li c. Ba or Mg

6.37 What family of elements has the lowest first ionization energies?

6.38 What family of elements has the highest first ionization energies?

MELTING POINT AND BOILING POINT TRENDS

6.39 Compare the trend in melting points for the alkali metals and the halogens.

6.40 Compare the trend in boiling points for the alkali metals and the halogens.

6.41 Which element in the second period has the highest melting point? (See Table 6.3.) What does this suggest about the attractions between these atoms?

6.42 Predict which element in each pair of metals can be expected to have a higher melting point.
a. W or Fe b. W or Pb c. Cr or K

DENSITY AND CONDUCTIVITY TRENDS

6.43 What is the trend in densities for alkali metals and for the halogens?

6.44 What is the trend in densities for period 5 elements?

6.45 Predict which element in each pair can be expected to have a greater density.
a. Mg or Al b. Au or Pb c. Ni or Pt

6.46 List the three metals that are the best conductors of heat and electricity. What other common metal is a good conductor?

A SURVEY OF ELEMENTS BY GROUPS

6.47 Give reasons why hydrogen could be placed in either Group IA or Group VIIA. Give reasons why neither group is exactly right for hydrogen.

6.48 Describe the expected reactivity of sodium and potassium in water and in mineral oil.
6.49 Of the alkaline earth metals, which one is the most common? List some common minerals that contain this element.
6.50 Which two ions from Group IIA are common in hard water and are responsible for white crystalline deposits around water faucets?
6.51 What element can be extracted from borax? Give two uses of the element or the mineral.
6.52 What element can be extracted from bauxite? Why is the extraction expensive?
6.53 List three allotropic forms of carbon.
6.54 Explain why a "tin can" is not really a tin can.
6.55 Describe the reactivity of phosphorus in water and in oxygen.
6.56 Give formulas of two allotropes of oxygen. Identify the one called ozone.
6.57 What is a fluorocarbon? Name a common fluorocarbon and describe its use.
6.58 Give the only nonmetallic liquid element at room temperature. Give some uses.
6.59 Iodine sublimes when heated. What does this mean?
6.60 Describe the discovery of helium.
6.61 How are iron and steel different?
6.62 What element is used in galvanizing? Why is this done?

INNER TRANSITION ELEMENTS

6.63 Distinguish between the lanthanides and the actinides.
6.64 What are the transuranium elements? What is special about these elements?

Additional Problems

6.65 From what you know about atomic and ionic sizes, compare the sizes of an oxygen atom and an oxide ion, O^{2-}.
6.66 Compare the sizes of a bromide ion, Br^-, and a rubidium ion, Rb^+. Are they isoelectronic?
6.67 A magnesium atom has two valence electrons. Do you expect the first ionization energy or the second ionization energy to be larger? Explain your reasoning.
6.68 Predict which element in each pair would be expected to have the lower first ionization energy. Which element in each pair has the greater tendency to form a positive ion? Explain your reasoning.
a. Cl or Cs b. Br or Zn c. C or Sn
6.69 Predict which element in each pair can be expected to have a greater density.
a. Au or Cu b. K or Fe c. W or Cr
6.70 What is the element present in both quartz sand and carborundum? List two important uses of this element.
6.71 Discuss both detrimental and beneficial aspects of ozone.
6.72 If an element with an atomic number of 118 is ever synthesized, what are some properties it can be expected to have? How is the periodic table useful in this regard?
6.73 Describe two allotropic forms of phosphorus.
6.74 What are "buckyballs"?

The hand-to-wrist bonding of one person to another is somewhat like the bonding of one atom to another within a molecule to give a specific arrangement.

7

Chemical Bonds

CONTENTS

In this chapter we will compare characteristics of chemical bonds—the attractive forces that hold atoms or ions together to form molecules or crystals. The types of bonds that are present in a substance are largely responsible for the physical and chemical properties of the substance. Bonding is also responsible for the attraction one substance has for another. For example, salt dissolves in water much better than in oil because of differences in bonding. Certain substances that are dissolved in water can conduct an electric current, but others cannot. Ethyl alcohol evaporates more quickly than water. Wax melts at a low temperature, but salt has a high melting point. These properties of substances, and many more, are closely related to their chemical bonding.

As you study the types of chemical bonding described in this chapter, you may also find it helpful to study the naming of chemical compounds as described in Chapter 8; you may even be asked to master all of Chapter 8 before discussing bonding. Either approach can be successful; both topics are fundamental to your success in subsequent topics.

7.1 Ionic Bonds

Sodium metal is a soft, silvery, solid metal that can be cut with a knife. It rapidly reacts with oxygen and water vapor in the air. Chlorine, a reactive nonmetal, is a pale greenish yellow gas with an irritating odor. The element is often used as a disinfectant for city water supplies and swimming pools. Perhaps you have been swimming in a pool containing so much chlorine that you could smell or taste it. Chlorine is quite irritating to the respiratory tract.

Chlorine in Swimming Pools Solid compounds that react with water to release chlorine are often used in swimming pools instead of the Cl_2 gas dispensed from tanks.

When a warm piece of sodium metal is dropped into a flask containing chlorine gas, a vigorous reaction occurs, and a stable white solid is formed. It is the familiar compound—sodium chloride—used as table salt (Figure 7.1). Chemists use the following chemical equation to represent this reaction of sodium with chlorine (a diatomic gas, Cl_2) to form sodium chloride, NaCl.

$$2\,Na + Cl_2 \longrightarrow 2\,NaCl$$

The Reaction of Sodium and Chlorine: Theory

We have described facts about the reaction that can be observed, but this does not explain *why* the reaction took place. We must look at reaction theory to understand what was taking place during the reaction. By losing an electron, a reactive sodium atom forms a sodium ion, Na^+. The loss of the valence electron reveals a complete octet of electrons. The electron configuration of the ion is like that of the noble gas neon. During ionization, a metal atom is said to be **oxidized** as it *loses* electrons. The oxidation can be represented with Lewis electron-dot symbols.

Oxidation always involves a *loss* of electrons.

$$\underset{\text{Sodium atom}}{Na\cdot} + \text{Energy} \longrightarrow \underset{\text{Sodium ion}}{Na^+} + \underset{\text{One electron}}{e^-}$$

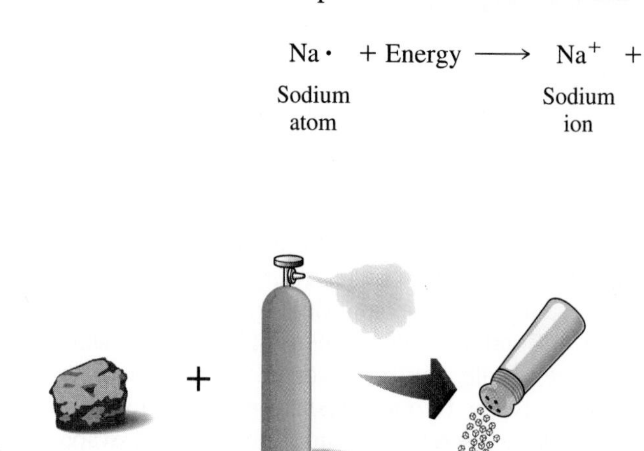

Figure 7.1 *Sodium, a soft silvery metal, reacts with chlorine, a greenish gas, to form sodium chloride (ordinary table salt).*

Chlorine atoms, on the other hand, tend to gain an electron to form chloride ions, Cl^-, with an octet of electrons like the nonreactive noble gas argon. During ionization, the nonmetal is said to be **reduced** when it *gains* an electron. Using Lewis symbols, this reduction can be represented as follows.

$$:\ddot{\underset{..}{Cl}}\cdot \ + \ e^- \longrightarrow \ :\ddot{\underset{..}{Cl}}:^- + \text{Energy}$$

Chlorine atom — One electron — Chloride ion

Notice that metal atoms *lose* electrons to form ions with completely empty outer energy levels. Nonmetal atoms *gain* electrons to form ions with completely filled outer energy levels. Only the valence electrons of an atom are involved in the electron transfer; the inner core of electrons is not changed. The tendency for nonmetal atoms to gain electrons until there are eight valence electrons is called the **octet rule**. (Because hydrogen atoms can accommodate only two electrons, they obey the *duet* rule instead of the octet rule.) Using Lewis electron-dot symbols, we can represent the ionic reaction of sodium with chlorine in terms of electron transfer between atoms as follows.

$$Na\cdot \ + \ \cdot\ddot{\underset{..}{Cl}}: \longrightarrow \ Na^+ \ + \ :\ddot{\underset{..}{Cl}}:^-$$

Sodium atom — Chlorine atom — Sodium ion — Chloride ion

The electron lost by a sodium atom is gained by a chlorine atom to produce a sodium ion, Na^+, and a chloride ion, Cl^-. While sodium atoms are oxidized, chlorine atoms are reduced. Oxidation and reduction always go together. Metals are oxidized by nonmetals; nonmetals are reduced by metals.

Study Hint
Remember that it is essential to say "ion" when referring to a charged particle such as a sodium ion. *Atoms* are neutral; they have an equal number of electrons and protons. *Ions* are not neutral; they carry a positive or negative charge. Sodium *ions* and sodium *atoms* are not the same.

Metals tend to lose their valence electrons to form positive ions (cations).

Nonmetals tend to gain electrons to form negative ions (anions).

When electrons are transferred, stable ions with an octet of electrons are formed.

When sodium metal reacts with chlorine gas, a sodium atom *transfers* an electron to a chlorine atom to form a sodium ion and a chloride ion. The sodium chloride that is formed is an **ionic compound.** The sodium ion, Na^+, and the chloride ion, Cl^-, not only have stable electronic structures like noble gases, but they also have opposite charges. These opposite charges attract. The attractive force between oppositely charged ions is called an **ionic bond.** Notice that the complete transfer of electrons is required to produce ionic bonds.

Actually, the greenish yellow chlorine gas is composed of chlorine molecules; each molecule consists of two atoms.

Ionic bonds are formed by the complete transfer of electrons.

In even a small amount of salt there are billions and billions of positive and negative ions. These ions arrange themselves in an orderly array (Figure 7.2). These arrangements are repeated in all directions—above and below, in front and behind, to the left and right—to make a three-dimensional cube-shaped **crystal** of

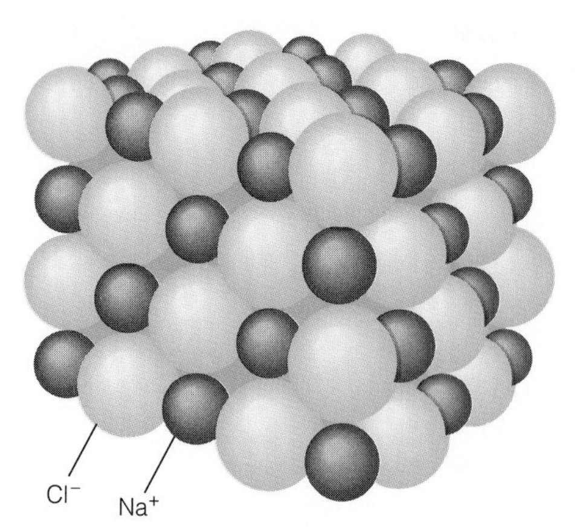

Figure 7.2 *The arrangement of ions in a sodium chloride crystal.*

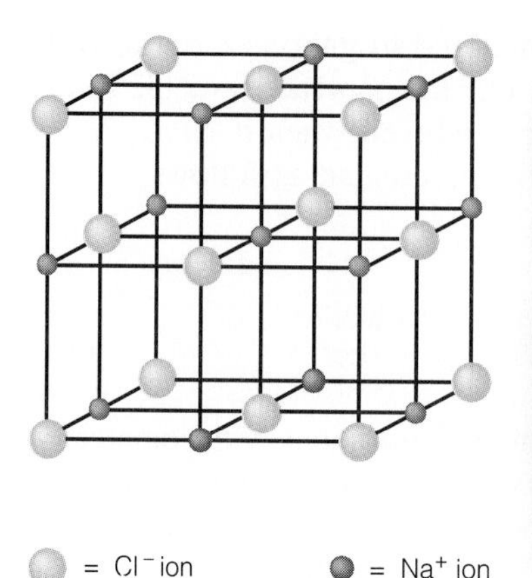

Figure 7.3 *Ball-and-stick model of a sodium chloride crystal. The diameter of a chloride ion is about twice that of a sodium ion.*

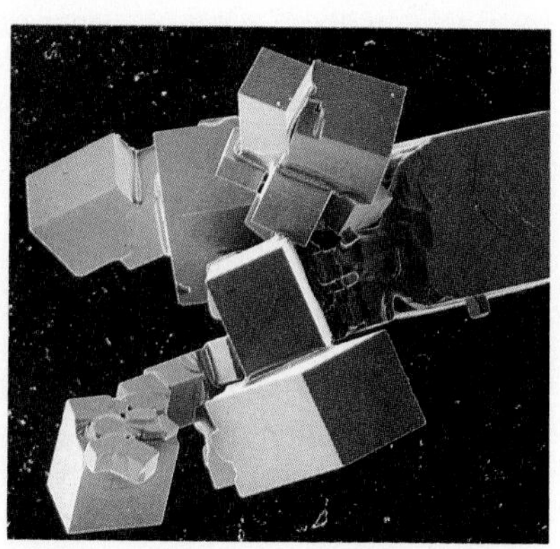

Figure 7.4 *Sodium chloride is a cubic-shaped crystalline solid that has a melting point of 808 °C.*

A CLOSER LOOK

Crystal Models

Scientists sometimes use different models to represent the same system. The model shown in Figure 7.2 is a space-filling model showing the relative sizes of the sodium and chloride ions. Sometimes a ball-and-stick model is used to show the geometry of the crystal (Figure 7.3). This model assists us in describing the cubic arrangement of the ions. Solid sodium chloride crystals themselves are cubic (Figure 7.4).

Each of the two models allows us to emphasize a different point; both are useful.

sodium chloride (Figure 7.3). Each sodium ion attracts (and is attracted by) the nearest six chloride ions. Similarly, each chloride ion attracts (and is attracted by) the surrounding six sodium ions. The attractive forces—the ionic bonds—hold the crystal together (Figure 7.4).

All chemical compounds are neutral in charge; their formulas do not show the charges of the individual ions.

With the ball-and-stick model (Figure 7.3), one can observe that there is one sodium ion for each chloride ion; the ratio of ions is 1 to 1. Thus, the simplest formula for the compound sodium chloride is written NaCl to represent this 1-to-1 ion ratio. There are many other ionic compounds besides sodium chloride. When metals react with nonmetals, they form compounds that are primarily ionic.

Other Ionic Compounds

Study Hint
Notice that the symbol of the metal ion is always placed before the symbol of the nonmetal ion in the chemical formula.

Potassium, K, another alkali metal, can react with chlorine, Cl, to produce the ionic compound called potassium chloride, KCl. This reaction involving electron transfer can be represented with Lewis electron-dot symbols.

$$\mathrm{K\cdot} + \mathrm{\cdot\ddot{\underset{..}{Cl}}:} \longrightarrow \mathrm{K^+} + \mathrm{:\ddot{\underset{..}{Cl}}:^-}$$

Potassium atom	Chlorine atom	Potassium ion	Chloride ion

The formula KCl does not show ion charges.

Potassium metal also reacts with bromine—a reddish brown liquid that is chemically similar to chlorine—to form a stable white crystalline solid called potassium bromide, KBr.

$$\mathrm{K\cdot + \cdot\ddot{\underset{\cdot\cdot}{Br}}: \longrightarrow K^+ + :\ddot{\underset{\cdot\cdot}{Br}}:^-}$$

Magnesium is a hard alkaline earth metal that is less reactive than potassium. When this metal reacts with chlorine gas, each magnesium atom gives up two valence electrons to form a Mg^{2+} ion. Each magnesium atom must, therefore, react with two chlorine atoms to produce the stable white crystalline solid called magnesium chloride, with the formula $MgCl_2$.

$$\mathrm{\cdot Mg\cdot + \begin{matrix}\cdot\ddot{\underset{\cdot\cdot}{Cl}}:\\ \cdot\ddot{\underset{\cdot\cdot}{Cl}}:\end{matrix} \longrightarrow Mg^{2+} + \begin{matrix}:\ddot{\underset{\cdot\cdot}{Cl}}:^-\\ :\ddot{\underset{\cdot\cdot}{Cl}}:^-\end{matrix}}$$

Magnesium atoms lose electrons while being oxidized to form magnesium ions. Chlorine, the nonmetal, gains electrons while being reduced to chloride ions. The total number of electrons lost by magnesium equals the total number of electrons gained by chlorine in the formation of magnesium chloride.

During the reaction of magnesium and nitrogen, each of three magnesium atoms loses two valence electrons (a total of six) while two nitrogen atoms gain a total of six electrons to make two nitride ions. The formula of the product, magnesium nitride, is written Mg_3N_2 to show the proper ratio of magnesium and nitrogen. In Lewis electron-dot symbols we can write

Chemical formula writing is described in detail in Chapter 8.

$$\mathrm{\begin{matrix}\cdot Mg\cdot\\ \cdot Mg\cdot\\ \cdot Mg\cdot\end{matrix} + \begin{matrix}\cdot\dot{\underset{\cdot}{N}}:\\ \cdot\underset{\cdot}{\dot{N}}:\end{matrix} \longrightarrow \begin{matrix}Mg^{2+}\\ Mg^{2+}\\ Mg^{2+}\end{matrix} + \begin{matrix}:\ddot{\underset{\cdot\cdot}{N}}:^{3-}\\ :\ddot{\underset{\cdot\cdot}{N}}:^{3-}\end{matrix}}$$

The formula of magnesium nitride is written Mg_3N_2.

Notice that the total positive and total negative charges on the ions produced are equal (6+ and 6−). All compounds are neutral in charge. During this ionic reaction, six electrons are transferred from magnesium to nitrogen; none are lost. Once again, metal atoms are oxidized to form positive ions, and nonmetal atoms gain electrons as they are reduced to form negative ions.

General Properties of Ionic Compounds

In general, metals on the left side of the periodic table react with nonmetallic elements on the right side of the periodic table (excluding the noble gases) to form stable, crystalline solids. The crystalline solids are held tightly together by oppositely charged ions; this is ionic bonding. The strong attractions within ionic solids are responsible for their high melting points—typically about 300 to 1000 °C.

All pure ionic compounds are solids at room temperature; none are liquids or gases.

Boiling points of ionic substances are very high—typically 1000 to 1500 °C.

Many ionic compounds are soluble in water. When they dissolve in water, they **dissociate**; that is, they break up into individual ions that move about freely. Ions

Electrolyte Balance
The *electrolyte balance* of sodium and potassium ions within the blood and other body fluids is essential to good health; a change in the electrolyte balance can lead to serious medical emergencies.

are held in solution by their attraction to water. The presence of dissociated ions allows a substance to conduct electricity. A substance that dissolves in water to give a solution that conducts electricity is called an **electrolyte**. (See Section 7.6 for additional information.)

EXAMPLE 7.1
Write Lewis electron-dot symbols to represent the reaction of calcium with chlorine that produces calcium chloride, $CaCl_2$. (Use the periodic table as a guide to valence electrons.)

SOLUTION

$$\cdot\text{Ca}\cdot + \begin{matrix}\cdot\ddot{\text{Cl}}: \\ \cdot\ddot{\text{Cl}}:\end{matrix} \longrightarrow \text{Ca}^{2+} + \begin{matrix}:\ddot{\text{Cl}}:^- \\ :\ddot{\text{Cl}}:^-\end{matrix}$$

The formula $CaCl_2$ does not show ion charges.

See Problems 7.1–7.8.

EXERCISE 7.1
Write Lewis electron-dot symbols to represent the reaction of calcium with oxygen to form calcium ions and oxide ions.

7.2 *Covalent Bonds*

You might expect a hydrogen atom, with its one electron, to acquire one more electron to make a stable electron structure like helium. Indeed, hydrogen atoms do just that in the presence of atoms of a very reactive metal, such as lithium, that readily gives up electrons. The negatively charged $H:^-$ ion produced is called a hydride ion. The compound LiH is named lithium hydride.

$$\text{Li}\cdot + \text{H}\cdot \longrightarrow \text{Li}^+ + \text{H}:^-$$

Metal hydrides are quite reactive.

If only hydrogen atoms are present, one hydrogen cannot grab an electron from another; all hydrogen atoms have an equal attraction for electrons. Instead, hydrogen atoms tend to acquire the desired noble gas configuration of helium by *sharing* electrons. Two hydrogen atoms—each with one electron—share an electron pair to make a *hydrogen molecule.*

$\text{H}\cdot$	+	$\cdot\text{H}$	$\longrightarrow$	$\text{H}:\text{H}$
Hydrogen atom		Hydrogen atom		Hydrogen molecule

The *shared pair* of electrons in the molecule is called a **covalent bond**.

A **molecule** is an electrically neutral cluster of two or more atoms joined by shared pairs of electrons (covalent bonds) that behaves as a single particle. A substance that is made up of molecules is called a *molecular substance.*

In forming a covalent bond, one can imagine two atoms approaching one another and getting their electron clouds or orbitals so enmeshed that they cannot easily be pulled apart. The molecule formed is more stable than the individual atoms. The covalent bond between two hydrogen atoms produces a *diatomic hydrogen molecule.*

Ionic and Covalent Compounds
Ionic compounds have crystal arrangements with ions packed in a three-dimensional array. *Molecular compounds* are made up of molecules that have a cluster of atoms joined by covalent bonds.

A chlorine atom—with seven valence electrons—will pick up an extra electron from another element if it can to form a chloride ion, Cl^-. But what if the only atom available is another chlorine atom? Chlorine atoms each lack one electron of having a completely filled energy level. Like hydrogen, chlorine atoms can share a pair of electrons to make a diatomic molecule that has one covalent bond.

$$:\ddot{\underset{\cdot\cdot}{Cl}}\cdot + \cdot\ddot{\underset{\cdot\cdot}{Cl}}: \longrightarrow :\ddot{\underset{\cdot\cdot}{Cl}}:\ddot{\underset{\cdot\cdot}{Cl}}:$$

Each chlorine atom in the chlorine molecule can use eight electrons to achieve a stable octet of electrons like that of the noble gas argon.

In the Cl_2 and H_2 molecules there is a covalent bond between the atoms. In Lewis electron-dot formulas of molecules, a covalent bond can be represented either by an electron pair between the atoms or by a dash in place of an electron pair. Lewis electron-dot formulas are shown here for hydrogen and for chlorine gas, but all halogens—as elements—are diatomic, with bonding that can be represented in the same way as shown for chlorine.

Study Hint
Recall that the diatomic hydrogen gas molecule is represented by the formula H_2, and the diatomic chlorine gas molecule is written as Cl_2.

$$H—H \text{ for } H_2 \qquad \text{and} \qquad :\ddot{\underset{\cdot\cdot}{Cl}}—\ddot{\underset{\cdot\cdot}{Cl}}: \text{ for } Cl_2$$

All diatomic elements have what are called **nonpolar covalent** bonds; that is, the electron pairs are shared equally between two atoms of the same kind of element (Figure 7.5).

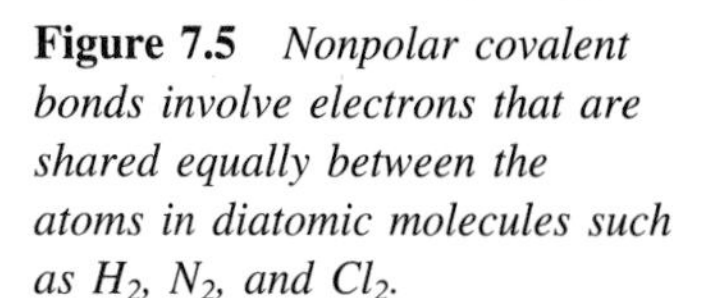

Figure 7.5 *Nonpolar covalent bonds involve electrons that are shared equally between the atoms in diatomic molecules such as H_2, N_2, and Cl_2.*

The sharing of electrons is not limited to one pair of electrons. Let's consider a nitrogen atom with five valence electrons. Its electron-dot symbol is

$$:\dot{\underset{\cdot}{N}}\cdot$$

We expect this atom to be reactive, based on what we know about the octet rule. If two nitrogen atoms share one pair of electrons, its Lewis structure would be

$$:\dot{\underset{\cdot}{N}}:\dot{\underset{\cdot}{N}}: \qquad \textit{(incorrect structure)}$$

In this arrangement, each atom has only six valence electrons around it—not enough for the desired octet. Each atom has two unpaired electrons "hanging out there" without partners, so, to become more stable, each nitrogen forms two more covalent bonds, to make a total of three covalent bonds. The nitrogen molecule can be represented by either of the following structures.

$$:N:::N: \qquad \text{or} \qquad :N\equiv N:$$

In both formulas for diatomic nitrogen, N_2, three pairs of electrons have been represented as being shared by the two atoms. Each nitrogen has achieved a stable, noble gas configuration. These three pairs of shared electrons between nitrogen atoms form a **triple bond**. Two pairs of shared electrons between atoms make a **double bond**, and a single pair of shared electrons make a **single bond**. For example, carbon atoms have four valence electrons that can be shared with other atoms to produce thousands of natural and synthetic molecules with single, double, and triple bonds as shown by the following structures for ethane (eth-ANE), ethene (eth-ENE), and ethyne (eth-INE). In each case, each carbon atom has four covalent bonds, and each hydrogen has one covalent bond.

Study Hint
Each pair of electrons can be represented by a pair of dots or one line. So a double bond is represented by a pair of lines.

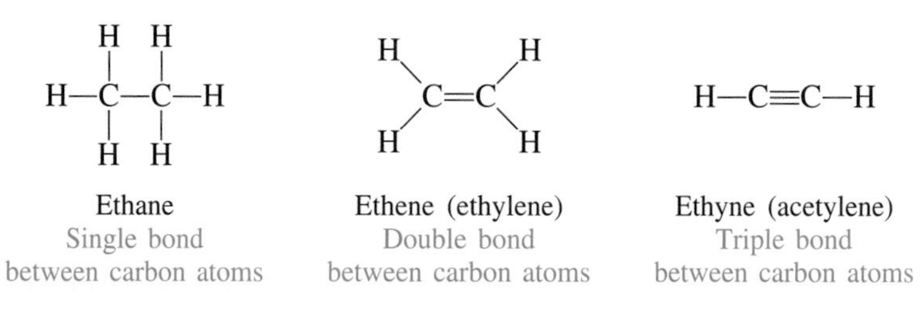

Ethane
Single bond between carbon atoms

Ethene (ethylene)
Double bond between carbon atoms

Ethyne (acetylene)
Triple bond between carbon atoms

EXAMPLE 7.2

Write electron-dot symbols for two bromine atoms. Show how they can form a covalently bonded molecule. Explain why bromine exists as diatomic molecules.

SOLUTION

$$:\ddot{\underset{\cdot\cdot}{Br}}\cdot + \cdot\ddot{\underset{\cdot\cdot}{Br}}: \longrightarrow :\ddot{\underset{\cdot\cdot}{Br}}:\ddot{\underset{\cdot\cdot}{Br}}:$$

Each bromine atom is quite reactive since it is short one electron of having an octet of electrons. Two bromine atoms can share an electron pair to form a covalent bond and thus achieve greater stability.

See Problems 7.9–7.16.

EXERCISE 7.2

Using electron-dot symbols show how two iodine atoms can come together to form a diatomic iodine molecule.

7.3 *Electronegativity*

Before studying the bonding in more compounds, we need to consider the importance of **electronegativity**, which is a measure of the tendency of an atom in a covalent bond to attract shared electrons to itself. Atoms of the most **electronegative elements** have the greatest attraction for electrons. They are the same elements—clustered in the upper right-hand corner of the periodic table—that have the greatest tendency to gain electrons to form negative ions.

Linus Pauling (Section 1.5) was the first chemist to develop a numerical scale of electronegativity. On the Pauling scale, fluorine, the most electronegative element, is assigned a value of 4.0 (see Figure 7.6). Oxygen is the second most electronegative element, followed by chlorine and nitrogen. Electronegativities decrease as

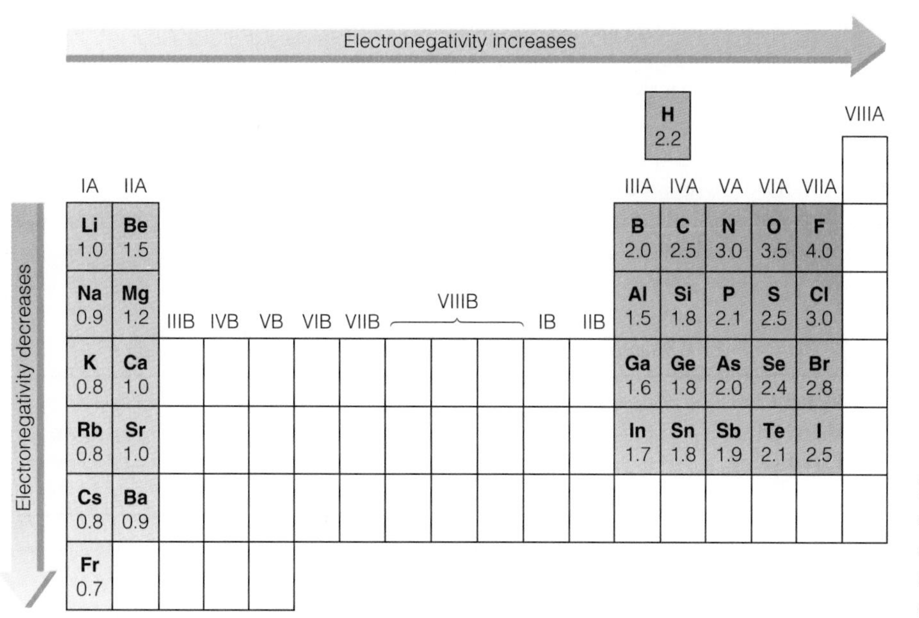

Figure 7.6 *Relative electronegativities of some representative elements. Electronegativities have no units; they are arbitrary numbers with relative values.*

metallic character increases. The most reactive metals (those in the lower left-hand corner of the periodic table) have the lowest electronegativity values (Figure 7.7). These trends are consistent with the trends in ionization energy discussed in Chapter 6.

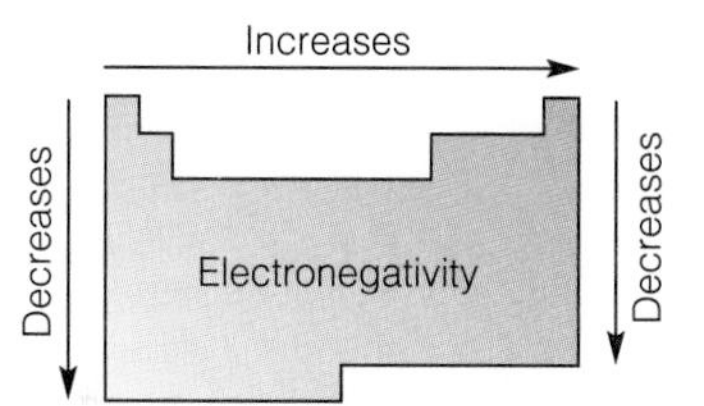

Figure 7.7 *General trends in electronegativities.*

Electronegativity increases in a period of elements as atomic number increases, but decreases within a group as atomic number increases.

Actually, it is the *difference* in electronegativities of atoms joined by a chemical bond that is important. When two atoms of the same element are joined in a covalent bond to make a diatomic molecule, both atoms have the same electronegativity values. Thus, the difference in electronegativity is zero, and the bond is nonpolar. Atoms joined by ionic bonds have large differences in electronegativity. When the difference in electronegativity is greater than about 1.7, the bond is considered to be mainly ionic. Polar covalent bonds (Section 7.4) have smaller differences in electronegativities. Thus, as the difference in electronegativity values decreases, the covalent character of the bond increases. For hydrogen chloride gas, the difference in electronegativity values (3.0 − 2.1) is 0.9. This difference is less than 1.7, so the bonding would be expected to have more covalent than ionic charcter. This prediction is in agreement with experimentation that shows HCl gas is a polar covalent molecule.

Study Hint

Whenever the difference in electronegativity is no greater than 0.4, the bond is essentially nonpolar covalent.

The electronegativity of hydrogen, 2.1, indicates, once again, that this element is not really like typical metallic elements in Group IA that have one valence electron, nor is it really like typical elements in Group VIIA that lack one valence electron of having a completely filled outer energy level. Instead, the electronegativity value for hydrogen is much closer to the electronegativity value of carbon, which is 2.5. Both hydrogen and carbon have an outer energy level that is half filled—hydrogen has one valence electron out of a possible two, and carbon has four valence electrons out of a possible eight. Both hydrogen and carbon tend to

An understanding of the concept of electronegativity is fundamental to a thorough understanding of chemical bonding.

form covalent bonds. In fact, covalent bonds joining hydrogen to carbon are present in nearly all organic compounds, as illustrated by the compounds ethane, ethene, and ethyne described in Section 7.2.

EXAMPLE 7.3

Determine differences in electronegativity for each of the following pairs of atoms and indicate whether a bond between the two is ionic, polar covalent, or nonpolar covalent: K and Cl, C and Cl, P and Cl, I and I.

SOLUTION

	Electronegativity	*Difference*	*Bonding*
K and Cl	3.0 − 0.8 = 2.2	Greater than 1.7	Ionic
C and Cl	3.0 − 2.5 = 0.5	Less than 1.7	Polar covalent
P and Cl	3.0 − 2.1 = 0.9	Less than 1.7	Polar covalent
I and I	2.5 − 2.5 = 0	No difference	Nonpolar covalent

See Problems 7.17–7.20.

EXERCISE 7.3

Are sulfur-to-oxygen bonds and oxygen-to-sodium bonds ionic, polar covalent, or nonpolar covalent?

7.4 Polar Covalent Bonds

Among types of bonding, there is—at one extreme—ionic bonding where electrons are transferred from one atom to another to form ions. At the other extreme is nonpolar covalent bonding of diatomic molecules where the electron distribution is perfectly balanced between atoms that have equivalent attractions for the electrons. In between these extremes is **polar covalent** bonding where electron pairs are shared *unequally* between atoms of different elements.

Most compounds contain *polar covalent* bonds; they fall between the extremes of nonpolar covalent and purely ionic.

Ionic bonding	**Polar covalent bonding**	**Covalent bonding**
Electrons are transferred	Electrons are shared unequally	Electrons are shared equally

Increasing ionic character

Hydrogen and chlorine react to form a colorless gas called hydrogen chloride. Using electron-dot symbols, the reaction can be represented by the equation

$$H\cdot + \cdot\ddot{\underset{\cdot\cdot}{Cl}}: \longrightarrow H:\ddot{\underset{\cdot\cdot}{Cl}}:$$

Both the hydrogen atom and the chlorine atom want an electron, so we can say they "compromise" by sharing an electron pair in a covalent bond. Because hydrogen and chlorine molecules are diatomic, the reaction is more accurately represented by the equation

$$\mathrm{H\!:\!H + :\underset{\cdot\cdot}{\ddot{Cl}}\!:\!\underset{\cdot\cdot}{\ddot{Cl}}\!: \longrightarrow 2\ H\!:\!\underset{\cdot\cdot}{\ddot{Cl}}\!:}$$

This reaction can also be written as

$$\mathrm{H_2 + Cl_2 \longrightarrow 2\ HCl}$$

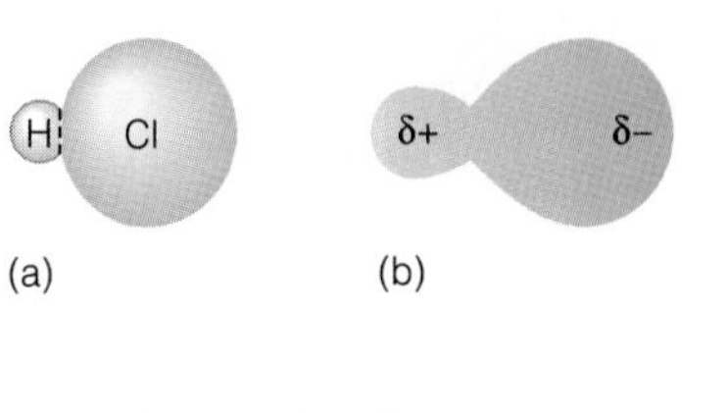

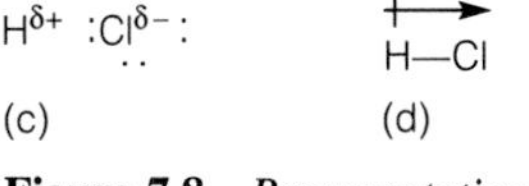

Figure 7.8 *Representations of the polar hydrogen chloride molecule. (a) A space-filling model of the molecule. (b) A diagram depicting the unequal distribution of electron density in the hydrogen chloride molecule. Electrons in the bond are attracted unequally toward chlorine, the more electronegative atom, so chlorine carries a partial negative charge, symbolized δ− chlorine atom. The symbols δ+ and δ− indicate partial positive and partial negative charges, respectively. (d) Polarity can also be symbolized by the crossed arrow that points toward the more electronegative atom in a covalent bond.*

Let us pause to consider an important question. One might reasonably ask why hydrogen molecules and chlorine molecules react at all. Have we not just explained that these diatomic molecules were formed to provide a more stable electron arrangement? Yes, we did say that. But there is stable, and there is more stable. Although a chlorine molecule is more stable than two separate chlorine atoms, the bond of a chlorine atom to a hydrogen atom has even greater stability. Nature always favors the formation of the more stable, stronger bonds.

When hydrogen and chlorine share an electron pair in the hydrogen chloride molecule, they do not share the electrons equally. In fact, any time two different kinds of atoms share electrons, one of the atoms will have a greater attraction for the electrons than will the other atom, so the bond is *polar covalent.*

The term *electronegativity* (see Section 7.3) is used to describe the relative attraction of an atom for electrons involved in a bond. Chlorine—which has greater attraction for electrons than does hydrogen—is more electronegative than hydrogen. Thus, in a hydrogen chloride molecule, shared electrons spend more time near the chlorine atom than they do near the hydrogen atom. If you think of an orbital as a fuzzy-looking electron cloud, then the electron cloud has greater electron density near the chlorine atom. The bond is *polar* (Figure 7.8). It is called a *polar covalent bond.* There is only one bond in the HCl molecule, and it is polar, so the entire HCl molecule is polar.

The following notation is often used to designate a polar covalent bond.

$$\mathrm{\overset{\delta+}{H}\!-\!\underset{\cdot\cdot}{\overset{\cdot\cdot\,\delta-}{Cl}}\!:} \quad \text{or} \quad \overset{+\!\!\longrightarrow}{\mathrm{H\!-\!Cl}}$$

The line between the two atoms represents the covalent bond. The δ+ and δ− (read "delta plus" and "delta minus") signify which end is partially positive and which end is partially negative. A crossed arrow (with a plus on its tail) pointed toward the more negative charge center can also be used to show the polarity of the bond. This unequal sharing of electrons in a covalent bond has a significant effect on the properties of a compound. For example, polar hydrogen chloride gas dissolves readily in water—which is also quite polar—to produce hydrochloric acid.

Study Hint
The atom that is "partially negative" is the one that is more electronegative and pulls electrons in the bond more closely to itself.

We can now make the following generalizations.

Chemical bonds

- between two identical nonmetal atoms are *nonpolar covalent.*
- between two different nonmetal atoms are *polar covalent.*
- between nonmetals and reactive metals are primarily *ionic.*

EXAMPLE 7.4

Draw Lewis electron-dot structures for Br_2, HBr, and NaBr. Discuss the ionic, polar covalent, and nonpolar covalent bonding of these molecules.

SOLUTION

$:\ddot{\underset{..}{Br}}-\ddot{\underset{..}{Br}}:$ Br_2 has one nonpolar covalent bond. Electrons in the single bond are shared equally.

$H-\ddot{\underset{..}{Br}}:$ The bond in HBr is polar covalent. Electrons in this single bond are shared unequally so the molecule is polar covalent.

$Na^+ :\ddot{\underset{..}{Br}}:^-$ NaBr is ionic. An electron is transferred from sodium to bromine to make sodium ions and bromide ions.

7.5 Metallic Bonding

The bonding of atoms in solid metallic crystals is referred to simply as **metallic bonding**. This type of bonding is distinctly different from the ionic and covalent bonding described in Sections 7.1 and 7.2. A model of a metallic solid can be pictured as a three-dimensional array of positive ions that remain fixed in the crystal lattice while the loosely held valence electrons move freely throughout the crystal (Figure 7.9). The fluidlike movement of these valence electrons through the crystal lattice makes metals good conductors of both heat and electricity. An important distinguishing characteristic of metals is that solid metals conduct electricity—ionic and covalently bonded solids do not.

Although metals are composed of a three-dimensional gridlike structure of *positive ions*, the metal remains *neutral* in charge. This is because the number of loosely held electrons that move through the crystal is exactly equal to the total positive charge of the ions.

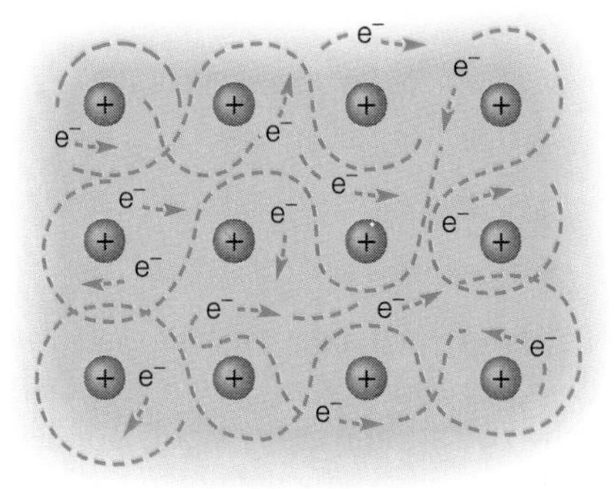

Figure 7.9 *Metallic bonding consists of positively charged metal ions in a three-dimensional lattice while loosely held valence electrons move freely in a fluidlike manner through the metal. Metals with the most loosely held electrons are the best conductors of electricity.*

7.6 Conductivity, Solubility, and Other Clues to Chemical Bonding

Can we test a substance to determine the type of bonding present? Yes, we can. Both conductivity tests (Figure 7.10) and the solubilities of substances can give important clues to their bonding characteristics. If the material being tested is a solid that conducts electricity and has a shiny appearance, we can expect the substance to be a metal.

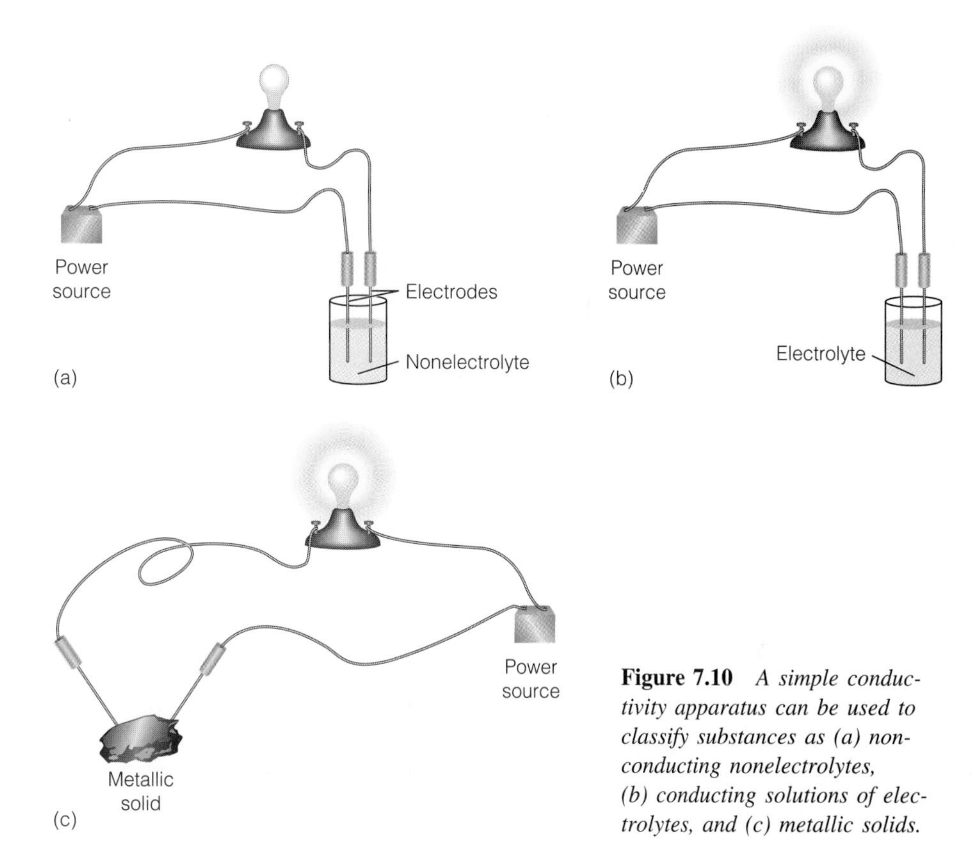

Figure 7.10 *A simple conductivity apparatus can be used to classify substances as (a) nonconducting nonelectrolytes, (b) conducting solutions of electrolytes, and (c) metallic solids.*

If a small amount of a solid material being tested will dissolve in water, and if the resultant solution will conduct an electric current, we can expect the material to be an ionic substance. Ionic substances all have high melting points (about 300 to 1000 °C) and, therefore, are solids at room temperature.

Although ionic compounds do not conduct electricity as solids, they are good conductors when fused (melted). The ions in the solid crystal are fixed—they cannot move about freely and cannot conduct a current. By contrast, an electric current can readily flow through a fused sample of the ionic compound because the individual ions are *dissociated* (they are separated from one another) and move about freely toward an electrode. An ionic compound that is *dissolved* in water will conduct an electric current for the same reason: the ions are dissociated and thus are free to move about. Any substance that gives ions in an aqueous solution and thereby conducts electricity is called an *electrolyte.* Pure water, by itself, is such an extremely poor conductor of electricity that it is not classified as an electrolyte.

The key terms **dissociate** and **electrolyte** are defined in Section 7.1.

The fact that many ionic compounds dissolve in water is, itself, an important characteristic property of substances that contain ions. Polar water molecules are attracted to both positive and negative ions. In fact, several water molecules surround and isolate individual ions as the compound becomes dissolved in water (Figure 7.11). Ionic compounds do not dissolve in nonpolar liquids such as gasoline and other similar petroleum products.

Ionic compounds dissolve in water.

If the substance being tested melts at a low temperature and is a nonconducting solid at room temperature, we can expect it to be a molecular substance; that is, it contains molecules with covalent bonding. Furthermore, a solid molecular substance also does not conduct an electric current when melted or when dissolved in

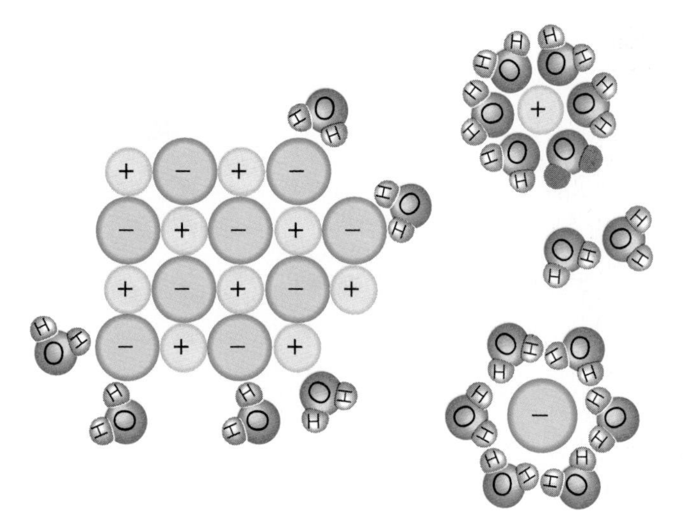

Figure 7.11 *As ionic solids dissolve in water, they dissociate. Several water molecules surround and associate with each positive and negative ion as the crystal dissolves. Notice that positive ions are attracted by the oxygen atom of water, while negative ions are attracted by the more positive hydrogen atoms of water. Nonpolar solvents do not dissolve ions; they lack the strong attractive forces required to hold ions in solution.*

a solvent. Glucose, a simple sugar, is a good example. It has a low melting point (83 °C), and it does not conduct electricity when melted or when dissolved in water. These properties are related to the fact that glucose is not ionic; it is a molecular substance with covalent bonds.

At room temperature, some molecular substances are solids, some are liquids, and some are gases.

If a substance being tested is a gas or a pure liquid, it cannot be ionic—all ionic substances have high melting points. If the substance is a *gas*, it is either one of the elements called noble gases (helium, neon, etc.) or else the gas is composed of molecules that have covalent bonding. Also, pure substances that are *liquid* at room temperature have covalent bonding.

There are two classifications of molecular liquids: polar and nonpolar. Water molecules are quite polar; the electron distribution in water molecules is unbalanced. Other polar liquids are miscible (soluble) in water, while most nonpolar liquids are immiscible (not soluble) in water. Oil and water are immiscible; they are not soluble in each other. Thus, oil must be nonpolar because it does not dissolve in the polar water.

Thinking It Through

Investigations with HCl Gas

Suppose we investigate the behavior of hydrogen chloride gas, HCl, in the lab. Since it is a gas at room temperature, we know it cannot be ionic, and, since it is chemically reactive, it cannot be a noble gas. Thus, it must be composed of covalent molecules. Bubbling the gas through hexane, C_6H_{14}, a nonpolar liquid (a petroleum product), gives a mixture that does not conduct electricity, but when HCl gas is bubbled through water, it dissolves to give a solution that conducts electricity.

What is going on here with the HCl? Since HCl in water conducts an electric current, we know ions are present at that time, but we also know that neither HCl gas nor any other gas is made up of ions. Thus, *ions must be formed* as the polar HCl dissolves in water. Here, the hydrogen-to-chlorine covalent bond is broken and hydrogen ions and chloride ions are formed. The dissociated ions in the water solution permit electrical conductivity.

Since the HCl gas in hexane did not conduct a current, we can conclude from the laboratory experiment that only molecules were present and that *no ions were formed* when HCl was bubbled through hexane.

Table 7.1 **Chemical Bonding Characteristics**

Characteristic	*Ionic Bonding*	*Covalent Bonding*	*Metallic Bonding*
Unit particles	Positive and negative ions	Molecules	Atoms
Physical state at room temperature	Solid	Some are solids Some are liquids Some are gases	All are solids except Hg
Melting point	High, about 300–1000 °C	Low; varies widely	Varies widely; >28 °C except Hg
Electrical conductivity			
As a solid	No	No	Yes
When melted	Yes, good	No	Yes
In water	Yes, good	No	Not applicable
Solubility	Soluble in polar solvents such as water	Nonpolar covalent compounds: soluble in nonpolar solvents Polar covalent compounds: soluble in polar solvents	Not soluble in nonpolar solvents; some react with acids, a few react with water
EXAMPLES	NaCl, $CaCl_2$	CH_4, CO_2, H_2O, I_2	Cu, Mg, Al, Fe

Laboratory tests involving electrical conductivity, melting point, and solubility can reveal a great deal about the chemical bonding of a substance. These properties and generalizations regarding bonding are summarized in Table 7.1.

EXAMPLE 7.5

The following samples were tested as described. Is the bonding in each case ionic, covalent, or metallic?

(a) Road oil splattered on a car does not dissolve in water, but it is soluble in charcoal lighter fluid and other nonpolar petroleum solvents. Also, the road oil does not conduct an electric current.
(b) Conductivity tests were performed on tap water. The tap water was found to be a rather poor conductor of electricity. Complete evaporation of 20 mL of tap water left a solid, white residue.

SOLUTION

(a) Road oil has nonpolar covalent bonds; its properties are typical of organic substances with high molecular masses.
(b) Conductivity confirms the presence of ions in the water sample. *Pure* water is not an electrical conductor, but the residue left after evaporation indicates that electrolytes—ionic substances—were dissolved in the tap water.

See Problems 7.21–7.28.

7.7 Writing Lewis Electron-Dot Formulas

Lewis electron-dot formulas (Lewis structures) of some simple nonpolar covalent molecules (Cl_2 and Br_2), polar covalent molecules (HCl and HBr), and ionic compounds (NaCl, $MgCl_2$, and Mg_3N_2) have been described in this chapter. After writing Lewis structures for these simple compounds, you should learn to write

Lewis structures for polyatomic molecules and ions. The following systematic procedure will be useful for writing electron-dot formulas for these more complex structures, especially those made up of four or more atoms.

Steps in Writing Electron-Dot Formulas

1. First write the symbol for the central atom of the structure (if three or more atoms are involved) and arrange the other atoms around the central atom. Common central atoms include the nonmetals (C, N, P, S, and sometimes O (in H_2O, N_2O, HOCl, and O_3).
2. Determine the total number of valence electrons by adding together the number of valence electrons for each atom in the molecule or ion.
   a. For a negative ion, add to this total a number of electrons equal to the negative charge of the ion.
   b. For a positive ion, subtract from this total a number of electrons equal to the positive charge of the ion.
3. Use a single bond (an electron pair) to connect each atom to the central atom. Arrange the remaining electrons around all atoms to make a complete octet of electrons around each atom except for hydrogen, which never has more than two electrons. (In large structures that contain hydrogen, such as H_2SO_4 or HSO_4^-, hydrogen bonds to oxygen, which, in turn, bonds to the central atom.)
4. If the total number of electrons available is fewer than the number needed to complete an octet, shift nonbonding (outer) electron pairs in the structure to make one or more double or triple bonds. (A double bond is present in the structure when there is a shortage of two electrons; two double bonds or one triple bond is signaled by a shortage of four electrons.)

Let us compare the differences between the polar covalent *bonding* in HCl and the polar covalent *bonding* in carbon dioxide, CO_2. To do this we will first construct a Lewis electron-dot formula and a structural formula for CO_2 using the rules that have been outlined.

The carbon atom has four valence electrons and each oxygen atom has six valence electrons. For CO_2 we have a total of $4 + (2 \times 6) = 16$ valence electrons. First use a single bond (an electron pair) to connect the central atom, carbon, to each of the two oxygen atoms. This gives

$$O{:}C{:}O \qquad \textit{(incomplete structure)}$$

We can distribute the remaining 12 electrons between the two oxygen atoms to make an octet of electrons around each oxygen atom.

$$:\underset{..}{\ddot{O}}{:}C{:}\underset{..}{\ddot{O}}: \qquad \textit{(incomplete structure)}$$

We have used all 16 valence electrons and each oxygen has an octet, but the carbon atom only has four electrons and needs four more electrons to make an octet. By shifting one pair of nonbonding electrons from each oxygen to each C—O bond, we can make double bonds between carbon and oxygen, thus producing an octet of electrons for the carbon as well as for each oxygen.

$$:\ddot{O}{:}{:}C{:}{:}\ddot{O}: \qquad \text{or} \qquad :\ddot{O}{=}C{=}\ddot{O}: \qquad \text{(correct structures)}$$

Oxygen is more electronegative than carbon. Thus, each carbon-to-oxygen double bond (C=O) is polar covalent with a partial negative charge on the oxygen atom. Yet, carbon dioxide—with two polar covalent bonds—is *not* polar! This is because CO_2 is a linear, symmetrical molecule. The two electron pairs on one side of the carbon atom are repelled by two electron pairs on the opposite side. One carbon-to-oxygen double bond has the same electron density as the other. This balance in electron density about the central atom makes the CO_2 molecule nonpolar. The polar covalent bond in HCl, however, is the only bond in the molecule, so it is responsible for the polar nature of the HCl molecule. In both examples polar bonds are present, but one of the molecules is nonpolar while the other is polar (Figure 7.12). You should practice writing electron-dot formulas for several molecules and ions.

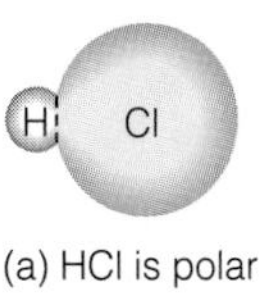

(a) HCl is polar

(b) CO_2 is nonpolar

Figure 7.12 *Both hydrogen chloride gas (a) and carbon dioxide gas (b) molecules have polar bonds. The electron density is balanced in CO_2 to give a nonpolar linear-shaped molecule, but the electron density in HCl is unbalanced so the molecule is quite polar.*

EXAMPLE 7.6

Write an electron-dot formula for the nitrate ion, NO_3^-.

SOLUTION

STEP 1 Arrange the three oxygen atoms around nitrogen, the central atom.

```
    O
O   N   O     (incomplete structure)
```

STEP 2 Determine the total number of valence electrons for these atoms.

1 nitrogen atom =	5 valence electrons
3 oxygen atoms = 3 × 6 =	18 valence electrons
1− charge =	1 additional electron
	24 total valence electrons

STEP 3 Connect the oxygen atoms to the nitrogen atom with single bonds and use the remaining electrons to complete octets of electrons around atoms. Enclose the ion in brackets and show the 1− charge of the ion.

```
 [      ..        ]−
 [     :O:        ]
 [ ..   |   ..    ]
 [:O — N — O:     ]     (incomplete structure)
 [ ..       ..    ]
```

STEP 4 Since the nitrogen atom appears to be short two electrons, shift a nonbonding electron pair from an oxygen atom to make a N=O double bond. The electrons in bonds between atoms are shared and can be used by each of the two bonded atoms. Thus, by sharing electrons, each atom can use an octet of eight electrons.

```
 [      ..        ]−
 [     :O:        ]
 [ ..   |   ..    ]
 [:O = N — O:     ]     (complete Lewis structure)
 [          ..    ]
```

For the electron-dot structure of the NO_3^- ion, all atoms have an octet of electrons. The N=O (double bond) was placed on the left side of the N atom as shown, but this double bond could have been drawn to the oxygen on the right side or the oxygen above the nitrogen. When the movement of electrons (not atoms) produces more than one equivalent structure, these are called **resonance structures.** There are three resonance structures for the nitrate ion. Double-headed arrows are placed between resonance structures to indicate that these structures are equivalent.

$$\left[\begin{array}{c} :\ddot{O}: \\ | \\ :\ddot{O}{=}N{-}\ddot{O}: \end{array} \right]^- \longleftrightarrow \left[\begin{array}{c} :\ddot{O} \\ \| \\ :\ddot{O}{-}N{-}\ddot{O}: \end{array} \right]^- \longleftrightarrow \left[\begin{array}{c} :\ddot{O}: \\ | \\ :\ddot{O}{-}N{=}\ddot{O}: \end{array} \right]^-$$

The actual bond distances in the nitrate ion are equivalent, indicating that the N-to-O bonds are all equivalent. There is an average of $1\frac{1}{3}$ bonds between the N and each O, but we cannot draw fractions of a bond; instead, we use the three resonance structures that are shown for the nitrate ion.

Now, practice writing the following Lewis electron-dot formula as well as others at the end of this chapter.

EXAMPLE 7.7

Write a Lewis electron-dot formula for the phosphate ion, PO_4^{3-}.

SOLUTION

STEP 1 Arrange the four oxygen atoms around phosphorus, the central atom.

$$\begin{array}{ccc} & O & \\ O & P & O \\ & O & \end{array} \quad \textit{(incomplete structure)}$$

STEP 2 Determine the total number of valence electrons for these atoms.

1 phosphorus atom =	5	valence electrons
4 oxygen atoms = 4 × 6 =	24	valence electrons
3− charge =	3	additional electrons
	32	total valence electrons

STEP 3 Connect the oxygen atoms to phosphorus with single bonds and use the remaining electrons to complete octets of electrons around atoms. Enclose the ion in brackets and show the 3− charge of the ion.

STEP 4 Now we check our work to make sure that all 32 electrons have been used properly and that the correct charge is shown.

See Problems 7.29–7.34.

7.8 Shapes of Molecules: Balloon Models and Electron Pair Repulsion

The shapes of molecules are similar to arrangements formed when different numbers of balloons of the same size are tied together.

The Linear Arrangement

Two balloons tied together tend to point in opposite directions so that the bond angle between them is 180°, a **linear** arrangement.

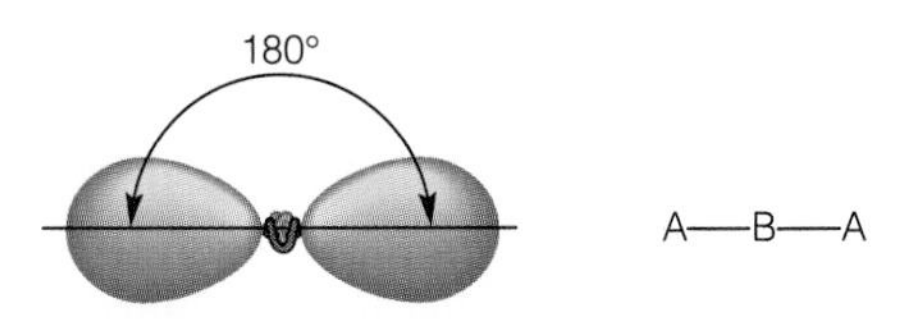

Examples of molecules with this linear (A—B—A) arrangement include CO_2 (described in Section 7.7), BeF_2, and similar molecules that have two atoms covalently bonded to a central atom. In BeF_2, the central atom, Be, has only two valence electrons to share with the two fluorine atoms to make two covalent bonds. Electron pair repulsion produces the linear shape. Beryllium atoms do not follow the octet rule.

The Trigonal Planar Arrangement

When three balloons are tied together, they all tend to lie in the same plane and assume positions such that the angles between them are 120°, a **trigonal planar** arrangement.

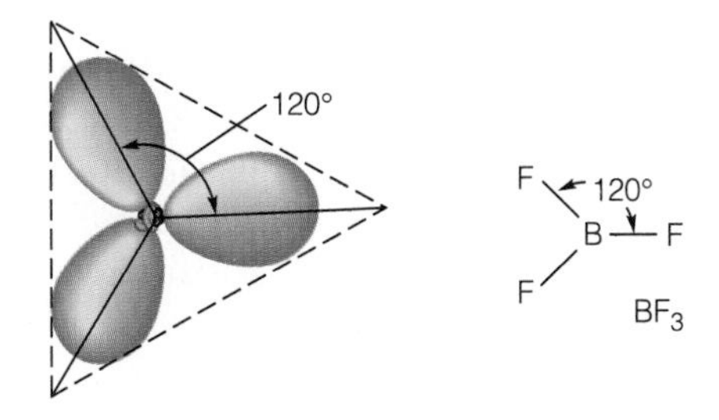

Examples of molecules with this trigonal planar shape include SO_3, BF_3, and BCl_3. In the boron compounds, the central atom (boron with three valence electrons) can have only three bonds (three electron pairs). Thus, boron is able to share only six electrons, so it does *not* follow the octet rule. These three electron pairs repel each other to give the molecule a trigonal planar shape.

The Tetrahedral Arrangement

Molecules with four atoms bonded to a central atom form structures having the same shape as four balloons tied together at a common center. They arrange themselves as far apart as possible to make 109.5° angles at the center. The cluster of four balloons, and also the molecule, would fit inside a *tetrahedron* (a structure with four faces, each of which is triangular in shape). Thus, the balloons and the molecule are said to have a **tetrahedral** arrangement, as shown here.

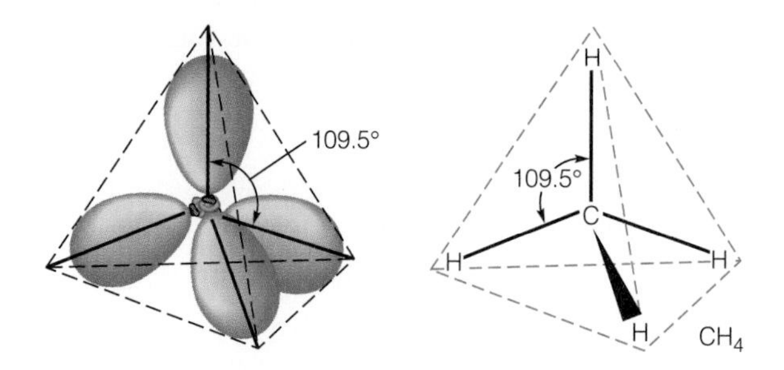

Examples of molecules that tend to form this tetrahedral arrangement include methane, CH_4, carbon tetrachloride, CCl_4, and other molecules with four atoms covalently bonded to a central atom. This is the arrangement that would be predicted, based on the **valence shell electron pair repulsion** (VSEPR) **theory**. The concept is quite simple—the negatively charged electron clouds surrounding the central atom in a molecule will stay as far apart as possible. That is, the electron pairs around the central atom stay as far apart as possible.

EXAMPLE 7.8

Predict the shapes of the BCl_3 and $BeCl_2$ molecules.

SOLUTION

First draw the electron-dot structures for each molecule.

:C̈l—B—C̈l: :C̈l—Be—C̈l:
|
:C̈l:

Boron has only three valence electrons, and the BCl_3 molecule has only three covalent bonds with no unshared electron pairs. The three bonds get as far apart as possible, so the molecule is trigonal and planar with 120° bond angles. In the $BeCl_2$ molecule, Be has two bonds and no additional unshared electron pairs. The two electron pairs get as far apart as possible, so the molecule is linear.

:C̈l:
B—C̈l: :C̈l—Be—C̈l:
:C̈l:

See Problems 7.35 and 7.36.

7.9 Water: A Bent Molecule

Passing a direct current through water can break up the water molecules to release two volumes of hydrogen gas for each volume of oxygen gas. (This process is called the electrolysis of water.) The 2-to-1 gas volume ratio is produced because water molecules contain twice as many hydrogen atoms as oxygen atoms. We might wonder why an oxygen atom in water bonds to two hydrogen atoms, and not to three or more atoms of hydrogen. This question can be answered in terms of bonding. An oxygen atom, with six valence electrons, needs two more electrons to have an octet. A hydrogen atom, however, needs only one more electron to fill its first energy level. Thus, the oxygen atom in water shares electron pairs with two hydrogen atoms to form two covalent (single) bonds. Lewis structures are shown here.

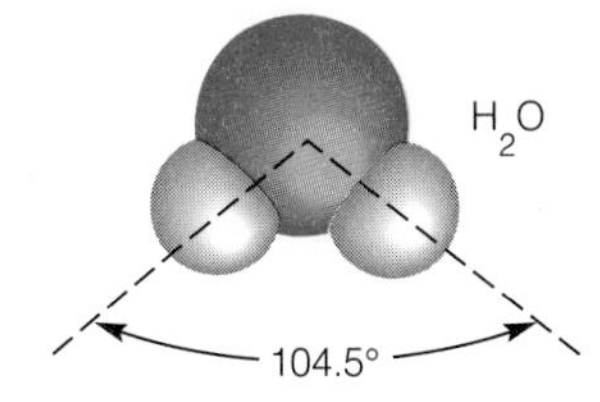

$$2\,\mathrm{H}\cdot + \cdot\ddot{\underset{\cdot}{\mathrm{O}}}: \longrightarrow \mathrm{H}:\ddot{\underset{\mathrm{\ddot{H}}}{\mathrm{O}}}:$$

Figure 7.13 *The water molecule has a nonlinear bent shape with a bond angle of 104.5°.*

This bonding arrangement gives oxygen a complete octet of electrons, like neon. Furthermore, it gives each hydrogen atom an electron structure like that of the noble gas helium.

Lewis structures cannot show precise bond angles, so the angular arrangement of electron-dot symbols is used to represent the **angular** or **bent shape** and the polar nature of the water molecule. The H—O—H bond angle in water is actually 104.5° (Figure 7.13). The polar nature of water can readily be shown experimentally. When we vigorously shake a mixture of water and oil, the two liquids separate into layers; they are immiscible. This is because oil is nonpolar and water has polar molecules. Determining the precise bond angle in the polar water molecule is not so simple; it requires sophisticated instrumentation.

The VSEPR theory offers a simple explanation for the shape of a water molecule. There are four pairs of electrons surrounding the oxygen atom in the water molecule—two bonding pairs and two nonbonding pairs. The **nonbonding electron pairs** are those that are not involved in covalent bonding with other atoms in the molecule. If we assume that all four electron pairs repel each other equally, then two nonbonding electron pairs are directed toward two corners of a tetrahedron, and the two bonds to hydrogen atoms are directed toward the other two corners of the tetrahedron (Figure 7.14).

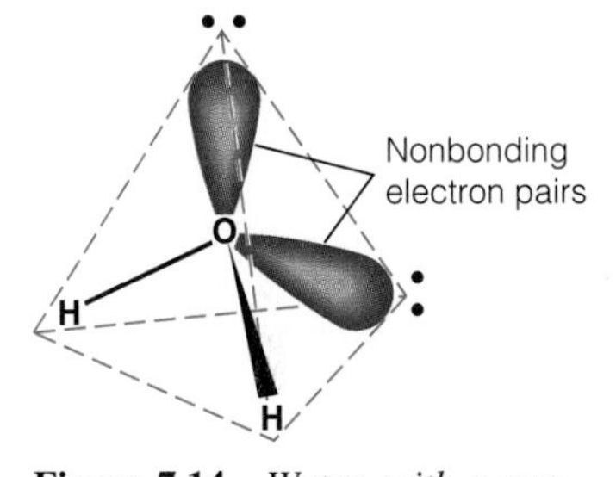

Figure 7.14 *Water, with a nonlinear bent shape, can be pictured inside a tetrahedron. The two nonbonding electron pairs are directed toward two corners of the tetrahedron, while the bonds to the two hydrogen atoms are directed toward two other corners of the tetrahedron. The true H—O—H bond angle is 104.5°, not the 109.5° angle present in a tetrahedral arrangement.*

The perfect tetrahedral angle of 109.5°—at the center—is a little greater than the actual H—O—H bond angle in water, which is 104.5°. According to the VSEPR theory, *nonbonding electron pairs* exert more repulsion than bonding pairs. Thus, the nonbonding electron pairs push the bonding pairs a little closer together to give a bond angle of 104.5°.

EXAMPLE 7.9

Write an electron-dot formula for hydrogen sulfide gas, H_2S. Predict its molecular shape using VSEPR theory.

Hydrogen sulfide gas smells like rotten eggs. It is present in most sewage, it is toxic, and it contributes to pollution.

SOLUTION

Write the electron-dot symbols for two hydrogen atoms and one sulfur atom.

$$\mathrm{H}\cdot \quad \mathrm{H}\cdot \quad \cdot\ddot{\underset{\cdot}{\mathrm{S}}}:$$

Sulfur is in the same group of the periodic table as oxygen and has the same number of valence electrons as oxygen, so we can expect the structure of H_2S to be similar to H_2O, a polar molecule with a bent shape. The H_2S has two pairs of bonding electrons covalently bonding H to S plus two nonbonding electron pairs that give the molecule a bent shape.

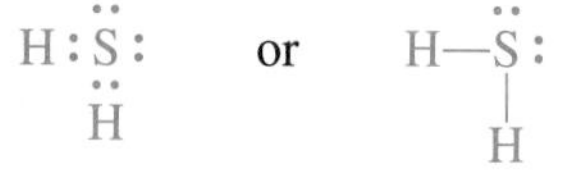

EXERCISE 7.4
Based on what you know about VSEPR theory and the shapes of H_2O and H_2S molecules, discuss their polarities. Should H_2S dissolve in water?

7.10 Ammonia: A Pyramidal Molecule

Ammonia, NH_3, is a gas at room temperature. It ranks among the top ten chemicals produced, with 30–35 billion pounds produced annually in the United States. Ammonia is liquefied under high pressure.

An atom of nitrogen, N, has five valence electrons. It can assume the neon configuration by sharing three pairs of electrons with *three* hydrogen atoms to form ammonia, NH_3.

$$3\ H\cdot + \cdot\ddot{N}\cdot \longrightarrow H:\ddot{N}:H \quad \text{or} \quad H—\ddot{N}—H$$

There are four electron pairs around the central nitrogen atom in the ammonia molecule. Using the VSEPR theory, one would expect a tetrahedral arrangement of these four electron pairs. The actual H—N—H bond angles of 107° are quite close to the 109.5° bond angles that would be present for a perfect tetrahedron. Presumably, the nonbonding electron pair exerts greater repulsion than do the bonding pairs, so the bond angle is decreased to 107° in the polar ammonia molecule. (This is closer to the angles in a tetrahedron than the 104.5° bond angle in water, which has two nonbonding electron pairs.)

The ammonia molecule has a **pyramidal shape** like that of a tripod with a hydrogen atom at the end of each ''leg'' and the nitrogen atom with its unshared pair of electrons sitting at the top (Figure 7.15). Each N—H bond is a polar bond since nitrogen is more electronegative than hydrogen. This, combined with the unsymmetrical pyramidal shape of ammonia, makes the entire molecule polar; the nitrogen end carries the partial negative charge.

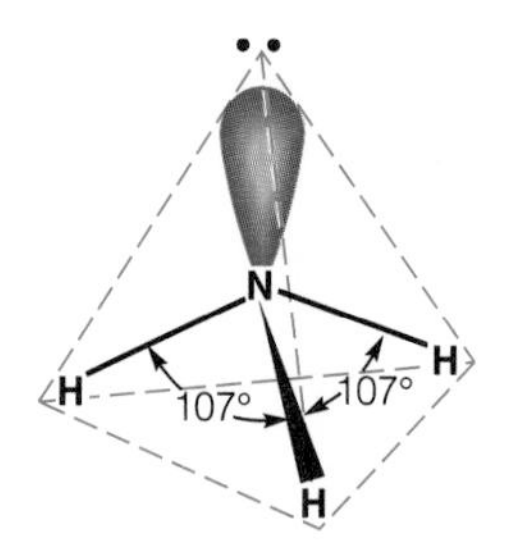

Figure 7.15 *The pyramidal ammonia molecule can be pictured inside a tetrahedron.*

Ammonia is quite soluble in water. Household ammonia and certain other cleaning agents contain dilute aqueous ammonia; these solutions are used to clean greasy surfaces and to remove wax from floors. Compressed ammonia is widely used as a fertilizer for crops such as corn that require large amounts of nitrogen (Figure 7.16).

7.11 Ammonium Ions and Coordinate Covalent Bonds

When ammonia gas, NH_3, is added to an acid, a chemical reaction occurs. All typical acids contain hydrogen ions, H^+; these ions give acids their characteristic sour taste. When an ammonia molecule reacts with a hydrogen ion from the acid, an *ammonium ion* is formed.

Figure 7.16 *Ammonia, NH_3, is a gas at room temperature. It ranks among the top ten chemicals with 30 to 35 billion pounds produced annually in the United States. Ammonia is liquefied under high pressure. Vast quantities of compressed ammonia, called anhydrous ammonia, are used each year as a fertilizer for crops that need large amounts of nitrogen.*

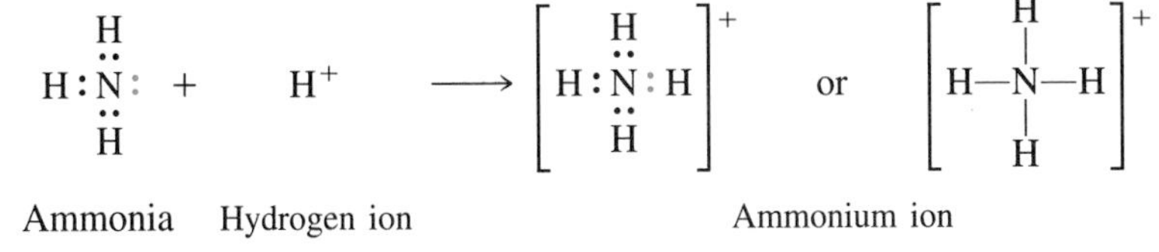

The ammonium *ion*—common in solid fertilizer (Figure 7.17)—is distinctly different from the ammonia gas *molecule* with its unique odor. For the ammonia *molecule* (Section 7.10), recall that each H—N covalent bond involves the sharing of one electron from a hydrogen atom with one electron from a nitrogen atom to

Figure 7.17 *Fertilizers containing ammonium ions also are widely used in agriculture.*

make a shared pair of electrons. For the *ammonium ion*, three of the N—H bonds are typical covalent bonds, but the fourth covalent bond is formed when the electron pair from nitrogen donates *both* of its electrons to form another N—H covalent bond. This bond is called a **coordinate covalent bond**, that is, a bond where one atom donates both of the electrons involved in the shared pair.

To help us remember the characteristics of the coordinate covalent bond, we could say that the nitrogen of ammonia, with one nonbonding electron pair, appears to say, ''Have pair, will share,'' and a coordinate covalent bond is formed. The properties of a H—N coordinate covalent bond, once formed, are not different from other H—N covalent bonds. Thus, we can't detect any difference in properties or bond lengths between the four covalent bonds in an ammonium ion.

EXAMPLE 7.10

Use the periodic table to assist you in predicting the shape of the PH_3 molecule. Write an electron-dot formula for PH_3. Would you expect this chemical to react with H^+ ions in acids? Explain.

SOLUTION

Since phosphorus has five valence electrons like nitrogen, we expect the bonding of PH_3 to be much like that of NH_3, so PH_3 should be pyramidal.

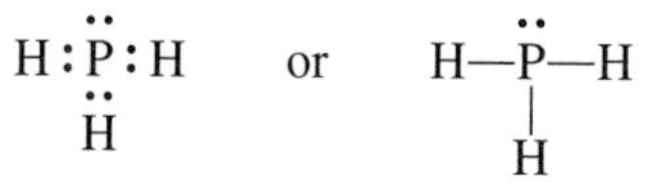

We should expect the polar PH_3 molecule with its lone electron pair to react with acids in a way similar to that described for ammonia. Again, the periodic table and bonding theory can help us make predictions about many compounds.

7.12 Methane: A Tetrahedral Molecule

An atom of carbon, C, has four valence electrons. It can obtain an octet of electrons by sharing pairs of electrons with four hydrogen atoms to form the gaseous compound methane, CH_4.

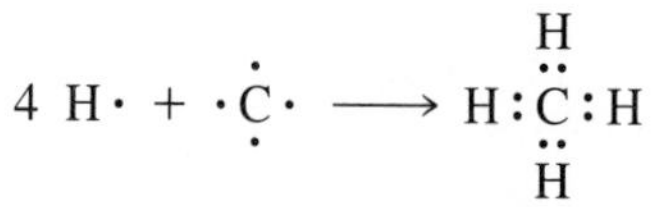

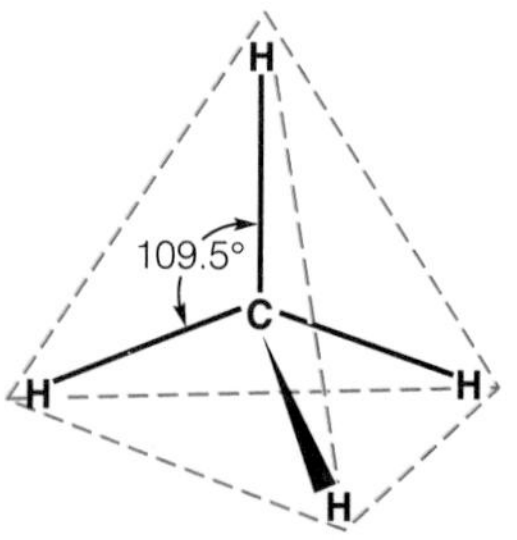

Figure 7.18 *The tetrahedral methane molecule is nonpolar.*

In methane, there are four shared pairs of electrons around the central carbon atom. Using the VSEPR theory, one would expect a tetrahedral arrangement and bond angles of 109.5°, which is in perfect agreement with actual findings (Figure 7.18). All four electron pairs are shared with hydrogen atoms, and all four pairs have equivalent repulsions. This tetrahedral arrangement makes methane a nonpolar molecule because polarity of the four equal H—C bonds is mutually canceled. Both methane and carbon dioxide (Section 7.7) have polar *bonds,* but both molecules are *nonpolar*. This is because the polar bonds in both compounds are perfectly balanced by other equivalent polar bonds.

Methane is the principal component of natural gas fuel. It burns with a hot flame, and, if sufficient oxygen is present, the principal products are carbon dioxide and water. Methane is produced during the decay of plant and animal material. It is often seen bubbling to the surface of swamps; hence its common name "marsh gas." Bacterial metabolism in the intestinal tract of all animals also produces methane, making it a component of intestinal gas.

Methane in the Environment Besides CO_2 and chlorofluorocarbons, methane gas can trap heat in the atmosphere. Each acre of rice produces about 480 lb of CH_4, and each cow about 77 lb, yearly. Overall, rice accounts for roughly 3.5% and cattle for about 2% of the heat-retaining capacity of the atmosphere.

EXAMPLE 7.11

Use the periodic table to assist you in predicting the shape of the carbon tetrachloride, CCl_4, molecule. Write an electron-dot formula for CCl_4. Would you expect this substance to dissolve in water? Explain.

SOLUTION

We would expect a CCl_4 molecule to be tetrahedral, like methane, and to have four equivalent C—Cl covalent bonds, making a nonpolar molecule.

$$4\ :\ddot{\underset{..}{Cl}}\cdot\ +\ \cdot\dot{\underset{\cdot}{C}}\cdot\ \longrightarrow\ \begin{array}{c} :\ddot{Cl}: \\ :\ddot{\underset{..}{Cl}}:\ddot{\underset{..}{C}}:\ddot{\underset{..}{Cl}}: \\ :\underset{..}{Cl}: \end{array}\quad \text{or}\quad \begin{array}{c} :\ddot{Cl}: \\ | \\ :\ddot{\underset{..}{Cl}}—C—\ddot{\underset{..}{Cl}}: \\ | \\ :\underset{..}{Cl}: \end{array}$$

We should not expect the nonpolar CCl_4 to dissolve in water, which is polar. This prediction is upheld experimentally; the two chemicals are immiscible.

7.13 Molecular Structures and the Periodic Table

In the preceding sections, we have shown that atoms consistently form a definite number of covalent bonds. Hydrogen forms one covalent bond, oxygen two, nitrogen three (four for the NH_4^+ ion), and carbon four. Table 7.2 shows the number of covalent bonds typically associated with various representative atoms. Notice that the number of unpaired valence electrons of an atom determines the number of covalent bonds that it can form.

We have seen that the location of an element in the periodic table allows us to predict the number of valence electrons for an atom of that element. This is also true for the number of covalent bonds it can form (Figure 7.19). Hydrogen, in Group IA, has only one valence electron and can form only one covalent bond. Other elements in this group tend to form ionic bonds. Beryllium, in Group IIA, forms a linear molecule when covalently bonded to two nonmetal atoms. Boron in Group IIIA can form three covalent bonds to give trigonal planar-shaped molecules. In Group IVA, the nonmetals carbon and silicon can form tetrahedral molecules by sharing four electrons to form four covalent bonds.

Nitrogen and phosphorus, nonmetallic elements with five valence electrons, tend to form compounds with three covalent bonds. Three of their five valence electrons are shared with other atoms to make covalent bonds; the other two valence electrons form a nonbonding electron pair. As predicted by the VSEPR theory, one nonbonding electron pair gives molecules a pyramidal shape. Molecules with this type of bonding are polar.

TABLE 7.2 **Representative Covalent Structures**

Electron-Dot Symbol	*Covalent Structure*	*Number of Covalent Bonds*	*Structures of Representative Molecules*
H·	H—	1	H—H H—C̈l:
·Be·	—Be—	2	:F̈—Be—F̈:
·B·	—B— (with one bond down)	3	:F̈—B—F̈: with :F̈: bonded below B
·Ċ·	—C— (with bonds up and down)	4	CH_4 (H—C—H with H above and below); :O=C=O:
·N̈·	—N̈— (with one bond down)	3	H—N̈—H with H below N; :N≡N:
·Ö:	—Ö: (with one bond down)	2	H—Ö: with H below O; H—Ö—Ö—H
·F̈:	—F̈:	1	H—F̈: :F̈—F̈:
·C̈l:	—C̈l:	1	H—C̈l: :C̈l—C̈l:

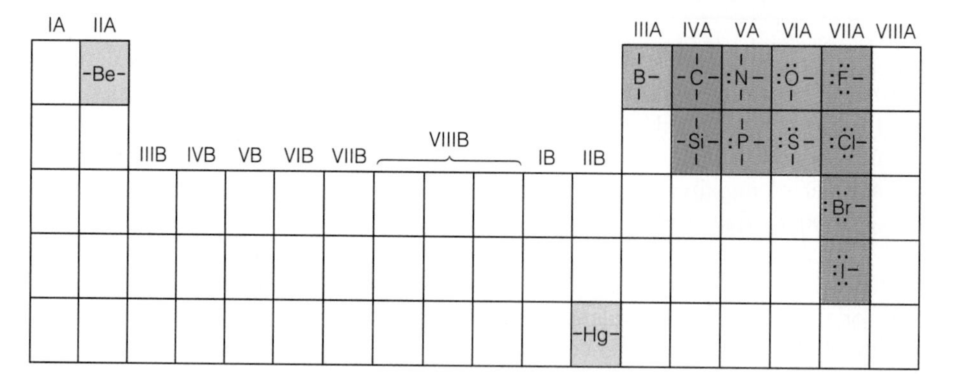

Figure 7.19 *Covalent bonds formed by certain elements in the periodic table.*

EXAMPLE 7.12

Hydrogen cyanide, HCN, is a poisonous gas that can cause the familiar cyanide poisoning. Determine the Lewis structure for HCN.

SOLUTION

We have

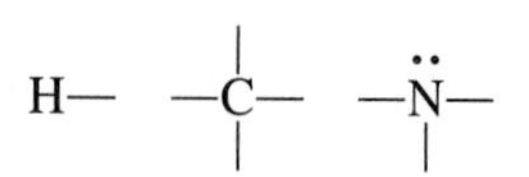

The carbon atom will form four bonds, the hydrogen atom will form one bond, and the nitrogen atom will form three bonds. The only way this combination can be put together is shown here.

$$H{-}C{\equiv}N{:}$$

Both H—C≡N: and :N≡C—H represent the same structure.

Oxygen and sulfur, with six valence electrons, need two valence electrons to complete an octet of electrons. When atoms of these elements form two covalent bonds with other atoms, the two pairs of nonbonding valence electrons on the central atom give these molecules a bent shape. This is in agreement with predictions based on the VSEPR theory.

Both water and hydrogen peroxide molecules contain oxygen atoms covalently bonded to hydrogen atoms. Both molecules have a bent shape. (Hydrogen peroxide is commonly used as a disinfectant and as a bleach for hair.) Each hydrogen peroxide molecule is composed of two hydrogen atoms and two oxygen atoms, as indicated by the molecular formula, H_2O_2. In only a few compounds do we find that an oxygen atom bonds to another oxygen atom, but such is the case for all peroxides, including hydrogen peroxide. Its structure is

Study Hint
In peroxides, two oxygen atoms are covalently bonded together, but in dioxides such as carbon dioxide, CO_2, and sulfur dioxide, SO_2, the oxygen atoms are separated by a different kind of atom.

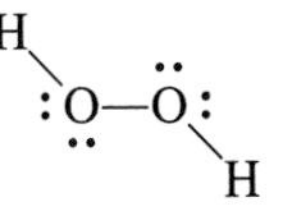

This structure represents the most stable arrangement of atoms in H_2O_2, but the molecule is free to rotate about the O—O bond, so both hydrogen atoms could be positioned, at least temporarily, on the same side of the O—O bond.

Finally, the halogens, with seven valence electrons, need only one additional electron to complete an octet of electrons. In most cases, a halogen atom forms one covalent bond by sharing its unpaired electron with an electron from another atom, but in certain compounds, halogens also share their electron pairs to form additional coordinate covalent bonds.

Table 7.3 **Bonding and the Shape of Molecules**

Number of Bonds	*Number of Unshared Pairs*	*Shape*	*Examples*
2	0	Linear	$BeCl_2$, $HgCl_2$
3	0	Trigonal planar	BF_3
4	0	Tetrahedral	CH_4, $SiCl_4$
3	1	Pyramidal	NH_3, PCl_3
2	2	Bent	H_2O, H_2S, SCl_2

As shown by the examples described here, the position of an element in the periodic table gives us valuable information about covalent bonding. Table 7.3 summarizes information about the number of covalent bonds, unshared (nonbonding) electron pairs, and the molecular shapes of representative compounds.

EXAMPLE 7.13

Describe the shape and polar or nonpolar nature of BCl_3 and PCl_3.

SOLUTION

We should first draw Lewis structures for each molecule.

:Cl:B:Cl: :Cl: :Cl:P:Cl: :Cl:

Boron has only three valence electrons, and BCl_3 has only three covalent bonds with no unshared electron pairs. Electron pair repulsion gives BCl_3 a trigonal planar shape, like BF_3, with 120° bond angles. The PCl_3 molecule has three covalent bonds and one unshared electron pair, like NH_3, so it has a pyramidal shape.

See Problems 7.37–7.42.

We expect BCl_3 to be nonpolar and PCl_3 to be polar, as is actually the case.

7.14 Hydrogen Bonding

There is one more very important type of chemical bonding that we will present here to allow comparison with ionic, covalent, and metallic bonding. This type of bonding, called **hydrogen bonding**, involves attractive forces *between* certain polar *molecules* containing hydrogen atoms.

The hydrogen bonding between water molecules (Figure 7.20) is responsible for the unusually high melting point and the unusually high boiling point of water. These high values are attributed to the especially strong hydrogen bonds between water molecules. In fact, both H_2S (34 g/mol) and H_2Se (81 g/mol) are gases at room temperature, but water (18 g/mol)—with a similar molecular shape and a lower molar mass—is a liquid.

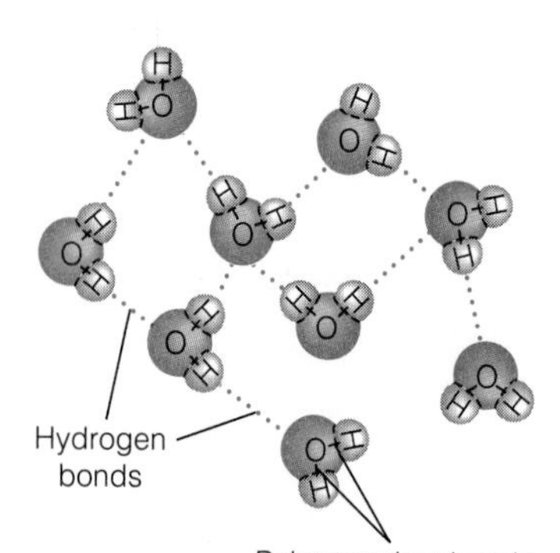

Figure 7.20 *Hydrogen bonding between water molecules is responsible for the unusually high melting point and boiling point of water compared to other molecules of similar molar mass.*

Not all molecules containing hydrogen become involved in hydrogen bonding.

> Only molecules where hydrogen atoms are attached to one of the highly electronegative elements fluorine, oxygen, and nitrogen can become involved in hydrogen bonding.

In molecules of this type, the bond between the electronegative element and the hydrogen is quite polar—the bonding electron pair is much closer to the electronegative atom. Since the hydrogen has rather low electron density, this nearly bare hydrogen proton is attracted to nonbonding electrons held by an electronegative element of a nearby molecule.

Hydrogen bonds between molecules are generally represented by dotted lines (Figure 7.21). In liquid hydrogen fluoride, hydrogen bonds are about 5% as strong as the hydrogen-to-fluorine covalent bonds. In ice, hydrogen bonds are also about 5% as strong as the covalent bonds between hydrogen and oxygen.

As water cools and approaches its freezing point of 0 °C, the hydrogen bonding between molecules causes the water molecules to become arranged in specific patterns with the oxygen of one molecule next to a hydrogen of a nearby water molecule. Because this is not the most compact arrangement possible, water expands as it freezes and forms ice crystals. As water freezes in the crevices of rocks, the expansion—due to hydrogen bonding—causes solid materials to crack and break. This is exactly what happens when water freezes in a glass bottle and causes it to break. Rocks are weathered away, and concrete or marble structures are broken down as this process occurs over and over again, all because of hydrogen bonding.

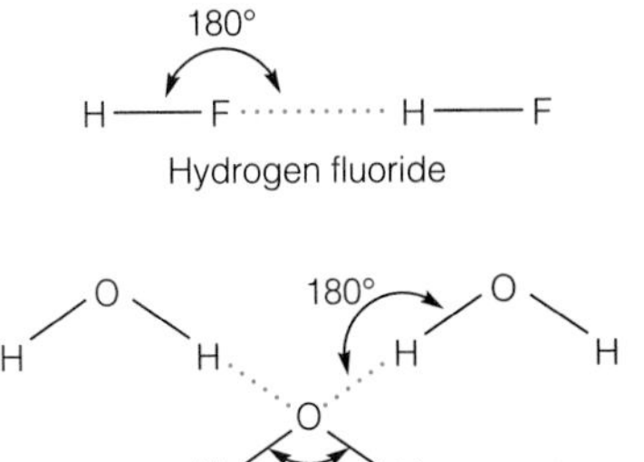

Figure 7.21 *Hydrogen bonding in hydrogen fluoride and in water.*

Chapter Summary

Chemical bonds are the forces of attraction that hold atoms or ions together in chemical compounds. The chemical bonding of a substance is closely related to such properties as electrical conductivity, melting point, boiling point, and solubility. Ionic bonds are formed when electrons are transferred, leaving positive cations and negative anions. In order to achieve a noble gas electron configuration, metal atoms become oxidized; they lose electrons to form cations. In contrast, nonmetal atoms are reduced as they gain electrons to form anions.

Compounds with ionic bonds exist as crystalline solids at room temperature. Those that dissolve in water give solutions that will conduct an electric current. This is due to the dissociated ions that are free to move about in the solution. A molecule, by contrast, is not ionic; it is a discrete group of atoms held together by one or more covalent bonds. All molecules are electrically neutral. The covalent bonds within molecules are formed when one, two, or three electron pairs are shared between atoms; these are called single, double, and triple bonds, respectively. In a typical single bond, each atom shares one electron with another atom to make a shared electron pair. A coordinate covalent bond is a special type of covalent bond that occurs when both electrons in the shared pair are donated by the same atom.

Electronegativity is the relative attraction of an atom in a molecule for a shared pair of electrons in a chemical bond. Elements in simple ionic compounds have the largest differences in electronegativity. When the difference in electronegativity between atoms is zero, the bond is covalent and nonpolar. Electronegativity differences that fall between these extremes are characteristic of polar covalent bonds. When polar bonds are distributed unequally about a central atom, the entire molecule is polar. Fluorine is the most electronegative element, followed by oxygen, nitrogen, chlorine, and other elements in the upper right-hand corner of the periodic table. Electronegativity decreases as metallic character increases.

In metallic bonding, positive metal ions remain fixed in a crystal lattice, while the loosely held valence electrons are free to flow through the crystal lattice. Thus, metals can conduct an electric current. Ionic compounds are all solids at room temperature. They have high melting points and do not conduct an electric current as solids, but do conduct an electric current when melted or dissolved in water. Molecules have covalent bonding with low melting points. At room temperature some molecular compounds are solids (like sugar and wax), some are liquids (like water, gasoline, and alcohol), and some are gases (like ammonia, methane, and CO_2). Molecular compounds do not conduct an electric current. Nonpolar molecules are miscible with nonpolar liquids but are immiscible with polar liquids.

Lewis electron-dot structures are useful in describing the bonding in both ionic and covalent compounds. Two or more valid Lewis structures for a molecule obtained by moving only electrons—not atoms—are called resonance structures. The shape of a molecule depends on the number of bonding electron pairs and nonbonding electrons associated with the central atom of a molecule. The repulsions of valence shell electron pairs is the key principle of the VSEPR theory, which allows us to predict the tetrahedral shape of molecules like methane, the pyramidal shape of molecules like ammonia, the bent shape of molecules like water, the trigonal planar shape of molecules like boron trifluoride, and the linear shape of molecules like carbon dioxide.

Hydrogen bonding is an extra strong attraction between molecules that have hydrogen atoms covalently bonded to fluorine, oxygen, or nitrogen atoms. The hydrogen bonding between water molecules is quite strong and is responsible for the unusually high melting point and boiling point of water.

Assess Your Understanding

1. Compare ionic, covalent, polar covalent, metallic, and hydrogen bonding. [7.1–7.5, 7.14]
2. Describe what happens to an atom as it becomes oxidized or reduced. [7.1]
3. Use Lewis electron-dot symbols to represent the reaction of metal atoms with nonmetal atoms to form compounds. [7.1]
4. Distinguish between polar covalent and nonpolar covalent bonds. [7.2, 7.4]
5. Use electronegativities to evaluate the covalent character of bonds. [7.3]
6. Compare models of metallic, ionic, and covalent bonding. [7.5]
7. Write Lewis electron-dot formulas for ionic and covalent compounds. [7.7]
8. Predict shapes of molecules and draw their structures using the VSEPR theory. [7.8–7.13]
9. Compare hydrogen bonding with the other types of bonding. [7.14]

Key Terms

angular (bent) shape [7.9]
chemical bonds [7.1]
coordinate covalent bond [7.11]
covalent bond [7.2]
crystal [7.1]
dissociate [7.1]
double bond [7.2]
electrolyte [7.1]
electronegative elements [7.3]
electronegativity [7.3]
hydrogen bonding [7.14]
ionic bond [7.1]

ionic compound [7.1]
linear [7.8]
metallic bonding [7.5]
molecule [7.2]
nonbonding electron pairs [7.9]
nonpolar covalent bond [7.2]
octet rule [7.1]
oxidized [7.1]
polar covalent bond [7.4]
pyramidal shape [7.10]
reduced [7.1]
resonance structures [7.7]
single bond [7.2]
tetrahedral [7.8]
trigonal planar [7.8]
triple bond [7.2]
valence shell electron pair repulsion theory [7.8]

Problems

IONIC AND COVALENT BONDS

7.1 Write a chemical equation to represent the ionization of a potassium atom using Lewis electron-dot symbols. Is this oxidation or reduction? Explain.

7.2 Write a chemical equation to represent the ionization of a magnesium atom using Lewis electron-dot symbols. Is this oxidation or reduction? Explain.

7.3 Write a chemical equation to represent the ionization of a bromine atom to form a bromide ion, Br^-, using Lewis electron-dot symbols. Is this oxidation or reduction? Explain.

7.4 Write a chemical equation to represent the ionization of a sulfur atom to form a sulfide ion, S^{2-}, using Lewis electron-dot symbols. Is this oxidation or reduction? Explain.

7.5 Use Lewis electron-dot symbols to represent the reaction of potassium with bromine to form potassium ions, K^+, and bromide ions, Br^-.

7.6 Use Lewis electron-dot symbols to represent the reaction of magnesium with sulfur to form magnesium ions, Mg^{2+}, and sulfide ions, S^{2-}.

7.7 What is an electrolyte? Is table salt, NaCl, an electrolyte? Explain.

7.8 What occurs during the *dissociation* of ions? How can one tell if ions are dissociated?

7.9 What is the difference between an ionic bond and a covalent bond?

7.10 What is the difference between a nonpolar covalent bond and a polar covalent bond?

7.11 What is meant by the term *coordinate covalent bond?*

7.12 Describe what happens when an atom is (a) oxidized and (b) reduced.

7.13 Criticize the use of the phrase "a sodium chloride molecule."

7.14 What is a single bond? a double bond? a triple bond?

7.15 Predict whether the following compounds would have ionic bonding, nonpolar covalent bonding, or polar covalent bonding. Explain your reasoning. *Hint*: Notice placement in the periodic table.
a. Cl_2 gas b. KCl solid c. CO_2 gas
d. HBr gas e. NH_3 gas f. CH_4 gas

7.16 Predict whether the following compounds would have ionic bonding, nonpolar covalent bonding, or polar covalent bonding. Explain your reasoning. *Hint*: Notice placement in the periodic table.
a. $MgBr_2$ solid b. H_2 gas c. CCl_4 liquid
d. O_2 gas e. HF gas f. PH_3 gas

ELECTRONEGATIVITIES

7.17 Carbon and bromine have different electronegativities. How can a CBr_4 molecule be nonpolar?

7.18 Draw a structure to represent the polarity of a gaseous HBr molecule by placing $\delta-$ on the more electronegative atom and $\delta+$ on the more electropositive atom.

7.19 Determine differences in electronegativity values for each of the following pairs of atoms. Also state whether a bond between the two is expected to be ionic, polar covalent, or nonpolar covalent.
a. Br and Br b. C and Br c. Na and Br

7.20 Determine differences in electronegativity values for each of the following pairs of atoms. Also state whether a bond between the two is expected to be ionic, polar covalent, or nonpolar covalent.
a. I and I b. S and O c. K and I

CONDUCTIVITY AND SOLUBILITY PROPERTIES

7.21 Based on all information presented in this chapter, which of the following compounds would you expect to be electrolytes? Why?
a. glucose sugar
b. N_2 gas

c. $CaCl_2$
d. CH_3CH_2OH (ethyl alcohol)

7.22 Based on all information presented in this chapter, which of the following compounds would you expect to be electrolytes?
a. NaBr b. HCl c. $CHCl_3$ (chloroform)
d. CCl_4 (carbon tetrachloride)

7.23 Unknown A was a liquid that did not conduct an electric current. It was immiscible with water and had a boiling point of 115 °C. Was the bonding of unknown A ionic, polar covalent, nonpolar covalent, or metallic?

7.24 Unknown B was a solid that did not conduct an electric current when molten. The unknown did dissolve in water, but the resultant solution would not conduct an electric current. Was the bonding of unknown B ionic, polar covalent, nonpolar covalent, or metallic?

7.25 Unknown C was a solid with a silvery shine. It did conduct an electric current, but the solid could not be melted with a laboratory Bunsen burner. The unknown did not dissolve in water or charcoal lighter fluid, but it did dissolve in acid. Was the bonding of unknown C ionic, polar covalent, nonpolar covalent, or metallic?

7.26 Unknown D was a white crystalline solid at room temperature. It dissolved in water to give a solution that conducted electricity. Was the bonding of unknown D ionic, polar covalent, nonpolar covalent, or metallic?

7.27 Candle wax melts at a low temperature, but it is not a conductor of electricity. Wax will not dissolve in water, but it is partially soluble when left in nonpolar solvents like gasoline and charcoal lighter fluid. What type of bonding is present in candle wax?

7.28 Battery acid is a solution of H_2SO_4 in water. It is a good conductor of electricity. What does this tell us about the bonding in H_2SO_4 when the acid is dissolved in water.

LEWIS ELECTRON-DOT STRUCTURES

7.29 Draw electron-dot formulas for the following atoms, molecules, and ions.
a. a chlorine atom
b. a chlorine molecule
c. a chloride ion

7.30 Draw electron-dot formulas for the following atoms, molecules, and ions.
a. an iodine atom
b. an iodine molecule
c. an iodide ion

7.31 Draw electron-dot formulas for the following molecules and ions.
a. NH_3 (ammonia gas)
b. NH_4^+ (ammonium ion)
c. $CHCl_3$ (chloroform)
d. CH_3OH (methyl alcohol or wood alcohol)

7.32 Draw electron-dot formulas for the following molecules and ions.
a. NO_3^- (nitrate ion)
b. NO_2^- (nitrite ion)
c. C_2H_2 (acetylene)
d. CH_3CH_2OH (ethyl alcohol or grain alcohol)

7.33 Draw two resonance structures for sulfur dioxide gas, SO_2.

7.34 Draw three resonance structures for the carbonate ion, CO_3^{2-}.

SHAPES OF MOLECULES AND STRUCTURAL FORMULAS

7.35 What is the shape of a methane, CH_4, molecule? What are its bond angles?

7.36 What is the shape of a carbon dioxide molecule? What is its bond angle?

7.37 Draw Lewis structures and describe the shape of each of the following.
a. PCl_3 b. BCl_3 c. H_2S d. BeF_2

7.38 Draw Lewis structures and describe the shape of each of the following.
a. CBr_4 b. C_2F_2 c. PBr_3 d. SCl_2

7.39 Draw the Lewis structure of water; describe the shape of a water molecule.

7.40 Draw the Lewis structure of ammonia; describe the shape of this molecule.

7.41 There are two different covalent molecules with the formula C_2H_6O. Draw Lewis structures for the two molecules.

7.42 Draw Lewis structures to show that both C_2H_6 and C_2H_4 are possible.

HYDROGEN BONDING

7.43 Describe the difference between the covalent bonds in water molecules and the hydrogen bonding for water.

7.44 What specific properties of water are related to hydrogen bonding?

7.45 How is hydrogen bonding responsible for the weathering of rocks?

7.46 Why does an ice cube float in a glass of water?

Additional Problems

7.47 Consider the hypothetical elements X, Y, and Z with electron-dot formulas

$$:\ddot{\underset{\cdot\cdot}{X}}\cdot \qquad :\ddot{\underset{\cdot}{Y}}\cdot \qquad :\dot{\underset{\cdot}{Z}}\cdot$$

a. To which group in the periodic table would each belong?
b. Show the Lewis structure for the simplest compound of each with hydrogen.
c. Write electron-dot formulas for the ions formed when X reacts with sodium and also when Y reacts with sodium.

7.48 Consider the hypothetical elements K, L, and M with electron-dot formulas

$$K\cdot \qquad \cdot L\cdot \qquad \cdot\dot{M}\cdot$$

a. To which group in the periodic table would each belong?
b. Use electron-dot symbols to write a chemical equation for the reaction of each "element" with chlorine to form ionic compounds.

7.49 Explain why hydrogen bromide, HBr, readily dissolves in water, but the liquid element bromine, Br_2, is only slightly soluble in water.

7.50 Explain why ammonia gas, NH_3, is quite soluble in water, but nitrogen gas, N_2, in the air is nearly insoluble in water.

7.51 Using the procedure described in Section 7.7, draw Lewis electron-dot structures for the following.
a. H_2SO_4 (sulfuric acid)
b. H_2SO_3 (sulfurous acid)

7.52 Using the procedure described in Section 7.7, draw Lewis electron-dot structures for the following.
a. H_3PO_4 (phosphoric acid)
b. H_3PO_3 (phosphorous acid)

7.53 Gasoline does not dissolve in water. What does this suggest about the molecules in gasoline?

7.54 Explain why ammonia gas readily dissolves in water but an open can of a soft drink quickly goes "flat" as it loses the CO_2 gas. *Hint*: Consider molecular shape.

7.55 Which of the following would you expect to dissolve in water?
a. H_2 gas
b. $CaCl_2$
c. chloride ions
d. CH_3CH_2OH (ethyl alcohol)

7.56 Which of the following compounds would you expect to dissolve in water?
a. NaBr
b. HCl gas
c. CH_4 (methane)
d. CCl_4 (carbon tetrachloride)

7.57 Which of the following compounds could participate in hydrogen bonding?
a. CH_3OH (methyl alcohol)
b. CH_3NH_2
c. HBr

7.58 Which of the following compounds could participate in hydrogen bonding?
a. CH_3CH_3 (ethane)
b. H_2S (hydrogen sulfide gas)
c. HF

7.59 Describe the type of bonding in each of the following.
a. O_2 gas
b. K_2O (potassium oxide)
c. CO (carbon monoxide gas)

7.60 Describe the type of bonding in each of the following.
a. H_2 gas
b. HCl gas
c. HCl dissolved in water

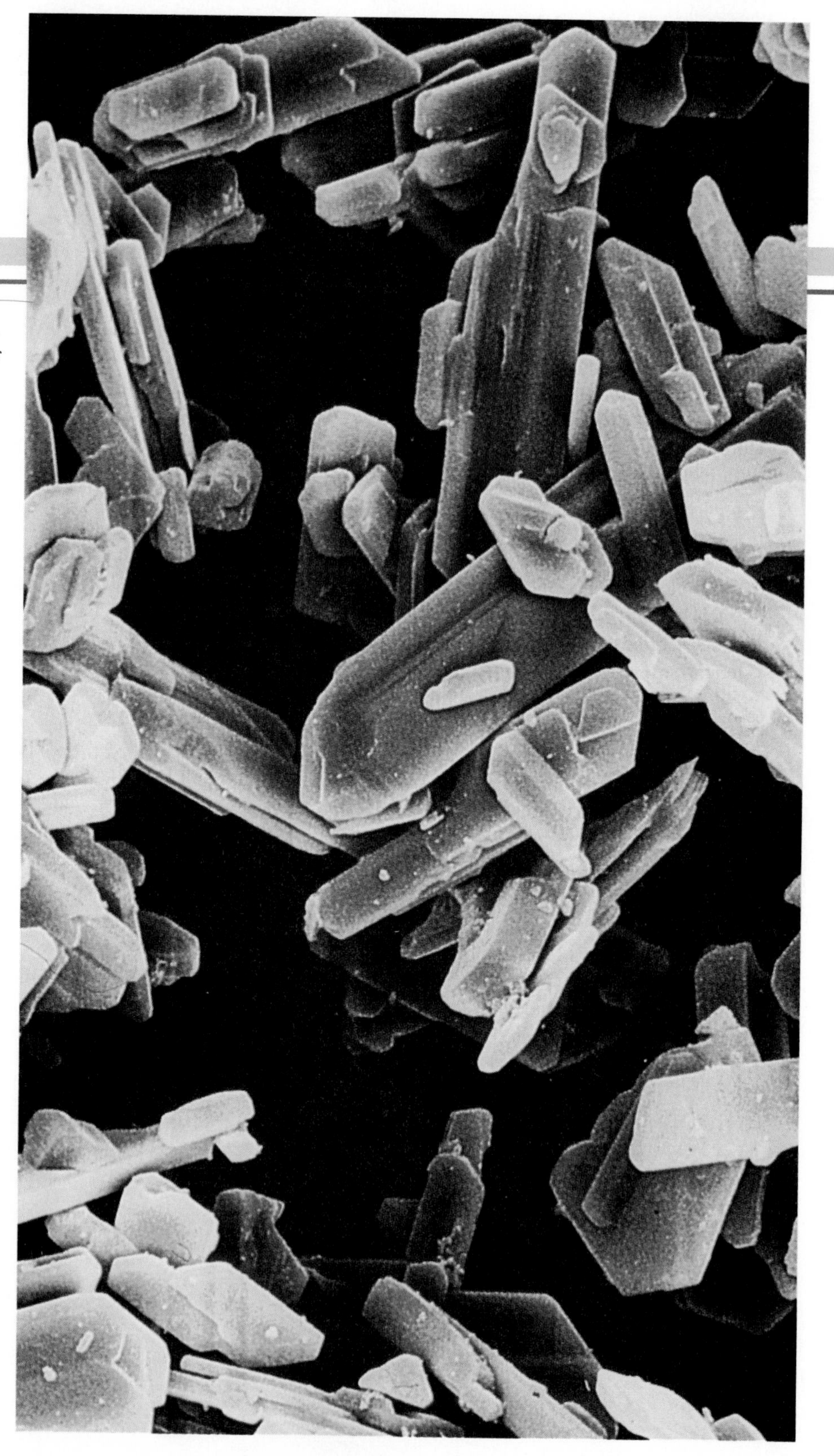

A color-enhanced scanning electron micrograph of crystals of calcium sulfate.

8

Names, Formulas, and Uses of Inorganic Compounds

CONTENTS

Chemists have identified over 10 million chemical compounds, and the list keeps growing. Each compound has a specific name and structure. With such a large number of chemicals, it is essential that we use an explicit and systematic method of naming these substances.

There are two major classifications of chemical compounds: organic and inorganic. Although carbon is not the most abundant element, more *different* compounds contain carbon than any other element except hydrogen. The large group of covalently bonded compounds that contain carbon atoms—the petroleum-based chemicals, plastics, synthetic fibers, carbohydrates, and many more—are classified as **organic chemicals**. All other compounds fall into the category of **inorganic chemicals**. The International Union of Pure and Applied Chemistry (IUPAC) has adopted an unambiguous system of names and formulas for both organic and inorganic substances.

In this chapter, we discuss **chemical nomenclature**—the system of names and formulas—for inorganic chemicals. Before considering subsequent topics, you should be able to write chemical formulas and names quickly and correctly for chemicals like those presented in this chapter. Consistent, daily practice is needed to develop your skills.

8.1 Monatomic Ions

Many compounds involve ionic bonding. Before we can write names and formulas for these compounds, we must be familiar with the names and formulas of individual ions. The names of simple positive ions are derived from those of the parent

Table 8.1 **Symbols and Names of Common Cations**

1+ Cations		*2+ Cations*		*3+ and 4+ Cations*	
FROM GROUP IA		FROM GROUP IIA		FROM GROUP IIIA	
Hydrogen	H^+	Magnesium	Mg^{2+}	Aluminum	Al^{3+}
Lithium	Li^+	Calcium	Ca^{2+}		
Sodium	Na^+	Strontium	Sr^{2+}		
Potassium	K^+	Barium	Ba^{2+}		
OTHERS		OTHERS		OTHERS	
Ammonium	NH_4^+	Zinc	Zn^{2+}		
Silver	Ag^+	Cadmium	Cd^{2+}		
Copper(I) or cuprous	Cu^+	Copper(II) or cupric	Cu^{2+}		
Mercury(I) or mercurous	Hg_2^{2+}	Mercury(II) or mercuric	Hg^{2+}		
		Chromium(II) or chromous	Cr^{2+}	Chromium(III) or chromic	Cr^{3+}
		Manganese(II) or manganous	Mn^{2+}	Manganese(III) or manganic	Mn^{3+}
		Iron(II) or ferrous	Fe^{2+}	Iron(III) or ferric	Fe^{3+}
		Cobalt(II) or cobaltous	Co^{2+}	Cobalt(III) or cobaltic	Co^{3+}
		Nickel(II) or nickelous	Ni^{2+}	Nickel(III)	Ni^{3+}
		Tin(II) or stannous	Sn^{2+}	Tin(IV) or stannic	Sn^{4+}
		Lead(II) or plumbous	Pb^{2+}	Lead(IV) or plumbic	Pb^{4+}

Table 8.2 **Symbols and Names of Common Anions**

1− Anions		*2− Anions*		*3− and 4− Anions*	
Peroxide*	O_2^{2-}	Oxide	O^{2-}	Nitride	N^{3-}
Hydride	H^-	Sulfide	S^{2-}	Phosphide	P^{3-}
Fluoride	F^-	Selenide	Se^{2-}	Arsenide	As^{3-}
Chloride	Cl^-	Telluride	Te^{2-}	Carbide	C^{4-}
Bromide	Br^-				
Iodide	I^-				
Hydroxide	OH^-				
Hydrogen carbonate (bicarbonate)	HCO_3^-	Carbonate	CO_3^{2-}		
Hydrogen sulfate (bisulfate)	HSO_4^-	Sulfate	SO_4^{2-}	Phosphate	PO_4^{3-}
Hydrogen sulfite (bisulfite)	HSO_3^-	Sulfite	SO_3^{2-}	Phosphite	PO_3^{3-}
Thiocyanate	SCN^-	Thiosulfate	$S_2O_3^{2-}$		
Cyanide	CN^-				
Acetate	CH_3COO^- or $C_2H_3O_2^-$	Oxalate	$C_2O_4^{2-}$		
Nitrate	NO_3^-	Chromate	CrO_4^{2-}		
Nitrite	NO_2^-	Dichromate	$Cr_2O_7^{2-}$		
Permanganate	MnO_4^-				
Perchlorate†	ClO_4^-				
Chlorate†	ClO_3^-				
Chlorite†	ClO_2^-				
Hypochlorite†	ClO^-				

*In peroxide each oxygen has an oxidation number of −1.
†Ions with bromine and iodine in place of chlorine are named similarly.

elements by addition of the word *ion*. For example, a sodium atom, Na, upon losing its lone valence electron, becomes a sodium *ion*, Na^+. Similarly, a magnesium atom, Mg, readily loses its two valence electrons to become a magnesium *ion*, Mg^{2+}. Names and symbols for the most common cations (positive ions) are given in Table 8.1.

Names of simple, negative ions that involve only one kind of nonmetal atom are derived from names of the parent elements by changing the ending to *-ide* and addition of the word ion. For example, a chlor*ine* atom, Cl, upon gaining an electron, becomes a chlor*ide* ion, Cl^-. A sulfur atom, S, with six valence electrons readily gains two electrons to form a sulf*ide* ion, S^{2-}. Names and symbols for the most common anions (negative ions) are given in Table 8.2.

Practice giving the specific names and symbols for each of the cations and anions listed in Tables 8.1 and 8.2.

It is essential that we use the precise name for an element or its ion in speaking or writing; their properties are entirely different. Sodium is a reactive, silver-colored metal, while sodium *ions*, Na^+, are stable and do not look or act like metallic sodium. Chlorine is a reactive gas with the characteristic odor of a swimming pool, but chloride *ions*, Cl^-, are common in saltwater and seawater and in the fluids of plant and animal cells. Be careful that you do not say that a particular solution contains chlor*ine* if you actually mean chlor*ide* ions—they are not the same at all! Also, avoid writing the symbol for the *element* when you intend to refer to the *ion*.

Metals lose valence electrons to form positive ions called *cations*. Nonmetals gain electrons to form negative ions called *anions*.

Periodic Relationships Among Simple Ions

The periodic relationship of some of the simple ions is shown in Figure 8.1. Notice that the alkali metals of Group IA form ions with a charge of 1+ (often written simply as a plus sign). Calcium and other Group IIA metals form ions with a charge of 2+. Similarly, aluminum and other Group IIIA metals form ions with a charge of 3+.

When nonmetallic elements gain electrons, negative ions are formed. The halogens in Group VIIA lack one electron of having an octet of electrons; they tend to gain one electron to form halide ions with a charge of 1−. Oxygen and sulfur in Group VIA lack two electrons of having an octet of electrons; they tend to gain two electrons to make oxide, O^{2-}, and sulfide, S^{2-}, ions. Similarly, nitrogen and phosphorus in Group VA tend to gain three electrons to make nitride, N^{3-}, and phosphide, P^{3-}, ions.

Most of the transition metals—the elements in the central region of the periodic table—can form more than one kind of ion; each has its unique charge. For example, iron can form two different ions: one is Fe^{2+} and the other is Fe^{3+}. There are two different methods of naming these ions that can have multiple charges. In one method, the name of the metal is followed immediately by a Roman numeral in parentheses to indicate the charge of the ion. For example, the Fe^{3+} ion is identified as iron(III); when speaking, we say, ''iron three.'' Similarly, the Fe^{2+} ion is identified as iron(II); in speaking, we say, ''iron two.'' The IUPAC recommends that compounds with ions like these be named by this method, called the **Stock system.**

The Stock system was developed by Alfred Stock (1876–1946), a German chemist who came to the United States and taught at Cornell University.

An older method of naming certain metal ions—still in use today—identifies

IA	IIA	IIIB	IVB	VB	VIB	VIIB	VIII	VIII	VIII	IB	IIB	IIIA	IVA	VA	VIA	VIIA	VIIIA
Li^+														N^{3-}	O^{2-}	F^-	
Na^+	Mg^{2+}											Al^{3+}		P^{3-}	S^{2-}	Cl^-	
K^+	Ca^{2+}						Fe^{2+} Fe^{3+}	Co^{2+} Co^{3+}	Ni^{2+} Ni^{3+}	Cu^+ Cu^{2+}	Zn^{2+}					Br^-	
Rb^+	Sr^{2+}									Ag^+			Sn^{2+} Sn^{4+}			I^-	
Cs^+	Ba^{2+}												Pb^{2+} Pb^{4+}				

Figure 8.1 *The periodic relationship of some simple ions.*

the ion with the lower charge number by the Latin name of the element followed with an *-ous* ending. The *-ic* ending is used to identify the ion with the higher charge number. Thus, the Fe^{2+} ion can be referred to as an iron(II) ion, or as a ferrous ion, while the Fe^{3+} ion can be called an iron(III) ion or a ferric ion.

It is suggested that you notice the special situation related to the ions of mercury. The mercuric or mercury(II) ion is written Hg^{2+} as one would expect, but the mercurous or mercury(I) ion is listed in Table 8.1 as Hg_2^{2+} to indicate that Hg^+ ions always exist in pairs.

Learn the names and symbols for the various transition metal ions and the other cations included in Table 8.1.

EXAMPLE 8.1

Use Tables 8.1 and 8.2 and the periodic table to assist you in writing symbols for the following.

(a) a copper atom, a copper(I) ion, and a cupric ion
(b) a bromine molecule, a bromine atom, and a bromide ion
(c) an oxygen molecule, an oxygen atom, and an oxide ion

SOLUTION

(a) Cu, Cu^+, Cu^{2+}. Atoms have no charge; ions do. See Table 8.1.
(b) Br_2, Br, Br^-. A bromine molecule is diatomic.
(c) O_2, O, O^{2-}. Oxygen molecules are diatomic, the atom is neutral, and the oxide ion has a 2− charge.

EXERCISE 8.1

Write symbols for (a) nitrogen gas, (b) a nitride ion, (c) a sulfide ion, and (d) a nickel(II) ion.

8.2 Polyatomic Ions

The **polyatomic ions** are those that involve a cluster of two or more atoms, covalently bonded, but possessing an overall charge. The ammonium ion, NH_4^+, and the nitrate ion, NO_3^-, are two examples. While there are many polyatomic anions listed in Table 8.2, notice that the ammonium ion is the only polyatomic cation listed in Table 8.1. The tetrahedral structure of the ammonium ion was described in Chapter 7. When a neutral ammonia molecule, NH_3, picks up an additional hydrogen ion, an ammonium ion, NH_4^+, is formed. The following information should be helpful as you learn the names and formulas of polyatomic anions shown in Table 8.2.

1. Ions with the suffixes *-ate* and *-ite* contain oxygen atoms.
   Examples: nitrate, NO_3^-, and nitrite, NO_2^-.
2. An ion with the suffix *-ite* has one fewer oxygen atoms than a corresponding ion with the suffix *-ate*.
   Examples: sulfate ion, SO_4^{2-}, and sulfite ion, SO_3^{2-}; nitrate ion, NO_3^-, and nitrite ion, NO_2^-.

3. Warning: All ions with the suffix *-ate* do not have the same specific number of oxygen atoms, nor the same charge. This is also true for ions with the suffix *-ite*.
   Examples: Compare a nitrate ion, NO_3^-, and a sulfate ion, SO_4^{2-}; compare a nitrite ion, NO_2^-, and a sulfite ion, SO_3^{2-}.
4. The prefix *bi-* for polyatomic ions means hydrogen—not the number 2.
   Examples: bicarbonate ion, HCO_3^-, and bisulfate ion, HSO_4^{2-}. The *bi-* prefix can also be read as hydrogen; bicarbonate is also hydrogen carbonate.
5. An anion with the prefix *per-* contains one more oxygen atom than the corresponding *-ate* ion without the prefix.
   Examples: perchlorate ion, ClO_4^-, and chlorate ion, ClO_3^-.
6. An anion with the prefix *hypo-* (Greek for "under") contains one less oxygen atom than the corresponding *-ite* ion without the prefix.
   Examples: hypochlorite ion, ClO^-, and chlorite ion, ClO_2^-.
7. The prefix *thio-* identifies the presence of sulfur in place of oxygen.
   Examples: sulfate ion, SO_4^{2-}, and thiosulfate ion, $S_2O_3^{2-}$; cyanate ion, OCN^-, and thiocyanate ion, SCN^-.

You will need to learn the names and formulas for ions listed in Tables 8.1 and 8.2. Learn to provide either the formula or the name quickly for each ion as shown by the following examples. Considerable practice may be required.

EXAMPLE 8.2

Provide either the formula or name as required for the following ions.

(a) ammonium ion (b) bicarbonate ion (c) sulfate ion
(d) sulfite ion (e) PO_4^{3-} (f) $Cr_2O_7^{2-}$

SOLUTION

Use Tables 8.1 and 8.2 for assistance now, but you need to learn to write these quickly without using the tables.

See Problems 8.1–8.8.

(a) ammonium ion, NH_4^+ (b) bicarbonate ion, HCO_3^-
(c) sulfate ion, SO_4^{2-} (d) sulfite ion, SO_3^{2-}
(e) PO_4^{3-}, phosphate ion (f) $Cr_2O_7^{2-}$, dichromate ion

EXERCISE 8.2

Give formulas for the following ions.
(a) chromate CrO_4^{2-} (b) perchlorate ClO_4^- (c) hypochlorite ClO^-

8.3 *Names and Formulas of Ionic Compounds*

For any ionic compound, the total charge of all positive ions (cations) must equal the total charge of all negative ions (anions).

All compounds are neutral; the overall charge is zero.

To *name* an ionic compound, name the cation first. Follow this with the name of the anion. For example, the compound consisting of magnesium ions and chloride ions is named magnesium chloride (Figure 8.2).

While learning to write chemical formulas for compounds, you will find it helpful to write the formula for the cation first with its charge followed by the anion with its charge, before writing the final formula. Again, remember that all compounds are neutral in charge; the sum of all positive and negative charges must be equal to zero. As long as the charge of a single cation exactly balances the charge of the anion (such as 2+ and 2−), the formula is simply written with one of each kind of ion. For example, the formula of calcium oxide is CaO with one ion of each kind, since the positive charge, 2+, balances the negative charge, 2−, to make an electrically neutral compound.

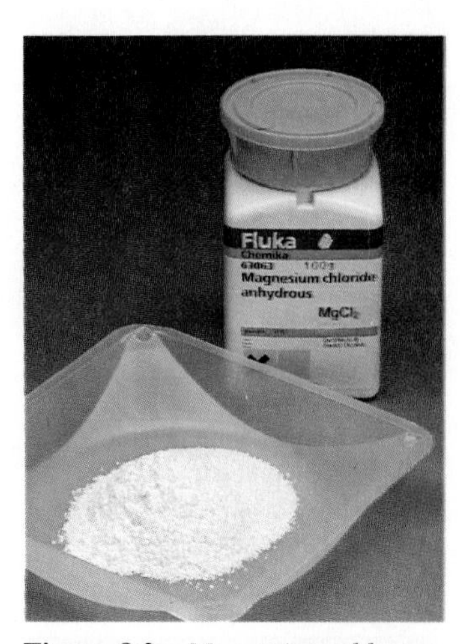

Figure 8.2 *Magnesium chloride, $MgCl_2$, is a white crystalline solid that is soluble in water. Uses include fireproofing wood, ceramic, thread lubricant, and paper manufacturing.*

One Ca^{2+} balances one O^{2-}, so the formula is CaO, calcium oxide.

Similarly, the formula for aluminum nitride is written AlN with one of each type, since the aluminum ion, Al^{3+}, charge exactly balances the nitride ion, N^{3-}, charge. The simplest whole-number ratio is used in writing formulas of ionic compounds. When there is no subscript, it is understood to be one. For example, in CaO and AlN the simplest ion ratios are 1 to 1. No subscripts are shown; they are understood to be one.

In the compound calcium chloride, one calcium ion, Ca^{2+}, combines with *two* chloride ions, Cl^-, to maintain electrical neutrality for the compound. This ratio is indicated in the formula $CaCl_2$ by use of the subscript 2 after Cl; this subscript applies to the Cl only, and not to the Ca. To summarize,

One Ca^{2+} ion balances two Cl^- ions; the formula is $CaCl_2$, calcium chloride.

Notice that the ionic charges are not shown in a chemical formula. Also, a subscript of 1 is understood when no other number appears in the position of the subscript. In the formula $CaCl_2$, the subscript 1 for the calcium ion is understood, but the 2 for chloride is explicitly written: $CaCl_2$ not Ca_1Cl_2. The formula $CaCl_2$ shows that a total of three ions are present, it gives the proportion of calcium and chloride ions (1 to 2), and it represents the compound, calcium chloride. Thus, the formula is a shorthand way of showing a compound with one Ca^{2+} ion and two Cl^- ions.

This same procedure for writing formulas can be used any time the charge of one ion is a simple multiple of the charge of the other ion. For example, the formula for potassium nitride is K_3N.

Three K^+ ions balance one N^{3-} ion, the formula is K_3N, potassium nitride.

A three-step process for writing chemical formulas is demonstrated in the following examples. It is especially useful when the charge of one ion is not a simple multiple of the other ion. It is essential that you learn the charges of common ions, but, until then, you may need to refer to Tables 8.1 and 8.2 for assistance in determining ion charges.

EXAMPLE 8.3

Write the formula for aluminum oxide (Figure 8.3).

SOLUTION

STEP 1 Write the symbols (including charge) of the cation and anion, with the cation listed first.

$$Al^{3+} \quad \text{and} \quad O^{2-}$$

The **least common multiple** is the smallest whole number into which both charge numbers can be divided. Exclude signs.

STEP 2 Determine the smallest number of each kind of ion needed to give an overall charge of zero. *Mentally*, determine the *least common multiple* (LCM) of ion charges—excluding signs. For Al^{3+} and O^{2-}, the LCM is $3 \times 2 = 6$.

We will need two Al^{3+} ions: $\dfrac{\text{LCM}}{\text{Ion charge}} = \dfrac{6}{3} = 2$

We will need three O^{2-} ions: $\dfrac{\text{LCM}}{\text{Ion charge}} = \dfrac{6}{2} = 3$

Disregard sign when determining the LCM.

Figure 8.3 *Aluminum oxide, Al_2O_3, is a white powder that melts at 2030 °C. Uses include the production of aluminum metal, abrasives, ceramics, paper, and artificial gems.*

STEP 3 Write the chemical formula, using the appropriate subscripts to make the compound neutral in charge. For aluminum oxide, the formula is written Al_2O_3. Then, check your work.

$$2\ Al^{3+} + 3\ O^{2-} = 2(+3) + 3(-2) = 0$$

In step 2 of Example 8.3, the LCM of 6 was used to determine that two aluminum ions would be used along with three oxide ions in writing the formula.

Now that you have seen *why* the subscripts 2 and 3 are used, you may find the following "shortcut" useful in writing formulas when one ion charge is *not* a simple multiple of the other ion. Notice that the subscript placed after one ion is the same as the numerical value (disregarding sign) of the charge on the other ion, as shown in the following examples.

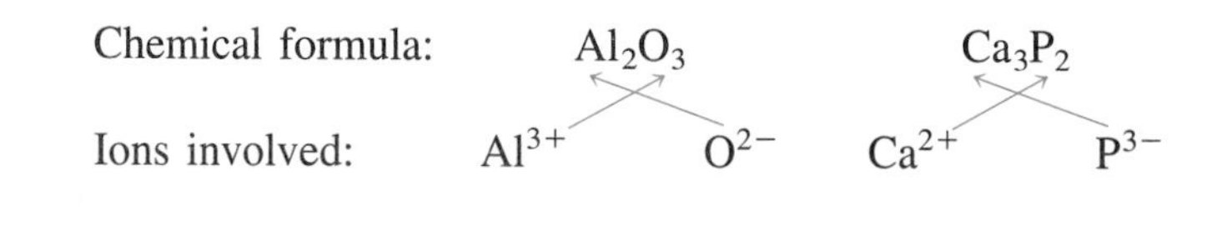

EXAMPLE 8.4

Write the formula for magnesium nitride.

SOLUTION

STEP 1 Write the symbols (including charge) of the cation and anion, with the cation listed first.

$$Mg^{2+} \quad \text{and} \quad N^{3-}$$

STEP 2 Determine the smallest number of each kind of ion needed to give an overall charge of zero. To do this, find the smallest whole number into which both charge numbers can be divided, the LCM. For Mg^{2+} and N^{3-} the LCM is $2 \times 3 = 6$. (Disregard signs.)

We will need three Mg^{2+} ions: $\dfrac{\text{LCM}}{\text{Charge}} = \dfrac{6}{2} = 3$

We will need two N^{3-} ions: $\dfrac{\text{LCM}}{\text{Charge}} = \dfrac{6}{3} = 2$

STEP 3 Write the chemical formula, using the appropriate subscripts to make the compound neutral in charge. The correct formula is written Mg_3N_2. Then, check your work.

$$3\,Mg^{2+} + 2\,N^{3-} = 3(+2) + 2(-3) = 0.$$

8.4 *The Use of Parentheses in Chemical Formula Writing*

When writing the chemical formula of a compound that contains a polyatomic ion, the ion is placed in parentheses before writing the subscript.

This rule for the use of parentheses is to be followed without variation, unless the subscript would be one (1), which is not written, so parentheses are not needed. For example, to write the formula for ammonium sulfate with NH_4^+ and SO_4^{2-} ions, we will need two ammonium ions with one sulfate ion to maintain neutrality.

Two NH_4^+ ions balance one SO_4^{2-} ion, so the formula is $(NH_4)_2SO_4$.

Similarly, to write the formula for magnesium phosphate with the ions Mg^{2+} and PO_4^{3-}, we will need three magnesium ions with two phosphate ions to achieve neutrality. Here, the least common multiple is 6.

Three Mg^{2+} ions balance two PO_4^{3-} ions, so the formula is $Mg_3(PO_4)_2$.

Notice that magnesium (a monatomic ion) was not placed in parentheses when followed by a subscript.

When writing chemical formulas, parentheses are *not* used where they are not needed.

1. Parentheses are not used unless they are followed by a subscript. For example, for Ca^{2+} and SO_4^{2-}, write $CaSO_4$, not $Ca(SO_4)$;
   for NH_4^+ and SO_4^{2-}, write $(NH_4)_2SO_4$, not $(NH_4)_2(SO_4)$.
2. Parentheses are not placed around monatomic ions in formulas. For example, for Ca^{2+} and Cl^-, write $CaCl_2$, not $Ca(Cl)_2$;
   for Al^{3+} and Cl^-, write $AlCl_3$, not $Al(Cl)_3$;
   for Al^{3+} and SO_4^{2-}, write $Al_2(SO_4)_3$, not $(Al)_2(SO_4)_3$.

Follow the steps outlined in the examples in this chapter, and practice writing chemical formulas as you do the problems included at the end of this chapter.

EXAMPLE 8.5

Write formulas for the following compounds.

(a) cobalt(III) hydroxide (b) cobalt(II) chloride

SOLUTION

(a) **STEP 1** Write the symbols (including charge) of the cation and anion, with the cation listed first.

$$Co^{3+} \quad \text{and} \quad OH^-$$

STEP 2 Determine the smallest number of each kind of ion needed to give an overall charge of zero. We will need one Co^{3+} ion and three OH^- ions.

STEP 3 Write the chemical formula, using the appropriate subscripts to make the compound neutral in charge. The correct formula is $Co(OH)_3$. Notice that the hydroxide ion (made up of one oxygen and one hydrogen) is enclosed in parentheses followed by the subscript 3. Thus, each formula unit contains one cobalt atom, three oxygen atoms, and three hydrogen atoms. Then, check your work: $1\ Co^{3+} + 3\ OH^- = 1(+3) + 3(-1) = 0$.

(b) The formula of cobalt(II) chloride is $CoCl_2$ with no parentheses; there are no polyatomic ions.

Study Hint

It may take considerable practice to become proficient at quickly writing chemical formulas and names of chemicals, but these skills will be needed as you proceed to the topics and the problems in subsequent chapters.

See Problems 8.9 and 8.10.

8.5 Determining the Name of an Ionic Compound from Its Formula

To write the name of an ionic compound when the formula is given, simply write the name of the positive ion first and then write the name of the negative ion as separate words. Follow through with the examples.

EXAMPLE 8.6

What is the name for the compound K_2S?

SOLUTION

The compound contains two potassium ions, K^+, and one sulfide ion, S^{2-}. The compound is potassium sulfide.

EXERCISE 8.3

Name the compounds (a) $CaCl_2$, (b) $Zn(NO_3)_2$, and (c) Na_2SO_4.

If one of the two ions has a variable charge, such as Fe(II) and Fe(III) or Sn(II) and Sn(IV), the name must explicitly identify which ion is present. The Stock system name, iron(II) chloride, or the older name, ferrous chloride, can be used, but it is incorrect to refer to the compound merely as iron chloride—an indefinite and therefore incorrect name.

When the cation can have more than one charge, its value may not be obvious. Remember that the sum of all positive charges and negative charges must be zero. If the charge of a positive ion is variable, first determine the total negative charge (this is the charge of the negative ion times the number of negative ions involved). The total positive charge must be equal to the total negative charge because compounds are neutral. Divide the total positive charge by the number of positive ions shown in the formula to obtain the charge of each cation. The number obtained is the charge of the ion with a variable charge.

For example, $Sn(SO_4)_2$ contains a tin ion, but there are two possible charges. To determine the charge of the tin ion, first determine the total negative charge for two sulfide ions: $2(-2) = -4$. The total positive charge must therefore be +4. Divide this total charge by the number of tin ions in the compound. Here,

$$\frac{+4 \text{ total positive charge}}{1 \text{ positive ion}} = +4$$

The cation is Sn^{4+}. The name of the compound is tin(IV) sulfate or stannic sulfate.

EXAMPLE 8.7

What is the name for the compound FeS?

See Problems 8.11 and 8.12.

SOLUTION

There are two kinds of iron ions, Fe^{2+} and Fe^{3+}. Since there is one sulfide ion, S^{2-}, for each iron ion in the formula, the iron ion in this compound must be Fe^{2+} to make the compound neutral. Thus, FeS has Fe^{2+} and S^{2-} ions. The name for FeS is iron(II) sulfide or ferrous sulfide

The name iron sulfide is unacceptable; it is not definite.

EXERCISE 8.4

What is the name for the compound $Fe_2(SO_4)_3$?

8.6 Names and Formulas of Binary Compounds of Nonmetals

A binary covalent compound is formed when atoms of two nonmetallic elements share one or more electron pairs. Carbon dioxide and carbon monoxide are examples. In naming covalent compounds, the IUPAC has established the following preferred sequence for determining which nonmetal to name first.

$$B > Si > C > P > N > H > S > I > Br > Cl > O > F$$

As shown in this sequence, the *least electronegative* element is usually named first. For example, in a compound of nitrogen and oxygen, the nitrogen would precede oxygen in the name and formula.

The number of atoms of each kind of element in the molecule is indicated by a Greek prefix (see Table 8.3). The prefix mono- is generally omitted as a prefix for the first element in a compound unless this would result in an ambiguous name. According to IUPAC guidelines, the last "o" in mono- and the last "a" in tetra,

Table 8.3 **Greek Prefixes Used in Chemical Names**

Number	*Prefix*
1	mono-
2	di-
3	tri-
4	tetra-
5	penta-
6	hexa-
7	hepta-
8	octa-
9	nona-
10	deca-
12	dodeca-
14	tetradeca-

Table 8.4 **Binary Compounds of Nonmetals**

	Formula	*Name*		*Formula*	*Name*
Carbon	CO	Carbon monoxide*	Nitrogen	NO	Nitrogen monoxide†
	CO_2	Carbon dioxide		NO_2	Nitrogen dioxide
	CS_2	Carbon disulfide		N_2O	Dinitrogen monoxide‡
	CCl_4	Carbon tetrachloride		N_2O_3	Dinitrogen trioxide
Sulfur	SO_2	Sulfur dioxide		N_2O_5	Dinitrogen pentoxide
	SO_3	Sulfur trioxide	Phosphorus	PBr_3	Phosphorus tribromide
	SF_6	Sulfur hexafluoride		PCl_5	Phosphorus pentachloride

*The last "o" of *mono-* is omitted when joined to oxide.
†The prefix mono is omitted for the first element. Nitric oxide is another name for NO.
‡Nitrous oxide is another name for N_2O.

penta, and so on are dropped when the prefix is joined to a name that begins with the letter "o". Thus, carbon monoxide is *not* written monocarbon monooxide.

For binary compounds—those with two kinds of atoms—the second element in the compound has the suffix *-ide*. For example, SO_2 is sulfur diox*ide*. Names and formulas for several binary covalent compounds are listed in Table 8.4. Two familiar binary molecular compounds that retain their traditional names are water, H_2O, and ammonia, NH_3. Others include nitric oxide, NO, and nitrous oxide, N_2O, also called laughing gas.

EXAMPLE 8.8

Write an appropriate name when the formula is given. Write the formula when the name is given.

(a) dinitrogen pentoxide (b) carbon tetrachloride
(c) NO_2 (d) SO_3

SOLUTION

The Greek prefixes indicate the number of atoms involved.

(a) N_2O_5 (b) CCl_4
(c) nitrogen dioxide (d) sulfur trioxide

See Problems 8.13 and 8.14.

8.7 Oxidation Numbers of Atoms in Polyatomic Compounds

The specific charges of monatomic ions such as Na^+, Mg^{2+}, and Cl^- have been discussed in Sections 8.1 and 8.3. In polyatomic compounds, these individual ions carry specific charges, but there is no clearly measurable ionic charge associated with certain other atoms. For example, in the compound sodium nitrate, $NaNO_3$, the sodium ion carries a 1+ charge and the nitrate ion carries a 1− charge, but there is no ionic charge on the nitrogen since it is covalently bonded to oxygen atoms. We can, however, determine "apparent charge" that appears to be associated with the nitrogen in the compound, since the compound is neutral in charge.

The simple ion charges and the "apparent charges" assigned to atoms within

Table 8.5 Oxidation Numbers of Manganese

Ion or Compound	Oxidation State of Manganese	Color
MnO_4^-	+7*	Purple
MnO_4^{2-}	+6	Dark green
MnO_3^-	+5	Light blue
MnO_2	+4	Brown (solid)
Mn^{3+}	+3	Rose
Mn^{2+}	+2*	Pink

*The most stable oxidation states.

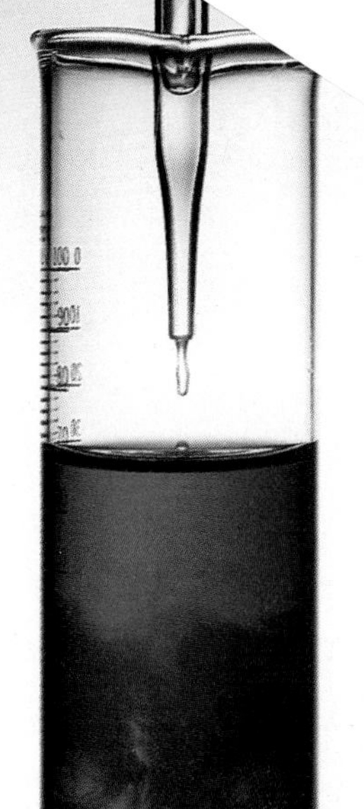

Figure 8.4 *Here green MnO_4^- ions with manganese in an oxidation state of +6 are reacting to form purple MnO_4^- ions with manganese in an oxidation state of +7.*

compounds are called **oxidation numbers**. The oxidation number, also called **oxidation state**, of an atom can be used in an electron bookkeeping system to keep track of electrons. It is sometimes useful to compare the oxidation states of an element like chlorine, for example, which can vary widely from one compound to another. Many transition metals also have several oxidation states. For example, manganese has six different oxidation states with six different colors, as shown in Table 8.5 and Figure 8.4.

Consider the following rules for assigning oxidation numbers to an element within an ion or a compound.

Rules for Assigning Oxidation Numbers

1. Any *element* not combined with a different element is assigned an oxidation number of zero. Examples: K, Fe, H_2, O_2.
2. For a *compound*, the sum of all oxidation numbers of all atoms is zero.
3. For a *polyatomic ion*, the sum of all oxidation numbers of all atoms is equal to the charge on the ion.
4. All *monatomic ions* are assigned oxidation numbers equal to the charge on their ions. Example: The oxidation number of a K^+ ion is +1.
5. When *oxygen* is present in a compound or ion, it usually has an oxidation number of −2. (Exceptions include peroxides, such as H_2O_2, in which oxygen has an oxidation number of −1.)
6. *Hydrogen* usually has an oxidation number of +1, except in metal hydrides, such as NaH and $LiAlH_4$, where H is −1.

As you determine the oxidation number of a particular kind of atom in a compound or polyatomic ion, it will be helpful to follow these steps.

Thinking It Through

Determining an Oxidation Number

1. Write down the known oxidation numbers for atoms represented in the formula.
2. Multiply the oxidation number of each element by the appropriate subscript shown in the formula. Write these total oxidation numbers below the corresponding symbols in the formula.
3. Write a simple equation where the sum of all oxidation numbers is set equal to the charge of the ion, or set equal to zero for a compound. (Use a symbol for the unknown oxidation number multiplied by the number of atoms of that element.) Solve for the missing oxidation number.

Practice using the rules and steps listed for determining oxidation numbers in the following examples.

EXAMPLE 8.9

Determine the oxidation number of chromium in potassium dichromate, $K_2Cr_2O_7$.

SOLUTION

STEP 1

Atom	*Oxidation Number*
K	+1
Cr	Cr (unknown = x)
O	−2

STEP 2 Total oxidation numbers

$$\underset{+2}{K_2}\underset{}{Cr_2}\underset{-14}{O_7}$$

STEP 3

$$+2 + 2x + (-14) = 0$$
$$2x = +12$$
$$x = +6$$

The sum of all oxidation numbers in a compound is zero.

The oxidation number of chromium in potassium dichromate is +6.

EXERCISE 8.5

See Problems 8.15 and 8.16.

Determine the oxidation number of nitrogen in $NaNO_3$.

EXAMPLE 8.10

Determine the oxidation number of chlorine in a chlorate ion, ClO_3^-.

SOLUTION

STEP 1

Atom	*Oxidation Number*
Cl	Cl (unknown = x)
O	−2

STEP 2 Total oxidation numbers

$$\underset{?}{Cl}\underset{-6}{O_3}^-$$

STEP 3

$$x + (-6) = -1$$
$$x = +5$$

The charge on the ion is −1.

With practice, you may learn to do step 3 in your head without writing it down.

The oxidation number of chlorine in a chlorate ion is +5.

8.8 Nomenclature of Acids and Their Salts

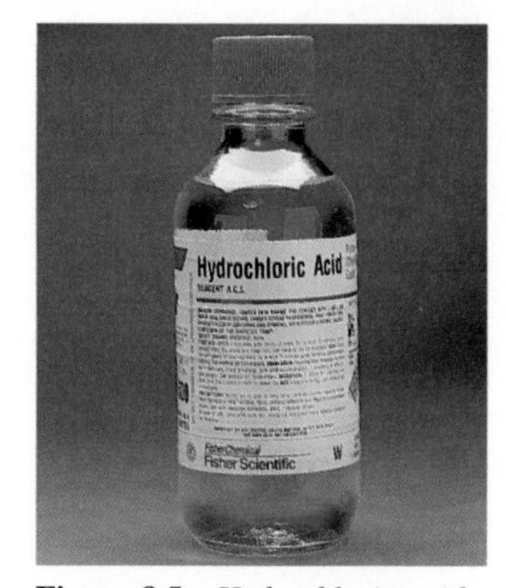

Figure 8.5 *Hydrochloric acid, HCl(aq), is produced by dissolving hydrogen chloride gas in water. The acid is widely used in petroleum refining, food processing, and metal cleaning, as well as for industrial processes.*

Certain compounds release hydrogen ions, H^+, when dissolved in water. These substances are called **acids**; they have a characteristic sour taste and react with certain metals to produce hydrogen gas. The traditional nomenclature (naming) of acids is different from the nomenclature of ionic compounds. For example, HNO_3 in a water solution is known as nitric acid, not hydrogen nitrate.

Hydrochloric Acid

Hydrogen chloride gas, HCl(g), is a polar covalent compound. However, an **aqueous** (abbreviated *aq*) solution of hydrogen chloride—a water solution of HCl—has entirely different properties; it is the familiar hydrochloric acid, HCl(aq) (Figure 8.5). As the HCl gas dissolves in water, ions are produced. This can be demonstrated by testing the electrical conductivity of the acid solution. Although distilled water is a nonconductor, the hydrochloric acid is an excellent conductor of electricity because it is completely ionized in water. Acids that are completely ionized are classified as *strong acids.*

When a hydrogen ion, H^+, is released by HCl, it combines with water to form a **hydronium ion**, H_3O^+. This reaction can be represented as follows.

$$\underset{\text{HCl gas}}{HCl(g)} + \underset{\text{Water}}{H_2O} \longrightarrow \underset{\text{Hydrochloric acid}}{H_3O^+ + Cl^-}$$

We can think of the hydronium ions in acid solutions as hydrogen ions riding "piggy back" on water molecules. All acidic solutions contain hydronium ions.

Acetic Acid and Carbonic Acid

The sour taste of vinegar is due to the presence of hydronium (hydrogen) ions in acetic acid. Unlike hydrochloric acid, acetic acid is a poor conductor of electricity; only a small fraction of its molecules are ionized when acetic acid dissolves in water. Acetic acid—like other carbon-containing acids—is classified as a weak acid. The formula of acetic acid is written $HC_2H_3O_2$ or CH_3COOH. The following structural formulas allow us to show the bonding and the ionization of acetic acid to give acetate and hydrogen ions.

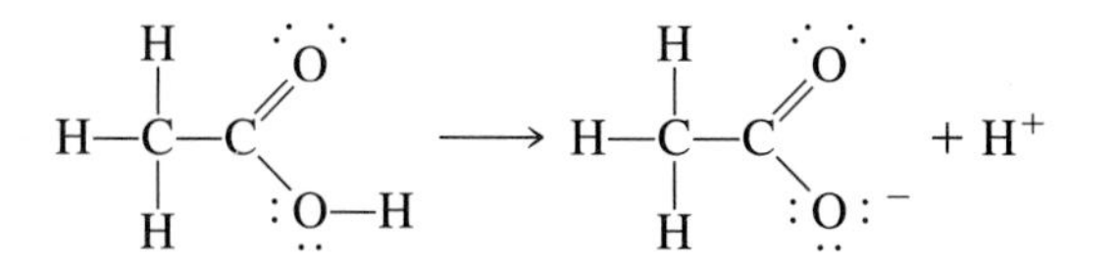

The ionization of acetic acid can also be shown with "line formulas" as follows.

$$\underset{\text{Acetic acid}}{CH_3COOH} \rightarrow \underset{\text{Acetate ion}}{CH_3COO^-} + \underset{\text{Hydrogen ion}}{H^+}$$

The Lewis structures show that the hydrogen atoms bonded to carbon do not ionize; the only hydrogen to break away and ionize is in the —COOH group of the

Figure 8.6 *All carbonated beverages contain carbonic acid, H_2CO_3, formed when carbon dioxide gas dissolves in water.*

acid. This —COOH group is called the **carboxyl group**. In addition to acetic acid, there are many other acids that contain the carboxyl group bonded to a string of carbons. In each case, the acidity of these organic acids—called **carboxylic acids**—is due to the hydrogen in the carboxyl group. Since the degree of ionization is small for these carboxylic acids, they are all classified as *weak acids*.

Another important weak acid is carbonic acid, H_2CO_3, formed when carbon dioxide dissolves in water. All carbonated beverages contain carbonic acid (Figure 8.6). You should learn the names and formulas of acetic acid and carbonic acid.

Other Acids, Their Salts, and Oxidation Numbers

When the hydrogen of an acid is replaced by a cation, the ionic compound formed is called a **salt**. For example, sodium chloride (common table salt) is the sodium salt of hydrochloric acid. Similarly, calcium chloride, $CaCl_2$, is the calcium salt of hydrochloric acid. The names and formulas of acids, their anions, and typical salts are listed in Table 8.6. You will need to learn the names and formulas of certain acids as directed by your instructor.

Several of the acids listed in Table 8.6 are **oxyacids**; their anions contain oxygen bonded to another nonmetal. Notice that oxyacids of nitrogen, phosphorus, sulfur, or chlorine with the *-ic acid* ending have corresponding anions with the *-ate* suffix. For example, sulfur*ic acid* releases sulf*ate* anions, and nitr*ic acid* releases nitr*ate* anions. Thus, sodium sulfate and calcium sulfate are salts of sulfuric acid; sodium nitrate and potassium nitrate are salts of nitric acid.

Acids that have the *-ous acid* ending have corresponding anions with the *-ite* suffix. For example, sulfur*ous acid* releases sulf*ite* anions, and nitr*ous acid* gives nitr*ite* anions. Thus, sodium sulfite is a salt of sulfurous acid, and sodium nitrite is a salt of nitrous acid.

The number of oxygen atoms in an oxyacid with the *-ous* ending is always one fewer than the number of oxygen atoms in the corresponding acid with the *-ic* ending (*-ous* acids also have a lower oxidation number for the central atom). For example, the number of oxygen atoms in sulfurous acid, nitrous acid, and phospho-

Table 8.6 **Important Acids and Their Salts**

Acid Formula	*Acid Name*	*Anion Formula*	*Anion Name*	*Typical Salt*	*Oxidation State of Atom Identified*
BINARY ACIDS OF HALOGENS					
HF	Hydrofluoric acid	F^-	Fluoride ion	NaF	F = −1
HCl	Hydrochloric acid	Cl^-	Chloride ion	NaCl	Cl = −1
HBr	Hydrobromic acid	Br^-	Bromide ion	NaBr	Br = −1
HI	Hydroiodic acid	I^-	Iodide ion	NaI	I = −1
ACIDS CONTAINING SULFUR					
H_2SO_4	Sulfuric acid	SO_4^{2-}	Sulfate ion	$CaSO_4$	S = +6
H_2SO_3	Sulfurous acid	SO_3^{2-}	Sulfite ion	Na_2SO_3	S = +4
H_2S	Hydrosulfuric acid	S^{2-}	Sulfide ion	Na_2S	S = −2
ACIDS CONTAINING NITROGEN OR PHOSPHORUS					
HNO_3	Nitric acid	NO_3^-	Nitrate ion	KNO_3	N = +5
HNO_2	Nitrous acid	NO_2^-	Nitrite ion	KNO_2	N = +3
H_3PO_4	Phosphoric acid	PO_4^{3-}	Phosphate ion	$Ca_3(PO_4)_2$	P = +5
H_3PO_3	Phosphorous acid	HPO_3^{2-}	Monohydrogen phosphite ion	Na_2HPO_3	P = +3
OXYACIDS CONTAINING CHLORINE					
$HClO_4$	Perchloric acid	ClO_4^-	Perchlorate ion	$KClO_4$	Cl = +7
$HClO_3$	Chloric acid	ClO_3^-	Chlorate ion	$KClO_3$	Cl = +5
$HClO_2$	Chlorous acid	ClO_2^-	Chlorite ion	$KClO_2$	Cl = +3
HClO	Hypochlorous acid	ClO^-	Hypochlorite ion	KClO	Cl = +1
IMPORTANT ACIDS CONTAINING CARBON					
H_2CO_3	Carbonic acid	HCO_3^-	Bicarbonate ion	$NaHCO_3$	C = +4
*CH_3COOH	Acetic acid	CH_3COO^-	Acetate ion	$NaCH_3COO$	C = 0

*The formula of acetic acid is often written $HC_2H_3O_2$, with acetate as $C_2H_3O_2^-$.

rous acid is one fewer than the number in the corresponding *-ic* acids, sulfuric acid, nitric acid, and phosphoric acid, respectively. Similarly, the number of oxygen atoms in a salt with the *-ite* ending is always one fewer than the number in the corresponding salt with the *-ate* ending. The compounds sodium nitrite, $NaNO_2$, and sodium nitrate, $NaNO_3$, illustrate this point.

When several different acids are formed from the same nonmetal, the oxidation number of the central nonmetal atom can vary widely, as shown in Table 8.6. For example, you will notice that five chlorine-containing acids are listed; each has a different oxidation number on its chlorine atom. There are three acids listed that contain sulfur; each has a different oxidation number for its sulfur atom. Two acids containing nitrogen and two acids containing phosphorus are also listed, along with the appropriate oxidation numbers.

CHEMISTRY IN OUR WORLD

Names Remembered and Forgotten

The names of chemicals—like the names of people—are easily forgotten unless you become familiar with them yourself and learn their unique characteristics. A name like sodium nitrite, for example, may have little more meaning than a name in a phone book until you meet it firsthand, work with it, and learn something interesting about it.

Sodium nitrite is a chemical name that you might recognize; it has been surrounded by considerable controversy. For many years this compound has been used to preserve meat. Besides inhibiting spoilage, it helps maintain the pink color of smoked hams, frankfurters, and luncheon meats. Nitrites, however, have been investigated as a possible cause of cancer of the stomach. In the presence of acid in the stomach, nitrites can be converted to nitrous acid, HNO_2. If the nitrous acid reacts with compounds classified as secondary amines, nitroso compounds can be formed. Research has shown that nitroso compounds are potent carcinogens—substances that can cause cancer. Also, there is concern that nitroso compounds may be produced when meats containing nitrites are heated to high temperatures during grilling or frying.

The amount of sodium nitrite allowed in foods has been reduced, but this action may not be enough. Some researchers claim that nitrates in tap water can be converted to nitrites in the lower intestine.

The search for more information and better products is endless; we must continually test products and replace those that are no longer desirable or effective. Like most things in life, the needs are always greater than the funds.

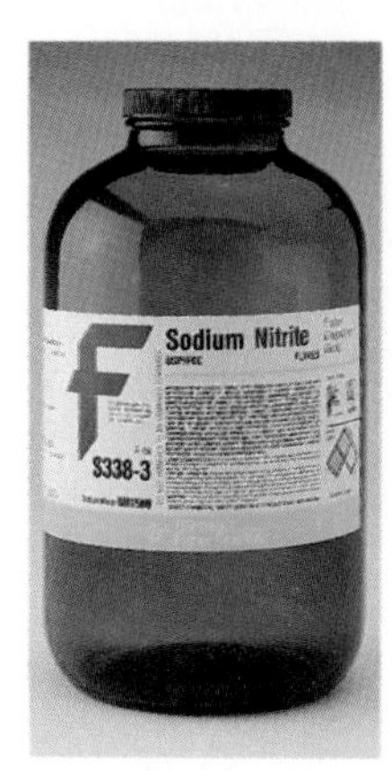

Sodium nitrite, $NaNO_3$.

Acids Containing Halogens and Their Salts

Let us compare the five different acids listed in Table 8.6 that contain chlorine. One of these, HCl, is listed with the binary acids—it does not contain oxygen. The other four acids of chlorine are oxyacids—they contain a chlorine atom bonded to various numbers of oxygen atoms. The oxyacids of chlorine are grouped together in Table 8.6. Notice that the oxidation number of chlorine is different in each of these acids, varying from -1 in HCl to $+7$ in $HClO_4$.

Suffixes for the oxyacids of chlorine and their corresponding anions are listed in Table 8.7. As described for other oxyacids, again notice that

1. Oxyacids with the suffic *-ic* form anions and salts that end in *-ate*.
2. Oxyacids with the suffix *-ous* form anions and salts that end in *-ite*.
3. An oxyacid with the suffix *-ous* always has a number of oxygen atoms that is one less than the corresponding acid that ends in *-ic*.

When there are more than two oxyacids of a particular nonmetal, the prefix *per-* is used along with the *-ic* suffix to indicate that the number of oxygens in an acid

Table 8.7 Suffixes of Acids and Their Salts

Acid Suffix	*Example*	*Ion Suffix*	*Example*	*Oxidation State of Cl*
-ic acid	Perchloric acid	-ate	Perchlorate	+7
	Chloric acid		Chlorate	+5
-ous acid	Chlorous acid	-ite	Chlorite	+3
	Hypochlorous acid		Hypochlorite	+1

and its corresponding anion has been increased by one (the *per-* also identifies a higher oxidation number for chlorine). The prefix *hypo-* (meaning "under") is used along with the *-ous* suffix to indicate that the number of oxygens in an acid and its corresponding anion has been decreased by one; the oxidation number of chlorine has been decreased. Lewis electron-dot structures of the four oxyacids of chlorine are shown here.

```
                                                 ..                 ..
                                                :O:                :O:
   .. ..           .. .. ..                 .. .. ..            .. .. ..
 H:O:Cl:         H:O:Cl:O:                H:O:Cl:O:           H:O:Cl:O:
   .. ..           .. .. ..                 .. .. ..            .. .. ..
                                                                   :O:
                                                                    ..
```

Hypochlorous acid | Chlorous acid | Chloric acid | Perchloric acid

Other halogen-containing acids and their salts are named similarly. For example, $HBrO_3$ and HIO_3 are named bromic acid and iodic acid; their anions, BrO_3^- and IO_3^-, are named bromate and iodate ions, respectively.

Salts of Polyprotic Acids

Sulfuric acid, H_2SO_4, is classified as a diprotic acid—it has two hydrogens that can be replaced by other cations. When both hydrogens of sulfuric acid are replaced by sodium ions, the salt sodium sulfate is formed. When all three hydrogens of phosphoric acid, H_3PO_4, are replaced by sodium ions, the salt sodium phosphate is formed. The names and formulas for salts like these are written in the way described for other ionic compounds in Section 8.3.

Polyprotic acids—those with more than one ionizable, acidic hydrogen—do not

Table 8.8 Salts and Ions That Contain Acidic Hydrogens

Acid	*Acidic Ion*	*Ion Name*	*Typical Salt*	*Name of Salt*
H_2CO_3	HCO_3^-	Hydrogen carbonate (bicarbonate)	$NaHCO_3$	Sodium hydrogen carbonate (sodium bicarbonate)
H_2SO_4	HSO_4^-	Hydrogen sulfate (bisulfate)	$NaHSO_4$	Sodium hydrogen sulfate (sodium bisulfate)
H_2SO_3	HSO_3^-	Hydrogen sulfite (bisulfite)	$NaHSO_3$	Sodium hydrogen sulfite (sodium bisulfite)
H_3PO_4	$H_2PO_4^-$	Dihydrogen phosphate	NaH_2PO_4	Sodium dihydrogen phosphate
	HPO_4^{2-}	Hydrogen phosphate	Na_2HPO_4	Sodium hydrogen phosphate*

*Some chemists would use disodium hydrogen phosphate for clarity.

necessarily lose all hydrogens in reactions. Salts that contain one or more hydrogens that can be replaced by other cations are called **acid salts**. For example, if only one hydrogen ion of sulfuric acid, H_2SO_4, is replaced by a sodium ion, the acid salt—sodium hydrogen sulfate, $NaHSO_4$ (or sodium bisulfate)—is formed. This acidic salt contains the hydrogen sulfate ion, HSO_4^-. Names and formulas of ions and salts that contain acidic hydrogens are listed in Table 8.8.

Sodium hydrogen sulfate is used in some toilet bowl cleaners.

EXAMPLE 8.11

For each of the following, write the chemical formula when the name is given; write the name when the formula is given.

(a) hypochlorous acid (b) sulfurous acid (c) bromic acid
(d) KCH_3COO (e) $Ca(ClO)_2$ (f) IO_3^-

SOLUTION

See Table 8.6 for the names and formulas of similar compounds.

(a) $HClO$ (b) H_2SO_3 (c) $HBrO_3$
(d) potassium acetate (e) calcium hypochlorite (f) iodate ion

See Problems 8.17–8.22.

8.9 *Hydrates*

A crystalline compound that contains a definite number of water molecules is called a **hydrate**. These crystalline solids have a definite composition and do not appear to contain moisture, but gentle heating releases a fixed amount of water from these inorganic compounds. For example, when the bright blue, crystalline solid hydrate of copper(II) sulfate is heated to about 100 °C in a test tube, water is released; it appears as droplets of moisture in the cool, upper portion of the test tube. After heating, an almost white solid residue remains in the test tube. The release of water by the hydrate can be represented as follows.

$$\underset{\text{Hydrate}}{CuSO_4 \cdot 5H_2O} \xrightarrow{\text{heat}} \underset{\text{Anhydrous salt}}{CuSO_4} + \underset{\text{Water}}{5\,H_2O}$$

When the **water of hydration** in the hydrate is removed, the resultant salt is called an **anhydrous salt**—it has no water of hydration. If the hydrate is colored,

Figure 8.7 *Copper(II) sulfate pentahydrate crystals have a distinctive bright blue color. The chemical is used in medicine, as a wood preservative, for electroplating process, and other uses.*

Table 8.9 Names and Formulas of Selected Hydrates

Formula of the Hydrate	*Chemical Name (Common Name)*	*Uses*
$CaSO_4 \cdot 2H_2O$	Calcium sulfate dihydrate (gypsum)	Dry wall sheets, plaster of Paris figurines
$CuSO_4 \cdot 5H_2O$	Copper(II) sulfate pentahydrate (blue vitriol)	Insecticide, algicide, wood preservative
$KAl(SO_4)_2 \cdot 12H_2O$	Potassium aluminum sulfate dodecahydrate (alum)	Aid in dyeing of fabrics
$MgSO_4 \cdot 7H_2O$	Magnesium sulfate heptahydrate (Epsom salt)	Medicine (cathartic), dyeing, tanning
$Na_2CO_3 \cdot 10H_2O$	Sodium carbonate decahydrate (washing soda)	Water softening, cleaning agents
$Na_2B_4O_7 \cdot 10H_2O$	Sodium tetraborate decahydrate (borax)	Laundry detergents, water-softening agent
$Na_2S_2O_3 \cdot 5H_2O$	Sodium thiosulfate pentahydrate (photographer's hypo)	Photographic developing

as described for the hydrate of copper(II) sulfate, a color change occurs along with the loss of water, but most hydrates are white, like their anhydrous salts.

To write the formula of a hydrate, place a centered dot between the formula of the anhydrous salt and the number of water molecules in the simplest unit of the hydrate. $CuSO_4 \cdot 5H_2O$ is the formula for the hydrate described in our investigation. In this compound, five formula units of H_2O are associated with one formula unit of $CuSO_4$ in the crystalline solid (Figure 8.7).

To write the name of a hydrate, write the name of the anhydrous salt followed by the word *hydrate* with the appropriate Greek prefix to indicate the number of water molecules in the formula unit. The name for $CuSO_4 \cdot 5H_2O$ is copper(II) sulfate pentahydrate. The names and formulas of selected hydrates are listed in Table 8.9. Notice that several hydrates have familiar common names and uses.

EXAMPLE 8.12

Write formulas for the following hydrates.

See Problems 8.23 and 8.24.

(a) sodium carbonate decahydrate, also called washing soda
(b) magnesium sulfate heptahydrate, also known as Epsom salt

SOLUTION

(a) $Na_2CO_3 \cdot 10H_2O$ (b) $MgSO_4 \cdot 7H_2O$

EXERCISE 8.6

Write chemical names for the following hydrates.

(a) $CaSO_4 \cdot 2H_2O$. This compound is also called gypsum. It is the chalky substance sandwiched between cardboard to make sheets of dry wall.
(b) $CoCl_2 \cdot 6H_2O$. As water is lost by this hydrate, the color changes from pink to blue. Paper test strips containing this hydrate can be used to check humidity in a room.

8.10 Some Chemicals and Their Uses

Many chemicals with practical uses are known to most people only by their common names. While systematic Stock chemical names identify precisely the makeup of the compound, common names do not. The general public, however, continues to use common names for familiar chemicals. The common names of some hydrates are given in Table 8.9. Table 8.10 lists common names of several more chemicals, along with their formulas, chemical names, and uses.

Among these chemicals you will find sodium bicarbonate, with the common name *baking soda* derived from its use in certain cooking recipes. It can also help eliminate odors from your refrigerator. Calcium oxide is another important chemical. It is often called *quicklime* or simply *lime*. It is a key ingredient in cement. When calcium oxide is mixed with water, considerable heat is liberated and calcium hydroxide is formed. The process is called *slaking*, and the product, calcium

Table 8.10 Some Common Chemicals, Their Names, Formulas, and Uses

Chemical Name	*Formula*	*Common Name*	*Uses*
Acetic acid	CH_3COOH	Vinegar	Pickling, salad dressing, manufacture of other chemicals
Calcium carbonate	$CaCO_3$	Limestone, marble, calcite	Antacids, tablet coatings, manufacture of cement
Calcium hydroxide	$Ca(OH)_2$	Slaked lime	Mortar, plaster, industrial neutralization of acids
Calcium oxide	CaO	Quicklime	Cement, mortar, steelmaking
Carbon dioxide solid	CO_2	Dry ice	Refrigeration of items mailed, fire extinguishers (liquefied CO_2)
Dinitrogen oxide (nitrous oxide)	N_2O	Laughing gas	Anesthesia, oxidant for high-energy fuel
Ethanol (ethyl alcohol)	C_2H_5OH	Grain alcohol	Liquor, beer, wine, fuel, industrial solvent, chemical manufacturing
Hydrochloric acid	HCl	Muriatic acid	Cleaning brick and metals, present in stomach acid
Lead(II) oxide	PbO	Litharge	Cement for metal tube fittings
Magnesium hydroxide	$Mg(OH)_2$	Milk of magnesia	Laxative, antacid
Methanol (methyl alcohol)	CH_3OH	Wood alcohol	Solvent, fuel, manufacture of adhesives, plastics, fibers
Potassium carbonate	K_2CO_3	Potash	Manufacture of glass, soap, other chemicals
Sodium bicarbonate	$NaHCO_3$	Baking soda	Baking soda, baking powder
Sodium carbonate	Na_2CO_3	Soda ash, soda	Manufacture of glass, paper, water softeners, and other chemicals
Sodium chloride	$NaCl$	Table salt	Seasoning, the melting of ice
Sodium hydroxide	$NaOH$	Caustic soda, lye	Neutralizing acids, cleaners, production of paper and other chemicals
Sodium thiosulfate	$Na_2S_2O_3$	Hypo	Photographic processing
Sucrose	$C_{12}H_{22}O_{11}$	Cane or beet sugar	Sweetener
Sulfuric acid	H_2SO_4	Battery acid	Manufacture of fertilizers and other chemicals, oil refining

hydroxide, is therefore known as *slaked lime*. Notice the common names and uses of several other common chemicals listed in Table 8.10.

Important Industrial Chemicals

Certain chemicals are especially important in the chemical industry today. You may be familiar with some of these, but not with others, since many of them become the raw materials for the production of other products that are more familiar. For example, each year the chemical industry produces tremendous quantities of a wide variety of chemicals used in the manufacture of metals and metal products, paper and paper products, food products, fertilizers, synthetic fibers, adhesives, plastics, paint and other coatings, hardware and building supplies, prescription and nonprescription drugs, cosmetics, cleaning products, and many more.

Virtually every product you buy involves chemical preparation, testing, or packaging—it was touched by the chemical industry. Some of the most widely used chemicals are listed in Table 8.11. They are arranged in order of annual production. Although the quantities produced in the United States vary from year to year, take a look at these quantities as well as the names, formulas, and uses of these chemicals. Try to think of products you buy or use. Which ones did not involve chemicals in some way? If you can name some, think again!

Sulfuric acid, H_2SO_4, is listed first in Table 8.11. With 81 billion pounds produced annually, it ranks first in chemical production in the United States. To get an

CHEMISTRY IN OUR WORLD

Sodium Carbonate, a Key Chemical

Sodium carbonate is a chemical that has several common names, including soda, soda ash, and washing soda. Sodium carbonate is used in laundry detergents and in cleansers as a water-softening agent. It is also used in swimming pools to control acidity. The United States produces about 9.5 million tons of sodium carbonate each year. Industrially, it is used in the manufacture of glass, soap, detergents, and paper, as well as in the manufacture of other chemicals.

Common detergents and household cleaners contain sodium carbonate, Na_2CO_3, which acts as a water-softening agent.

Table 8.11 **Some Industrial Chemicals, Their Production, and Uses***

Chemical Name Formula	*Annual U.S. Production (billion lb)*	*Approximate Price per Unit*	*Source*	*Some Uses*
Sulfuric acid H_2SO_4	81	\$75/ton	Oxidation of sulfur, SO_3 plus water	Fertilizers (70%), metals processing, manufacture of other chemicals
Nitrogen N_2 (liquid)	65 (870 bcf)†	\$0.50/100 ft^3	Liquefied air	Metals, chemical manufacturing; refrigeration (−196 °C)
Oxygen O_2	46 (560 bcf)†	\$0.55/100 ft^3	Liquefied air	Metals processing and fabrication (60%), chemicals (20%)
Ethylene $CH_2{=}CH_2$	41	\$0.15/lb	Petroleum refining	Polyethylene plastics (75%), antifreeze (10%), fibers (5%)
Slaked lime $Ca(OH)_2$	37	\$55/ton	CaO (from limestone) plus water	Metals manufacturing, neutralizing acids, pollution control
Ammonia NH_3	35	\$110/ton	Reaction of $N_2 + H_2$	Fertilizers (80%), fibers, plastics, explosives (5%)
Sodium hydroxide NaOH	26	\$175/ton	Electrolysis of NaCl	Manufacture of chemicals (50%), paper, soap, cleansers
Chlorine Cl_2	24	\$200/ton	Electrolysis of NaCl	Chemicals and plastics (65%), paper products (15%)
Phosphoric acid H_3PO_4	23	\$250/ton	Phosphate rock + acid	Fertilizers (95%), detergents, animal feeds
Sodium carbonate Na_2CO_3	19	\$85/ton	Minerals or brine	Glass and chemicals (75%), detergents and cleansers (10%)
Nitric acid HNO_3	17	\$275/ton	$NH_3 \rightarrow NO_2$, $NO_2 + H_2O$	Fertilizers (70%), plastics, explosives, manufacture of chemicals
Ammonium nitrate NH_4NO_3	16	\$135/ton	Ammonia + HNO_3	Fertilizers, explosives, organic chemicals
Urea $(NH_2)_2C{=}O$	12	\$100/ton	$NH_3 + CO_2$ reaction	Fertilizers (80%), animal feeds, plastics
Carbon dioxide CO_2	10	\$70/ton	By-product of natural gas, and NH_3 manufacturing	Refrigeration (50%), beverages (20%), metals and chemical production (15%)
Hydrochloric acid HCl	6	\$70/ton	By-product of chemical manufacturing	Petroleum refining, chemicals and metals processing

*Chemicals are listed in order of annual U.S. production for 1993.
†bcf = billions of cubic feet.

idea of the impact that this one chemical has on our economy, we can readily calculate the total annual sales using figures from Table 8.11.

$$\frac{81{,}000{,}000{,}000 \text{ lb}}{\text{year}} \times \frac{1 \text{ ton}}{2000 \text{ lb}} \times \frac{\$75.00}{\text{ton}} = \$3.0 \text{ billion/year}$$

Sulfuric acid, the leading chemical in U.S. production, is employed in the manufacture of products used by virtually everyone.

Annual sales of other major chemicals can be calculated similarly.

Since sulfuric acid is the least expensive acid, it is used extensively in industry for many purposes. About 70% of all sulfuric acid is used in the production of fertilizers. When sulfuric acid is reacted with minerals rich in calcium phosphate, phosphoric acid can be produced. The phosphoric acid is used in producing tremendous quantities of phosphate-based fertilizers.

Another 10% of the sulfuric acid production goes into the processing of metal ores and the fabrication of metal products. Some sulfuric acid is used in automobile storage batteries. The chemical industry uses large quantities of this acid in the production of other chemicals and other products, including petroleum refining, plastics, detergents, dyes, medicines, insecticides, herbicides, and many more. Sulfuric acid remains the most important industrial chemical; its price greatly affects the prices of many other products.

See Problems 8.25–8.28.

Chapter Summary

With over 10 million different chemicals, unambiguous names and formulas are essential to any system of nomenclature. Each chemical species has a definite formula and name. Before naming ionic compounds, one must first know the precise names and symbols for the cations and anions involved. Several ions are listed in Tables 8.1 and 8.2. In compounds, cations with differing oxidation numbers are named by the Stock system, using Roman numerals to identify the oxidation state, or sometimes by an older system with *-ous* and *-ic* suffixes.

Anions containing oxygen (oxyanions) end with the suffix *-ate* or *-ite*. The *-ate* suffix is used to identify the oxyanion with the higher oxidation number (the oxyanion with more oxygen atoms). When more than two oxyanions of the same nonmetal exist, the prefix *per-* is used with the *-ate* ending to identify an ion with a higher oxidation number, and more oxygen atoms. Similarly, the prefix *hypo-* is used with the anion having the *-ite* ending to identify an ion with a lower oxidation number, and fewer oxygen atoms.

Binary compounds—those with only two kinds of atoms—all end with the suffix *-ide*. Any ion that contains more than one kind of atom is placed in parentheses before writing the subscript. Rules were provided for naming ionic compounds when formulas are given and for writing formulas when names are given. The oxidation number for a particular kind of atom in a polyatomic ion or compound can be determined by difference, using fixed oxidation numbers of certain species.

Binary compounds of nonmetals are named using Greek prefixes (Table 8.3) to identify the number of atoms of a particular kind in the compound. The more electropositive element is generally written first in both the name and formula.

Aqueous acids contain hydrogen ions and anions. Salts are formed when ionizable hydrogens of acids are replaced by cations. Examples of names and formulas for acids and their salts are listed in Table 8.6. An acid with the *-ic acid* ending has a salt with the *-ate* suffix. An acid with the *-ous acid* ending has a salt with the *-ite* suffix.

Hydrates are crystalline compounds with a definite number of water molecules per formula unit. To write the name of a hydrate, the name of the anhydrous salt is followed by the word hydrate with the appropriate Greek prefix to indicate the number of water molecules in the formula unit. See Table 8.9.

Common names for many chemicals continue to be used (Table 8.10). Some of the chemicals produced in greatest quantities are listed in Table 8.11, beginning with sulfuric acid, the number one chemical in terms of annual production. Virtually every consumer product you buy involves chemical production, testing, and packaging. You will need to write chemical names and formulas as you work problems in Chapter 9 and as you study chemical reactions in Chapter 10.

Assess Your Understanding

1. Write symbols and names for cations and anions. [8.1, 8.2]
2. Write formulas for ionic compounds when names are given. [8.3, 8.4]
3. Write names for ionic compounds when formulas are given. [8.5]
4. Write the formulas and names for binary compounds of nonmetals. [8.6]
5. Determine oxidation numbers of all elements in compounds or ions. [8.7]
6. Write names and formulas for acids and their salts. [8.8]
7. Write names and formulas for hydrates. [8.9]
8. Identify the common names and uses of selected chemicals. [8.10]
9. Identify names and uses of major industrial chemicals. [8.10]

Key Terms

acid [8.8]
acid salt [8.8]
anhydrous salt [8.9]
aqueous [8.8]
carboxyl group [8.8]
carboxylic acid [8.8]
chemical nomenclature [intro]
hydrate [8.9]
hydronium ion [8.8]
inorganic chemicals [intro]
organic chemicals [intro]
oxidation numbcr [8.7]
oxidation state [8.7]
oxyacid [8.8]
polyatomic ions [8.2]
salt [8.8]
Stock system of nomenclature [8.1]
water of hydration [8.9]

Problems

SYMBOLS OF INDIVIDUAL IONS

8.1 What element is present in ions having both the *-ate* and *-ite* suffixes?

8.2 In chemical nonmenclature, describe what the prefixes *bi-*, *thio-*, *hypo-*, and *per-* tell us.

8.3 Write symbols, including charge, for each of the following ions. Also determine the total number of atoms present in each ion represented.
a. bicarbonate ion b. carbonate ion
c. thiosulfate ion d. chromate ion
e. acetate ion f. phosphate ion
g. phosphite ion h. phosphide ion

8.4 Write symbols, including charge, for each of the following ions. Also determine the total number of atoms present in each ion represented.
a. ammonium ion b. bisulfate ion
c. bisulfite ion d. permanganate ion
e. hydroxide ion f. dichromate ion
g. hypochlorite ion h. perchlorate ion

8.5 Write symbols and names using the *-ous* and *-ic* suffixes for the following ions. For each pair, describe how the *-ous* and *-ic* suffixes are used.
a. copper(I) and copper(II)
b. iron(II) and iron(III)
c. tin(II) and tin(IV)
d. mercury(I) and mercury(II)

8.6 Write symbols for a peroxide ion and a mercurous ion. How are they similar?

8.7 Write names for each of the following ions.
a. ClO_3^- b. ClO_4^- c. CO_3^{2-}
d. SCN^- e. SO_4^{2-}

8.8 Write names for each of the following ions.
a. Br^- b. N^{3-} c. O^{2-}
d. P^{3-} e. OH^-

NOMENCLATURE OF IONIC COMPOUNDS

8.9 Show individual ions and formulas for each of the following compounds.
a. aluminum chloride
b. aluminum hydroxide
c. ammonium sulfate
d. ferrous phosphate
e. antimony(III) sulfide
f. calcium hydroxide (used in mortar)
g. stannous fluoride (used in toothpaste)
h. ammonium phosphate (used in fertilizers)
i. calcium carbonate (used in antacids)
j. calcium hypochlorite (used in swimming pools)

8.10 Show individual ions and formulas for each of the following compounds.
a. sodium peroxide
b. ferric oxide
c. cobalt(III) nitrate
d. chromium(III) sulfate
e. potassium permanganate
f. potassium carbonate (used in making glass)
g. calcium acetate (used in canned heat)
h. magnesium hydroxide (used in laxatives)
i. sodium thiosulfate (used in photography)
j. sodium carbonate (used in detergents)

8.11 Write Stock (IUPAC) names along with *-ous* and *-ic* names for these compounds.
a. $SnCl_4$ b. Hg_2Cl_2
c. FeO d. $MnCl_2$
e. CuS f. $Co(NO_3)_3$

8.12 Write Stock (IUPAC) names along with *-ous* and *-ic* names for these compounds.
a. Cr_2O_3 b. $CuCH_3COO$
c. $Pb(NO_3)_2$ d. HgS
e. $Fe_2(SO_4)_3$ f. $Cu(OH)_2$

NONMENCLATURE OF BINARY COMPOUNDS OF NONMETALS

8.13 Write formulas when names are given; write names using Greek prefixes when formulas are given.
a. nitrous oxide b. carbon tetrachloride
c. SF_6 d. N_2O_5
e. NO_2 f. PCl_3

8.14 Write formulas when names are given; write names using Greek prefixes when formulas are given.
a. nitric oxide b. carbon disulfide
c. N_2O_3 d. N_2O
e. SO_3 f. P_2O_5

DETERMINATION OF OXIDATION NUMBERS

8.15 Determine the oxidation number of the element specified in each of the following compounds and ions.
a. P in Na_3PO_4 b. P in PO_3^{3-}
c. Mn in MnO_4^- d. Mn in MnO_2
e. N in HNO_3 f. N in HNO_2

8.16 Determine the oxidation number of the element specified in each of the following compounds and ions.
a. S in K_2SO_4 b. S in SO_3^{2-}
c. S in SO_3 gas d. I in HIO_3
e. I in HIO_2 f. I in HIO

NONMENCLATURE OF ACIDS AND THEIR SALTS

8.17 Write formulas for each of the following acids and their salts.
a. hydrobromic acid b. nitric acid
c. nitrous acid d. carbonic acid
e. silver bromide f. silver nitrate
g. sodium nitrite h. potassium carbonate

8.18 Write formulas for each of the following acids and their salts.
a. hypochlorous acid b. chlorous acid
c. chloric acid d. perchloric acid
e. sodium hypochlorite (bleach)
f. sodium chlorite
g. potassium chlorate (used in fireworks)
h. potassium perchlorate

8.19 Write appropriate names for the following acids and their salts.
a. H_3PO_4 b. K_3PO_4
c. K_2HPO_4 d. KH_2PO_4
e. H_2SO_4 f. $KHSO_4$
g. HNO_3 h. KNO_3

8.20 Write appropriate names for the following acids and their salts.
a. H_2CO_3 b. $KHCO_3$
c. K_2CO_3 d. HF(aq)
e. H_2SO_3 f. $NaHSO_3$
g. Na_2SO_3 h. CH_3COOH

8.21 Draw Lewis electron-dot structures for H_2SO_4 and a sulfate ion.

8.22 Draw Lewis electron-dot structures for H_2SO_3 and a sulfite ion.

DESCRIPTIVE CHEMISTRY, COMMON NAMES OF CHEMICALS, AND INDUSTRIAL CHEMICALS

8.23 Write the chemical formula, name, and a common use for each of the following chemicals. (See Tables 8.9 and 8.10 for formulas.)

a. the dihydrate called gypsum
b. Epsom salt, a heptahydrate
c. borax, a decahydrate
d. lye, caustic soda
e. quicklime
f. limestone
g. the key ingredient in vinegar
h. cane sugar

8.24 Write the chemical formula, name, and a common use for each of the following chemicals. (See Tables 8.9 and 8.10 for formulas.)
a. the decahydrate called washing soda
b. the pentahydrate called photographer's hypo
c. slaked lime d. milk of magnesia
e. baking soda f. battery acid
g. dry ice h. laughing gas

8.25 Write the chemical formula, the industrial source, the approximate annual production, and two uses for each of these important industrial chemicals.
a. sulfuric acid b. phosphoric acid
c. sodium hydroxide d. sodium carbonate

8.26 Write the chemical formula, the industrial source, the approximate annual production, and two uses for each of these important industrial chemicals.
a. slaked lime b. ammonia
c. urea d. nitric acid

8.27 Calculate the market value of all ethylene produced in the United States in one year. (See Table 8.11.) How is this ethylene used?

8.28 Calculate the market value of all ammonia produced in the United States in one year. What other industrial chemicals are heavily dependent on the production of ammonia?

Additional Problems

8.29 Which ion is shown first when writing the formula of an ionic compound?

8.30 What is always true about the sum of all oxidation numbers in a compound?

8.31 What kinds of substances contain ionizable hydrogen? What does this hydrogen ion become in a water solution?

8.32 What is a carboxyl group? Do formulas of all acids contain carboxyl groups?

8.33 What is the difference between CO and Co?

8.34 What is the difference between HF and Hf?

8.35 What is the difference between the structure of a peroxide and a dioxide?

8.36 What is the difference between a chemical formula and a Lewis structure?

8.37 What is the difference between NO_2 and NO_2^-?

8.38 What is the difference between SO_3 and SO_3^{2-}?

8.39 When are parentheses used in writing formulas?

8.40 Which of the following formulas should not have the parentheses?
a. $(NH_4)_2SO_4$ b. $(NH_4)NO_3$
c. $Ca(Cl)_2$ d. $Ca(NO_3)_2$

8.41 What is a hydrate?

8.42 How are formulas of hydrates written?

8.43 How many ions will be formed as each formula unit of the following compounds becomes dissolved in water?
a. Na_2SO_4 b. $(NH_4)_2CO_3$ c. NaOH

8.44 How many ions will be formed as each formula unit of the following compounds becomes dissolved in water?
a. K_3PO_4 b. $Ca(NO_3)_2$ c. NH_4NO_3

8.45 Write names and formulas of compounds formed when chlorine combines with each element in period 3 of the periodic table—Na through Cl. (Consider the number of valence electrons available.)

8.46 Using the periodic table, predict formulas for compounds formed with some ions that are not listed among the common ions.
a. rubidium iodide (This is not intended to be "runs batted in," RBI.)
b. radium nitride (This is not the past tense of "run".)
c. hydrogen combined with astatine (Would you wear it?)
d. plutonium(III) nitride (This is the way this problem was to be taken.)

8.47 Write formulas for the following compounds or ions.
a. hydrogen cyanide gas
b. oxalic acid
c. sodium peroxide
d. sodium potassium sulfate
e. gold(III) oxide
f. iodine monochloride (a disinfectant)
g. potassium aluminum sulfate dodecahydrate (one of several alum compounds)
h. table salt

8.48 Write formulas for the following compounds or ions.

a. lead(IV) oxide
b. silicon dioxide (quartz)
c. ammonium dichromate
d. sodium dihydrogen phosphate
e. hydrogen sulfide gas (rotten egg odor)
f. silver sulfide (tarnished silverware)
g. magnesium chloride hexahydrate
h. marble and calcite

8.49 Write names for the following compounds or ions.

a. $NaSCN$ b. NO_2 c. NO_2^-
d. $(NH_4)_2C_2O_4$ e. $Mg(HCO_3)_2$ f. H_3PO_3

8.50 Write names for the following compounds or ions.

a. SF_6 b. $Ca(ClO)_2$ c. LiH
d. $Ba(OH)_2$ e. KCN f. $SbCl_3$

8.51 Categorize each of the following compounds as an acid, an ionic salt, a hydrate, or a binary covalent compound of nonmetallic elements.

a. $HCl(g)$ b. $HCl(aq)$ c. $CaCl_2$

8.52 Categorize each of the following compounds as an acid, an ionic salt, a hydrate, or a binary covalent compound of nonmetallic elements.

a. Na_2CO_3 b. H_2CO_3 c. NO_2

8.53 Name the following compounds.

a. $NiCl_2$ b. $NiCl_3$ c. PCl_3

8.54 Give the oxidation number of manganese in each compound.

a. K_2MnO_4 b. $KMnO_3$ c. MnO_2

What mass of copper sulfate pentahydrate—the blue crystals—should be used to prepare a solution with a certain concentration? What volume of a particular solution of sodium chloride is needed to deliver a certain number of grams of NaCl—the white crystals? Questions of this type can be answered when you understand how to calculate chemical quantities, as described in this chapter.

9

Chemical Quantities

CONTENTS

Calculations involving chemical quantities are fundamental to a beginning chemistry course. Quantities of chemicals present in the blood and in food and drug products are important to all of us. It's not the fact that cholesterol is present in a blood sample that's most important (we all produce some cholesterol); it's the quantity of cholesterol present that makes the difference.

One person may need to know the quantity of caffeine in a pain reliever tablet, or can of soda, while another person may need to know the quantity of sodium in a serving of potato chips or some other food. The potato chip package and cereal box, for example, list quantities of several chemicals, including sodium, potassium, iron, zinc, calcium, vitamin A, vitamin C, vitamin B_1 (thiamine), and numerous other ingredients present in a single serving (Figure 9.1).

NUTRITION INFORMATION PER SERVING

SERVING SIZE: 1 OZ. (ABOUT 2/3 CUP) (28.35g)
SERVINGS PER PACKAGE: 16

	1 OZ. (28.35g) CEREAL	WITH 1/2 CUP VITAMIN A&D SKIM MILK
CALORIES	90	130*
PROTEIN	3g	7g
CARBOHYDRATE	23g	29g
FAT	0	1g*
CHOLESTEROL	0	0*
SODIUM	210mg	270mg
POTASSIUM	190mg	400mg

PERCENTAGES OF U.S. RECOMMENDED DAILY ALLOWANCES (U.S. RDA)

PROTEIN	4%	10%
VITAMIN A	25%	30%
VITAMIN C	**	2%
THIAMINE	25%	30%
RIBOFLAVIN	25%	35%
NIACIN	25%	25%
CALCIUM	**	15%
IRON	45%	45%
VITAMIN D	10%	25%
VITAMIN B6	25%	25%
FOLIC ACID	25%	25%
VITAMIN B12	25%	35%
PHOSPHORUS	15%	30%
MAGNESIUM	15%	20%
ZINC	10%	15%
COPPER	10%	10%

*2% MILK SUPPLIES AN ADDITIONAL 20 CALORIES, 2g FAT AND 5mg CHOLESTEROL.
**CONTAINS LESS THAN 2% OF THE U.S. RDA OF THESE NUTRIENTS.

INGREDIENTS

WHOLE GRAIN WHEAT, WHEAT BRAN, SUGAR, NATURAL FLAVORING, SALT AND CORN SYRUP.

VITAMINS AND MINERALS

IRON, VITAMIN A PALMITATE, NIACINAMIDE, ZINC OXIDE (SOURCE OF ZINC), VITAMIN B6, RIBOFLAVIN (VITAMIN B2), THIAMINE MONONITRATE (VITAMIN B1), VITAMIN B12, FOLIC ACID AND VITAMIN D.

KRAFT GENERAL FOODS, INC.
BOX BF-16
WHITE PLAINS, NY 10625.

CARBOHYDRATE INFORMATION

	1 OZ. CEREAL	WITH 1/2 CUP SKIM MILK
DIETARY FIBER	5g	5g
SOLUBLE FIBER	.5g	.5g
INSOLUBLE FIBER	4.5g	4.5g
COMPLEX CARBOHYDRATE	13g	13g
SUCROSE AND OTHER SUGARS	5g	11g
TOTAL CARBOHYDRATE	23g	29g

Figure 9.1 *Quantities of sodium, potassium, cholesterol, and certain other chemicals present in a typical serving are required for various packaged food products. This information is provided to assist persons with a variety of dietary needs and to supply consumers with information about nutrition.*

In this chapter we introduce only a few new terms and concepts, but the mathematical calculations described here will interconnect several topics presented previously. For example, we use metric measurements, conversion factors, and dimensional analysis presented in Chapter 3. We will also use atomic masses presented in Chapter 4 and the names and formulas of chemicals that were presented in Chapter 8. An understanding of chemical bonding (Chapter 7) is also required. Thus, the framework of chemistry continues to expand as we build on a solid foundation of fundamental principles.

9.1 Formula Weights and Molecular Weights

The atomic mass (also called atomic weight) for an element shown on the periodic table is actually the average mass (in atomic mass units, amu) of its natural isotopes, based on the abundances of isotopes. Atomic masses were also discussed in Chapter 4. For any compound, the sum of the atomic masses in amu is called the **formula weight** (F.W.). When the compound is molecular—not ionic—this sum can also be called the **molecular weight** (M.W.). If a compound has ionic bonding, it is not technically correct to use the term *molecular weight*, but formula weight can always be used. Review formula weights of compounds (Section 4.11) and work through the following example.

EXAMPLE 9.1

Use atomic masses shown in the periodic table to determine the formula weight of ammonium sulfate, $(NH_4)_2SO_4$.

SOLUTION

There are $2 \times 1 = 2$ N atoms, $2 \times 4 = 8$ H atoms, 1 S atom, and 4 O atoms. Multiply the number of atoms represented by their atomic masses and add.

$$\begin{aligned} 2 \times \text{atomic mass of N} &= 2 \times 14 = 28 \text{ amu} \\ 8 \times \text{atomic mass of H} &= 8 \times 1 = 8 \text{ amu} \\ 1 \times \text{atomic mass of S} &= 1 \times 32 = 32 \text{ amu} \\ 4 \times \text{atomic mass of O} &= 4 \times 16 = 64 \text{ amu} \\ \text{Formula weight} &= 132 \text{ amu} \end{aligned}$$

See Problems 9.1–9.4.

9.2 Moles and Molar Masses

The mole is one of the seven basic SI units of measurement. As explained in Section 4.11, a **mole** (abbreviated **mol** and used with numerical quantities) is defined as the amount of a substance containing as many *formula units* as there are atoms in exactly 12 g of the carbon-12 isotope, ${}^{12}_{6}C$. The formula units can be small molecules (such as O_2 or CO_2) or large molecules (such as caffeine, $C_8H_{10}N_4O_2$). They can be ionic compounds (such as $NaNO_3$), or atoms (such as Na, N, or O), or ions (such as Na^+ and NO_3^-).

One mole of carbon-12 (exactly 12 g) has 6.022×10^{23} atoms of carbon—rounded to four significant figures. This is known as Avogadro's number and is symbolized by the letter N. Thus, a mole of *any* substance contains 6.022×10^{23}

Mole Day

Each year Mole Day is celebrated from 6:02 A.M. to 6:02 P.M. on October 23 (the tenth month, the twenty-third day). This celebration is promoted annually by the National Mole Day Foundation to recognize Amedeo Avogadro and also the contributions of chemistry.

formula units. While a dozen (12) and a gross (144) are counting units used for typical, visible quantities, the mole is the standard unit used for counting extremely large numbers of small particles. With the mole, we count Avogadro's number of molecules, atoms, ions, electrons, protons, or any other kind of formula units specified.

The chemical formula gives the ratio for the numbers of different kinds of atoms present in the compound. These ratios are the same ratios in atoms, dozens of atoms, millions of atoms, or moles of atoms. For example, the atom ratios in Na_2CO_3 are as follows.

	Consists of		
Na_2CO_3 Formula Units	*Number of Na Atoms*	*Number of C Atoms*	*Number of O Atoms*
1	2	1	3
1 doz	2 doz	1 doz	3 doz
1 mol	2 mol	1 mol	3 mol
$1(6.02 \times 10^{23})$	$2(6.02 \times 10^{23})$	$1(6.02 \times 10^{23})$	$3(6.02 \times 10^{23})$

Ratios of ions are also provided by formulas of ionic compounds. Again, for Na_2CO_3, an ionic compound, compare the following ion ratios.

	Consists of	
Number of Na_2CO_3 Formula Units	*Number of Na^+ Ions*	*Number of CO_3^{2-} Ions*
1 Na_2CO_3 formula unit	2 Na^+ ions	1 CO_3^{2-} ion
1 doz Na_2CO_3 formula units	2 doz Na^+ ions	1 doz CO_3^{2-} ions
1 mol Na_2CO_3 formula units	2 mol Na^+ ions	1 mol CO_3^{2-} ions
1 N* Na_2CO_3 formula units	2 N Na^+ ions	1 N CO_3^{2-} ions

*N = Avogadro's number, 6.02×10^{23}.

Thus, the same whole-number ratios apply, whether dealing with individual particles, dozens, or moles. Use these ratios as you consider Example 9.2.

EXAMPLE 9.2

How many moles of sodium ions, Na^+, are contained in 3.84 mol of Na_2CO_3?

SOLUTION

The number of moles of Na_2CO_3 given must be converted to moles of sodium ions.

$$\text{Plan: } 3.84 \text{ mol } Na_2CO_3 \longrightarrow \text{? mol } Na^+ \text{ ions}$$

We can use the mole ratios shown in this section. One mole of Na_2CO_3 contains 2 mol of sodium ions. Thus, we can write the following two conversion factors.

$$\frac{2 \text{ mol } Na^+ \text{ ions}}{1 \text{ mol } Na_2CO_3} \quad \text{or} \quad \frac{1 \text{ mol } Na_2CO_3}{2 \text{ mol } Na^+ \text{ ions}}$$

To find the number of moles of sodium ions, multiply the original number of moles of Na_2CO_3 by the appropriate conversion factor, the one with moles of

Na_2CO_3 in the denominator. Thus, moles of Na_2CO_3 cancels, leaving the answer in moles of Na^+ ions, as shown here.

$$3.84 \text{ mol } Na_2CO_3 \times \frac{2 \text{ mol } Na^+ \text{ ions}}{1 \text{ mol } Na_2CO_3} = 7.68 \text{ mol } Na^+ \text{ ions}$$

EXERCISE 9.1
How many moles of carbonate ions, CO_3^{2-}, are contained in 3.84 mol of Na_2CO_3?

To help you understand the meaning of Avogadro's number, *N*, several examples are given in "How Big Is Avogadro's Number?" On the one hand, it takes 6.02×10^{23} molecules of water, a tremendous number, to make a mole of water. On the other hand, a mole of liquid water has a mass of only 18.0 g and a volume of only 18.0 mL (less than 4 teaspoons). It takes a tremendous number of molecules to make even a small visible mass and volume.

Admittedly, 6.02×10^{23} is not a nice round number. A million is a nice number; so, also, is a million million. However, the real beauty of 6.02×10^{23} lies in the fact that it is very easy to calculate the mass of this number of particles, that is, the mass of one mole of a substance. To calculate the mass of one mole, called the **molar mass** of a compound, simply add up the atomic masses of each of the elements (multiplied by the appropriate subscript) shown in the formula and express this quantity in *grams* instead of atomic mass units. Since the formula weight of H_2O is $2(1.0) + 16.0 = 18.0$ amu, one *mole* of water has a molar mass of 18.0 g. If you measure out 18.0 g of water, you have a mole of water, 6.02×10^{23} molecules.

EXAMPLE 9.3

(a) Determine the mass of 1.00 mol (the molar mass) of Cl_2 gas.
(b) How many molecules of chlorine, Cl_2, are present in 1.00 mol of chlorine gas?

SOLUTION

(a) Recall that chlorine gas is composed of diatomic molecules. Thus, 1.00 mol of chlorine gas, $Cl_2(g)$, contains 2.00 mol of Cl atoms.

$$\text{Plan: } 1.00 \text{ mol } Cl_2 \rightarrow ? \text{ g } Cl_2$$

$$1.00 \text{ mol } Cl_2 = 2.00 \text{ mol of Cl atoms} = 2 \times 35.5 \text{ g} = 71.0 \text{ g}$$

The molar mass is also written 71.0 g/mol.
(b) One mole of chlorine gas has Avogadro's number of Cl_2 molecules—6.02×10^{23} molecules.

EXAMPLE 9.4

Determine the molar mass of the hydrate $Na_2CO_3 \cdot 10H_2O$, called washing soda. It is used in washing powders as a water-softening agent.

A CLOSER LOOK

How Big Is Avogadro's Number?

Avogadro's number is such a large number that it may have little meaning unless we use some examples. Similarly, an automobile trip of 2000 miles may have little meaning until you have experienced it by driving the distance. We trust that at least one of the following examples will help you comprehend the enormous number of particles represented by Avogadro's number, 6.02×10^{23}.

1. Avogadro's number of snowflakes would cover the entire United States to a depth of about 1000 m (3400 ft).
2. If atoms were the size of ordinary glass marbles, Avogadro's number of these marble-sized atoms would cover the entire United States to a depth of about 110 km (70 mi).
3. If atoms were the size of peas, Avogadro's number of these pea-sized atoms would cover the surface of the Earth to a depth of about 15 m.
4. If you had a fortune worth 6.02×10^{23} dollars, Avogadro's number of dollars, you could spend a billion dollars each second of your entire life and have used only about 0.001% of your money.
5. To count Avogadro's number of marbles, peas, dollars, or anything else at a rate of one per second (that's 6.02×10^{23} s) it would take 51,000 Earths full of people, with each person counting ceaselessly for one lifetime of 75 years. Incredible! See the calculation.

$$6.02 \times 10^{23}\ \text{s} \times \frac{1\ \text{min}}{60\ \text{s}} \times \frac{1\ \text{hr}}{60\ \text{min}} \times \frac{1\ \text{day}}{24\ \text{hr}}$$

$$\times \frac{1\ \text{yr}}{365\ \text{days}} \times \frac{1\ \text{person}}{75\ \text{yr}} \times \frac{1\ \text{planet Earth}}{5 \times 10^{9}\ \text{persons}}$$

$= 51{,}000$ Earths full of people, all counting for 75 years

One mole of a substance contains 6.02×10^{23} particles—a tremendous number; yet

One mole of table salt has a mass of 58.5 g; 1 mole of water (18.0 g) has a volume of 18.0 mL; 1 mole of any gas occupies 22.4 L, enough to fill a balloon to a 35-cm diameter.

- 1 mole of water has a mass of only 18.0 g and a volume of 18.0 mL—less than 4 teaspoonfuls.
- 1 mole of any gas occupies only 22.4 L, enough to fill a balloon to a diameter of 35 cm (14 in.), at standard temperature and pressure.
- 1 mole of salt, NaCl, has a mass of 58.5 g—a quantity you could hold in the palm of your hand.

Have you "experienced" the size of Avogadro's number? Do you know what is meant by a mole of a substance? You do if you can explain it to someone else. Try it!

SOLUTION

Plan: 1.00 mol $Na_2CO_3 \cdot 10H_2O \longrightarrow$? g $Na_2CO_3 \cdot 10H_2O$

For the hydrate, we will add together the mass of 1.00 mol of Na_2CO_3 and 10.0 mol of H_2O. (The dot in the hydrate does *not* mean to multiply; it indicates the number of water molecules present in each formula unit of the crystalline compound.) For 1 mol of $Na_2CO_3 \cdot 10H_2O$ we will add

$$\begin{aligned}
&\text{2 mol of Na atoms} = 2 \times 23.0\text{ g} = 46.0\text{ g}\\
&\text{1 mol of C atoms} = 1 \times 12.0\text{ g} = 12.0\text{ g}\\
&\text{3 mol of O atoms} = 3 \times 16.0\text{ g} = 48.0\text{ g}\\
&\text{10 mol of } H_2O = 10 \times 18.0\text{ g} = \underline{180.0\text{ g}}\\
&\text{1 mol } Na_2CO_3 \cdot 10H_2O = 286.0\text{ g}
\end{aligned}$$

See Problems 9.5–9.10.

The molar mass is written 286.0 g/mol

EXERCISE 9.2
Determine the mass of 2.63 mol of washing soda.

9.3 Composition Calculations

One snack food package lists the amount of sodium chloride (table salt) in a serving as 0.300 g of NaCl. A particular bag of potato chips, however, shows that a 1-ounce serving contains 0.200 g of sodium. These two quantities pertain to different chemicals (NaCl and Na^+ ion) and should not be compared directly. A person concerned about total dietary sodium (actually sodium ion, Na^+) intake, however, may wish to compare the quantity of sodium present in a single serving of the two products. The following calculation involves a one-step conversion to determine the quantity of sodium ions in 0.300 g of NaCl.

A mole of sodium atoms or sodium ions is 23.0 g. Masses of electrons are negligible in our calculations.

$$0.300\text{ g NaCl} \times \frac{23.0\text{ g Na}^+}{58.5\text{ g NaCl}} = 0.118\text{ g Na}^+\text{ ions}$$

Thus, one serving of the snack food with 0.300 g of NaCl actually contains 0.118 g of sodium as Na^+ ions. This is less sodium than is present in a single serving of the potato chips with 0.200 g of sodium ions.

In setting up the problem, we began with the quantity of NaCl given in grams. The conversion factor needed must relate the mass of NaCl to the mass of sodium. Since one mole of NaCl (58.5 g) is made up of one mole of sodium ions (23.0 g) and one mole of chloride ions (35.5 g), these masses are all proportional. Thus, we can use these masses to develop conversion factors that relate the mass of a mole of Na^+ to a mole of NaCl as shown here.

$$\frac{23.0\text{ g Na}^+}{58.5\text{ g NaCl}} \quad \text{or} \quad \frac{58.5\text{ g NaCl}}{23.0\text{ g Na}^+}$$

Since the original quantity given was in grams of NaCl, we need to use the conversion factor that has g NaCl in the denominator. Thus, g NaCl cancels, and the resultant answer is in grams of sodium ions.

The molar masses of any compound and its component parts are interrelated by conversion factors similar to the factors described here. In Example 9.4, the molar mass of sodium carbonate decahydrate was determined. The quantity of water present in a specific sample of this hydrate can be determined, as shown in Example 9.5. This calculation is similar to the determination of the sodium ion content in a given quantity of NaCl.

EXAMPLE 9.5

Determine the quantity of water (in grams) that can be released by moderately heating 8.00 g of the hydrate $Na_2CO_3 \cdot 10H_2O$.

SOLUTION

$$\text{Plan: } 8.00 \text{ g } Na_2CO_3 \cdot 10H_2O \longrightarrow ?\text{ g } H_2O$$

STEP 1 Determine the mass of 1.00 mol of the hydrate, $Na_2CO_3 \cdot 10H_2O$. (This mass was determined in Example 9.4.)

STEP 2 Multiply the quantity given by the appropriate conversion factor that relates molar masses of the two components involved.

The mass of the hydrate and the mass of 10 mol of water are determined as follows.

$$1 \text{ mol } Na_2CO_3 \cdot 10H_2O \text{ (determined in Example 9.4)} = 286 \text{ g hydrate}$$

$$10.0 \text{ mol of water} = 10.0 \text{ mol} \times 18.0 \text{ g/mol} = 180. \text{ g water}$$

We can now write the following two conversion factors.

$$\frac{286 \text{ g } Na_2CO_3 \cdot 10H_2O}{180 \text{ g } H_2O} \quad \text{or} \quad \frac{180 \text{ g } H_2O}{286 \text{ g } Na_2CO_3 \cdot 10H_2O}$$

For grams of the hydrate to cancel, we will need to use the second conversion factor. Thus,

$$8.00 \text{ g } Na_2CO_3 \cdot 10H_2O \times \frac{180 \text{ g water}}{286 \text{ g } Na_2CO_3 \cdot 10H_2O} = 5.03 \text{ g water}$$

Therefore, 8.00 g of the hydrate contains 5.03 g of water

See Problems 9.11–9.16.

Percent Composition

Percentages are often used to express the proportion by weight of elements present in a particular compound. We would be more precise to use the term *mass* rather than *weight*, but the term **percent by weight** is widely used. The percent by weight of an element present in a particular compound is equivalent to the number of grams of an element present in 100 g of the compound.

A listing of the percents by weight of each element in a compound is called the **percent composition** for that compound. If the sum of the percentages is not exactly 100%, any deviation is due either to rounding off numbers or to errors in computations. To minimize errors due to rounding off numbers, it is recommended that either three or four significant figures be used in all calculations.

When the chemical formula is known, the determination of percent composition for a compound can be broken down into two steps.

Thinking It Through

Determining Percent Composition

1. Determine the mass of 1 mol of the substance (the molar mass), as shown in Examples 9.3 and 9.4.
2. Divide the mass of each element in the formula by the molar mass and multiply each decimal fraction obtained by 100%. These percentages are usually rounded to the nearest 0.1%.

Figure 9.2 *The set of three numbers on a bag of fertilizer, such as the 3-10-6 shown on this bulb fertilizer, give the percentages of nitrogen, phosphorus, and potassium—in that order—in the fertilizer. Flowering plants need a high phosphorus fertilizer, while a typical lawn fertilizer is high in nitrogen.*

We can write the mathematical operation for step 2 as follows.

$$\frac{\text{Total mass of an element in a compound}}{\text{Molar mass of the compound}} \times 100\% = \text{Percent of the element}$$

The determination of percent composition for a compound is illustrated in Example 9.6. It is recommended that you closely observe the format used in setting up this sample problem. This strategy can be used in setting up and solving other similar problems.

EXAMPLE 9.6

Determine the percent composition for ammonium phosphate, $(NH_4)_3PO_4$, a compound that is sometimes used as a fertilizer.

SOLUTION

Plan: Mass of 1 mol of the compound $\longrightarrow$ Percent of each element

STEP 1 Determine the mass of 1 mol of $(NH_4)_3PO_4$.

$$\begin{aligned} 3 \text{ mol of N atoms} &= 3 \times 14.0 \text{ g} = 42.0 \text{ g} \\ 12 \text{ mol of H atoms} &= 12 \times 1.01 \text{ g} = 12.1 \text{ g} \\ 1 \text{ mol of P atoms} &= 1 \times 31.0 \text{ g} = 31.0 \text{ g} \\ 4 \text{ mol of O atoms} &= 4 \times 16.0 \text{ g} = \underline{64.0 \text{ g}} \\ 1 \text{ mol of } (NH_4)_3PO_4 &= 149.1 \text{ g} \end{aligned}$$

STEP 2 Determine the percentages of each element present.

$$\text{N:} \quad \frac{42.0 \text{ g N}}{149.1 \text{ g } (NH_4)_3PO_4} \times 100\% = 28.2\% \text{ N}$$

$$\text{H:} \quad \frac{12.1 \text{ g H}}{149.1 \text{ g } (NH_4)_3PO_4} \times 100\% = 8.1\% \text{ H}$$

$$\text{P:} \quad \frac{31.0 \text{ g P}}{149.1 \text{ g } (NH_4)_3PO_4} \times 100\% = 20.8\% \text{ P}$$

Study Hint

Recall from math that multiplying by 100% is equivalent to multiplying by a factor of 1 because 100% of anything is one complete unit. A score on an exam of 90 out of 100 (meaning 90/100) can be expressed as 0.90 in decimal form. Multiplying 0.90 × 100% = 90%. The fraction of hydrogen in the compound described is 0.081 in decimal form. Multiplying 0.081 × 100% = 8.1%. The quantities 0.081 and 8.1% are equivalent.

$$\text{O:} \quad \frac{64.0 \text{ g O}}{149.1 \text{ g } (NH_4)_3PO_4} \times 100\% = \underline{42.9\% \text{ O}}$$

$$\text{Sum of percentages} = 100.0\%$$

See Problems 9.17–9.22.

The percent by mass for a single element in a formula can be calculated without determining the entire percent composition for all elements represented by the formula. For example, we might wish to determine only the percentage of nitrogen in the compound described in Example 9.6. In this example, the formula shows that a mole of ammonium phosphate contains 3 mol of nitrogen. In determining the percentage of N, be sure to divide the mass of *3 mol* of nitrogen atoms by the molar mass of the compound. Similarly, when determining the percentage of hydrogen, all hydrogens must be counted, and for oxygen, all oxygens in the formula must be counted when the percent by weight is desired.

9.4 *Mass and Mole Conversions*

A specific number of moles of any substance can be expressed in grams. We must think through a plan for the conversion

This type of conversion involving moles and mass is fundamental; it will be used quite often in solving problems.

$$\text{Moles} \longrightarrow \text{Grams}$$

Here, our reasoning can be broken down into the following two steps.

Thinking It Through

Converting Moles to Mass

1. Determining the mass of 1 mol of the substance (the molar mass) by expressing the formula weight of the substance in grams per mole (written g/mol).
2. Multiply the original quantity of the substance given in moles by the molar mass determined in step 1.

These steps are illustrated in Example 9.7 and are discussed in the explanation that follows.

EXAMPLE 9.7

Calculate the mass of 0.500 mol of carbon dioxide.

SOLUTION

$$\text{Plan:} \quad 0.500 \text{ mol } CO_2 \longrightarrow ? \text{ g } CO_2$$

STEP 1 Determine the mass of 1 mol of CO_2. This is called the molar mass. One mole of CO_2 has a mass equal to that of 1 mol of carbon atoms and 2 mol of oxygen atoms. (Locate atomic masses on the periodic table.)

$$\begin{aligned} \text{1 mol of C atoms} &= 1 \times 12.0 \text{ g} = 12.0 \text{ g} \\ \text{2 mol of O atoms} &= 2 \times 16.0 \text{ g} = \underline{32.0 \text{ g}} \\ 1 \text{ mol } CO_2 &= 44.0 \text{ g} \end{aligned}$$

This is written 44.0 g/mol.

STEP 2 Calculate the mass of 0.500 mol of CO_2 using the molar mass determined in step 1.

$$0.500 \text{ mol } CO_2 \times \frac{44.0 \text{ g } CO_2}{1 \text{ mol } CO_2} = 22.0 \text{ g } CO_2$$

EXERCISE 9.3
Calculate the mass of 1.64 mol of carbon dioxide.

In step 2 of Example 9.7, a specific number of moles of carbon dioxide was converted into grams. This conversion can be represented as

$$\text{Moles A} \longrightarrow \text{Grams A}$$

where A represents a specific chemical—the same chemical. To perform this conversion, the given quantity (in moles) was written first. The conversion factor that is used to convert grams to moles or moles to grams for a particular substance is always the molar mass (the formula weight expressed in grams). The molar mass of carbon dioxide used in Example 9.7 could be written in the following two ways.

$$\frac{44.0 \text{ g } CO_2}{1 \text{ mol } CO_2} \quad \text{or} \quad \frac{1 \text{ mol } CO_2}{44.0 \text{ g } CO_2}$$

If we wish to convert grams of a substance to moles, we will need to use the second conversion factor, with grams in the denominator and moles in the numerator. Thus, grams of the quantity given will cancel with grams in the denominator of the conversion factor. The conversion of grams of a substance to moles is shown in Example 9.8.

EXAMPLE 9.8

Convert 28.6 g of carbon dioxide into moles.

SOLUTION

$$\text{Plan:} \quad 28.6 \text{ g } CO_2 \longrightarrow ? \text{ mol } CO_2$$

Always start with the *quantity* given, and develop a plan or pathway to solve the problem using conversion factors. The molar mass is the conversion factor needed to convert grams A → moles A. If we invert the molar mass to put *moles* in the numerator and *grams* in the denominator, notice that g CO_2 cancels, and the answer is in moles.

See Problems 9.23–9.26.

$$28.6 \cancel{\text{g } CO_2} \times \frac{1 \text{ mol } CO_2}{44.0 \cancel{\text{g } CO_2}} = 0.650 \text{ mol } CO_2$$

EXERCISE 9.4
Convert 91.6 g of carbon dioxide into moles.

9.5 Calculations Involving Avogadro's Number

As mentioned in Section 9.2, a mole is often compared to a dozen; both are units that represent a specific number of items. A dozen nickels, or a dozen dimes, or a dozen quarters always represents 12 of the item specified. Similarly, a mole of a substance also represents a specific number; it always represents 6.02×10^{23} of whatever you're talking about. A dozen nickels, a dozen dimes, and a dozen quarters, however, all have specific but different masses. The mass of a specific number of each coin is *constant*—it does not change. Similarly, a mole of CO_2 and a mole of $NaNO_3$ each has Avogadro's number, N, of formula units, but a mole of each compound has a different and unique mass, like the nickels, dimes, and quarters. The two compounds are also made up of different numbers of different kinds of atoms.

Study the comparison of a mole of CO_2 and a mole of $NaNO_3$; then follow through with the example that follows.

A Comparison of One Mole of Two Different Compounds

1 mol of CO_2, a covalently bonded molecule, has a mass of 44.0 g; it contains 6.02×10^{23} *molecules* made up of 6.02×10^{23} *atoms* of carbon and $2(6.02 \times 10^{23})$ *atoms* of oxygen.

1 mol of $NaNO_3$, an ionic compound, has a mass of 85.0 g; it consists of 6.02×10^{23} *formula units* made up of 6.02×10^{23} Na^+ *ions* and 6.02×10^{23} NO_3^- *ions.*

EXAMPLE 9.9

For 1 mol each of CO_2 and $NaNO_3$ (two compounds we have discussed), compare (a) the total number of moles of atoms and (b) the total atoms present.

SOLUTION

One mole of each substance has Avogadro's number of formula units, but

(a) One mole of CO_2 has 1 mol of carbon atoms + 2 mol of oxygen atoms = 3 mol of atoms. One mole of $NaNO_3$ has 1 mol of Na + 1 mol of N + 3 mol of O = 5 mol of atoms.

(b) One mole of CO_2 has

$$3 \text{ mol of atoms} \times \frac{6.02 \times 10^{23} \text{ atoms}}{1 \text{ mol of atoms}} = 18.1 \times 10^{23} \text{ atoms} \quad \text{(total)}$$

One mole of $NaNO_3$ has

$$5 \text{ mol of atoms} \times \frac{6.02 \times 10^{23} \text{ atoms}}{1 \text{ mol of atoms}} = 30.1 \times 10^{23} \text{ atoms} \quad \text{(total)}$$

We have shown calculations for determining the total number of *moles* of atoms and the total number of individual atoms for both compounds.

The Mass of a Unit Particle

As described in Section 4.11, we can calculate the number of sugar crystals in a bowl of sugar by dividing the total mass of sugar in the bowl by the mass of one sugar crystal. Similarly, the number of thumbtacks in a large box of thumbtacks can be determined by dividing the total mass of thumbtacks by the mass of a single thumbtack as shown here.

$$\frac{\text{Total mass/box}}{\text{Individual mass/tack}}$$

When we invert the denominator of this complex fraction and multiply, we obtain

$$\frac{\text{Total mass}}{\text{Box}} \times \frac{\text{1 Tack}}{\text{Individual mass}} = \frac{\text{Total tacks}}{\text{Box}}$$

Notice that mass in the numerator of the first factor and in the denominator of the second factor cancels, giving the total tacks per box. Using the same ratios arranged differently—so ''box'' cancels—we can determine the individual mass of one tack, as shown here.

$$\frac{\text{Total mass}}{\text{Box}} \times \frac{\text{1 Box}}{\text{Total tacks}} = \frac{\text{Individual mass}}{\text{Tack}}$$

This approach is precisely what we must use to calculate masses of individual atoms, ions, molecules, or other unit particles. We will use mass per mole—the molar mass—in place of mass per box in the equation. We will multiply this factor by 1 mole/(6.02×10^{23}), which is the inverted form of Avogadro's number of particles per mole. This type of calculation is shown in Example 9.10.

EXAMPLE 9.10

Determine the mass of 1 molecule of water.

SOLUTION

Plan: 1 molecule $H_2O \longrightarrow$? g H_2O

Begin with the molar mass—the mass per mole—and multiply by the factor 1 mol/Avogadro's number. In words, the setup is

$$\frac{\text{Mass}}{\text{Mol water}} \times \frac{\text{1 mol water}}{\text{Avogadro's number of molecules}} = \text{Mass/molecule}$$

Thus,

$$\frac{18.0\text{ g}}{\text{1 mol water}} \times \frac{\text{1 mol water}}{6.02 \times 10^{23}\text{ molecules}} = 2.99 \times 10^{-23}\text{ g/molecule}$$

See Problems 9.27–9.32.

Conversions summarized: Mass/mol → Mass/molecule.

EXERCISE 9.5

Determine the mass of 1 molecule of CO_2.

Determination of the Number of Particles in a Given Quantity

When the quantity of a particular substance is given in either moles or grams, we can calculate the number of unit particles present. Again, starting with the quantity given, outline a pathway involving conversion of units. Then set up the problem using the appropriate conversion factors. Example 9.11 demonstrates a conversion where the original quantity in moles is known. Example 9.12 demonstrates calculations needed when the original quantity is given in grams.

EXAMPLE 9.11

How many hydroxide ions, OH^-, can be released into solution as 1×10^{-4} mol of $Ca(OH)_2$ is dissolved in water?

SOLUTION

Each formula unit of $Ca(OH)_2$ can release two OH^- ions. Since the original quantity is given in moles, we will perform the following conversions.

Plan: Moles $Ca(OH)_2$ $\longrightarrow$ Moles OH^- ions $\longrightarrow$ Total number of OH^- ions

$$1 \times 10^{-4} \text{ mol } Ca(OH)_2 \times \frac{2 \text{ mol } OH^- \text{ ions}}{1 \text{ mol } Ca(OH)_2} \times \frac{6.02 \times 10^{23}\ OH^- \text{ ions}}{1 \text{ mol } OH^- \text{ ions}}$$

$$= 1.20 \times 10^{20}\ OH^- \text{ ions}$$

Conversions summarized: Moles given $\rightarrow$ Moles of OH^- ions $\rightarrow$ Total number of OH^- ions.

EXAMPLE 9.12

How many hydroxide ions, OH^-, can be released into solution as 1.00 mg of $Ca(OH)_2$ is dissolved in water?

SOLUTION

Each formula unit of $Ca(OH)_2$ can release two OH^- ions. Express the original quantity in grams and carry out the following conversions.

Plan: Mass Ca $(OH)_2$ $\longrightarrow$ Moles $Ca(OH)_2$ $\longrightarrow$ Moles OH^- ions $\longrightarrow$ Number of OH^- ions

$$1 \times 10^{-3} \text{ g } Ca(OH)_2 \times \frac{1 \text{ mol } Ca(OH)_2}{74.1 \text{ g } Ca(OH)_2} \times \frac{2 \text{ mol } OH^- \text{ ions}}{1 \text{ mol } Ca(OH)_2}$$

$$\times \frac{6.02 \times 10^{23}\ OH^- \text{ ions}}{1 \text{ mol } OH^- \text{ ions}} = 1.62 \times 10^{19}\ OH^- \text{ ions}$$

Conversions summarized: Milligrams $\rightarrow$ Grams $\rightarrow$ Moles $\rightarrow$ Moles of OH^- ions $\rightarrow$ Number of OH^- ions.

See Problems 9.33–9.36.

9.6 Molarity

A solution is obtained when one chemical becomes completely dissolved in another. A solution of table salt in water is a good example. The NaCl that becomes dissolved is called the **solute**, while the water is the **solvent**. The **concentration** of a solution is a measure of the quantity of solute dissolved in the solution. In Chapter 14 several methods will be described that are used to express the concentration of a solution. One of the most useful methods of expressing concentration is called **molarity**, M, which is defined as the number of moles of solute in a solution divided by the total number of liters of solution.

$$\text{Molarity} = \frac{\text{Moles of solute}}{\text{Total liters of solution}}$$

When 1.00 mol of NaCl is dissolved in enough water to make a total volume of 1.00 L, the solution is called a 1.00 **molar** solution, written 1.00 M, with an uppercase ''M.'' We certainly do not always need to work with 1.00-L volumes. For example, when 0.300 mol of NaCl is dissolved in enough water to make 400. mL (0.400 L) of solution, we have a concentration of

$$\frac{0.300 \text{ mol}}{0.400 \text{ L}} = 0.750 \text{ mol/L NaCl}$$
$$= 0.750 \text{ M NaCl}$$

Study Hint
When solving a problem that involves a solution with a specific molarity, M, substitute moles per liter for M when setting up the calculations. For example, use 3.0 mol/L in place of 3.0 M when solving a problem in order for units to cancel.

The following example shows how to calculate the number of grams of a substance that must be used to make up a given volume of solution with a specified molarity.

EXAMPLE 9.13
How many grams of NaCl must be used to make 250.0 mL of a 0.125 M solution?

SOLUTION

STEP 1 Start with the *known quantity*—the volume—in milliliters and convert this to liters.

STEP 2 Convert liters to moles using molarity as a conversion factor.

STEP 3 Convert moles to grams using the formula weight as the conversion factor.

The series of conversions can be summarized as follows.

Plan: Milliliters ⟶ Liters ⟶ Moles ⟶ Grams

$$250.0 \text{ mL} \times \frac{1 \text{ L}}{1000 \text{ mL}} \times \frac{0.125 \text{ mol}}{\text{L}} \times \frac{58.5 \text{ g NaCl}}{\text{mol}} = 1.83 \text{ g NaCl}$$

To prepare the solution, 1.83 g of NaCl must be dissolved in enough water to give a total volume of 250. mL. The solution should be labeled 0.125 M NaCl.

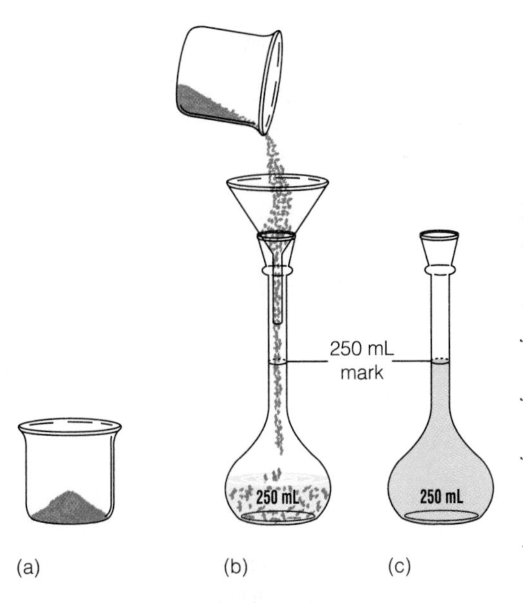

Figure 9.3 *To prepare 250 mL of a solution of specified molarity, (a) obtain the appropriate mass of the chemical, (b) transfer the chemical to a 250-mL volumetric flask that is partly filled with distilled water, and (c) add enough distilled water to fill the flask to the appropriate mark. Once the solution is thoroughly mixed, it is ready to be placed in an appropriately labeled bottle for laboratory use.*

Once the solution with a specific molarity is prepared as described in Example 9.13 and Figure 9.3, we can measure out the specific volume needed to obtain a given number of moles or grams of the chemical desired. Again, start with the known quantity (the number of moles or grams) and use the molarity as a conversion factor to determine the volume needed, as described in Example 9.14.

EXAMPLE 9.14

How many milliliters of 0.125 M NaCl solution (prepared as described in Example 9.13) must be used to obtain 0.0250 mol of NaCl?

SOLUTION

STEP 1 Start with the known quantity—the number of moles of NaCl.

STEP 2 Convert moles to liters using molarity as a conversion factor.

STEP 3 Convert liters to milliliters since the quantity is to be reported in milliliters.

The series of conversions can be summarized as follows.

Plan: Moles ⟶ Liters ⟶ Milliliters

$$0.0250 \text{ mol} \times \frac{1 \text{ L}}{0.125 \text{ mol}} \times \frac{1000 \text{ mL}}{1 \text{ L}} = 200. \text{ mL}$$

Notice that the molarity, M, in moles per liter is a conversion factor written with liters in the numerator (on top) so moles will cancel to give a quantity in liters. Then, in the second factor, liters are converted to milliliters.

See Problems 9.37–9.40.

EXERCISE 9.6

How many milliliters of 0.125 M NaCl solution (prepared as described in Example 9.13) must be used to obtain 1.36 g of NaCl? *Hint*: Start with grams of NaCl, convert to moles, and proceed as shown in Example 9.14.

Solution Preparation by Dilution

A solution with a known concentration can be diluted with water to prepare a solution with any desired concentration that is lower than the original solution. For example, concentrated hydrochloric acid, which can be purchased as a 12.0 M HCl solution, can be diluted to prepare a specified volume of a 2.0 M HCl solution. The number of moles of solute in the original solution can be determined by multiplying the volume of the original solution in liters (V_1) by the molarity of the original solution (M_1 in moles per liter).

$$\text{Volume of original solution in liters} \times \frac{\text{Moles}}{\text{Liter}} = \text{Moles of solute before dilution}$$

$$V_1M_1 = \text{Moles}_1$$

Similarly, multiplying the volume of the diluted solution in liters (V_2) by the molarity of the diluted solution (M_2 in moles per liter) gives the number of moles of solute in the diluted solution.

$$\text{Volume of diluted solution in liters} \times \frac{\text{Moles}}{\text{Liter}} = \text{Moles of solute after dilution}$$

$$V_2M_2 = \text{Moles}_2$$

During the dilution, the number of moles of solute does not change.

$$\text{Moles}_1 = \text{Moles}_2$$

Thus, the volume times molarity (V_1M_1) before dilution is equal to the volume times molarity (V_2M_2) after dilution. This gives the following equation, which is quite useful in making calculations involving dilutions.

$$V_1M_1 = V_2M_2$$

The volume units used in this equation do not have to be in liters, but we must use the same units for both V_1 and V_2. We could use volume units of milliliters, quarts, pints, teaspoonfuls, fluid ounces, or any other volume units desired as long as we use the same units for both volumes. Similarly, the concentrations (C_1 and C_2) can be in molarity, in percentage, or in any other appropriate concentration units. Again, we must use the same concentration units for both the original and final solutions. Thus, any appropriate volume units and concentration units can be used for calculations involving dilutions as long as we do not switch units during the calculations. We can write the following general equation.

$$V_1C_1 = V_2C_2$$

Examples 9.15 and 9.16 demonstrate calculations involving dilutions.

EXAMPLE 9.15

Calculate the number of milliliters of concentrated hydrochloric acid, 12.0 M HCl, that must be used to prepare 500. mL of a 2.00 M HCl solution. Explain how to carry out the dilution.

SOLUTION

STEP 1 This is a dilution problem with molar concentrations. The equation is

$$V_1M_1 = V_2M_2$$

STEP 2 Identify original and final (after dilution) values.

Original	*Final*
$V_1 = ?$	$V_2 = 500.\ \text{mL}$
$M_1 = 12.0\ \text{M}$	$M_2 = 2.00\ \text{M}$

STEP 3 Substitute the appropriate values into the equation.

$$V_1 \times 12.0\ \text{M} = 500.\ \text{mL} \times 2.00\ \text{M}$$

$$V_1 = \frac{(500.\ \text{mL})(2.00\ \text{M})}{(12.0\ \text{M})} = 83.3\ \text{mL}$$

STEP 4 To prepare the 2.00 M solution, dilute 83.3 mL of the 12.0 M HCl to give a *total* volume of 500. mL and mix thoroughly.

See Problems 9.41–9.42.

EXAMPLE 9.16

How many gallons of a 16.0% insecticide solution would a commercial lawn care company use to prepare 800. gal of a 0.0450% solution? Explain how to carry out the dilution.

SOLUTION

STEP 1 For this dilution, concentrations are in percentages. We can use the general equation for dilutions.

$$V_1C_1 = V_2C_2$$

STEP 2 Identify original and final values; they must be in the same units.

Original	*Final*
$V_1 = ?$	$V_2 = 800.\ \text{gal}$
$C_1 = 16.0\%$	$C_2 = 0.0450\%$

STEP 3 Substitute the appropriate values into the equation.

$$V_1 \times 16.0\% = 800.\ \text{gal} \times 0.0450\%$$

$$V_1 = \frac{(800.\ \text{gal})(0.0450\%)}{(16.0\%)}$$

$$V_1 = 2.25\ \text{gal} \qquad \text{(This quantity is equivalent to 9 qt.)}$$

STEP 4 To prepare the 0.0450% solution, dilute 2.25 gal (9 qt) of the 16.0% solution to a *total* volume of 800. gal and mix thoroughly.

9.7 Empirical Formulas and Molecular Formulas

Empirical formulas and molecular formulas are two distinctly different types of chemical formulas that are used for different purposes. Let us compare acetylene gas, C_2H_2, which is used in simple welding applications, and benzene, C_6H_6, which is a liquid used as a solvent and in the industrial production of many organic chemicals. The percent compositions of the two chemicals can be determined by the procedure described in Section 9.3. Although the molecular formulas are different, notice that benzene has three times as many atoms of C and H as are present in acetylene. The following calculations show that both compounds have the same percentages of carbon and hydrogen.

One Mole Acetylene, C_2H_2 = 26.0 g	*One Mole Benzene, C_6H_6 = 78.0 g*
C: $\frac{24.0\text{ g C}}{26.0\text{ g }C_2H_2} \times 100\% = 92.3\%\text{ C}$	C: $\frac{72.0\text{ g C}}{78.0\text{ g }C_6H_6} \times 100\% = 92.3\%\text{ C}$
H: $\frac{2.0\text{ g H}}{26.0\text{ g }C_2H_2} \times 100\% = 7.7\%\text{ H}$	H: $\frac{6.0\text{ g H}}{78.0\text{ g }C_6H_6} \times 100\% = 7.7\%\text{ H}$
Sum of percentages = 100.0%	Sum of percentages = 100.0%

The **molecular formula** for a compound gives the *actual number* of atoms of each element present in a molecule. The molecular formulas of acetylene and benzene are C_2H_2 and C_6H_6, respectively.

The **empirical formula**, also called the **simplest formula**, gives the simplest whole-number ratio of atoms of each element present in a compound. For acetylene, C_2H_2, the simplest ratio of C atoms to H atoms is 1 : 1. Thus, the empirical formula for acetylene is CH. The simplest ratio of C atoms to H atoms in benzene, C_6H_6, is also 1 : 1. The empirical formula of benzene is CH. Thus, both compounds have the same empirical formula, which is CH.

To summarize, acetylene and benzene have different molecular formulas, C_2H_2 and C_6H_6, but these compounds have the same empirical formula, CH, and contain the same percentages of C and H.

EXAMPLE 9.17

For each molecular formula, give the appropriate empirical formula.

(a) glucose, $C_6H_{12}O_6$
(b) water, H_2O
(c) ethylene glycol in antifreeze, $C_2H_6O_2$

SOLUTION

(a) Divide subscripts by 6 to obtain CH_2O.
(b) The empirical formula is also H_2O.
(c) Divide subscripts by 2 to obtain CH_3O.

See Problems 9.43–9.46.

EXERCISE 9.7

Give the empirical formulas for (a) hydrogen peroxide, H_2O_2, and (b) butane, C_4H_{10}.

9.8 Determination of Empirical Formulas

The empirical formula of a compound can be determined from experimental data. This is possible if we know either the number of grams of each element that combines to give a particular compound or the percentage of each element in the compound (the percent composition).

Recall that the empirical formula shows the simplest ratio of atoms present in the compound. The atom ratios are proportional to the number of moles of each kind of atom present. If we can determine the number of moles of each element present, we can also determine the simplest whole-number ratios of atoms present. This simplest set of whole numbers corresponds to the subscripts in the empirical formula for the compound. Study the steps outlined in the following examples. Examples 9.18 and 9.19 illustrate how to determine empirical formulas when either the number of grams of each element or the percentages of each element are known.

EXAMPLE 9.18

A 6.50-g sample of powdered chromium was heated in pure oxygen. The compound produced had a mass of 9.50 g. What is its empirical formula?

SOLUTION

STEP 1 Determine the mass of each element present in the compound.

$$\text{Mass of Cr} = 6.50\text{ g} \quad \text{(given)}$$

$$\text{Mass of O} = 9.50\text{ g combined mass} - 6.50\text{ g Cr}$$

$$= 3.00\text{ g oxygen}$$

The mass of oxygen is obtained, here, by "difference."

STEP 2 Use molar masses to convert grams of each element to moles.

$$\text{Cr:}\quad 6.50\text{ g Cr} \times \frac{1\text{ mol Cr}}{52.0\text{ g Cr}} = 0.125\text{ mol Cr}$$

$$\text{O:}\quad 3.00\text{ g O} \times \frac{1\text{ mol O}}{16.0\text{ g O}} = 0.188\text{ mol O}$$

The mole ratios obtained here are the same as the atom ratios. Therefore, we could write the empirical formula as $Cr_{0.125}O_{0.188}$, but this is *not* the accepted form. We must determine the smallest set of *whole* numbers.

STEP 3 Obtain the smallest set of whole-number atom ratios. First, divide the number of moles of each element by the smaller of the two values.

$$\text{Cr:}\quad \frac{0.125\text{ mol}}{0.125\text{ mol}} = 1.00 \qquad \text{O:}\quad \frac{0.188\text{ mol}}{0.125\text{ mol}} = 1.50$$

If a decimal value appears at this point, as occurs in this problem, we have not obtained the smallest set of whole numbers—the atom ratios. To eliminate the decimals, multiply both values by the smallest possible integer (2, 3,

Table 9.1 **Decimal and Simple Fraction Equivalencies**

Decimal Value	*Simple Fraction*	*Multiplied by This Integer*	*Gives This Whole Number*	*Decimal Value*	*Simple Fraction*	*Multiplied by This Integer*	*Gives This Whole Number*
0.500	$= \frac{1}{2}$	$\times 2$	1	0.200	$= \frac{1}{5}$	$\times 5$	1
0.333	$= \frac{1}{3}$	$\times 3$	1	0.400	$= \frac{2}{5}$	$\times 5$	2
0.667	$= \frac{2}{3}$	$\times 3$	2	0.600	$= \frac{3}{5}$	$\times 5$	3
0.250	$= \frac{1}{4}$	$\times 4$	1	0.800	$= \frac{4}{5}$	$\times 5$	4
0.750	$= \frac{3}{4}$	$\times 4$	3				

4, or 5) that will give the smallest set of whole numbers. For this problem, we need to multiply both values by 2 since the decimal for oxygen is in halves.

Cr: $1.00 \times 2 = 2$ O: $1.50 \times 2 = 3$

The empirical formula is Cr_2O_3.

In step 3 of Example 9.18, the decimal value for oxygen was 1.50. In this case we multiplied by 2. That is because 1.50 is equivalent to $1\frac{1}{2}$ or $\frac{3}{2}$. Whenever the decimal value appears as a multiple of 0.50 or $\frac{1}{2}$, we can eliminate fractions by multiplying by 2. Similarly, if the decimal value is a multiple of 0.333 or 0.666, the fractional equivalent is in thirds. We can eliminate multiples of thirds by multiplying by 3. Also, if the decimal value is a multiple of 0.250 or 0.750, the fractional equivalent is in fourths. We can eliminate multiples of fourths by multiplying by 4. Finally, if the decimal value is a multiple of 0.20, including 0.40, 0.60, 0.80, and so on, the fractional equivalent is in fifths. We can eliminate multiples of fifths by multiplying by 5. Decimal values with simple fraction equivalents are summarized in Table 9.1. It should be understood that a calculated value of 0.498 or 0.499 presumably is due to small errors in rounding off values, and should be treated as 0.500. Similarly, values of 1.32 and 1.65 should be treated as 1.333 and 1.666, respectively.

All calculations to determine empirical formulas should be carried out to three or four significant figures. Rounding off too soon (using fewer than three significant figures) can result in errors in the simplest whole-number ratios.

Percentages can also be used in the determination of empirical formulas instead of the individual masses of the elements present in the compound, as shown in Example 9.19. We will simply use a 100.0-g sample of the compound and multiply each percentage by 100.0 g. For example, 82.7% of 100.0 g is 82.7 g, and 17.3% of 100.0 g is 17.3 g. Thus, we can use percentages as we do masses.

EXAMPLE 9.19

Glycerol, a chemical used in hand lotions, is 39.10% carbon, 8.77% hydrogen, and 52.13% oxygen. Determine the empirical formula of glycerol.

SOLUTION

STEP 1 When the quantities are given as percentages, assume you have a 100.0-g sample of the compound. To determine the mass of each element in the sample, multiply the percentage of each element by 100 g. For 100.0 g of glycerol, we have 39.10 g of carbon, 8.77 g of hydrogen, and 52.13 g of oxygen.

STEP 2 Use molar masses to convert grams of each element (step 1) to moles.

$$\text{C:}\quad 39.10 \text{ g C} \times \frac{1 \text{ mol C}}{12.0 \text{ g C}} = 3.258 \text{ mol C}$$

$$\text{H:}\quad 8.77 \text{ g H} \times \frac{1 \text{ mol H}}{1.01 \text{ g H}} = 8.683 \text{ mol H}$$

$$\text{O:}\quad 52.13 \text{ g O} \times \frac{1 \text{ mol O}}{16.0 \text{ g O}} = 3.258 \text{ mol O}$$

The mole ratios obtained here are the same as the atom ratios. We must simply determine the smallest set of *whole* numbers.

STEP 3 Divide the number of moles of each element by the smallest of the three quantities calculated. Here, we will divide by 3.258 mol.

$$\text{C:}\quad \frac{3.258 \text{ mol}}{3.258 \text{ mol}} = 1.00 \qquad \text{H:}\quad \frac{8.683 \text{ mol}}{3.258 \text{ mol}} = 2.66$$

$$\text{O:}\quad \frac{3.258 \text{ mol}}{3.258 \text{ mol}} = 1.00$$

One of the values, 2.66, is not a whole number; it is in thirds. We can multiply by 3 to obtain the smallest set of whole numbers.

$$\text{C:}\quad 1.00 \times 3 = 3 \qquad \text{H:}\quad 2.66 \times 3 = 7.98 \text{ or } 8.0$$

$$\text{O:}\quad 1.00 \times 3 = 3$$

The empirical formula is $C_3H_8O_3$.

See Problems 9.47–9.50.

9.9 Determination of Molecular Formulas

If the molar mass of a compound and its empirical formula are known, the molecular formula can be determined. As explained in Section 9.7, both acetylene, C_2H_2, with a molar mass (or molecular weight) of 26.0 g/mol and benzene, C_6H_6, with a molar mass of 78.0 g/mol have the same empirical formula, CH.

Acetylene, C_2H_2, contains two empirical formula units. Its molecular formula contains two times as many C and H atoms as are present in the empirical formula, and its molar mass is two times the empirical formula mass.

$$\frac{\text{Molar mass of acetylene}}{\text{Empirical formula mass for CH}} = \frac{26.0 \text{ g}}{13.0 \text{ g}}$$

$$= 2 \text{ empirical formula units}$$

Acetylene, with two empirical formula units, has a molecular formula of C_2H_2.

Benzene has a molecular formula that contains six times as many C and H atoms as are present in the empirical formula, and its molar mass is six times the empirical formula weight.

$$\frac{\text{Molar mass of benzene}}{\text{Empirical formula mass for CH}} = \frac{78.0\text{ g}}{13.0\text{ g}}$$
$$= 6 \text{ empirical formula units}$$

Benzene, with six empirical formula units, has a molecular formula of C_6H_6.

Thus, when we divide the molar mass of a compound by its empirical formula mass, we can determine the number of empirical formula units present in the molecular formula. The molecular formula can be determined by multiplying each subscript in the empirical formula by the number of empirical formula units.

$$\text{Empirical formula} \times \text{Number of empirical formula units} = \text{Molecular formula}$$

Examples:

CH	×	2	=	C_2H_2	(acetylene)
CH	×	6	=	C_6H_6	(benzene)
CH_3O	×	2	=	$C_2H_6O_2$	(ethylene glycol)

EXAMPLE 9.20

By using a mass spectrometer in an analytical chemistry laboratory, one of the compounds in gasoline was found to have a molar mass of 114.0 g/mol. By further analysis, the percentages of C and H in the compound were determined, and the compound was found to have the empirical formula C_4H_9. What is the molecular formula for this compound?

SOLUTION

STEP 1 Determine the number of empirical formula units in the compound.

$$\frac{\text{Molar mass of compound}}{\text{Empirical formula mass for } C_4H_9} = \frac{114.0\text{ g}}{57.0\text{ g}}$$
$$= 2 \text{ empirical formula units}$$

STEP 2 Determine the molecular formula as follows.

$$\text{Empirical formula} \times \text{Number of empirical formula units} = \text{Molecular formula}$$
$$C_4H_9 \times 2 = C_8H_{18}$$

The molecular formula is C_8H_{18}.

See Problems 9.51–9.54.

In the next chapter, we will work with several kinds of chemical reactions. Then, in Chapter 11, we will combine the calculations described in this chapter with information regarding specific chemical reactions (Chapter 10) to determine specific quantities of chemicals involved in chemical reactions.

Chapter Summary

Our ability to make use of chemical information is greatly limited until we learn to make calculations involving the chemical mole. The sum of the atomic masses in atomic mass units for a compound gives the formula weight that is also called the molecular weight when the compound is covalently bonded.

With the mole, we can count Avogadro's number (6.02×10^{23}) of molecules, atoms, ions, electrons, or any other kind of formula units. The chemical formula gives the ratio of atoms and the ratio of moles of atoms in the compound. The percent composition of a compound is a listing of the percentages, by mass, of each kind of element in a compound.

The molar mass of a compound is equivalent to the formula weight expressed in grams. The molar mass is used as a conversion factor when converting grams to moles, and vice versa. When the mass of a mole of particles is known, the mass of any unit particle can be determined by dividing the mass of a mole of the particles by Avogadro's number.

The molarity, M, of a solution is a measure of the number of moles of a substance dissolved in enough water to make a liter of solution. Once a solution with a specific molarity is prepared, the volume of solution needed to deliver a specific number of moles or grams of solute can be calculated.

$$\text{Volume (in liters)} \times \frac{\text{Moles}}{\text{Liter}} = \text{Moles of solute}$$

$$\text{Volume (in liters)} \times \frac{\text{Moles}}{\text{Liter}} \times \frac{\text{Grams}}{\text{Mole}} = \text{Grams of solute}$$

The volume, V_1, of a solution of known concentration, C_1, required to prepare a solution with a specific volume, V_2, and concentration, C_2, by dilution, can be determined by using the equation

$$V_1C_1 = V_2C_2$$

The simplest whole-number ratio of atoms of each element present in a compound is called the empirical formula. The molecular formula gives the actual number of atoms of each element present in a molecule. The percent composition of a compound can be used to calculate the empirical formula for a compound. When the molar mass of the compound is known, the number of empirical formula units and the actual molecular formula can be determined. The molecular formula can be the same as the empirical formula or can be a simple integer multiple (1, 2, 3, etc.) of the empirical formula.

The best way to learn to work problems like these is to *practice* working the problems found at the end of this chapter.

Assess Your Understanding

1. Determine formula weights and molecular weights for compounds. [9.1]

2. Describe the chemical mole and Avogadro's number. [9.2]
3. Define molar mass and determine molar masses for compounds. [9.2]
4. From a chemical formula, calculate the percent composition. [9.3]
5. Convert grams of a substance to moles, and vice versa. [9.4]
6. Calculate the masses of individual atoms and molecules. [9.5]
7. Interconvert mass, moles, and number of atoms or ions. [9.5]
8. Describe how to prepare solutions with molar concentrations. [9.6]
9. Distinguish between empirical and molecular formulas. [9.7]
10. Determine empirical and molecular formulas from data. [9.7–9.9]

Key Terms

concentration [9.6]
empirical formula [9.7]
formula weight [9.1]
molar [9.6]
molarity [9.6]
molar mass [9.2]
mole (mol) [9.2]
molecular formula [9.7]
molecular weight [9.1]
percent composition [9.3]
percent by weight [9.3]
simplest formula [9.7]
solute [9.6]
solvent [9.6]

Problems

FORMULA WEIGHTS AND MOLECULAR WEIGHTS

9.1 Describe when to use the terms "formula weight" and "molecular weight."

9.2 Criticize this statement: "The molecular weight of KCl is 74.6 amu."

9.3 Determine formula weights for the following compounds.
a. $Mg(OH)_2$ (present in "milk of magnesia")
b. $(NH_4)_3PO_4$ (used in fertilizers)
c. calcium hydroxide
d. CH_3COOH, acetic acid (in vinegar)

9.4 Determine formula weights for the following compounds.
a. $Ca_3(PO_4)_2$ (present in bones)
b. C_2H_5OH, ethanol (ethyl alcohol)
c. sulfuric acid
d. magnesium nitrate

MOLES AND MOLAR MASSES

9.5 The formula of oxygen gas is O_2. Determine
a. the mass of one mole (the molar mass) of oxygen gas
b. the number of atoms in one molecule of oxygen gas
c. the total number of atoms in one mole of oxygen gas
d. the mass of 2.50 mol of oxygen gas
e. the number of moles of oxygen gas in 70.0 g of oxygen gas

9.6 The formula of carbon dioxide gas is CO_2. Determine
a. the mass of one mole (the molar mass) of CO_2 gas
b. the number of atoms in one molecule of CO_2 gas
c. the total number of atoms in one mole of CO_2 gas
d. the mass of 2.50 mol of CO_2 gas
e. the number of moles of CO_2 gas in 70.0 g of CO_2 gas

9.7 The formula of caffeine is $C_8H_{10}N_4O_2$. Determine
a. the mass of one mole of caffeine
b. the number of atoms in one molecule of caffeine
c. the total number of atoms in one mole of caffeine
d. the mass of 0.125 mol of caffeine
e. the number of moles of caffeine in 50.0 g of caffeine

9.8 The formula of glucose is $C_6H_{12}O_6$. Determine
a. the mass of one mole of glucose
b. the number of atoms in one molecule of glucose
c. the total number of atoms in one mole of glucose
d. the mass of 0.125 mol of glucose
e. the number of moles of glucose in 50.0 g of glucose

9.9 For calcium hydroxide, $Ca(OH)_2$, used in mortar, determine
a. the number of calcium ions, Ca^{2+}, in one formula unit of $Ca(OH)_2$
b. the number of hydroxide ions, OH^-, in one formula unit of $Ca(OH)_2$
c. the number of moles of calcium ions, Ca^{2+}, in 2.50 mol of $Ca(OH)_2$
d. the number of moles of hydroxide ions, OH^-, in 2.50 mol of $Ca(OH)_2$

9.10 For $(NH_4)_3PO_4$, used in many fertilizers, determine
a. the number of ammonium ions, NH_4^+, in one formula unit of $(NH_4)_3PO_4$
b. the number of phosphate ions, PO_4^{3-}, in one formula unit of $(NH_4)_3PO_4$
c. the number of moles of ammonium ions, NH_4^+, in 0.240 mol of $(NH_4)_3PO_4$
d. the number of moles of phosphate ions, PO_4^{3-}, in 0.240 mol of $(NH_4)_3PO_4$

COMPOSITION CALCULATIONS

9.11 For each 1000-g quantity of $(NH_4)_3PO_4$ in a lawn fertilizer, determine the number of grams of nitrogen present in the fertilizer sample.

9.12 For each 1000-g quantity of $(NH_4)_3PO_4$ in a lawn fertilizer, determine the number of grams of phosphorus present in the fertilizer sample.

9.13 For a 10.0-g NaCl (table salt) sample, how many grams of Na are present?

9.14 For each 10.0 kg of Cu_2S ore mined, how many kilograms of Cu can be obtained?

9.15 For each 10.0 kg of PbS ore (called galena), how many kilograms of Pb can be obtained?

9.16 For each 10.0 kg of Zn_2SiO_4 ore mined, how many kilograms of Zn can be obtained?

9.17 Determine the percent composition (by mass) of ammonia gas, NH_3.

9.18 Determine the percent composition (by mass) of ammonium nitrate, NH_4NO_3.

9.19 Determine the percent composition (by mass) of ammonium sulfate, $(NH_4)_2SO_4$.

9.20 Determine the percent composition (by mass) of urea, N_2H_4CO.

9.21 Each of the compounds listed in Problems 9.17–9.20 is used as a fertilizer. Which of these compounds has the highest percentage (by mass) of nitrogen?

9.22 Which compound, ammonium phosphate or calcium phosphate, has the greater percent (by mass) of phosphorus?

MASS AND MOLE CONVERSIONS

9.23 Convert each of the following quantities to moles.
a. 10.0 g of Fe
b. 10.0 g of Fe_2O_3
c. 92.0 g of ethanol, C_2H_5OH
d. 92.0 g of gold

9.24 Convert each of the following quantities to moles.
a. 44.0 g of H_2O
b. 44.0 g of CO_2
c. 90.0 g of glucose, $C_6H_{12}O_6$
d. 90.0 g of H_2 gas

9.25 Determine the number of grams present in each of the following samples.
a. 0.800 mol of Fe
b. 0.800 mol of Fe_2O_3
c. 1.50 mol of ethanol, C_2H_5OH
d. 1.50 mol of gold

9.26 Determine the number of grams present in each of the following samples.
a. 1.50 mol of H_2O
b. 1.50 mol of CO_2
c. 0.750 mol of glucose, $C_6H_{12}O_6$
d. 0.750 mol of H_2 gas

CALCULATIONS WITH AVOGADRO'S NUMBER

9.27 Determine the mass (in grams) of one atom of carbon.

9.28 Determine the mass (in grams) of one atom of nitrogen.

9.29 Determine the mass (in grams) of one molecule of carbon dioxide.

9.30 Determine the mass (in grams) of one molecule of dinitrogen pentoxide, N_2O_5.

9.31 The mass of a single drop of water was determined to be 0.0500 g. How many water molecules are present in this drop?

9.32 Two crystals of sucrose, $C_{12}H_{22}O_{11}$ (table sugar), have a mass of 0.0012 g. How many sucrose molecules are present in this sample?

9.33 How many hydroxide ions can be released into solution from 1.00 mg of $Al(OH)_3$?

9.34 How many hydroxide ions can be released into solution from 1.00 mg of $Mg(OH)_2$?

9.35 How many chloride ions are present in a sample of 1.50 g of $CaCl_2$?

9.36 How many molecules are present in 1.00 mg of vitamin C, $C_6H_8O_6$?

MOLARITY

9.37 How many grams of glucose, $C_6H_{12}O_6$, are needed to prepare 250 mL of a 0.150 M solution? Describe how to prepare the solution.

9.38 How many grams of $Mg(OH)_2$ are needed to prepare 500 mL of a 1.25 M solution? Describe how to prepare the solution.

9.39 To obtain 2.00 g of glucose, how many milliliters of the 0.150 M glucose solution (Problem 9.37) should be used?

9.40 To obtain 5.00 g of $Mg(OH)_2$, how many milliliters of the 1.25 M $Mg(OH)_2$ solution (Problem 9.38) should be used?

9.41 Determine the number of milliliters of a 12.0 M concentrated hydrochloric acid solution needed to prepare 2.00 L of a 0.100 M solution. Describe how to perform the dilution.

9.42 Determine the amount of a 6.00 M sulfuric acid solution needed to prepare 500 mL of a 1.50 M solution. Describe how to perform the dilution.

EMPIRICAL AND MOLECULAR FORMULAS

9.43 Write the empirical formula for each of the following.
a. C_8H_{18}, octane (in gasoline)
b. $C_{12}H_{22}O_{11}$, sucrose (table sugar)
c. Hg_2Cl_2 d. $CaCl_2$

9.44 Write the empirical formula for each of the following.
a. $C_{20}H_{42}$, present in paraffin
b. C_2H_4, ethene (used to make polyethylene)
c. $C_3H_8O_3$, glycerol
d. $C_{10}H_{22}$, decane

9.45 From the following empirical formulas and the formula weights for each compound, determine the correct molecular formulas.
a. CH_3, F.W. = 30.0 amu
b. CH_2, F.W. = 56.0 amu
c. C_5H_7N (nicotine), F.W. = 81.0 amu
d. P_2O_5, F.W. = 284 amu

9.46 From the following empirical formulas and the formula weights for each compound, determine the correct molecular formulas.
a. CH_2, F.W. = 84.0 amu
b. CH_2O, F.W. = 60.0 amu
c. $C_3H_4O_3$, (vitamin C), F.W. = 176 amu
d. BH_3, F.W. = 27.7 amu

9.47 Determine the empirical formula for a compound sample that contained 18.6 g of phosphorus and 14.0 g of nitrogen.

9.48 Determine the empirical formula for a compound sample that contained 18.6 mg of phosphorus and 12.6 mg of nitrogen.

9.49 Determine the empirical formula of a compound that is 35.6% phosphorus and 64.4% sulfur.

9.50 Determine the empirical formula of a compound that is 43.7% phosphorus and 56.3% sulfur.

9.51 Hydrazine is a chemical sometimes used as a rocket fuel. It is 87.5% nitrogen and 12.5% hydrogen and has a molecular weight of 32.0. Determine
a. the empirical formula of hydrazine
b. the molecular formula of hydrazine

9.52 A chemical present in vinegar was found to be 40.0% carbon, 6.67% hydrogen, and 53.3% oxygen and to have a molecular weight of 60.0. Determine
a. the empirical formula of the compound
b. the molecular formula of the compound

9.53 A compound with a molecular weight of 98.0 was determined to be 24.49% carbon, 4.08% hydrogen, and 72.43% chlorine. Determine
a. the empirical formula of the compound
b. the molecular formula of the compound

9.54 An organic acid with a molecular weight of 88.0 was determined to be 54.55% carbon, 9.09% hydrogen, and 36.36% oxygen. Determine
a. the empirical formula of the compound
b. the molecular formula of the compound

Additional Problems

9.55 Determine the mass of a single mole of water and a single molecule of water.

9.56 Determine the mass of a single mole of lead and a single atom of lead.

9.57 Determine the mass of a mole of ammonium phosphate and the number of ammonium ions present in a 0.100-g sample of the compound.

9.58 How many sodium ions are present in a 0.100-g sample of NaCl?

9.59 How many milliliters of a 10.0% NaOH solution should be used to prepare 250 mL of a 2.0% solution?

9.60 How many gallons of a 12.0% disinfectant solution should be used to prepare 4.0 gal of a 2.0% solution?

9.61 How many grams of copper(II) sulfate pentahydrate must be used to prepare 1.00 L of a 0.200 M solution?

9.62 How many grams of NaOH must be used to prepare 500. mL of a 6.00 M solution?

9.63 How many milliliters of a 0.200 M copper(II) sulfate solution must be used to obtain 6.00×10^{-3} mol of Cu^{2+} ions?

9.64 How many milliliters of 6.00 M sodium hydroxide solution must be used to obtain 3.0×10^{-4} mol of OH^- ions?

9.65 How many Cl^- ions are present in 1.00 mL of a 0.100 M $CaCl_2$ solution?

9.66 How many hydrogen ions, H^+, can be obtained from 1.0 mL of a 6.00 M H_2SO_4 solution?

9.67 Calculate the number of grams of nitrogen in 50.0 g of the amino acid glycine, CH_2NH_2COOH.

9.68 Calculate the number of pounds of iron that could be obtained from 500. lb of Fe_2O_3.

9.69 A Freon gas sample with a molecular weight of 121 was found to be 9.92% carbon, 58.68% chlorine, and 31.40% fluorine. What is the molecular formula of this Freon gas?

9.70 Determine the number of sucrose (cane sugar) molecules present in 30.0 g (1 teaspoon) of $C_{12}H_{22}O_{11}$.

9.71 How many grams of $NaHCO_3$ are present in 15.0 mL of a 0.200 M $NaHCO_3$ solution?

9.72 How many milliliters of 0.200 M $NaHCO_3$ solution must be used to obtain 500. mg of $NaHCO_3$?

9.73 Determine the number of milliliters of a 6.00 M NaOH solution needed to prepare 500. mL of a 0.100 M solution?

9.74 How many milligrams of NaOH are present in 12.0 mL of a 0.100 M NaOH solution?

This vigorous reaction of iron(III) oxide with aluminum powder, called the thermite reaction, produces small quantities of molten iron while releasing a considerable amount of heat. The reaction was used to weld rails in building railroads of the old West.

10

Chemical Reactions

CONTENTS

As you drive your car, gasoline combines explosively with oxygen gas to give carbon dioxide, water vapor, and a specific amount of energy. This is an example of a common, but exceedingly important, chemical reaction (Figure 10.1a). During a complex series of reactions within the cells of your body, glucose and other carbohydrates in food are consumed (metabolized) as they react with oxygen to produce carbon dioxide and water vapor that are exhaled as you breathe (Figure 10.1b). Both of these examples demonstrate that some substances disappear and other substances are produced during chemical reactions.

The two reactions are similar in several ways. In both cases, a compound containing carbon reacts with oxygen to produce carbon dioxide and water. Whether inside the human body, or in an automobile engine, or in the open air, substances

(a) (b)

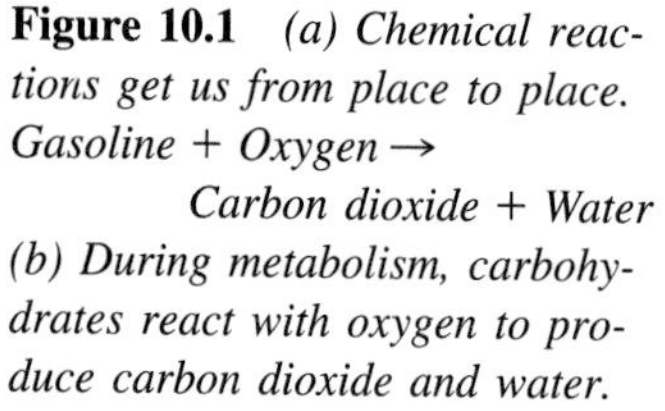

Figure 10.1 *(a) Chemical reactions get us from place to place. Gasoline + Oxygen → Carbon dioxide + Water (b) During metabolism, carbohydrates react with oxygen to produce carbon dioxide and water.*

react to produce new and different substances. Once we understand what is happening, we find that reactions are predictable. In each chemical reaction no atoms are created or destroyed; they are rearranged to form different substances. In each case, matter is conserved. There is no change in total mass.

In order to be able to write chemical equations for reactions like the ones described, you must first be quite familiar with chemical formulas and what they represent. Chemical formulas were presented in Chapter 8; they will be used extensively in this chapter. If that topic is difficult for you, it may be necessary to review before attempting to write chemical equations.

10.1 Chemical Reactions and Chemical Equations

Chemical reactions occur when substances undergo fundamental changes in identity; one or more substances are used up while one or more substances are formed. Examples include the burning of gasoline and of glucose. The substances present at the beginning of a reaction, the starting materials, are called **reactants**. The substances *produced* by the reaction are called **products**.

Chemical equations are used to represent, symbolically, what is taking place during the reaction. The *reaction* of glucose (a sugar) with oxygen gas during metabolism to produce carbon dioxide and water can be written as a *chemical equation* in words or in symbols (chemical formulas).

$$\text{Glucose} + \text{Oxygen} \longrightarrow \text{Carbon dioxide} + \text{Water}$$

$$C_6H_{12}O_6 + 6\,O_2(g) \longrightarrow 6\,CO_2(g) + 6\,H_2O(g)$$

The reactants, or starting materials, are shown on the left side of the equation and are separated by the plus sign (+). The products are shown on the right side of the equation. Reactants and products are separated by an arrow (→), which is read **yields** (or produces). The reaction of glucose with oxygen "yields" carbon dioxide and water. While the chemical equation can be written in words, the equation written in chemical formulas tells us much more.

The chemical reactions that have been described, thus far, were chosen because they pertain to familiar processes. We will also write chemical equations for reactions that are much less familiar, but no less important.

Special symbols are often used in chemical equations to give specific information about the substances involved, or the conditions for the reaction. Substances that exist as a gas at the time of the reaction may be designated by a (**g**) immediately after a formula, as was shown with the gases in the equation for the combustion of glucose. The symbols (**s**) and (**l**) may be used to identify solids and liquids, respectively. A substance that is dissolved in water, to make what we call an **aqueous solution**, may be identified by the symbol (**aq**) in the equation. The Greek capital letter delta (Δ) is sometimes shown over or under the arrow to indicate that heat is required to initiate the reaction.

Sometimes a substance is added to speed up a reaction that otherwise would take place at a slow rate, or not at all. This substance, called a **catalyst**, is shown above the arrow; it is not used up during the reaction. Enzymes are special catalysts that are manufactured and used by the human body for each and every reaction that occurs during metabolism.

10.2 What Balanced Chemical Equations Tell Us

A reaction can be represented by a word equation to tell the identities of reactants and products, but the chemical equation, shown with symbols, gives this information as well as the formulas and the proportional amounts of each substance involved. As shown by the equation for the combustion of glucose (Section 10.1), the chemical equation summarizes, in a symbolic way, what is taking place.

The chemical equation, unfortunately, does not tell us everything we may wish to know about a reaction. It does not tell us how quickly the reaction will occur. It could proceed, for example, as slowly as the rusting of iron or fast enough to be an explosion. The basic equation does not show how much heat, light, or any other form of energy is produced or is required to cause the reaction to occur. It also does not describe the appearance of reactants or products during the reaction, or even tell us if the reaction will in fact occur at all.

On the Atomic Level

In order to understand what is actually taking place during a reaction, it is often helpful to visualize what is occurring at the molecular or atomic level. A chemical equation contains a great deal of this type of information. For example, magnesium metal burns in oxygen to produce a white powder, magnesium oxide (Figure 10.2). At the atomic and molecular level, this equation

Figure 10.2 *Magnesium metal burns in oxygen to produce magnesium oxide.*

$$2\,Mg + O_2(g) \longrightarrow 2\,MgO$$

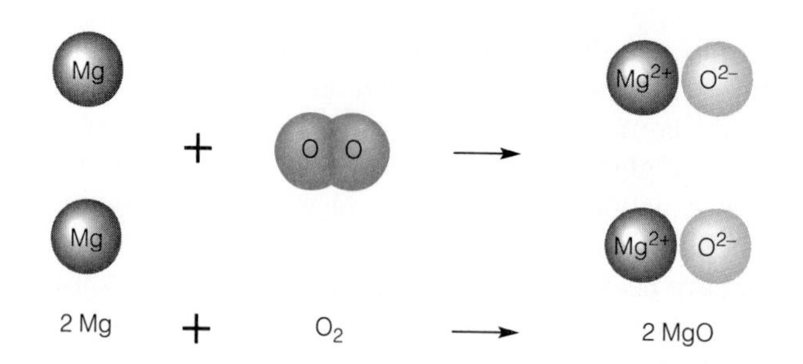

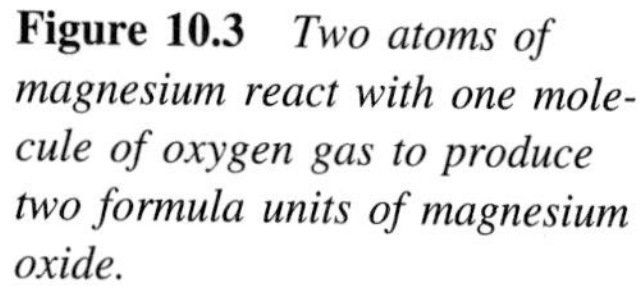

Figure 10.3 *Two atoms of magnesium react with one molecule of oxygen gas to produce two formula units of magnesium oxide.*

indicates with formulas that two atoms of magnesium react with one molecule of oxygen gas to produce two formula units of magnesium oxide. (We will not say molecules of magnesium oxide because the bonding is more ionic than covalent.)

$$2 \text{ atoms Mg} + 1 \text{ molecule } O_2 \longrightarrow 2 \text{ formula units MgO}$$

Figure 10.3 can help us visualize what is happening on the atomic level. In a chemical reaction, no atoms are gained, none are lost; matter is conserved. Thus, in a **balanced chemical equation**, the number of atoms of each kind of element represented as reactants and products must be equal; atoms are *balanced.*

Now, let us scale up the proportions by a factor of 12. Twelve items make a dozen, so we can use dozens in writing the quantities of reactants and products.

$$2 \text{ doz atoms Mg} + 1 \text{ doz molecules } O_2 \longrightarrow 2 \text{ doz formula units MgO}$$

These quantities are far too small to see, so let's use a million as the factor.

$$2 \text{ million atoms Mg} + 1 \text{ million molecules } O_2 \longrightarrow 2 \text{ million formula units MgO}$$

Even this large number of atoms or formula units is too small to see with your eyes, so let's scale up the reaction with 6.02×10^{23} as the factor; the proportions are the same. Because 6.02×10^{23} is the number of unit particles in one mole—just as 12 is the number of items in a dozen—we can write

$$2 \text{ mol Mg} + 1 \text{ mol } O_2 \longrightarrow 2 \text{ mol MgO}$$

Now we have quantities we can see and weigh. Although the numbers of unit particles—or moles of unit particles—are in simple, whole-number ratios, the ratios of masses are not. Just as a dozen Ping-Pong balls has a different mass than a dozen golf balls, a mole of Mg has a different mass than a mole of O_2. These masses are the molar masses we discussed in Chapter 9. Since a mole of Mg has a mass of 24.3 g, and a mole of O_2 has a mass of 2×16.0 or 32.0 g, and a mole of MgO has a mass of $24.3 + 16.0$ or 40.3 g, we can write the following equations, showing that the mass before the reaction is equal to the mass after the reaction.

$$
\begin{aligned}
2\,\text{Mg} + O_2 &\longrightarrow 2\,\text{MgO} \\
2 \times 24.3 \text{ g} + 32.0 \text{ g} &\longrightarrow 2 \times 40.3 \text{ g} \\
48.6 \text{ g} + 32.0 \text{ g} &\longrightarrow 80.6 \text{ g} \\
80.6 \text{ g} &= 80.6 \text{ g}
\end{aligned}
$$

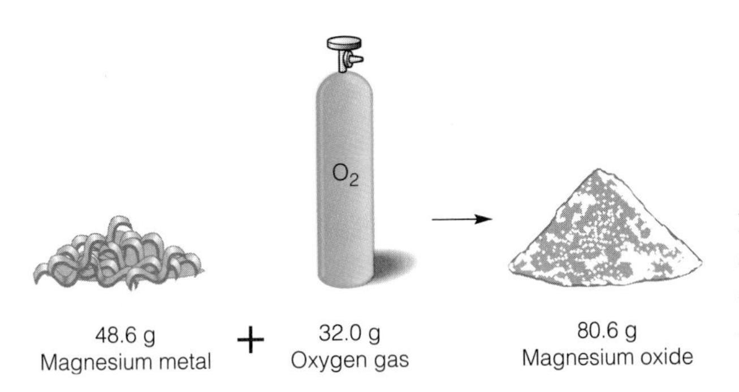

Figure 10.4 *Two moles of magnesium (48.6 g) react with one mole of oxygen gas (32.0 g) to give two moles of magnesium oxide (80.6 g).*

The equations tell us the proportions in numbers of unit particles and in mass. The total mass of the reactants is equal to the total mass of the products (Figure 10.4). This agrees with the *law of conservation of mass*: Atoms are neither created nor destroyed, and mass is neither created nor destroyed. All of these relations are represented by the balanced equation.

10.3 Writing and Balancing Chemical Equations

Let us start with a chemical equation for the reaction that occurs when a mixture of hydrogen gas and oxygen gas is ignited with a spark to produce water and enough energy to make a very loud bang (Figure 10.5). We *first write the correct formulas* for the reactants and products. The hydrogen and oxygen may be identified as gases (g) if you wish. This is optional.

Recall that H_2 and O_2 are diatomic gases.

$$H_2(g) + O_2(g) \longrightarrow H_2O(l) \quad \textit{(not balanced)}$$

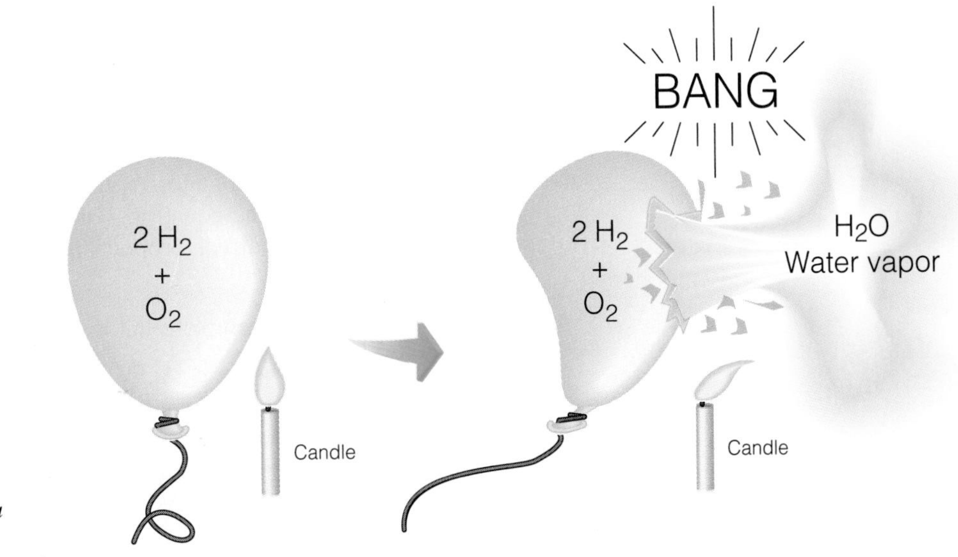

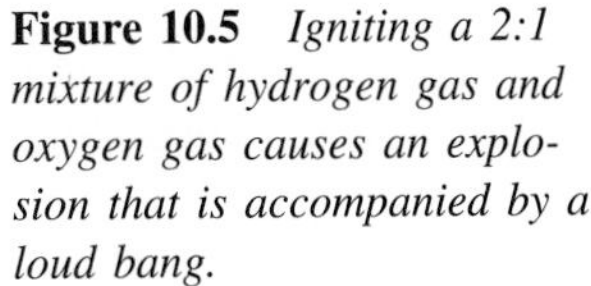

Figure 10.5 *Igniting a 2:1 mixture of hydrogen gas and oxygen gas causes an explosion that is accompanied by a loud bang.*

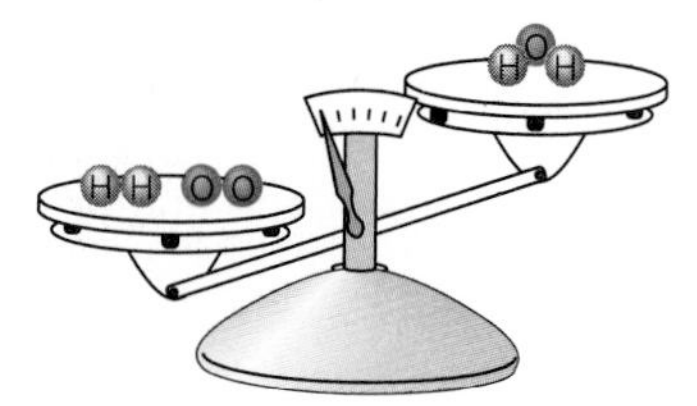

Figure 10.6 $H_2 + O_2 \rightarrow H_2O$ *(not balanced)*

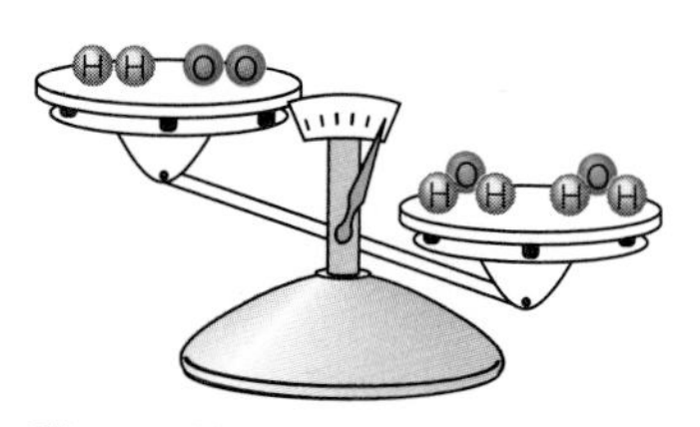

Figure 10.7 $H_2 + O_2 \rightarrow 2\ H_2O$ *(not balanced)*

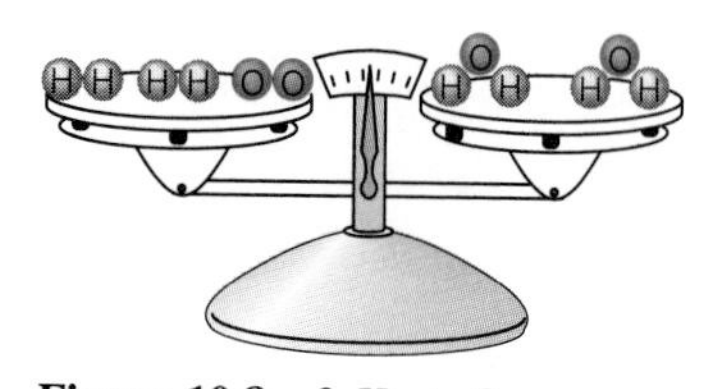

Figure 10.8 $2\ H_2 + O_2 \rightarrow 2\ H_2O$ *The equation is balanced. Mass is conserved.*

The numbers of atoms in the equation we have just written are not balanced (Figure 10.6). Two oxygen atoms are shown among the reactants (as O_2) and only one oxygen atom appears among the products (in H_2O). To balance the oxygen atoms, we place the number 2 in front of the formula for water (Figure 10.7).

$$H_2(g) + O_2(g) \longrightarrow 2\ H_2O(l) \qquad \textit{(not balanced)}$$

This number indicates that *two molecules* of water are produced for each molecule of oxygen that reacts. The number that we place in front of a chemical formula in an equation is called a **coefficient**. A coefficient of 1 is understood where no other number appears. A coefficient preceding a formula is an instruction to multiply everything in the formula by that number. When the H_2O is doubled, we have represented two oxygen atoms and four hydrogen atoms. But the equation in our example is still not balanced.

To balance hydrogen atoms, we place the coefficient 2 in front of H_2.

$$2\ H_2(g) + O_2(g) \longrightarrow 2\ H_2O(l) \qquad \text{(balanced)}$$

Now, there are four hydrogen atoms as well as two oxygen atoms represented on both sides of the equation. The equation is *balanced* (Figure 10.8). Always check to see if the equation is balanced by counting and comparing atoms of each element present as reactants and as products.

Atoms of Reactants	*Atoms of Products*
4 H	4 H
2 O	2 O

The balanced equation shows that atoms are neither created nor destroyed.

If we had attempted to balance the equation by changing the subscript for oxygen in water, the equation would *appear* to be balanced

$$H_2(g) + O_2(g) \longrightarrow H_2O_2(l) \qquad \textit{(a change in meaning!)}$$

but it would not mean "hydrogen reacts with oxygen to form *water*." The formula H_2O_2 represents *hydrogen peroxide*, an entirely different compound. Thus, when balancing a chemical equation, *we cannot change the subscripts* of formulas because that would change the compounds represented and also the meaning of the equation. Instead, we balance the equation—as shown earlier—by changing the coefficients to represent the correct proportions of substances in the reaction.

A chemical equation is not complete until it is balanced.

One of the best ways of learning something new is actually to try it for yourself. Work through each of the following examples to make sure you follow the thought processes involved in balancing equations like these. Hints will be provided along the way to help you with balancing several types of equations.

EXAMPLE 10.1

Balance the following equation for the reaction of nitrogen gas with hydrogen gas to produce ammonia gas, NH_3.

$$N_2(g) + H_2(g) \longrightarrow NH_3(g) \qquad \textit{(not balanced)}$$

SOLUTION

The equation shows two hydrogen atoms on the left and three on the right. The least common multiple of 2 and 3 is 6. Thus, six will be the smallest number of hydrogen atoms that we can have as reactants and products. We will need three molecules of hydrogen and two of NH_3.

$$N_2(g) + 3\ H_2(g) \longrightarrow 2\ NH_3(g) \qquad \text{(balanced)}$$

We should now check to see if the equation is balanced by counting atoms of each kind present as reactants and as products.

Atoms of Reactants	*Atoms of Products*
2 N	2 N
6 H	6 H

While balancing the number of hydrogen atoms, we have also balanced the number of nitrogen atoms. The entire equation is balanced (Figure 10.9).

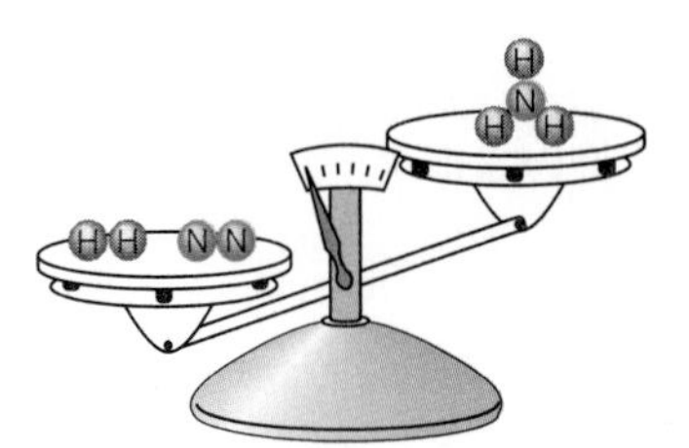

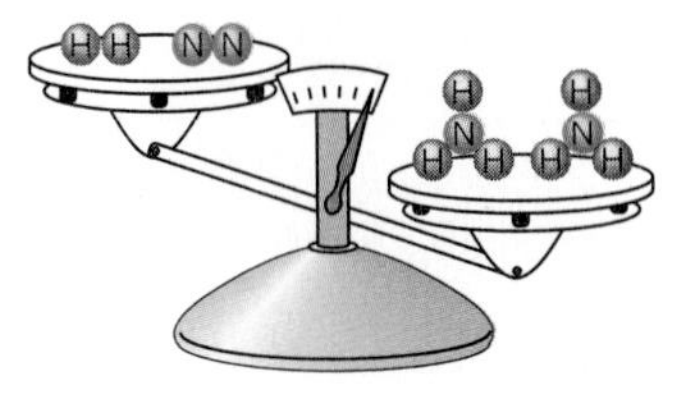

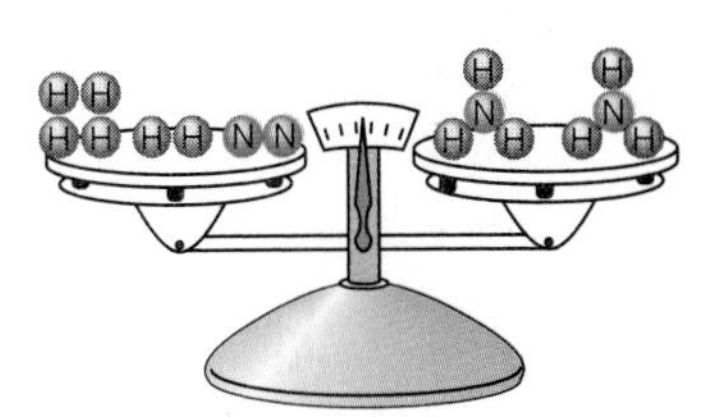

Figure 10.9 *To balance the equation for the reaction of $H_2(g)$ with $N_2(g)$, first place the correct formulas for reactants on the left and the correct formula for ammonia the product, on the right. If we double the amount of $NH_3(g)$ on the right and triple the $H_2(g)$ on the left, the equation will be balanced with two nitrogen atoms and six hydrogen atoms on each side. The balanced equation is written $N_2 + 3\ H_2 \rightarrow 2\ NH_3$.*

EXAMPLE 10.2

Write and balance an equation for the reaction involving the rusting of iron, showing that iron reacts with oxygen to produce iron(III) oxide, iron rust (Figure 10.10).

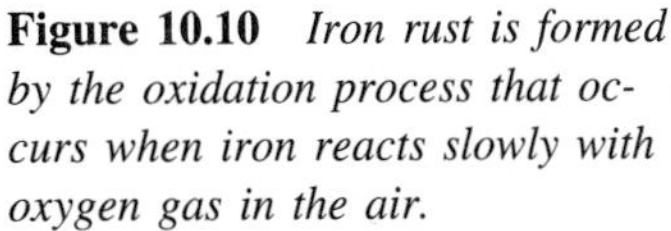

Figure 10.10 *Iron rust is formed by the oxidation process that occurs when iron reacts slowly with oxygen gas in the air.*

The chemical formulas were not given this time!

SOLUTION

STEP 1 Write the unbalanced equation with correct formulas of reactants and products.

$$Fe + O_2 \longrightarrow Fe_2O_3 \qquad \textit{(not balanced)}$$

STEP 2 Balance the equation by using coefficients where appropriate. It is recommended that you not begin with a single element standing alone like the iron shown here, since it will easily be taken care of later, after other kinds of elements are balanced. Instead, count the number of atoms of oxygen on each side. With two atoms of oxygen on the left and three on the right, the lowest multiple is six. Thus, we can balance the oxygen by using a coefficient of 3 for the O_2 and a coefficient of 2 for the Fe_2O_3.

$$Fe + 3\,O_2 \longrightarrow 2\,Fe_2O_3 \qquad \textit{(not balanced)}$$

The oxygen is now balanced, but atoms of iron are not. The four atoms of iron represented on the right can be balanced by placing a coefficient of four in front of the Fe on the left.

$$4\,Fe + 3\,O_2 \longrightarrow 2\,Fe_2O_3 \qquad \text{(balanced)}$$

An ''atoms count'' shows that there are equal numbers of atoms of each kind on both sides of the equation. The equation is balanced.

EXAMPLE 10.3

Write and balance the equation for the combustion of methane, CH_4, the main chemical in natural gas. During combustion, or burning, a chemical combines with oxygen. Combustion of a substance containing C and H atoms always gives carbon dioxide and water when we have ''complete'' combustion.

SOLUTION

STEP 1 Write the unbalanced equation using correct chemical formulas.

$$CH_4 + O_2 \longrightarrow CO_2 + H_2O \qquad \textit{(not balanced)}$$

Study Hint
Balance C, H, and O in that order in equations involving combustion of carbon-containing substances.

STEP 2 Balance the equation. Since the oxygen appears in two different products, we will leave it for last. For reactions involving combustion, it is generally best to balance the carbon, then the hydrogen, and leave the oxygen for last. (The oxygen appears as an individual element on the left, so placing any coefficient we may need in front of the O_2 will not interfere with other elements that have been balanced.) In this example, carbon is already balanced, with one atom on each side of the equation. To balance the hydrogen, with four atoms already on the left and two on the right, we place the coefficient 2 in front of H_2O on the right. We now have four hydrogen atoms on each side.

$$CH_4 + O_2 \longrightarrow CO_2 + 2\,H_2O \qquad \textit{(not balanced)}$$

The last element to balance is oxygen. There are four oxygen atoms represented on the right. Placing a 2 in front of O_2 balances oxygen atoms.

$$CH_4 + 2\,O_2 \longrightarrow CO_2 + 2\,H_2O \quad \text{(balanced)}$$

After counting to make sure there are equal numbers of atoms of each kind on both sides of the equation—an ''atoms count''—we conclude that the equation is now balanced.

EXAMPLE 10.4

Write and balance an equation for the reaction of aluminum sulfate with barium nitrate to produce aluminum nitrate and a solid, chalky white precipitate of barium sulfate (Figure 10.11). Use (s) to identify the precipitate.

Application
The barium sulfate described here is the substance used in the ''cocktail'' given to a patient during an upper or lower GI medical exam to make the intestinal tract show up on the X-ray.

SOLUTION

STEP 1 Write the unbalanced equation using correct chemical formulas.

$$Al_2(SO_4)_3 + Ba(NO_3)_2 \longrightarrow Al(NO_3)_3 + BaSO_4(s) \quad \textit{(not balanced)}$$

STEP 2 Balance the equation. Here we have an equation that involves compounds with polyatomic ions, the sulfate and the nitrate. These ions should be treated as individual units and balanced as a whole. When the reaction occurs, the Ba^{2+} ions from one compound react with the SO_4^{2-} ions from the other compound to produce the $BaSO_4$ precipitate, leaving the aluminum nitrate in solution.

The technique of ''following through'' as you would use in tennis, bowling, or golf can help in balancing the equation. Pick one key component appearing in only one compound on each side of the equation, such as the Al in this equation, and ''follow through'' with it. To balance the Al, we will place a coefficient of 2 in front of $Al(NO_3)_3$.

$$Al_2(SO_4)_3 + Ba(NO_3)_2 \longrightarrow 2\,Al(NO_3)_3 + BaSO_4(s) \quad \textit{(not balanced)}$$

But the coefficient of 2 on $Al(NO_3)_3$ also doubles the number of nitrate ions, producing $2 \times 3 = 6$ nitrate ions, so ''follow through'' with the nitrate by balancing it next. To get six nitrate ions on the left, we will need to place a 3 in front of the $Ba(NO_3)_2$.

$$Al_2(SO_4)_3 + 3\,Ba(NO_3)_2 \longrightarrow 2\,Al(NO_3)_3 + BaSO_4(s) \quad \textit{(not balanced)}$$

With the 3 in front of $Ba(NO_3)_2$, we now have three barium ions on the left side of the equation. We will therefore ''follow through'' by balancing the barium. To do this, place a 3 in front of the $BaSO_4$ on the right side of the equation.

$$Al_2(SO_4)_3 + 3\,Ba(NO_3)_2 \longrightarrow 2\,Al(NO_3)_3 + 3\,BaSO_4(s) \quad \text{(balanced)}$$

With the barium now balanced, we should ''follow through'' by checking the sulfate that is combined with the barium. As we ''follow through'' with sulfate, we find that it is already in balance. We have now gone full circle back to the aluminum sulfate where we started.
The equation is balanced. Check it by doing an ''atoms count.''

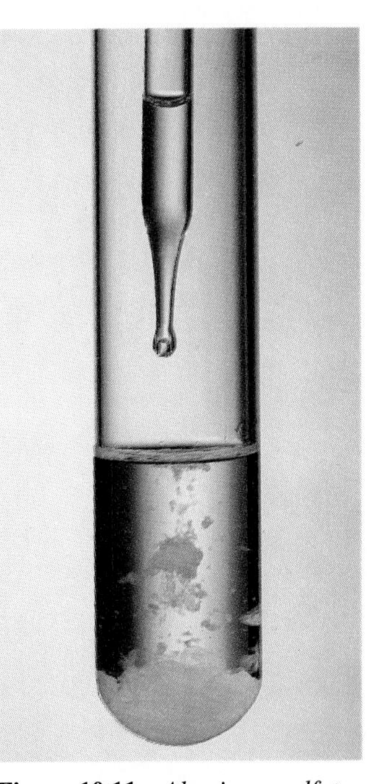

Figure 10.11 *Aluminum sulfate, $Al_2(SO_4)_3$, reacts with barium nitrate, $Ba(NO_3)_2$, to form a white precipitate of barium sulfate, $BaSO_4$.*

Exercise 10.1
Write and balance an equation for the reaction of sodium hydroxide with iron(III) chloride to produce sodium chloride and a precipitate of iron(III) hydroxide.

Now that the chemical equations in the examples have been written and balanced, let us summarize what is really ''balanced'' and what is not.

1. *Atoms.* The number of *atoms* of each kind of element is the same in reactants as in products. This is in agreement with the law of conservation of matter.
2. *Mass* Since the atoms do not change in mass, and since there is no change in numbers of atoms present before and after the reaction, we can be assured there is also *no mass change* during the reaction.
3. *Molecules.* The total numbers of *molecules* (or formula units) represented as reactants and products are not necessarily equal. Compare reactants and products for balanced equations in the preceding examples.
4. *Moles.* Since a mole is a specific quantity—it's Avogadro's number of unit particles—the total numbers of *moles* of reactants and products are not necessarily equal. To verify this, look again at the balanced equation at the end of Example 10.4: there are four formula units of reactants for every five formula units of products. Thus, there are 4 *moles* of reactants for every 5 *moles* of products. Do a similar comparison for Examples 10.1, 10.2, and 10.3. (Moles of reactants equals moles of products only for Example 10.3.)

Study Hint
The more equations you balance, the quicker you will become proficient at it. It may look easy as you watch someone else, but only after you have practiced thinking through the process will you be sure you can do it quickly and with confidence.

Thus, during chemical reactions, the atoms in compounds are rearranged to form different compounds, but no atoms are created or lost in the process. Atoms and mass are conserved; molecules and moles are not.

The example equations that were balanced here are only a beginning. To acquire skill, you need to practice balancing a large number of equations such as those at the end of this chapter.

See Problems 10.1–10.12.

10.4 *Classifying Reactions*

Now that we are familiar with balancing chemical equations, we should look more closely at several different types of reactions and how they can be classified. Most chemical reactions can be placed into one or more of the following five categories.

1. *Combustion Reactions.* During **combustion**, compounds containing carbon, hydrogen, and sometimes oxygen burn in air (consuming oxygen) to produce carbon dioxide and water. Example 10.3 involving the burning of methane, CH_4, was a typical combustion reaction. This type of reaction will be described in more detail in Section 10.5.
2. *Combination (Synthesis) Reactions.* When one element reacts or combines with another element to produce a compound, we can say that a new substance is synthesized. Reactions of this type are classified as **combination** or **synthesis** reactions. They can be represented in a general way as follows.

$$A + B \longrightarrow AB$$

The synthesis of ammonia (Example 10.1) illustrates this type of reaction. More synthesis reactions are described in Section 10.6.

3. *Decomposition Reactions.* A **decomposition** reaction is one in which a single compound, symbolized as AB, is broken down into two or more simple substances. This type of reaction can be represented as follows.

$$AB \longrightarrow A + B$$

Examples of decomposition reactions are provided in Section 10.7.

4. *Single-Replacement Reactions.* In **single-replacement reactions**, an element, symbolized as A, reacts with a compound, BC, to take the place of one of the components of the compound. This type of reaction can be represented by the following general equation.

$$A + BC \longrightarrow AC + B$$

Single-replacement reactions for metals are discussed in Section 10.8. Single-replacement reactions for nonmetals are discussed in Section 10.9.
Many of the reactions that fit into one of these first four categories involve an oxidation–reduction process that will be described later in this chapter.

5. *Double-Replacement Reactions.* In **double-replacement reactions**, two compounds, AB and CD, can be thought of as ''exchanging partners'' to produce two different compounds, AD and CB.

$$AB + CD \longrightarrow AD + CB$$

The positive ion, A, in the first compound combines with the negative ion, D, in the second compound while the positive ion, C, of the second compound combines with the negative ion, B, in the first compound. Double-replacement reactions are described in Sections 10.10 and 10.12.

EXAMPLE 10.5

Classify these reactions in terms of the five categories described.

(a) From Example 10.2, $4\ Fe + 3\ O_2 \rightarrow 2\ Fe_2O_3$.
(b) From Example 10.4, $Al_2(SO_4)_3 + 3\ Ba(NO_3)_2 \rightarrow 2\ Al(NO_3)_3 + 3\ BaSO_4$.

SOLUTION

(a) This is a combination (or synthesis) reaction: $A + B \rightarrow AB$.
(b) This is a double-replacement reaction. The ions ''exchange partners.''

See Problems 10.13 and 10.14.

EXERCISE 10.2

Classify the reactions described in Examples 10.1 and 10.3.

Most reactions fit into one of the five categories described here, but this is not the only way reactions can be classified. As will be shown in the following sections, one can become familiar with many similar reactions by grouping them together into several smaller categories.

10.5 Combustion

When a substance containing carbon and hydrogen (a hydrocarbon) undergoes complete **combustion**, or burning, oxygen is consumed as carbon dioxide and water are produced. The general unbalanced equation becomes

---

Chemistry in Our World

Air Pollution and the Greenhouse Effect

The burning of petroleum products and other fossil fuels is believed to contribute to global warming because of the greenhouse effect, whereby energy from the sun warms the Earth and is trapped by carbon dioxide and other gases in the atmosphere.

Companies—such as Viacom featured here—are working with the U.S. Department of Transportation to drastically curb the ever-increasing atmospheric carbon dioxide levels and to reduce all forms of air pollution.

As one of the tri-state area's biggest employers, Viacom is responsible for developing ways 2 minimize air pollution during peak commuting hours. So we've developed CO2, a program designed 2 find COMMUTER OPTIONS 2 CLEAN THE AIR.

All Viacom employees will receive Department of Transportation Commuter Surveys in the next few weeks, and it's really important that we all fill them out.

First, because we have 2 find out how we are commuting 2 work. Second, because we can use this information 2 reduce the amount of pollution we're putting in2 the air every day.

AND THAT'S A GOOD OPTION FOR EVERYONE.

Reducing the number of cars used by commuters is expected to lead to cleaner air.

$$\text{Hydrocarbon} + O_2 \longrightarrow CO_2 + H_2O \qquad \textit{(not balanced)}$$

Example 10.3 demonstrated the balancing of the equation for the combustion of methane, CH_4, a simple hydrocarbon. The quickest way to get this done is to

1. Balance the carbon atoms first.
2. Balance the hydrogen atoms second.
3. Balance the oxygen atoms last.

If it is necessary to use fractional coefficients, especially multiples of $\frac{1}{2}$, go ahead and do so. Then, to eliminate the fractional coefficients, multiply the entire equation by the lowest common denominator of the fractions.

The use of fractional coefficients is shown in Example 10.6.

At the beginning of the chapter, the reaction for the combustion of gasoline was mentioned. Let us balance the equation for that reaction.

EXAMPLE 10.6

Write and balance the equation for the combustion of octane, C_8H_{18}, present in gasoline.

SOLUTION

STEP 1 Write the unbalanced equation, with correct formulas.

$$C_8H_{18} + O_2 \longrightarrow CO_2 + H_2O \qquad \textit{(not balanced)}$$

STEP 2 Balance the equation, using the three steps outlined in this section.

$$C_8H_{18} + 12\tfrac{1}{2}\,O_2 \longrightarrow 8\,CO_2 + 9\,H_2O \qquad \text{(balanced)}$$

While the equation is actually balanced, the fractional coefficient for oxygen should be eliminated by multiplying all coefficients by 2, to give

$$2\,C_8H_{18} + 25\,O_2 \longrightarrow 16\,CO_2 + 18\,H_2O \qquad \text{(balanced with whole numbers)}$$

EXERCISE 10.3

Write and balance the equation for the combustion of ethanol, C_2H_5OH. (See Figure 10.12.)

Figure 10.12 *The combustion of ethanol (also called ethyl alcohol or grain alcohol), shown here, and the metabolism of ethanol by the human body have the same overall reaction and release the same amount of energy.*

Your skill in writing and balancing chemical equations will increase with practice. Watching your instructor balance equations or reading what is printed here will not provide enough experience to learn the process. Study the example and the exercise; then develop your skill and speed by doing problems at the end of the chapter.

See Problems 10.15 and 10.16.

10.6 Synthesis (Combination) Reactions

The production of a single compound from the reaction of two or more substances can be called a **combination reaction** or a **synthesis reaction**. Reactions of this type have the general form

$$A + B \longrightarrow AB$$

CHEMISTRY AT WORK

Hydrogen as a Fuel

Hydrogen fuel was used by the Saturn V rocket that carried the first astronauts to the moon. Hydrogen is also the main fuel used by space shuttle rockets. Separate tanks with a diameter of 8 meters and a length of 40 meters carry the liquid hydrogen and liquid oxygen for shuttle rockets.

Nuclear-powered electric energy has been proposed as a means of decomposing ocean water to release hydrogen, but the process is not yet economical.

Hydrogen is not an original energy source but, rather, an energy carrier. This is because the energy to release the hydrogen must be obtained from another energy source such as solar power or a fossil fuel. Thus, hydrogen should be thought of as a means of storing or transporting energy.

Hydrogen is carried in the space shuttle's large fuel tank. The main engines of the shuttle burn a mixture of liquid hydrogen and liquid oxygen.

The synthesis of ammonia gas, $NH_3(g)$, from $N_2(g)$ and $H_2(g)$ is an important industrial process. The ammonia can be used directly as a fertilizer or to produce other chemicals used in fertilizers, explosives, and by industry.

$$N_2(g) + 3\,H_2(g) \longrightarrow 2\,NH_3(g)$$

Another synthesis reaction is that of $H_2(g)$ with $O_2(g)$ to produce water. The equation for this reaction was balanced in Section 10.3.

$$2\,H_2(g) + O_2(g) \xrightarrow{\text{spark}} 2\,H_2O(l)$$

Hydrogen can burn in oxygen with a steady flame and can be used as a fuel, but a mixture of the two gases, ignited by a spark, reacts rapidly, with an explosion. See the box ''Hydrogen as a Fuel.'' Other synthesis reactions include

$$N_2(g) + O_2(g) \longrightarrow 2\,NO(g)$$

Nitrogen monoxide — NO produced by automobile engines increases pollution.

$$2\,NO(g) + O_2(g) \longrightarrow 2\,NO_2(g)$$

Nitrogen dioxide — Colorless NO readily reacts with oxygen in the air to form reddish brown NO_2 gas.

See Problems 10.17 and 10.18.

More synthesis reactions are included in problems at the end of this chapter.

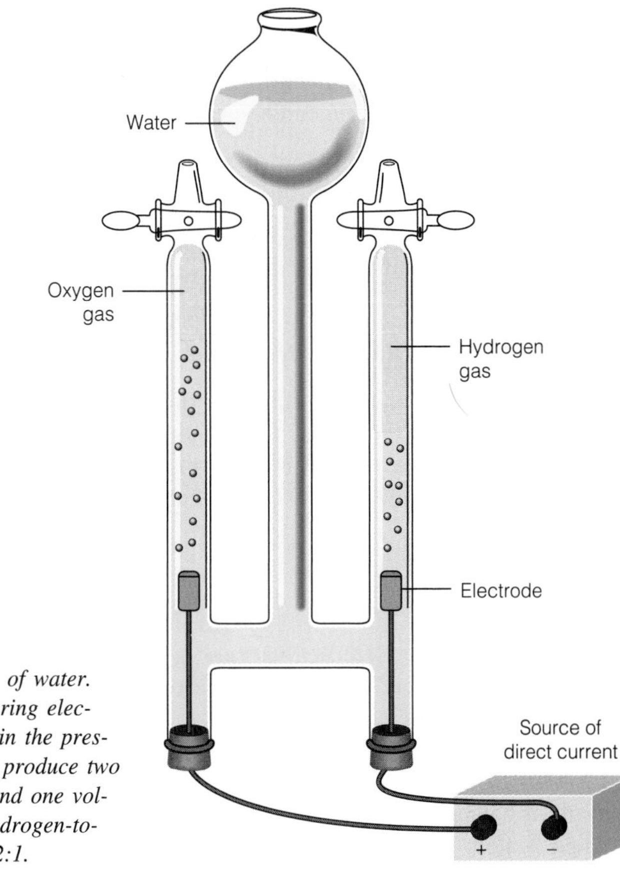

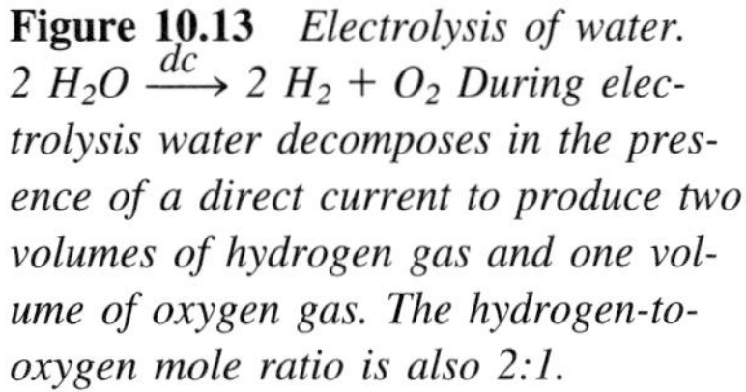
Figure 10.13 *Electrolysis of water.* $2\ H_2O \xrightarrow{dc} 2\ H_2 + O_2$ *During electrolysis water decomposes in the presence of a direct current to produce two volumes of hydrogen gas and one volume of oxygen gas. The hydrogen-to-oxygen mole ratio is also 2:1.*

10.7 Decomposition Reactions

When a single compound breaks down into two or more simpler substances, the reaction involves **decomposition**, as indicated by the general equation

$$AB \longrightarrow A + B$$

The synthesis reaction of $H_2(g)$ with $O_2(g)$ to produce water also releases a specific amount of energy. The reverse reaction—decomposition—must take up energy (Figure 10.13). The energy to carry out the reaction must be supplied continually from a battery or another source of direct current (dc). The process is called **electrolysis** (Greek for separating by electricity). When the energy source is turned off, the reaction ceases.

$$2\ H_2O(l) \xrightarrow{dc} 2\ H_2(g) + O_2(g)$$

Decomposition of Metal Oxides

Antoine Lavoisier, a French chemist (see Section 2.6), was able to decompose red-orange mercuric oxide powder to give liquid mercury and a gas, which he named oxygen. He focused sunlight through a lens to provide the heat for this decomposition. Lavoisier performed the investigation in a closed system and found that there was no mass change during the reaction. His investigations helped establish chemistry as an experimental science. The reaction is

$$2\ HgO(s) \longrightarrow 2\ Hg(l) + O_2(g)$$

Some heavy metal oxides like HgO and PbO_2 can easily be decomposed by heat, while other metal oxides, like Al_2O_3, are very difficult to decompose.

Decomposition of Chlorates and Nitrates

When compounds containing chlorates are heated, they decompose to give the metal chloride and oxygen gas. Chlorates are used in fireworks and road flares (Figure 10.14). A catalyst like MnO_2 can be used to speed up the reaction. The catalyst does not change; it is written over the arrow rather than as a reactant or product.

$$2\ KClO_3(s) \xrightarrow[\text{heat}]{MnO_2} 2\ KCl(s) + 3\ O_2(g)$$

When metal nitrates are heated they do not readily release all of the oxygen atoms in the compound. They decompose to give the metal nitrite and oxygen gas, as shown in the following example.

EXAMPLE 10.7

Write a balanced chemical equation for the decomposition of sodium nitrate when heated.

SOLUTION

STEP 1 Write the unbalanced equation using correct formulas for all substances.

Warning
An equation cannot correctly represent a reaction if formulas are written incorrectly.

$$NaNO_3 \longrightarrow NaNO_2 + O_2(g) \qquad \textit{(not balanced)}$$

STEP 2 Balance the equation using appropriate coefficients. Begin by counting and comparing oxygen atoms on each side of the equation. Since $NaNO_3$ contains three oxygen atoms and $NaNO_2$ contains two, double the

Figure 10.14 *Safety flares, shown here, and many types of fireworks use potassium chlorate.*

$NaNO_3$; it has an odd number of oxygen atoms. Also double the $NaNO_2$ to keep the Na atoms in balance.

$$2\,NaNO_3 \longrightarrow 2\,NaNO_2 + O_2(g) \qquad \text{(balanced)}$$

Counting oxygen atoms on the left (six) and oxygen atoms on the right (six) shows us that oxygen atoms are balanced. The coefficient in front of the O_2 does not need to be adjusted.

See Problems 10.19 and 10.20.

Decomposition of Hydrogen Peroxide

One more decomposition reaction of special interest involves hydrogen peroxide, H_2O_2, which decomposes in the presence of a catalyst to produce oxygen gas and water (Figure 10.15). The iodide ion or MnO_2 catalyzes this reaction.

$$2\,H_2O_2 \xrightarrow{MnO_2 \text{ or } I^- \text{ ion}} 2\,H_2O + O_2(g)$$

You may have a bottle of 3% hydrogen peroxide in the medicine cabinet. A 6% solution of hydrogen peroxide can be used to bleach hair, while a 90% solution has been used as a source of oxygen in some rockets.

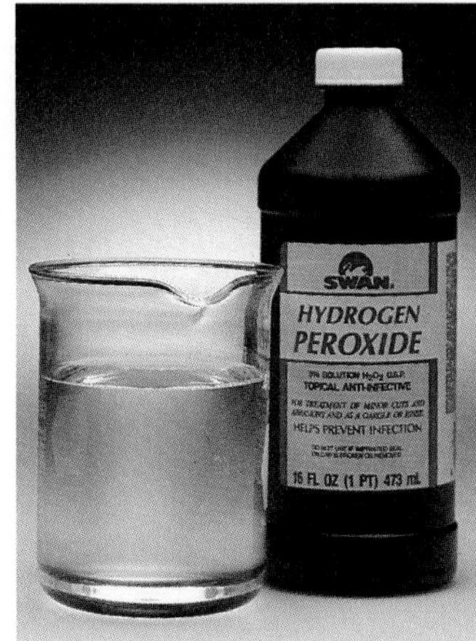

Figure 10.15 *Hydrogen peroxide decomposes in the presence of a variety of catalysts to give oxygen gas and water.*

10.8 *Reactions of Metals*

Very reactive metals in Group IA of the periodic table react rapidly with oxygen in the air to form metal oxides, with the general formula M_2O. These metals are stored under mineral oil or kerosene to prevent reaction with the atmosphere or with water. Other metals react less rapidly with oxygen. Some examples follow.

$$\text{Metal} + \text{Oxygen} \longrightarrow \text{Metal oxide}$$

Examples:

$$\begin{array}{llll} 4\,Na & + O_2(g) & \longrightarrow 2\,Na_2O(s) & \text{Rapid.} \\ 2\,Mg & + O_2(g) & \xrightarrow{\text{burn}} 2\,MgO(s) & \text{Mg ignites in a flame.} \\ 4\,Fe & + 3\,O_2(g) & \longrightarrow \underset{\text{Rust}}{2\,Fe_2O_3(s)} & \text{Slow, unless heated.} \end{array}$$

Some Group IA metals also form peroxides with the general formula M_2O_2 or superoxides with the general formula MO_2, but we shall not attempt to present these special cases here.

Reactions of Metal Oxides with Water

The last set of reactions produced metal oxides. Metal oxides that are water soluble dissolve in water to produce metal hydroxides that are **basic**. Basic solutions contain OH^- ions that can neutralize H^+ ions of acids. Acid–base indicators can be used to detect the presence of bases. We will discuss acids and bases in more detail later. For the following reactions, we shall start with the metal oxides produced by the reactions shown in the previous set of examples.

$$\text{Metal oxide} + \text{Water} \longrightarrow \text{Metal hydroxide}$$

$$\begin{array}{llll} Na_2O & + H_2O & \longrightarrow & 2\,NaOH \\ MgO & + H_2O & \longrightarrow & Mg(OH)_2 \end{array}$$

Reactions of Metals with Various Oxidizing Agents

Fine copper wire mesh becomes oxidized and glows as it is heated in a flame, but it also glows when it is heated and thrust into a bottle of chlorine gas, producing a greenish solid. The equation is

$$Cu(s) + Cl_2(g) \longrightarrow CuCl_2(s)$$

Reactions of metals with oxygen were described at the beginning of this section. They could be classified as combination reactions, but they can also be classified as **oxidation–reduction** reactions. During the reaction with oxygen and also with chlorine, the neutral metal atom loses electrons as it is "oxidized" by the nonmetal to become a positively charged metal ion. In the general equation

$$M \longrightarrow M^{n+} + n\,e^- \quad \text{(In oxidation, electrons are lost.)}$$

M represents the metal and *n* represents the ion charge as well as the number of electrons lost. A *loss of electrons* by the metal always occurs during **oxidation**. It is accompanied by a **reduction**, a *gain of electrons*, by another chemical called the **oxidizing agent**.

Whenever one substance is oxidized, another is reduced.

In the earlier examples, oxygen was the oxidizing agent, so it is easy to see why the process is called oxidation, but chlorine can also cause copper to lose electrons (to be oxidized). Thus, chlorine is also an oxidizing agent. Any substance that causes the metal to lose electrons can be an oxidizing agent. The substance that loses electrons is the **reducing agent**; it gets oxidized. Good examples of spontaneous oxidation—where the reaction occurs without addition of heat—include the following.

$$2\,Sb(s) + 3\,Cl_2(g) \longrightarrow 2\,SbCl_3$$

$$2\,Al(s) + 3\,I_2(s) \longrightarrow 2\,AlI_3$$

The nonmetals (chlorine and iodine) change from neutral elements to negative ions. This involves a gain of electrons, so they are reduced. For example,

$$3\,Cl_2 + 6\,e^- \longrightarrow 6\,Cl^-$$

The nonmetal (symbolized by X) is reduced as it gains electrons. A general equation for reduction is shown here.

$$X + n\,e^- \longrightarrow X^{n-} \quad \text{(In reduction, electrons are gained.)}$$

We could call the nonmetal an electron grabber. As the nonmetal is reduced, it acts as an oxidizing agent for the metal. To summarize what is happening, the oxidizing and reducing agents are labeled for the following equation.

Reducing agent		Oxidizing agent		
$2\,Al(s)$	$+$	$3\,I_2(s)$	$\longrightarrow$	$2\,AlI_3$
loses electrons and is oxidized.		gains electrons and is reduced.		

Study Hint
When metals react with nonmetals, the metal is the reducing agent and the nonmetal is the oxidizing agent.

Single-Replacement Reactions of Metals

When a piece of copper wire is submerged in a solution of silver nitrate, a chemical reaction occurs. Shiny needlelike crystals of silver form on the copper wire, as shown in Figure 10.16. The reaction is

$$Cu(s) + 2\,AgNO_3(aq) \longrightarrow Cu(NO_3)_2(aq) + 2\,Ag(s) \qquad \text{(balanced)}$$

In this reaction, the copper is oxidized to Cu^{2+} ions and has *replaced* the Ag^+ ions of $AgNO_3$. The Ag^+ ions are reduced to solid silver. Reactions like these fit the single-replacement equation form

$$A + BC \longrightarrow AC + B$$

We know that the alkali metals are quite reactive, but a look at the periodic table does not clearly reveal which metals will replace others. Instead, experimental investigations that are relatively easy to perform in the lab provide information that allows us to make a list with metals arranged in order of reactivity. This list is called an **activity series**. Table 10.1 shows the activity series for some of the more common metals in order of decreasing reactivity, with the most reactive metal at the top of the list.

The most reactive metals, the alkali (Group IA) metals, readily replace hydrogen from cold water to produce the metal hydroxide and hydrogen gas

Figure 10.16 *A copper wire placed in a silver nitrate solution reacts to produce shiny needlelike crystals of metallic silver along with copper(II) ions.*

Table 10.1 **An Activity Series of Metals**

Reduced Form (metal atom) → *Oxidized Form (metal ion)*

$$\text{Metal atom} \xrightarrow{\text{oxidation}} \text{Metal ion} + n\,e^-$$

$$\text{Metal atom} \xleftarrow{\text{reduction}} \text{Metal ion} + n\,e^-$$

(Arrow pointing up: Relative ease of oxidation)

Reduced Form (metal atom)		Oxidized Form (metal ion)		
Li	$\longrightarrow$	Li^+	$+ e^-$	React with cold water, steam, or acids releasing hydrogen gas
K	$\longrightarrow$	K^+	$+ e^-$	
Ca	$\longrightarrow$	Ca^{2+}	$+ 2e^-$	
Na	$\longrightarrow$	Na^+	$+ e^-$	
Mg	$\longrightarrow$	Mg^{2+}	$+ 2e^-$	React with steam or acids releasing hydrogen gas
Al	$\longrightarrow$	Al^{3+}	$+ 3e^-$	
Zn	$\longrightarrow$	Zn^{2+}	$+ 2e^-$	
Cr	$\longrightarrow$	Cr^{3+}	$+ 3e^-$	
Fe	$\longrightarrow$	Fe^{2+}	$+ 2e^-$	
Cd	$\longrightarrow$	Cd^{2+}	$+ 2e^-$	React with acids releasing hydrogen gas
Ni	$\longrightarrow$	Ni^{2+}	$+ 2e^-$	
Sn	$\longrightarrow$	Sn^{2+}	$+ 2e^-$	
Pb	$\longrightarrow$	Pb^{2+}	$+ 2e^-$	
H_2	$\longrightarrow$	$2H^+$	$+ 2e^-$	
Cu	$\longrightarrow$	Cu^{2+}	$+ 2e^-$	Do not react with acids to release hydrogen gas
Ag	$\longrightarrow$	Ag^+	$+ e^-$	
Hg	$\longrightarrow$	Hg^{2+}	$+ 2e^-$	
Au	$\longrightarrow$	Au^{3+}	$+ 3e^-$	

Figure 10.17 *Potassium and other alkali metals react vigorously with water to give hydrogen gas and the metal hydroxide. Here the reaction released enough heat energy to ignite the hydrogen.*

(Figure 10.17). These reactions are quite rapid, even though water has strong covalent bonds that must be broken. Two alkaline earth metals, calcium and barium, react at a moderate rate with water, but water must be in the form of steam for it to react with magnesium. Some examples follow.

$$\text{Active metal} + \text{Water} \longrightarrow \text{Metal hydroxide} + \text{Hydrogen gas}$$

$$2\,Na + 2\,H_2O \longrightarrow 2\,NaOH + H_2(g)$$

$$Ca + 2\,H_2O \longrightarrow Ca(OH)_2 + H_2(g)$$

Compare these reactions with the reactions described earlier where metal oxides placed in water produce metal hydroxides alone, with no hydrogen gas.

EXAMPLE 10.8

Write an equation for the reaction of potassium with water.

SOLUTION

STEP 1 Write the unbalanced equation with correct formulas for all substances.

$$K + H_2O \longrightarrow KOH + H_2(g) \qquad \textit{(not balanced)}$$

STEP 2 Balance the equation using the smallest set of integer coefficients. Begin by counting and comparing numbers of atoms of each type for reactants and products. With an odd number of H atoms on the right and an even number on the left, double the KOH on the right to give an even number of H's. Now, with four H's on the right, double the H_2O on the left, to balance the H's. Balance the K last; it stands alone on the left. Count all atoms to check your work.

$$2\,K + 2\,H_2O \longrightarrow 2\,KOH + H_2(g) \qquad \text{(balanced)}$$

The most reactive metals readily replace $H_2(g)$ from *cold water* or acids. The next group of metals in the activity series can replace $H_2(g)$ from *steam* or from acids. Next in the series are metals like nickel that will replace hydrogen from acids, but not from steam or cold water. Silver and other metals that are below hydrogen in the list will not replace hydrogen, even from acids.

The reason for differences in reactivity is related to the tendency to give up electrons to form ions. Metals higher in the activity series give up electrons (they are oxidized) more readily than do metals that are lower in the series. Learn to use this activity series when predicting which metals will react with acids and with other compounds.

EXAMPLE 10.9

Will magnesium react with hydrochloric acid, HCl? If so, write a balanced equation for the reaction.

SOLUTION

STEP 1 Check the activity series to see if the Mg metal is more reactive than the H^+ ions of the acid. It is, so we can write the unbalanced equation. (If no reaction is predicted, we simply write "no reaction" after the arrow.)

$$Mg + HCl(aq) \longrightarrow MgCl_2(aq) + H_2(g) \qquad \textit{(not balanced)}$$

STEP 2 Balance the equation using the smallest set of whole-number coefficients.

$$\text{Mg} + 2\,\text{HCl(aq)} \longrightarrow \text{MgCl}_2\text{(aq)} + \text{H}_2\text{(g)} \qquad \text{(balanced)}$$

EXERCISE 10.4
Will mercury or gold react with an acid? (See Table 10.1.)

Now we can explain why the copper wire reacted with the silver nitrate. A reactive metal in the activity series will replace the *ion* of any metal lower on the list. Thus, the Cu metal would be expected to react with Ag^+ ions placed lower on the list. Similarly, zinc metal reacts with $Pb(NO_3)_2$ in a solution. As the Zn—which is more reactive than lead—is oxidized to Zn^{2+} ions, the Pb^{2+} ions are reduced to lead metal crystals. The conversion of Pb^{2+} ions to Pb is the reverse reaction of the one given in the activity series because it involves reduction rather than oxidation.

Locate these metals and their ions in the activity series as you read this paragraph.

$$\text{Zn(s)} + \text{Pb(NO}_3)_2\text{(aq)} \longrightarrow \text{Zn(NO}_3)_2\text{(aq)} + \text{Pb(s)}$$

Zinc can react with $Pb(NO_3)_2$; the reverse reaction does not occur.

EXAMPLE 10.10
Will silver metal react with a solution containing Zn^{2+} ions? If so, write a balanced equation for the reaction.

SOLUTION
Check the activity series to see if the Ag *metal* is more reactive (is higher on the list) than Zn^{2+} *ions*. This is not the case; no reaction is expected.

Since there is no reaction, we write

$$\text{Ag(s)} + \text{Zn}^{2+}\text{(aq)} \longrightarrow \text{No reaction}$$

See Problems 10.21–10.28.

With the aid of the activity series (Table 10.1), you should be able to predict many reactions similar to these.

10.9 Reactions of Nonmetals

Nonmetals burn in air—react with oxygen—to form nonmetal oxides. For example, light yellow sulfur burns in pure oxygen gas with a pale blue flame to produce sulfur dioxide, an irritating, colorless gas (Figure 10.18). The sulfur has the formula S_8 because sulfur atoms form eight-membered rings.

Figure 10.18 *Sulfur burns in oxygen to give sulfur dioxide gas, which has a distinctive but irritating odor somewhat like that of a burning match.*

$$\text{S}_8\text{(s)} + 8\,\text{O}_2\text{(g)} \longrightarrow 8\,\text{SO}_2\text{(g)}$$

Further oxidation of SO_2 can produce sulfur trioxide, SO_3. Carbon dioxide is produced as charcoal and other forms of carbon are burned. The burning of phosphorus, P_4, yields tetraphosphorus decoxide, P_4O_{10}, also called diphosphorus pentoxide because its simplest formula is P_2O_5.

$$P_4(s) + 5\,O_2(g) \longrightarrow P_4O_{10}(s)$$

Phosphorus can also be oxidized by chlorine and bromine to produce halides such as PCl_5 and PBr_3.

Reactions of Nonmetal Oxides with Water

Nonmetal oxides react with water to produce acids. Some examples are

$$SO_2 + H_2O \longrightarrow H_2SO_3$$
Sulfurous acid

$$SO_3 + H_2O \longrightarrow H_2SO_4$$
Sulfuric acid

$$CO_2 + H_2O \longrightarrow H_2CO_3$$
Carbonic acid

EXAMPLE 10.11

Write a balanced equation for the reaction of P_4O_{10} with water to produce phosphoric acid, H_3PO_4.

SOLUTION

STEP 1 Write the unbalanced equation with correct formulas for all substances.

$$P_4O_{10} + H_2O \longrightarrow H_3PO_4 \quad \textit{(not balanced)}$$

STEP 2 Balance the equation using the smallest set of integer coefficients. Since oxygen is in each of the compounds, balance it last. Start with the P, and follow through with H before checking the oxygen.

$$P_4O_{10} + 6\,H_2O \longrightarrow 4\,H_3PO_4 \quad \text{(balanced)}$$

Single-Replacement Reactions of Nonmetals

Some halogens are more reactive than others. More active halogens react with compounds containing ions of less active halogens. In these reactions, the active halogen—as an element—is the oxidizing agent, and it is reduced in the process. The order of reactivity follows, with the most reactive halogen listed first in the series: F_2, Cl_2, Br_2, and I_2. This order is the same as the top-to-bottom order of the halogens in the periodic table, with the most electronegative and most nonmetallic, fluorine, first in the series. Compare the following.

Activity Series for Halogens

F_2
Cl_2
Br_2
I_2

$$Cl_2(g) + 2\,HI(g) \longrightarrow I_2(s) + 2\,HCl(g)$$
greenish colorless purple colorless

$$Cl_2(g) + NaF(aq) \longrightarrow \text{No reaction}$$

As shown, the chlorine replaces iodide *ions*, I^-, but does not replace fluoride *ions*, F^-. Look carefully at the activity series of halogens as you repeat that last statement. Now try this example.

EXAMPLE 10.12

Write a balanced equation for what you expect to happen when bromine, a red liquid, is mixed with a colorless NaCl solution.

SOLUTION

STEP 1 Check the periodic table and compare the positions of bromine and chlorine. Bromine is below fluorine and chlorine in the table, so it would be expected to replace iodide ions but *not* fluoride or chloride ions.

STEP 2 Write down your prediction.

$$Br_2 + NaCl(aq) \longrightarrow \text{No reaction}$$

See Problems 10.29–10.32.

EXERCISE 10.5

Write a balanced equation for what you expect to happen when bromine is mixed with a solution of sodium iodide.

10.10 Double-Replacement Reactions

Double-replacement or **metathesis** reactions take the form

$$AB + CD \longrightarrow AD + CB$$

In other words, ions in compounds AB and CD switch partners. This type of reaction takes place in aqueous solution when at least one of the products is

1. An insoluble or nearly insoluble solid, called a **precipitate**.
2. A covalent compound, including water and common gases.

The driving force behind these reactions and many others is the formation of a stable product.

Formation of a Gas

Heating sulfuric acid, H_2SO_4, with sodium sulfide releases hydrogen sulfide, H_2S, a pungent gas with the odor of rotten eggs. The equation for the reaction is

$$H_2SO_4 + Na_2S \longrightarrow H_2S(g) + Na_2SO_4$$

The driving force for this reaction is the formation of hydrogen sulfide gas.

Hydrogen chloride gas is prepared commercially, and also in the lab, by the reaction of concentrated sulfuric acid with sodium chloride, as follows.

$$H_2SO_4 + NaCl \longrightarrow HCl(g) + NaHSO_4$$

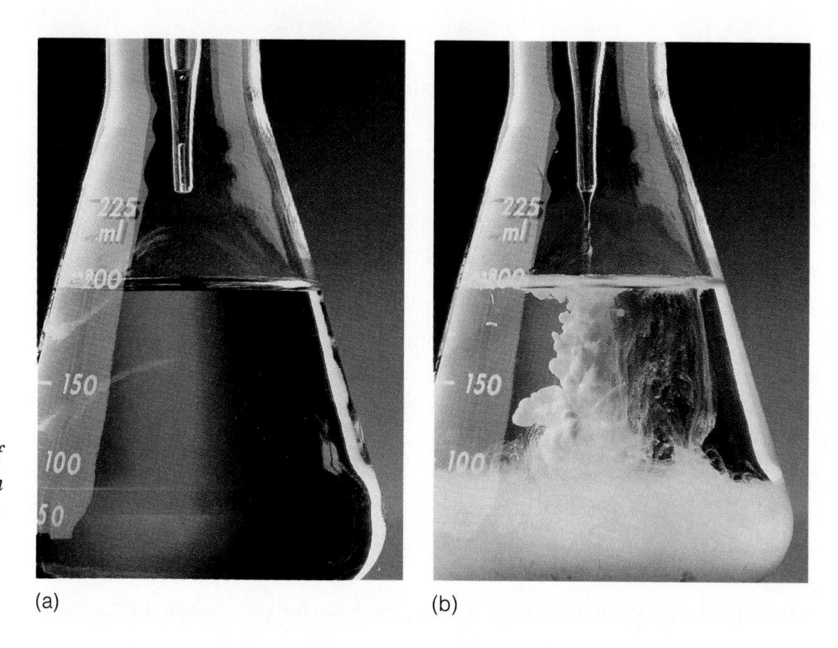

(a) (b)

Figure 10.19 *Clear solutions of (a) lead(II) nitrate and potassium iodide react (b) to form a yellow precipitate of lead(II) iodide. Potassium ions and nitrate ions remain in solution.*

The HCl gas can then be dissolved in water to make hydrochloric acid. Also, poisonous hydrogen cyanide gas, HCN(g), is produced as acids react with cyanides.

$$HCl(aq) + NaCN \longrightarrow HCN(g) + NaCl$$

Precipitation Reactions

Mixing an aqueous (water) solution of colorless lead nitrate with a colorless solution of potassium iodide gives an opaque, bright yellow product that, upon standing, settles to the bottom of the flask (Figure 10.19). This insoluble product is called a *precipitate* (abbreviated ppt). The precipitate can be identified in an equation by (s) for solid, as shown here.

$$Pb(NO_3)_2(aq) + 2\ KI(aq) \longrightarrow PbI_2(s) + 2\ KNO_3(aq)$$

During the reaction, the Pb^{2+} ions from $Pb(NO_3)_2$(aq) and I^- ions from KI(aq) form the precipitate of PbI_2(s). The ions have switched partners.

Check with your instructor about which solubility rules you should know.

The formation of a precipitate is the driving force for this reaction and for many others. Table 10.2 gives several general solubility rules to help you decide which,

Table 10.2 **Solubility Rules for Ionic Compounds***

1. Alkali metal compounds, nitrates, and ammonium compounds are soluble.
2. Hydroxides of alkali metals and some alkaline earths (Ca^{2+}, Sr^{2+}, and Ba^{2+}) are soluble. All other hydroxides are insoluble.
3. All chlorides are soluble except AgCl, $PbCl_2$, and Hg_2Cl_2.
4. Most sulfates are soluble; exceptions include $BaSO_4$, $PbSO_4$, and $CaSO_4$.
5. Most phosphates, carbonates, chromates, and sulfides are insoluble except those of the alkali metals.*

*The terms *insoluble* and *slightly soluble* are used interchangeably; only extremely small amounts dissolve in water. For chemicals that are *soluble*, solutions of 0.10 M or greater can be prepared.

if any, products of a reaction are precipitates. The following metathesis reactions illustrate the formation of precipitates.

$$NaCl(aq) + AgNO_3(aq) \longrightarrow NaNO_3(aq) + AgCl(s)$$

Silver halides are insoluble.

$$K_2SO_4(aq) + BaCl_2(aq) \longrightarrow 2\ KCl(aq) + BaSO_4(s)$$

Ba, Sr, and Pb sulfates are insoluble.

$$3\ KOH(aq) + FeCl_3(aq) \longrightarrow 3\ KCl(aq) + Fe(OH)_3(s)$$

All transition metal hydroxides are insoluble.

EXAMPLE 10.13

Write a balanced chemical equation for the reaction of aqueous solutions of potassium chromate and silver nitrate to form a precipitate.

SOLUTION

STEP 1 Write the unbalanced equation with correct formulas for all substances.

$$K_2CrO_4(aq) + AgNO_3(aq) \longrightarrow KNO_3 + Ag_2CrO_4 \qquad \textit{(not balanced)}$$

To decide which of the two products is responsible for the formation of the precipitate, we will use a process of elimination. Since the first product listed is a nitrate, and we know all nitrates are soluble, the KNO_3 cannot be the precipitate. Therefore, the precipitate formed has to be the other product, Ag_2CrO_4, which should be identified in the equation with an (s).

STEP 2 Balance the equation using the smallest set of integer coefficients.

$$K_2CrO_4(aq) + 2\ AgNO_3(aq) \longrightarrow 2\ KNO_3(aq) + Ag_2CrO_4(s) \qquad \text{(balanced)}$$

See Problems 10.33 and 10.34.

10.11 Ionic and Net Ionic Equations

The equation that we balanced in Example 10.13 is a typical, standard chemical equation. It gives a great deal of information, including formulas of all substances involved in the reaction and the proportions in which they react. It also shows that solid silver chromate is produced as a precipitate.

There is more that can be easily shown for reactions that occur in an aqueous solution. The original potassium chromate solution contains both potassium ions, K^+, and chromate ions, CrO_4^{2-}, dissolved in water. These ions move around freely in the water solution. Similarly, the silver ions, Ag^+, and the nitrate ions, NO_3^-, are also dissolved in water and move around freely and independently. The same is true for the potassium and nitrate ions that show up as products. In fact, the only component for which this is not true is the $Ag_2CrO_4(s)$, which is a precipitate. We can write an **ionic equation**; it shows all dissolved ions written separately.

$$2\ K^+(aq) + CrO_4^{2-}(aq) + 2\ Ag^+(aq) + 2\ NO_3^-(aq) \longrightarrow 2\ K^+(aq) + 2\ NO_3^-(aq) + Ag_2CrO_4(s)$$

To keep the equation balanced, each of the ions is multiplied by the same coefficient involved in the standard balanced equation shown earlier. Notice that potassium chromate dissolves to give *two* separate K^+ ions, shown as $2\ K^+$, not as K_2^+. Study this ionic equation, as a model for others.

An ionic equation shows more accurately what is going on during a reaction, but there is a disadvantage to writing this type of equation. It is rather long and looks complicated. However, as we examine this ionic equation, we see two K^+ ions and two NO_3^- ions as reactants and as products. These ions that appear to "stand around" to "watch" as the reaction takes place are called **spectator ions**. The reaction occurs because silver ions react with chromate ions to form silver chromate, a precipitate, so these ions are taken out of solution.

Omitting the spectator ions gives the following equation.

Study Hint
Be sure you can distinguish among these three types of equations and that you can represent a metathesis reaction using any one of the three.

$$CrO_4^{2-}(aq) + 2\ Ag^+(aq) \longrightarrow Ag_2CrO_4(s)$$

This equation is called a **net ionic equation**. It includes only the ions that react to form the precipitate. Spectator ions are left out. Thus, the net ionic equation tells—in a concise way—which ions react, and why the reaction occurs.

EXAMPLE 10.14

Rewrite this equation as an ionic and then as a net ionic equation.

$$3\ NaOH(aq) + FeCl_3(aq) \longrightarrow 3\ NaCl(aq) + Fe(OH)_3(s)$$

SOLUTION

See Problems 10.35–10.38.

The ionic equation must show all dissolved ions written separately.

$$3\ Na^+(aq) + 3\ OH^-(aq) + Fe^{3+}(aq) + 3\ Cl^-(aq) \longrightarrow 3\ Na^+(aq) + 3\ Cl^-(aq) + Fe(OH)_3(s)$$

Whether the OH^- ions are listed before or after the Fe^{3+} ions in the net ionic equation is arbitrary. Here, they were listed in the order they appear in the ionic equation.

The following net ionic equation does not show spectator ions.

$$3\ OH^-(aq) + Fe^{3+}(aq) \longrightarrow Fe(OH)_3(s)$$

10.12 *Neutralization: A Double-Replacement Reaction*

An **acid** (containing H^+ ions) neutralizes a **base** (containing OH^- ions) to produce water and a *salt*. The **salt** contains the cation from the base and the anion from the acid. The general equation and a sample equation follow.

$$\text{Acid} + \text{Base} \longrightarrow \text{Water} + \text{Salt}$$

$$\text{Sample:}\quad HCl(aq) + NaOH(aq) \longrightarrow H_2O(l) + NaCl(aq)$$

The driving force behind this **neutralization** reaction is the reaction of H^+ ions with OH^- ions to form covalent molecules of water.

Ionic equations were described in the previous section. Let us now write an ionic equation for the neutralization of hydrochloric acid by sodium hydroxide.

$$H^+(aq) + Cl^-(aq) + Na^+(aq) + OH^-(aq) \longrightarrow H_2O(l) + Na^+(aq) + Cl^-(aq)$$

If we omit the Na^+ and Cl^- spectator ions, we have the net ionic equation

$$H^+(aq) + OH^-(aq) \longrightarrow H_2O(l)$$

All acid–base neutralization reactions have this same net ionic equation. One more acid–base reaction is included here. Chapter 16 provides more information about acids and bases.

See Problems 10.39 and 10.40.

$$H_2SO_4(aq) + Ca(OH)_2(aq) \longrightarrow 2\,H_2O + CaSO_4(s)$$

$CaSO_4$ is a component of cement, wallboard, stucco, plaster, paper, and paint.

10.13 Miscellaneous Reactions: Calcium Compounds

Carbonates yield carbon dioxide gas when reacted with an acid, or when heated strongly, as shown by these reactions for calcium carbonate, $CaCO_3$.

$$CaCO_3(s) + 2\,HCl(aq) \longrightarrow CO_2(g) + H_2O(l) + CaCl_2(aq)$$

$$\underset{\text{(limestone)}}{\underset{\text{Calcium carbonate}}{CaCO_3(s)}} \xrightarrow{\text{heat}} CO_2(g) + \underset{\text{(quicklime)}}{\underset{\text{Calcium oxide}}{CaO(s)}}$$

CHEMISTRY IN OUR WORLD

Chemical Reactions in Limestone Caves

Limestone caves can be formed as water containing CO_2 gas comes in contact with and dissolves the calcium carbonate, $CaCO_3$, present in limestone rock.

$$CaCO_3(s) + H_2O + CO_2(g) \longrightarrow Ca^{2+}(aq) + 2\,HCO_3^-(aq)$$

The formation of $CaCO_3$ in *stalactites* found hanging from the ceiling of limestone caves, and in *stalagmites* that form on the floor of caves, occurs when the last reaction is reversed.

$$Ca^{2+}(aq) + 2\,HCO_3^-(aq) \longrightarrow CaCO_3(s) + H_2O + CO_2(g)$$

The reversal of the reaction is made possible by the fact that water slowly evaporates from the damp surfaces in caves and, as the carbon dioxide gas and water leave, the slightly soluble $CaCO_3$ is left behind as a precipitate.

Limestone formations in Lost Caverns in Pennsylvania.

Calcium hydroxide, $Ca(OH)_2$ (called lime or slaked lime), is prepared by adding quicklime, CaO, to water. The reaction releases a considerable amount of heat.

$$CaO(s) + H_2O \longrightarrow Ca(OH)_2(s)$$

Calcium hydroxide is widely used to neutralize acids remaining after various manufacturing processes. Calcium oxide is also used in pollution control to take up sulfur dioxide, SO_2, gas or sulfur trioxide, SO_3, gas produced by coal-burning electric power plants.

$$CaO(s) + SO_2(g) \longrightarrow CaSO_3(s)$$

There are many ''miscellaneous'' reactions. Only reactions of calcium compounds were described here. They were chosen because of their common natural source, their commercial uses, and their importance in pollution control. Heating limestone to give CaO is a decomposition reaction. The conversion of CaO to $Ca(OH)_2$ is a synthesis reaction, as is the conversion of CaO to $CaSO_3$. Finally, the reaction of $CaCO_3$ with HCl(aq) involves both double replacement and the formation of stable covalent molecules, CO_2 and H_2O.

Important chemical reactions are all around us—in nature, in industry, and in our homes—even though they may often go unnoticed.

Chapter Summary

One of the most important topics in all of chemistry is the study of chemical reactions and the symbolic notation used to write chemical equations. The equation summarizes information about the identities (formulas) of reactants and products. It provides an accounting system for each atom involved in the reaction, with all atoms being accounted for in both reactants and products. The chemical equation must be in agreement with the law of conservation of mass, verifying that no matter is created or lost in the process.

The coefficients placed in front of reactants and products provide the key to determining the proportions in which substances react, whether it is in terms of atoms or moles of atoms. The description of a chemical reaction must first be translated from words into formulas and equations. In order to balance an equation, it is crucial that each formula is written correctly. During the process of balancing the equation, only the *coefficients* are adjusted. Ionic equations and net ionic equations can be used to represent more completely what is taking place during the reaction. Admittedly, this all takes practice.

To help you learn general types of reactions, they have been grouped into categories. Sections 10.4–10.13 provide numerous samples and worked-out examples of reactions involving combustion reactions; synthesis and decomposition reactions; and single- and double-replacement (metathesis) reactions involving metals, nonmetals, acids, bases, and salts.

Besides knowing the terminology, you should be able to complete an equation and balance it when either reactants or products are given for reactions similar to those presented in this chapter. For reactions that occur in solution, you should also be able to write ionic and net ionic equations. Learning is deceptive. Watching someone else does not mean you can balance the equations or predict the products. The best way to learn to write and balance chemical equations is to work through the examples in the chapter and then do the problems at the end of the chapter.

Assess Your Understanding

1. Distinguish between chemical reactions and chemical equations. [10.1]
2. Describe at the molecular level what a given equation means. [10.2]
3. Describe what is ''balanced'' in a ''balanced equation.'' [10.3]
4. Balance chemical equations for which all formulas are given. [10.3]
5. Use descriptions of chemical reactions to write word equations and balanced chemical equations. [10.3]
6. Classify reactions by the following categories: combustion, synthesis, decomposition, single replacement, double replacement, oxidation and reduction, and neutralization. [10.4–10.13]
7. Predict the products and balance equations for reactions similar to those in each category presented. [10.4–10.13]
8. Write ionic and net ionic equations when reactants or products are given. [10.11]

Key Terms

acid [10.12]
activity series [10.8]
aqueous solution [10.1]
base [10.2]
catalyst [10.1]
chemical equation [10.1]
chemical reaction [10.1]
coefficient [10.3]
combination reaction [10.6]
combustion [10.5]
decomposition reaction [10.7]
double-replacement reaction [10.10]
electrolysis [10.7]
ionic equation [10.11]
metathesis reaction [10.10]
net ionic equation [10.11]
neutralization [10.12]
oxidation [10.8]
oxidation–reduction [10.8]
oxidizing agent [10.8]
precipitate [10.10]
products [10.1]
reactants [10.1]
reducing agent [10.8]
reduction [10.8]
salt [10.12]
single-replacement reaction [10.8]
spectator ions [10.11]
synthesis reaction [10.6]
yields [10.1]

Problems

CHEMICAL EQUATIONS

10.1 Use the law of conservation of mass to explain why we balance equations.

10.2 How does the law of conservation of mass pertain to the number of atoms of reactants and products?

10.3 Before balancing a chemical equation, what must one have written down?

10.4 In a balanced chemical equation, which of the following must be balanced: atoms, mass, molecules, moles? Explain.

BALANCING EQUATIONS

10.5 Balance these chemical equations.

a. $Al + O_2 \rightarrow Al_2O_3$

b. $N_2 + O_2 \rightarrow NO_2$

c. $H_2O_2 \rightarrow H_2O + O_2(g)$
d. $LiOH + CO_2 \rightarrow Li_2CO_3 + H_2O$
e. $Fe_2(SO_4)_3 + NaOH \rightarrow Fe(OH)_3 + Na_2SO_4$

10.6 Balance these chemical equations.
a. $Cr + O_2 \rightarrow Cr_2O_3$
b. $SO_2 + O_2 \rightarrow SO_3$
c. $PbO_2 \rightarrow PbO + O_2(g)$
d. $NaOH + CO_2 \rightarrow Na_2CO_3 + H_2O$
e. $Al_2(SO_4)_3 + NaOH \rightarrow Al(OH)_3 + Na_2SO_4$

10.7 For a balanced equation, which statements are true and which are false?
a. Atoms of reactants must equal atoms of products.
b. Moles of atoms of reactants must equal moles of atoms of products.
c. Grams of reactants must equal grams of products.
d. Total molecules of reactants must equal total molecules of products.
e. Total moles of reactants must equal total moles of products.

10.8 Explain your answers for each part of Problem 10.7, using an example.

10.9 Balance these chemical equations.
a. $Mg + H_2O(g) \rightarrow Mg(OH)_2 + H_2(g)$
b. $NaHCO_3 + H_3PO_4 \rightarrow Na_2HPO_4 + H_2O + CO_2$
c. $Al + H_2SO_4(aq) \rightarrow Al_2(SO_4)_3 + H_2(g)$
d. $C_3H_8 + O_2(g) \rightarrow CO_2 + H_2O$
e. $CH_3OH + O_2(g) \rightarrow CO_2 + H_2O$

10.10 Balance these chemical equations.
a. $Ca + H_2O \rightarrow Ca(OH)_2 + H_2(g)$
b. $KHCO_3 + H_3PO_4 \rightarrow K_2HPO_4 + H_2O + CO_2$
c. $(NH_4)_2Cr_2O_7 \rightarrow Cr_2O_3 + H_2O + N_2(g)$
d. $C_2H_2 + O_2(g) \rightarrow CO_2 + H_2O$
e. $C_3H_7OH + O_2(g) \rightarrow CO_2 + H_2O$

10.11 Balance these chemical equations.
a. $CaCO_3 + HCl \rightarrow CaCl_2 + H_2O + CO_2(g)$
b. $PCl_5 + H_2O \rightarrow H_3PO_4 + HCl$
c. $KClO_3 \rightarrow KCl + O_2(g)$
d. $Ba(OH)_2 + H_3PO_4 \rightarrow Ba_3(PO_4)_2 + H_2O$
e. $C_2H_6(g) + O_2(g) \rightarrow CO_2(g) + H_2O(g)$

10.12 Balance these chemical equations.
a. $CaCO_3 + H_2SO_4 \rightarrow CaSO_4 + CO_2 + H_2O$
b. $Ag + H_2S + O_2 \rightarrow Ag_2S + H_2O$
c. $NaClO_3 \rightarrow NaCl + O_2(g)$
d. $Cu(OH)_2 + H_3PO_4 \rightarrow Cu_3(PO_4)_2 + H_2O$
e. $C_4H_9OH(g) + O_2(g) \rightarrow CO_2(g) + H_2O(g)$

REACTION CLASSIFICATIONS

10.13 Classify each reaction specified as being combustion, synthesis, decomposition, single replacement, or double replacement.
a. the reaction in Problem 10.10c
b. the reaction in Problem 10.9a
c. the reaction in Problem 10.9d
d. the reaction in Problem 10.11d
e. the reaction in Problem 10.12c

10.14 Classify each reaction specified as being combustion, synthesis, decomposition, single replacement, or double replacement.
a. the reaction in Problem 10.10e
b. the reaction in Problem 10.11c
c. the reaction in Problem 10.9e
d. the reaction in Problem 10.10a
e. the reaction in Problem 10.12d

COMBUSTION REACTIONS

10.15 When butane gas, C_4H_{10}, from a small pocket cigarette lighter burns in air, carbon dioxide and water are produced. Write a balanced equation for this reaction.

10.16 Acetone, C_3H_6O, is a main component of fingernail polish remover. Write a balanced equation for the complete combustion of acetone to produce carbon dioxide and water.

SYNTHESIS REACTIONS

10.17 Complete and balance the following equations.
a. Hydrogen gas and chlorine gas explode to make hydrogen chloride gas.
b. Phosphorus, P_4, reacts with bromine spontaneously to give phosphorus tribromide.
c. Hydrogen gas and oxygen gas in a pop bottle will explode, when ignited with a spark, to produce water vapor.
d. $SO_2(g) + O_2(g) \rightarrow$ Sulfur trioxide
e. ______ + ______ $\rightarrow AlCl_3$

10.18 Complete and balance the following equations.
a. Carbon monoxide gas reacts with oxygen gas to produce carbon dioxide.
b. Zinc heated with sulfur powder (S_8) produces zinc sulfide.
c. Nitrogen gas can be reacted with hydrogen gas at a high pressure and a moderate temperature to produce ammonia, used heavily as a fertilizer.
d. $NO(g) + O_2(g) \rightarrow$ Nitrogen dioxide
e. ______ + ______ $\rightarrow FeBr_3$

DECOMPOSITION REACTIONS

10.19 Complete and balance equations for these decomposition reactions.

a. $Al_2O_3 \xrightarrow{dc}$
b. $PbO_2 \rightarrow PbO +$ ______
c. $NaClO_3 \rightarrow$
d. $KNO_3 \rightarrow$
e. $H_2O_2 \xrightarrow{I^-}$

10.20 Complete and balance equations for these decomposition reactions.

a. $H_2O \xrightarrow{dc}$
b. $BaO_2 \rightarrow BaO +$ ______
c. $KClO_3 \rightarrow$
d. $NaNO_3 \rightarrow$
e. $HgO(s) \rightarrow$

OXIDATION OF METALS

10.21 Complete and balance equations for these reactions involving the oxidation of metals.

a. ______ + ______ $\rightarrow Fe_2O_3$
b. ______ + ______ $\rightarrow SbCl_3$
c. Calcium metal reacts with oxygen in air to give calcium oxide.

10.22 Complete and balance equations for these reactions involving the oxidation of metals.

a. ______ + ______ $\rightarrow Li_2O$
b. ______ + ______ $\rightarrow AlI_3$
c. Iron metal is oxidized by liquid bromine to give iron(III) bromide.

ACTIVE METALS AND METAL OXIDES WITH WATER

10.23 Complete and balance equations for these reactions involving metals and metal oxides with water. If no reaction occurs, show reactants, and write "no reaction" after the arrow.

a. $Li_2O + H_2O \rightarrow$ b. $Na + H_2O \rightarrow$
c. $Mg + H_2O(g) \rightarrow$ d. $Ag + H_2O \rightarrow$
e. $SrO + H_2O \rightarrow$

10.24 Complete and balance equations for these reactions involving metals and metal oxides with water. If a reaction should not occur, show reactants, and write "no reaction" after the arrow.

a. $MgO + H_2O \rightarrow$ b. $K + H_2O \rightarrow$
c. $Pb + H_2O \rightarrow$ d. $Ca + H_2O \rightarrow$
e. $Fe_2O_3 + H_2O \rightarrow$

10.25 Complete and balance equations for these single-replacement reactions. If a reaction should not occur, show reactants and write "no reaction" after the arrow.

a. $Zn + HCl(aq) \rightarrow$
b. $Cu + HCl(aq) \rightarrow$
c. $Mg + Fe(NO_3)_3(aq) \rightarrow$
d. $AgNO_3(aq) + Al \rightarrow$
e. $Fe + MgCl_2(aq) \rightarrow$

10.26 Complete and balance equations for these single-replacement reactions. If a reaction should not occur, show reactants and write "no reaction" after the arrow.

a. $Al + HNO_3(aq) \rightarrow$
b. $Ag + HNO_3(aq) \rightarrow$
c. $Pb + CuCl_2(aq) \rightarrow$ d. $H_2SO_4 + Au \rightarrow$
e. $Zn + Ni(NO_3)_2(aq) \rightarrow$

10.27 Complete and balance equations for these single-replacement reactions. If a reaction should not occur, show reactants and write "no reaction" after the arrow.

a. Calcium metal was placed in hydrochloric acid.
b. A gold ring was dropped into sulfuric acid.
c. A solution of copper(II) sulfate was left in a galvanized (zinc-coated) bucket.
d. Copper metal was left in a solution of sodium chloride.
e. Aluminum metal was cleaned with a solution containing hydrochloric acid.

10.28 Complete and balance equations for these single-replacement reactions. If a reaction should not occur, show reactants and write "no reaction" after the arrow.

a. Sulfuric acid was poured into an aluminum pan.
b. A gold ring was dropped into a vat of hydrochloric acid.
c. An iron nail was left in a solution of copper(II) sulfate.
d. After writing an appropriate equation, explain why apple butter, containing acetic acid, can be made in a copper kettle.
e. Write an equation and describe whether there would be a problem with shipping lead(II) nitrate in an aluminum container.

REACTIONS INVOLVING NONMETALS

10.29 Complete and balance the following equations involving nonmetals. If no reaction is predicted, write "no reaction" after the arrow.

a. $S_8 + O_2(g) \rightarrow$
b. $SO_2 + H_2O \rightarrow$ (sulfurous acid, only)
c. $N_2O_5 + H_2O \rightarrow$ (nitric acid, only)
d. $KBr + Cl_2 \rightarrow$
e. $KCl + I_2 \rightarrow$

10.30 Complete and balance the following equations involving nonmetals. If no reaction is predicted, write "no reaction" after the arrow.
a. $P_4 + O_2 \rightarrow$
b. $P_4O_{10} + H_2O \rightarrow$ (phosphoric acid, only)
c. $CO_2 + H_2O \rightarrow$ (carbonic acid, only)
d. $Br_2 + KCl \rightarrow$
e. $KI + Cl_2 \rightarrow$

10.31 Describe what would happen if sodium iodide is added to a swimming pool containing chlorine. Write an equation to explain your reasoning.

10.32 What would you expect to happen if HI(g) is mixed with chlorine gas? Write an equation to explain your reasoning.

DOUBLE-REPLACEMENT REACTIONS

10.33 Write equations for each of the following double-replacement reactions. In each case a precipitate is formed, which you should identify by an (s). Also, underline the precipitate so it will be clear which product is the precipitate. Use the solubility rules in this chapter to help you.
a. Silver nitrate and potassium chloride solutions are mixed.
b. Iron(III) chloride and sodium hydroxide solutions are mixed.
c. Aluminum sulfate and barium nitrate solutions are mixed.
d. Lead(II) nitrate and potassium dichromate solutions are mixed.
e. Silver nitrate and potassium chromate solutions are mixed.

10.34 Write equations for each of the following double-replacement reactions. In each case a precipitate is formed, which you should identify by an (s). Also, underline the precipitate so it will be clear which product is the precipitate. Use the solubility rules in this chapter to help you.
a. Sodium bromide and silver nitrate solutions are mixed.
b. Nickel(II) chloride and potassium hydroxide solutions are mixed.
c. Potassium chromate and lead(II) nitrate solutions and mixed.
d. Barium chloride and aluminum sulfate solutions are mixed.
e. Silver nitrate and potassium iodide solutions are mixed.

IONIC AND NET IONIC EQUATIONS

10.35 Write complete ionic equations for the reactions in Problem 10.33.

10.36 Write complete ionic equations for the reactions in Problem 10.34.

10.37 Write net ionic equations for the reactions in Problem 10.33.

10.38 Write net ionic equations for the reactions in Problem 10.34.

NEUTRALIZATION REACTIONS

10.39 Give the balanced equation, or provide the information requested.
a. Sulfuric acid neutralizes potassium hydroxide.
b. Identify the acid, base, salt, and water for the chemicals in part (a).
c. Write a net ionic equation for the reaction in part (a).
d. What ion is present in all acids?
e. Write the equation for what happens when hydrochloric acid neutralizes magnesium hydroxide.

10.40 Give the balanced equation, or provide the information requested.
a. Hydrochloric acid neutralizes calcium hydroxide.
b. Identify the acid, base, salt, and water for the chemicals in part (a).
c. Write a net ionic equation for the reaction in part (a).
d. What does neutralization mean, at the ionic level?
e. Write an equation for the reaction of carbonic acid with NaOH.

Additional Problems

10.41 Why is a neutralization reaction really a double-replacement reaction?

10.42 What type of reaction was involved when precipitates were formed?

10.43 Balance this equation.

$$Cu + H_2SO_4 \longrightarrow SO_2 + CuSO_4 + H_2O$$

10.44 Complete and balance $HCl + MgCO_3 \rightarrow$

10.45 Classify an acid and base neutralization reaction as one of the following: combustion, synthesis, decomposition, single replacement, or metathesis (double replacement).

10.46 Classify the reaction of a metal plus an acid to give a salt plus H_2 gas as one of the following: combustion, synthesis, decomposition, single replacement, or metathesis (double replacement).

10.47 Classify the reaction (five categories).

$$NH_3 + HNO_3 \rightarrow NH_4NO_3$$

10.48 Classify the reaction (five categories).

$$Zn + CuCl_2 \rightarrow ZnCl_2 + Cu$$

10.49 Balance the following equation used to produce pure iron from its ore.

$$Fe_2O_3 + CO(g) \rightarrow Fe + CO_2(g)$$

10.50 Balance the following equation used to make sodium thiosulfate, $Na_2S_2O_3$, called ‘‘hypo’’ and used in photographic developing.

$$Na_2CO_3 + Na_2S + SO_2(g) \rightarrow CO_2(g) + Na_2S_2O_3$$

10.51 Write a balanced chemical equation for the reaction of hydrochloric acid with a marble gravestone. (Marble and limestone contain calcium carbonate.)

10.52 Write a balanced chemical equation to show that sulfur dioxide gas (produced by burning coal) can be removed from smokestack gases by using calcium oxide (quicklime).

How much of a particular chemical must be used to react completely with a certain quantity of another chemical? Questions of this type can be answered once you understand how to carry out the calculations described in this chapter.

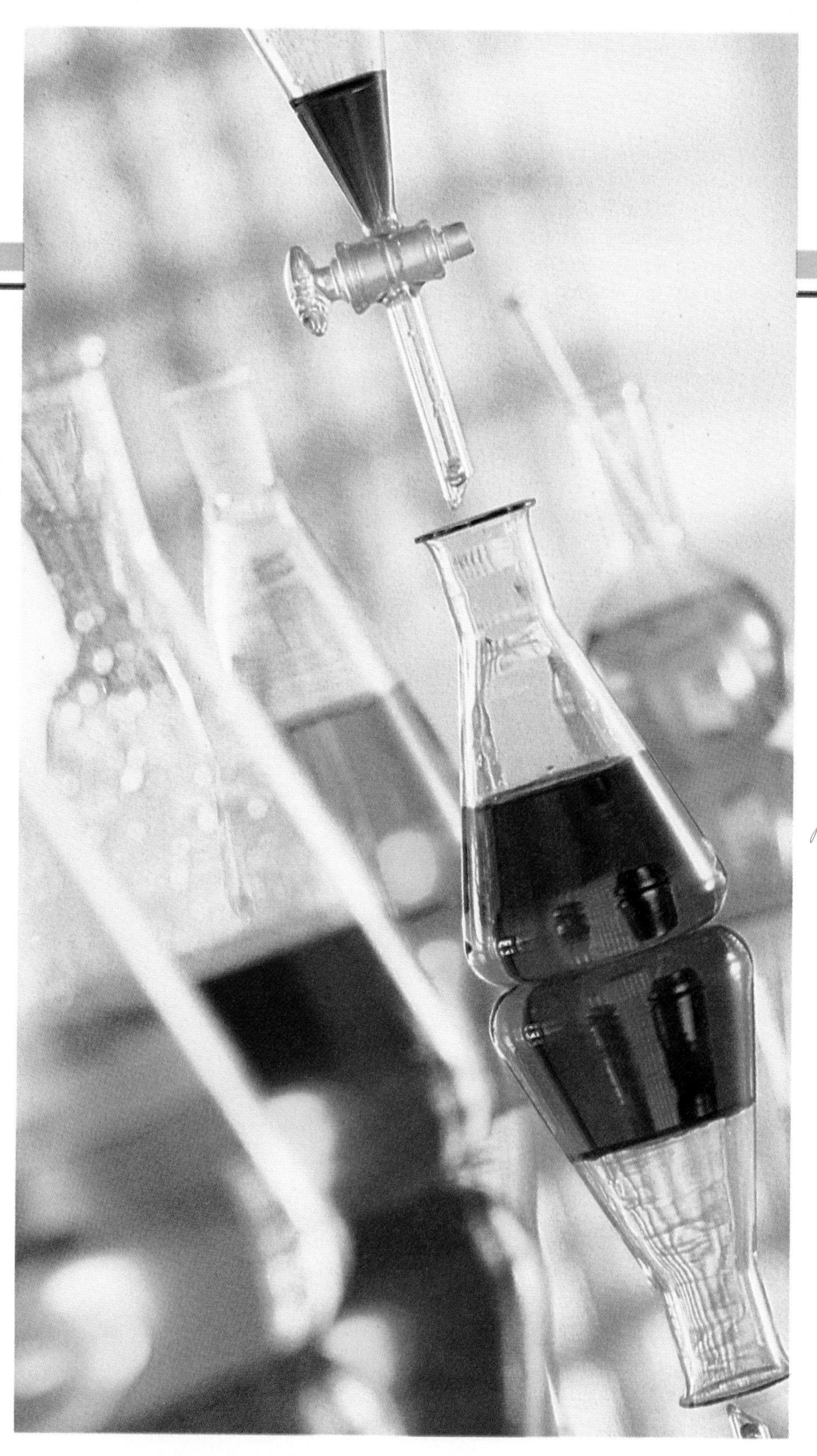

11

Stoichiometry: Calculations Based on Chemical Equations

CONTENTS

The term **stoichiometry** is used to designate the calculation of quantities of substances involved in chemical reactions. When we know the amount of one substance involved in a chemical reaction, and when we have the balanced chemical equation (Chapter 10), we can determine the quantities of the other reactants and products. These quantities (Chapter 9) can be in moles, masses (grams), or volumes (liters). Calculations of this type are at the heart of chemistry; they are routinely used during chemical analysis and during the production of every chemical used by industry or sold to consumers.

Stoichiometry (STOY-key-AAH-muh-tree) is derived from the Greek *stoicheion,* element + *metron,* to measure.

As an example, we can use stoichiometric calculations to determine the quantity of oxygen required to burn a specific quantity of ethyl alcohol. Incidentally, the number of moles—or the number of grams—of oxygen gas required to burn the alcohol sample is the same whether the alcohol is being burned in an open flame, in an engine, or within the cells of the human body; that is, the stoichiometry is the same.

11.1 *Mole Ratios from Chemical Equations*

Recall from Chapter 10 that the numerical coefficients in a balanced chemical equation give the simplest whole-number ratios of moles of each chemical involved in the reaction. For example, consider the balanced equation for the reaction of nitrogen gas with hydrogen gas to produce ammonia gas.

$$N_2(g) + 3\ H_2(g) \longrightarrow 2\ NH_3(g)$$

The numerical coefficients used to balance the equation indicate that 1 mole of nitrogen gas reacts with 3 moles of hydrogen gas to produce 2 moles of ammonia. With these numerical coefficients, we can write a **mole ratio** for any two substances represented by the equation. Since there are three different chemicals involved in the chemical equation for the synthesis of ammonia, we can write three pairs of mole ratios. The following pair of mole ratios relates the two reactants, nitrogen gas and hydrogen gas.

$$\frac{1\ \text{mol}\ N_2}{3\ \text{mol}\ H_2} \quad \text{or} \quad \frac{3\ \text{mol}\ H_2}{1\ \text{mol}\ N_2}$$

The mole ratios for nitrogen gas (a reactant) and ammonia gas (the only product of the reaction) are written as follows.

$$\frac{1\ \text{mol}\ N_2}{2\ \text{mol}\ NH_3} \quad \text{or} \quad \frac{2\ \text{mol}\ NH_3}{1\ \text{mol}\ N_2}$$

Similarly, the mole ratios for hydrogen gas (a reactant) and ammonia gas are written as follows.

$$\frac{3\ \text{mol}\ H_2}{2\ \text{mol}\ NH_3} \quad \text{or} \quad \frac{2\ \text{mol}\ NH_3}{3\ \text{mol}\ H_2}$$

Coefficients
The coefficients in chemical equations are exact numbers, so they do not limit the number of significant figures in calculations.

We can use the numerical coefficients in any balanced chemical equation to write mole ratios for each pair of chemicals involved. We will later show how these mole ratios are used as conversion factors in determining reacting quantities of substances.

Study the following worked-out example involving mole ratios for the combustion of ethyl alcohol before setting up mole ratios for problems included at the end of this chapter.

EXAMPLE 11.1

Write the mole ratios for each pair of chemicals involved in the balanced equation for the combustion of ethyl alcohol, C_2H_5OH.

$$C_2H_5OH + 3\ O_2 \longrightarrow 2\ CO_2 + 3\ H_2O$$

SOLUTION

Since there are four different chemicals involved in the equation, there are six pairs of mole ratios, making a total of 12 ratios, as follows.

C_2H_5OH and O_2: $\dfrac{1\text{ mol } C_2H_5OH}{3\text{ mol } O_2}$ or $\dfrac{3\text{ mol } O_2}{1\text{ mol } C_2H_5OH}$

C_2H_5OH and CO_2: $\dfrac{1\text{ mol } C_2H_5OH}{2\text{ mol } CO_2}$ or $\dfrac{2\text{ mol } CO_2}{1\text{ mol } C_2H_5OH}$

C_2H_5OH and H_2O: $\dfrac{1\text{ mol } C_2H_5OH}{3\text{ mol } H_2O}$ or $\dfrac{3\text{ mol } H_2O}{1\text{ mol } C_2H_5OH}$

O_2 and CO_2: $\dfrac{3\text{ mol } O_2}{2\text{ mol } CO_2}$ or $\dfrac{2\text{ mol } CO_2}{3\text{ mol } O_2}$

O_2 and H_2O: $\dfrac{3\text{ mol } O_2}{3\text{ mol } H_2O}$ or $\dfrac{3\text{ mol } H_2O}{3\text{ mol } O_2}$

CO_2 and H_2O: $\dfrac{2\text{ mol } CO_2}{3\text{ mol } H_2O}$ or $\dfrac{3\text{ mol } H_2O}{2\text{ mol } CO_2}$

See Problems 11.1 and 11.2.

11.2 Mole–Mole Calculations

When the balanced chemical equation is known, and the number of moles of any one of the reactants or products is also known, the proportional number of moles of any other reactant or product can be determined by using an appropriate mole ratio. For an example, let us again use the reaction for the production of ammonia described in Section 11.1.

$$N_2(g) + 3\ H_2(g) \longrightarrow 2\ NH_3(g)$$

Assuming that we are provided with a sufficient supply of nitrogen, we can easily determine how many moles of ammonia gas, $NH_3(g)$, can be produced from 10.8 mol of hydrogen gas. Consider the following sequence of steps used in this type of determination.

1. Obtain a balanced chemical equation. (This equation is provided.)
2. Write down the given quantity of one chemical (expressed here in moles) as the starting point. This can be any reactant or product of the reaction. (For this problem we are given 10.8 mol of hydrogen gas as the starting point.)

3. Also write down the appropriate mole ratio from the balanced chemical equation. This mole ratio must be in the following form.

$$\frac{\text{Moles of desired chemical}}{\text{Moles of starting chemical}}$$

For this problem we start with moles of H_2 to determine moles of NH_3. From the equation, the conversion is $3\ H_2(g) \rightarrow 2\ NH_3(g)$. Thus, the appropriate mole ratio is written

$$\frac{2\ \text{mol}\ NH_3}{3\ \text{mol}\ H_2}$$

4. Multiply the given number of moles of the starting chemical (step 2) by the appropriate mole ratio (step 3) to obtain moles of the desired chemical.

$$\text{Given moles of starting chemical} \times \frac{\text{Moles of desired chemical}}{\text{Moles of starting chemical}}$$

For this problem, the conversion is written

$$10.8\ \cancel{\text{mol}\ H_2} \times \frac{2\ \text{mol}\ NH_3}{3\ \cancel{\text{mol}\ H_2}} = 7.20\ \text{mol}\ NH_3 \qquad \text{(answer)}$$

Notice that moles of the starting chemical cancel, giving 7.20 mol of NH_3. In this problem, we were actually converting between the moles of a given substance, which could be called "moles of A," and the moles of the desired substance, which could be called "moles of B," with the aid of the mole ratio as the conversion factor. The conversion could be symbolized as follows.

$$\text{Moles A} \longrightarrow \text{Moles B}$$

Now, learn to follow through with this stepwise approach in converting "moles of A" to "moles of B" as illustrated in Example 11.2.

Study Hint
You may find it helpful to write 1.20 mol below the C_2H_5OH in the equation and ? mol below the O_2.

EXAMPLE 11.2

Determine the number of moles of oxygen gas required to burn 1.20 mol of ethyl alcohol, C_2H_5OH. The balanced chemical equation is shown here.

$$C_2H_5OH + 3\ O_2(g) \longrightarrow 2\ CO_2(g) + 3\ H_2O(g)$$

SOLUTION

This problem follows the form

$$\text{Plan: Moles A} \longrightarrow \text{Moles B}$$

STEP 1 Obtain a balanced chemical equation. (This equation is provided.)

STEP 2 Write down the given quantity of one chemical (expressed here as moles A) for the starting point. (For this problem we are given 1.20 mol of C_2H_5OH as the starting point.)

STEP 3 Also write down the appropriate mole ratio from the balanced chemical equation. This mole ratio must be in the form

$$\frac{\text{Moles of desired chemical}}{\text{Moles of starting chemical}}$$

For this problem, start with moles of C_2H_5OH to determine moles of O_2. From the equation, the moles A → moles B conversion is

$$1 \text{ mol } C_2H_5OH \longrightarrow 3 \text{ mol } O_2$$

Thus, the appropriate mole ratio is written

$$\frac{3 \text{ mol } O_2}{1 \text{ mol } C_2H_5OH}$$

STEP 4 Multiply the given number of moles of the starting chemical (step 2) by the appropriate mole ratio (step 3) to obtain moles of the desired chemical.

$$1.20 \cancel{\text{ mol } C_2H_5OH} \times \frac{3 \text{ mol } O_2}{1 \cancel{\text{ mol } C_2H_5OH}} = 3.60 \text{ mol } O_2 \qquad \text{(answer)}$$

See Problems 11.3 and 11.4.

EXERCISE 11.1
How many moles of carbon dioxide can be produced by burning 1.20 mol of ethyl alcohol, C_2H_5OH?

11.3 Stoichiometric Calculations Involving Moles and Masses

Stoichiometric calculations involving quantities of reacting substances and reaction products can be carried out in terms of moles, as described in Section 11.2, but quantities of chemicals are often expressed as masses (in grams or kilograms). Conversions between masses and moles were described in Chapter 9. We can now combine these mass-to-mole conversions with the "moles of A" to "moles of B" conversions described in Section 11.2.

To perform this series of conversions, the mass of one chemical, "grams of A," is first converted to "moles of A" using the molar mass (grams per mole) of chemical A. Then, a mole ratio is used to convert "moles of A" to "moles of B." The moles of the second chemical, "moles of B," can be converted back to "grams of B" by using the molar mass of chemical B. This series of conversions can be symbolized as follows.

$$\text{Grams A} \longrightarrow \text{Moles A} \longrightarrow \text{Moles B} \longrightarrow \text{Grams B}$$

See Figure 11.1 for a "road map" type of diagram that shows the series of conversions to be made along with the conversion factors needed in carrying out these stoichiometric calculations.

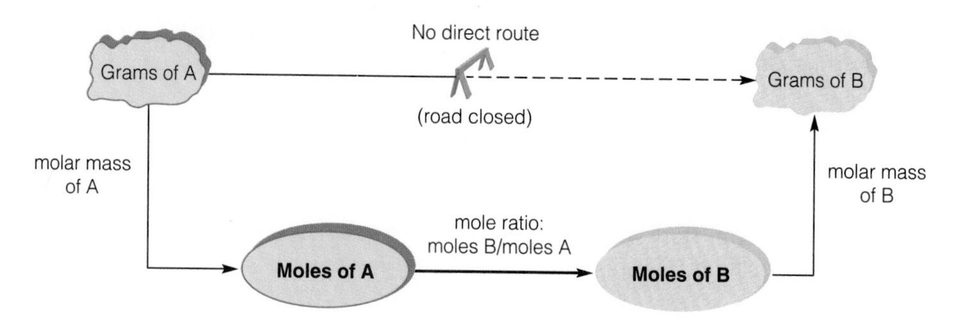

Figure 11.1 *A road map to stoichiometric conversions. Conversion factors are shown alongside arrows.*

The "grams of A" to "grams of B" conversion sequence can be accomplished by following through with this series of steps.

1. Obtain a balanced chemical equation.
2. Write down the given quantity of one chemical labeled as "grams of A" for the starting point. (This can be any reactant or product of the reaction.)
3. Convert "grams of A" to "moles of A" by using the molar mass of A as the conversion factor.
4. Convert "moles of A" to "moles of B" by using the mole ratio (mol B/mol A) from the balanced chemical equation.
5. Convert "moles of B" to "grams of B" by using the molar mass of B as the conversion factor.

The conversions in a typical problem of this type would take the general form

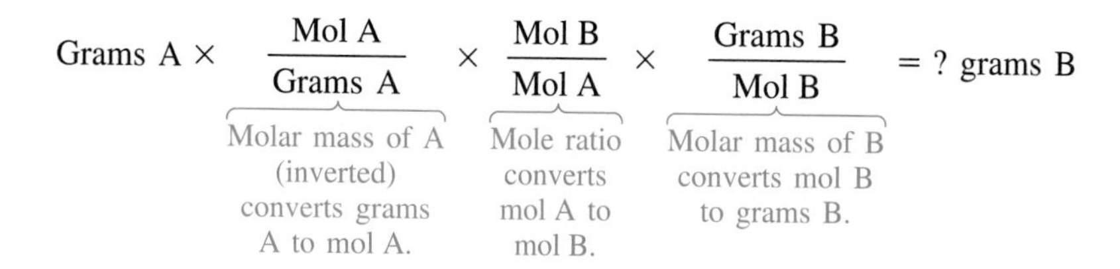

$$\text{Grams A} \times \underbrace{\frac{\text{Mol A}}{\text{Grams A}}}_{\substack{\text{Molar mass of A} \\ \text{(inverted)} \\ \text{converts grams} \\ \text{A to mol A.}}} \times \underbrace{\frac{\text{Mol B}}{\text{Mol A}}}_{\substack{\text{Mole ratio} \\ \text{converts} \\ \text{mol A to} \\ \text{mol B.}}} \times \underbrace{\frac{\text{Grams B}}{\text{Mol B}}}_{\substack{\text{Molar mass of B} \\ \text{converts mol B} \\ \text{to grams B.}}} = \text{? grams B}$$

Example 11.3 illustrates the use of this series of steps in solving stoichiometric calculations involving "grams A" to "grams B" conversions.

EXAMPLE 11.3

Determine the number of grams of oxygen gas required to burn 10.0 g of ethyl alcohol, C_2H_5OH. The balanced chemical equation is shown here.

$$C_2H_5OH + 3\,O_2(g) \longrightarrow 2\,CO_2(g) + 3\,H_2O(g)$$

SOLUTION

This problem follows the form

$$\text{Plan: } \underset{\text{(alcohol)}}{\text{Grams A}} \longrightarrow \text{Moles A} \longrightarrow \underset{\text{(oxygen)}}{\text{Moles B}} \longrightarrow \text{Grams B}$$

STEP 1 Obtain a balanced chemical equation. (This equation is provided.)

STEP 2 Write down the given quantity of one chemical labeled as "grams of

A'' for the starting point. (For this problem we are given 10.0 g of C_2H_5OH as the starting point.)

STEP 3 Set up the conversion from ''grams of A'' to ''moles of A'' by using the molar mass of A as the conversion factor.

$$10.0 \text{ g } C_2H_5OH \times \frac{1 \text{ mol } C_2H_5OH}{46.0 \text{ g } C_2H_5OH} \quad (\textit{incomplete})$$

STEP 4 Set up the conversion from ''moles of A'' to ''moles of B'' by using the mole ratio (mol B/mol A) from the balanced chemical equation.

$$10.0 \text{ g } C_2H_5OH \times \frac{1 \text{ mol } C_2H_5OH}{46.0 \text{ g } C_2H_5OH} \times \frac{3 \text{ mol } O_2}{1 \text{ mol } C_2H_5OH} \quad (\textit{incomplete})$$

STEP 5 To convert ''moles of B'' to ''grams of B,'' use the molar mass of B (oxygen) as the conversion factor. Now, follow through with the mathematical calculations using a calculator.

$$10.0 \text{ g } C_2H_5OH \times \frac{1 \text{ mol } C_2H_5OH}{46.0 \text{ g } C_2H_5OH} \times \frac{3 \text{ mol } O_2}{1 \text{ mol } C_2H_5OH} \times \frac{32.0 \text{ g } O_2}{1 \text{ mol } O_2} = 20.9 \text{ g } O_2 \quad \text{(answer)}$$

Thus, 20.9 g of oxygen gas is required to completely burn 10.0 g of ethyl alcohol. We could also determine the grams of CO_2 and H_2O produced.

EXERCISE 11.2
How many grams of carbon dioxide can be produced by burning 10.0 g of ethyl alcohol?

For Example 11.3, the quantity of the starting chemical was given in grams, but if the starting quantity is given in moles (''moles of A'' instead of ''grams of A''), we can simply omit the first conversion step in the sequence

$$\text{Grams A} \longrightarrow \text{Moles A} \longrightarrow \text{Moles B} \longrightarrow \text{Grams B}$$

Similarly, if ''moles of B'' is desired instead of ''grams of B,'' we can simply omit the final conversion step in the sequence.

Notice that regardless of the units used for starting materials and final products, the quantities of two chemicals can be converted only when they are first expressed as moles.

Example 11.4 illustrates a conversion sequence where the quantity of the starting chemical is given in moles and the quantity of the second chemical is to be determined in grams.

EXAMPLE 11.4

Determine the number of grams of lead(II) sulfide, PbS, that can be oxidized by 5.22 mol of oxygen gas according to the following equation.

$$2 \text{ PbS} + 3 \text{ O}_2(g) \longrightarrow 2 \text{ PbO} + 2 \text{ SO}_2(g)$$

Galena is the principal lead ore and is primarily PbS. The PbS is converted to PbO, which is placed in a blast furnace and reduced to lead metal, Pb.

SOLUTION
This problem follows the form

Plan: Moles A ⟶ Moles B ⟶ Grams B
(oxygen) (PbS) (PbS)

STEP 1 Obtain a balanced chemical equation. (This equation is provided.)
STEP 2 Write down the given quantity of one chemical labeled as "moles of A" for the starting point. (For this problem we are given 5.22 mol of oxygen as the starting point.)
STEP 3 Set up the conversion from "moles of A" to "moles of B" by using the mole ratio (mol B/mol A) from the balanced chemical equation.

$$5.22 \text{ mol } O_2 \times \frac{2 \text{ mol PbS}}{3 \text{ mol } O_2} \quad \textit{(incomplete)}$$

STEP 4 Convert "moles of B" to "grams of B" (PbS in this problem) by using the molar mass of B as the conversion factor. Now, follow through with the mathematical calculations using a calculator.

$$5.22 \text{ mol } O_2 \times \frac{2 \text{ mol PbS}}{3 \text{ mol } O_2} \times \frac{239 \text{ g PbS}}{1 \text{ mol PbS}} = 832 \text{ g PbS} \quad \text{(answer)}$$

Using the chemical equation provided, 832 g PbS can be oxidized by 5.22 mol O_2.

See Problems 11.5–11.10.

11.4 *Stoichiometric Calculations Involving Molar Solutions*

It has been shown that stoichiometric calculations of reacting quantities can be carried out quickly when all quantities are expressed in moles (Section 11.2). Stoichiometric calculations can also be carried out when chemical quantities are expressed in grams (Section 11.3), but more steps are required.

For many chemical reactions, either one or both of the reacting chemicals is dissolved in water (especially acids, bases, and water-soluble salts) to give an aqueous (water) solution. In this section, stoichiometric calculations involving aqueous solutions will be described. We can determine the number of moles of the dissolved chemical when the concentration of the solution is given in molarity (M, in moles per liter) and the volume of the solution is also given (Section 9.6). In the following calculations the volume and molarity of one solution are given and the quantity of a second chemical is to be determined in grams. The conversion sequence becomes

$$\underset{\text{(solution)}}{\text{Volume A}} \xrightarrow{\text{molarity of A}} \text{Moles A} \longrightarrow \text{Moles B} \longrightarrow \text{Grams B}$$

On the other hand, if the mass of one chemical is given ("grams of A") and the volume of a second solution (with a known molarity) is to be determined, the conversion sequence becomes

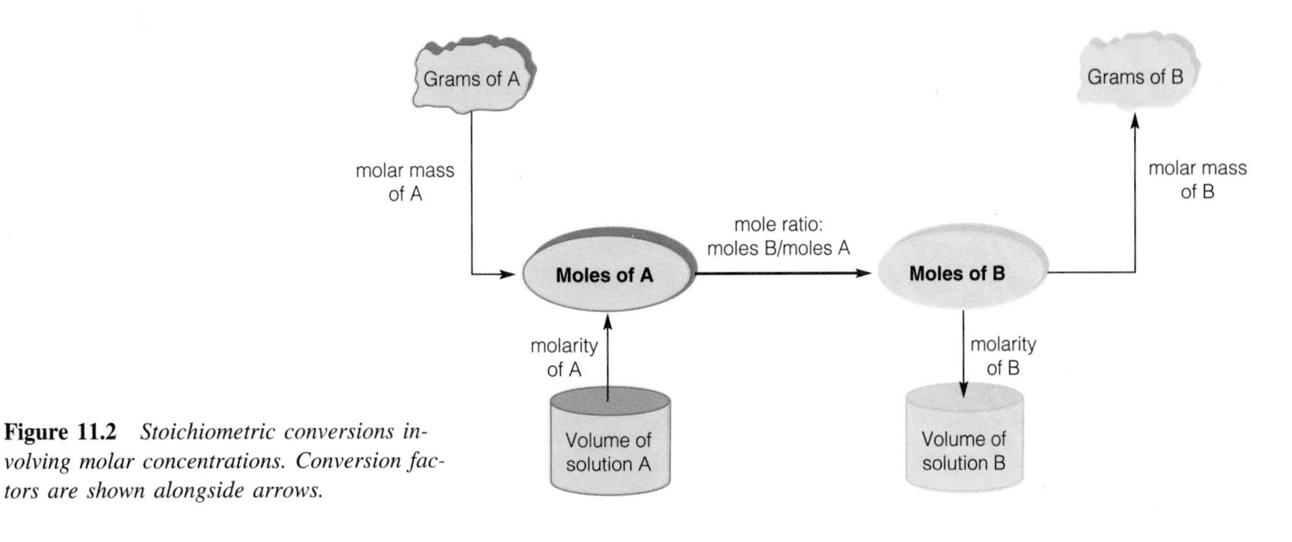

Figure 11.2 *Stoichiometric conversions involving molar concentrations. Conversion factors are shown alongside arrows.*

$$\text{Grams A} \longrightarrow \text{Moles A} \longrightarrow \text{Moles B} \xrightarrow{\text{molarity of B}} \underset{\text{(solution)}}{\text{Volume B}}$$

Figure 11.2 shows an expanded road map of stoichiometric conversions with solution quantities measured by volume and with concentration in molarities. Calculations of this type are carried out in Example 11.5.

EXAMPLE 11.5

Determine the number of milliliters of 0.125 M hydrochloric acid, HCl(aq), required to react completely with 0.500 g of $CaCO_3$ contained in one antacid tablet according to the following chemical equation. (See Figure 11.3.)

$$2\,HCl + CaCO_3 \longrightarrow CaCl_2 + H_2O + CO_2(g)$$

SOLUTION

Since the known chemical quantity is given in grams and the volume of solution for another chemical (with known molarity) is to be determined, we will use the following conversion sequence.

Plan: $\text{Grams A} \longrightarrow \text{Moles A} \longrightarrow \text{Moles B} \xrightarrow{\text{molarity of B}} \underset{\text{(solution)}}{\text{Volume B}}$

STEP 1 Obtain a balanced chemical equation. (This equation is provided.)

STEP 2 Write down the given quantity of one chemical labeled as "grams of A" for the starting point. (For this problem we are given 0.500 g of $CaCO_3$ as the starting point.)

STEP 3 Set up the conversion from "grams of A" to "moles of A" by using the molar mass of A as the conversion factor.

$$0.500\text{ g }CaCO_3 \times \frac{1\text{ mol }CaCO_3}{100.0\text{ g }CaCO_3} \qquad \text{(incomplete)}$$

STEP 4 Set up the conversion from "moles of A" to "moles of B" by using the mole ratio (mol B/mol A) from the balanced chemical equation.

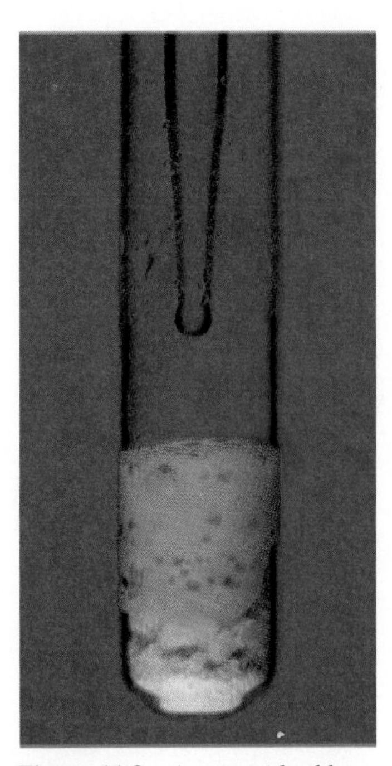

Figure 11.3 *An antacid tablet containing calcium carbonate reacts with hydrochloric acid—or any acid in the stomach—to release carbon dioxide gas.*

$$0.500 \text{ g } CaCO_3 \times \frac{1 \text{ mol } CaCO_3}{100.0 \text{ g } CaCO_3} \times \frac{2 \text{ mol HCl}}{1 \text{ mol } CaCO_3} \quad \textit{(incomplete)}$$

STEP 5 For this problem, "moles of B" is to be converted to a volume of solution, "volume of B." The molarity of the solution (0.125 M, meaning 0.125 mol/L) is used for this conversion, but the inverted form 1 L/ 0.125 mol HCl is used so terms will cancel. Then, follow through with these calculations using a calculator.

$$0.500 \cancel{\text{ g } CaCO_3} \times \frac{1 \cancel{\text{ mol } CaCO_3}}{100.0 \cancel{\text{ g } CaCO_3}} \times \frac{2 \cancel{\text{ mol HCl}}}{1 \cancel{\text{ mol } CaCO_3}} \times \frac{1 \text{ L solution}}{0.125 \cancel{\text{ mol HCl}}}$$

$$= 0.0800 \text{ L solution}$$

Since the volume of HCl(aq) is to be stated in milliliters, we need to make one more conversion.

$$0.0800 \cancel{\text{ L solution}} \times \frac{1000 \text{ mL}}{1 \cancel{\text{L}}} = 80.0 \text{ mL HCl(aq)} \quad \text{(answer)}$$

See Problems 11.11–11.14.

Thus, 80.0 mL of the hydrochloric acid solution specified is needed to react with 0.500 g of $CaCO_3$ in the antacid tablet.

EXERCISE 11.3

Determine the number of milliters of 0.100 M HCl that are needed to react with 243 mg of $CaCO_3$.

11.5 *Gas Stoichiometry*

Some of the earliest chemical reactions investigated were between gases. In 1809, Joseph Louis Gay-Lussac found that when all measurements were made at the same temperature and pressure, the volumes of gaseous reactants and products were in a small whole-number ratio. For example, when he allowed hydrogen to react with oxygen to form steam at 100°C, it took two volumes of hydrogen (i.e., two volumes of equal size) to unite with one volume of oxygen to make two volumes of water vapor (Figure 11.4). The volume ratio was 2:1:2. It was later

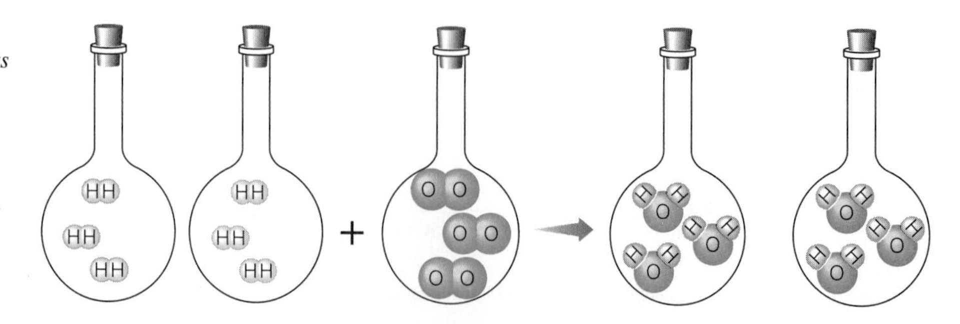

Figure 11.4 *The reaction of hydrogen gas with oxygen gas. Two volumes of H_2 react with one volume of O_2 to give two volumes of water vapor. The total volume for reactants is 3, and the total volume for products is 2. Each volume contains the same number of molecules. The same numbers of hydrogen and oxygen atoms have been represented as reactants and as products. Atoms are neither created nor destroyed in chemical reactions.*

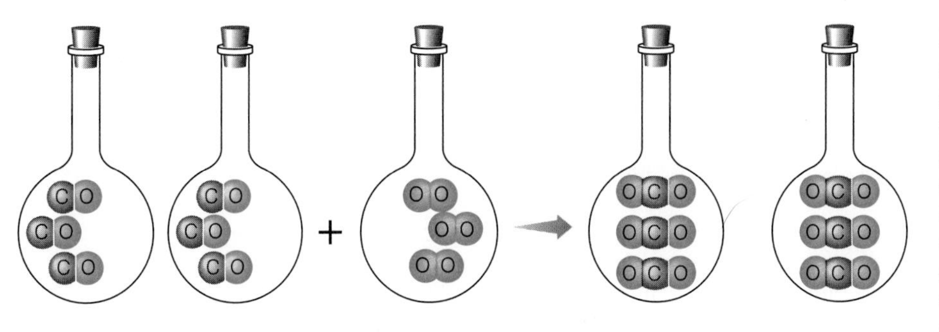

Figure 11.5 *The reaction of carbon monoxide gas with oxygen gas. Two volumes of CO react with one volume of O_2 to give two volumes of CO_2. The total volume for reactants is 3, and the total volume for products is 2. Each volume contains the same number of molecules. Also, the same numbers of carbon and oxygen atoms have been represented as reactants and as products. Atoms are neither created nor destroyed in chemical reactions.*

understood that the volume ratios are the same as the mole ratios, 2:1:2, for the reaction, as indicated by the coefficients in the equation

$$2\ H_2(g) + O_2(g) \longrightarrow 2\ H_2O(g)$$

In another experiment (Figure 11.5), Gay-Lussac found that two volumes of carbon monoxide, CO, combined with one volume of oxygen to form two volumes of carbon dioxide. Again, the 2:1:2 volume ratio corresponds to the mole ratios indicated by the coefficients in the equation.

$$2\ CO(g) + O_2(g) \longrightarrow 2\ CO_2(g)$$

The ratio is not always 2:1:2. When nitrogen gas reacts with hydrogen gas (Figure 11.6) to form ammonia, the combining volume ratio is one volume of nitrogen with three volumes of hydrogen to give two volumes of ammonia (1:3:2). Once more, the volume ratio corresponds to the mole ratios indicated by the coefficients in the equation.

$$N_2(g) + 3\ H_2(g) \longrightarrow 2\ NH_3(g)$$

Although Gay-Lussac found that volumes of gaseous reactants and products are in small whole-number ratios, it was the Italian chemist Amadeo Avogadro who

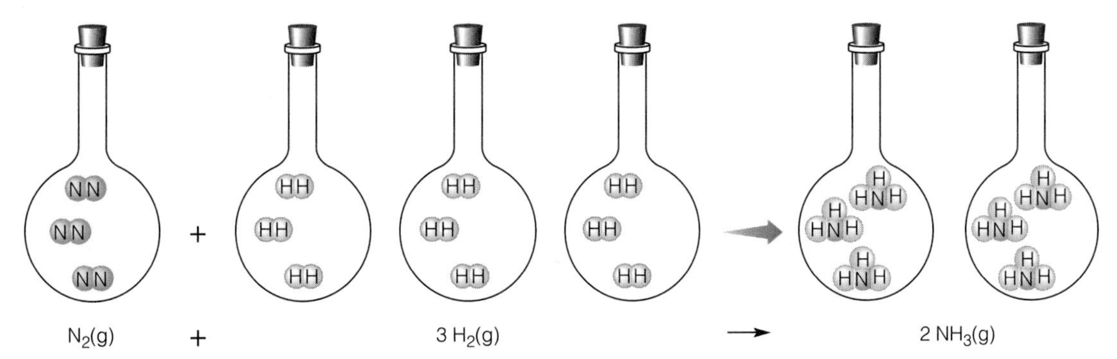

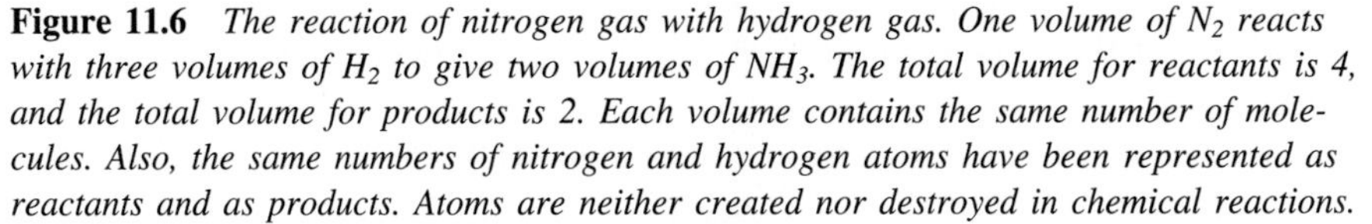

Figure 11.6 *The reaction of nitrogen gas with hydrogen gas. One volume of N_2 reacts with three volumes of H_2 to give two volumes of NH_3. The total volume for reactants is 4, and the total volume for products is 2. Each volume contains the same number of molecules. Also, the same numbers of nitrogen and hydrogen atoms have been represented as reactants and as products. Atoms are neither created nor destroyed in chemical reactions.*

It was Avogadro who first suggested that certain elements such as hydrogen, oxygen, and nitrogen were made up of diatomic molecules, that is, molecules containing two atoms each.

first explained the **law of combining volumes**. According to this well-known **Avogadro hypothesis**,

Equal volumes of all gases (at the same temperature and pressure) contain the same number of molecules.

Since equal volumes of gases at the same pressure and temperature contain the same number of molecules, and since a specific number of molecules (6.02×10^{23}) are present in a mole, we can conclude that

Equal volumes of gases at the same conditions contain the same number of moles of gas.

Thus, two *volumes* of a gas will contain twice as many molecules and also twice as many moles of the gas as are contained in a single *volume* of the gas. Furthermore, two *moles* of gas will contain twice as many molecules and have a volume twice as large as one *mole* of gas at the same conditions. And three moles of gas will occupy a volume three times as large as one mole of gas.

The fact that the mole ratios are the same as the volume ratios—at the same conditions—for gases in a chemical reaction is quite useful. For example, in the ammonia synthesis reaction, the gases are in a simple 1:3:2 mole ratio; the volume ratio is also 1:3:2, regardless of the volume units used.

	$N_2(g)$	$+ 3\ H_2(g)$		$2\ NH_3(g)$
Moles of gas:	1 mol	3 mol	$\longrightarrow$	2 mol
Volume in liters:	1 L	3 L	$\longrightarrow$	2 L
Volume in milliliters:	1 mL	3 mL	$\longrightarrow$	2 mL
	50 mL	150 mL	$\longrightarrow$	100 mL
Any volume units:	1 volume	3 volumes	$\longrightarrow$	2 volumes

EXAMPLE 11.6

What volume of $H_2(g)$ is required to produce 120. L of ammonia at the same conditions?

SOLUTION

Since both chemicals are gases, the volume ratio is the same as the mole ratio. Again, the reaction is

$$N_2(g) + 3\ H_2(g) \longrightarrow 2\ NH_3(g)$$

We will start the calculation with ammonia, NH_3, because this volume is given.

See Problems 11.15 and 11.16.

$$120.\ \text{L}\ NH_3 \times \frac{3\ \text{mol}\ H_2}{2\ \text{mol}\ NH_3} = 180.\ \text{L}\ H_2\ \text{gas} \qquad \text{(answer)}$$

EXERCISE 11.4

Determine the volume of hydrogen that can be obtained from 3000. L of methane, CH_4, at the same conditions, based on the following equation.

$$CH_4(g) \longrightarrow C(s) + 2\ H_2(g)$$

11.6 Limiting Reactant Calculations

When the quantity of one reactant is known, the mole ratios obtained from the chemical equation (Section 11.1) can be used to determine the stoichiometric quantities of any and all other chemicals involved in a particular reaction. When an excess of one of the reactants is present, as is often the case in many chemical processes, a portion of that chemical remains after the reaction is completed. The reactant that is fully consumed during the reaction is called the **limiting reagent.**

When quantities for both reactants are given, the quantities of products for the reaction are based on the quantity of the limiting reagent. Thus, if we do not know which reagent is the *limiting reagent,* this must be determined first.

The **reagents** are the chemicals used to carry out the reaction. The **limiting reagent** is the reactant that is totally consumed during a chemical reaction.

For an illustration, let us go back to the reaction of nitrogen and hydrogen to produce ammonia.

$$N_2(g) + 3\ H_2(g) \longrightarrow 2\ NH_3(g)$$

In this reaction 1 mol of nitrogen requires 3 mol of hydrogen. Consider the following case where 2.5 mol of nitrogen is used in a reaction with 8.0 mol of hydrogen. Since the ratio of moles of nitrogen to hydrogen to ammonia is 1:3:2, the change in moles during the reaction is summarized as follows.

Equation	$N_2(g)$ +	$3\ H_2(g)$	$\longrightarrow$ $2\ NH_3(g)$
Moles at start:	2.5	8.0	0
Change in moles:	−2.5	−2.5 × 3	+2.5 × 2
Moles remaining after reaction:	0	0.5	5

When reacting quantities are given in grams, these values must first be converted to moles and then compared to determine which chemical is in excess and which is the limiting reagent. Quantities of products must be based on the quantity of the limiting reagent. You must think through each problem, but the following steps can help you solve problems that involve limiting reagents.

Thinking It Through

Limiting Reagent Calculations

1. Use the quantities given in the problem to determine the number of moles of each reactant.
2. Determine which chemical is the *limiting reagent.* To do this, use the number of moles given for one reactant—we'll call it reactant A—to determine the number of moles of the other reactant—reactant B—needed to consume all of reactant A. (Use the appropriate mole ratio.)

   If more moles of reactant B are available than are actually needed (see step 1), then reactant A is the limiting reagent.

   If there are not enough moles of reactant B available, relative to the number of moles of reactant A, then reactant B is the limiting reagent.
3. Use the number of moles of the limiting reagent to determine the quantity of any reaction product as required.

Now that you have the idea behind solving problems involving a limiting reagent, follow through with calculations shown in Example 11.7.

EXAMPLE 11.7

If 55.0 g of nitrogen gas is placed in a reaction container with 55.0 g of hydrogen gas, determine which chemical is the limiting reagent and the number of grams of ammonia gas that can be produced by this reaction.

$$N_2(g) + 3\,H_2(g) \longrightarrow 2\,NH_3(g)$$

SOLUTION

The information given lets you know this is a limiting-reagent problem.

STEP 1 Use the quantities given in the problem to determine the number of moles of each reactant.

$$\text{For nitrogen:}\quad 55.0\text{ g } N_2 \times \frac{1\text{ mol } N_2}{28.0\text{ g } N_2} = 1.96\text{ mol } N_2$$

$$\text{For hydrogen:}\quad 55.0\text{ g } H_2 \times \frac{1\text{ mol } H_2}{2.02\text{ g } H_2} = 27.2\text{ mol } H_2$$

STEP 2 Determine which chemical is the limiting reagent. Use the mole ratio from the chemical equation to determine the number of moles of hydrogen needed to react completely with 1.96 mol N_2. The mole ratio is 3 mol H_2 to 1 mol N_2.

$$1.96\text{ mol } N_2 \times \frac{3\text{ mol } H_2}{1\text{ mol } N_2} = 5.88\text{ mol } H_2\text{ needed}$$

With 27.2 mol H_2 available, we have an *excess* of H_2. Thus, N_2 is the limiting reagent.

The quantity of the limiting reagent, N_2, is used to determine the quantity of ammonia (in grams) that can be produced by the reaction, as shown by these calculations.

$$1.96\text{ mol } N_2 \times \frac{2\text{ mol } NH_3}{1\text{ mol } N_2} \times \frac{17.0\text{ g } NH_3}{1\text{ mol } NH_3} = 66.6\text{ g } NH_3 \qquad \text{(answer)}$$

11.7 *Percent Yield*

In Example 11.7, we determined that 66.6 g of ammonia gas can be produced when 55.0 g of nitrogen, the limiting reagent, reacts with a sufficient quantity of hydrogen gas. This calculated quantity of ammonia, called the theoretical yield, is the maximum yield (the maximum amount) that can be produced by the given reaction. The **theoretical yield** can be more accurately defined as the maximum quantity of a substance that can be produced by complete reaction of all of the limiting reagent in accordance with the chemical equation.

When chemical reactions are carried out in laboratory and manufacturing operations, it is very difficult to actually obtain 100% of the theoretical yield. This is caused by several factors, including incomplete reaction of the limiting reagent, less than ideal reacting conditions, reversible reactions, the formation of unwanted reaction products, and loss of product in transferring from one vessel to another. The quantity of product (usually in grams) that is finally obtained from the reaction is called the **actual yield**.

The **percent yield** for a reaction is obtained by dividing the actual yield by the theoretical yield and multiplying by 100% as shown here.

$$\frac{\text{Actual yield}}{\text{Theoretical yield}} \times 100\% = \text{Percent yield}$$

For Example 11.7 we calculated a theoretical yield of 66.6 g of ammonia. If the actual amount of ammonia obtained after the reaction was 56.9 g, the percent yield is

$$\frac{56.9\text{ g (actual yield)}}{66.6\text{ g (theoretical yield)}} \times 100\% = 85.4\%$$

The following example illustrates the calculation of a theoretical yield and percent yield.

EXAMPLE 11.8

(a) Calculate the theoretical yield of sodium chloride produced by the reaction of 20.0 g of $NaHCO_3$ with 50.0 mL of 6 M hydrochloric acid based on the following chemical equation.

$$NaHCO_3 + HCl(aq) \longrightarrow NaCl + H_2O + CO_2(g)$$

(b) What is the percent yield if 12.3 g of NaCl was actually obtained?

SOLUTION

For part (a) we will follow through with the following steps.

STEP 1 Determine the number of moles of each reactant.

$$\underset{\text{(mass given)}}{\text{Moles } NaHCO_3} = 20.0\text{ g } NaHCO_3 \times \frac{1\text{ mol } NaHCO_3}{84.0\text{ g } NaHCO_3} = 0.238\text{ mol } NaHCO_3$$

$$\underset{\text{(volume given)}}{\text{Moles } HCl(aq)} = 0.050\text{ L } HCl(aq) \times \frac{6\text{ mol}}{1\text{ L}} = 0.300\text{ mol } HCl(aq)$$

STEP 2 Determine which chemical is the limiting reagent. According to the chemical equation, equal numbers of moles of $NaHCO_3$, and HCl(aq) are consumed by the reaction. Thus, our 0.238 mol $NaHCO_3$ requires 0.238 mol HCl. Since we have an excess of HCl (0.300 mol present), the $NaHCO_3$ is the limiting reagent.

STEP 3 Use the number of moles of $NaHCO_3$, the limiting reagent, to determine the theoretical yield of NaCl.

$$0.238\text{ mol } NaHCO_3 \times \frac{1\text{ mol NaCl}}{1\text{ mol } NaHCO_3} \times \frac{58.5\text{ g NaCl}}{1\text{ mol NaCl}} = \underset{\text{(theoretical yield)}}{13.9\text{ g NaCl}}$$

(b) Determine the percent yield.

$$\frac{12.3\text{ g (actual yield)}}{13.9\text{ g (theoretical yield)}} \times 100\% = 88.5\%$$

See Problems 11.17–11.20.

EXERCISE 11.5
What would be the percent yield if the actual yield in the example had been 13.3 g (just 1 g greater)?

11.8 Energy Changes in Chemical Reactions

The energy released during chemical reactions is a major driving force behind these reactions. This energy change helps account for why reactions occur—reactions tend to proceed toward products that are at a lower energy state. We take advantage of this fundamental principle when we burn natural gas or any other fuel. The main component of natural gas is methane, CH_4. As the methane burns, it consumes oxygen from the air and releases heat energy while producing carbon dioxide gas and water vapor (a gas). One way that we can show that heat is released during the reaction is to write the word ''heat'' on the right-hand side of the equation, as follows.

$$CH_4(g) + 2\,O_2(g) \longrightarrow CO_2(g) + 2\,H_2O(g) + \text{Heat}$$

This equation, however, does not indicate the amount of energy released. Quantities of heat energy are sometimes expressed in the familiar **calorie** (cal) or the **kilocalorie** (kcal), but the SI unit preferred is the **joule** (J) or the **kilojoule** (kJ). These units can be readily interconverted. By definition.

$$1 \text{ cal} = 4.184 \text{ J}$$

$$1 \text{ kcal} = 4.184 \text{ kJ}$$

The amount of energy change that takes place during a reaction depends upon the chemicals present and also upon the conditions under which the reaction is carried out. We can rewrite the equation to show the specific amount of energy released when the combustion of methane is carried out in the atmosphere (where pressure is constant) at 25°C and where the water produced is a vapor. Under these conditions, the combustion of 1 mol $CH_4(g)$ requires 2 mol $O_2(g)$ to produce 1 mol $CO_2(g)$ and 2 mol $H_2O(g)$ while releasing 802 kJ (192 kcal).

$$CH_4(g) + 2\,O_2(g) \longrightarrow CO_2(g) + 2\,H_2O(g) + 802 \text{ kJ}$$

The change in energy is different when the water vapor produced is in the liquid state because water vapor contains more energy than does liquid water.

$$CH_4(g) + 2\,O_2(g) \longrightarrow CO_2(g) + 2\,H_2O(l) + 890 \text{ kJ}$$

Energy is released during the combustion of methane, gasoline, alcohol, and all other fuels. A reaction that involves the release of heat energy is called an **exothermic reaction**.

Endothermic reactions are those that absorb or take up heat energy. Although most reactions are exothermic, some reactions are endothermic. Endothermic reactions require that energy be continually supplied during the reaction. The decomposition of water by an electric current from a battery or other direct current source is a good example of an endothermic reaction. This is known as the **electrolysis** of water. It is one method of producing oxygen gas and hydrogen gas of high purity. To decompose 1 mol of water requires 283 kJ (67.6 kcal) of energy. The 283 kJ is shown on the left side of the equation.

The fractional coefficient is used here because we are emphasizing the energy change for *one* mole of water.

$$H_2O(l) + 283 \text{ kJ} \longrightarrow H_2(g) + \tfrac{1}{2}\, O_2(g)$$

The reverse reaction, however, is exothermic; it involves the release of precisely the same amount of energy. When a mixture of hydrogen gas and oxygen gas is ignited, water is formed, and a loud explosion accompanies the release of energy. For the combustion of one mole of hydrogen we can write

$$H_2(g) + \tfrac{1}{2}\, O_2(g) \longrightarrow H_2O(l) + 283 \text{ kJ}$$

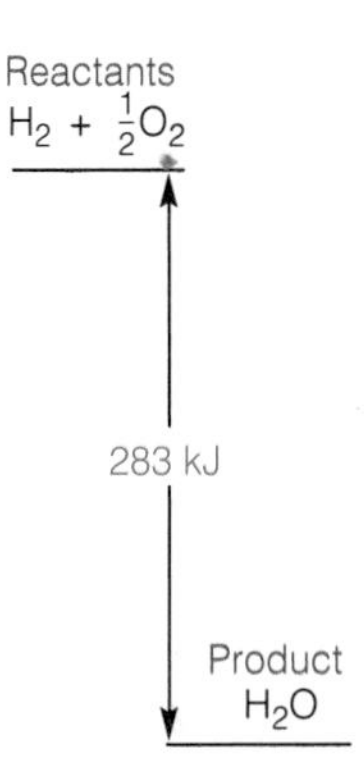

Figure 11.7 *The exothermic reaction of hydrogen gas with oxygen gas. The combustion of hydrogen is an exothermic reaction: $\Delta H = -283$ kJ. The energy possessed by the product, water, is less than the energy of the reactants, so ΔH is negative.*

The heat energy released in an exothermic reaction is directly related to the difference between the *chemical energy* present in the products and the reactants. We are not as interested in the chemical energy of the reactants or products as we are interested in the *heat change* for a reaction, called the change in **enthalpy** (when pressure is constant), symbolized ΔH (Δ for change, H for enthalpy). The enthalpy change for a reaction, ΔH, is equal to the sum of the enthalpies of the products minus the sum of the enthalpies of the reactants. For an exothermic reaction, the enthalpies of the products are smaller than the enthalpies of the reactants. Thus, ΔH is a negative number. The exothermic reaction for the combustion of hydrogen

$$H_2(g) + \tfrac{1}{2}\, O_2(g) \longrightarrow H_2O(l) + 283 \text{ kJ}$$

is often written in the following alternate form, showing the negative ΔH value.

$$H_2(g) + \tfrac{1}{2}\, O_2(g) \longrightarrow H_2O(l) \qquad \Delta H = -283 \text{ kJ}$$

The ΔH value applies to the reaction as it is written in moles, that is, 1 mol H_2, 0.5 mol O_2, and 1 mol H_2O. Figure 11.7 shows an energy level diagram for this exothermic reaction. Notice that the energy level of the products is shown to be lower than that of the reactants; the ΔH for the reaction is negative.

Since the electrolysis of water is an endothermic reaction, ΔH is positive.

$$H_2O(l) \xrightarrow{\text{electrolysis}} H_2(g) + \tfrac{1}{2}O_2(g) \qquad \Delta H = +283 \text{ kJ}$$

An energy level diagram for this endothermic reaction is shown in Figure 11.8. For this reaction, the energy level of the reactants is shown to be lower than that of the products since ΔH for the reaction is positive.

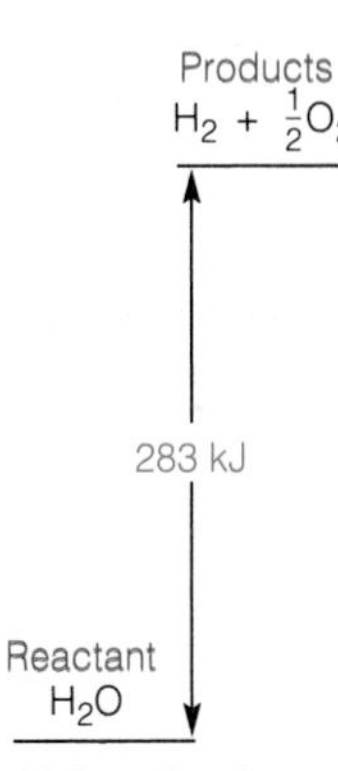

Figure 11.8 *The decomposition of water is an endothermic reaction: $\Delta H = +283$ kJ. The energy possessed by the products is greater than the energy of the reactants, so ΔH is positive. When a direct current is used to supply the energy, the process is called electrolysis.*

As an important biological application, consider the exothermic reaction for the combustion of glucose.

$$\underset{\text{glucose}}{C_6H_{12}O_6(s)} + 6\, O_2(g) \longrightarrow 6\, CO_2(g) + 6\, H_2O(g) \qquad \Delta H = -2540 \text{ kJ}$$

Whether one mole of glucose (180 g) is burned in the open air or is metabolized within the cells of the human body (or any other organism), the same amount of energy is liberated, 2540 kJ or 607 kcal (or 607 dieter's Calories).

The reverse reaction occurs during photosynthesis in green plants. By a series of reactions, carbon dioxide and water vapor react in the presence of chlorophyll to produce glucose and oxygen. The plant obtains energy from the sun for this endothermic process. Again, notice that the ΔH value has the opposite sign when the reaction is reversed.

$$6\,CO_2(g) + 6\,H_2O(g) \longrightarrow \underset{\text{glucose}}{C_6H_{12}O_6} + 6\,O_2(g) \qquad \Delta H = +2540\text{ kJ}$$

EXAMPLE 11.9

Determine the energy released (in kilojoules) by the metabolism of 10.0 g of glucose.

SOLUTION

Write the equation and the change in enthalpy for the combustion of glucose as printed in this section.

$$C_6H_{12}O_6 + 6\,O_2(g) \longrightarrow 6\,CO_2(g) + 6\,H_2O(g) \qquad \Delta H = -2540\text{ kJ}$$

According to this equation, the combustion (or metabolism) of 1 mol of glucose (180. g) releases 2540 kJ. We can write this as a conversion factor.

$$\frac{2540\text{ kJ released}}{\text{Mol glucose}}$$

Now, begin with the quantity of glucose in grams and multiply by the appropriate conversion factors to obtain energy *released.*

See Problems 11.21–11.26.

$$10.0\text{ g glucose} \times \frac{1\text{ mol glucose}}{180.\text{ g glucose}} \times \frac{2540\text{ kJ released}}{\text{Mol glucose}} = 141\text{ kJ} \qquad \text{(answer)}$$

Chapter Summary

Stoichiometric calculations are used to determine the proportional quantities of chemicals involved in a reaction when a balanced chemical equation is given and the quantity of one of the reactants or products is known.

The numerical coefficients from the balanced chemical equation can be used to write mole ratios for all pairs of chemicals involved in the reaction. The number of moles of one substance can be used to determine the number of moles of a second substance, symbolized as follows.

$$\text{Moles A} \longrightarrow \text{Moles B}$$

To perform this conversion, the number of moles of a given substance is multiplied by the appropriate mole ratio to obtain the number of moles of a second substance. This conversion operation takes the general form

$$\underset{\text{(given)}}{\text{Moles A}} \times \frac{\text{Mol B}}{\text{Mol A}} = \text{? moles B}$$

When the mass of any reactant or product is given (grams A), the conversion sequence for determining the mass of any other chemical involved in the reaction (grams B) can be written

$$\text{Grams A} \longrightarrow \text{Moles A} \longrightarrow \text{Moles B} \longrightarrow \text{Grams B}$$

The series of conversions takes the general form

$$\text{Grams A} \times \frac{\text{Mol A}}{\text{Grams A}} \times \frac{\text{Mol B}}{\text{Mol A}} \times \frac{\text{Grams B}}{\text{Mol B}} = \text{? grams B}$$

If the quantity of the first chemical is given in moles instead of grams, we simply omit the first conversion step of the sequence. Similarly, if the final quantity is desired in moles, we omit the final conversion step of the sequence. For gases involved in chemical reactions, the volume ratios are the same as the mole ratios (at constant pressure and temperature).

When the volume and molarity are given for an aqueous solution of the starting chemical, the sequence takes the following general form.

$$\text{Volume A} \xrightarrow{\text{molarity of A}} \text{Moles A} \longrightarrow \text{Moles B} \longrightarrow \text{Grams B}$$

Furthermore, ''moles of B'' can be converted to ''volume of B'' when the molarity of solution B is given.

$$\text{Grams A} \longrightarrow \text{Moles A} \longrightarrow \text{Moles B} \xrightarrow{\text{molarity of B}} \text{Volume B}$$

The reactant that is fully consumed in a reaction is called the limiting reagent. The theoretical yield is the maximum quantity of the product that can be produced by a complete reaction based on stoichiometric calculations. The actual yield is the quantity of the desired product that is finally obtained when a reaction is carried out. The percent yield for a reaction is calculated as follows.

$$\frac{\text{Actual yield}}{\text{Theoretical yield}} \times 100\% = \text{Percent yield}$$

The heat of reaction, or change in enthalpy for a reaction, is dependent upon the quantities of reacting chemicals. For an exothermic reaction, heat energy is released; the change in enthalpy is a negative value. For an endothermic reaction, heat energy is absorbed; the change in enthalpy is a positive value.

Regardless of whether we are concerned with the manufacture of plastics, medicines, cleaning agents, or synthetic fibers; the production of metals from ores; the formation of polluting gases in the atmosphere; or the metabolism of food, all of these processes are stoichiometric. For these processes and for all other chemical changes, there are definite quantities of chemicals and definite energy changes involved. Matter and energy are neither created nor destroyed during chemical changes; thus, we can use stoichiometric calculations to determine the specific quantities involved.

Assess Your Understanding

1. Write mole ratios for all pairs of substances involved in given chemical reactions. [11.1]
2. Make "moles of A" to "moles of B" stoichiometric calculations. [11.2]
3. Make "moles of A" to mass of B" and "mass of A" to "moles of B" calculations. [11.3]
4. Make stoichiometric calculations that involve molar solutions. [11.4]
5. Apply the Avogadro hypothesis to stoichiometric calculations involving gases ("volume of gas A" to "volume of gas B"). [11.5]
6. Use the quantities of chemicals given to determine the limiting reagent and calculate the quantities of reaction products. [11.6]
7. Calculate theoretical yields and percentage yields for reactions. [11.7]
8. Calculate energy changes for exothermic and endothermic reactions. [11.8]

Key Terms

actual yield [11.7]
Avogadro hypothesis [11.5]
electrolysis [11.8]
endothermic reaction [11.8]
enthalpy [11.8]
exothermic reaction [11.8]
law of combining volumes [11.5]
limiting reagent [11.6]
mole ratio [11.1]
percent yield [11.7]
stoichiometry [intro]
theoretical yield [11.7]

Problems

MOLE RATIOS FROM CHEMICAL EQUATIONS

11.1 For the following chemical equation

$$Al_2(SO_4)_3 + 3\ Ba(NO_3)_2 \rightarrow 3\ BaSO_4 + 2\ Al(NO_3)_3$$

give the mole ratio needed to convert
a. from moles of $Al_2(SO_4)_3$ to moles of $BaSO_4$.
b. from moles of $BaSO_4$ to moles of $Al_2(SO_4)_3$.
c. from moles of $Ba(NO_3)_2$ to moles of $Al(NO_3)_3$.
d. from moles of $Ba(NO_3)_2$ to moles of $Al_2(SO_4)_3$.

11.2 For the following chemical equation

$$3\ NO_2(g) + H_2O(l) \rightarrow 2\ HNO_3(aq) + NO(g)$$

give the mole ratio needed to convert
a. from moles of NO_2 to moles of HNO_3.
b. from moles of NO_2 to moles of NO.
c. from moles of HNO_3 to moles of H_2O.
d. from moles of NO to moles of HNO_3.

MOLE–MOLE CALCULATIONS

11.3 For the reaction given in problem 11.1, determine the number of
a. moles of $BaSO_4$ that can be prepared from 0.225 mol $Al_2(SO_4)_3$.
b. moles of $Al_2(SO_4)_3$ needed to prepare 3.30 mol $BaSO_4$.
c. moles of $Al(NO_3)_3$ that can be prepared from 0.750 mol $Ba(NO_3)_2$.
d. moles of $Al_2(SO_4)_3$ needed to react with 9.33 mol $Ba(NO_3)_2$.

11.4 For the reaction given in problem 11.2, determine the number of
a. moles of HNO_3 that can be produced from 63.3 mol NO_2.
b. moles of NO that can be produced from 12.3 mol NO_2.
c. moles of water that will be needed to produce 6.44 mol HNO_3.
d. moles of HNO_3 that can be formed along with 7.25 mol NO.

STOICHIOMETRIC CALCULATIONS INVOLVING MOLES AND MASSES

11.5 For the acid–base neutralization reaction represented here

$$2\ HCl(aq) + Ca(OH)_2 \rightarrow 2\ H_2O + CaCl_2(aq)$$

and 0.684 mol HCl(aq), determine the
a. maximum number of grams of $Ca(OH)_2$ consumed during the reaction.
b. maximum number of grams of water that can be formed.
c. maximum number of grams of $CaCl_2$ that can be formed.
d. number of grams of HCl consumed during the reaction.
e. total grams of reactants and total grams of products. Discuss.

11.6 For the reaction represented here

$$3\ Mg + 2\ H_3PO_4(aq) \rightarrow Mg_3(PO_4)_2(aq) + 3\ H_2(g)$$

and 0.482 mol H_3PO_4(aq), determine the
a. maximum number of grams of Mg consumed during the reaction.
b. maximum number of grams of $Mg_3(PO_4)_2$ that can be formed.
c. maximum number of grams of hydrogen gas that can be produced.
d. number of grams of H_3PO_4 consumed during the reaction.
e. total grams of reactants and total grams of products. Discuss.

11.7 For the combustion of methanol, CH_3OH (also called methyl alcohol),

$$2\ CH_3OH + 3\ O_2 \rightarrow 2\ CO_2 + 4\ H_2O$$

calculate the number of
a. moles of oxygen required to burn 50.0 g of methanol.
b. grams of oxygen required to burn 50.0 g of methanol.
c. moles of methanol burned to produce 45.0 g of water vapor.
d. grams of methanol burned to produce 45.0 g of water vapor.

11.8 For the combustion of octane, C_8H_{18}, present in gasoline

$$2\ C_8H_{18} + 25\ O_2 \rightarrow 16\ CO_2 + 18\ H_2O$$

calculate the number of
a. moles of oxygen required to burn 50.0 g of octane.
b. grams of oxygen required to burn 50.0 g of octane.
c. moles of carbon dioxide produced while burning 50.0 g of octane.
d. grams of carbon dioxide produced while burning 50.0 g of octane.

11.9 When 20.0 g of Mg was placed in a beaker of hydrochloric acid, a vigorous reaction produced hydrogen gas, magnesium chloride, and enough heat to make the beaker hot to the touch. The Mg metal was completely reacted with an excess of the acid.
a. Write a balanced chemical equation for the reaction.
b. How many moles of $H_2(g)$ were produced?
c. How many grams of HCl were consumed during the reaction?

11.10 When 20 g of Zn was placed in a beaker of sulfuric acid, a vigorous reaction produced hydrogen gas and zinc sulfate while the beaker became quite warm. The Zn metal was completely reacted.
a. Write a balanced chemical equation for the reaction.
b. How many moles of $H_2(g)$ were produced?
c. How many grams of H_2SO_4 were consumed during the reaction?

STOICHIOMETRIC CALCULATIONS INVOLVING MOLAR SOLUTIONS

11.11 In problem 11.9c the number of grams of HCl for the reaction with Mg were determined. How many milliliters of 6.00 M hydrochloric acid would have been needed to react with all of the magnesium metal?

11.12 In problem 11.10c the number of grams of H_2SO_4 for the reaction with Zn were determined. How many milliliters of 6.00 M sulfuric acid would have been needed to react with all of the zinc metal?

11.13 How many milliliters of 1.00 M hydrochloric acid would be needed to react completely with a milk of magnesia tablet that contains 310 mg of $Mg(OH)_2$? The neutralization reaction produces magnesium chloride, $MgCl_2$, and water.

11.14 How many milliliters of 3.00 M hydrochloric acid would be needed to react completely with 10.0 g of lime, $Ca(OH)_2$? The neutralization reaction produces calcium chloride, $CaCl_2$, and water.

GAS STOICHIOMETRY (APPLYING THE AVOGADRO HYPOTHESIS)

11.15 Acetylene gas, C_2H_2, is the fuel used for ordinary welding applications.

a. If tanks of acetylene and oxygen are used that have the same pressure, volume, and temperature, how many tanks of oxygen would be needed to supply enough oxygen to burn one tank of acetylene? The equation is

$$2\,C_2H_2(g) + 5\,O_2(g) \rightarrow 4\,CO_2(g) + 2\,H_2O(g)$$

b. Give an explanation for why welders often use a cart that carries a small tank of acetylene and a large tank of oxygen.

11.16 During the complete combustion of methane from natural gas both carbon dioxide and water vapor are produced. If the two gases are compared at the same temperature and pressure, how many liters of water vapor are produced during the formation of 12.0 L of CO_2? The equation is

$$CH_4(g) + 2\,O_2(g) \rightarrow CO_2(g) + 2\,H_2O(g)$$

CALCULATIONS INVOLVING LIMITING REAGENTS AND PERCENT YIELDS

11.17 The reaction of aluminum metal with bromine, a liquid nonmetal, is spontaneous—no outside energy is necessary to start the reaction. The quantities of chemicals mixed together are shown below the reactants.

$$\underset{4.00\text{ g}}{2\,Al} + \underset{42.0\text{ g}}{3\,Br_2(l)} \rightarrow 2\,AlBr_3$$

a. Which chemical is the limiting reagent? (Show calculations.)

b. What is the theoretical yield of $AlBr_3$?

c. For an actual yield of 32.2 g of $AlBr_3$, what is the percent yield?

11.18 When a lump of phosphorus is added to liquid bromine, the reaction is spontaneous and liberates heat. The quantities of chemicals mixed together are shown below the reactants.

$$\underset{5.00\text{ g}}{P_4} + \underset{40.5\text{ g}}{6\,Br_2(l)} \rightarrow 4\,PBr_3$$

a. Which chemical is the limiting reagent? (Show calculations.)

b. What is the theoretical yield of PBr_3?

c. For an actual yield of 37.5 g of PBr_3, what is the percent yield?

11.19 Methanol, CH_3OH, has the common name methyl alcohol. Most methanol is produced commercially by the reaction of carbon monoxide, CO, with $H_2(g)$ at a high temperature and pressure. For 72.0 kg of CO reacted with 5.50 kg of H_2, make calculations based on the following equation.

$$CO(g) + 2\,H_2(g) \rightarrow CH_3OH(l)$$

a. Which chemical is the limiting reagent? (Show calculations.) *Hint:* Work the problem using grams first. The quantities in kilograms can be treated as if they are grams—they are proportional.

b. What is the theoretical yield of methanol?

c. For an actual yield of 39.5 kg of methanol, what is the percent yield?

11.20 Ethanol, C_2H_5OH, has the common name ethyl alcohol. Much of the ethanol produced commercially is by the reaction of ethene, C_2H_4, with water. For 80.0 kg of ethene reacted with 55.0 kg of water, make calculations based on the following chemical equation.

$$C_2H_4 + H_2O \rightarrow C_2H_5OH$$

a. Which chemical is the limiting reagent? (Show calculations.)

b. What is the theoretical yield of ethanol?

c. For an actual yield of 125 kg of ethanol, what is the percent yield?

STOICHIOMETRY AND ENERGY CHANGES

11.21 What is the difference between an exothermic reaction and an endothermic reaction.

11.22 Define enthalpy. Does the change in enthalpy for an exothermic reaction have a positive or a negative value? Explain your answer.

11.23 Determine the change in enthalpy for the combustion of 10.0 g of methane, CH_4, in natural gas. Why is the enthalpy change negative?

$$CH_4(g) + 2\,O_2(g) \rightarrow CO_2(g) + 2\,H_2O(g) \qquad \Delta H = -802\text{ kJ}$$

Compare this energy with the value calculated for 10.0 g of glucose (Example 11.9 in the chapter).

11.24 Determine the change in enthalpy for the combustion of 10.0 g of octane, C_8H_{18}, in gasoline. (The enthalpy change given is for the equation as written, with 2 mol of octane.)

$$2\ C_8H_{18}(l) + 25\ O_2(g) \rightarrow 16\ CO_2(g) + 18\ H_2O(g) \quad \Delta H = -5460\ kJ$$

Compare this energy with the value calculated for 10.0 g of methane (Problem 11.23).

11.25 Calcium carbonate, $CaCO_3$, from limestone is heated to a high temperature in a kiln to produce calcium oxide, CaO. The CaO, called quicklime, is present in cement. Rewrite the equation shown here in the alternate form that shows the change in enthalpy. Also, determine the enthalpy change required to produce 1000. g of CaO.

$$CaCO_3(s) + 178\ kJ \rightarrow CaO(s) + CO_2(g)$$

11.26 Calcium oxide, CaO, called quicklime, is mixed with water to produce calcium hydroxide, $Ca(OH)_2$, called slaked lime. The slaked lime is used in mortar and plaster. Rewrite the equation shown here in the alternate form that shows the change in enthalpy. Also, determine the enthalpy change involved when 1000. g of CaO is converted to slaked lime.

$$CaO(s) + H_2O(l) \rightarrow Ca(OH)_2(s) + 65.5\ kJ$$

Additional Problems

11.27 For centuries, ethanol, C_2H_5OH, has been produced by the fermentation of glucose from starch in grain (especially corn and barley). Thus, ethanol has long been known as grain alcohol. The reaction is symbolized

$$\underset{\text{glucose}}{C_6H_{12}O_6} \xrightarrow{\text{fermentation}} 2\ \underset{\text{ethanol}}{C_2H_5OH} + 2\ CO_2$$

If we start with 454 g (1 pound) of glucose, what is the maximum number of grams of ethanol that can be produced during fermentation?

11.28 Hydrogen peroxide, H_2O_2, decomposes to give water and oxygen gas in the presence of a catalyst—a substance that increases the rate of the reaction without being consumed during the reaction. The equation is

$$2\ H_2O_2 \xrightarrow{\text{catalyst}} 2\ H_2O + O_2(g)$$

How many grams of oxygen will be produced by decomposing 51.0 g of H_2O_2?

11.29 The fuel in disposable cigarette lighters is butane, C_4H_{10}.

a. Write a balanced chemical equation for the combustion of butane. *Hint*: Start with a coefficient of 2 for butane.
b. Calculate the number of moles of oxygen required to burn the 5.00 g of butane in one disposable lighter.
c. Calculate the number of grams of O_2 required to burn 5.00 g of butane.

11.30 Propane, C_3H_8, in a tank that is a little larger than a basketball is often used as fuel for outdoor cooking grills.

a. Write a balanced chemical equation for the combustion of propane.
b. Calculate the number of moles of oxygen required to burn 5000. g (11.0 lb) of propane.
c. Calculate the number of kilograms of O_2 required to burn the propane.

11.31 Sodium peroxide, $Na_2O_2(s)$, can be used in a breathing apparatus to take up exhaled CO_2. Based on the following chemical equation, calculate the number of grams of CO_2 that can be taken up by 75.0 g of $Na_2O_2(s)$.

$$2\ Na_2O_2(s) + 2\ CO_2(g) \rightarrow 2\ Na_2CO_3(s) + O_2(g)$$

11.32 Sulfuric acid reacts with NaCl (common salt) by a series of reactions that can be combined to give the following chemical equation.

$$H_2SO_4 + 2\ NaCl \rightarrow Na_2SO_4 + 2\ HCl(g)$$

The HCl(g) is used to prepare hydrochloric acid, and the Na_2SO_4 is used in the manufacture of paper, glass, soap, and several other chemicals. For every 10.0 kg of HCl produced, determine the number of kilograms of sodium sulfate produced as a by-product.

11.33 Baking soda is actually sodium bicarbonate. When the $NaHCO_3$ reacts with any acid, CO_2 gas is produced. Based on the following equation, how many grams of sodium bicarbonate will react with 10. mL of 2.00 M HCl(aq)?

$$NaHCO_3 + HCl(aq) \rightarrow NaCl + H_2O + CO_2(g)$$

11.34 Acid rain can destroy exposed surfaces of marble used in buildings and sculptures. Sulfuric acid is one of the components of acid rain that reacts with marble, $CaCO_3$, as shown by the following equation.

$$H_2SO_4(aq) + CaCO_3(s) \rightarrow CaSO_4(s) + H_2O(l) + CO_2(g)$$

How many liters of 0.0100 M sulfuric acid will be used in reacting completely with 500. g (slightly over 1 lb) of $CaCO_3$? The figures in this calculation may surprise you!

11.35 Coal varies in sulfur content, but coal mined east of the Mississippi River has a higher sulfur content than coal mined in western states. A certain coal-fired power plant burns 100 train carloads of coal each day. Each carload averages 100 tons of coal with a 2.00% sulfur content.

a. Calculate the number of tons of sulfur burned each day by the plant.

b. Based on the following equation for the combustion of sulfur, S_8,

$$S_8(s) + 8\ O_2(g) \rightarrow 8\ SO_2(g)$$

determine the number of tons of sulfur dioxide, SO_2, produced daily. *Hint*: Make calculations as you would for grams; tons and grams are proportional.

11.36 Over 13 billion lb of nitric acid, HNO_3, is produced annually in the United States. The Ostwald process for the commercial production of nitric acid involves the reaction of nitrogen dioxide gas, NO_2, with water.

$$3\ NO_2(g) + H_2O \rightarrow 2\ HNO_3 + NO(g)$$

How many pounds of NO_2 are needed to produce 1000 lb of nitric acid?

11.37 Nitric acid, HNO_3, can be prepared in a laboratory by heating sodium nitrate, $NaNO_3$, with concentrated (18 M) sulfuric acid. The equation is

$$NaNO_3 + H_2SO_4 \rightarrow NaHSO_4 + HNO_3$$

a. Determine the theoretical yield of nitric acid for the reaction of 30.0 g of $NaNO_3$ and 22.0 mL of 18 M H_2SO_4.

b. A student obtained 17.0 g of HNO_3. What is the percent yield?

11.38 Ammonium sulfate, $(NH_4)_2SO_4$, is considered to be the most important solid fertilizer in the world. It can be produced by the reaction of aqueous ammonia $NH_3(aq)$ with sulfuric acid. The equation is

$$2\ NH_3(aq) + H_2SO_4(aq) \rightarrow (NH_4)_2SO_4$$

a. Determine the theoretical yield of $(NH_4)_2SO_4$ for a mixture of 40.0 mL of 15 M $NH_3(aq)$ and 20.0 mL of 18 M (concentrated) $H_2SO_4(aq)$.

b. A theoretical yield of 1200. kg was calculated for an industrial batch process, but 1170. kg was actually recovered. What is the percent yield?

11.39 The concentration of ozone, O_3, present in the air after morning rush-hour traffic in a large city later drops as the ozone is converted to ordinary oxygen, O_2, in the presence of ultraviolet light.

$$2\ O_3(g) \xrightarrow{\text{ultraviolet light}} 3\ O_2(g)$$

Assuming the gases are at the same pressure and temperature, what volume of oxygen is produced by 1000 L of ozone?

11.40 Ultraviolet light in the stratosphere can split chlorine atoms from Freon molecules. Once formed, the free chlorine atoms can react with and destroy the ozone, O_3, layer that protects people on the Earth from the ultraviolet radiation that can cause sunburn and skin cancer. One of the reactions involved is represented here.

$$Cl + O_3 \rightarrow ClO + O_2$$

For every 10,000 L of ozone, how many liters of O_2 are formed during this reaction?

11.41 Sodium hypochlorite, NaClO, is used in laundry bleaches such as Clorox. The NaClO can be prepared by reacting sodium hydroxide with chlorine gas.

$$2\ NaOH + Cl_2 \rightarrow NaClO + NaCl + H_2O$$

How many kilograms of chlorine gas will be used in the reaction with 50.0 kg of NaOH?

11.42 One method of obtaining copper metal is to heat a mixture of copper(I) oxide, Cu_2O, with copper(I) sulfide, Cu_2S.

$$2\ Cu_2O(s) + Cu_2S(s) \rightarrow 6\ Cu + SO_2(g)$$

How many kilograms of $Cu_2O(s)$ must be mixed with 10.0 kg of $Cu_2S(s)$ to achieve a complete reaction for both reactants?

11.43 Calcium hydroxide, $Ca(OH)_2$, called lime or slaked lime, is used in many manufacturing processes to neutralize acid solutions. How many kilograms of slaked lime are needed to neutralize 1000. L of a 0.500 M solution of discarded hydrochloric acid? The equation is

$$Ca(OH)_2(s) + 2\ HCl(aq) \rightarrow CaCl_2(aq) + 2\ H_2O$$

11.44 Barium sulfate, $BaSO_4$, has a very low solubility in water—it is virtually insoluble in water. The chalky white powder mixed with water looks milky; it is used in the gastrointestinal tract prior to making an X-ray of the specified region. A reaction to prepare a precipitate of $BaSO_4$ is symbolized

$$Ba(NO_3)_2(aq) + Na_2SO_4(aq) \rightarrow BaSO_4(s) + 2\ NaNO_3$$

How many milliliters of 0.1 M $Na_2SO_4(aq)$ will be needed to react completely with 20.0 mL of a 0.150 M solution of $Ba(NO_3)_2(aq)$?

11.45 When laundry bleach with NaClO is mixed with hydrochloric acid, HCl(aq), present in some liquid toilet bowl cleaners, a vigorous reaction occurs. The equation is

$$NaClO(aq) + HCl(aq) \rightarrow Cl_2(g) + NaOH(aq)$$

a. How many moles of chlorine gas would be produced by mixing 250 mL (about 1 cup) of 2.00 M HCl(aq) with an excess of bleach?
b. How many grams of chlorine gas would be produced?
c. Why should these two household chemicals never be used together?

11.46 A sterling silver spoon becomes tarnished when placed in certain foods, especially foods containing sulfides, such as cooked eggs. The tarnish is actually Ag_2S. Based on the following equation

$$4\ Ag + 2\ H_2S(g) + O_2(g) \rightarrow 2\ Ag_2S + 2\ H_2O$$

a. How many grams of Ag would be converted to Ag_2S by 2.00 g of H_2S?
b. Why would a silver spoon placed in a dish of food prepared with cooked eggs be more likely than usual to become tarnished?

11.47 Aluminum oxide, Al_2O_3, is a very hard, white, crystalline solid called alumina. Because of the extremely strong bonding, it is difficult to free the aluminum from the oxide. An electric current—electrolysis—is required. Based on the equation given, determine the energy required to obtain 500. g (a little more than a pound) of aluminum.

$$2\ Al_2O_3(s) + 3340\ kJ \rightarrow 4\ Al + 3\ O_2(g)$$

11.48 Determine the change in enthalpy for the combustion of 10.0 g of ethanol, C_2H_5OH, called ethyl alcohol.

$$C_2H_5OH(l) + 3\ O_2(g) \rightarrow 2\ CO_2(g) + 3\ H_2O(g) \quad \Delta H = 1235\ kJ$$

You may wish to compare this enthalpy change with values previously calculated for the combustion of 10.0 g of methane and 10.0 g of octane.

11.49 How many milliliters of stomach acid (considered to be 0.100 M HCl for our problem) will be consumed during a reaction with one antacid tablet that contains 500. mg of $CaCO_3$? The reaction produces CO_2 gas according to the equation

$$CaCO_3(s) + 2\ HCl(aq) \rightarrow CaCl_2 + CO_2(g) + H_2O(l)$$

11.50 How many milliliters of stomach acid (considered to be 0.100 M HCl for our problem) will be consumed in a neutralization reaction with one milk of magnesia tablet that contains 330. mg of $Mg(OH)_2$?

11.51 Quicklime (calcium oxide) is used in Portland cement. How many kilograms of CaO can be obtained when 8.64 kg of limestone, $CaCO_3$, is decomposed by heating?

$$CaCO_3(s) \rightarrow CaO(s) + CO_2(g)$$

11.52 What is the minimum number of tons of quicklime, CaO, that must be used to take up 200 tons of SO_2 gas produced by burning coal at an electric power plant? The equation is

$$CaO(s) + SO_2(g) \rightarrow CaSO_3(s)$$

The beauty of this New Zealand sunset is made possible by the interaction of light with molecules of gases in the atmosphere.

12

Gases

CONTENTS

The gases that surround planet Earth make it unique in our solar system. Astronauts have seen firsthand Earth's barren, airless moon. Our spaceships have photographed the desolation of Mercury from a few kilometers up and have measured the high temperatures of Venus, which are hot enough to melt lead. Photographs taken from space have even given us close-up portraits of the crushing, turbulent atmospheres of Jupiter and Saturn. Atmospheric data from Mars and experiments on the harsh Martian surface failed to detect the presence of life there.

After these investigations and more, it is becoming increasingly clear that the planet Earth—a small island of green and blue in the vastness of space—is uniquely equipped to serve the needs of the life that inhabits it. One of the essentials to the life-support system of Spaceship Earth is our atmosphere, a relatively thin layer of gases that surrounds and envelops our planet (Figure 12.1).

Figure 12.1 *Spaceship Earth with the desolate moon in the foreground.*

All gases, air included, obey certain physical laws. In this chapter, we will examine some of these laws as we consider certain unique properties of gases. In particular, we will look closely at relationships involving pressure, volume, temperature, and the quantity of gas in a sample.

12.1 The Atmosphere

Earth's atmosphere, composed of about 5.2×10^{15} metric tons of air, appears to be unique in its ability to support life.

The thin blanket of gases that surrounds our planet is called the **atmosphere**. It is difficult to measure the depth of the atmosphere; it does not end abruptly. The number of molecules per volume unit gradually decreases as the distance from the surface of the Earth increases. While the boundary is indefinite, we do know that 99% of the atmosphere lies within 30 km of the surface of the Earth. If we compare the Earth and its atmosphere with an apple, the thin layer of air would be proportionally thinner than the apple peel. That thin layer of air is all that separates us from the emptiness of space. Our supply of air, once thought of as inexhaustible, now appears to be limited.

Air is so familiar, and yet so nebulous, that it is difficult to think of it as matter. But it is matter—matter in the gaseous state. All gases, air included, have mass and occupy space. We can reach out and touch solids and liquids, but we cannot feel the air unless the wind is blowing. Furthermore, since air and most gases are colorless, we do not see them.

Air is made up of a mixture of gases. Dry air is (by volume) about 78% nitrogen (N_2), 21% oxygen (O_2), and 1% argon (Ar). The amount of water vapor varies up to about 4%. There are a number of minor constituents, the most important of which is carbon dioxide, CO_2. The concentration of carbon dioxide in the atmosphere is believed to have increased from 296 parts per million (ppm) in 1900 to its present value of more than 360 ppm. It will most likely continue to rise as we burn more fossil fuels (coal, oil, and gas). The composition of the atmosphere is summarized in Table 12.1.

12.2 The Kinetic Molecular Theory

We say air is a mixture of gases, but what are gases? Several general properties of each of the three states of matter were compared briefly in Chapter 2, but five important physical properties of gases can be summarized as follows.

Table 12.1 **Composition of Earth's Dry Atmosphere**

Substance	*Formula*	*Number of Molecules per 10,000 Molecules*	*Percent by Volume*
Nitrogen	N_2	7,800	78.00%
Oxygen	O_2	2,100	21.00
Argon	Ar	93	0.93
Carbon dioxide	CO_2	3	0.03
All others	—	4	0.04
Total		10,000	100.00%

1. Gases have no definite shape or volume; they expand to fill the entire volume of the container and conform to the shape of the container.
2. Gases are compressible; by increasing the pressure, gases can be made to occupy a much smaller volume.
3. Gases have low densities compared to liquids and solids. The density of dry air, for example, is about 0.00117 g/cm^3 at room pressure and temperature. As pressure increases, gas density also increases.
4. Gases that are confined to a container exert uniform pressure on all walls of the container. This is not the case with liquids; the force exerted on the walls of a container by a liquid is greater at greater depths due to gravitational forces.
5. Gases mix spontaneously and completely with each other at constant pressure, provided no chemical reaction is involved. This is called **diffusion** (Figure 12.2). For example, if ammonia gas is released in one corner of a room, it diffuses throughout the room until all areas of the room finally have the same concentration of the gas.

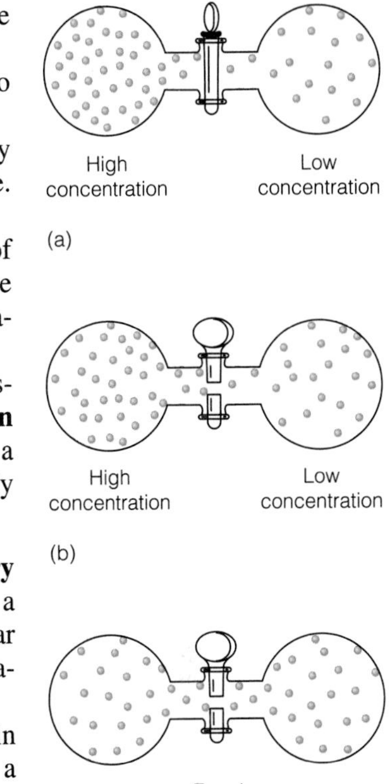

Figure 12.2 *Gases diffuse; they tend to flow from a region of high concentration to a region of low concentration. (a) With the stopcock closed, no flow is possible. (b) With the stopcock open, there is a net flow of gases from the region of higher concentration on the left to the region of lower concentration on the right. (c) With the stopcock open, there is no net flow of gas once concentrations are equal.*

Perhaps gases can best be understood in terms of the **kinetic molecular theory** (KMT)—or simply the **kinetic theory**. This theory, like atomic theory, provides a model with broad generalizations about properties of matter. The kinetic molecular theory enables us to visualize and understand the behavior of gases, but the fundamentals of this theory also apply to liquids and solids.

The kinetic molecular theory treats gases as collections of individual particles in rapid motion (hence the term *kinetic*). The word *particle* can mean an atom or a molecule, depending upon the gas involved. For example, particles of argon gas, Ar, are atoms, but particles of nitrogen gas, N_2, and carbon dioxide, CO_2, are molecules. The primary assumptions of this model for explaining gas behavior are described as follows.

Kinetic Theory of Gases

1. Gas particles move continuously, rapidly, and randomly in straight lines in all directions.

The motions of gas particles could be compared to billiard balls in constant motion on a pool table, bouncing ceaselessly off the sides and even off each other (Figure 12.3). Molecules of air have an average velocity (rate of movement) of about 500 m/s. That is equivalent to about 1100 miles per hour. The random mo-

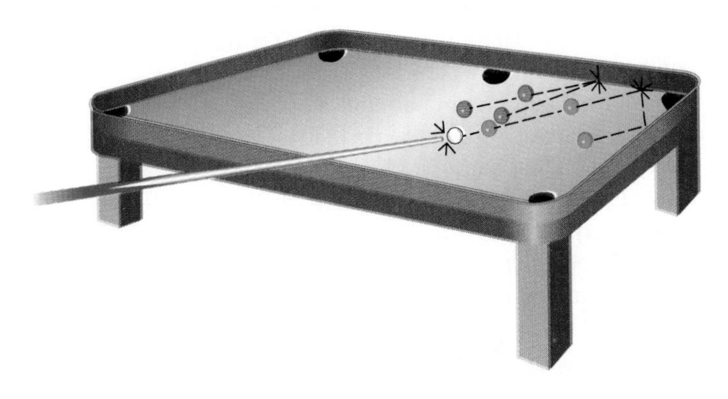

Figure 12.3 *Gas molecules can be compared to billiard balls in random motion, bouncing off each other and off the sides of the pool table.*

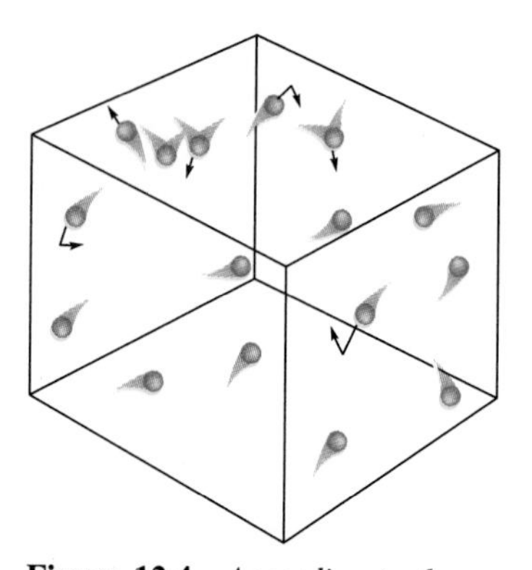

Figure 12.4 *According to the kinetic molecular theory, particles of a gas are in constant motion, occasionally bouncing off one another and off the walls of their container.*

tion of molecules allows us to explain the observation that gases expand to fill a closed container, which was listed as the first physical property for gases.

2. Gas particles are extremely tiny and distances between them are large.

Most of the gas sample is actually empty space (Figure 12.4). To get a mental picture of this fact, consider a mole of water (18.0 g). A mole of liquid water occupies 18.0 mL or 18.0 cm^3. When a mole of water at its boiling point (100°C or 373 K) is heated enough to become steam (a gas), the volume increases to 30,600 cm^3. The volume has expanded by a factor of 1700. The number of molecules involved did not change—the volume did. If the water molecules were touching, originally, and then the volume expands by a factor of 1700, the molecules of steam must be widely separated.

The concept of small particles with large distances between them makes it possible for us to explain the observation that gases can be easily compressed. This was listed as the second physical property of gases. The fact that gases have low densities—listed as the third physical property of gases—can also be understood when we realize that the particles are small and take up a small fraction of the total volume. In addition, the fact that gases diffuse or mix spontaneously—the fifth physical property of gases—can be explained in terms of the rapid movement of small particles that have sufficient room in which to mix.

Besides allowing us to explain properties of gases, the kinetic molecular theory also allows us to understand the noncompressible nature of liquids and solids. Unlike gas particles, individual particles of solids and liquids are in contact with one another. Since they are already touching, very little compression is possible.

3. For gases, both gravitational forces and forces of attraction between gas particles are negligible.

Since particles are separated by relatively large distances and are moving rapidly, attractive forces are negligible. Gravity has relatively little effect on gases. Gas particles move up, down, and sideways with ease. Unlike liquids and solids, they do not fall to the bottom of a container. Any container of gas is completely filled. By *filled*, we do not mean that the gas is packed tightly, but rather that it is distributed throughout the container's entire volume.

4. When gas particles collide with each other, or with the walls of the container, no energy is lost; all collisions are perfectly elastic.

The term **perfectly elastic** means that the particles continue to collide with no loss of energy. There is no tendency for the particles, taken collectively, to slow down and eventually stop.

A billiard ball moving around a pool table can be used to illustrate movements that are relatively—but not perfectly—elastic. The ball may bounce off several sides of the table and off other balls before coming to rest. If the collisions were truly elastic—with no energy loss—the ball would never come to rest. Collisions of gas particles are perfectly elastic. If they were not, the pressure inside a sealed container of gas at a given temperature would continually decrease.

The continual bombardment of all container walls by gas particles allows us to understand that gases can exert uniform pressure on all walls of the container—the fourth physical property listed for gases.

5. The average kinetic energy is the same for all gases at the same temperature; it varies proportionally with the temperature in kelvins.

Kinetic energy is the energy that molecules possess because of their motion. Light gases, such as hydrogen and helium, have the same kinetic energy as heavier gases like chlorine and carbon dioxide at the same temperature. Gases with smaller masses move more rapidly than do gases with greater masses.

Potential energy, unlike kinetic energy, is the stored energy possessed by molecules as a result of attractive and repulsive forces within molecules, especially those due to chemical bonding.

As the temperature of a particular gas increases, the kinetic energy of the gas increases. Thus, according to the theory, temperature is actually a measure of the average kinetic energy of the gas particles. The higher the temperature of the gas particles, the faster the particles are moving. In any single sample, however, some particles are moving faster than others; the Kelvin temperature is proportional to the *average* kinetic energy of the gas particles.

The kinetic energy of particles can be described more precisely in a mathematical equation as being equal to one-half the mass (m) times the velocity squared (v^2). This equation is written

$$\text{K.E.} = \tfrac{1}{2}mv^2$$

As indicated by the equation, if mass is fixed, any change in kinetic energy must be related to a change in the velocity of the gas particles. Oxygen, for example, has a fixed molar mass of 32.0 g/mol. The kinetic energy of the gas is proportional to the square of the velocity of the gas particles.

Let us now put the pieces of the puzzle together. As the temperature of a gas increases, the particles move faster. The increase in speed, as shown by the equation, results in an increase in the kinetic energy of the particles—they bombard the walls of the container more frequently. When the walls of the container are bombarded more frequently, the pressure increases; that is, the force per unit area increases. For a fixed number of gas particles in a closed container, pressure increases as temperature increases.

See Problems 12.9–12.22.

12.3 Atmospheric Pressure

Molecules of air are constantly bouncing off each of us. Since these molecules are so tiny, we do not feel their individual impact. In fact, at ordinary altitudes we don't feel the gas molecules pushing on our bodies because gas molecules inside our bodies are also pushing out with the same force. However, when we increase our altitude rapidly by driving up a mountain or by riding an elevator up to the top of a tall building, our ears pop because there are fewer molecules on the outside pushing in than there are on the inside pushing out. Once we are at the top of the mountain or building, the pressures are soon equalized inside and outside the ear, and the popping stops.

We sometimes say that the air is "thinner" at high attitudes, but we actually mean that there are fewer molecules per unit volume.

Marine life at the bottom of the ocean has become accustomed to living in an environment where the pressure is quite high. Even those of us on the surface of the Earth live at the bottom of an ocean—an ocean of air. We, too, are accustomed to living in an environment with considerable pressure—called atmospheric pressure—due to this blanket of air. **Atmospheric pressure** can be defined as the total force exerted by the molecules of air on each unit of area. This force is due to the

(a)

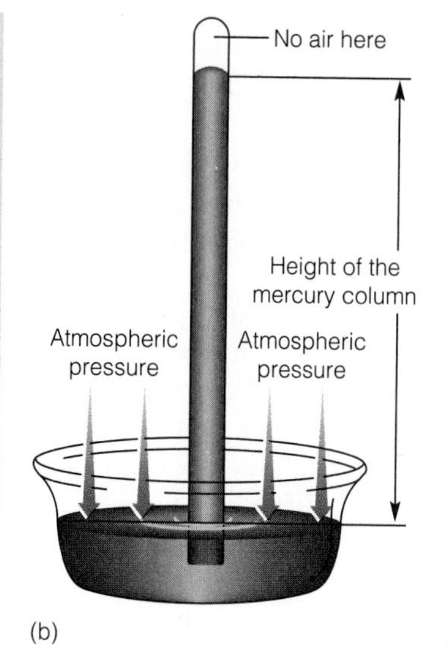

Figure 12.5 *(a) A scientific laboratory barometer. (b) Schematic representation of a mercury barometer.*

attraction of the Earth for the column of air that extends from the surface of the Earth to the outermost molecules in our atmosphere.

Pressure is defined as the force exerted per unit area.

$$\text{Pressure} = \frac{\text{Force}}{\text{Area}}$$

The pressure of the atmosphere is measured by a device called a **barometer**. The simplest type of barometer is made from a glass tube that is about 1 m long and closed at one end. The tube is filled completely with mercury and inverted in a shallow dish containing mercury (Figure 12.5). Initially, some of the mercury will drain out of the tube, but *not all* of it. The mercury will drain out only until the pressure exerted by the mercury remaining in the tube exactly equals the pressure exerted by the atmosphere on the surface of the mercury in the dish. We could say that the mercury in the tube is trying to flow out under the influence of gravity while the air pressure is pushing it back in. At some point they reach *equilibrium* (a stalemate with both forces in balance).

The air pressure can be determined by measuring the standing height of mercury in the tube (i.e., the difference between the mercury levels inside and outside the tube). This height, or difference, is often expressed in millimeters of mercury or inches of mercury, but other equivalent pressure units are also used.

Mercury is a liquid that is quite dense (13.6 g/cm^3); it is 13.6 times as dense as water. On the average, at sea level, a column of mercury 760. mm high has a force (per unit area) on the surface of the liquid in the dish that is equal to the force of the atmosphere on the surface of the liquid. Thus, the average atmospheric pressure at sea level is said to be 760. mm Hg. This pressure is defined as 1 **atmosphere** (atm) and is also the pressure used as the reference standard; it is called **standard pressure**. The pressure unit **millimeters of mercury** is also called **torr** (after the seventeenth-century Italian physicist Evangelista Torricelli, the inventor of the mercury barometer).

See Problems 12.1–12.8.

$$1 \text{ atm} = 760. \text{ mm Hg} = 760. \text{ torr}$$

A CLOSER LOOK

Gas Pressure Units

Several different pressure units are widely used. Each gas unit will be defined in terms of *exactly* 1 atmosphere of pressure.

Weather reports in the United States often include atmospheric pressure in inches of mercury.

$$1 \text{ atm} = 29.9 \text{ in. Hg} = 760. \text{ torr}$$

Engineers generally use pounds (of air) per square inch (psi).

$$1 \text{ atm} = 14.7 \text{ psi}$$

The preferred SI unit for pressure—now gaining wider usage—is the pascal (Pa), a unit that is interpreted in terms of newtons (N). (A newton is the force that, when applied to a mass of 1 kg, will give that mass an acceleration of 1 m/s during each second.)

$$1 \text{ atm} = 101{,}325 \text{ Pa} \quad \text{or} \quad 101{,}325 \text{ N/m}^2$$
$$= 101.325 \text{ kPa}$$

12.4 *Boyle's Law: The Pressure–Volume Relationship*

The relationship between the volume and the pressure of a gas was first determined in 1662 by Robert Boyle (Figure 12.6), an Irish chemist and physicist. Using a J-tube apparatus similar to that shown in Figure 12.7, Boyle found that the volume

Figure 12.6 *Robert Boyle (1627–1691) was born in Ireland. He developed a pump that he used in producing evacuated containers and showed that sound cannot be produced in a vacuum. He is especially remembered for his experiments demonstrating that gas pressure times volume equals a constant (Boyle's law).*

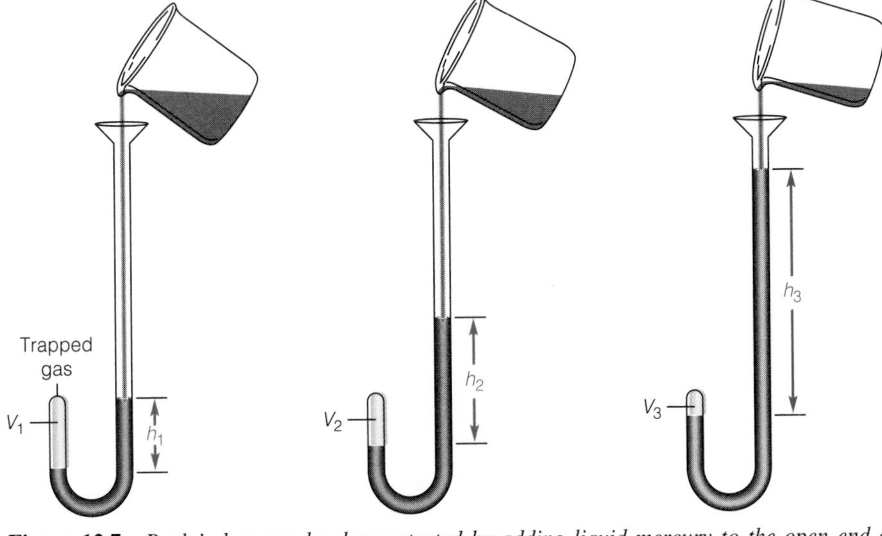

Figure 12.7 *Boyle's law can be demonstrated by adding liquid mercury to the open end of a J-tube. As the pressure is increased by addition of mercury, the volume of the sample of trapped gas decreases. Gas pressure and volume are inversely related; one increases when the other decreases.*

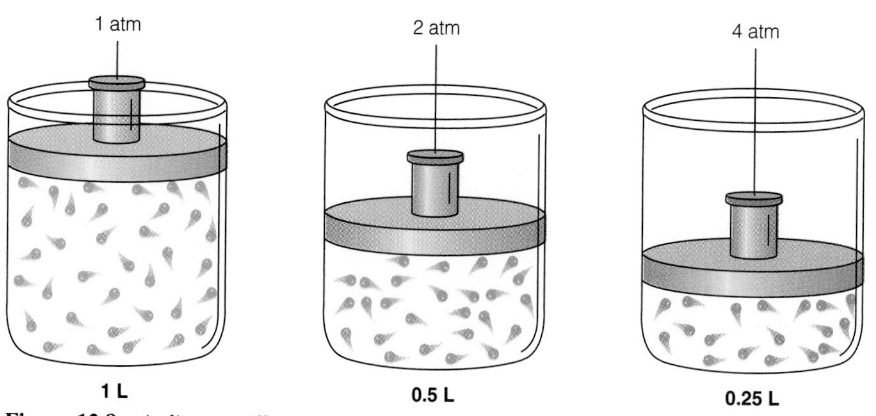

Figure 12.8 *A diagram illustrating the effect of different pressures on the volume of a gas.*

of a sample of trapped gas decreases as the external pressure increases when temperature is held constant (unchanged).

Each time more mercury is added to the open end of the J-tube (Figure 12.7), the pressure exerted on the small gas sample increases and the volume of the gas sample decreases. Eventually, after enough mercury has been added to cause the pressure to be doubled, the volume of gas is found to be compressed to one-half of its original volume. Furthermore, when the external pressure is tripled, the volume of a gas sample is decreased to one-third, and when the pressure is quadrupled, the volume of a gas is decreased to one-fourth of its original volume (Figure 12.8). Volume and pressure are said to be inversely related; one component decreases while the other increases. A graph of experimental data involving pressure and volume changes for a fixed sample of gas at constant temperature is shown in Figure 12.9. Notice the shape of the curve for this inverse relationship.

As Robert Boyle showed, the inverse relationship involving pressure and volume applies to all gases. The law stating this relationship is named in his honor.

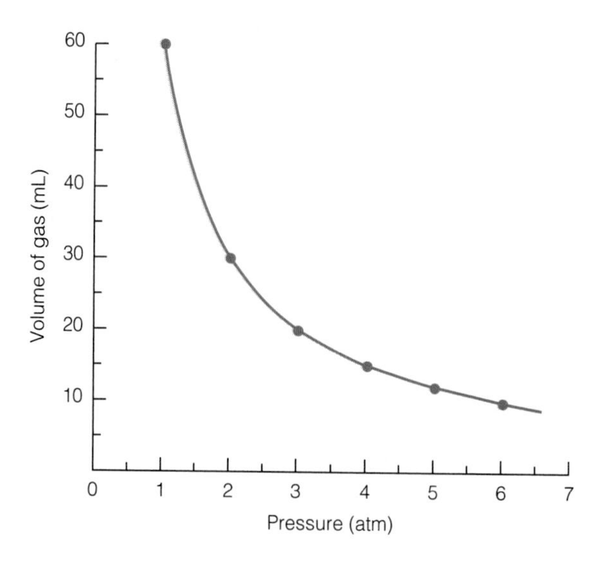

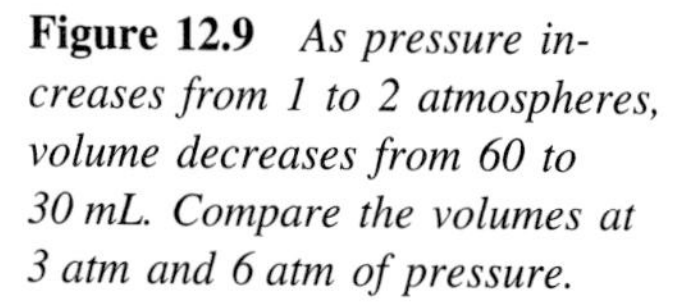

Figure 12.9 *As pressure increases from 1 to 2 atmospheres, volume decreases from 60 to 30 mL. Compare the volumes at 3 atm and 6 atm of pressure.*

Boyle's law: At constant temperature, the volume, V, occupied by a gas sample is inversely proportional to the pressure, P.

$$V \propto \frac{1}{P} \qquad \text{(with constant temperature and mass)}$$

The $\propto$ symbol means "is proportional to." The mathematical expression for Boyle's law is read "V is proportional to 1 over P." This means that V is *inversely proportional* to P. The inverse proportion can be changed to an equation by including a *proportionality constant, k.*

$$V = \frac{k}{P}$$

Multiplying both sides by P gives

$$PV = k$$

This is an elegant and precise, if somewhat abstract, way of summarizing a lot of experimental data. If the product of V times P is to be a constant, then V must decrease as P increases, and vice versa. We can actually make greater use of this equation if we consider a case where a specific sample of gas with a volume V_1 is compressed to a smaller volume, V_2. The initial pressure, P_1, is increased to a new value, P_2. For the initial and final conditions we can write two equations.

$$P_1V_1 = k_1$$

$$P_2V_2 = k_2$$

As long as we are using the *same* sample of trapped gas at a constant temperature, k_1 is equal to k_2, and the product of the initial pressure times the initial volume is equal to the product of the final pressure times the final volume. Thus, the following useful equation representing Boyle's law can be written.

$$P_1V_1 = P_2V_2$$

Solving this equation for the final volume, V_2, gives

$$V_2 = V_1 \times \frac{P_1}{P_2}$$

That is, when the pressure of a sample of gas is changed, the new volume can be calculated by multiplying the initial volume by a ratio of the two pressure values. If the final pressure is greater than the initial pressure, the volume must decrease. Thus, the original volume must be multiplied by a ratio of the pressures with a value that is less than 1—the numerator, P_1, is *smaller* than the denominator, P_2. However, if the final pressure is less than the initial pressure, the volume must increase—the pressure ratio must have a value that is greater than 1—the numerator, P_1, is *greater* than the denominator, P_2.

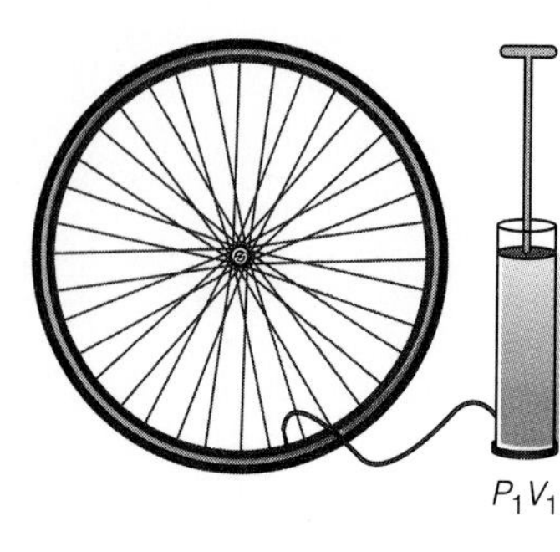

When both volumes are known, the final pressure can be calculated by multiplying the initial pressure by the appropriate volume ratio.

$$P_2 = P_1 \times \frac{V_1}{V_2}$$

For all calculations involving Boyle's law, remember that the pressure or volume ratio needed is the one that maintains the inverse proportion.

A Boyle's Law Model

It is generally easier to understand a particular principle if you can think of a familiar example, a model, that demonstrates the principle. A simple bicycle tire pump effectively demonstrates Boyle's law—when the volume of gas in the cylinder of the tire pump is decreased by pushing on the piston, the gas pressure increases (Figure 12.10).

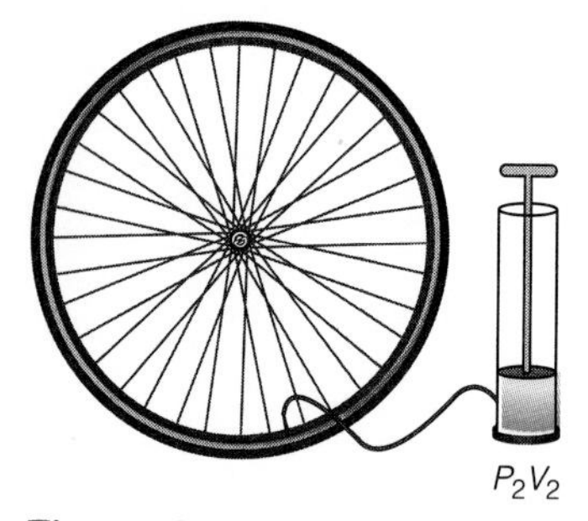

Figure 12.10 *A simple bicycle tire pump demonstrates Boyle's law. A pressure increase from* P_1 *to* P_2 *causes a volume decrease from* V_1 *to* V_2*. Pressure and gas volume are inversely related.*

Boyle's Law Theory

To get a mental picture of what is happening—at the molecular level—during a change in pressure and volume for a fixed sample of gas, think of gases as described by the kinetic molecular theory. A sample of gas in a container exerts a certain pressure because the particles are bouncing against the walls at a certain rate with a certain force. If the volume of the container is decreased, the particles will have to travel shorter distances before they strike the walls. Also, the surface area of the walls decreases as the volume decreases, so each unit of area is struck by more particles per unit of time. To make it simple, when the number of strikes increases, the pressure increases.

Boyle's law has a number of practical applications that can be illustrated by a series of examples. Any appropriate pressure and volume units can be used in working problems involving Boyle's law, as long as the same units are used for both initial and final conditions.

EXAMPLE 12.1

A cylinder of oxygen has a volume of 2.00 L. The pressure of gas is 1470 psi at 20 °C. What volume will the oxygen occupy at standard atmospheric pressure (14.7 psi), assuming no temperature change?

SOLUTION

First, make a list of the original and final conditions.

Initial	*Final*
$P_1 = 1470$ psi	$P_2 = 14.7$ psi
$V_1 = 2.00$ L	$V_2 = ?$ L

Direct substitution of these values into the Boyle's law equation is one method of determining the final volume. You can also use the reasoning approach, as follows. You are asked for the volume that results when the pressure is changed from 1470 psi to 14.7 psi. The pressure has been decreased; therefore, the volume must increase. Write down the initial volume, 2.00 L, and multiply this volume by a fraction made up of the two pressures. To make the fraction greater than 1 (so the volume will increase), place the larger pressure value in the numerator (on top).

$$V_2 = 2.00 \text{ L} \times \frac{1470 \text{ psi}}{14.7 \text{ psi}} = 200. \text{ L} \qquad \text{(answer)}$$

Gases are stored for use under high pressure, even though they will be used at atmospheric pressure. Using this arrangement, much oxygen, or any other gas, can be stored in a small tank.

EXAMPLE 12.2

A space capsule is equipped with a tank of air that has a volume of 0.100 m^3. The air is under a pressure of 100. atm. After a space walk, during which the cabin pressure is reduced to zero, the cabin is closed and filled with the air from the tank. What will be the final pressure if the volume of the capsule is 12.5 m^3?

SOLUTION

First, make a list of the original and final conditions.

Initial	*Final*
$P_1 = 100.$ atm	$P_2 = ?$ atm
$V_1 = 0.100$ m^3	$V_2 = 12.5$ m^3

Since the volume in which the air is confined increases, the pressure must decrease. Write down the initial pressure, 100. atm, and multiply this pressure by a fraction made up of the two volumes. Use the volumes, in cubic meters, to make a multiplier that is a fraction with a value less than 1 (pressure decreases).

$$P_2 = 100. \text{ atm} \times \frac{0.100 \text{ m}^3}{12.5 \text{ m}^3} = 0.800 \text{ atm} \qquad \text{(answer)}$$

Double-check your work. For this problem, P_2 must be less than P_1.

See Problems 12.23–12.30.

EXERCISE 12.1

A weather balloon is partially filled with helium gas. On the ground, where the atmospheric pressure is 740. torr, the volume of the balloon is 10.0 m^3. What will the volume be when the balloon reaches an altitude of 5300 m, where the pressure is 370. torr, assuming the temperature is constant?

12.5 Charles's Law: The Volume–Temperature Relationship

In 1787, the French physicist Jacques A. C. Charles studied the relationship between the temperature and volume of gases. At that particular time hot-air ballooning was getting a lot of attention in France, and Charles was one of the pioneer balloonists. He is credited with being the first to use hydrogen gas to inflate a balloon for carrying people. Charles, however, did more about his interests than other balloon enthusiasts—he proceeded to carry out scientific investigations involving the effect of temperature on the volume of a gas.

When gas is cooled at constant pressure, its volume decreases. When the gas is heated, its volume increases. Temperature and volume are directly proportional; that is, they increase or decrease together. The relationship, however, requires a bit more thought. If 1 L of gas is heated from 100. °C to 200. °C at constant pressure, the volume does not double but only increases to about 1.3 L. This volume and temperature relationship is not as simple as we might like.

Remember how temperature scales are defined? While a pressure of zero—no matter what units are used—means there is no pressure to be measured, and a volume of zero means there is no volume to be measured, a temperature of zero degrees Celsius, you will recall, is simply a mark arbitrarily placed on the thermometer at the freezing point of water (0 °C). The Celsius scale does not end at this designated zero point; it extends in both directions to allow for the measurement of colder as well as warmer temperatures.

Figure 12.11 shows a graph of data similar to the data obtained by Charles during his experiments with the effect of temperature on the volume of a gas sample at a fixed pressure. The fact that the plotted points lie in a straight line indicates that gas volume and temperature are *directly proportional*. Compare this graph for a directly proportional relationship with the pressure–volume graph that shows an inversely proportional relationship.

The line through the points plotted in Figure 12.11 can be extended, in theory, to lower temperatures, even to the point at which the volume of the gas hits zero. Before a gas actually reaches this point it liquefies, so this is an exercise that we can only imagine. (When the line through the points on a graph is extended beyond the limits of actual experimental data, it is said to be *extrapolated* and is shown on the graph as a dashed line.) From the extrapolated line on the volume–temperature

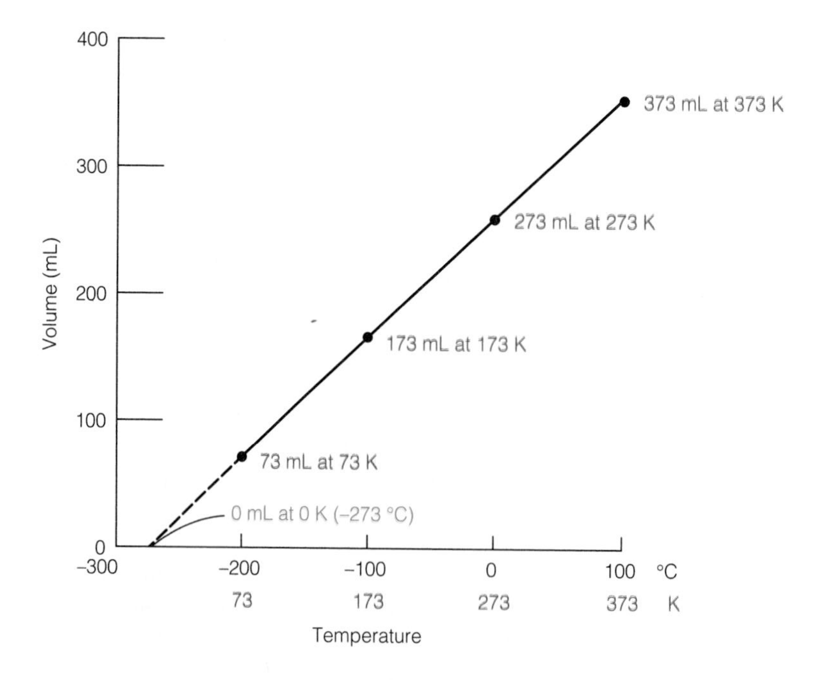

Figure 12.11 *The volume–temperature relationship for gases at constant pressure.*

graph, you can determine the temperature at which the volume of the gas would reach zero—assuming the gas does not liquefy. That temperature is -273.15 °C and is known as absolute zero; it was used as the zero point on the absolute scale with temperatures measured in kelvins, K (see Section 3.12).

Charles compared changes in the volume of a gas sample with changes in temperature on the *absolute scale*. When he did this, he found that causing the temperature to be doubled (from 100. K to 200. K, for example) also caused the volume to be doubled (from 1 L to 2 L, for example). Here is the simple relationship we were looking for. The law stating this relationship is known as Charles's law.

Charles's law: At constant pressure, the volume occupied by a gas sample is directly proportional to its Kelvin temperature.

$$V \propto T \qquad \text{(with constant pressure and mass)}$$

The direct proportion shown in Charles's law can be changed to an equation by including a proportionality constant, k.

$$V = kT$$

Dividing both sides by T gives

$$\frac{V}{T} = k$$

From this equation we can write two equations, one for initial conditions using V_1 and T_1, and the other for the final conditions using V_2 and T_2.

$$\frac{V_1}{T_1} = k_1 \qquad \text{and} \qquad \frac{V_2}{T_2} = k_2$$

As long as we are working with the same sample of gas at constant pressure, k_1 is equal to k_2, and the V/T ratios are, therefore, also equal. Thus, we can write the following useful equation representing Charles's law.

$$\frac{V_1}{T_1} = \frac{V_2}{T_2}$$

Solving this equation for the final volume, V_2, gives

$$V_2 = V_1 \times \frac{T_2}{T_1}$$

As indicated here, when the temperature of a specific gas sample at constant pressure is changed, the new volume can be calculated by multiplying the initial volume by a ratio of the two Kelvin temperatures. Remember that volume and temperature are proportional—when the Kelvin temperature increases, the volume also increases and vice versa. Thus, when temperature increases, we use the ratio of Kelvin temperatures that is greater than 1, and when temperature decreases, we use the ratio of Kelvin temperatures that is less than 1.

Kelvin temperatures (not °C or °F) must be used for all gas law calculations.

Figure 12.12 *As liquid nitrogen (−196 °C) is poured over a balloon, the gas trapped in the balloon is cooled and the volume decreases while pressure remains constant. This is in agreement with Charles's law.*

When we know both the initial and final volumes, but wish to calculate the final temperature, T_2, we can multiply the initial temperature by the appropriate volume ratio needed to maintain the direct proportion.

$$T_2 = T_1 \times \frac{V_2}{V_1}$$

For all calculations involving Charles's law, remember that the volume or temperature ratio needed is the one that maintains the direct proportion.

A Charles's Law Model

A simple balloon filled with any gas can serve as a simple model of Charles's law. The external pressure exerted on the balloon by the air is constant. When the balloon of gas is heated, it expands—volume increases. When the balloon of gas is cooled, it gets smaller—volume decreases. Pouring liquid nitrogen (−196 °C) over a balloon filled with air is a dramatic way to show that the volume decreases when the temperature decreases (Figure 12.12).

Charles's Law Theory

The kinetic molecular theory gives us a model of what is happening at the molecular level during a change in temperature and volume for a specific gas sample at constant pressure. When we heat a gas, we supply it with energy (K.E. = $\frac{1}{2}mv^2$), and the particles of the gas begin moving faster—mass *(m)* does not change, but velocity *(v)* increases. These speedier particles strike the walls of their container more frequently. If the *pressure* is to remain constant, the volume of the container must increase. Increased volume means the particles have more space to move around in; they will take longer to travel from one wall to another. In addition, the increased wall surface area means each unit of area will be hit less frequently.

For gas pressure to remain constant, the volume of the container must expand just enough to compensate for the extra energy of warmer gas particles. Within a balloon, for example, the pressure exerted by slowly moving (low temperature) particles confined within a small volume is the same as the pressure of faster moving (higher temperature) particles contained within a larger volume.

EXAMPLE 12.3

A balloon, indoors at a temperature of 27 °C, has a volume of 2.00 L. What will its volume be outside where the temperature is −23 °C? (Assume no change in pressure—the atmospheric pressure is constant.)

SOLUTION

First make a list of the original and final conditions.

Initial	*Final*
$V_1 = 2.00$ L	$V_2 = ?$ L
$T_1 = 27°$ C	$T_2 = -23$ °C

All temperatures must be changed to kelvins by adding 273 to the Celsius temperature.

$$T_1 = 27 + 273 = 300.\ \text{K}$$

$$T_2 = -23 + 273 = 250.\ \text{K}$$

Since the temperature decreases, the volume must also decrease.

$$V_2 = 2.00\ \text{L} \times \frac{250.\ \text{K}}{300.\ \text{K}} = 1.67\ \text{L} \qquad \text{(answer)}$$

See Problems 12.31–12.38.

Exercise 12.2
What would be the final volume of the balloon in the example if it were measured where the temperature was 47 °C? (Assume no change in pressure.)

12.6 *Gay-Lussac's Law: The Pressure–Temperature Relationship*

At about the same time that Charles was doing experiments involving gas temperature and volume, Joseph Gay-Lussac, a French chemist, was investigating the relationship between gas pressure and temperature. The law stating this relationship is known as Gay-Lussac's law.

Gay-Lussac's law: At constant volume, the pressure exerted by a specific gas sample is directly proportional to its Kelvin temperature.

$$P \propto T \qquad \text{(with constant volume and mass)}$$

As an equation, the direct proportion shown in Gay-Lussac's law becomes

$$P = kT$$

Dividing both sides by T gives

$$\frac{P}{T} = k$$

For a fixed volume and mass of gas, we can now write the following equation that relates the initial conditions, P_1 and T_1, with the final conditions, P_2 and T_2. Thus, Gay-Lussac's law can also be written

$$\frac{P_1}{T_1} = \frac{P_2}{T_2}$$

When a gas with an initial pressure, P_1, is involved in a temperature change from T_1 to T_2, the new pressure, P_2, can be determined by multiplying the initial pressure, P_1, by the ratio of Kelvin temperatures. Since pressure and temperature are proportional, the pressure increases when the Kelvin temperature increases, and vice versa.

$$P_2 = P_1 \times \frac{T_2}{T_1}$$

Also, multiplying the given Kelvin temperature by the appropriate pressure ratio allows us to calculate the required temperature when volume is constant.

$$T_2 = T_1 \times \frac{P_2}{P_1}$$

For all calculations involving Gay-Lussac's law, the pressure or Kelvin temperature ratio needed is the one that maintains the direct proportion.

A device to measure pressures at various temperatures can be made by attaching a pressure gauge to a rigid, closed container. The device can be placed in an oven at various temperatures or submerged in liquids at various temperatures to obtain data over a wide range of conditions. After recording and plotting the data on a graph, a straight line can be drawn through the plotted points. When this line is extended (extrapolated), the temperature corresponding to a pressure of zero can be determined. This temperature, called absolute zero, is found to be −273 °C, which is the same value obtained by using the Charles's law plot of gas volume and temperature.

A Gay-Lussac's Law Model

Examples that illustrate Gay-Lussac's law are all around us. Simply pick up a pressurized can of paint, hair spray, deodorant, bathroom cleaner, or any other product sold in a pressurized can. A typical *Caution* statement on this type of can is likely to be worded

Contents under pressure. Do not incinerate container.
Do not expose to heat or store at temperatures above 120 °F.

The reason for such caution statements lies in the relationship of temperature and pressure, as described by Gay-Lussac's law.

Since an increase in temperature causes an increase in pressure for a gas sample at a constant volume, an explosion can result from the heating of a closed container—especially one that is already under considerable pressure at room temper-

HAIR SPRAY
Contents under pressure. Do not incinerate container. Do not expose to heat or store at temperatures above 120° F.
(a)

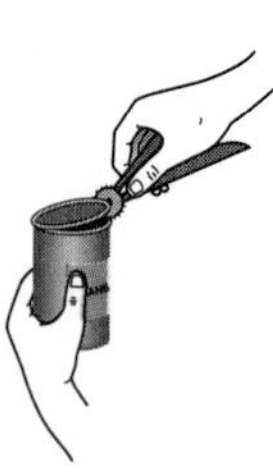
(b)

(c)

Figure 12.13 *A pressurized can of hair spray or deodorant has a constant volume. Pressure increases as temperature increases. Heating the can could cause it to explode or rupture. A can of food must be punctured or opened before heating directly over a campfire. These examples illustrate Gay-Lussac's law; pressure and temperature are directly proportional.*

ature. The same principle is learned by the Girl Scout or Boy Scout who is instructed to put a hole in the lid of a can of food before placing it directly over a campfire. Again, the pressure of the trapped gas within the can increases with temperature (Figure 12.13).

Gay-Lussac's Law Theory

The kinetic molecular theory again gives us a model of what is happening at the molecular level during the heating of a specific quantity of gas with a fixed volume. As the temperature increases at constant volume, the average kinetic energy of the particles increases, and the molecules move faster. The fast-moving particles bombard the walls of the container more frequently, and with greater force. Thus, pressure—the force per each unit area—increases as temperature increases when volume is held constant.

EXAMPLE 12.4

When a 17-ounce aerosol spray can with a pressure of 850. torr at 21 °C is thrown into a fire with a temperature of 450. °C, what pressure can be reached if the can does not burst?

SOLUTION

First, make a list of the original and final conditions.

Initial	*Final*
V_1 = 17 oz.	V_2 = 17 oz. (unchanged)
P_1 = 850. torr	P_2 = ? torr
T_1 = 21 °C	T_2 = 450. °C

All temperatures must be expressed in kelvins. Add 273 to the Celsius temperature.

$$T_1 = 21 + 273 = 294 \text{ K}$$

$$T_2 = 450. + 273 = 723 \text{ K}$$

Because the temperature increases, the pressure must also increase.

$$P_2 = 850. \text{ torr} \times \frac{723 \text{ K}}{294 \text{ K}} = 2090 \text{ torr} \qquad \text{(answer)}$$

The air pressure within an automobile tire is also related to its temperature, as described by Gay-Lussac's law. If we assume that the volume of air within an automobile tire remains constant, an increase in temperature must result in an increase in pressure. A simple tire pressure gauge can be used to measure these pressures, but the gauge pressure only measures the difference between the internal pressure and the external (atmospheric) pressure. Thus, the actual pressure in psi (pounds per square inch) is equal to the gauge pressure plus 14.7 psi. Be sure to make this pressure adjustment when working problems that provide tire pressure gauge measurements.

EXAMPLE 12.5

Before taking a trip, an automobile tire had a gauge pressure of 32.0 psi at 20. °C. After several hours of driving, the pressure was again checked. The new gauge pressure was 36.0 psi. How hot (°C) was the air in this tire?

SOLUTION

First, make a list of the original and final conditions.

Initial	*Final*
$P_1 = 32.0 + 14.7 = 46.7$ psi	$P_2 = 36.0 + 14.7 = 50.7$ psi
$T_1 = 20. + 273 = 293$ K	$T_2 = ?$ °C

Caution: Tire gauges measure the difference between the pressures inside and outside of the tire. You must add atmospheric pressure (14.7 psi) to a tire gauge pressure to obtain the total pressure shown in our table of values. Multiply the initial temperature (in kelvins) by the ratio of pressures that results in a temperature increase.

$$T_2 = 293 \text{ K} \times \frac{50.7 \text{ psi}}{46.7 \text{ psi}} = 318 \text{ K}$$

See Problems 12.39–12.48.

$T_2 = 318 \text{ K} - 273 = 45$ °C (answer valid to two significant figures)

EXERCISE 12.3

If the gauge pressure on the same tire was later found to be 34.0 psi, how hot was the air in the tire?

12.7 *Standard Temperature and Pressure*

Since the volume of a gas is sensitive to both pressure and temperature, a small volume of a gas at a low temperature and high pressure may have the same mass as a large volume of the same gas at a higher temperature and lower pressure. Thus, if the volumes of any two gas samples are to be compared, they must be at the same set of conditions. The conditions used for reference are called **standard conditions**. Standard temperature, by definition, is 273 K (0 °C). Standard pressure is defined as 1 atm (760. torr). Together, these standard conditions are referred to as **standard temperature and pressure** (or **STP**).

EXAMPLE 12.6

Give the appropriate value for each of the following units at STP.

(a) pressure in atmospheres
(b) pressure in torr
(c) pressure in psi
(d) temperature in kelvins
(e) temperature in degrees Celsius

SOLUTION

Each of the pressures must be equivalent to 1 atm, standard pressure. Each of the temperatures must be equivalent to 273 K, standard temperature.

(a) 1 atm (b) 760 torr (c) 14.7 psi (d) 273 K (e) 0 °C

12.8 The Combined Gas Law

For each gas law that has been described in this chapter, pressure, volume, or temperature was held constant along with the amount of gas, but in many actual situations this is not practical. For a specific gas sample, it is possible to relate initial and final pressure, volume, and temperature variations by using a single equation. Since the following equation combines all variables represented in Boyle's law, Charles's law, and Gay-Lussac's law, it is known as the **combined gas law**.

Combined gas law: $$\frac{P_1V_1}{T_1} = \frac{P_2V_2}{T_2}$$

In this equation six different variables are represented. As long as values are known for five of the six variables, the missing quantity can be determined by solving the equation. Notice that when the temperatures T_1 and T_2 are equal (constant), the temperatures actually drop out of the equation—we have reduced the equation to Boyle's law. When the pressures P_1 and P_2 are equal, the pressures drop out of the equation—we have reduced the equation to Charles's law. Also, when the volumes V_1 and V_2 are equal, the volumes drop out of the equation—we have reduced the equation to Gay-Lussac's law. Thus, when we are dealing with a *specific gas sample*—with no change in mass or number of moles—we can use the combined gas law equation to determine any one of the six variables if the other five values either are known or are held constant.

If the final volume, V_2, is to be determined, we can solve the equation for V_2, as shown here, and substitute the known values into the equation.

Study Hint
Regardless of which variable is to be determined, it is highly recommended that you solve the equation for the unknown *before* substituting values into the equation. Rearranging a few terms takes less time than rearranging and rewriting complex terms.

$$V_2 = V_1 = \frac{P_1}{P_2} \times \frac{T_2}{T_1}$$

We have a choice. Instead of using the combined gas law equation approach for determining the unknown pressure, volume, or temperature, we can use a two-step reasoning approach. To illustrate, let us again assume the final volume, V_2, is to be determined. Start with the known volume, V_1, and multiply by the appropriate ratio of pressures and also multiply by the appropriate ratio of temperatures. Keep in mind the inverse and direct proportions involved. We arrive at the same final equation for V_2 regardless of whether we use this reasoning approach or whether we solve the combined gas law equation for V_2.

EXAMPLE 12.7

A balloon is partially filled with helium on the ground at 22 °C at a pressure of 740. torr. At these conditions the volume is 10.0 m^3. What would be the volume (in cubic meters) at an altitude of 5300 m, where the pressure is 370. torr and the temperature is −23 °C?

SOLUTION

First, make a list of the original and final conditions.

Initial	*Final*
P_1 = 740. torr	P_2 = 370. torr
V_1 = 10.0 m^3	V_2 = ? m^3
T_1 = 22 + 273 = 295 K	T_2 = −23 + 273 = 250 K

Using the reasoning approach, begin with the initial volume and multiply by the ratio of pressures. (The pressure decreases; the ratio must lead to a volume increase.) Set up this part of the problem, but wait to do the calculations.

$$V_2 = 10.0\ \text{m}^3 \times \frac{740.\ \text{torr}}{370.\ \text{torr}} \qquad \text{(incomplete)}$$

Now, multiply by a ratio of the Kelvin temperatures. How will the initial volume be affected by the decrease in temperature? Since volume and temperature are directly proportional, the ratio of temperatures must lead to a volume decrease.

$$V_2 = 10.0\ \text{m}^3 \times \frac{740.\ \text{torr}}{370.\ \text{torr}} \times \frac{250.\ \text{K}}{295\ \text{K}} = 16.9\ \text{m}^3 \qquad \text{(answer)}$$

The order in which you use the pressure and temperature ratios does not matter. You could also substitute the appropriate values into the equation

The order in which terms are multiplied does not matter.

$$V_2 = V_1 \times \frac{T_2}{T_1} \times \frac{P_1}{P_2}$$

EXAMPLE 12.8

Determine the volume at STP for a sample of carbon dioxide that has a volume of 10. L at 25 °C and a pressure of 4.0 atm.

SOLUTION

The list of the original and final conditions is as follows.

Initial	*Final*
P_1 = 4.0 atm	P_2 = 1.0 atm at STP
V_1 = 10. L	V_2 = ? L
T_1 = 25 + 273 = 298 K	T_2 = 273 K at STP

Multiply the initial volume by the ratio of pressures. The pressure decrease from 4.0 atm to 1.0 atm (standard pressure) should cause a volume increase. Also, multiply by the ratio of temperatures. The temperature decrease should be accompanied by a volume decrease.

See Problems 12.45–12.48.

$$V_2 = 10.\ \text{L} \times \frac{4.0\ \text{atm}}{1.0\ \text{atm}} \times \frac{273\ \text{K}}{298\ \text{K}} = 37\ \text{L} \qquad \text{(answer)}$$

EXERCISE 12.4

Determine the volume at STP for a sample of helium that has a volume of 4.50 L at 21 °C and 744 torr pressure.

12.9 *Molar Volume and Gas Density at STP*

Densities of gases usually are reported in the literature in grams per liter at STP. Recall (Section 9.5) that there are 6.02×10^{23} molecules in a mole of any chemical—solid, liquid, or gas. Recall also that the mass of a mole (the molar mass) of gas is the formula weight expressed in grams. If we begin with the molar mass (grams per mole) and use the gas density at STP (grams per liter) as a conversion factor, we can determine the number of liters occupied by a mole of the gas at STP. For example, the molar mass of nitrogen gas is 28.0 g/mol. Its density at STP is 1.25 g/L. We determine that its molar volume is

$$\frac{28.0\ \text{g}}{\text{mol}} \times \frac{1\ \text{L}}{1.25\ \text{g}} = 22.4\ \text{L/mol for nitrogen at STP}$$

Notice that the density term is inverted to put liters in the numerator and grams in the denominator. Multiplying eliminates grams.

For oxygen, O_2, the density at STP is 1.43 g/L. The molar mass is 32.0 g/mol. Its molar volume is

$$\frac{32.0\ \text{g}}{\text{mol}} \times \frac{1\ \text{L}}{1.43\ \text{g}} = 22.4\ \text{L/mol for oxygen at STP}$$

In fact, the volume occupied by a mole of *any* gas at STP is quite close to 22.4 L. This value can also be verified by laboratory investigations. Thus, 22.4 L is known as the **molar volume** of a gas at STP.

For a rough comparison, the box in which a basketball fits snugly has a volume close to 22.4 L. To be more precise, the box should be 28.2 cm (11.1 in.) on each edge to have a total volume of 22.4 L (Figure 12.14). No matter what gas we are

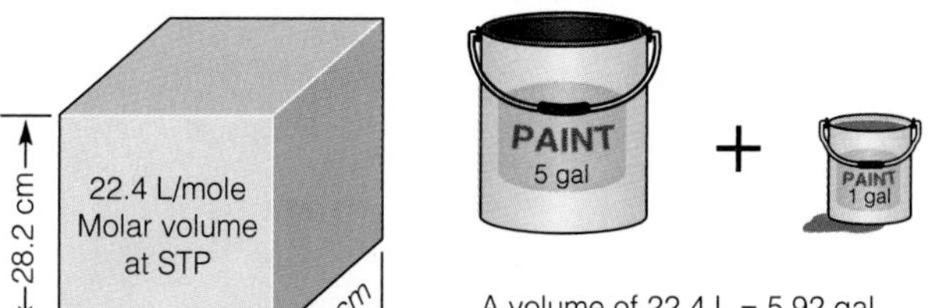

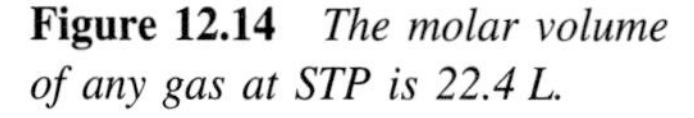

Figure 12.14 *The molar volume of any gas at STP is 22.4 L.*

considering, a mole of the gas occupies the same volume as a mole of any other gas under the same conditions. We have arrived at the following very useful constant.

The molar volume of any gas at STP is 22.4 L/mol.

Now, we can explain Gay-Lussac's law of combining volumes (Section 11.5). which states that a mole of one gas has the same volume as a mole of any other gas at the same conditions. If we begin with the molar mass (grams per mole) and use the volume of one mole (the molar volume) at STP as a conversion factor, we can determine the density of any gas at STP, as shown in the example.

EXAMPLE 12.9

Calculate the density of methane gas, CH_4, at STP.

SOLUTION

Begin with the molar mass of methane (16.0 g/mol) and use the molar volume (22.4 L/mol) as a conversion factor to find the gas density at STP.

See Problems 12.49–12.56.

$$\frac{16.0 \text{ g}}{\text{mol}} \times \frac{1 \text{ mol}}{22.4 \text{ L}} = 0.714 \text{ g/L density at STP}$$

EXERCISE 12.5

Calculate the density of chlorine gas, Cl_2, at STP.

Since the worked-out example shows that the density of any gas at STP can be obtained by multiplying the molar mass by a fixed value (1 mol/22.4 L), we can conclude that the molar mass is proportional to the density of the gas. Thus, a gas with high molar mass will also have a high density, and vice versa.

12.10 *The Ideal Gas Law*

Our gas calculations have all dealt with a fixed quantity of gas, where pressure, volume, and temperature are the only variables, but it is also possible to calculate the number of grams, moles, or the molar mass for a gas sample. Let us go back to the basic gas laws. From Boyle's law we know that $PV = k$, where k is a constant for a fixed mass and temperature. Also, from the combined gas law we can write

$$\frac{PV}{T} = k \qquad \text{(for a fixed mass of a gas)}$$

Again, we have a constant, k, for a fixed sample of gas. But if pressure and temperature are held constant while we put in twice as many molecules, the volume (left-hand side) must double, and the value on the right-hand side of the equation must also be doubled; we need a multiplier on the right-hand side of the equation.

Recall from the Avogadro hypothesis (Section 11.5) that equal volumes contain the same number of molecules and moles of gas at the same conditions (constant P and T). Thus, two volumes of a gas have twice as many moles and molecules as are contained in one volume of the gas (constant P and T). The **Avogadro hypothesis** can be stated as follows: volume, V, is proportional, $\propto$, to the number of moles, n, of gas at constant pressure and temperature.

$$V \propto n \quad \text{or} \quad V = kn \quad \text{(constant pressure and temperature)}$$

The multiplier we need on the right-hand side of our earlier equation ($PV/T = k$) is n, representing the number of moles. When volume doubles on the left side, the number of moles on the right side also doubles. We can now rewrite the equation with n on the right-hand side along with the **universal gas constant**, R, in place of k.

$$\frac{PV}{T} = nR$$

This equation is called the **ideal gas law** and is usually written in linear form.

Ideal gas law: $PV = nRT$

With this equation, we can determine the number of moles, n, when P, V, and T are given, but we must also know the value of R, the gas constant. We could simply tell you the value of R, but you will understand it more fully if we show you how the value of R can be readily determined. First, rearrange the equation to solve for R, and then simply use the values that we know for one mole of gas at STP, as follows.

$$R = \frac{PV}{nT}$$

$$= \frac{1 \text{ atm} \times 22.4 \text{ L}}{1 \text{ mol} \times 273 \text{ K}} \quad \text{(We use known values for 1 mol of gas at STP.)}$$

$$R = 0.0821 \frac{\text{L-atm}}{\text{mol-K}} \quad \text{(L-atm means liters} \times \text{atm, and mol-K means moles} \times \text{K.)}$$

The R value is read as 0.0821 liter-atmospheres per mole-kelvin.

Now that we have a value for R, we can use the equation to determine P, V, T, or n for any specific gas sample at a fixed set of conditions.

While the equation is quite useful, there are limitations—calculations must be based on the assumption that we are dealing with an ideal gas.

An **ideal gas** is defined as a gas that perfectly conforms to the ideal gas law—and the other gas laws—under all conditions.

Real gases (gases that actually exist) do not conform perfectly to these gas laws because they are composed of molecules that actually have volume as well as small attractive forces that are not factored into the equations. The deviation from ideal conditions becomes quite significant at high pressures or very low temperatures, where molecules are close together. At these conditions, molecular attractions are increased and the volume of the molecules becomes a significant fraction of the total volume. However, at ordinary pressures the ideal gas law is quite adequate for most calculations.

EXAMPLE 12.10

Use the ideal gas law to calculate the volume occupied by 1.00 mol of nitrogen gas at 25 °C and 1.00 atm pressure.

SOLUTION

Rearrange the ideal gas law equation ($PV = nRT$) to solve for V and substitute appropriate values into the equation.

$$V = \frac{nRT}{P} \qquad \text{or, broken into factors:} \qquad V = \frac{n}{P} \times R \times T$$

$$V = \frac{1.00\ \text{mol}}{1.00\ \text{atm}} \times \frac{0.0821\ \text{L-atm}}{\text{mol-K}} \times 298\ \text{K} = 24.5\ \text{L} \qquad \text{(answer)}$$

Notice that we can cancel all units except liters. The answer is a volume in liters. The units for the gas constant, R, may look complex, but this is necessary if all units are to cancel properly.

The calculations for Example 12.10 show that a mole of nitrogen—or any gas—has a volume of 24.5 L at 25 °C and 1 atm. In other words, the molar volume of a gas at 25 °C and 1 atm is 24.5 L. Since the molar volume of any gas at STP is 22.4 L, we have again shown that a gas at a higher temperature must occupy a greater volume when other factors are held constant.

We can apply the ideal gas law to more situations if we make a substitution for the number of moles, n. Since the number of moles, n, can be determined by dividing the number of grams, g, by the molar mass, M_m, we can write

$$n = \frac{\text{Mass in grams}}{\text{Grams/mole}} = \frac{\text{Mass in grams}}{\text{Molar mass}} = \frac{g}{M_m}$$

Substituting g/M_m for n in the ideal gas law gives

$$PV = \frac{g}{M_m}RT$$

or, written as a linear equation.

$$M_m PV = gRT$$

With this modified ideal gas law equation, we can calculate the molar mass, M_m, pressure, P, volume, V, grams, g, or temperature, T, for any gas sample if all units are chosen appropriately to cancel with the units of the gas constant, R.

EXAMPLE 12.11

A 10.0-g quantity of liquid nitrogen was allowed to warm to room temperature. What volume will be occupied by the nitrogen gas at 20. °C and 1 atm?

SOLUTION

Solve the modified ideal gas law equation for V.

$$V = \frac{gRT}{M_m P} = \frac{g}{M_m} \times R \times \frac{T}{P}$$

Then, substitute the values into the equation. The nitrogen gas has a molar mass of 28.0 g/mol, and the temperature of 20. °C is 293 K.

$$V = \frac{10.0\text{ g}}{28.0\text{ g/mol}} \times \frac{0.0821\text{ L-atm}}{\text{mol-K}} \times \frac{293\text{ K}}{1\text{ atm}} = 8.59\text{ L} \quad \text{(answer)}$$

See Problems 12.57–12.64.

Exercise 12.6
What is the pressure (in atm) of a 10.0-g sample of nitrogen gas in a 2.00-L rigid container at 22.0 °C?

12.11 *Dalton's Law of Partial Pressures*

John Dalton is most renowned for his atomic theory (Section 4.7), but his wide range of interests also included the field of meteorology. While trying to understand the weather, he did a number of experiments on water vapor in the air. He found that when he added water vapor to dry air, the pressure exerted by the air would increase by an amount equal to the pressure of the water vapor. After numerous investigations, Dalton concluded that each of the gases in a mixture behaves independently of the other gases. Each gas exerts its own pressure. The total pressure of the mixture is equal to the sum of the **partial pressures** exerted by the separate gases (Figure 12.15). This is known as **Dalton's law of partial pressures**. Mathematically, it can be expressed as

$$P_{\text{total}} = P_1 + P_2 + P_3 + \cdots$$

where the terms on the right side refer to the partial pressures of gases 1, 2, 3, and so on.

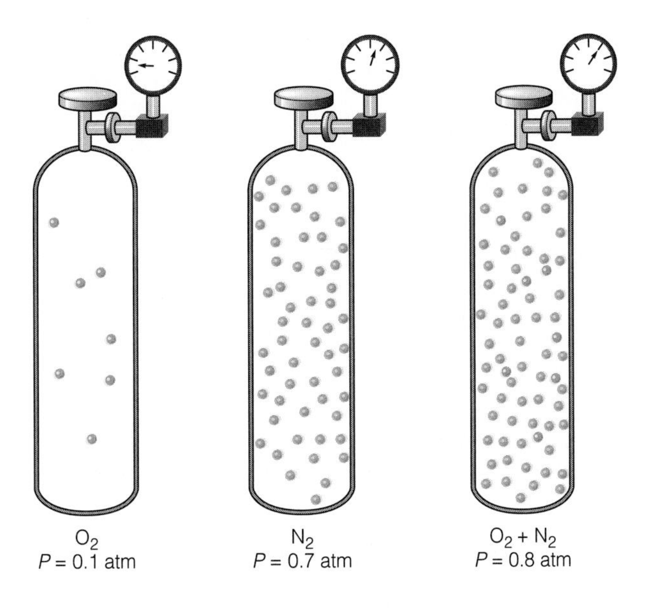

Figure 12.15 *Dalton's law of partial pressures states that the total pressure of a mixture of gases is equal to the sum of the partial pressures exerted by each gas.*

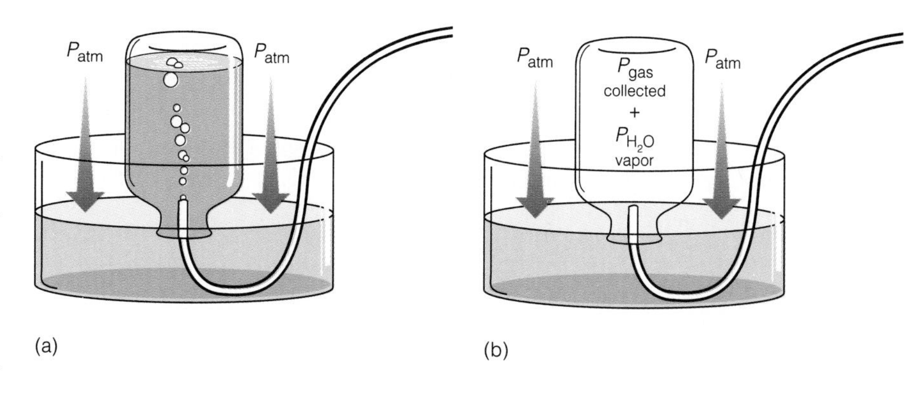

Figure 12.16 *To collect a gas by displacement of water, the bottle is filled completely with water and inverted in a large vessel of water (a). As the gas to be collected is released into the bottle, gas bubbles rise to fill the bottle as water in the bottle is displaced. When the water levels inside and outside the bottle are equal, pressures inside and outside the bottle are also equal (b). At this point* $P_{atm} = P_{gas\ coll'd} + P_{water\ vapor}$.

Gases such as oxygen, nitrogen, and hydrogen consist of nonpolar molecules. Since they are only slightly soluble in water, they are often collected over water by a technique called displacement of water (Figure 12.16). Gases collected in this manner are said to be "wet"—they contain water vapor. The total pressure in the collection vessel is equal to the sum of the pressure of the gas being collected plus the pressure of the water vapor. In mathematical form

$$P_{total} = P_{gas\ collected} + P_{water\ vapor}$$

The **vapor pressure** of a substance is the partial pressure exerted by the molecules of the substance that are in the gas phase above the liquid phase of the substance. The vapor pressure of water depends on the temperature of the water—the quantity of liquid water doesn't matter. The hotter the water, the higher is its vapor pressure. As long as the temperature of the water is known, we can use a table of vapor pressures (Table 12.2) to find the appropriate vapor pressure for the water. The partial pressure of the gas being collected is obtained by subtracting the vapor pressure of water from the total pressure within the collecting vessel.

$$P_{gas\ collected} = P_{total} - P_{water\ vapor}$$

Table 12.2 **Vapor Pressures of Water at Various Temperatures**

Temperature (°C)	*Vapor Pressure (torr)*	*Temperature (°C)*	*Vapor Pressure (torr)*
0	4.6	24	22.4
5	6.5	25	23.8
10	9.2	30	31.8
15	12.8	35	42.2
17	14.5	40	55.3
18	15.5	50	92.5
19	16.5	60	149.4
20	17.5	70	233.7
21	18.7	80	355.1
22	19.8	90	525.8
23	21.2	100	760.0

EXAMPLE 12.12

Oxygen gas is collected over water at 20. °C and at a barometric pressure of 744 torr. The water levels inside and outside the collecting bottle are equalized. What is the pressure of the dry oxygen (O_2 alone)?

SOLUTION

Since the water levels inside and outside the collecting bottle are equalized, the total pressure inside the bottle is the same as the barometric pressure. From Table 12.2 we find the vapor pressure of water at 20. °C is 17.5 torr (or 18 torr). We will apply Dalton's law of partial pressures.

$$P_{\text{gas collected}} = P_{\text{total}} - P_{\text{water vapor}}$$
$$= 744 - 18 = 726 \text{ torr}$$

The composition of dry air is listed in Table 12.1. According to Dalton's law of partial pressures, it doesn't matter what gases we are mixing (as long as they don't react); their partial pressures are added to give the total pressure. Since 78.0% of dry air is nitrogen, then 78.0% of the air pressure is due to the nitrogen in the air. If the total air pressure is 700. torr, then the partial pressure of nitrogen in the air is 78.0% times 700. torr or 546 torr.

EXAMPLE 12.13

Using the percentage of oxygen in dry air given in Table 12.1, determine the partial pressure of oxygen in an air sample at 738 torr.

SOLUTION

From Table 12.1 we obtain the percentage of oxygen in air (21.0%). Since the total pressure is 738 torr, the partial pressure of oxygen is

$$P_{\text{oxygen}} = 0.210 \times 738 \text{ torr} = 155 \text{ torr}$$

See Problems 12.65–12.70.

EXERCISE 12.7

Determine the partial pressure of nitrogen in air at 738 torr. (Air is 78.0% nitrogen.)

12.12 *Gas Stoichiometry: Putting It All Together*

In Chapter 11 we dealt with numerous types of stoichiometric calculations including the conversions of grams of A to moles of A to moles of B to grams of B, and stoichiometric relations where both the reactant and product in question were gases at the same temperature and pressure. Now, we can also handle problems involving conversions from liters of gas to moles. If the gas is at STP we can handle conversions from liters to moles by using the molar volume of a gas, 22.4 L/mole, but, if the gas is not at STP, the ideal gas law, $PV = nRT$, is used for such conversions.

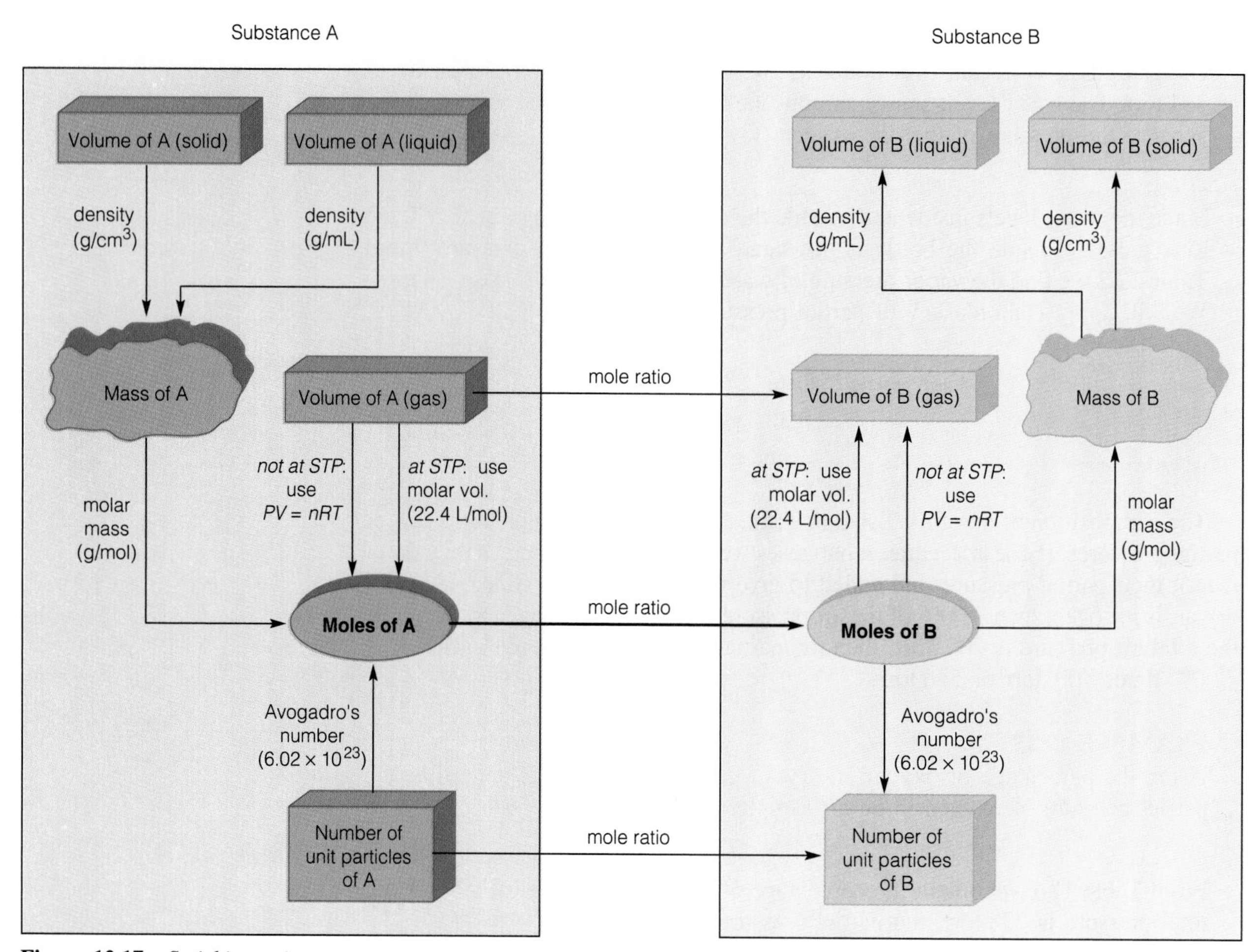

Figure 12.17 *Stoichiometric conversions for gases, liquids, and solids. Conversion factors are shown alongside arrows.*

Some important stoichiometric conversions are summarized in Figure 12.17. When you use appropriate conversions and equations, you can work problems involving reactants and products that are gases, liquids, solids, or solutions.

EXAMPLE 12.14

Some drain cleaners contain mixtures of NaOH (lye) with aluminum metal flakes. When this mixture of solids is added to water, hydrogen gas is produced. The drain-unclogging action is due partly to the pressure of the H_2 gas and partly to the agitating action of the H_2 gas, which mixes the NaOH with the grease.

$$2\,NaOH(s) + 2\,Al(s) + 6\,H_2O(l) \longrightarrow 2\,NaAl(OH)_4(aq) + 3\,H_2(g)$$

Determine the mass of aluminum in a drain cleaner sample that reacted with an excess of NaOH and water to produce 195 mL of hydrogen gas collected over water at a barometric pressure of 740. torr and 19 °C. The water levels inside and outside the collecting bottle were equalized.

SOLUTION

We can first determine the pressure of dry hydrogen by using Dalton's law of partial pressures. At 19 °C the vapor pressure of water is 16 torr.

$$P_{\text{outside}} = P_{\text{inside}} \quad \text{(when water levels are equal)}$$

$$740.\ \text{torr} = P_{\text{hydrogen}} + 16\ \text{torr}$$

Solve for P_{hydrogen}.

$$P_{\text{hydrogen}} = 740. - 16 = 724\ \text{torr}$$

Convert torr to atm.

$$P_{\text{hydrogen}} = 724\ \text{torr} \times \frac{1\ \text{atm}}{760.\ \text{torr}} = 0.953\ \text{atm}$$

Use the ideal gas law to find moles of H_2.

$$n = \frac{PV}{RT} = \frac{0.953\ \text{atm} \times 0.195\ \text{L}}{0.0821\ \text{L-atm/mol-K} \times 292\ \text{K}} = 7.75 \times 10^{-3}\ \text{mol}\ H_2$$

Use moles of H_2 to determine grams of Al.

Plan: Mol H_2 ⟶ Mol Al ⟶ Grams Al

$$7.75 \times 10^{-3}\ \text{mol}\ H_2 \times \frac{2\ \text{mol Al}}{3\ \text{mol}\ H_2} \times \frac{27.0\ \text{g Al}}{1\ \text{mol Al}} = 0.139\ \text{g Al} \quad \text{(answer)}$$

In Chapter 11, we were limited in our ability to solve stoichiometry problems; all gas quantities had to be expressed in grams or in number of moles. Now, with the ideal gas law, we can convert from moles to gas volume (and vice versa) at any specified conditions. Thus, besides using gas laws to deal with changes in P, V, and T, our ability to deal with a wide range of stoichiometry problems has also been expanded.

Chapter Summary

Gases have no definite shape. They are compressible. They have low densities. They exert uniform pressure. They mix spontaneously. Each of these physical properties of gases can be explained by the kinetic molecular theory, which states that gases are composed of tiny particles (separated by large distances) that move continuously, rapidly, and randomly. Gas particles have no significant attractive forces, their collisions are perfectly elastic, and all gases at the same temperature have the same kinetic energy (K.E. = $\frac{1}{2}mv^2$).

We can summarize some facts about gases by comparing two familiar gases, H_2 and O_2, in separate tanks with the same volume, pressure, and temperature. There is an equal number of molecules in both tanks (Avogadro hypothesis), and their kinetic energies are also equal, but the lighter H_2 molecules must move more rapidly in order to exert the same pressure as the heavier O_2 molecules.

Key facts about the gas laws are summarized in the table below.

Law	*Equations*		*Constant*	*Type of Relationship*
Boyle's	$PV = k$	$P_1V_1 = P_2V_2$	n and T	Inverse: P up, V down
Charles's	$\frac{V}{T} = k$	$\frac{V_1}{T_1} = \frac{V_2}{T_2}$	n and P	Direct: T up, V up
Gay-Lussac's	$\frac{P}{T} = k$	$\frac{P_1}{T_1} = \frac{P_2}{T_2}$	n and V	Direct: T up, P up
Combined	$\frac{PV}{T} = k$	$\frac{P_1V_1}{T_1} = \frac{P_2V_2}{T_2}$	n	Both direct and inverse
Avogadro's	$\frac{V}{n} = k$	$\frac{V_1}{n_1} = \frac{V_2}{n_2}$	T and P	Direct: n up, V up
Ideal gas	$\frac{PV}{nT} = k = R$	$PV = nRT$	R	P, V, n, and T are all variables that are interdependent
Dalton's	$P_{\text{total}} = P_1 + P_2 + P_3 + \cdots$		T and V	Additive

A standard temperature of 273 K (0 °C) and a standard pressure of 1 atm (760 torr) are collectively defined as standard conditions (called STP). At STP, the molar volume of any gas is about 22.4 L/mol. The density of a gas at STP can be determined by dividing its molar mass by the molar volume. As shown by Dalton's law, gas pressures are additive—the total pressure is equal to the sum of the individual partial pressures.

Variations in P, V, and T for a fixed sample of gas can be determined in accordance with Boyle's law, Charles's law, Gay-Lussac's law, or the combined gas law. For a fixed sample (n moles) of gas at any fixed P, V, and T, the ideal gas law, $PV = nRT$, can be used to determine any single missing value when all other required values are known. The units used for P, V, and T in this equation must coincide with the units in the universal gas constant, R, which is 0.0821 L-atm/mol-K. By applying the equations and relationships described in this chapter, we can begin to understand important properties of gases and how to handle problems that involve pure gases or mixtures of gases.

Assess Your Understanding

1. Describe five physical properties that characterize gases. [12.2]
2. Describe gas behavior in terms of five assumptions of the kinetic molecular theory (KMT). [12.2]
3. Define pressure, and explain how a gas exerts pressure [12.3]
4. Describe the use and operation of a mercury barometer. [12.3]
5. Describe and use three different units for expressing pressure. [12.3]

6. Describe (mathematically and in words) and apply Boyle's law, Charles's law, Gay-Lussac's law, Dalton's law, and Avogadro's hypothesis. [12.4–12.11]
7. Interpret graphs involving gas pressure, volume, and temperature. [12.4–12.6]
8. Work problems involving changes in pressure, volume, and temperature. [12.4–12.8]
9. Give the standard temperature and pressure and molar volume at STP. [12.7–12.9]
10. Use the molar volume to determine the density of a gas at STP. [12.9]
11. Explain and use the ideal gas law in working problems. [12.10]
12. Use Dalton's law to find the pressure of a gas collected over water. [12.11]
13. Work stoichiometric problems involving gases, liquids, and solids. [12.12]

Key Terms

atmosphere (pressure unit) [12.3]
atmospheric pressure [12.3]
Avogadro hypothesis [12.10]
barometer [12.3]
Boyle's law [12.4]
Charles's law [12.5]
combined gas law [12.8]
Dalton's law of partial pressures [12.11]
diffusion [12.2]
Gay-Lussac's law [12.6]
ideal gas [12.10]
ideal gas law [12.10]
kinetic energy [12.2]
kinetic molecular theory (kinetic theory) [12.2]
molar volume [12.9]
partial pressure [12.11]
perfectly elastic collisions [12.2]
real gases [12.10]
standard conditions [12.7]
standard pressure [12.3]
standard temperature and pressure, STP [12.7]
universal gas constant [12.10]
vapor pressure [12.11]

Problems

THE ATMOSPHERE AND ATMOSPHERIC PRESSURE

12.1 What do we mean when we speak of the Earth's atmosphere?

12.2 List the four major gases found in our atmosphere.

12.3 Define the terms *pressure* and *atmospheric pressure*.

12.4 Why is atmospheric pressure greater at sea level than at high altitudes?

12.5 How does a mercury barometer work?

12.6 Why doesn't the mercury all drain out of a mercury barometer?

12.7 Carry out the following gas pressure conversions.

a. 1.00 atm = ? torr
b. 912 torr = ? atm
c. 0.500 atm = ? mm Hg
d. 1200 psi = ? atm
e. 2.00 atm = ? kPa
f. 3.0 in. Hg = ? atm

12.8 Carry out the following gas pressure conversions.

a. 10 atm = ____ psi
b. 646 torr = ____ atm
c. 3.5 atm = ____ torr
d. 35 psi = ____ atm
e. 35 psi = ____ torr
f. 14.7 psi = ____ kPa

THE KINETIC MOLECULAR THEORY (KMT)

12.9 Gases have no definite shape or volume; they expand to fill the container. Explain this physical property in terms of the kinetic molecular theory (KMT).

12.10 Gases are compressible, they have low densities, and they diffuse (mix). Explain these physical properties in terms of one key assumption of the KMT.

12.11 Using the KMT, explain why decreasing the volume of a gas is accompanied by an increase in pressure at constant temperature.

12.12 Using the KMT, explain how the kinetic energies of helium and nitrogen particles can be the same in tanks of the gases at the same pressure, volume, and temperature.

12.13 According to the KMT, what kind of change in temperature occurs when the average speed of the particles decreases?

12.14 According to the KMT, what kind of change in pressure occurs when the walls of a container are struck less often by gas particles?

12.15 A tank of helium and a tank of krypton have equal pressures, volumes, and temperature. Compare (without numbers) their kinetic energies, masses, rates of movement, and number of particles in each tank. Explain your reasoning.

12.16 Tank A has a volume of 10. L and contains 800. g of O_2. Tank B has a volume of 5.0 L and contains 400. g of O_2. Using the KMT as basis for your judgment, compare the pressures in the two tanks (constant temperature).

12.17 Density is defined as mass per unit volume. Two tanks of oxygen gas, A and B, have the same volume and temperature, but tank A is at a higer pressure. Which gas sample has a higher density? Explain in terms of particles per equivalent volume.

12.18 Two hot-air balloons, A and B, have the same volume and gas pressure, but the gas temperature in balloon A is much higher than it is in balloon B. Which gas has a higher density?

12.19 Use the KMT in explaining how the distinctive aroma of coffee can be detected in a different room from where it is being prepared.

12.20 Use the KMT in explaining why sleeping on an air mattress is more comfortable than sleeping on the ground.

12.21 Use the KMT in explaining why additional air can be added to an automobile tire that already appears to be fully pressurized.

12.22 Use the KMT in explaining why gas bubbles in boiling water get larger as they come closer to the surface. Why do the bubbles rise to the top?

PRESSURE AND VOLUME RELATIONSHIPS

12.23 What *specific* relationship about gases did Robert Boyle discover? Explain why a bicycle tire pump can be used as a model for Boyle's law.

12.24 What does "inversely proportional" mean? How can you recognize a graph that represents an inverse proportion?

12.25 Use the following P and V data to determine the value of k for a Boyle's law experiment. Express your answer in scientific notation.

Pressure (torr)	*Volume (mL)*	k
828	90.0	?
980	76.0	?
1263	59.0	?
1656	45.0	?

12.26 Under what circumstances is P_1V_1 equal to P_2V_2 for a gas? How does this apply to the data supplied in problem 12.25.

12.27 A tank contains 500. mL of compressed air at 1800. torr. What volume will the compressed air occupy at 750. torr, assuming no temperature change?

12.28 An automobile engine cylinder with a volume of 400. cm^3 is compressed to a volume of 100. cm^3 at constant temperature. If the initial gas pressure was 1 atm, what was the final pressure?

12.29 A 13.0-L tank used in scuba diving was filled with air at a pressure of 115 atm. What volume (in liters) will the gas occupy at a pressure of 775 torr? (First, express all pressures in the same units, e.g., atm.)

12.30 Oxygen used in a particular hospital is stored at a pressure of 2200. psi in gas cylinders with a volume of 60.0 L. What volume would the gas from one cylinder occupy at atmospheric pressure of 14.7 psi?If gas flow to a patient is adjusted to 8.00 L/min, how long (in hours) will the tank last?

VOLUME AND TEMPERATURE RELATIONSHIPS

12.31 What specific relationship about gases did Charles discover? What type of proportion is involved? What type of graph is obtained from plotting gas volume against temperature?

12.32 Explain why a rubber balloon filled with air at room temperature and cooled to a much lower temperature can be used as a model for Charles's law.

12.33 Determine a volume/temperature ratio for each of the data points plotted on the graph shown in Figure 12.11. Why must temperatures be in kelvins?

12.34 Explain how absolute zero can be determined from Charles's law experimental data.

12.35 A helium-filled balloon had a volume of 5.00 L at 27 °C. What will be its volume at −93 °C assuming no change in pressure?

12.36 A helium-filled balloon had a volume of 400. mL when cooled to −120. °C. What will be the volume if the balloon is warmed in an oven to 100. °C assuming no change in pressure?

12.37 If a 1500.-mL sample of air at 20. °C is heated enough to expand the volume to 1750. mL at constant pressure, what is the final Celsius temperature required? What was the temperature change?

12.38 If a 1500.-mL sample of air at 22 °C is cooled enough to cause the volume to decrease to 750. mL at constant pressure, what is the final Celsius temperature required? What was the temperature change?

PRESSURE AND TEMPERATURE RELATIONSHIPS

12.39 Name two items that can serve as models for Gay-Lussac's law. Explain.

12.40 Describe in terms of the kinetic molecular theory the pressure–temperature relationship for a fixed sample of gas at a constant volume.

12.41 A light bulb with an internal pressure of 720. torr at 20. °C is thrown into an incinerator operating at 750. °C. What internal pressure must the light bulb be able to withstand if it does not break?

12.42 The air in a jar of home-canned green beans is heated to the boiling point of water at 100. °C and the jar is sealed at that temperature. Assume that the pressure is 1.00 atm when the jar is sealed. What will be the pressure in the jar when it is cooled to room temperature at 20. °C?

12.43 An automobile tire had a gauge pressure of 32 psi on a fall day when the temperature was 22 °C. In the winter when the temperature dropped to −23 °C (which is −10 °F) what was the tire *gauge* pressure, assuming no change in volume? *Hint*: Gauge pressure is not *total* pressure; you must add 14.7 psi to gauge pressures when doing the math.

12.44 An automobile tire had a gauge pressure of 30. psi at 20. °C. After racing the car, the gauge pressure was found to be 34 psi. Assuming no change in volume, what was the Celsius temperature of the air in the tire?

THE COMBINED GAS LAW

12.45 What volume will 150. mL of a gas at 23 °C and 710. torr occupy at STP?

12.46 At STP one mole of a gas has a volume of 22.4 L. What volume will the gas occupy at 100. °C when the pressure is 1.2 atm? Also, determine the percent increase in volume by dividing the volume *increase* by the original volume and expressing the quotient as a percentage.

12.47 If a gas has a volume of 800. mL at 10. °C and 1.00 atm, what will be its pressure at a temperature of 100. °C if the volume increases to 850. mL?

12.48 A helium-filled balloon has a volume of 8.50 L on the ground at 20. °C and at a pressure of 750. torr. After the balloon was released, it rose to an altitude where the temperature was −20. °C and the pressure was 425 torr. What was the volume of gas in the balloon at these conditions?

MOLAR VOLUME AND GAS DENSITY

12.49 What is the volume of 0.200 g of hydrogen at STP?

12.50 What is the volume of 0.200 g of oxygen gas at STP?

12.51 How many milligrams of oxygen are in a 4.00-L bottle of the gas at STP?

12.52 How many milligrams of CO_2 are in a 4.00-L bottle of the gas at STP?

12.53 How many moles of helium can be in a balloon with a volume of 6.00 L at STP?

12.54 How many moles of N_2 can be in a balloon with a volume of 6.00 L at STP?

12.55 Determine the density of CO_2 gas at STP.

12.56 Determine the density of nitrogen gas at STP.

IDEAL GASES

12.57 What is an ideal gas? Can you name one?

12.58 How do real gases differ from ideal gases?

12.59 When the universal gas constant, R, has a value of 0.0821, what are the specific units that must be used?

12.60 Using appropriate values for the volume of one mole of gas at STP, and using a method similar to that shown in the text for calculating R, determine the appropriate value of R with the units L-torr/mol-K.

12.61 What pressure, in atmospheres, is exerted by 0.120 mol of steam (water vapor) at its boiling point of 100. °C if the volume is contained in a 2.00-L teakettle?

12.62 What pressure, in atmospheres, is exerted by 44.0 mol of propane in a 36-L tank at 22 °C?

12.63 What volume will 10.0 g of oxygen gas occupy at a temperature of 22 °C and a pressure of 740. torr? *Hint*: Change pressure units to atmospheres.

12.64 How many grams of helium are in a balloon filled with 8.50 L of the gas at a temperature of 20. °C and a pressure of 800. torr?

PARTIAL PRESSURES

12.65 A container holds oxygen at a partial pressure of 0.25 atm, nitrogen at a partial pressure of 0.50 atm, and helium at a partial pressure of 0.20 atm. What is the pressure inside the container?

12.66 A container is filled with equal numbers of nitrogen, oxygen, and carbon dioxide molecules. The total pressure is 750. torr. What is the partial pressure of nitrogen in the container?

12.67 Atmospheric pressure on the surface of Mars is about 6.0 torr. The partial pressure of carbon dioxide is 5.7 torr. What percentage of the Martian atmosphere is carbon dioxide?

12.68 The pressure on the surface of Venus is about 100. atm. Carbon dioxide gas makes up about 97.0% of the atmospheric gases. What is the partial pressure of CO_2 in the atmosphere of Venus?

12.69 Oxygen gas is collected over water at a temperature of 25 °C. If the collected sample has a pressure of 740. torr, what is the partial pressure of oxygen in the container? (You may refer to Table 12.2.) If the gas sample has a volume of 95.0 mL, how many moles of $O_2(g)$ are in the sample?

12.70 Hydrogen gas is collected over water. If the 50.0-mL sample of gas collected has a temperature of 22 °C and a pressure of 744 torr, what is the partial pressure of hydrogen? How many moles of $H_2(g)$ were collected?

GAS STOICHIOMETRY

12.71 Calcium carbide, CaC_2, reacts with water to produce acetylene gas, C_2H_2, which is burned in miner's lamps by spelunkers while cave exploring.

$$CaC_2(s) + 2\ H_2O(l) \rightarrow Ca(OH)_2 + C_2H_2(g)$$

How many milliliters of acetylene can be produced at a temperature of 20. °C and at a pressure of 740. torr by the complete reaction of 2.50 g of CaC_2?

12.72 Ethanol, C_2H_5OH, is produced by the fermentation of sugars, but CO_2 is produced at the same time. The equation for the fermentation of glucose, $C_6H_{12}O_6$, is

$$C_6H_{12}O_6 \rightarrow C_2H_5OH + CO_2(g) \qquad \text{(unbalanced)}$$

What volume of $CO_2(g)$ is produced at a temperature of 20. °C and a pressure of 1.00 atm by the fermentation of 500. g of glucose?

12.73 Magnesium metal reacts with hydrochloric acid, HCl(aq), to produce hydrogen gas according to the following equation.

$$Mg(s) + 2\ HCl(aq) \rightarrow MgCl_2(aq) + H_2(g)$$

If 2.15 g of magnesium reacts completely, what volume of hydrogen can be produced at STP? (Disregard the vapor pressure of water.) What will be the volume if the gas is at 735 torr and a temperature of 25 °C?

12.74 When 500. g of gasoline, C_8H_{18}, is completely burned, what volume of $CO_2(g)$ is produced at standard atmospheric pressure and a temperature of 18 °C? The unbalanced equation for the reaction is shown here.

$$C_8H_{18}(l) + O_2(g) \rightarrow CO_2(g) + H_2O(g) \qquad \text{(unbalanced)}$$

Additional Problems

12.75 What is a common item that can serve as a model for Boyle's law?

12.76 The pressure in a tire decreases when temperature decreases, at constant volume. What gas law is demonstrated here?

12.77 Determine the density of ammonia gas, NH_3, at STP.

12.78 What is the molar mass for a gas with a density of 1.96 g/L at STP?

12.79 Calculate the volume occupied by 0.600 mol of a gas at a temperature of 310. K and a pressure of 0.800 atm.

12.80 What volume is occupied by 25.0 mol of natural gas (mostly CH_4) at a temperature of 15 °C and under a pressure of 15.0 atm?

12.81 What volume of water vapor is produced by burning 500. g of gasoline at the conditions specified in problem 12.74.

12.82 What volume of $H_2(g)$ is produced at STP by electrolysis of 26.0 g of H_2O?

12.83 Determine the density of sulfur dioxide gas at STP.

12.84 The density of a gas was determined to be 0.759 g/L at STP. What is its molar mass? Do you know of a gas with this molar mass?

The beauty of water in three states is captured in this photograph of Paradise Bay, Antartica. Water in the gaseous state—water vapor—is present in the clouds and the air. Water present in the bay and frozen water in the form of ice and snow are water in the two condensed states, the focus of this chapter.

13

Liquids and Solids

CONTENTS

Our presentation of chemical bonding in Chapter 7 focused on how atoms combine to form molecules and why elements react to form compounds. Now, we will look more closely at why some substances are solids, others are liquids, and still others are gases at the same temperature. The particles of a gas fly about at random, but the particles of liquids and solids cling together. Forces of attraction must keep the particles of liquids and solids from flying about. Gas particles interact so little with one another that a collection of them retains neither a specific volume nor a specific shape. In the liquid state, particles are held together with enough force to cause them to conform to a specific volume even though the shape is not rigid. In the solid state, particles are held together so rigidly that both the shape and volume of the sample are fixed.

These mysterious forces of attraction are responsible for why water droplets stand up on glass and on flat surfaces of a polished automobile. They give shape and volume to liquids and solids—the condensed forms of matter. They are also responsible for the relatively high boiling point of water and the relatively low boiling points of diethyl ether (36 °C) and hexane (69 °C), the inability of water to dissolve in oil, and the cooling effects that occur during evaporation.

In this chapter we will study changes in the states of matter, including what happens when a solid is converted to a liquid, or a liquid to a gas. Is that important to us? Well, consider perspiration. From the amount of advertising directed against this lowly liquid, you would think it was an unnecessary annoyance. However, were it not for the cooling effect that occurs during conversion of liquid to a vapor on skin surfaces, we would find it difficult just to survive in warm climates, let alone to carry out vigorous physical activity. We can explain these effects, and other properties of liquids and solids, in terms of the same kinetic molecular theory we used to describe physical properties of gases.

13.1 *Gases, Liquids, and Solids: Some Generalizations*

Tiny, individual gas particles are far apart, so gases can be compressed significantly, but in liquids and solids, the individual particles are in contact with one another, so very little compression is possible. To illustrate the vast difference in the spacing between molecules in the gaseous state and the condensed states (liquids and solids), let's consider a specific volume of water vapor (a gas) and compare it with the same volume of liquid water. At a temperature of 20. °C and a pressure of 1 atm, 1 mol of water vapor (6.02×10^{23} molecules) occupies 24.0 L, as shown by the following calculation based on the ideal gas law.

$$V = \frac{nRT}{P} + \frac{1 \text{ mol} \times 0.0821 \text{ L-atm} \times 293 \text{ K}}{1 \text{ atm} \times \text{mol-K}} = 24.0 \text{ L}$$

If the same volume (24.0 L) is filled by water in the liquid state, the number of molecules required would be many times greater. To determine how many times greater, we'll start by assuming that 24.0 L of water has a mass of 24,000 g, since the density of liquid water is approximately 1.00 g/mL. We will also use the molar mass of water (18.0 g/mol). With these values and Avogadro's number, we can calculate the number of molecules of liquid water required to fill a 24.0-L container, as shown here.

$$24{,}000 \text{ g} \times \frac{1 \text{ mol}}{18.0 \text{ g}} \times \frac{6.02 \times 10^{23} \text{ molecules}}{1 \text{ mol}} = 8027 \times 10^{23} \text{ molecules}$$

Now, when we divide this number of molecules of liquid water in 24.0 L by the number of molecules of water vapor (a gas) in 24.0 L,

$$\frac{\text{Number of liquid water molecules}}{\text{Number of gaseous water molecules}} = \frac{8027 \times 10^{23}}{6.02 \times 10^{23}} = 1333$$

we notice that there are roughly 1300 times as many molecules of liquid as there are of a gas in the same volume. Therefore, molecules of a liquid must be in a rather crowded situation (Figure 13.1).

Intermolecular forces are forces of attraction between a molecule and neighboring molecules. **Intramolecular forces** exist between atoms within a molecule due to chemical bonding.

Even in the gaseous state there is some attraction between molecules. These forces of attraction *between* molecules are called **intermolecular forces**. The attractions can be great enough to cause deviation from "ideal gas" behavior, as

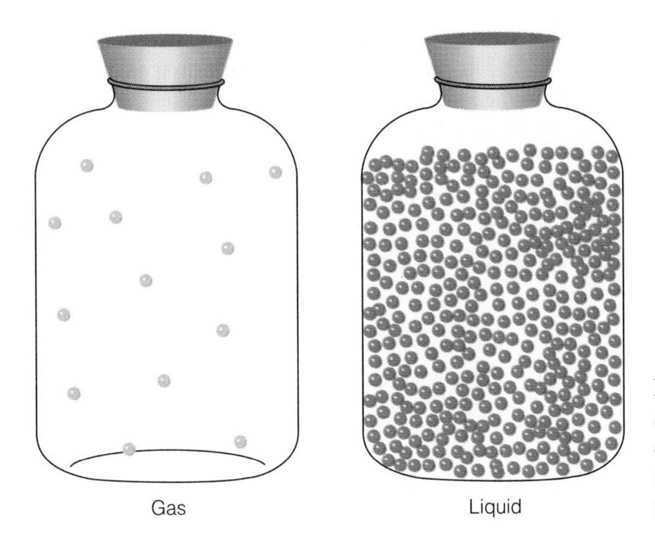

Figure 13.1 *There are over a thousand times as many molecules in a given volume of a liquid as there are in the same volume of a gas.*

stated in Chapter 12. Generally, however, these intermolecular forces are much weaker in the gaseous state than in the liquid and solid states.

Before studying specific types of intermolecular forces in detail, let's review some generalizations about gases, liquids, and solids.

1. All *gases* are either monatomic (consisting of individual atoms such as He and Ar) or molecules with covalent bonding (such as N_2 and CO_2).
2. All pure *liquids* at room temperature (such as ethanol, C_2H_5OH, and water) are composed of molecules with covalent bonding. Mercury is an exception; it is the only metal that is a liquid at room temperature. Liquids that are nonpolar are miscible with other nonpolar liquids. Liquids that are polar are immiscible (not miscible) with nonpolar liquids.
3. Although we cannot make a third generalization to cover all solids, we can say that *all ionic compounds* are crystalline solids at toom temperature. They have relatively high melting points (about 300 °C to 1200 °C) because their ionic bonds are very strong. When molten or when dissolved in water, the dissociated ions can conduct an electric current. Ionic compounds have very low solubilities in nonpolar liquids.
4. All *metals* are solids at room temperature (about 22 °C), except for mercury, which is liquid at room temperature. Metals conduct both heat and electricity in the solid state. Some metals can be melted at low temperatures, but others melt only at high temperatures. The range is quite large: cesium and gallium have melting points of 29 °C and 30 °C, respectively, silver and iron melt at 961 °C and 1536 °C, respectively, but tungsten tops the list with a melting point of 3407 °C.
5. Many *molecular* (covalently bonded) *substances* are solids at room temperature. Examples include glucose, $C_6H_{12}O_6$, sulfur, S_8, iodine, I_2, and numerous organic molecules such as the carbohydrates and plastics.

A **polar** molecule has an unbalanced molecular shape and electron distribution, so there are negative and positive poles. A **nonpolar** molecule has a balanced and symmetrical molecular shape and electron distribution.

The physical state of a molecular substance at a given temperature is related to its molar mass—particles with greater mass have greater attractions—and to various *intermolecular forces*. If one of these two variables can be eliminated—or held constant—then simple generalizations are possible. For example, the halogens, all of which exist as nonpolar diatomic molecules, have similar intermolecular forces, so variations in properties can be attributed to variations in molar mass. Fluorine, F_2, and chlorine, Cl_2, with molar masses of 38 g/mol and 71 g/mol, respectively, are gases at room temperature. Bromine, Br_2, with a molar mass of 160 g/mol is a liquid, and iodine, I_2, with a molar mass of 254 g/mol is a solid. The melting point increases as molar mass increases.

For another example, we can compare a series of compounds containing carbon and halogens that have similar intermolecular forces.

Compound	CF_4	CCl_4	CBr_4	CI_4
Molar mass (g/mol)	88	154	332	520
Physical state at 20 °C	Gas	Liquid	Solid	Solid

Once again, variations in physical state (at a given temperature) parallel changes in molar mass when intermolecular forces are similar. We will consider the various types of forces in detail in the following sections.

EXAMPLE 13.1

Based on the information given, predict the type of bonding and molecular polarity for these substances.

(a) A white waxy solid melts at 80 °C. It does not conduct an electric current and does not dissolve in water, but dissolves in hexane.
(b) A liquid at room temperature is miscible with both water and hexane.

SOLUTION

(a) The solid is molecular and nonpolar. Molecular solids have low melting points. It dissolves in a nonpolar liquid so it must be nonpolar.
(b) The liquid is molecular and intermediate in polarity. All pure liquids (at room temperature) are molecular. Since this liquid is miscible with both polar and nonpolar liquids, it is intermediate in polarity.

See Problems 13.1–13.8.

13.2 *Interionic and Intermolecular Forces*

As we proceed to compare forces that exist in liquids and solids, we need to distinguish clearly between *intramolecular forces* and *intermolecular forces*.

Intramolecular forces are those that exist between atoms *within* a molecule or compound due to bonding (Chapter 7), such as the carbon-to-oxygen covalent bonds in a carbon dioxide molecule or the ionic attractions between positive and negative ions. The **interionic forces** (forces between ions) are—with few exceptions—the strongest of all the forces that hold solids and liquids together, as indicated by the high melting points for ionic compounds. This is because ions are electrically charged, and ions with opposite charges attract one another. Each ion in a crystalline solid is attracted by several ions that surround it in the crystal lattice. Attractive forces are greater between ions of higher charge. For example, the attraction between a calcium ion, Ca^{2+}, and an oxide ion, O^{2-}, is greater than that between a sodium ion, Na^+, and a chloride ion, Cl^-. The aluminum ion, Al^{3+}, to

Certain covalent network solids such as SiC and SiO_2 (Section 13.5) have melting points that are even higher than those of ionic solids.

nitride ion, N^{3-}, attraction is greater still. Again, the greater the force of attraction, the higher will be the melting point.

Intermolecular forces—the attractions of molecules for one another—are much weaker than intramolecular forces, but they help determine physical properties of molecular substances. Intermolecular forces, as a group, are also called **van der Waals forces** in honor of Johannes van der Waals, a Dutch physicist who first emphasized their importance. We will study three types of intermolecular forces.

Intermolecular Forces

- Dipole forces
- Hydrogen bonds
- London forces

Dipole Forces

We have seen (Section 7.4) that polar covalent bonds occur when electron pairs are shared unequally between atoms of different kinds of elements. A hydrogen chloride, HCl, molecule is polar; the electron pair in its bond is attracted more strongly toward chlorine, the more electronegative atom, than it is toward hydrogen, the less electronegative atom. Such molecules, with separate centers of partial negative and partial positive charges, are called **dipoles**.

When molecules that are dipoles are brought close enough together, the positive end of one molecule attracts the negative end of another molecule. This is similar to the way opposite poles of magnets attract one another. Dipole forces may exist throughout the structure of a liquid or a solid (Figure 13.2).

In general, attractive forces between dipoles are fairly weak. However, they are stronger than the forces between nonpolar molecules having comparable molar masses. Examples of compounds with dipolar molecules include hydrogen chloride, HCl, hydrogen bromide, HBr, and angular V-shaped molecules such as hydrogen sulfide, H_2S, and sulfur dioxide, SO_2. Dipole forces are only about 1% as strong as ionic bonds.

Attractions between dipolar molecules result from forces between centers of *partial* charge, while the much stronger attraction between ions involves fully charged particles.

Hydrogen Bonds

The intermolecular forces between polar molecules containing hydrogen atoms bonded to fluorine, oxygen, or nitrogen are stronger than would be expected on the basis of dipole attractive forces alone. These intermolecular forces are strong enough to be given a special name, the **hydrogen bond**. The term *hydrogen bond*

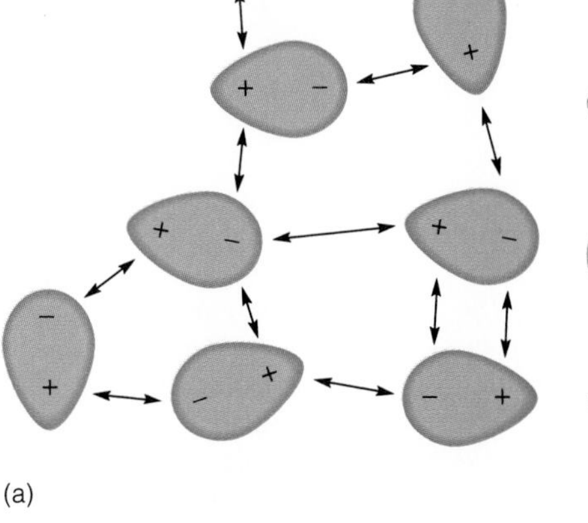

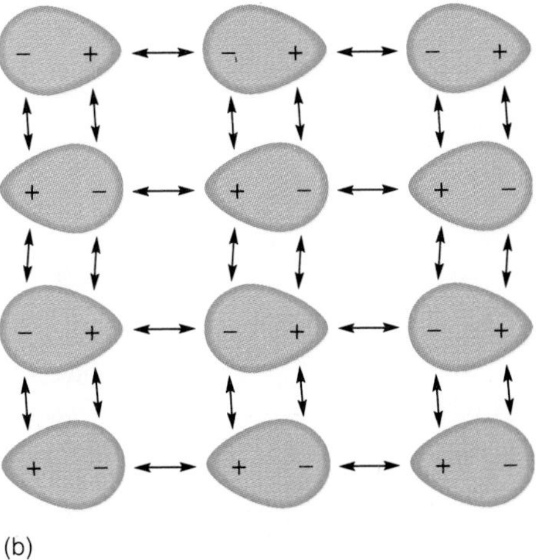

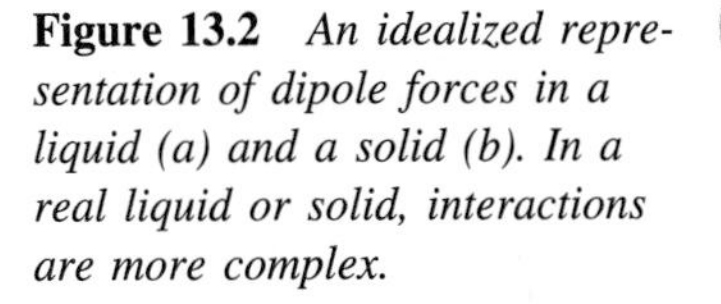

Figure 13.2 *An idealized representation of dipole forces in a liquid (a) and a solid (b). In a real liquid or solid, interactions are more complex.*

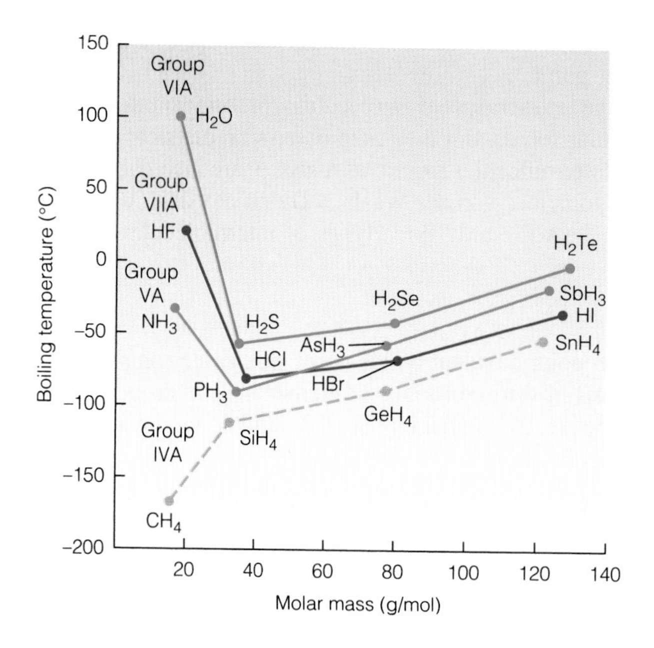

Figure 13.3 *Boiling point trends for compounds of hydrogen with elements in Group IVA, Group VA, Group VIA, and Group VIIA. Hydrogen bonding accounts for the unusually high boiling points of H_2O, HF, and NH_3, which violate the general trend that smaller molecules have lower boiling points.*

is a somewhat misleading name because it emphasizes only the hydrogen component.

The compounds HF, H_2O, and NH_3 all have much higher boiling points than would be expected, based on boiling point trends for compounds of hydrogen with elements in the same families (Figure 13.3). Their unusually strong forces of intermolecular attraction can be attributed to hydrogen bonding. Figure 13.4 shows that water molecules in ice crystals are held in hexagonal arrangements by hydrogen bonds, represented here by dotted lines between molecules. Molecules involved in hydrogen bonding have at least three common characteristics.

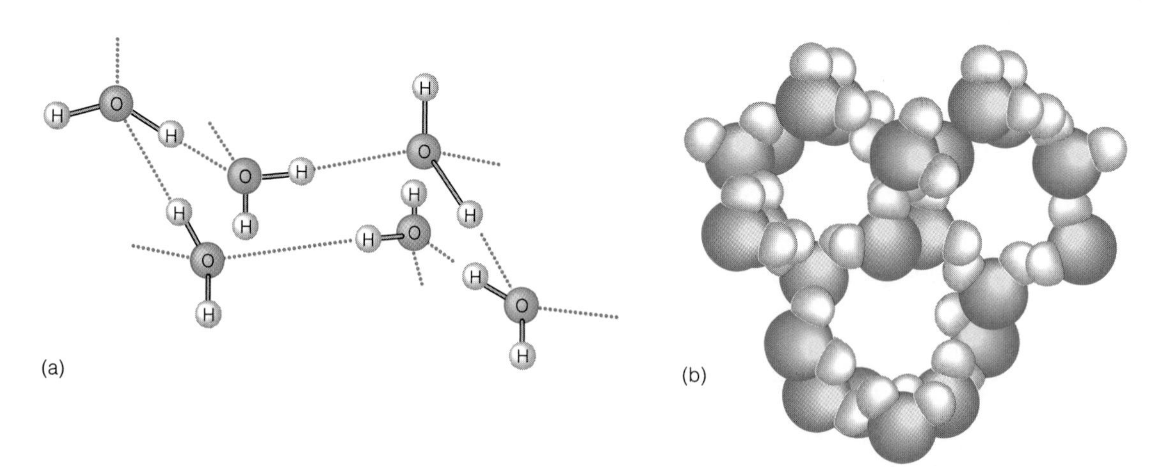

Figure 13.4 *(a) The hydrogen bonding in ice is indicated here by dotted lines between ball-and-stick structures of water molecules. Each water molecule has two hydrogen atoms and two nonbonding electron pairs that can participate in hydrogen bonding with other molecules. These intermolecular attractions cause water molecules to form hexagonal crystal arrangements. (b) The hexagonal arrangement present in ice is less compact than liquid water, shown by the open crystal structure.*

1. Each molecule involved in hydrogen bonding has a hydrogen atom covalently bonded to a highly electronegative atom of fluorine, oxygen, or nitrogen. The electron pair in this covalent bond is attracted more strongly toward the highly electronegative atom that takes on a partial negative charge. This leaves the hydrogen with a partial positive charge.
2. The hydrogen of one molecule is attracted toward the nonbonding electron pair of a fluorine, oxygen, or nitrogen atom on a neighboring molecule that carries a partial negative charge. We can picture the tiny hydrogen as being sandwiched between two highly electronegative atoms.
3. Hydrogen bonds tend to form within a cluster of molecules somewhat like the attractions formed in a cluster of several small magnets—all arranged with the positive end of one directed toward the negative end of another.

The **hydrogen bond** may seem merely to be a piece of chemical theory, but its importance to life and health is immense. The structure of proteins—chemicals essential to life—is determined, in part, by hydrogen bonding. Also, the heredity that one generation passes on to the next is dependent on an elegant application of hydrogen bonding.

Since fluorine forms only one covalent bond, there is only one pure fluorine-containing compound, hydrogen fluoride, HF, capable of intermolecular hydrogen bonding. Oxygen-containing compounds that can form hydrogen bonds include not only water but also methanol, CH_3OH, ethanol, C_2H_5OH, and some other alcohols. Nitrogen-containing compounds that can form hydrogen bonds include ammonia, NH_3, and a number of organic compounds such as the amines. More examples are shown in the next chapter (Figure 14.4).

Dispersion Forces (London Forces)

If one understands that positive attracts negative, then it is easy enough to understand how polar molecules are held together. But how can we explain the fact that nonpolar substances such as bromine, Br_2, and iodine, I_2, exist in the liquid and solid states? Even hydrogen can exist as a liquid or solid if the temperature is low enough. Something must hold these molecules together.

The answer arises from the fact that the electron cloud pictures (Sections 7.2 and 7.4) are only *average* positions. For example, in the tiny, nonpolar hydrogen molecule, the two electrons are—on the average—between (and equidistant from) the two nuclei where they are shared equally. At any given instant, however, the electrons may be at one end of the molecule; at some other instant the electrons may be at the other end of the molecule. Such electron motions give rise to momentary dipoles (Figure 13.5). One dipole, however momentary, can induce a similar momentary dipole in a neighboring molecule. (When the electrons of one molecule are at one end, the electrons in the next molecule move away from that end.) This results in an attractive force between the electron-rich end of one molecule and the electron-poor end of the next. These small, transient, attractive forces between nonpolar molecules are called **dispersion forces** or **London forces**.

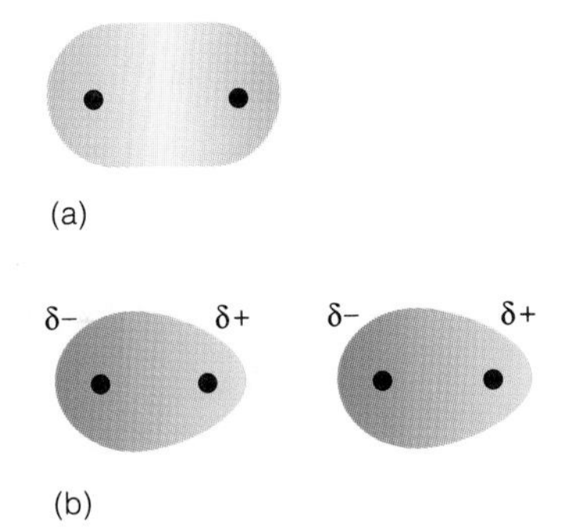

Figure 13.5 *Electron cloud shapes on hydrogen molecules. (a) Average picture with no net dipole. (b) Instantaneous pictures with momentary dipoles. The transient dipoles are constantly changing, but the net result is attraction.*

Dispersion forces are stronger for larger nonpolar molecules than for smaller ones. These forces are greater in bromine, Br_2, and iodine, I_2, than in fluorine, F_2. Larger atoms have larger electron clouds, and their outermost electrons are farther from the nucleus than those of smaller atoms. The electrons that are farther from the nucleus are also more loosely bound and can shift toward another atom more readily than can the tightly bound electrons of a smaller atom. This makes molecules with larger atoms more polarizable than small ones. Iodine molecules are attracted to one another more strongly than bromine molecules are attracted to one another. Bromine molecules have stronger dispersion forces than chlorine molecules, and chlorine molecules, in turn, have stronger dispersion forces than fluorine molecules. Compare this order (iodine to fluorine) with the order in which these

London forces were named after Fritz London, professor of chemical physics at Duke University.

halogens appear in the periodic table; iodine is the largest and fluorine is the smallest. As will be further described in Section 13.4, dispersion forces determine, to a large extent, the physical properties of nonpolar compounds.

London dispersion forces are important even when other types of forces are also present. Even though London forces are individually weaker than dipolar attractions or ionic bonds, their cumulative effect can be considerable in substances composed of large molecules.

EXAMPLE 13.2

Arrange the following types of forces in order of increasing strength (weakest first): hydrogen bonding, dispersion forces, dipole forces, and covalent bonding. Identify each force as being intermolecular or intramolecular.

SOLUTION

In order of increasing strength we have dispersion forces, dipole forces, hydrogen bonding, and covalent bonding. Covalent bonding is an intramolecular force; the others listed are intermolecular forces.

See Problems 13.9–13.22.

EXERCISE 13.1

What intermolecular forces are present for iodine, I_2, and hydrogen iodide, HI?

13.3 The Liquid State

The molecules of a liquid are in constant motion, but their movements are greatly restricted by neighboring molecules. They can be compressed only slightly. One liquid can diffuse into another, but this diffusion is much slower than in gases because of the restricted molecular motion of liquids.

Viscosity

Viscosity (Section 2.2) is a measure of the resistance of a liquid to flow—the higher the viscosity, the slower is its rate of flow.

The **viscosity** of a liquid is related to the shape of the molecules that make up the liquid. Liquids that have low viscosity—those that flow readily—generally consist of small, symmetrical molecules with weak intermolecular forces.

Two types of intermolecular forces lead to high viscosities. The relatively weak London dispersion forces are responsible for the high viscosities of large nonpolar molecules such as octadecane (Figure 13.6). Certain small, unsymmetrical mole-

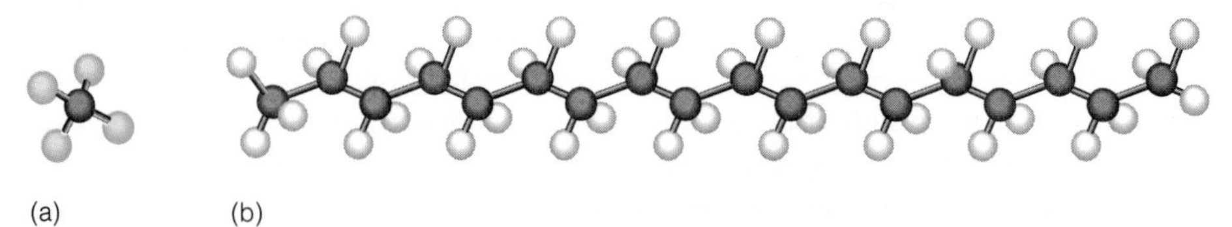

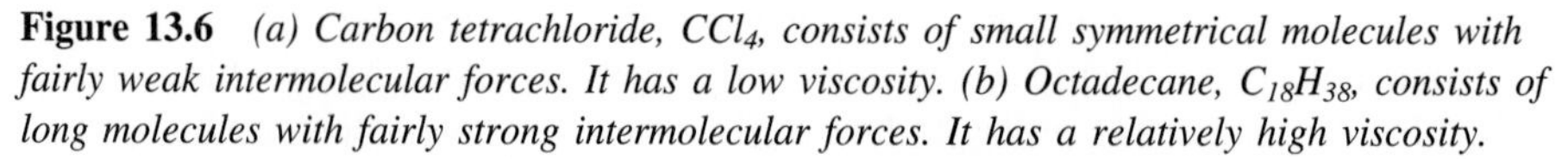

(a) (b)

Figure 13.6 *(a) Carbon tetrachloride, CCl_4, consists of small symmetrical molecules with fairly weak intermolecular forces. It has a low viscosity. (b) Octadecane, $C_{18}H_{38}$, consists of long molecules with fairly strong intermolecular forces. It has a relatively high viscosity.*

cules with strong hydrogen bonds also have high viscosities. For example, ethylene glycol, $HOCH_2CH_2OH$, the key ingredient in antifreeze, has a viscosity similar to that of syrup. The —OH group on both ends of each ethylene glycol molecule can be involved in hydrogen bonding with other molecules.

Viscosity generally decreases with increasing temperature. Increased kinetic energy partially overcomes the intermolecular forces. For example, cooking oil poured from a bottle is rather viscous; it is thick and "oily." During heating, its viscosity decreases; it becomes thinner and more like the consistency of water.

Automotive oils carry a viscosity rating identified by an SAE number. Oil with an SAE number of 30 is more viscous than an SAE number 10 oil. The SAE abbreviation is for "Society of Automotive Engineers," which sets the standards for oil viscosities.

Surface Tension

Another property of liquids is **surface tension**. A glass of water can be slightly overfilled with water before it spills over. A small needle, carefully placed, can be made to float horizontally on the surface of water—even though steel is about eight times more dense than water. A variety of insects can walk on water or skate across the surface of a pond with ease. These phenomena indicate something quite unusual about the surface of water. There is a special force or tension at the surface that resists being disrupted by the penetration of a needle or a water bug (Figure 13.7).

Figure 13.7 *The surface tension of water enables this water strider to walk across the surface of a pond. Notice how the water is indented, but not penetrated, by the insect's legs.*

These surface forces can be explained in terms of the intermolecular forces that have been described. In general, liquids with strong intermolecular forces have higher surface tension than liquids with weak intermolecular forces. A molecule in the center of a liquid is attracted equally in all directions by the molecules surrounding it. A molecule on the surface, however, is attracted by molecules at its sides and below it only (Figure 13.8). There is no corresponding upward attraction. These unequal forces tend to exert a force inward at the surface of the liquid and cause it to contract. Thus, a small amount of liquid will "bead" and a liquid drop will tend to be spherical to minimize surface area.

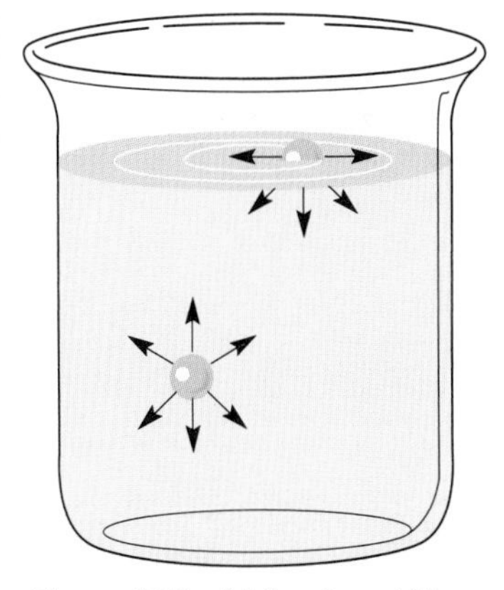

Figure 13.8 *Molecules within the body of a liquid are attracted equally in all directions. Those at the surface, however, are pulled downward and sideways, but not upward.*

The surface tension of water can be lowered by using detergents. This enables the water to *wet* a solid surface. The **wetting** action of a liquid, especially water, enables it to spread evenly over a surface as a thin film. Chemicals that reduce the surface tension of water are called **surfactants**. They are widely used as ingredients in laundry and dishwashing products and industrial cleaners.

EXAMPLE 13.3

Based on characteristics that affect viscosity, compare molecular sizes for SAE number 40 motor oil with SAE number 5 motor oil. Which oil would have greater surface tension? *Hint:* See the margin note on automotive oils.

SOLUTION

A higher SAE number indicates a higher viscosity. The higher the viscosity, the greater are the intermolecular forces, which, in turn, are greater for larger molecules. Thus, SAE 40 oil is expected to have larger molecules (with higher molar masses) than SAE number 5 oil.

The SAE number 40 oil would also be expected to have greater surface tension, again based on intermolecular attractions.

See Problems 13.23–13.30.

13.4 *Vaporization and Condensation*

Molecules of a liquid are in constant motion; some move fast, some move more slowly. Occasionally one of the molecules has enough kinetic energy to escape from the liquid's surface and become a molecule of vapor. This is called **vaporization**—the process by which molecules of a liquid break away and go into the gas phase.

Volatile liquids, such as ethanol and isopropyl alcohol, are those that evaporate readily at room temperature. Most of them have low boiling points and high vapor pressures.

If a small amount of a volatile liquid such as water or isopropyl alcohol (rubbing alcohol) is placed in an open container, it will soon disappear as it is converted from a liquid to a gas (a vapor). The process is called **evaporation** (Figure 13.9). As molecules of vapor become dispersed throughout the atmosphere, more liquid molecules escape into the vapor state until, eventually, all the liquid has evaporated. The rate of evaporation of a liquid depends on the temperature of the liquid and the amount of exposed surface area.

The vapor exerts a partial pressure (Section 12.11) that is constant at a given temperature. Vapor pressure increases as temperature increases.

If the same volatile liquid is placed in a closed container, it does not go away. Some of the liquid is converted to vapor, but the vapor molecules are trapped within the container. Eventually the air above the liquid becomes saturated and vaporization appears to stop. It may appear that nothing further is happening within the closed container, but molecular motion has not ceased. Some molecules of liquid are still escaping into the vapor state. Vapor molecules in the space above the liquid occasionally strike the liquid's surface where they are captured, and thus return to the liquid state. This conversion of vapor to liquid—the reverse of vaporization—is called **condensation**.

At the start, there are many liquid molecules but no vapor molecules. Then, vaporization begins to occur as some of the liquid is converted to vapor. As more molecules pass into the vapor state, the rate of condensation increases. Eventually, the rate of condensation becomes equal to the rate of vaporization in the closed container; there appears to be no change in the amount of liquid or the amount of vapor. The system is at *equilibrium*. On the molecular level, the number of molecules of liquid that change to vapor in a given unit of time equals the number of molecules of vapor that change back to liquid—the number of molecules of vapor remains constant and the number of molecules of liquid remains constant. A condition called a *dynamic equilibrium* has been achieved.

$$\text{Liquid} \underset{\text{condensation}}{\overset{\text{vaporization}}{\rightleftharpoons}} \text{Vapor}$$

For any **dynamic equilibrium** two opposing processes are occurring at equal rates. This is one of the most important concepts in all of chemistry. Visibly, the quantities of chemicals of each kind and state appear to be static or fixed—they are at equilibrium. On the molecular level, there is a continuous movement of particles, but the number of particles that move in one direction is equal to the number that move in the opposite direction. Because activity has not actually ceased, the equilibrium is said to be *dynamic.*

If the closed container of liquid and vapor is heated, more molecules of the liquid would have enough energy to escape from the liquid. Thus, the vapor pressure would increase, but equilibrium would soon be reestablished at the higher temperature. While the rate of vaporization would be greater at the higher temperature, so also would be the rate of condensation. At the newly established equilibrium, the rates would once again be equal (Figure 13.10), but the equilibrium vapor pressure would be higher at the higher temperature.

Figure 13.9 *A volatile liquid evaporates from an open container. The rate of evaporation of a particular liquid depends on the temperature of the liquid and the amount of exposed surface area.*

The Boiling Point

When a liquid is placed in an open container, the escape of molecules of the liquid is opposed by atmospheric pressure. When the liquid is heated, the vapor pressure increases. Continued heating will eventually result in a vapor pressure equal to atmospheric pressure. At that temperature, the liquid will begin to boil. As the liquid boils, vaporization takes place not only at the surface of the liquid but also in the body of the liquid, with vapor bubbles forming and rising to the surface. The **boiling point** of a liquid is the temperature at which its vapor pressure becomes

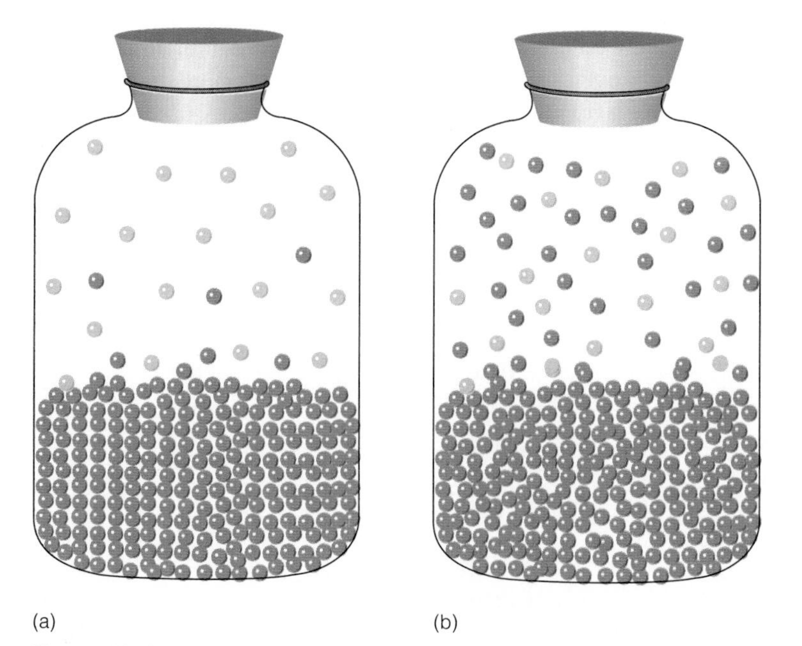

Figure 13.10 *(a) Diagram of a liquid with its vapor at equilibrium in a closed container at a given temperature. (b) The same system at a higher temperature.*

Figure 13.11 *The boiling point of water at different altitudes.*

equal to the atmospheric pressure. Since the atmospheric pressure varies with altitude and weather conditions, boiling points of liquids also vary (Figure 13.11).

The cooking of foods requires that they be supplied with a certain amount of energy. When the pressure is 1 atm, water boils at 100 °C, and an egg can be placed in the water and soft-boiled in 3 minutes. At reduced atmospheric pressure, water boils at a lower temperature, so it has less heat energy to cook the egg. It would take longer to cook an egg in boiling water on top of Mount Everest.

The boiling point is increased when external pressure is increased. The operation of pressure cookers and hospital autoclaves is based on this principle. We can achieve a higher temperature at the higher pressures attained in these closed vessels. (Heat added to a liquid at its boiling point in an open container would simply convert liquid to vapor. No increase in temperature would occur until all the liquid had vaporized.) Bacteria—even resistant spores—are killed more rapidly in a pressure cooker or an autoclave, not directly by the increased pressure but, rather, by the higher temperatures attained. Table 13.1 gives the temperatures attainable with pure water at various pressures.

The boiling point of a liquid is a useful physical property; it is often used as an

Table 13.1 **Boiling Points of Pure Water at Various Pressures**

Boiling Point (°C)	*Pressure (mm Hg)*
80	355
85	434
90	526
95	634
98	707
100	760
102	816
104	875
106	938
110	1075

Table 13.2 **Boiling Points of Several Liquids at 1 atm Pressure**

Compound	*Boiling Point (°C)*
Diethyl ether, $C_2H_5OC_2H_5$	34.6
Acetone, C_3H_6O	56.2
Methanol, CH_3OH	64.5
Ethanol, C_2H_5OH	78.3
Water, H_2O	100.0
Mercury, Hg	356.6

aid in identifying compounds. Since boiling point varies with pressure, it is necessary to define the **normal boiling point** as that temperature at which a liquid boils under standard pressure (1 atm, or 760 torr). As an alternative, one can specify the pressure at which the boiling point was determined. For example, the *Handbook of Chemistry and Physics* lists the boiling point of antipyrine (a pain reliever and fever reducer) as 319^{741}. This means that the compound boils at 319 °C under a pressure of 741 torr. Table 13.2 gives the normal boiling points of some familiar liquids.

Distillation

Liquids can be purified by a process called **distillation**. Imagine a mixture of water and some nonvolatile material such as table salt. If the mixture is heated until it boils, the water will vaporize, but the nonvolatile material will not. The water vapor can then be condensed back to the liquid state and collected in a separate container as shown in Figure 13.12. The condensed vapor sample obtained is called

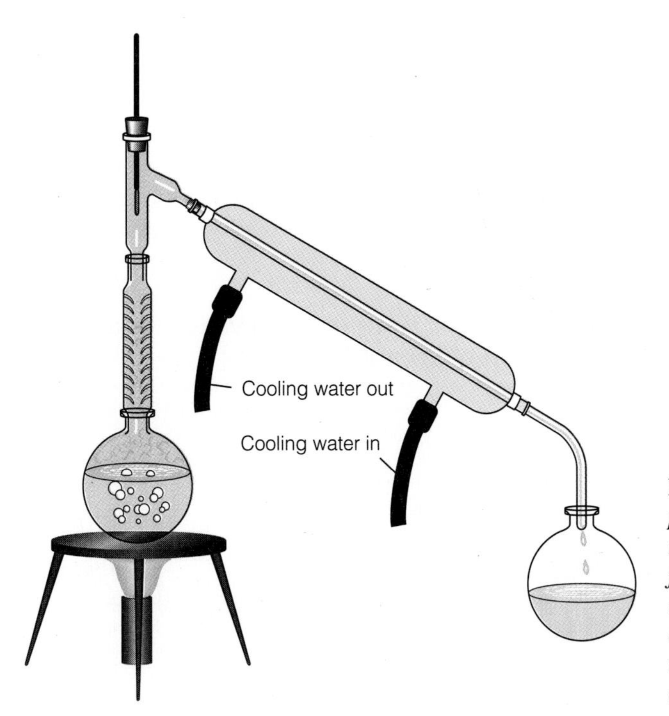

Figure 13.12 *A distillation apparatus. A mixture is heated in the flask at the left. The vapors formed travel up the vertical column, are then condensed in the cooled tube angled downward toward the right, and are finally collected in the flask at the right.*

the **distillate**. Thus, the distillate, water in this case, is separated from the other component of the mixture and is thereby purified.

Even if a mixture contains two or more volatile components, purification by distillation is possible. Let's consider a mixture of two components, one of which is somewhat more volatile than the other. At the boiling point of such a mixture, both components will contribute some molecules to the vapor. The more volatile component, because it is more easily vaporized, will have a larger fraction of its molecules in the vapor state than will the less volatile component. When this vapor is condensed into another container, the resulting liquid (the distillate) will be richer in the more volatile component than was the original mixture. As distillation continues, the boiling point rises until, finally, the boiling point of the less volatile liquid is reached. The first portion of the distillate has the greatest concentration of the more volatile component. The concentration of the less volatile component in the distillate increases as distillation proceeds. Thus, the more volatile liquid in the mixture can be obtained by collecting only the first portion of the distillate.

Heat of Vaporization

Heat is required to convert a liquid to a vapor. A liquid that evaporates at room temperature absorbs heat from its surroundings, so it has a cooling effect on the surroundings. Even on a warm day, we feel cool after a swim because the water evaporating from our skin removes heat from the skin. Evaporation of perspiration also produces a cooling effect. Bathing the skin with alcohol also acts to cool the skin by evaporation. At the molecular level, it is the molecules with higher than average kinetic energy that evaporate first. This takes energy, so the average kinetic energy of the remaining molecules decreases. Evaporating 1 g of water takes 2.26 kJ (540 cal).

Certain liquids (such as ethyl chloride) can evaporate from the skin rapidly enough to freeze a small area and cause it to be insensitive to pain.

The quantity of heat required to vaporize 1 mol of a liquid at constant pressure and temperature is called the **molar heat of vaporization**. This value is a characteristic property of a given liquid that largely depends on the types of intermolecular forces in the liquid. Water, with molecules strongly associated through hydrogen bonding, has a heat of vaporization of 40.7 kJ/mol. Methane, with molecules held by weak dispersion forces, has a heat of vaporization of only 0.971 kJ/mol. Table 13.3 gives heats of vaporization of several liquids.

The following examples demonstrate the conversion of molar heat of vaporization to heat of vaporization per gram and the determination of the amount of heat required to vaporize a given mass of a specific liquid sample.

Table 13.3 **Molar Heats of Vaporization (at the Boiling Point) of Several Liquids**

	Molar Heat of Vaporization	
Compound	*kJ/mol*	*kcal/mol*
Diethyl ether, $C_2H_5OC_2H_5$	26.0	6.21
Methanol, CH_3OH	38.0	9.08
Ethanol, C_2H_5OH	39.3	9.39
Water, H_2O	40.7	9.72
Mercury, Hg	59.2	14.20

EXAMPLE 13.4

The molar heat of vaporization of ammonia, NH_3, is 2.33 kJ/mol. What is the heat of vaporization in joules per gram?

SOLUTION

Use the molar mass of NH_3, 17.0 g/mol, to convert kilojoules per mole to joules per gram.

$$\frac{2.33\ \text{kJ}}{\cancel{\text{mol}}} \times \frac{1\ \cancel{\text{mol}}}{17.0\ \text{g}} = 0.137\ \text{kJ/g or } 137\ \text{J/g}$$

See Problems 13.31–13.48.

EXERCISE 13.2

How much heat (in kilojoules) would be required to vaporize 400. g of water at its boiling point? The heat of vaporization of water is 2.26 kJ/g.

When a vapor condenses to a liquid, it gives up exactly the same amount of heat energy as was used in converting the liquid to a vapor. During the operation of a refrigerator, a fluid is alternately vaporized and condensed. The heat required to vaporize the fluid is taken from the refrigerated compartment. The heat is then released to the outside surroundings when the fluid is condensed back to the liquid state.

13.5 The Solid State

In solids, the particles (atoms, molecules, or ions) are so close together that very little compression is possible. The motions of particles in the liquid and solid states differ significantly. In liquids, particles are in constant—but somewhat restricted—motion. In solids, there is little motion of particles other than gentle vibration about a fixed point. Consequently, diffusion in solids is extremely slow. An increase in temperature increases the vigor of the vibrations in a solid. If the vibrations become sufficiently violent, the solid turns to a liquid—it melts (Section 13.6).

Noncrystalline Solids

Glass, rubber, wax, and many plastics are examples of **amorphous solids**. This term is derived from a Greek word that means "without shape." The particles in amorphous solids do not have a definite or regular order; the pattern does not repeat itself throughout the solid. These solids are not crystalline; they shatter in irregular patterns to give pieces with jagged edges and with irregular angles. In many amorphous solids there is some freedom of motion of the particles. They do not have a definite melting point; they soften gradually when heated.

Additional information on various crystalline arrangements can be obtained by consulting more advanced texts. Much research is currently being done with crystalline arrangements in superconducting materials.

Crystalline Solids

In **crystalline solids**, the particles are arranged in regular, systematic patterns called a **crystal lattice**. Three types of crystal lattices based on the cube are shown

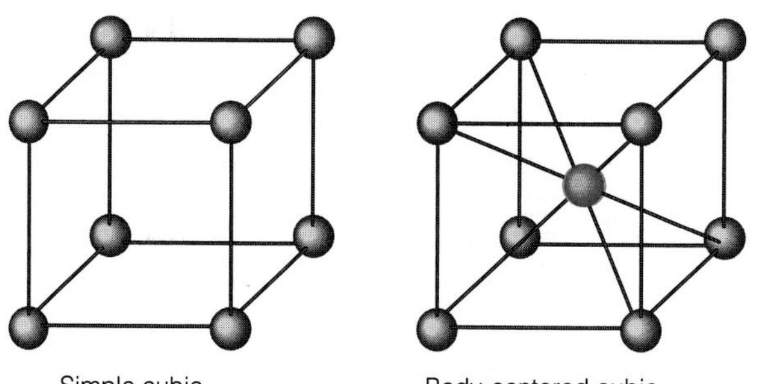

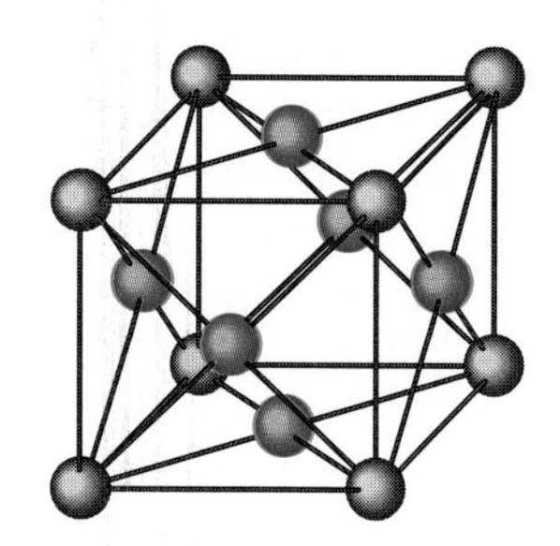

Figure 13.13 *Three types of crystal lattices based on the cube.*

Table 13.4 **Some Characteristics of Crystalline Solids**

Crystal Type	*Particles in Crystal*	*Principal Attractive Force Between Particles*	*Melting Point*	*Electrical Conductivity of Liquid*	*Characteristics of the Crystal*	*Examples*
Ionic	Positive and negative ions	Interionic attraction between ions (very strong)	High	High	Hard, brittle, most dissolve in polar solvents	NaCl, CaF_2, K_2S, MgO
Molecular						
Hydrogen-bonded	Molecules with H on N, O, or F	Hydrogen bonds (intermediate)	Intermediate	Very low	Fragile, soluble in other H-bonding liquids	H_2O, HF, NH_3, CH_3OH
Polar	Polar molecules (no H bonds)	Electrostatic attraction between dipoles (rather weak)	Low	Very low	Fragile, soluble in other polar and many nonpolar solvents	HCl, H_2S, $CHCl_3$, ICl
Nonpolar	Atoms or nonpolar molecules	Dispersion forces only (weak)	Very low	Extremely low	Soft, soluble in nonpolar or slightly polar solvents	S_8, I_2, CH_4, CO_2, CCl_4
Covalent network	Atoms	Covalent bonds (very strong)	Generally do not melt	—	Very hard, insoluble	Diamond (C), SiC, AlN
Metallic	Positive ions plus mobile electrons	Metallic bonds (strong)	Most are high	Very high	Most are hard, malleable, ductile, good conductors, insoluble unless a reaction occurs	Cu, Ca, Al, Pb, Zn, Fe, Na, Ag

in Figure 13.13. In one type of lattice structure, atoms (or other unit particles) are positioned directly above and below each other to form a *simple cubic* arrangement. This arrangement is not common; closer packing is possible with other arrangements. Chromium, manganese, iron, and the alkali metals—notice their locations in the periodic table—have a *body-centered cubic* arrangement that is like the simple cubic arrangement except one additional atom is at the center of the cube. Copper, silver, gold, nickel, palladium, and platinum have *face-centered cubic* arrangements. (Notice their locations in the periodic table.) Certain ionic compounds including sodium chloride, potassium chloride, and calcium oxide also have the face-centered arrangement with particles positioned at each corner and at the center of each of the six faces of a cube.

Bonding Classifications for Solids

Crystalline solids may also be classified on the basis of the types of forces holding the particles together. The four classes are ionic, molecular, covalent network (macromolecular), and metallic. Characteristics of these bonding types are described in the following paragraphs and are summarized in Table 13.4.

Ionic solids have ions at each lattice point in the crystal. Sodium chloride, NaCl, is a typical ionic solid. It has a face-centered cubic arrangement; each chloride ion is surrounded by six sodium ions, and each sodium ion is surrounded by six chloride ions. Because interionic forces are quite strong, ionic solids are quite hard. They have high melting points and low vapor pressures. They also tend to be water soluble but are insoluble in nonpolar solvents.

Molecular solids have individual, discrete covalent molecules at the lattice points of crystals. These solids are held together by several types of rather weak forces. Examples include the London dispersion forces present in crystalline iodine, I_2, the dispersion forces plus dipolar forces in iodine chloride, ICl, and the hydrogen bonds present in ice. Dry ice (solid CO_2), sulfur crystals (made up of S_8 molecules), and many organic chemicals are molecular. Molecular solids are typically soft, have low melting points, and are generally—not always—insoluble in water but are more soluble in nonpolar solvents. Many huge molecules, such as viruses, can also have definite crystalline arrangements, as shown in Figure 13.14.

One nonpolar molecular solid, *para*-dichlorobenzene, $C_6H_4Cl_2$, is used as moth flakes. It has a low melting point of 53 °C and dissolves in nonpolar solvents. It has

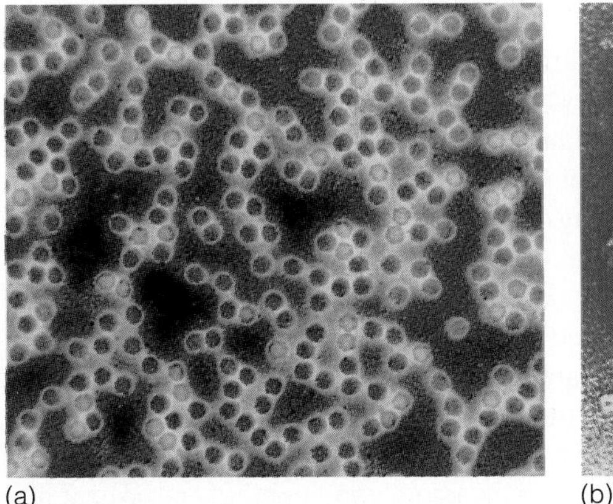

(a)

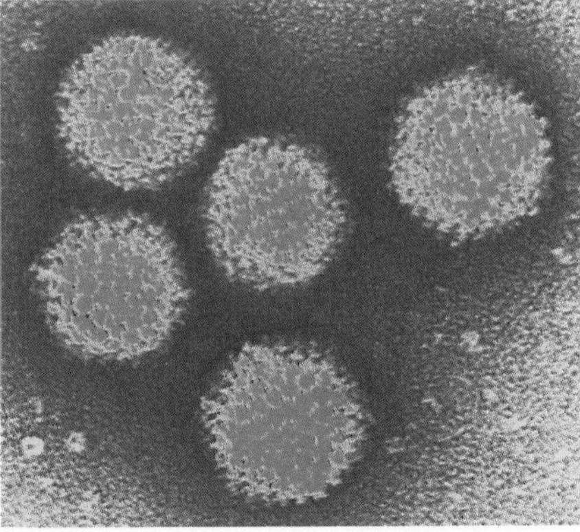

(b)

Figure 13.14 *(a) Color-enhanced human polio virus magnified about 20,000 times. (b) Adenovirus magnified about 210,000 times. Viruses like this cause upper respiratory tract infections.*

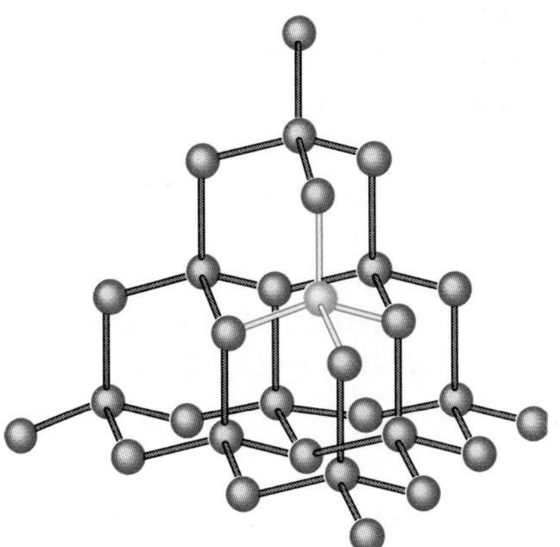

Figure 13.15 *The cyrstal structure of diamond, a covalent network solid. Each corner or intersection of lines represents a carbon atom. Notice the tetrahedral arrangement at each carbon atom.*

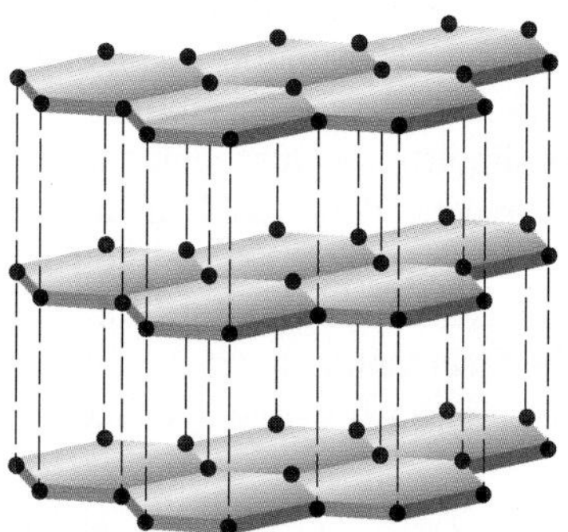

Figure 13.16 *In graphite the carbon atoms form a network of hexagonal rings within each plane. The planes are stacked in layers that can slide over one another.*

a relatively high vapor pressure and a distinctive odor. Camphor, $C_{10}H_{16}O$, is a nonpolar molecular solid that is used in certain medical preparations. It has a distinctive odor and a relatively low melting point of 179 °C.

Covalent network or **macromolecular solids** have atoms at the lattice points that are joined by covalent bonds to make networks that extend throughout a sample. Thus, each crystal is in essence one enormous molecule—a macromolecule. These solids are generally extremely hard and nonvolatile. They melt (sometimes with decomposition) at extremely high temperatures.

Silicon carbide, SiC, also called carborundum, has an extensive network of covalent bonds. The hardness of this compound makes it useful for many applications, including grinding stones, grinding wheels, and abrasive products.

Diamond conducts heat about six times better than the best conducting metals.

Diamond is a crystalline form of carbon with atoms in a covalent network. Each carbon atom is covalently bonded in a tetrahedral arrangement to four other carbon atoms as shown in Figure 13.15 to form an extremely hard crystal. But pure carbon is not always diamond; carbon can have other arrangements of atoms. When there are two or more different physical forms of an element in the same state, they are called **allotropes**. Diamond, graphite, and buckminsterfullerene molecules are allotropes of carbon. Graphite (the ''lead'' of pencils) is black and slippery; it is used as a dry lubricant. The carbon atoms in graphite form a network of hexagonal rings joined together within a plane, somewhat like traditional chicken wire. In the total graphite structure (Figure 13.16), the planes of carbon atoms are stacked in layers. These layers can slide over one another because they are not joined by covalent bonds. This accounts for the lubricating properties of graphite. The loosely held electrons between graphite layers account for its ability to conduct electricity. *Coke* (from coal heated in an absence of air), *charcoal* (from wood heated in an absence of air), and *carbon black* (finely divided soot from a smoky flame) are forms of carbon that are thought to have the graphite structure.

The **buckminsterfullerene molecule,** an allotropic form of carbon with 60 carbon atoms in a soccer-ball-shaped spherical arrangement, is described in a special box in Section 6.7.

Metallic solids have positive ions at the lattice points. The valence electrons are distributed throughout the lattice, almost like a fluid. These electrons, which can move freely about the lattice, make the metals act as good conductors of heat and electricity. Some metals, such as sodium or potassium, are fairly soft and have low melting points. Others, such as manganese and iron, are hard and have high melting points. (Their extra electrons seem to lead to stronger forces between atoms.) Most metals are *malleable*: they can be hammered or rolled into sheets. Many metals are *ductile*: they can be drawn or pulled into wire.

EXAMPLE 13.5

Describe differences and give an example for each of the following.

(a) Crystalline and amorphous solids
(b) Molecular and network solids.

SOLUTION

(a) Crystalline solids such as NaCl have discrete arrangements of atoms, ions, or molecules in a lattice system. Particles in amorphous solids such as glass or rubber have no definite arrangement.
(b) Both molecular and network solids have covalent bonds, but there are no discrete, individual molecules in network solids. Molecular solids such as camphor and iodine have low melting points. Network solids such as diamond are extremely hard and have high melting points.

See Problems 13.49–13.54.

13.6 Melting and Freezing

When a crystalline solid is heated, its particles vibrate more vigorously. As the temperature increases, the substance turns from a solid to a liquid when the attractive forces within the solid are overcome by vigorous vibration. This process is called **melting**. The temperature at which the solid and liquid exist in dynamic equilibrium is called the **melting point**. That is, at the melting point, particles move from the solid to the liquid state at the same rate they move from the liquid to the solid state. This equilibrium is often represented as follows.

$$\text{Solid} \underset{\text{freezing}}{\overset{\text{melting}}{\rightleftharpoons}} \text{Liquid}$$

As the temperature of a liquid drops, the substance turns from a liquid to a solid. This process is called **freezing**. The **freezing point** is the temperature at which the liquid and solid are in dynamic equilibrium.

The temperature at the melting point of a solid is precisely the same as the temperature at the freezing point of the liquid—the distinction is the direction from which we approach this equilibrium temperature. We would give the melting point for iron or lead; they are solids at room temperature. On the other hand, we would give the freezing point of water or alcohol; they are liquids at room temperature. The freezing point of water is the same as the melting point of ice; the distinction depends on the original state.

Although water is an exception, nearly all substances expand when they melt. Substances that have weak intermolecular forces have low melting points; substances with strong intermolecular forces have higher melting points. Pure crystalline solids have sharp melting points while amorphous materials like glass and plastic soften over a wider temperature range.

EXAMPLE 13.6

The melting point of tin is 232 °C. What is the freezing point of tin? The freezing point of mercury is −39 °C. What is its melting point?

SOLUTION

The melting point is the same temperature as the freezing point. The freezing point (or melting point) of tin is 232 °C The melting point (or freezing point) of mercury is −39 °C

Heat of Fusion

Fusion means melting. Thus, the molar heat of fusion is the heat required to melt 1 mole of the sample.

The amount of heat required to convert 1 mol of a solid to a liquid at its melting point is called the **molar heat of fusion**. The molar heat of fusion of water is 5.98 kJ. The intermolecular forces of attraction within liquids are not as great as those present in solids; the difference in energy between the two is the heat of fusion. For comparison, the intermolecular forces of attraction within gases are quite small compared to those present in liquids; the difference in energy between the two is the heat of vaporization (Section 13.4). Water has a molar heat of vaporization of 40.7 kJ, a value that is considerably larger than its molar heat of fusion, 5.98 kJ (see Figure 13.17).

At the molecular level we can explain why the molar heat of fusion is much smaller than the molar heat of vaporization for a substance: When a sample is melted, energy is required to disrupt the crystal lattice to allow the particles to move around freely in the liquid state, but the particles remain in contact with one another under mutual attractions. This is much less energy than is required during vaporization where nearly all intermolecular forces must be overcome. Table 13.5 gives the molar heat of fusion in kilojoules per mole and in kilocalories per mole for several solids.

EXAMPLE 13.7

How much heat (kilojoules) would be required to melt 425 g of ice?

SOLUTION

Begin with the quantity of ice in grams, convert this to moles, and use the molar heat of fusion (Table 13.5) to determine the heat in kilojoules.

See Problems 13.55–13.58.

$$425\text{ g} \times \frac{1\text{ mol}}{18.0\text{ g}} \times \frac{5.98\text{ kJ}}{\text{mol}} = 141\text{ kJ}$$

Table 13.5 **Molar Heats of Fusion (at the Melting Point) of Several Substances**

Substance	*Melting Point (°C)*	*Molar Heat of Fusion*	
		kJ/mol	*kcal/mol*
Ethanol, C_2H_5OH	−117	5.02	1.20
Water, H_2O	0	5.98	1.44
Sodium chloride, NaCl	804	30.2	7.22
Copper, Cu	1083	13.0	3.11
Iron, Fe	1530	14.9	3.56
Tungsten, W	3407	35.4	8.05

Exercise 13.3
The molar heat of fusion of naphthalene, $C_{10}H_8$ (used in moth balls), is 19.3 kJ. What is the heat of fusion in kilojoules per gram? *Hint*: Convert kJ/mol to kJ/g.

13.7 *Heating and Cooling Curves*

In the laboratory, we can start with a solid sample of a pure substance and record temperatures every half-minute as the sample is heated at a constant rate. The data can then be used to plot a graph of temperature on the ordinate (the vertical axis) versus the time of heating—or energy—on the abscissa (the horizontal axis) as shown in Figure 13.17. This is called a **heating curve**.

As a solid sample is heated, we observe that the temperature rises steadily until it begins to melt. The temperature then remains constant until the sample is melted.

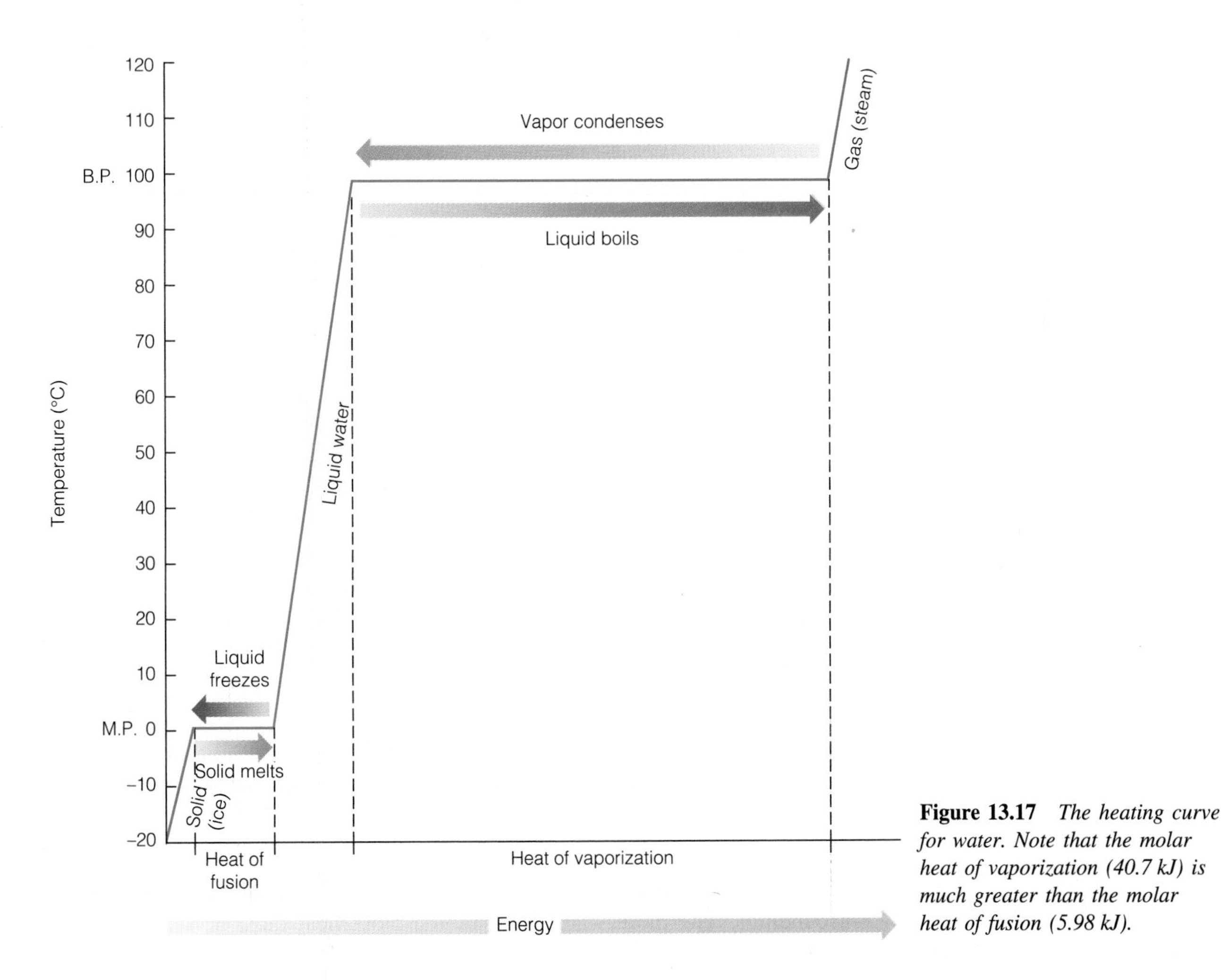

Figure 13.17 *The heating curve for water. Note that the molar heat of vaporization (40.7 kJ) is much greater than the molar heat of fusion (5.98 kJ).*

Table 13.6 **Specific Heats of Some Common Substances**

	Specific Heat	
Substance	*J/g-°C*	*cal/g-°C*
Water, H_2O, liquid	4.18	1.000
Ethanol, C_2H_5OH	2.46	0.588
Water, H_2O, ice	2.09	0.500
Water, H_2O, steam	1.97	0.471
Wood	1.8	0.43
Aluminum, Al	0.900	0.215
Glass	0.502	0.120
Iron, Fe	0.444	0.106
Copper, Cu	0.385	0.0920
Silver, Ag	0.225	0.0538
Gold, Au	0.128	0.0306

As heating continues, the temperature once again begins to climb. Eventually, the sample begins to boil, and the temperature once again stops climbing. It remains constant until all the sample has been vaporized if we are using a container that is open to the atmosphere. If the sample is in a sealed vessel, vapor cannot escape, so the temperature of the vapor once again begins to climb. **Superheated steam** is produced when water is heated above its boiling point in a closed vessel such as a pressure cooker or a hospital autoclave.

For portions of the heating curve where the temperature is rising, the increase in heat energy causes an increase in the average kinetic energy of the molecules. While the substance remains in one particular state, the heat energy required to warm the sample is equal to the mass of the sample multiplied by the temperature change and by its specific heat. (Specific heats of several substances are listed in Table 13.6.) When a substance is melting or boiling, the curve remains horizontal (flat)—the temperature and the average kinetic energy remain constant. What is happening here? Recall that the heat of fusion is the energy taken in by the molecules during melting. This energy is needed to break up the crystal lattice. The curve also is flat during boiling; energy equal to the heat of vaporization is needed to overcome nearly all attractive forces between the molecules.

A cooling curve for a substance looks like a mirror image of the heating curve. If we cool a vapor, its temperature decreases steadily until the boiling point is reached. At the boiling point, the curve levels off until the vapor is condensed to a liquid. Upon further cooling, the temperature of the liquid decreases steadily until the freezing point is reached. After freezing is complete, further cooling decreases the temperature of the solid. Energy is released by the substance during each stage of the cooling process.

EXAMPLE 13.8

Describe why the temperature of a container of water placed in a freezer drops consistently to the freezing point, but then stays right at the freezing point for an extended period of time before freezing completely.

SOLUTION

As the liquid turns to a solid, a definite amount of energy must be released to the surroundings. The more water there is, the more heat must be released. The substance cannot freeze until this energy has been removed from the liquid. The energy that must be removed during freezing is equal to the mass of the sample in grams multiplied by the heat of fusion.

13.8 Water: A Most Unusual Liquid

Water is probably the most familiar liquid compound to everyone, but it is a most unusual chemical. For example, at room temperature, it is the only liquid compound with a molar mass as low as 18.0 g/mole. Other special properties are described here.

Densities of Solid and Liquid Water

The solid form of water (ice) is less dense than the liquid, a very rare situation. The consequences for life on this planet due to this peculiar characteristic are immense. Ice forms on the surface of lakes when the temperature drops below freezing; it insulates the water below, enabling fish and other aquatic organisms to survive the winters of the temperate zones. If ice were more dense than liquid water, it would sink to the bottom as it is formed. This would permit more water to freeze and also sink to the bottom. As a result, even the deeper lakes of the northern latitudes would freeze solid in winter.

CHEMISTRY IN OUR WORLD

Flash Freezing

Many foods we buy are ''flash frozen''; that is, they are frozen so rapidly that any ice crystals formed are kept very small. Thus, any damage resulting from expansion and rupturing of food cells is kept to a minimum.

Perhaps you have heard of plans to preserve the body of a person who dies of cancer and then to revive the body after a cure for cancer has been found. This would require ''flash freezing'' the body almost instantly after death. The freezing portion of the scheme has been attempted on a number of bodies. To date, no revivals have succeeded.

Many packaged fruits and vegetables that we buy at the supermarket are flash frozen to minimize rupturing the tiny cells in the food.

The same property—the relative densities of ice and water—has dangerous consequences for living cells. Since ice has a lower density than liquid water, 1.00 g of ice occupies a larger volume than 1.00 g of liquid water. As ice crystals form in living cells, the expansion ruptures and kills the cells. The slower the cooling, the larger the crystals of ice and the more damage to the cell.

The Specific Heat of Water

Another unusual property of water is its relatively high specific heat. It takes precisely 1 cal (4.184 J) of heat to raise the temperature of 1 g of liquid water 1 °C. That's roughly 10 times as much energy as is required to raise the temperature of the same amount of iron 1 °C. (Compare the specific heats of several common substances that are given in Table 13.6.)

Cooking utensils made of iron, copper, aluminum, or glass have low specific heats, so they heat up very quickly. Handles of many frying pans are made of plastic or other insulating materials because they have high specific heats. When heated, their temperatures increase slowly.

Energy Changes: Ice to Steam

To determine the amount of heat required to raise the temperature of a solid sample at a temperature below its freezing point to a gas at its boiling point, we must work the problem in stages, as shown in the following example.

EXAMPLE 13.9

How much energy (in kilojoules) is required to change 10.0 g of ice at −10.0 °C to steam at 100 °C?

SOLUTION

This problem involves phase changes; it is worked in several stages.

STEP 1 First calculate the energy required to raise the temperature of 10.0 g of ice from −10.0 °C to 0 °C, a change of 10.0 °C. The specific heat of ice (Table 13.6) is 2.03 J/g-°C.

$$10.0 \text{ g} \times 10.0 \text{ °C} \times \frac{2.03 \text{ J}}{\text{g-°C}} = 203 \text{ J}$$

STEP 2 Use the heat of fusion of water, 335 J/g, to obtain the energy required to melt the given mass of ice at 0 °C.

$$10.0 \text{ g} \times \frac{335 \text{ J}}{\text{g}} = 3350 \text{ J}$$

STEP 3 Calculate the heat required to raise the temperature of water 100 °C (that is, 100 °C − 0 °C). The specific heat of water is 4.18 J/g-°C.

$$10.0 \text{ g} \times 100.0 \text{ °C} \times \frac{4.18 \text{ J}}{\text{g-°C}} = 4.18 \text{ kJ}$$

STEP 4 Calculate the amount of heat required to change the water at 100 °C to steam at 100 °C. The heat of varporization of water is 40.7 kJ/mol or 2.26 kJ/g.

$$10.0 \text{ g} \times \frac{2.26 \text{ kJ}}{\text{g}} = 22.6 \text{ kJ}$$

STEP 5 Finally, total all the calculated values in kJ.

To raise the temperature of ice from −10 °C to 0 °C	0.20 kJ
To change the ice to liquid water	3.35 kJ
To raise the temperature of water from 0 °C to 100 °C	4.18 kJ
To change the water to steam	22.6 kJ
	30.3 kJ

Notice that almost 75% of the total energy is used to vaporize the water.

See Problems 13.59–13.62.

The Specific Heat of Water: Environmental Implications

Because water has a relatively high specific heat, a relatively large amount of heat is required to raise the temperature of a water sample, and a large amount of heat is given off during a small drop in temperature. The vast amounts of water on the surface of the Earth alternately store and release heat. This tends to moderate daily temperature variations. To appreciate this important property of water, we need only consider the extreme temperature changes on the surface of the moon where no bodies of water are present. The temperature of the moon varies from just above the boiling point of water (100 °C) to about −175 °C, a range of 275 °C. In contrast, temperatures on the Earth rarely fall below −50 °C or rise above 50 °C, a range of only 100 °C.

Heat of Vaporization of Water: Its Implications

Another unusual property of water is its relatively high heat of vaporization. Thus, a large amount of heat is required to evaporate a small amount of water. This is of great importance to animals. Large amounts of body heat, produced as a by-product of metabolic processes, can be dissipated by the evaporation of small amounts of water (perspiration) from the skin. The heat to vaporize this water is obtained from the body, so the body is cooled by the process. Conversely, when steam condenses, a considerable amount of heat is released. For this reason, steam causes serious burns when it contacts the skin.

Along with the temperature-modifying effect of the high specific heat of water, the high heat of vaporization of water also has a temperature-moderating effect. A large portion of the heat that would otherwise warm the land is used, instead, to vaporize water from lakes and oceans. Thus, summer daytime temperatures do not tend to rise as high in areas that are near large bodies of water.

Hydrogen Bonding in Water: The Implications

The special properties of water that have been described here result from the unique structure of polar water molecules. In the liquid state, water molecules are strongly associated by hydrogen bonding (Section 13.2). These forces must be overcome if vaporization is to take place, so a large amount of energy—equal to the heat of vaporization—must be supplied to convert liquid water to a vapor.

In the solid state, water molecules—in the form of ice—are in an ordered arrangement as shown in Figure 13.18. This orderly, crystalline arrangement in ice is unique; the molecules are less compact than in the liquid state. As a result of hydrogen bonding, large hexagonal holes are incorporated into the ice lattice. It is this empty space that makes ice less dense than liquid water.

Figure 13.18 *A three-dimensional model of ice showing large hexagonal holes formed by six water molecules. Hydrogen bonds betwen water molecules are shown in yellow.*

Water as a Solvent

Because of its highly polar nature, water is a good solvent for many ionic substances. Although high temperatures are required to break the strong bonds of sodium chloride, an ionic solid, these bonds are readily broken by simply placing the sodium chloride in water at room temperature. The ionic bonds are broken as the salt dissolves. In the melting process, we are simply putting in enough heat energy to disrupt the crystal lattice. In the dissolving process, the ions are attracted away from the crystal lattice.

Here is what is happening. Water molecules surround the lattice. As they approach a negative ion, they align themselves so that the *positive* water dipole ends point toward the ion. When water molecules approach a positive ion, however, they align themselves so that the *negative* dipole ends point toward the ion. The attraction between a single water dipole and an ion is less than the attraction between two ions, but when several water molecules surround each ion, the ion-to-ion attractions are overcome by the attractions of several water dipoles. Once the ion is taken into solution (and is dissolved), it remains surrounded by several water molecules (Figure 13.19). Solutions are quite important; we shall devote the entire next chapter to this subject.

See Problems 13.63–13.68.

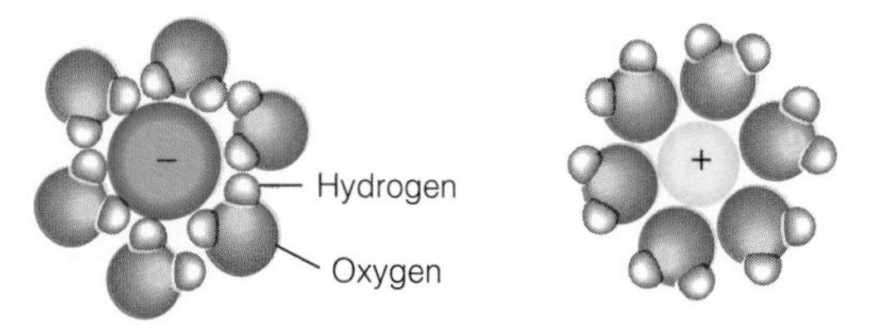

Figure 13.19 *The interaction of polar water molecules with ions.*

Chapter Summary

Gases have neither a definite shape nor volume. Liquids have a definite volume, but no definite shape. Solids have both a definite shape and a definite volume. The particles in a gas are much farther apart than in a liquid. We have shown, by calculation, that the volume occupied by a gas is roughly 1300 times as great as the volume occupied by the same number of molecules of the liquid.

All gases at room temperature are either covalent molecules or monatomic noble gases. Except for metallic mercury, all pure liquids at room temperature are also covalent molecules. Differences in viscosity and surface tension are related to intermolecular forces and molecular size. Vaporization occurs whenever molecules of a liquid break away from the surface of the liquid and go into the gaseous state. In an open container, the quantity of liquid diminishes during vaporization; it evaporates. In a closed container, dynamic equilibrium is established when the rate of evaporation is equal to the rate of condensation.

As a liquid is heated, the vapor pressure of the liquid increases. The boiling point is reached when the vapor pressure of the liquid is equal to the atmospheric pressure; the lower the atmospheric pressure, the lower the boiling point. During distillation, the vapor is condensed and collected. If a mixture of liquids is distilled, the more volatile liquid tends to vaporize first, allowing it to be separated from the less volatile component in the mixture.

Metallic solids conduct an electric current. Molecular solids are nonconductors and have low melting points. They are held together by several types of intermolecular forces, called van der Waals forces, which include dipole forces, hydrogen bonds, and dispersion (London) forces.

Noncrystalline solids are also called amorphous solids. Crystalline solids have definite lattice structures such as simple cubic, body-centered cubic, and face-centered cubic arrangements. Ionic substances form crystalline solids with high melting points. Their water solutions conduct an electric current. Insoluble, nonreactive solids with extremely high melting points have covalent network bonding; they are also called macromolecular solids.

Different forms of the same element are called allotropes. They either have different crystalline arrangements or no crystalline arrangement at all. Diamond, graphite, and buckminsterfullerene molecules are allotropes of carbon.

A heating curve for a substance shows that as a solid is heated, the temperature climbs consistently until it reaches its melting point where the temperature remains constant during melting. Here, energy equal to the heat of fusion is required to convert the solid to a liquid. Similarly, the temperature of the liquid climbs consistently until it reaches the boiling point where the temperature once again stays constant. Here, energy equal to the heat of vaporization is taken in by the molecules as they are converted from liquid to gas.

The special structure and bonding characteristics of water give it unique properties. As a result of hydrogen bonding, molecules in ice are less compact than in liquid water; the density of ice is less than that of the liquid. The expansion of water on freezing is responsible for weathering of rocks and for the formation of ice on the surface of a lake. Because of a high specific heat, much energy is absorbed in raising the temperature of a water sample only a few degrees, but, conversely, much heat is released by water when its temperature drops only a few degrees. With a high heat of vaporization, much energy is taken up in evaporating a sample of water. This is why perspiration is an effective means of cooling the body. Of all liquids and solids we have considered, water is unique—its properties mean the difference between life and death.

Assess Your Understanding

1. Use physical properties of substance to make generalizations about bonding in gases, liquids, and solids. [13.1]
2. Describe and identify intramolecular and intermolecular forces. [13.2]
3. Describe dipole forces, hydrogen bonds, and dispersion forces. [13.2]
4. Describe reasons for variances in viscosity and surface tension. [13.3]
5. Describe dynamic equilibrium, evaporation, and boiling point. [13.4]
6. Calculate energy changes that occur during changes of state. [13.4, 13.6]

7. Describe the bonding in crystalline and noncrystalline solids. [13.5]
8. Describe the different portions of heating and cooling curves. [13.7]
9. List and explain implications of the unique properties of water. [13.8]

Key Terms

allotropes [13.5]
amorphous solids [13.5]
boiling point [13.4]
condensation [13.4]
covalent network (macromolecular) solids [13.5]
crystal lattice [13.5]
crystalline solids [13.5]
dipoles [13.2]
distillate [13.4]
distillation [13.4]
dynamic equilibrium [13.4]
evaporation [13.4]
freezing [13.6]
freezing point [13.6]
heating curve [13.7]
hydrogen bond [13.2]
interionic forces [13.2]
intermolecular forces [13.1, 13.2]
intramolecular forces [13.2]
ionic solids [13.5]
London (dispersion) forces [13.2]
melting [13.6]
melting point [13.6]
metallic solids [13.5]
molar heat of fusion [13.6]
molar heat of vaporization [13.4]
molecular solids [13.5]
normal boiling point [13.4]
superheated steam [13.7]
surface tension [13.3]
surfactants [13.3]
van der Waals forces [13.2]
vaporization [13.4]
viscosity [13.3]
wetting [13.3]

Problems

GENERALIZATIONS RELATED TO GASES, LIQUIDS, AND SOLIDS

13.1 Why are liquids and solids virtually incompressible?

13.2 Using the kinetic theory, describe differences between particles of liquids and of solids. In what ways are liquids and solids similar and different?

13.3 With gas law calculations we can determine that 1 mol of water molecules at 1 atm and at 20 °C occupies 24.0 L. Assume that a container with this volume is filled with liquid water. Using the density of water as 1.00 g/mL, and other factual information, determine the number of molecules that would be present in this 24.0 L of liquid water.

13.4 Using information in Problem 13.3, determine the ratio for the number of gaseous versus liquid water molecules in 24.0 L. Explain.

13.5 For molecules with similar intermolecular forces at a fixed temperature, describe the relationship of molar mass and physical state.

13.6 For the halogens, discuss physical state in terms of molar mass.

13.7 From the information given, determine whether a substance described is ionic, polar covalent, nonpolar covalent, or metallic.
a. A liquid (a vegetable oil) is miscible with hexane but not with water.
b. A solid (in a fertilizer) dissolves in water and the solution will conduct electricity.

13.8 From the information given, determine whether a substance described is ionic, polar covalent, nonpolar covalent, or metallic.
a. A pure liquid (methanol) is miscible with water but not with hexane.
b. A solid that melts at 150 °C will not dissolve in water but dissolves in hexane. The solid does not conduct electricity.

INTERMOLECULAR AND INTERIONIC FORCES

13.9 Distinguish between *intermolecular* and *intramolecular* forces.

13.10 Distinguish between *interionic* and *intermolecular* forces.

13.11 How are van der Waals forces and intermolecular forces related?

13.12 In addition to interionic forces, give three different intermolecular forces. Rank all four types of forces in order of increasing strength.

13.13 For the following compounds, classify the types of forces present as London forces, dipole forces, hydrogen bonds, or combinations of these.

a. Cl_2 b. HF c. CO d. NO_2
e. CH_3CH_2OH

13.14 For the following compounds, classify the types of forces present as London forces, dipole forces, hydrogen bonds, or combinations of these.
a. NH_3 b. PH_3 c. Br_2 d. SO_2
e. $CH_3CH_2—O—CH_2CH_3$

13.15 Rank N_2, H_2O, NaCl, and HCl in order of increasing strength of intermolecular and interionic forces (weakest first). What forces are present for each?

13.16 Rank NH_3, HI, $NaNO_3$, and I_2 in order of increasing strength of intermolecular and interionic forces (weakest first). What forces are present for each?

13.17 Compare the dispersion (London) forces for F_2 and Br_2.

13.18 Compare the dispersion (London) forces for I_2 and Cl_2.

13.19 Butane, $CH_3CH_2CH_2CH_3$, has a boiling point (b.p.) of 0.4 °C and hexane, $CH_3CH_2CH_2CH_2CH_2CH_3$, has a b.p. of 69 °C. Explain.

13.20 The compounds CCl_4 and CF_4 have boiling points of 77 °C and −129 °C, respectively. Explain. (Both molecules are nonpolar.)

13.21 For which of the following compounds, if any, would hydrogen bonding be an important intermolecular force?
a. CH_3OH b. CH_3OCH_3 c. CH_3Cl

13.22 For which of the following compounds, if any, would hydrogen bonding be an important intermolecular force?
a. CH_3CH_3 b. CH_3NH_2 c. H_2S

SURFACE TENSION

13.23 Describe, in terms of intermolecular forces, the cause of the phenomenon of surface tension.

13.24 What is a surfactant, and what does a surfactant do to the surface tension of water?

13.25 Why does water "bead" up on the hood of a newly waxed car? Why does this not occur on a dirty car?

13.26 Describe why a person—using great care—can fill a glass with water even above the brim?

VISCOSITY

13.27 Which is more viscous, motor oil or gasoline? Describe the effect of temperature on viscosity.

13.28 Explain why a lower viscosity motor oil, such as SAE 10, is more desirable in very cold winters. Why don't you use this oil in hot summers?

13.29 Describe the effects of molecular size and symmetrical shape on viscosity.

13.30 Without actually knowing the chemical composition of SAE number 10 motor oil or SAE number 30 motor oil, which of the two do you expect to have larger molecules?

DYNAMIC EQUILIBRIUM AND BOILING

13.31 Describe the dynamic equilibrium in a closed bottle that contains only one cup of water.

13.32 The bottle of water (described in the previous problem) is moved from the kitchen table to the refrigerator where it is left for several hours. Is the dynamic equilibrium changed? If so, how is it changed; if not, why not?

13.33 Give a clear definition of boiling point.

13.34 Describe the effect of atmospheric pressure on boiling point.

13.35 Explain why foods can be cooked in water in a pressure cooker more quickly than when boiled in water in an open pan.

13.36 At high altitudes it takes longer to boil potatoes, but there is no difference in the time required to fry the potatoes. Explain.

13.37 The normal boiling point of a substance depends on the molecular mass *and* on the type of intermolecular attractions. Rank the following compounds in order of increasing boiling points (lowest boiling first).
a. H_2O b. CO c. O_2

13.38 The normal boiling point of a substance depends on the molecular mass *and* on the type of intermolecular attractions. Rank the following compounds in order of increasing boiling points (lowest boiling first).
a. H_2Se b. H_2S c. H_2Te

13.39 Based on molecular structure and intermolecular forces, which of the following compounds would have a higher boiling point, ethyl alcohol, CH_3CH_2OH, or ethyl fluoride, CH_3CH_2F? (Their molar masses are similar.)

13.40 Using intermolecular forces, explain why water (18.0 g/mol) is a liquid at room temperature but ammonia (17.0 g/mol) is a gas—its boiling point is below room temperature.

13.41 Explain why it takes longer to prepare a boiled egg at a very high altitude.

13.42 Potatoes are boiling in a pan of water. Will turning up the heat cause the potatoes to cook any faster, or is energy being wasted by this practice?

DISTILLATION

13.43 A mixture of water and ethyl alcohol was distilled by a student. Several milliliters of distillate were collected at 79 °C before the temperature slowly began to climb—finally leveling off at 99 °C. The student turned in the thermometer to the instructor and asked for a new one, saying that the thermometer wasn't any good—it got stuck at two different temperatures. What do you recommend that the instructor do here?

13.44 When distilling a mixture of ethyl alcohol and water, will the first portion of the distillate, the middle portion, or the final portion have the highest alcohol content?

HEAT OF VAPORIZATION

13.45 The heat of vaporization of bromine, Br_2, is 188 J/g. What is the molar heat of vaporization of bromine in kilojoules per mole?

13.46 The heat of vaporization of ammonia, NH_3, is 1368 J/g. What is the molar heat of vaporization of ammonia in kilojoules per mole?

13.47 Acetone, used in fingernail polish remover, is quite volatile. The molar heat of vaporization of acetone, C_3H_6O, is 7.23 kcal/mol. How much heat (in kilocalories) is absorbed as 7.50 g of acetone evaporates?

13.48 The molar heat of vaporization of acetic acid, CH_3COOH (in vinegar), is 5.81 kcal/mol. How much heat (in kilocalories) is needed to vaporize 6.75 g of acetic acid?

SOLIDS

13.49 Compare the bonding for diamond and graphite. Are these crystalline or amorphous solids? In what way are these materials alike?

13.50 Compare the bonding for charcoal and diamond. In what ways are these materials alike and how are they different?

13.51 Compare the types of bonding present in S_8 molecules and in graphite.

13.52 Describe the type of bonding present in sand, SiO_2. It does not melt at high temperatures, it does not conduct an electric current, and it does not dissolve in water or hexane. It is nonreactive.

13.53 Aspirin, acetylsalicylic acid, has a melting point (m.p.) of 135 °C. It is moderately soluble in ethyl alcohol and nonpolar solvents. Categorize the type of bonding for aspirin.

13.54 Iodine, I_2, crystals dissolve in ethyl alcohol to make a solution called "tincture of iodine." Some years ago this solution was used as a common disinfectant. When pure iodine crystals are heated, they do not appear to melt. Instead, they appear to go directly from a solid to a gas when heated; they *sublime*. What types of intermolecular forces are present in solid iodine?

ENERGY CHANGES (Refer to Tables 13.3, 13.5, and 13.6 for data.)

13.55 Is energy absorbed or released during melting? Is energy absorbed or released during freezing? How are the processes related?

13.56 Is energy absorbed or released during condensation? Is energy absorbed or released during vaporization? How are the processes related?

13.57 The heat of fusion of water is 80. cal/g. How much heat (in kilocalories) would be required to melt a 15.0-kg block of ice?

13.58 How much heat (in kilocalories) must be removed from 15.0 kg of water at its freezing point to convert it to ice?

13.59 How much energy (in kilojoules) is required to change 75.0 g of ice at −5 °C to steam at 100 °C?

13.60 How much energy (in kilojoules) is released as 50.0 g of water in the form of steam at 100 °C is converted to ice at −10 °C?

13.61 To obtain water each day, a bird in winter eats 5.00 g—we'll say—of snow at 0 °C. How many kilocalories (these are the food Calories—capital "C") of energy does it take to melt this snow and warm the liquid to the body temperature of the bird at 40 °C?

13.62 A person lost in a snowstorm decides to eat snow to obtain water. How many extra kilocalories (food Calories) of food would the person have to take in each day to raise 1500 g of snow from −10 °C to body temperature of 37 °C?

WATER: AN UNUSUAL COMPOUND

13.63 What is unusual about the density of water as a solid and as a liquid?

13.64 Describe some important implications related to the densities of ice and liquid water.

13.65 Describe important implications related to the specific heat of water.

13.66 How does perspiring cool the body? Why is this important?

13.67 Describe the process in which water dissolves an ionic solid.

13.68 How is it possible for water to dissolve an ionic compound when the attraction between a single water dipole and an ion is less than the attraction between two ions?

Additional Problems

13.69 For the following compounds, classify the types of forces present as London forces, dipole forces, hydrogen bonds, or combinations of these.
a. ICl b. HF c. HBr d. I_2

13.70 For the following compounds, classify the types of forces present as London forces, dipole forces, hydrogen bonds, or combinations of these.
a. F_2 b. H_2O
c. $CH_3CH_2CH_2OH$ d. N_2O

13.71 Account for the fact that oxygen can be liquefied at low temperatures even though O_2 molecules are nonpolar.

13.72 Which gas, HCl or HI, would you expect to require less pressure to liquefy? Use two reasons in your explanation.

13.73 Explain why ethylene glycol, $HOCH_2CH_2OH$, which is used in antifreeze, has a high boiling point and is slow to evaporate, but ethanol, CH_3CH_2OH, has a lower boiling point and evaporates quickly.

13.74 Would you expect ethane, CH_3CH_3, or methanol, CH_3OH, to have a higher boiling point? Explain.

13.75 A dish half filled with water is placed beside another dish half filled with gasoline. What does the fact that the gasoline evaporates first tell us about surface tension, volatility, and intermolecular forces?

13.76 Two tablespoonsful of motor oil and two tablespoonsful of gasoline are poured onto a concrete floor. The gasoline soon evaporates, but the oil remains, leaving an oily spot on the concrete weeks later. What does this tell us about differences in intermolecular forces?

13.77 Which noble gas, neon or xenon, would you expect to have a higher boiling point? Explain. (Both are monatomic gases.)

13.78 A chemist for a sugar-refining company suggests that reduced pressure (a vacuum) be used to remove water from a sugar and water solution without risking the chance of burning the sugar? You are the boss. Explain why you should use or should not use the employee's idea.

13.79 Dry ice is a solid form of carbon dioxide, CO_2, which is a nonpolar compound. At room temperature it does not melt, but goes directly from a solid to a gas without passing through the liquid state—it sublimes. When the vapor is cooled, it forms orthorhombic crystals. What types of intermolecular forces are present in the solid?

13.80 ''Crystal goblets'' have a much higher PbO content than common glass. The goblet shatters when dropped, and pieces can be found with irregular angles. How would you classify this solid?

13.81 The refrigerant for many air conditioners is Freon-12; it is actually dichlorodifluoromethane, CCl_2F_2, with a molar heat of vaporization of 35.0 kJ/mol. How much energy (in kilojoules) is absorbed in vaporizing 500. g of Freon-12?

13.82 How much heat (in kilojoules) is absorbed as 200. g of water is vaporized at its boiling point? The heat of vaporization of water is 2.26 kJ/g.

13.83 Phenol is an organic chemical present in certain antiseptic throat sprays. It has a melting point of 42 °C and a boiling point of 180 °C. Sketch a cooling curve—gas to solid—for this compound. Explain the various parts of the curve.

13.84 Naphthalene is an organic chemical used in moth balls. It has a melting point of 80. °C. Sketch a cooling curve—liquid to solid—for naphthalene.

13.85 Why does steam at 100 °C cause more severe burns than liquid water at the same temperature?

13.86 How could hot cooking oil cause more severe burns than boiling water?

13.87 How much heat (in kilojoules) is absorbed when 50.0 g of ammonia, NH_3, is vaporized? The heat of vaporization of ammonia is 1.37 kJ/g.

13.88 How much heat (in kilojoules) is absorbed when 50.0 g of diethyl ether, $C_2H_5OC_2H_5$, is vaporized? The heat of vaporization is 351 J/g.

Oceans, rivers, and lakes are solutions of minerals and nutrients dissolved in water, the most common solvent on Earth.

14

Solutions

CONTENTS

Solutions of gases, liquids, and solids are almost everywhere. The air is a solution of gases. The water you drink is not pure H_2O; it is a solution that contains calcium ions, magnesium ions, and traces of many other ions. Evaporation of tap water leaves white deposits of substances that were dissolved in the water—they were in solution. A colorless sample of ocean water is a solution that is rich in chloride ions, Cl^-, sodium ions, Na^+, sulfate ions, SO_4^{2-}, magnesium ions, Mg^{2+}, and many more. Water has the ability to mix with and to dissolve many substances to form solutions.

Numerous commercial products are sold as solutions. Examples include bottled soft drinks and other beverages, mouthwash, cough syrup, colognes, throat sprays, nasal sprays, vinegar, flavorings, liquid bleach, insecticides, glass cleaners and other household cleaning solutions, and many other household and industrial chemicals (Figure 14.1). Many life processes take place in solution and at the interfaces between solutions and membranes. Solutions are essential to all forms of life.

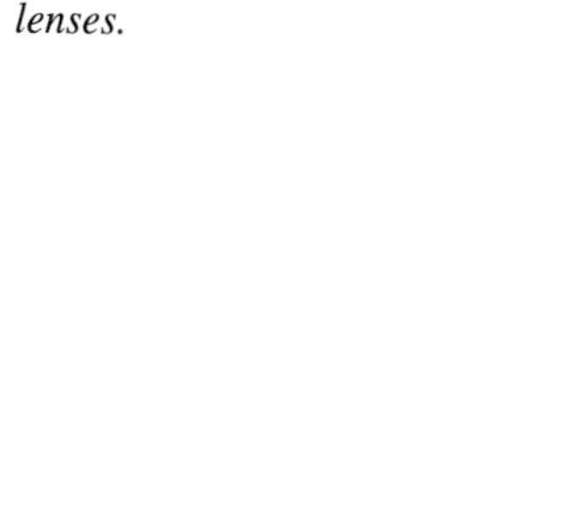

Figure 14.1 *Common commercial products sold as solutions include throat sprays, after shave lotions, bleaches, glass cleaners, fragrances, nasal sprays, and saline solutions for contact lenses.*

In this chapter, we will study different types of solutions and several ways of reporting the amount of a substance that is dissolved in a given quantity of solution; that is, the concentration of the solution. To be ready for this chapter, you should be familiar with intermolecular forces (Chapter 13), homogeneous and heterogeneous substances (Chapter 2), and chemical quantities (Chapter 9). Be sure to review any of these topics when needed.

14.1 *What Is a Solution?*

Put a teaspoonful of sugar in a cup of water. Stir until no more sugar crystals are visible. Using a straw, taste the sweetened water at the top, the bottom, and near the sides of the cup. If the sugar is completely *dissolved*, the sweetness is the same throughout, making a **homogeneous mixture** (Section 2.4). You could add more sugar to make the water sweeter, or you could use less sugar to make it less sweet. Further, you could evaporate or boil the water away and recover the sugar. The sugar and water do not react chemically; the sugar *dissolves* in the water.

When the sugar is **dissolved** in the water, the mixture is called a *solution*. A **solution** can be defined as an intimate, *homogeneous* mixture of two or more substances. The substances can be in the form of atoms (such as the copper and zinc in brass), ions (such as sodium chloride dissolved in water), or molecules (such as table sugar dissolved in water). In true solutions, the mixture is intimate right down to the level of individual atoms, ions, and molecules. For example, in a solution of salt in water or sugar in water, there are no solid lumps of crystals floating around. Instead, there are individual ions of the salt and molecules of the sugar randomly distributed among the water molecules. On a much larger scale we can picture a solution as marbles of one color distributed randomly among marbles of another color (Figure 14.2).

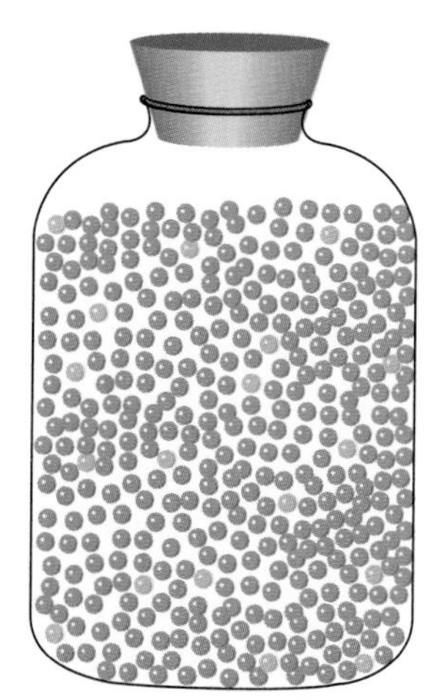

Figure 14.2 *In a solution, the molecules of solute (such as sugar) are randomly distributed among the solvent (often water) molecules.*

Table 14.1 **Types of Solutions**

Solute	*Solvent*	*Solution*	*Example*
Gas	Gas	Gas	Air (O_2 in N_2)
Gas	Liquid	Liquid	Carbonated beverages (CO_2 in H_2O) Swimming pool (Cl_2 in H_2O)
Liquid	Liquid	Liquid	Wine (ethanol in H_2O) Vinegar (acetic acid in H_2O)
Liquid	Solid	Solid	Dental amalgam for fillings (liquid mercury in solid silver)
Solid	Liquid	Liquid	Saline (NaCl in H_2O) Sugar in water
Solid	Solid	Solid	14-karat gold (Ag in Au) Steel (carbon in iron)

The components of a solution are given special names. The substance being dissolved (or the substance with the lesser amount) is called the **solute**. The component whose physical state is retained (or the substance present in the greater amount) is called the **solvent**. Water is undoubtedly the most familiar solvent. Its physical state is retained as it dissolves familiar substances such as table sugar (sucrose), table salt (NaCl), and ethyl alcohol (ethanol). But there are many other solvents. Gasoline and charcoal lighter fluid dissolve grease. Certain medications are dissolved in ethanol. Hexane and toluene are solvents used in rubber cement and in permanent marking pens.

Solvents and the solutions formed are not always liquids, as shown by the examples given in Table 14.1. Air is a gaseous solution of oxygen, argon, water vapor, and other gases in nitrogen gas. Steel is a solution of one solid dissolved in another solid; carbon (the solute) is dissolved in iron (the solvent). Brass, coinage metals, and most other alloys are solutions. Dental amalgam is an unusual solution; the solute, mercury, is a liquid and the solvent, silver, is a solid. There are many types of solutions, but in this chapter we will deal primarily with **aqueous solutions**, those in which the solvent is water.

Alloys are made by mixing together two or more molten metals or a metal with a nonmetal. For example, 14-karat and 18-karat gold contain different percentages of gold, silver, and copper (see Table 2.4).

EXAMPLE 14.1

For the following solutions, which chemical is the solvent?

(a) 2 ounces of oil and 2 gallons of gasoline
(b) carbon dioxide and water (in carbonated water)
(c) 70 mL of isopropyl alcohol and 30 mL of water (in rubbing alcohol)
(d) 25% Ni and 75% Cu (in coinage nickel)

SOLUTION

(a) Gasoline is the solvent; there is more of it present in the solution.
(b) Water is the solvent; it is present in the greater amount and its physical state is retained.
(c) Isopropyl alcohol is the solvent; it is present in the greater amount.
(d) Copper is the solvent in coinage nickel, which is 75% Cu.

14.2 Solubility Terminology

We say that table sugar is *soluble* in water. Just what does that mean? Can we dissolve a teaspoonful of sugar in a cup of water? How about dissolving 10 teaspoonfuls or 100 teaspoonfuls of sugar? We know from everyday experience that there is a limit to the amount of table sugar we can dissolve in a given quantity of water. Nevertheless, we still find it convenient to say that table sugar is **soluble** in water. We mean that an appreciable amount of the sugar will dissolve. We sometimes also use terms such as ''moderately soluble'' and ''slightly soluble'' to indicate the extent of the solubility.

Study Hint
You may wish to review solubility rules given in Table 10.2 (Section 10.10).

The terminology used in the preceding paragraph was useful but not precise. In scientific terminology, the **solubility** of a substance is a measure of how much solute will dissolve in a given quantity of solvent at a specific temperature. For example, the solubility of sodium chloride, NaCl, is 36 g per 100. g of water at 20. °C.

Two liquids that dissolve in each other are said to be **miscible**. If they do not dissolve in each other when mixed, they are **immiscible**. A few substances are miscible regardless of the proportions in which they are mixed; water and ethanol (ethyl alcohol) are familiar examples. We say that such substances are *completely* miscible. For most liquids, though, there is a limit to the amount of one that will dissolve in another. These liquids are *partially* miscible.

When a substance does not appear to dissolve in a solvent, it is said to be **insoluble**. Put an iron nail in a beaker of water. There is no apparent change. We say that the iron is *insoluble* in water. However, even insolubility is relative. If we had an instrument sensitive enough, we would find that some iron had dissolved. The amount might well be regarded as insignificant, and that is the sense in which the term *insoluble* is used. Thus, terms such as soluble and insoluble are useful, but they are imprecise and must be used with care.

Consider the terms ''large'' and ''small.'' They are relative terms that allow us to make rough comparisons. Similarly, the terms *concentrated* and *dilute* are imprecise but useful. They allow us to express, in a general way, the relative amount of solute in a solution. A **concentrated solution** is one that contains a relatively large amount of the solute. A **dilute solution** is one that contains a relatively small amount of the solute. The terms concentrated and dilute are used in a more quantitative way for solutions of acids and bases. These meanings will be specified in Chapter 16.

EXAMPLE 14.2

Copper water pipes in a house were joined with solder (67% Pb and 33% Sn). Solder is said to be insoluble in water, but chemical analysis revealed that water standing in the pipes overnight picked up 5 parts per billion of Pb. Relate this information to the statement that ''insoluble'' is an imprecise term.

SOLUTION

Lead appears to be insoluble in water, but traces of Pb dissolve in water when left in contact for long periods of time, as shown by the chemical analysis. (Is the Pb in this sample over the federal limit of 0.015 mg/L?)

See Problems 14.1–14.8.

14.3 The Solubility of Ionic Compounds

The unique structure of water not only results in relatively strong forces between water molecules but also enables water to dissolve ionic compounds, as described in Chapter 13. Intermolecular forces that exist between identical molecules in pure liquids were also described in Chapter 13. Now, we will look at attractions between solute and solvent particles. The solubility of a given solute is dependent upon the relative attractions between particles in the pure substances and particles in the solution (Figure 14.3).

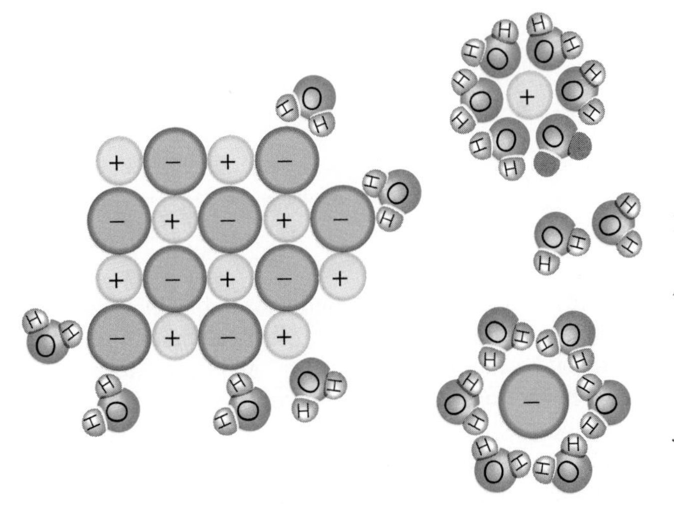

Figure 14.3 *When sodium chloride—or any other ionic compound—dissolves in water, each ion becomes completely separated from the solid crystal and becomes hydrated (surrounded by water molecules). Notice the difference in orientations of water molecules that surround the hydrated ions.*

Solubility rules for ionic compounds were given in Section 10.10. Nearly all compounds of alkali metals are quite soluble in water. Furthermore, nearly all compounds containing the nitrate ion or the ammonium ion are soluble. Why do these compounds dissolve in water? At the end of the last chapter (Section 13.8), we described—at the molecular level—what happens when water dissolves a salt such as sodium chloride. Essentially three things must happen when a salt goes into solution.

1. The attractive forces holding the salt ions together must be overcome.
2. The attractive forces holding at least some of the water molecules must be overcome.
3. The solute and solvent molecules must interact; that is, they must attract one another.

The process in which water molecules surround the ions of the solute is called **hydration**. (The more general term **solvation** is used when a solute is dissolved by a liquid other than water.) In order for hydration to occur, the energy released by the interaction of the solute with the solvent must be greater than the sum of the energy needed to overcome the forces holding the ions together in the crystal lattice and the energy needed to separate the solvent molecules.

When forces holding the ions together are sufficiently great, they cannot be overcome by the hydration of the ions. Many solids that have ions that are doubly

Table 14.2 **Solubilities of Solid Ionic Compounds in Pure Water***

	NO_3^-	CH_3COO^-	Cl^-	SO_4^{2-}	OH^-	S^{2-}	CO_3^{2-}	PO_4^{3-}
NH_4^+	S	S	S	S	N	S	S	S
Na^+	S	S	S	S	S	S	S	S
K^+	S	S	S	S	S	S	S	S
Ba^{2+}	S	S	S	I	S	D	I	I
Ca^{2+}	S	S	S	P	P	P, D	I	I
Mg^{2+}	S	S	S	S	I	D	I	I
Cu^{2+}	S	S	S	S	I	I	I	I
Fe^{2+}	S	S	S	S	I	I	I	I
Fe^{3+}	S	N	S	P	I	D	N	I
Zn^{2+}	S	S	S	S	I	I	I	I
Pb^{2+}	S	S	P	I	I	I	I	I
Ag^+	S	P	I	I	N	I	I	I
Hg_2^{2+}	S, D	P	I	I	N	I	I	I
Hg^{2+}	S	S	S	D	N	I	N	I

*S, is soluble in water; P, is partially soluble in water; I, is insoluble in water; D, decomposes; N, does not exist as an ionic solid.

or triply charged are essentially insoluble in water. Examples include calcium carbonate with Ca^{2+} and CO_3^{2-} ions, aluminum phosphate with Al^{3+} and PO_4^{3-} ions, and barium sulfate with Ba^{2+} and SO_4^{2-} ions. The large electrostatic forces between the ions hold the particles together despite the attraction of polar water molecules.

The solubilities of several ionic compounds are summarized in Table 14.2. Use this table—or solubility rules in Table 10.2—along with information provided in this section as you answer questions pertaining to solubility.

EXAMPLE 14.3

Indicate the extent of water solubility for these compounds using the following categories: soluble, partially soluble, insoluble, and decomposes.

(a) ammonium nitrate (used in fertilizer)
(b) magnesium hydroxide (used in ''milk of magnesia'')
(c) calcium sulfate (used in plaster of paris and drywall)
(d) calcium carbonate (present in limestone and marble)

SOLUTION

Use Table 14.2 and information summarized in Section 14.3.

(a) NH_4NO_3 is soluble in water.
(b) $Mg(OH)_2$ is insoluble in water.
(c) $CaSO_4$ is partially soluble in water.
(d) $CaCO_3$ is insoluble in water. (However, it does react with acids.)

EXERCISE 14.1

Using the terminology presented here, describe what happens when table salt dissolves in water.

14.4 The Solubility of Covalent Compounds

An old but helpful rule states that

> Like dissolves like.

This means that nonpolar (or only slightly polar) solutes dissolve best in nonpolar solvents and that highly polar solutes dissolve best in polar solvents such as water. The rule works well for nonpolar substances. Fats, oils, and greases (which are nonpolar or slightly polar) dissolve in nonpolar solvents such as hexane, C_6H_{14}. The forces that hold nonpolar molecules together are generally weak. Thus, only a small amount of energy is needed to separate nonpolar solute molecules and to disrupt the attractive forces between solvent molecules.

The rule that like dissolves like is not as helpful for polar substances and, in particular, for aqueous solutions. Water solubility for molecules such as table sugar is dependent upon the formation of hydrogen bonds between water and the solute. Thus, molecules containing a high proportion of nitrogen or oxygen atoms usually dissolve in water because these are the elements that can form hydrogen bonds. Examples include methanol, CH_3OH (methyl alcohol), which is completely miscible with water, and methylamine, CH_3NH_2, which is also quite soluble in water. Figure 14.4 gives the structures of these molecules (and others mentioned in this section) and shows how they interact with water by forming *hydrogen bonds.*

Many water-soluble molecules have a hydrogen atom bonded to an oxygen or nitrogen atom. These polar O—H and N—H bonds can enter into hydrogen bonding with water molecules (Figure 14.4). In the drawing, hydrogen bonds between molecules are indicated by dotted lines.

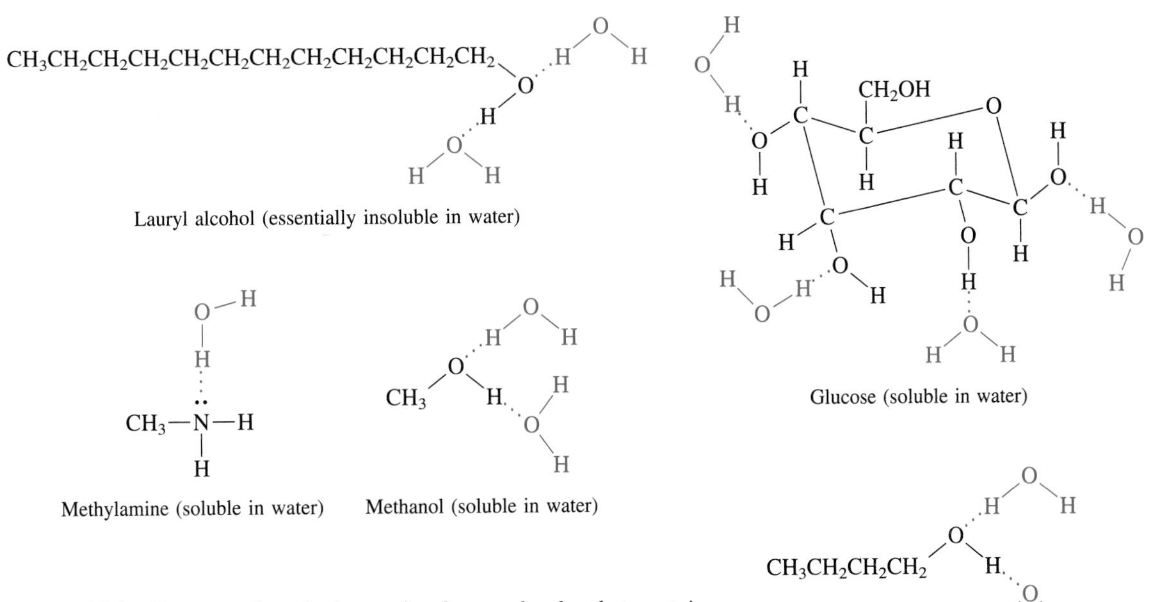

Figure 14.4 *Water can form hydrogen bonds to molecules that contain nitrogen or oxygen atoms. Those molecules with no more than four carbon atoms per nitrogen or oxygen atom are usually soluble in water.*

Each nitrogen or oxygen atom in a solute molecule can carry along into solution about four attached carbon atoms. Alcohols (compounds containing the —OH group covalently bonded to carbon atoms) that have 3 or fewer carbon atoms are completely miscible with water. Butyl alcohol, with 4 carbon atoms, is only partially soluble, but lauryl alcohol, with 12 carbon atoms and only one —OH group, is essentially insoluble. Some rather complex molecules are soluble in water if they contain several —OH groups. Glucose, $C_6H_{12}O_6$, is quite soluble in water; it has six —OH groups that form hydrogen bonds to water.

In general, hydrogen bonding is more important than polarity alone in determining the water solubility of a molecular substance. Methyl chloride, CH_3Cl, and methanol, CH_3OH (methyl alcohol), have about the same polarity, yet methyl chloride is essentially insoluble in water while methanol is completely miscible with water. Methyl chloride does not engage in hydrogen bonding, but methanol does. A few polar compounds, such as hydrogen chloride gas, HCl(g), dissolve in water because they react to form ions. We will look more closely at reactions of this type in Chapter 16, as we deal with acids.

EXAMPLE 14.4

Which of these compounds would not be expected to be water-soluble: glucose; methyl chloride, CH_3Cl; methylamine, CH_3NH_2; hexane, $CH_3CH_2CH_2CH_2CH_2CH_3$?

SOLUTION

Methyl chloride and hexane are essentially insoluble in water. They do not form hydrogen bonds with water. Both glucose and methylamine are water-soluble.

See Problems 14.9–14.22.

14.5 *Solubility Equilibria*

For most substances, there is a limit to how much can be dissolved in a given volume of solvent. This limit varies with the nature of the solute and the solvent. The solubilities of solids in liquids also vary with temperature; generally—but not always—solubilities increase with increasing temperature.

The water solubilities of lithium carbonate, Li_2CO_3, and calcium hydroxide, $Ca(OH)_2$, for example, actually decrease with increasing temperature.

Solubilities are often expressed in terms of grams of solute per 100. g of solvent. Since solubility varies with temperature, it is necessary to indicate the temperature at which the solubility is measured. For example, 100. g of water will dissolve up to 109 g of sodium hydroxide, NaOH, at 20. °C, and up to 145 g of NaOH at 50. °C. In a shorthand method, the solubility of sodium hydroxide is expressed as 109^{20} and 145^{50} (the 100. g of water is understood).

We need not restrict solubility calculations to 100-g quantities. At a specific temperature, the quantity of a given solute that will dissolve is proportional to the quantity of solvent used. We can determine the quantity of a given solute that will dissolve in a specific quantity of water by multiplying the quantity of water to be used by a conversion factor made up of the solubility information. For example, to determine the quantity of NaOH that can be dissolved in 175 g of water at 50. °C, start with the quantity of water given and multiply by the appropriate solubility factor.

$$175 \text{ g water} \times \frac{145 \text{ g NaOH}}{100. \text{ g water}} = 254 \text{ g NaOH can be dissolved}$$

The solubility of sodium chloride, NaCl, is 36 g per 100. g of water at 20. °C. What happens if we place 40. g of NaCl in 100. g of water at 20. °C? Initially, many of the sodium ions, Na^+, and chloride ions, Cl^-, leave the crystal surfaces and wander about at random through the water. Some of the ions, however, return to the crystal surfaces to become, once more, a part of the crystal lattice. As more salt dissolves, there are also more "wanderers" that recrystallize out of solution. Eventually, when 36 g of NaCl has dissolved, the number of ions leaving the surface of the undissolved crystal will just equal the number of returning ions. A condition of *dynamic equilibrium* will have been established.

Here, the *net* quantity of sodium chloride in solution will remain the same even though ions come and go from the crystal surfaces. The net quantity of undissolved crystals also remains constant. Some small crystals may even disappear as others grow larger, but the net amount of undissolved salt will not change. The rate at which ions dissolve will just equal the rate at which they crystallize. The dynamic solubility equilibrium can be represented as follows.

$$\text{Solute} + \text{Solvent} \underset{\text{crystallizing}}{\overset{\text{dissolving}}{\rightleftharpoons}} \text{Solution}$$

A solution at a given temperature that exists in dynamic equilibrium with undissolved solute is said to be **saturated**. When a solution contains less solute than the solubility limit, it is said to be **unsaturated**. A solution with 24 g of NaCl dissolved in 100. g of water at 20. °C would be unsaturated because it could dissolve an additional 12 g at that temperature before becoming saturated.

EXAMPLE 14.5

How many grams of NaCl can be dissolved in 220. g of water at 20. °C? The solubility of NaCl at 20. °C is 36 g NaCl per 100. g water.

SOLUTION

Start with the quantity of water (the solvent) to be used and multiply by the solubility factor.

$$220.\ \cancel{\text{g water}} \times \frac{36\ \text{g NaCl}}{100.\ \cancel{\text{g water}}} = 79\ \text{g NaCl can be dissolved}$$

See Problems 14.23–14.26.

EXERCISE 14.2

How many grams of KCl can be dissolved in 375 g of water at 20. °C? The solubility of KCl at 20. °C is 23.8 g KCl per 100. g water.

14.6 *Effects of Temperature and Pressure on Solubility*

Solubility of Solids in Liquids

A specific mass of a solid solute will stay in solution at a given temperature. That is, solubility equilibrium is established at a given temperature. Pressure has little effect on the solubilities of solids in liquids because solids and liquids are virtually

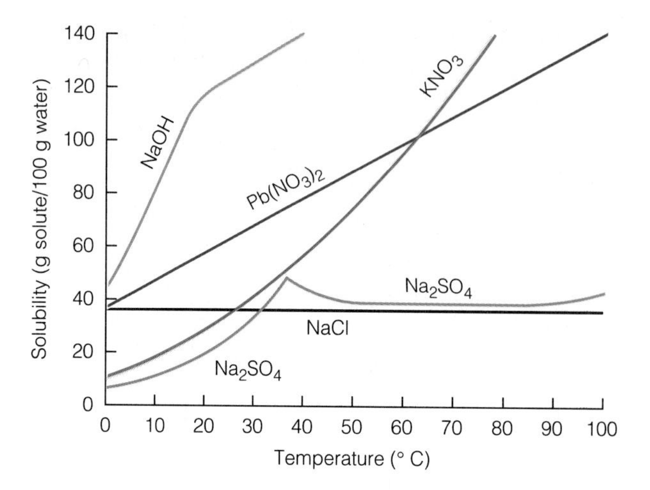

Figure 14.5 *The effect of temperature on the solubility of several solids in water.*

incompressible. However, if the temperature changes, more solute will either dissolve or else precipitate out of solution until equilibrium is reestablished at the new temperature.

Most solid compounds are increasingly soluble as the temperature is raised (see solubility curves in Figure 14.5). As the temperature goes up, the motion of all the particles increases. More ions are knocked loose from the lattice and are free to go into solution. Further, it is more difficult for the crystal to recapture the ions that return to its surface because they are moving at higher speeds. There are a few exceptions to this general rule of increased solubility at higher temperatures; see the curves for NaCl and Na_2SO_4 in Figure 14.5.

If a saturated solution of lead(II) nitrate—in equilibrium with an excess of solid lead(II) nitrate—is cooled, more solute precipitates until equilibrium is once again established at the lower temperature. For example, 100. g of water will dissolve 120. g of lead(II) nitrate, $Pb(NO_3)_2$, at 90. °C. When the solution is cooled to 20. °C, the solution at equilibrium can contain only 54 g of $Pb(NO_3)_2$. The excess 66 g will precipitate out, increasing the amount of undissolved solute.

Now consider what would happen if one started to cool a saturated solution of lead(II) nitrate with *no* excess solute present. Would the $Pb(NO_3)_2$ precipitate out of solution? It might. Then again, it might not. There is no equilibrium; no crystals are present to capture the wandering ions. One might be able to cool the solution to 20 °C without precipitation. Such a solution, containing solute in excess of the amount it could contain *if* it were at equilibrium is said to be **supersaturated**. This system is not stable because it is not at equilibrium. Solute may precipitate when the solution is stirred or when the inside of the container is scratched with a glass rod. This is not magic—scratching the container walls does facilitate the formation of small crystals. Also, addition of a "seed" crystal of solute (Figure 14.6) will nearly always result in the precipitation of all the excess solute. Equilibrium is often established rather rapidly when there is a crystal to which the ions can attach themselves.

Wine Chemistry
Some wines contain high concentrations of potassium hydrogen tartrate, $KHC_4H_4O_6$. When chilled, the solution becomes supersaturated. Crystals often form and settle out when the wine is stored in the consumer's refrigerator. Some wineries solve the problem—while also making the wine less acidic—by chilling the wine to −3 °C and adding tiny seed crystals of $KHC_4H_4O_6$. Precipitation is complete in 2 or 3 hr, and the crystals are filtered out.

Supersaturated solutions can be found in nature. Honey is one example; the solute is sugar. If honey is left to stand, the sugar crystallizes. We say—not very scientifically—that the honey has "turned to sugar." Some prepared foods, such as jellies, also contain supersaturated sugar solutions. The sugar often crystallizes from jelly that has been standing for a long time. Also, old-fashioned "rock candy" is prepared by suspending a string in a supersaturated sugar solution. Once

Figure 14.6 *Addition of a seed crystal induces rapid crystallization of excess solute from a supersaturated solution.*

small crystals of the sugar begin to form on the string, large angular-shaped rock candy crystals are formed as additional molecules of sugar attach themselves to the small crystals.

Solubility of Gases in Liquids

Solutions in which gases are dissolved in water are common. For example, bottled soft drinks are solutions; they contain carbon dioxide, flavoring, and sweetening agents in water. Some glass-cleaning solutions contain ammonia gas, $NH_3(g)$, dissolved in water. Blood contains dissolved oxygen and carbon dioxide. Water in lakes and rivers also contains small amounts of dissolved oxygen gas. For example, 0.0043 g of O_2 will dissolve in 100. g of water at 20. °C; this dissolved oxygen is vital to the survival of fish and other aquatic species.

Unlike most solid solutes, gases become *less* soluble as the temperature increases. That is because heat increases the molecular motion of both solute and solvent particles, and gaseous solute molecules can escape from the solution when they reach the surface of a liquid in an open container.

If you watch carefully when water is being heated, you will notice bubbles of gas appear long before the water begins to boil (Figure 14.7). As the solution of gas in liquid is heated, the solubility of gases decreases and gas bubbles are formed. Because gas bubbles have low density, they soon rise to the surface and escape into the atmosphere. A graph showing the effect of temperature on the solubility of oxygen gas in water is presented in Figure 14.8.

Figure 14.7 *Bubbles of air form when a beaker of water is heated.*

The solubility of gases in water also varies with the pressure of the gas. At higher pressures, more gas will dissolve in a given amount of water. Figure 14.9 shows how the solubility of oxygen gas in water at 25 °C varies with pressure. For another example, picture a can or bottle of carbonated beverage being opened; the bubbling liquid may overflow the container. This common occurrence is a demonstration of the change in gas solubility with pressure. Carbonated beverages are canned or bottled under a pressure that is slightly greater than one atmosphere. This increases the solubility of the carbon dioxide gas. Once the can or bottle is opened, the pressure immediately drops back to atmospheric pressure and gas solubility decreases. As gas bubbles escape from solution, some of the foaming

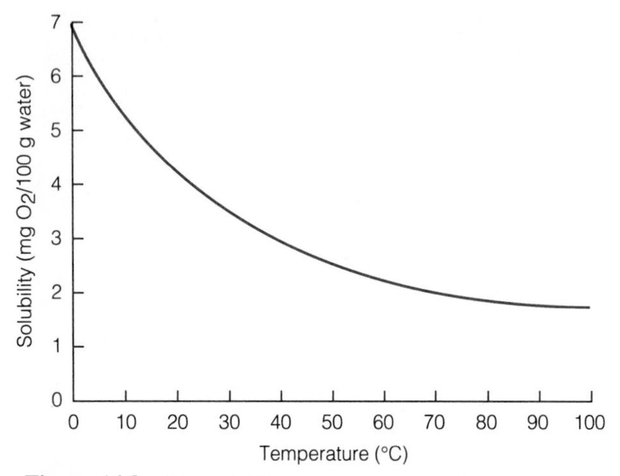

Figure 14.8 *The solubility of oxygen at various temperatures at 1 atm of pressure.*

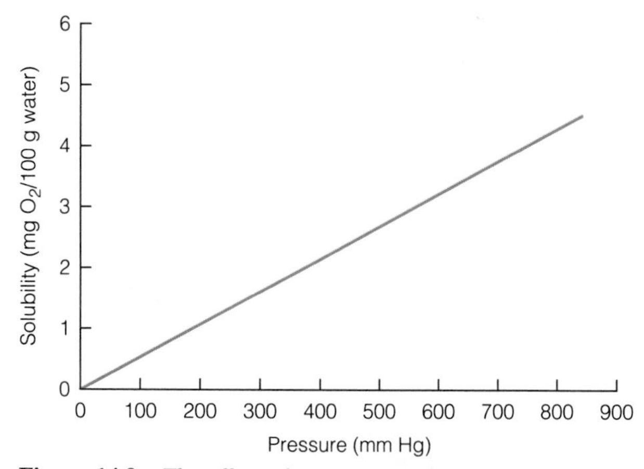

Figure 14.9 *The effect of pressure on the solubility of oxygen in water at 25 °C.*

Figure 14.10 *When the carbonated beverage is opened, the pressure immediately drops back to atmospheric pressure and gas solubility decreases. As gas bubbles escape from the solution, some of the foaming liquid may gush from the container.*

See Problems 14.27–14.34.

liquid may overflow the container (Figure 14.10). Even after the initial rapid release of gas, carbon dioxide continues to escape. Eventually, the tangy taste is gone (most of the gas is gone); the beverage has gone ''flat.'' If the liquid is warm, the beverage goes ''flat'' even faster because gases are less soluble at higher temperatures, as described earlier in this section.

EXAMPLE 14.6

Figure 14.5 shows that 30. g of KNO_3 dissolves in 100. g of water at 20. °C, but 108 g of it will dissolve in 100. g of water at 60. °C. When 150. g of KNO_3 was stirred with 100. g of water at 70 °C, some did not dissolve.

(a) After cooling to 60. °C, the mixture of solid and solution was filtered. What mass of solid KNO_3 could be recovered at 60. °C?
(b) The remaining solution was then cooled further to 20. °C. How much more solid KNO_3 can be recovered at 20. °C?

SOLUTION

(a)

Total mass of KNO_3 present = 150 g
Mass of KNO_3 soluble at 60. °C = 108 g
Mass of solid KNO_3 recovered at 60. °C = 42 g

(b)

Total mass of KNO_3 in solution after first filtering = 108 g
Mass of KNO_3 soluble at 20. °C = 30 g
Additional mass of KNO_3 recovered upon cooling to 20. °C = 78 g

14.7 Solution Concentration Expressions

In the previous section, the solubility of a solute in a given solvent was expressed in grams of solute per 100. g of solvent at a given temperature. We can dilute saturated solutions to prepare solutions of varying *concentrations*. The **concentra-**

tion of a solution is a measure of the amount of solute in a specific quantity of solvent. Having described what happens when solutions form, and what factors affect solubility, we are now ready to consider several methods of expressing solution concentrations.

Molarity

The most common method that chemists use to express solution concentrations is **molarity** (abbreviated M), which is defined as the number of moles of solute per liter of *solution*.

Study Hint
Remember that molarity is the concentration of a solution measured in moles per liter of *solution*, not per liter of solvent.

$$\text{Molarity (M)} = \frac{\text{Moles of solute}}{\text{Liters of solution}}$$

This is a very convenient method of expressing the concentration of a solution because we often are interested in measuring out a specific number of moles of solute. Calculations involving molarities of solutions were presented in Section 9.6. Then, in Section 11.4, we did stoichiometric calculations to determine the volume of a solution with a specific molarity needed to react with a given quantity of another substance.

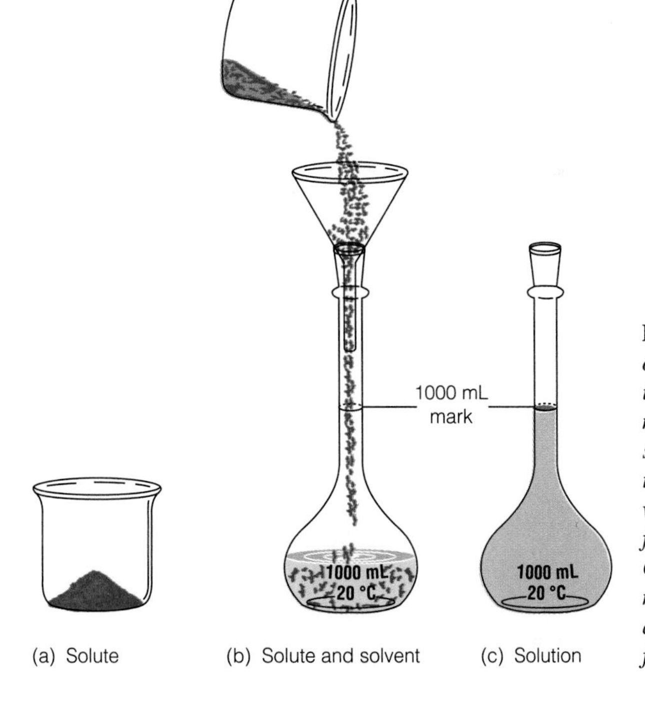

Figure 14.11 *To prepare 1.00 L of a solution of specified molarity, (a) obtain the appropriate mass of solute, (b) transfer the solid to a 1.00-L volumetric flask that is partly filled with distilled water, (c) add enough water to fill the flask to the 1.00-L mark. Once the solution is thoroughly mixed, it is ready to be placed in an appropriately labeled bottle for laboratory use.*

Directions for preparing a solution with a given molarity are shown in Figure 14.11. Work through the following example for a review of molarity calculations.

EXAMPLE 14.7

Describe how to prepare 500.0 mL of a 6.00 M solution of NaOH.

SOLUTION

1. Start with the *known quantity*—the volume—in mL and convert this to L.
2. Convert liters to moles using molarity as the conversion factor.
3. Convert moles to grams using the molar mass as the conversion factor.

The series of conversions can be summarized as follows.

Plan: mL ⟶ Liters ⟶ Moles ⟶ Grams

$$\frac{0.500\ \text{L}}{1} \times \frac{6.00\ \text{mol}}{\text{L}} \times \frac{40.0\ \text{g}}{\text{mol}} = 120.\ \text{g}$$

With the calculations completed, prepare the solution following the three steps outlined in Figure 14.11. (a) Weigh out 120. g of NaOH in an appropriate container. (b) Transfer the NaOH to a 500.0-mL volumetric flask partly filled with pure (distilled) water and shake to dissolve. (The solution will get quite warm due to the heat of solution.) (c) When the solution has cooled, add enough water to fill the flask to the 500.0-mL mark. Put the stopper in the flask and invert repeatedly to mix thoroughly.

We can quickly measure out the specific volume of a solution needed to obtain a given number of moles (or grams) of the chemical desired. For a review of calculations required to determine the volume of a specific solution needed to provide a given number of moles of a chemical, work through Example 14.8. If the number of grams (instead of moles) is given, convert grams to moles first.

EXAMPLE 14.8

How many milliliters of a 12.0 M reagent HCl solution would be used to obtain 0.480 mol of HCl?

SOLUTION

1. Start with the known quantity—the number of moles of HCl.
2. Convert moles to liters using molarity as a conversion factor.
3. Convert liters to milliliters.

The series of conversions can be summarized as follows.

Plan: Moles ⟶ Liters ⟶ mL

$$\frac{0.480\ \text{mol}}{1} \times \frac{1\ \text{L}}{12.0\ \text{mol}} \times \frac{1000\ \text{mL}}{1\ \text{L}} = 40.0\ \text{mL}$$

See Problems 14.35–14.40

EXERCISE 14.3

How many grams of HCl are present in 15.0 mL of 12.0 M reagent HCl solution?

Percent by Volume

Pick up a bottle of rubbing alcohol and read the label. It is likely to state, ''Isopropyl alcohol, 70% by volume.'' *Percent by volume* is another common method of expressing concentrations. For many purposes, it is not necessary to know the number of moles of solute but only the relative amounts of solute and solvent. If both the solute and solvent are liquids, concentrations can be expressed as **percent by volume**. To make up 100. mL of a 70.0% (by volume) isopropyl alcohol solution, 70.0 mL of the alcohol is added to enough water to make a *total* volume of 100. mL. Mathematically, we can write

$$\text{Percent by volume} = \frac{\text{Volume of solute}}{\text{Total volume of solution}} \times 100\%$$

To save printing space, the label of the isopropyl alcohol bottle may simply show 70% (v/v). It is important that we understand that the solution is made by diluting the alcohol with enough water to make a *total* volume of 100. mL. This is *not* the same as mixing 70.0 mL of alcohol with 30.0 mL of water because the liquid volumes do not add up to 100. mL when mixed. Try it for yourself. As shown in Example 14.9, we are not limited to 100.-mL batches of solution. Volumes can be in liters, fluid ounces, or any other volume unit as long as all volumes are measured in the same manner. The units cancel.

EXAMPLE 14.9

What is the percent by volume of an isopropyl alcohol solution prepared by mixing 25.0 mL of the alcohol with enough water to make a total volume of 125 mL of solution?

SOLUTION

Use the equation

$$\text{Percent by volume} = \frac{\text{Volume of solute}}{\text{Total volume of solution}} \times 100\%$$

$$= \frac{25.0\ \cancel{\text{mL}}}{125\ \cancel{\text{mL}}} \times 100\% = 20.0\%\ (\text{v/v})$$

(Notice that volume units cancel.)

See Problems 14.41–14.46.

Percent by Mass

Concentrations of solutions can also be expressed as percent by mass. This method is used most often for solids dissolved in liquids, but this is not always the case. For example, concentrated hydrochloric acid is commercially sold in solutions labeled as 38.0% by mass, but some labels are printed in the form 38.0% (w/w), where w/w is an abbreviation for weight/weight. The percent by mass or w/w notation indicates—for our example—that for every 100. g of *solution* there are 38.0 g of HCl and 62.0 g of water. Notice that the total amount of solution is 100. g, not 100. mL. Mathematically, the definition can be written

$$\text{Percent by mass} = \frac{\text{Mass of solute}}{\text{Total mass of solution}} \times 100\%$$

For masses in grams, the definition can be written

$$\text{Percent by mass} = \frac{\text{g solute}}{\text{g solute} + \text{g solvent}} \times 100\%$$

When the solvent is water—and it usually is—we can measure out the specified quantity in milliliters instead of grams, if we assume 100. mL of water has a mass of 100. g.

EXAMPLE 14.10

How would you prepare 250. g of 6.0% by mass NaCl aqueous solution?

SOLUTION

STEP 1 Calculate the mass of solute, NaCl, needed.

$$6.0\% \times 250.\ \text{g} = 0.060 \times 250.\ \text{g} = 15\ \text{g NaCl (the solute)}$$

STEP 2 Subtract the mass of solute from the total mass of solution to obtain the mass of the solvent needed.

$$250.\ \text{g total} - 15\ \text{g solute} = 235\ \text{g water (the solvent)}$$

STEP 3 To prepare the solution, dissolve 15 g of NaCl in 235 g of water. Since 1.0 g of water has a volume of 1.0 mL (the density of water is 1.0 g/mL), we can simply mix 235 mL of water and 15 g of NaCl to make the solution.

See Problems 14.47–14.50.

EXERCISE 14.4

How would you prepare 250. g of 6.0% (w/w) table sugar solution?

A 10.0% by mass solution of NaOH, or HCl, or any other compound contains 10.0 g of the specified solute per 100. g of solution. Although both the salt and sugar solutions described in these examples have the same mass of solute per 100. g of solution, the numbers of *moles* of salt and of sugar differ.

Solutions of two chemicals with the same molarity have an equal number of moles of solute in a specified volume, but solutions with the same percent by mass do not contain the same number of moles. Instead, they contain an equal number of grams of solute per 100 g of solution.

Concentrations of Very Dilute Solutions

Concentrations of very dilute solutions are often expressed in **parts per million** (ppm), in **parts per billion** (ppb), or even in **parts per trillion** (ppt). These units are frequently used to report extremely low levels of toxic substances. A concentration of 1 ppm means 1 part (using any unit) is present in a million parts (with the same unit). For example, depending on units selected, 1 ppm could be 1 gram in 1 million grams, 1 drop in a million drops, or 1 mL in a million milliliters. We can convert ppm to ppb using the relationship

$$1\ \text{ppm} = 1000\ \text{ppb}$$

In units of time, 1 ppm is equivalent to 1 second in 1 million seconds (11.5 days). One ppb is equivalent to 1 second in 1 billion seconds (31.7 yr).

For a comparison using money, 1 ppm is equivalent to the value of 1 cent (a penny) in 1 million cents, that is, 1 cent in $10,000. One ppb is equivalent to the value of 1 cent in 1 billion cents (1 cent in $10 million).

CHEMISTRY IN OUR WORLD

Lead: How Much Is Too Much?

Lead is a toxic metal that has been linked to impaired mental abilities, especially in children. Concentrations of lead in about one out of every five of the largest municipal water systems in the United States exceed the federal health guideline of 15 parts per billion (ppb), according to the U.S. Environmental Protection Agency (EPA). The acceptable limit of lead in U.S. municipal water supplies was lowered from 0.050 ppm (50 ppb) to 15 ppb in 1993, but some public health experts believe the limit should be set even lower.

Households are most likely to have high levels of lead in drinking water when they are served by municipal water lines made of lead or have interior water pipes made of lead or copper sealed with lead solder. EPA tests conducted in 660 large public water systems revealed that, in ten cities, some homes had water with lead concentrations that exceed 70 ppb. In one city, lead concentrations of 211 ppb were found. In certain other cities, lead concentrations were found to be 175, 163, 100 (2 cities), and 84 ppb (3 cities).

Consumers in areas with relatively high levels of lead are encouraged to let water run from faucets for several minutes in the morning before using the water for drinking or cooking, or to buy bottled water.

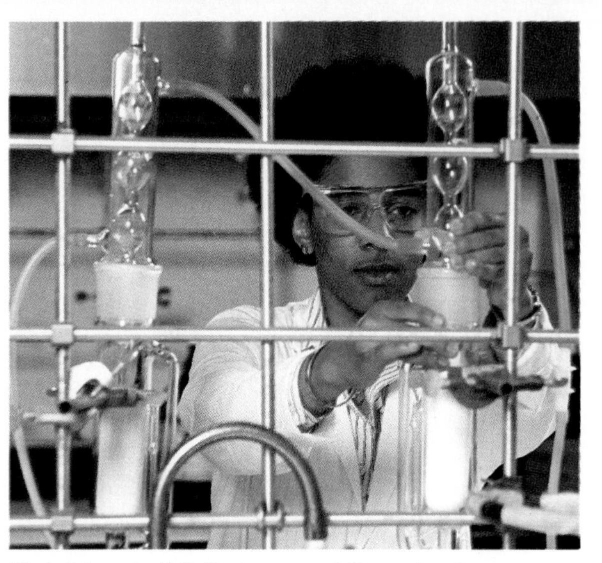

Technicians in U.S. Environmental Protection Agency (EPA) laboratories routinely check water samples for levels of lead and other contaminants.

For aqueous solutions: 1 ppm = 1 milligram/liter (mg/L)
1 ppb = 1 microgram/liter (μg/L)

Table 14.3 lists the U.S. Environmental Protection Agency (EPA) standards for trace element limits in drinking water. Notice that the values are given in milligrams per liter or parts per million.

Table 14.3 **Some Trace Element Limits in Drinking Water***

Element	*EPA Limit (ppm)*
Arsenic	0.05
Cadmium	0.01
Chromium	0.05
Copper	1.00
Lead	0.015
Manganese	0.05
Mercury	0.002
Zinc	5.00

*U.S. Environmental Protection Agency standards.

EXAMPLE 14.11

The federal limit of lead in tap water is 0.015 ppm. Express this concentration in (a) mg/L and (b) ppb.

SOLUTION

(a) The relation 1 ppm = 1 mg/L gives us the appropriate conversion factor. Multiply the quantity given (0.015 ppm) by the conversion factor.

$$0.015\ \cancel{\text{ppm}} \times \frac{1\ \text{mg/L}}{1\ \cancel{\text{ppm}}} = 0.015\ \text{mg/L}$$

(b) Using the relation 1 ppm = 1000 ppb, we can make the conversion to ppb.

See Problems 14.51–14.58.

$$0.015\ \text{ppm} \times \frac{1000\ \text{ppb}}{1\ \text{ppm}} = 15.\ \text{ppb}$$

Preparation of Solutions by Dilution

A sample from a solution with a known concentration can be diluted with water to prepare a solution with any desired concentration lower than that of the original solution. This method of preparing solutions was described in Section 9.6. Any volume units for V_1 and V_2 and any concentration units for C_1 and C_2 can be used for calculations involving dilutions, but we must not switch units during the calculations. The general equation for dilutions can be written

$$V_1C_1 = V_2C_2$$

Thus, the volume times the concentration before dilution equals the volume times the concentration after dilution.

EXAMPLE 14.12

How many milliliters of a 5.00% salt solution should be used to prepare 750. mL of a 1.00% solution of the salt?

SOLUTION

V_1 is unknown, C_1 is 5.00%, V_2 is 750. mL, and C_2 is 1.00%. Use the dilution equation.

$$V_1C_1 = V_2C_2$$

$$V_1 \times 5.00\% = 750.\ \text{mL} \times 1.00\%$$

$$V_1 = 150.\ \text{mL}$$

See Problems 14.59–14.64.

Dilute 150. mL of the 5.00% salt solution to make 750. mL of solution.

14.8 *Colligative Properties of Solutions*

Solutions have higher boiling points and lower freezing points than the corresponding pure solvent. The antifreeze in automobile cooling systems is there precisely because of these effects. Water used by itself as an engine coolant would boil away in the heat of summer and freeze in the cold temperatures of northern winters. Addition of antifreeze to the water raises the boiling point of the coolant and also prevents the coolant from freezing when the temperature drops below 0 °C. Salt is spread on city streets and sidewalks for the same reason: to lower the freezing point of water. When salt is used, the water remains liquid at temperatures below the normal freezing point of water.

The extent to which freezing points and boiling points are affected by solutes is related to the number of solute particles in solution. The higher the concentration of solute particles, the greater the effect. **Colligative properties** of solutions are those, like boiling point elevation and freezing point depression, that depend directly on the number of solute particles present in solution. Colligative properties are related to the *concentration* of dissolved solute particles, regardless of their identity. For living systems, perhaps the most important colligative property is osmotic pressure, a phenomenon described in Section 14.10.

See if you can answer the following questions related to the number of particles in various solutions. How many solute particles are there in 1.00 L of a 1 M glucose, $C_6H_{12}O_6$, solution? How many solute particles are in 1.00 L of 1 M sodium chloride, NaCl, solution? Finally, how many solute particles are in 1.00 L of 1 M calcium chloride, $CaCl_2$ solution? Should the answer be 6.02×10^{23}? All the solutions contain 1 mol of their respective solute compounds, but the questions did *not* ask for the number of *formula units*; they asked for the number of *solute particles*.

Glucose is composed of covalent molecules; its atoms are firmly bonded together in molecules. In the glucose solution, each solute particle is a glucose molecule, and there are 6.02×10^{23} of these. But in the sodium chloride solution, each formula unit of NaCl consists of a separate sodium ion, Na^+, and chloride ion, Cl^-, in solution. When sodium chloride dissolves in water, the individual ions are carried off into solution by solvent molecules. Thus, 6.02×10^{23} formula units of NaCl produce 12.04×10^{23} particles in solution (6.02×10^{23} sodium ions plus 6.02×10^{23} chloride ions). Each calcium chloride, $CaCl_2$, formula unit produces one calcium ion plus two chloride ions, so there are 18.06×10^{23} ions.

The effect of a 1 M NaCl solution on colligative properties is nearly twice that of a 1 M glucose solution—each NaCl formula unit produces two particles in solution while each glucose formula unit only produces one. Because each $CaCl_2$ unit produces three particles in solution, a calcium chloride solution has almost three times the effect of a glucose solution of the same molarity.

EXAMPLE 14.13

Qualitatively, compare the effect of the following covalent (nonionized) compounds on the freezing point of water.

(a) 1 mol of sucrose (342 g, table sugar) dissolved in 5 kg of water
(b) 1 mol of ethylene glycol (62.0 g) dissolved in 5 kg of water

SOLUTION

The freezing points will be lowered by the same number of degrees in both cases. Although the solute masses are different (342 g versus 62 g), we have 1 mol of particles dissolved in the same mass of water. Ethylene glycol is used as an antifreeze, but sugar is not. (Can you think of any factors that are important in selecting components for antifreeze other than the effect on the freezing and boiling points?)

See Problems 14.65–14.68.

14.9 Colloids

If table sugar is dissolved in water, the molecules become intimately mixed. The solution is homogeneous; that is, it has the same properties throughout. The sugar cannot be filtered out by ordinary filter paper, nor does it settle out on standing. On the other hand, if one tries to dissolve sand in water, the two substances may momentarily appear to be mixed, but the sand rapidly settles to the bottom. The

Table 14.4 **Properties of Solutions, Colloids, and Suspensions**

Property	*Solution*	*Colloid*	*Suspension*
Particle size	0.1–1.0 nm	1–100 nm	>100 nm
Settles on standing?	No	No	Yes
Filter with paper?	No	No	Yes
Separate by dialysis?	No	Yes	Yes
Homogeneous?	Yes	Borderline	No

temporary dispersion of sand in water is called a **suspension**. By allowing the water to pass through filter paper, one can recover the sand. The mixture is obviously heterogeneous because part of it is clearly sand with one set of properties and part of it is water with another set of properties.

Is there anything in between true solutions, with particles similar to the size of ordinary molecules and ions, and suspensions with large chunks of insoluble matter? Yes, there is what is called a *colloidal dispersion*, or simply *colloid*.

Colloidal dispersions are defined not by the kind of matter they contain, but by the *size of the particles* involved. True solutions have particles of about 0.1 to about 1.0 nanometers (nm) in diameter (1 nm = 10^{-9} m). Suspensions have particles with diameters of 100 nm or more. Particles intermediate between these are said to be colloidal. Fine dust and soot particles in air and corn starch dispersed in water are colloidal dispersions.

(a)

(b)

Figure 14.12 *(a) A beam of light passing through a colloidal solution is clearly visible because the light is scattered by the colloidal particles. This phenomenon is called the Tyndall effect. (b) The beacon light path through fog, which is a dispersion of water in air, also shows the Tyndall effect.*

The properties of a *colloidal dispersion* are different from those of a true solution and also are different from those of a suspension (Table 14.4). Colloidal dispersions usually appear milky or cloudy. Even those that may appear to be clear will show the path of a beam of light which passes through the dispersion (Figure 14.12). This phenomenon, called the **Tyndall effect**, is not observed in true solutions. Colloidal particles, unlike tiny molecules, are large enough to scatter and reflect light off to the side. You have probably observed the Tyndall effect in a movie theater. The shaft of light that originates in the projection booth and ends at the movie screen is brought to you through the courtesy of colloidal dust particles in the air.

There are eight different types of colloids, based on the physical state of the colloidal particles themselves (the dispersed phase) and the state of the "solvent" (the dispersing phase). A **foam** is one type of colloid. It is produced when a gas is dispersed in a liquid or solid. Examples include shaving cream and marshmallows (Figure 14.13). Another important type of colloid system is the **emulsion** that consists of a liquid dispersed (not dissolved) in a liquid or solid. Examples include milk, mayonnaise, and butter. Examples of all eight types of colloidal dispersions are listed in Table 14.5.

Some colloids are stabilized by addition of a material that provides a protective coating. Oil is ordinarily insoluble in water, but it can be emulsified by soap. The soap molecules form a negatively charged layer at the surface of each tiny oil droplet. These negative charges keep the oil particles from coming together and separating out. In a similar manner, bile salts are excreted into the upper intestine where they emulsify the fats we eat, keeping them dispersed as tiny particles during digestion. Milk is an emulsion in which fat droplets are stabilized by a coating of casein, a protein. Casein, soap, and bile salts are examples of **emulsifying agents**—substances that stabilize emulsions.

In some colloidal dispersions, the particles are charged. This charge is often due

Table 14.5 **Types of Colloidal Dispersions**

Type	*Particle Phase*	*Medium Phase*	*Example*
Foam	Gas	Liquid[†]	Whipped cream
Solid foam	Gas	Solid	Floating soap
Aerosol	Liquid	Gas	Fog, hair sprays
Liquid emulsion	Liquid	Liquid	Milk, mayonnaise
Solid emulsion	Liquid	Solid	Butter
Smoke	Solid	Gas	Fine dust or soot in air
Sol*	Solid	Liquid	Starch solutions, jellies
Solid sol	Solid	Solid	Pearl

*Sols that set up in semisolid, jellylike form are called gels.
†By their very nature, gas mixtures always qualify as solutions. The homogeneity and size of particles in gas mixtures fulfill the requirements of solutions.

to ions being picked up on the surface of a particle. A given colloid will preferentially adsorb (not absorb) or pick up only one kind of ion—either positive or negative—on its surfaces; thus, all the particles of a given colloidal dispersion bear like charges. Since like charges repel, the particles tend to stay away from one another. They cannot come together and form particles large enough to settle out of solution. However, by adding ions of opposite charge, particularly those doubly or triply charged, colloidal particles can be made to come together—to coalesce—and separate out of solution. Aluminum chloride, $AlCl_3$, with its Al^{3+} ions is quite good at breaking up colloids with negatively charged particles.

EXAMPLE 14.14

For each of the following colloids, identify the ''type'' of colloidal dispersion involved.

(a) whipped cream (b) hair spray (c) grape jelly

SOLUTION

Several examples are provided in Table 14.5.

(a) Whipped cream is a foam (a gas dispersed in a liquid).
(b) Hair spray is an aerosol (a liquid dispersed in a gas).
(c) Grape jelly is a sol (a solid dispersed in a liquid).

See Problems 14.69–14.72.

Figure 14.13 *Shaving cream, Ivory soap, and marshmallows are common examples of foams. Foams are colloids formed by the dispersion of a gas in a liquid or solid.*

14.10 Osmosis and Dialysis (optional)

Put a paper coffee filter or a piece of laboratory filter paper into a kitchen strainer or in a funnel to act as a holder for the paper. When grape juice, red wine, cranberry juice, or another solution is poured into the filter, it passes through, but undissolved solid particles do not pass through the filter. We say the filter is **permeable** to water and other solvents and to solutions.

Everyday experience tells us that certain materials are **impermeable**. Water and solutions will not pass through the metal walls of an aluminum can or through the plastic and glass walls of bottles and other containers. Are there, perhaps, materials with intermediate properties? Do some materials allow solvent molecules to pass through while holding back the solute? Are some materials permeable to some solutes but not to others? The answer is an emphatic yes! Many natural membranes are **semipermeable**. Cell membranes, the lining of the digestive tract, and the walls of the blood vessels are all semipermeable; they allow certain substances to go through while holding others back. If the semipermeable membrane allows only the solvent molecules to pass through, the process is called **osmosis**. If the membrane allows the selective passage of small ions and small molecules along with the solvent, but retains large molecules and colloidal particles, the process is called **dialysis**, and the membrane is called a *dialyzing membrane*.

Semipermeable membranes can be pictured as having extremely small pores. The size of these pores is such that tiny water molecules can pass through, but larger particles, such as sugar and protein molecules, cannot pass through. If a membrane with these characteristics is used to separate one compartment containing pure water from another compartment containing a sugar solution, something interesting happens. The volume of the liquid in the compartment containing sugar increases while the volume in the pure water compartment decreases. This is due to the process called osmosis.

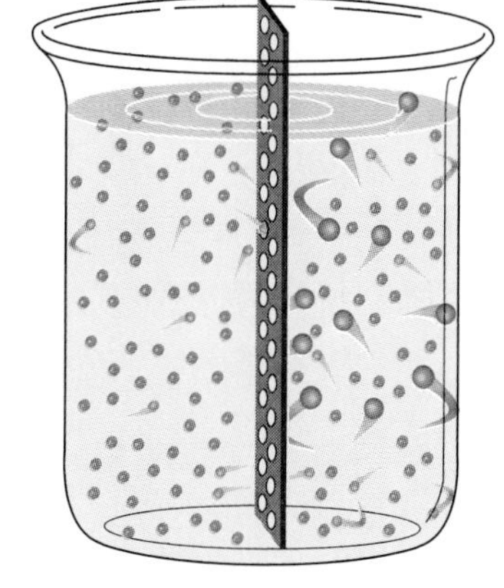

Figure 14.14 *The sieve model of osmosis holds that the semipermeable membrane has pores large enough to permit the passage of small solvent molecules such as water but too small to permit the passage of larger molecules.*

In order to understand what is going on during osmosis, we need to look at what is happening at the molecular level, as pictured in Figure 14.14. All molecules on both sides of the membrane move about at random, so occasionally they bump against the membrane. When a water molecule happens to hit one of the pores, it can pass (diffuse) through the membrane into the other compartment. However, when the much larger sugar molecule strikes a pore, it bounces back off the membrane; it cannot pass through. The more sugar molecules there are in solution (i.e., the more concentrated the solution), the smaller the chance that a water molecule will strike a pore. In other words, as time passes, more water molecules will diffuse through the membrane from the compartment that has more water molecules per unit volume than will diffuse in the reverse direction. Thus, in our example pictured in Figure 14.14, there is a *net* flow of water from the left compartment (the compartment with pure water) into the right compartment (the compartment with the sugar solution). This *net* diffusion of the solvent (water) through a semipermeable membrane is called *osmosis*.

During osmosis, there is always a spontaneous *net* flow of solvent across a membrane in one direction—from the compartment containing dilute solution into the compartment containing the more concentrated solution (Figure 14.15). For another example, the net diffusion of water across a membrane would be from a 5% sugar solution into a 10% sugar solution.

As water flow continues into the compartment with the more concentrated solution, the liquid level in that compartment rises while the liquid in the other com-

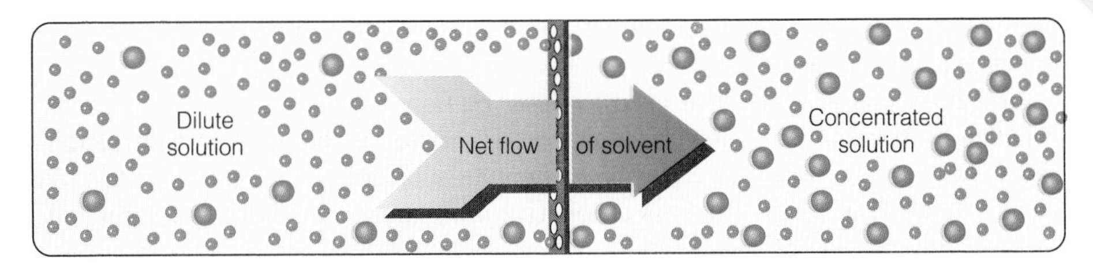

Figure 14.15 *Net solvent flow through a semipermeable membrane occurs spontaneously in only one direction, from the compartment containing dilute solution (or pure solvent) into the compartment of concentrated solution. Remember—ordinarily the terms* dilute *and* concentrated *are used to describe the concentration of* solute. *The net flow of solvent is from where the solvent is more concentrated to where the solvent is less concentrated.*

partment drops. Pressure gradually builds up in the compartment in which the liquid level rises. This increase in pressure makes it more difficult for additional water molecules to flow into that compartment. (Think of how a barometer works; see Section 12.3.) Eventually the pressure becomes great enough to prevent further net flow of water into that compartment. However, solvent molecules do not stop crossing the membrane; the *rates* at which they diffuse back and forth across the membrane become equal.

Instead of waiting for the liquid level to build up and stop the net flow of water, we can apply an external pressure to the compartment containing the more concentrated solution and accomplish the same thing. The precise amount of pressure needed to prevent the net flow of solvent from the dilute solution to the more concentrated solution is called the **osmotic pressure**. The magnitude of the osmotic pressure depends on the concentration of all dissolved particles in the solution. An **isotonic solution** is one that exhibits the same osmotic pressure as that of the fluid inside a living cell. When body fluids are replaced intravenously, it is important that the fluid be isotonic. For example, a 0.92% (or 0.16 M) NaCl solution, called physiological saline, and a 5.5% (or 0.31 M) glucose solution are isotonic with the fluid inside red blood cells.

Solutions and colloidal dispersions are within us and all around us in our environment. Without them, life itself would be impossible.

Chapter Summary

A solution is a *homogeneous* mixture made up of a solute dissolved in a solvent. In aqueous solutions, the solvent is water. A substance that dissolves in a given solvent is said to be soluble, but its *solubility* more precisely gives the maximum quantity of solute that will dissolve in a specific quantity of solvent at a given temperature.

A solution that contains a relatively large amount of solute is said to be concentrated, while a dilute solution contains a relatively small amount of solute, but these are imprecise, relative terms. A substance that does not appear to dissolve in a given solvent is said to be insoluble, but this is another imprecise term; chemical analysis may show that trace amounts actually do dissolve. The extent to which two liquids will dissolve in each other is indicated by the terms miscible, partially miscible, and immiscible.

Solvation is a general term used to describe how particles of a solute are surrounded by solvent molecules in a solution. Hydration is the more restrictive term that identifies the solvent as water. For nonpolar substances, the general rule "like dissolves like" is useful, but it is an oversimplification, especially in reference to the solubility of compounds in water where hydrogen bonding is quite important.

The relative amounts of solute and solvent in a solution can vary within solubility limits, depending on temperature and the specific substances present in the solution. When there is more solute than will dissolve in a given quantity of solvent (i.e., a saturated solution), a dynamic equilibrium is established with some solute particles being continually dissolved while other solute particles crystallize out of solution. A solution containing less than the maximum amount of solute at a given temperature is said to be unsaturated, and a solution containing more than its solubility limit is said to be supersaturated. Although most solid substances are more soluble at higher temperatures, this is not characteristic of all solids. Pressure has little effect on the solubility of solids, but the solubilities of gases are affected by both pressure and temperature. The solubility of a gas in a liquid decreases with increasing temperature, but increases at higher pressure.

Quantitatively, the concentrations of solutions can be expressed by several methods, including molarity, M; percent by volume, % (v/v); percent by mass % (w/w); parts per million, ppm (or mg/L); and parts per billion, ppb (or μg/L).

Colligative properties of solutions are those that depend directly on the number of solute particles in solution rather than on their identity. Increasing the number of solute particles in solution lowers the freezing point but elevates the boiling point and the osmotic pressure.

Finally, colloids have particles larger than those found in true solutions, but smaller than those present in suspensions. Aerosols, foams, emulsions, and sols are all colloidal dispersions.

Assess Your Understanding

1. Describe the nature of a solution at the molecular level. [14.1]
2. Identify the solute and the solvent for specific solutions. [14.1]
3. Explain why the terms soluble and insoluble, miscible and immiscible, concentrated and dilute are useful but imprecise. [14.2]
4. List three things that happen when a salt goes into solution. [14.3]
5. Predict what types of compounds will dissolve in given solvents. [14.4]
6. Describe dynamic solubility equilibria. [14.5]
7. Describe the effects of temperature and pressure on solubility. [14.6]
8. Use solubility curves or data to determine solubilities at specific temperatures. [14.6]
9. Make calculations involving concentration expressions and describe the steps followed in preparing solutions. [14.7]
10. Describe practical examples involving colligative properties. [14.8]
11. Describe foams and emulsions and give examples of each. [14.9]
12. Describe osmosis and dialysis, and differences in the membranes required for each process. [14.10]

Key Terms

aqueous solution [14.1]
colligative properties [14.8]
colloidal dispersion [14.9]
concentrated solution [14.2]
concentration [14.7]
dialysis [14.10]
dilute solution [14.2]
dissolve [14.1]
emulsifying agents [14.9]
emulsion [14.9]
foam [14.9]
homogeneous mixture [14.1]
hydration [14.3]
immiscible [14.2]
impermeable [14.10]
insoluble [14.2]
isotonic solution [14.10]
miscible [14.2]
molarity [14.7]
osmosis [14.10]
osmotic pressure [14.10]
parts per billion [14.7]
parts per million [14.7]
part per trillion [14.7]
percent by mass [14.7]
percent by volume [14.7]
permeable [14.10]
saturated solution [14.5]
semipermeable [14.10]
solubility [14.2]
soluble [14.2]
solute [14.1]
solution [14.1]
solvation [14.3]
solvent [14.1]
supersaturated [14.6]
suspension [14.9]
Tyndall effect [14.9]
unsaturated solution [14.5]

Problems

SOLUTIONS AND SOLUTION TERMINOLOGY

14.1 Explain the difference between the solute, the solvent, and the solution.

14.2 Describe the nature of a solution at the molecular level.

14.3 Identify the solute, the solvent, and the physical state of each for these solutions.
a. Chlorine gas dissolved in water in a swimming pool.
b. Vinegar (5% acetic acid and 95% water).
c. Brass (60 to 82% Cu and 18 to 40% Zn).

14.4 Identify the solute, the solvent, and the physical state of each for these solutions.
a. Carbon dioxide dissolved in water to make carbonated beverages.
b. The outer layer of a quarter (since 1965), which is 75% Cu and 25% Ni.
c. Scotch whiskey that is 86 proof is 43% ethanol (primarily) in water.

14.5 Explain why the term ''soluble'' is imprecise. Give an example.

14.6 Explain why the term ''insoluble'' is imprecise. Give an example.

14.7 Isopropyl alcohol (rubbing alcohol) and water dissolve in each other regardless of the proportions. What term describes the solubilities of these chemicals in each other?

14.8 Butyl alcohol and water dissolve in each other to a limited extent. What term best describes the solubilities of these chemicals?

SOLUBILITIES OF IONIC AND COVALENT COMPOUNDS

14.9 In the rule ''like dissolves like,'' what does the ''like'' mean?

14.10 Explain why $I_2(s)$ crystals dissolve readily in carbon tetrachloride, CCl_4, and hexane, $CH_3CH_2CH_2CH_2CH_2CH_3$, but not in water.

14.11 Account for the fact that glucose, $C_6H_{12}O_6$, unlike most six-carbon molecules, is water soluble.

14.12 Account for the fact that although hexane, C_6H_{14}, is a liquid, it is not miscible with water.

14.13 Without referring to Table 14.2, indicate whether each compound listed here would be expected to be soluble or insoluble in water. Explain each answer. You may use a periodic table.
a. CH_3CH_2OH
b. CH_3CH_2Cl
c. KCl
d. $BaSO_4$
e. $HOCH_2CH_2OH$ (ethylene glycol)
f. $Ca(NO_3)_2$

14.14 Without referring to Table 14.2, indicate whether each compound listed here would be expected to be soluble or insoluble in water. Explain each answer. You may use a periodic table.
a. $CH_3CH_2NH_2$
b. CH_3OH
c. $(NH_4)_2SO_4$
d. NaBr
e. $CH_3CH_2CH_2CH_2CH_2CH_3$
f. K_2CO_3

14.15 Explain why ammonia gas, NH_3, is quite soluble in water but not in hexane.

14.16 Explain why chlorine gas, Cl_2, which is added to swimming pools does not tend to stay in solution and must be continually replaced.

14.17 Certain ionic solids dissolve in water; others do not. Explain these differences in terms of energy changes.

14.18 Account for the fact that NaCl is water soluble but $AlPO_4$ is essentially insoluble in water.

14.19 What is the difference between "hydration" and "solvation"?

14.20 Describe, at the molecular level, what is happening as NaCl dissolves in water.

14.21 For the following compounds, indicate the extent of their solubilities in water using the categories soluble, partially soluble, insoluble, and decomposes. (See Table 14.2.)
- a. Calcium acetate (used in "canned heat").
- b. The mineral "dolomite" contains both calcium carbonate and magnesium carbonate. Describe its solubility in pure water. (Considering reactions of carbonates, how might caves be formed in dolomite?)
- c. Compounds of nitrates.
- d. Hydroxides of transition metals.

14.22 For the following compounds, indicate the extent of their solubilities in water using the categories soluble, partially soluble, insoluble, and decomposes. (See Table 14.2.)
- a. Silver chloride (light sensitive and used in photography).
- b. Lead(II) sulfate (present in deposits on car battery terminals).
- c. Carbonates and phosphates (except for those of ammonium and the alkali metal compounds).
- d. Alkali metal sulfides.

SOLUBILITY EQUILIBRIA

14.23 In a dynamic equilibrium, two processes are occurring at the same rate. In the equilibrium involving a saturated solution, which processes occur at equal rates?

14.24 Assume that additional NaCl is added to a saturated salt solution. What is observed? What actually happens?

14.25 The solubility of KNO_3 at 70. °C is 135 g per 100. g of water. At 10. °C the solubility drops to 20. g per 100. of water. A 200.-g quantity of KNO_3 was stirred into 200. mL of water.
- a. At 70. °C how many grams (if any) of crystals will settle out of solution?
- b. After cooling the solution to 10. °C, how many grams (if any) of crystals can be recovered?

14.26 Based on the information given in the previous problem, how much KNO_3 will dissolve in 80.0 g of water at 70. °C? If this saturated solution is cooled in an ice bath to 10. °C, how many grams (if any) of crystals can be recovered?

EFFECTS OF TEMPERATURE AND PRESSURE ON SOLUBILITY

14.27 Use the kinetic molecular theory to explain why most solid solutes become more soluble with increasing temperature but gases become less soluble.

14.28 Fish require the oxygen dissolved in water. Would it be a good idea to thoroughly boil the water you place in a fish aquarium? Explain.

14.29 What actually causes a bottle of soft drink to overflow while being opened?

14.30 Would a glass of warm or of cold soft drink go "flat" more quickly? Explain.

14.31 Could 50. g of NaCl be dissolved in 175 g of water? (See Figure 14.5.)

14.32 Could 50. g of Na_2SO_4 be dissolved in 150. g of water? (See Figure 14.5.)

14.33 When a seed crystal is placed in a supersaturated solution at constant temperature, precipitation is induced. When no more solid appears to precipitate, is the solution saturated, unsaturated, or supersaturated?

14.34 Which of the following involves a supersaturated solution?
- a. honey
- b. grape jelly
- c. grape juice

CONCENTRATIONS: MOLARITY (a review from Section 9.6)

14.35 You are asked to prepare 500. mL of a 0.10 M sucrose, $C_{12}H_{22}O_{11}$, solution. How many grams of sucrose would you use? What steps would you follow in making up the solution?

14.36 You are asked to prepare 250. mL of a 3.00 M NaOH solution, by starting with solid NaOH. How many grams of NaOH would you use? What steps would you follow in making up the solution?

14.37 A procedure calls for 0.040 mol of sucrose. How many milliliters of a 0.100 M sucrose solution (prepared in Problem 14.35) would you use?

14.38 A procedure calls for 0.150 mol of NaOH. How many milliliters of a 3.00 M NaOH solution (prepared in Problem 14.36) would you use?

14.39 To obtain 8.00 g of sucrose, how many milliliters of 0.100 M sucrose solution (prepared in Problem 14.35) would you use?

14.40 To obtain 8.00 g of NaOH, how many milliliters of 3.00 M NaOH solution (prepared in Problem 14.36) would you use?

CONCENTRATIONS: PERCENT BY VOLUME

14.41 What is the percent by volume concentration of a solution containing 300. mL of isopropyl alcohol and enough water to give 400. mL of solution? Which component is the solute and which is the solvent?

14.42 What is the percent by volume concentration of a solution containing 200. mL of ethanol and enough water to give 500. mL of solution? Which component is the solute and which is the solvent?

14.43 How many milliliters of isopropyl alcohol would you use to prepare 500. mL of a 60.0% (v/v) solution of isopropyl alcohol? How would you make the solution?

14.44 How many milliliters of acetone would you use to prepare 500. mL of a 60.0% (v/v) solution of acetone? How would you prepare the solution?

14.45 If a wine is 12.0% (v/v) ethanol, how many milliliters of ethanol are present in a glass containing 120. mL of the wine?

14.46 If a wine is 10.0% (v/v) ethanol, how many milliliters of ethanol are present in a bottle containing 750. mL of the wine?

CONCENTRATIONS: PERCENT BY MASS

14.47 Give the number of grams of solute and number of grams of water that must be used to prepare each of the following aqueous solutions.
a. 500. g of 5.0% (w/w) glucose solution
b. 500. g of 5.0% (w/w) $NaHCO_3$ solution
c. 2.0 kg of 3.0% by mass sodium carbonate solution

14.48 Give the number of grams of solute and number of grams of water that must be used to prepare each of the following aqueous solutions.
a. 800. g of 0.25% (w/w) sucrose solution
b. 800. g of 0.25% (w/w) KCl solution
c. 5.0 kg of 0.92% by mass NaCl solution used as ''saline solution''

14.49 Concentrated nitric acid is 70.0% HNO_3 by mass. How many grams of HNO_3 are in 500. g of the acid?

14.50 Concentrated hydrochloric acid used in the laboratory is 38.0% HCl by mass. How many grams of HCl are in 500. g of the acid?

CONCENTRATION CONVERSIONS

14.51 The maximum concentration of lead permitted in drinking water is 0.015 ppm. Express this concentration in parts per thousand using scientific notation.

14.52 The maximum concentration of zinc permitted in drinking water is 5.00 ppm. Express this concentration in parts per thousand using scientific notation.

14.53 Express 0.840 ppm in parts per billion.

14.54 Express 5.0 ppb in parts per million.

14.55 The U.S. Environmental Protection Agency limit for mercury in drinking water is 0.002 ppm. What is this value in milligrams per liter and in parts per billion?

14.56 The U.S. Environmental Protection Agency limit for arsenic in drinking water is 0.05 ppm. What is this value in milligrams per liter and in parts per billion?

14.57 A single dose of ethanol, C_2H_5OH, raises the intracellular level of calcium ions. The Ca^{2+} ions released by 0.10 M ethanol kill rat liver cells [*Science*, **224**, 1361 (1984)]. Express this concentration in grams per liter. What is the percent by volume? (The density of ethanol is 0.785 g/mL.)

14.58 A cyanide solution used to leach gold from its ore is made by adding 1 lb of sodium cyanide, NaCN, to 1 ton of water. Give the NaCN concentration in (a) percent by mass and (b) parts per million. (1 ton = 2000 lb.)

DILUTION: A REVIEW (see Section 9.6)

14.59 What volume of a 70.0% (w/w) antiseptic solution would be used to prepare 350. mL of 40.0% solution? How much water should be used?

14.60 What volume of a 10.0% (w/w) $NaHCO_3$ solution would be used to prepare 275 mL of 2.00% (w/w) solution? How much water should be used?

14.61 How many milliliters of 12.0 M HCl aqueous solution would be used to prepare 500. mL of

0.100 M HCl solution? How should the solution be prepared?

14.62 How many milliliters of 15.0 M NH_3 aqueous solution would be used to prepare 100. mL of 6.00 M NH_3 solution? How should the solution be prepared?

14.63 What would be the molarity of a solution obtained when 50.0 mL of 3.00 M HCl acid solution is mixed with 70.0 mL of H_2O?

14.64 What would be the concentration of a solution obtained when 250. mL of 6.00 M NaOH solution is mixed with 150. mL of H_2O?

COLLIGATIVE PROPERTIES

14.65 Would 5 kg of ethanol, C_2H_5OH, or 5 kg of methanol, CH_3OH, be more effective as an antifreeze when added to a radiator? *Hint*: Consider molar mass and number of particles.

14.66 Would an equal number of moles of sodium chloride, NaCl, or calcium chloride, $CaCl_2$, do a better job of keeping streets from freezing? Explain. Why do most highway departments use both chemicals?

14.67 How many solute particles does each formula unit of the following compounds give in aqueous solution? (Consider ionic and covalent compounds described in Chapter 7.)

a. KCl
b. CH_3OH
c. $(NH_4)_2SO_4$
d. $C_2H_6O_2$ (ethylene glycol)

14.68 How many solute particles does each formula unit of the following compounds give in aqueous solution?

a. $CaCl_2$
b. NaOH
c. $(NH_4)_3PO_4$
d. $C_{12}H_{22}O_{11}$ (sucrose)

COLLOIDS

14.69 Compare particle sizes for solutions, colloids, and suspensions. What is the range in particle sizes for colloids?

14.70 Why can one sometimes see the "shaft" of light coming through the window into a room? Explain.

14.71 For each of the following examples, give the *type* of colloidal dispersion (foam, aerosol, emulsion, etc.), the particle phase, and the phase of the dispersing medium.

a. fog
b. jelly
c. Ivory soap (it floats)
d. shaving cream

14.72 For each of the following examples, give the *type* of colloidal dispersion (foam, aerosol, emulsion, etc.), the particle phase, and the phase of the dispersing medium.

a. butter
b. mayonnaise
c. marshmallow
d. hair spray

Additional Problems

14.73 How does a dilute solution differ from a saturated solution? Explain.

14.74 What is meant by this statement? Antifreeze (primarily ethylene glycol) is completely miscible with water.

14.75 Why does an opened bottle of cola or other soft drink go "flat" more quickly when left at room temperature than when stored in a refrigerator?

14.76 Water from a lake or river is often used by an industry or electric power plant for cooling purposes and then returned to the lake. Why is the temperature of the water being returned important? (See Figure 14.8.)

14.77 Identify the solute, the solvent, and the state of each for the following.

a. a one-cent coin (95% Cu and 5% Zn)
b. glass cleaner solution that contains ammonia gas in water
c. vinegar, a 5% acetic acid in water solution

14.78 Identify the solute, the solvent, and the state of each for the following.

a. a 5% saline solution, which is a 5% (w/w) NaCl solution
b. sterling silver (92.5% Ag and 7.5% Cu)
c. tincture of iodine (a small amount of I_2 in alcohol)

14.79 Without referring to Table 14.2, indicate whether each compound listed here would be expected to be soluble or insoluble in water. Explain each answer. You may use a periodic table.

a. $RbCl$ b. $(NH_4)_3PO_4$
c. $C_{10}H_{20}O$ d. Na_2CO_3

14.80 Without referring to Table 14.2, indicate whether each compound listed here would be expected to be soluble or insoluble in water. Explain each answer. You may use a periodic table.
a. $(NH_4)_2CO_3$
b. CS_2
c. $CH_3CH_2CH_2CH_2CH_2CH_2CH_2CH_2OH$
d. K_3PO_4

14.81 Concentrated sulfuric acid is 95% H_2SO_4 by mass. How many grams of H_2SO_4 are in 800. g of the acid? How many milliliters of the acid can be used to supply 800. g of H_2SO_4? (The density of the acid is 1.84 g/mL.)

14.82 A 10.0% (w/w) NaOH solution is available in the laboratory. How many grams of this solution would be needed to deliver 12.0 g of NaOH. How many milliliters of the solution should be used? (The density of the solution is 1.11 g/mL.)

14.83 If 100.0 g of glucose is mixed with 400.0 g of water, what is the percent by mass concentration of the solution?

14.84 If 2.0 g of $NaHCO_3$ is mixed with 38 mL of water, what is the percent by mass concentration of the solution?

14.85 The U.S. Environmental Protection Agency limit for lindane (an insecticide) in drinking water is 0.004 mg/L. What is this value in parts per million and in parts per billion?

14.86 The U.S. Environmental Protection Agency limit for 2,4-D (a herbicide) in drinking water is 0.1 mg/L. What is this value in parts per million and in parts per billion?

14.87 What is the difference between osmosis and dialysis?

14.88 What is the difference between an osmotic membrane and a dialyzing membrane?

14.89 Celery left in a saltwater solution becomes limp, but the crispness can be restored by placing the celery in water. Explain.

14.90 Boiling vegetables in water to which salt has been added causes them to lose crispness. Explain.

14.91 What is the Tyndall effect? Give an example.

14.92 How do detergents and soaps act as emulsifying agents?

Dynamic equilibrium is established between two levels of a building when the number of people going up the escalator equals the number of people going down the escalator every hour.

15

Reaction Rates and Chemical Equilibrium

CONTENTS

Some reactions, such as the combustion of gasoline in an engine, are explosively fast. Other reactions, such as the rusting of iron, are exceedingly slow. Although several types of chemical reactions and examples of each type were described in Chapter 10, little was said about the varying rates of those reactions, or the extent to which they occur. We will begin this chapter by looking at factors that affect reaction rates.

The extent to which chemical reactants are converted to products is also quite important. In certain reactions, virtually all of the reactants are converted to products. (An automobile engine is ''tuned'' to maximize the complete conversion of gasoline to carbon dioxide and water.) Many other reactions are reversible; the products react to re-form the original chemicals. A reaction carried out in a closed container eventually appears to stop because the concentrations of reactants and products stop changing. At this point, an *equilibrium* is established between the forward and the reverse reactions. The conditions that affect reaction equilibria are important in industrial operations, environmental concerns, and biochemical processes.

In this chapter we will deal with factors that affect reaction rates and chemical equilibria. Although we will work mainly with simple reactions, the principles described here also apply to more complex systems, including the chemistry of living cells.

15.1 *Reaction Rates: Collision Theory*

The term **chemical kinetics** refers to the study of **rates of reaction** and factors that affect reaction rates. Before atoms, molecules, or ions can react (i.e., transfer or share valence electrons), they must first get together; they must *collide.* Second, they must come together in the proper *orientation* unless the particles involved are single atoms or small, symmetrical molecules. Third, the collision must provide a certain minimum energy, called the *activation energy.* Let us look more closely at each of these factors.

Collision Frequency

The first factor, **collision frequency,** is controlled by concentration and temperature. The more concentrated the reactants, the more frequently the particles will collide, simply because there are more of them in a given volume. An increase in temperature also increases the frequency of collision because particles move faster at higher temperatures, so they come in contact more frequently. These effects will be discussed in more detail in Section 15.2.

Orientation

Now, let's consider **orientation,** that is, the position of the particles relative to each other (their geometry) at the time of collision. If one perfectly symmetrical particle, such as a billiard ball, collides with another perfectly symmetrical particle, the collision is equivalent regardless of which "face" of the symmetrical particle is struck. However, to catch a baseball the orientation of the ball glove relative to the ball is quite important. Similarly, if two people approach each other for a kiss, their orientation is obviously important.

With chemical reactions, there are a few instances for which orientation is not important. When two hydrogen atoms react to form a hydrogen molecule,

$$2\,H \longrightarrow H_2$$

their orientation is not important because the hydrogen atoms behave like symmetrical (spherical) electron clouds. There is no front, back, top, or bottom; all "faces" of hydrogen atoms are alike. However, for most particles, proper orientation during the collision is essential. For example, consider the chemical equation for the reaction of nitrogen dioxide, NO_2, with CO to form NO and CO_2.

$$NO_2(g) + CO(g) \longrightarrow NO(g) + CO_2(g)$$

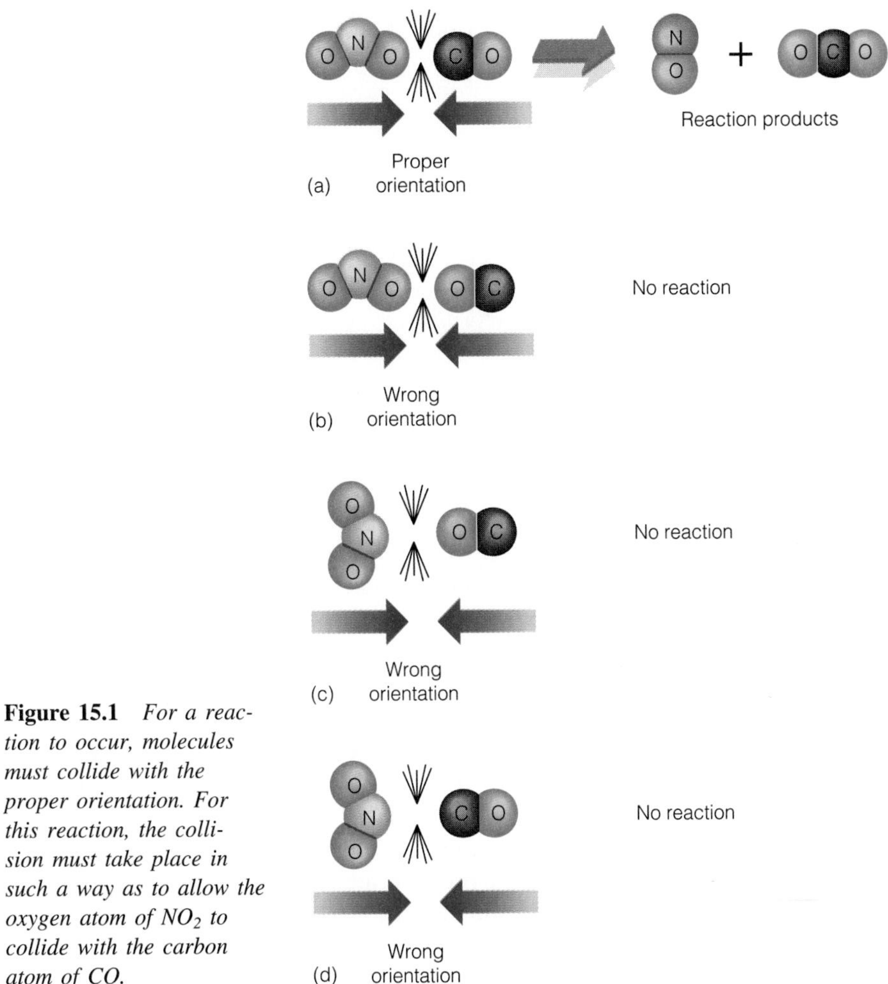

Figure 15.1 *For a reaction to occur, molecules must collide with the proper orientation. For this reaction, the collision must take place in such a way as to allow the oxygen atom of NO_2 to collide with the carbon atom of CO.*

During this reaction, an oxygen atom is transferred from NO_2 to CO to produce NO and CO_2. For this transfer to take place, an oxygen atom of NO_2 must collide with the carbon atom of CO. Figure 15.1 shows the required orientation for this reaction as well as some possible collisions that are ineffective.

Activation Energy

A reaction does not necessarily occur even when colliding molecules have proper orientation. Two molecules that bump each other gently are likely to bounce back without reacting. When particles collide, they must also possess a specific amount of kinetic energy before an *effective collision*—a reaction—can occur. This minimum kinetic (collision) energy that reacting molecules must have is called the **activation energy,** abbreviated E_a. The activation energy for a reaction depends on the specific kinds of molecules involved. For reactions that occur instantaneously when reactants are mixed, the activation energy is low. For reactions that are slow to get started, the activation energy is high.

Exergonic is the term used to describe any reaction that releases heat, light, or any form of energy.

Potential energy changes for chemical reactions were described in Section 11.8. Recall that *exothermic* reactions are those for which heat is released during the

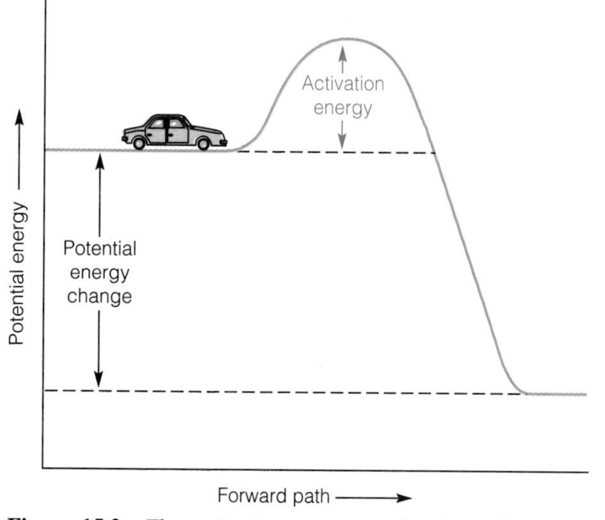

Figure 15.2 *The activation energy can be pictured as an "energy hill" that must be overcome in order to get to the "valley of stability" on the other side.*

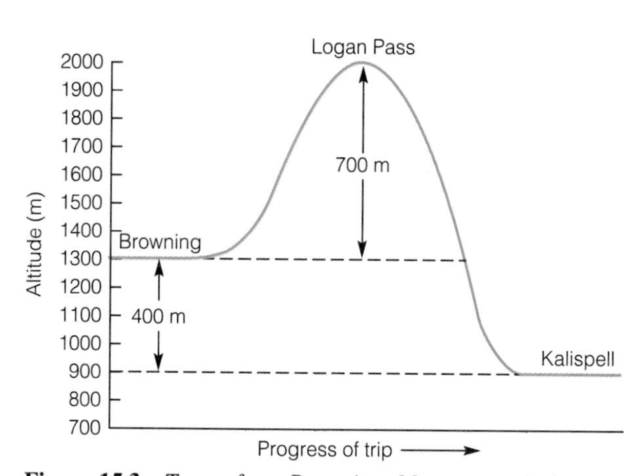

Figure 15.3 *To get from Browning, Montana, to Kalispell via Going-to-the-Sun Highway, we would first have to climb to Logan Pass, even though Kalispell is 400 m lower in elevation than Browning.*

reaction to give products that are more stable and lower in energy than the reactants. One might therefore think all exothermic reactions occur instantaneously when reactants are mixed. They do not. What, then, is the barrier to their instantaneous reaction? The barrier is the activation energy that must be overcome in order to get to the products, even though the products are at a lower energy. As pictured in Figure 15.2, overcoming the activation energy can be compared to pushing a car to the top of a hill so it can roll downhill to a valley that lies at a lower level (a lower potential energy).

Study Hint
This is an analogy; it is used to help you picture a complex process by comparing it to one that can be visualized, but an analogy should not be expected to be accurate in all respects.

For another analogy, imagine you were in Browning, Montana (elevation 1300 m), and wished to travel to Kalispell (elevation 900 m). One route you could take would be to drive the scenic Going-to-the-Sun Highway through the beautiful Glacier National Park. If you take this route, you would have to cross the continental divide at Logan Pass (elevation 2000 m). First, you would have to climb 700 m, but then it would be downhill the rest of the way (Figure 15.3).

Now, for a specific chemical reaction, consider the burning of charcoal (carbon), which has a lot of chemical energy. Carbon reacts with oxygen to form carbon dioxide with the release of considerable heat energy. But charcoal doesn't react very rapidly with oxygen at ordinary temperatures. In fact, one can store charcoal indefinitely without it reacting with the oxygen in air. Before charcoal will ignite and burn steadily, its energy must be elevated to a specific value that is equal to the activation energy (Figure 15.4). Once this activation energy level is reached, the heat energy released by the reaction will keep the reaction going. Overall, more energy is released by the reaction than is put in. The reaction is *exothermic.* (For an *endothermic* reaction, more energy must be put in than is released.)

Glucose—one type of sugar—burns in oxygen and releases energy. Initially, the reactants (glucose and oxygen) are at a higher energy level than are the products (carbon dioxide and water), as shown in Figure 15.5. But this does not mean that the reaction path is straight downhill from reactants to products. Indeed, glucose can be in an open container exposed to oxygen in the air, but no perceptible reaction occurs. As with the reaction of charcoal and oxygen, a certain amount of energy—the activation energy—must be supplied before glucose will burn. The overall reaction releases energy; it is exothermic.

We can read an activation energy diagram (such as Figure 15.5) from left to right or right to left. With the burning of glucose, we are considering an exothermic reaction in which higher energy reactants are converted to lower energy products. When this diagram is read in the opposite direction, the energy changes represented are those that occur during the endothermic reaction called photosynthesis. Notice that the climb to the top of the ''energy hill'' is longer from one side than from the other. This climb (the difference between the top of the energy hill and the starting materials) corresponds to the activation energy, E_a. Thus, the activation energy for photosynthesis, the reverse reaction, is greater than the activation energy for the oxidation of glucose, the forward reaction (Figure 15.5).

Certain substances, called *catalysts* (they will be described in the next section), may substantially lower the activation energy and thus increase the rate of a reaction without being consumed.

The potential energy diagram showing the activation energy as an ''energy hill'' is actually an oversimplification of the energy changes. Indeed, one seldom crosses a mountain by going straight up one side and straight down the other side. Similarly, reactions seldom proceed by a smooth, one-bump potential energy change.

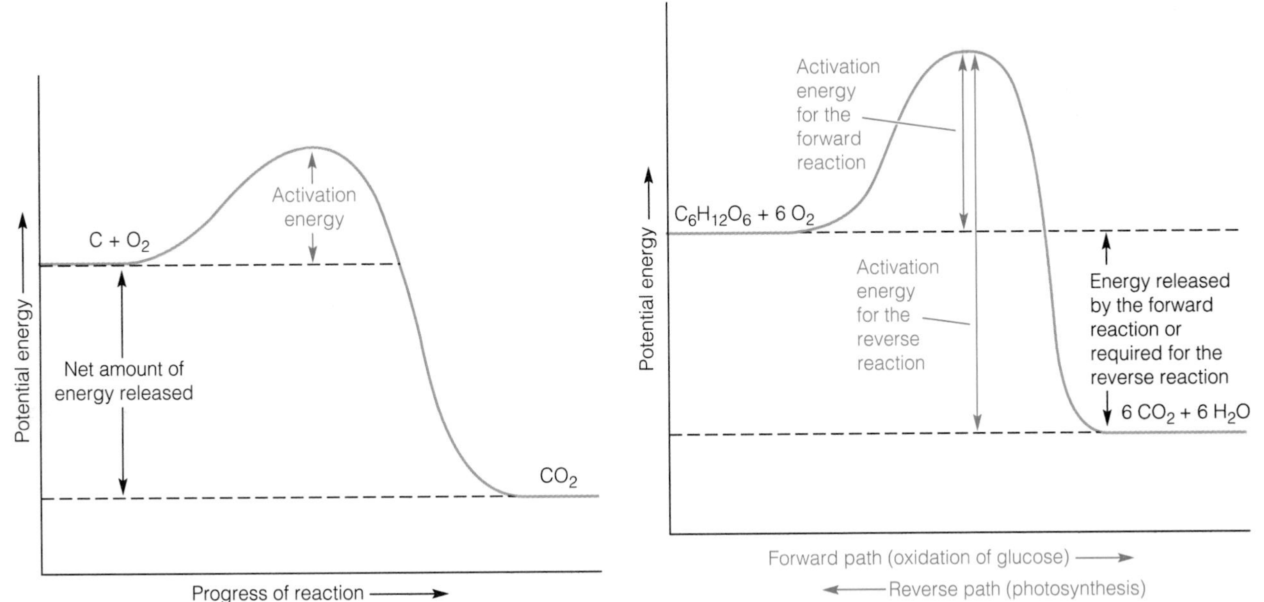

Figure 15.4 *To get from reactants (carbon and oxygen) to product (carbon dioxide), we must first put some energy into the system.*

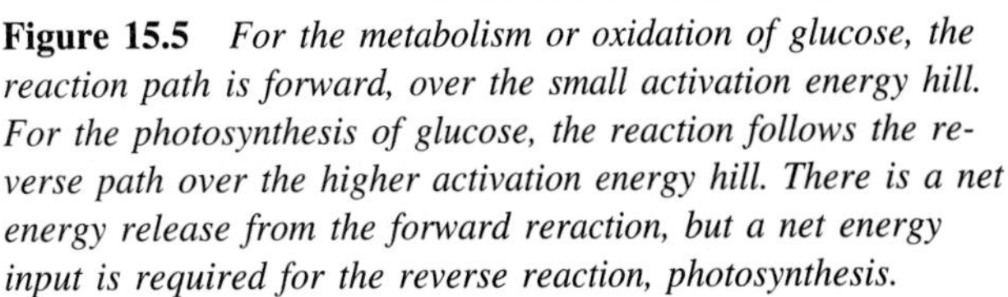

Figure 15.5 *For the metabolism or oxidation of glucose, the reaction path is forward, over the small activation energy hill. For the photosynthesis of glucose, the reaction follows the reverse path over the higher activation energy hill. There is a net energy release from the forward reraction, but a net energy input is required for the reverse reaction, photosynthesis.*

Multiplying the three factors described in this section gives the overall rate of a chemical reaction.

$$\text{Reaction rate} = \begin{matrix}\text{Collision}\\ \text{frequency}\\ \text{factor}\end{matrix} \times \begin{matrix}\text{Collision}\\ \text{orientation}\\ \text{factor}\end{matrix} \times \begin{matrix}\text{Collision}\\ \text{energy}\\ \text{factor}\end{matrix}$$

When we actually measure a reaction rate in the laboratory, it is expressed as a *change in concentration per unit time.*

EXAMPLE 15.1

Refer to Figure 15.5 while answering these questions.

(a) Compare the activation energy for photosynthesis with the activation energy for the metabolism (oxidation) of glucose.
(b) Justify or criticize this statement: The activation energy has no net effect on the energy of reaction.

SOLUTION

(a) The activation energy for the oxidation of glucose (the forward reaction) is small compared to the activation energy for photosynthesis (the reverse reaction).
(b) The statement is correct. The energy of reaction is the difference between the potential energies of the reactants and the products. (The height of the ''energy hill'' does not affect the net energy change.)

See Problems 15.1–15.8

EXERCISE 15.1

Describe three factors that determine the overall rate of a chemical reaction.

15.2 *Factors That Control Reaction Rates*

Three factors that affect reaction rates were given in the previous section. The frequency of collision—the first factor mentioned—is controlled by both temperature and concentration. We will now describe these effects in some detail and also discuss the effects of catalysts and of surface area.

The Effect of Temperature on Reaction Rate

Reactions generally take place at a faster rate when temperatures are high. For example, a roast in the oven or a steak on the grill cooks more rapidly at higher temperatures. On the other hand, lowering the temperature generally slows down the reaction rate. We place food in the refrigerator or freezer to slow down chemical reactions that take place when food spoils. Similarly, a cold automobile battery may not put out enough energy to crank an engine on a cold morning. Even the insects move around more slowly in the fall of the year when temperatures are cooler. This is because the temperature of the surroundings controls the body temperature of these cold-blooded creatures, and their biochemical reactions are slowed down at the lower temperatures.

Hibernation and Chemistry
When animals are hibernating, their body temperatures are low, so their metabolism rates are low. With a decreased rate of metabolism, the need for food is decreased. This allows the hibernating animal to survive the winter.

CHEMISTRY IN OUR WORLD

Body Temperatures: Fever and Hypothermia

The chemical reactions that occur in our bodies normally do so at a constant temperature of 37.0 °C (98.6 °F). A few degrees' rise in temperature—a fever—leads to an increase in pulse rate, respiration rate, and biochemical reaction rate. A drop in body temperature of a few degrees slows these same processes considerably. When there is a significant drop in inner body temperature, symptoms of hypothermia can be observed.

Inner Body Temperature (°C)	*Symptom*
37	Normal body temperature
35–36	Uncontrollable shivering
33–35	Confusion, amnesia
30–32	Muscular rigidity and stiffness
26–30	Unconsciousness, erratic heartbeat
Below 26	Heart failure, death

For heart surgery, the body temperature of the patient is sometimes lowered to about 28 to 30 °C. Ordinarily, the brain is permanently damaged when its oxygen supply is interrupted for more than 5 minutes, but, at the lower temperature, metabolic processes—biochemical reactions—are slowed and the oxygen requirement is reduced. The surgeon can stop the heartbeat, perform an hour-long surgical procedure on the heart, and then restart the heart and bring the patient's temperature back to normal.

Heart bypass surgery can be performed when the patient's heart is isolated from the general circulation and cooled to 4 °C to slow metabolic processes and minimize possible tissue damage.

Temperature effects in living systems are vital to life, but the manipulation of reaction rates through changes in temperature is severely restricted. An increase in temperature may destroy enzymes that control chemical reactions within cells—enzymes that are essential to life. (One way of killing germs is by sterilization.) For living cells, there is a rather narrow range of optimum temperatures. Both higher and lower temperatures can be disabling, if not deadly.

There is a general rule of thumb.

For many reactions, the rate of reaction roughly doubles for a temperature increase of 10 °C.

This can be explained in terms of the kinetic molecular theory (Section 12.2), which states that molecules move more rapidly at higher temperatures. When they move more rapidly, they collide more frequently, so the reaction rate is increased.

At higher temperatures, the rapidly moving molecules also have more kinetic energy; temperature is a measure of the average kinetic energy of molecules. As described in the previous section, particles must collide with at least the minimum activation energy to be effective (reaction-producing) collisions. At higher temperatures, a greater fraction of the colliding molecules possess sufficient energy—the activation energy—needed to break the covalent bonds of reacting molecules. For the reaction of hydrogen gas with chlorine gas.

This equation says nothing about the sequence in which the bonds are broken. The detailed sequence of small reaction steps is called the mechanism of the reaction.

$$H_2 + Cl_2 \longrightarrow 2\,HCl$$

the H—H and Cl—Cl bonds must be broken before the reaction can occur. This is an exothermic reaction; once it has started, the energy released during the formation of hydrogen-to-chlorine bonds more than compensates for the energy required to break H—H and Cl—Cl bonds. As a result, there is a net conversion of some chemical energy to heat energy.

The Effect of Concentration on Reaction Rate

The frequency of collision, and therefore the reaction rate, is also affected by the concentration of reactants. The more reactant molecules there are in a specific volume of liquid or gas, the more collisions will occur per unit of time. For example, if you light a wood splint (a small, thin stick) and then blow out the flame, the splint will continue to glow as the wood reacts slowly with the oxygen of the air. If the glowing splint is placed in pure oxygen, it will burst into flame, indicating a much more rapid reaction. This more rapid reaction can be interpreted in terms of the concentration of oxygen, and air is about one-fifth oxygen. The concentration of O_2 molecules in pure oxygen at the same temperature and pressure is, therefore, about five times as great as it is in air.

For reactions in solution, the concentration of a reactant can be increased if more of it is dissolved. One of the first studies of reaction rates was done by Ludwig Wilhelmy in 1850. He studied the rate of reaction of sucrose (cane or beet sugar) with water in an acid solution. The products are two simpler sugars, glucose and fructose.

$$\text{Sucrose} + H_2O \xrightarrow{HCl(aq)} \text{Glucose} + \text{Fructose}$$

We will not be concerned with the chemical formulas for these sugars at this time. Hydrochloric acid, HCl, speeds up the reaction, but it is not consumed by the reaction; it is a *catalyst,* so its formula is often written above the arrow. (Catalysts will be described in more detail under the next topical heading.) Wilhelmy found that the rate of the reaction was proportional to the concentration of sucrose. If he doubled the concentration of sucrose, the reaction rate also doubled.

In general, when the temperature is constant, the rate of the reaction can be related quantitatively to the amounts of reacting substances. The relationship is not necessarily a simple one, however. In order to know *which* reacting substances are involved in determining the rate, one must know the **reaction mechanism,** that is, the step-by-step detail of how the molecules collide, come apart, and recombine. See the box ''Reaction Mechanisms.''

The Effect of Catalysts on Reaction Rate

The reaction involving the decomposition of hydrogen peroxide, H_2O_2, to give oxygen gas and water proceeds very slowly at room temperature in the dark. However, if you add a little manganese(IV) oxide, the hydrogen peroxide decomposition immediately quickens; you will observe bubbling as oxygen gas is released. The manganese(IV) oxide serves as a catalyst for this reaction; its formula

The catalyst for this reaction, manganese(IV) oxide, MnO_2, is also called manganese dioxide.

CHEMISTRY AT WORK

Reaction Mechanisms: The Little Pieces of Big Puzzles

The *mechanism* of a reaction is the series of individual small steps required to produce the overall reaction that is observed. This step-by-step sequence can be determined by studying reaction rates as various conditions—temperature, concentrations, and catalysts—are changed individually.

Investigations of mechanisms are quite important for industrial processes, but they are also important in determining the chemistry of living cells. For example, many types of cancer are thought to be induced by chemicals in the environment. A lot of research is under way to work out the mechanism by which the chemicals act—or are acted upon—in the induction of cancer.

Certain diseases of genetic origin involve the disruption of a single step in the mechanism of a reaction essential to good health. Knowledge of such mechanisms has permitted treatment of what might otherwise have been a fatal defect. Many research investigations are now under way to find reaction mechanisms that will lead to an understanding and effective treatment of the common cold, AIDS, and many other diseases.

is written above the arrow. A **catalyst** is a substance that speeds up the rate of a chemical reaction without itself being consumed in the reaction.

$$2\,H_2O_2 \xrightarrow{MnO_2} 2\,H_2O + O_2(g)$$

In the laboratory, we can show that the solid manganese(IV) oxide does not get used up during the reaction. First, we determine the mass in grams of a small lump of MnO_2. (We could also use granules of the black MnO_2 powder in a small cloth bag.) The MnO_2 sample can then be tied to a string and lowered into a beaker containing hydrogen peroxide (Figure 15.6). When the rapid reaction ceases, the

(a) (b)

Figure 15.6 *The decomposition of hydrogen peroxide, H_2O_2, at room temperature is so slow it cannot be detected with the naked eye (a). When a lump of manganese(IV) oxide, MnO_2, is lowered into 30% H_2O_2, bubbles of oxygen gas are instantly produced (b). The MnO_2 serves as a* catalyst; *it speeds up the reaction but is not itself used up. There is no mass change for MnO_2 during the reaction. Several other substances, such as platinum metal, Pt, and the iodide ion, I^-, also act as catalysts for this reaction.*

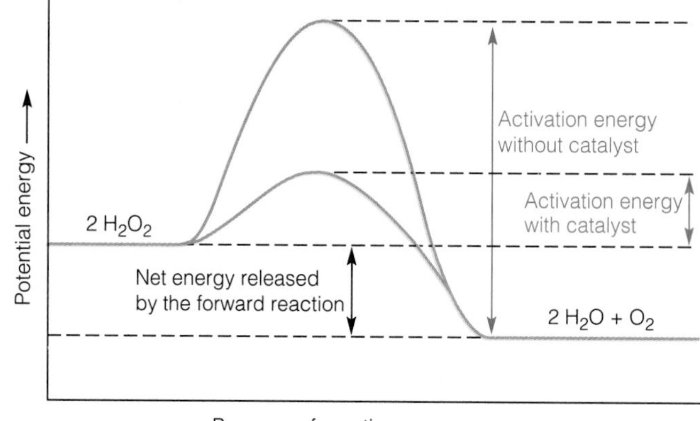

Figure 15.7 *The decomposition of hydrogen peroxide, H_2O_2, proceeds very slowly at ordinary temperatures. The use of a catalyst considerably lowers the activation energy. As Figure 15.6 shows, when MnO_2 is the catalyst, the reaction proceeds rapidly at room temperature.*

sample of MnO_2 can be removed and allowed to dry before the mass is again determined. As with all catalysts, the mass of MnO_2 does not change; it is not consumed during the reaction and can be used over and over again.

In general, catalysts act by lowering the activation energy required for the reaction to occur, as shown in the potential energy diagram for the decomposition of hydrogen peroxide (Figure 15.7). If the activation energy is lowered, then more of the slowly moving molecules possess enough energy for effective collisions, as indicated by the lower temperature required for a catalyzed reaction. When a catalyst is present, the energy of activation is lowered because the catalyst changes the *path* of the reaction.

Let us return to our analogy of the trip from Browning to Kalispell. It is possible to take an alternate route: U.S. Highway 2 crosses the continental divide through Marias Pass. This route involves a climb of only 300 m, compared with 700 m via

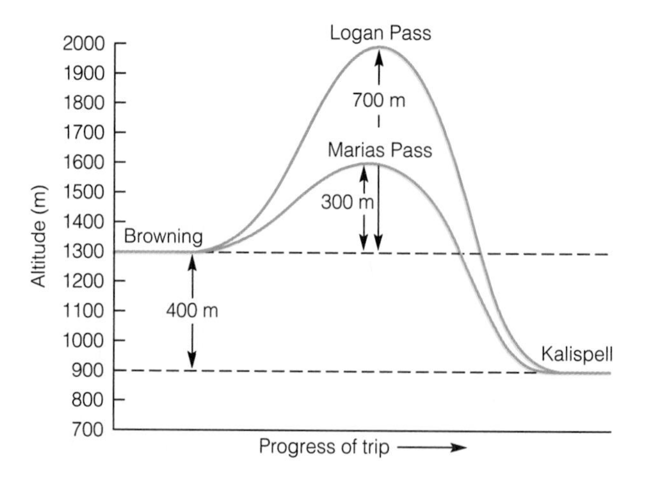

Figure 15.8 *To get from Browning to Kalispell via Highway 2 involves a lower energy barrier than the route via Logan Pass.*

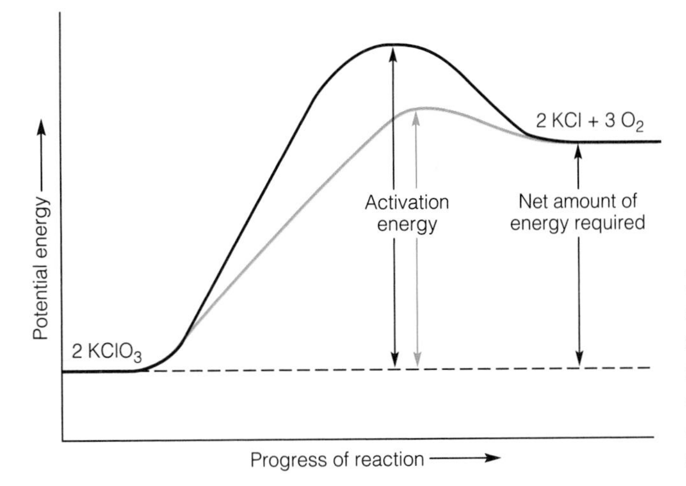

Figure 15.9 *The decomposition of potassium chlorate is endothermic. A catalyst acts to lower the energy of activation (colored arrow) compared with that required for the uncatalyzed reaction.*

Logan Pass (Figure 15.8). This alternate route is analogous to that provided by a catalyst in a chemical reaction.

The decomposition of potassium chlorate, $KClO_3$, to produce potassium chloride and oxygen is an endothermic reaction. For the oxygen to be produced at a useful rate, the potassium chlorate must be heated to over 400 °C. However, if we add a small amount of manganese(IV) oxide, MnO_2, the decomposition will deliver oxygen gas at the same rate when the reactant is heated to just 250 °C. The potential energy diagram for this endothermic reaction is shown in Figure 15.9. When the reaction is complete, the catalyst can be completely recovered, unchanged. The balanced equation is

$$2\ KClO_3 + \text{Heat} \xrightarrow{MnO_2} 2\ KCl + 3\ O_2(g)$$

Catalysts are of great importance in the chemical industry and in automobile exhaust catalytic converters (see the box ''Automobile Catalytic Converters''). A reaction that would otherwise be so slow as to be impractical can be made to proceed at a reasonable rate with the proper catalyst.

Catalysts are even more essential to the biochemical reactions that occur in living organisms where reactions are carried out at a constant temperature of 37 °C. Biological catalysts are called **enzymes;** they catalyze nearly all the chemical reactions that take place in living systems.

The Effect of Surface Area on Reaction Rate

A large log in the fireplace burns slowly. If the log is chopped into kindling (small pieces), the wood burns much faster. If the same wood is converted to dry sawdust, it can burn so rapidly as to cause a forceful explosion when ignited by a spark. Many grain elevators in midwestern states have been leveled by a single explosion for the same reason. In each of these reactions, the molecules at the surface of the cellulose in the wood or grain must react with the oxygen in air. The smaller the wood chips or grain dust, the greater is the surface area. This makes it possible for more collisions to occur between the reactants, so reaction rate increases. In these examples, one of the reactants was a solid and the other reactant was a gas. As shown here, when reactants are in two different physical states, an increase in surface area allows an increase in the frequency of collisions. The result is an increase in reaction rate.

EXAMPLE 15.2

A lump of table sugar (sucrose) can be held in the hand (at body temperature) without it reacting with oxygen. Holding a match under the lump of sugar does not ignite it. If the sugar is swallowed and metabolized by the body, carbon dioxide and water are produced, and energy is released.

CHEMISTRY AT WORK

Automobile Catalytic Converters

Catalysts are used to reduce the air pollution resulting from automobile emissions. Catalytic converters installed on automobiles convert combustion by-products such as carbon monoxide, CO, and oxides of nitrogen into carbon dioxide, CO_2, and nitrogen gas, N_2.

The catalyst provides a surface—something like a workbench—upon which the molecules meet and react. Platinum and certain transition metal oxides have been found to be the best catalysts for these reactions. The heat released by these exothermic reactions causes the catalyst to become quite hot. Then, as carbon monoxide and unreacted molecules of fuel strike these hot surfaces, they gain the activation energy needed for reaction. The fact that the catalytic converter is quite hot is one indication that the catalyst is working properly.

One of the catalyzed reactions that occurs in the catalytic converter is the oxidation of carbon monoxide to carbon dioxide.

$$2\,CO(g) + O_2(g) \xrightarrow{Pt} 2\,CO_2(g) + \text{Heat}$$

Small amounts of nitrogen oxides, such as NO and NO_2, are also produced in an automobile engine. These oxides, often symbolized NO_x, are also major pollutants. In a converter, the appropriate catalyst can be used to promote the decomposition of these oxides to the elements. Two examples are included here.

$$2\,NO(g) \xrightarrow{\text{catalyst}} N_2(g) + O_2(g) + \text{Heat}$$

$$2\,NO_2(g) \xrightarrow{\text{catalyst}} N_2(g) + 2\,O_2(g) + \text{Heat}$$

Catalytic converters are but one more example of how important chemical knowledge is to the products we use daily. Research continues; effective catalysts are needed for all kinds of industrial processes.

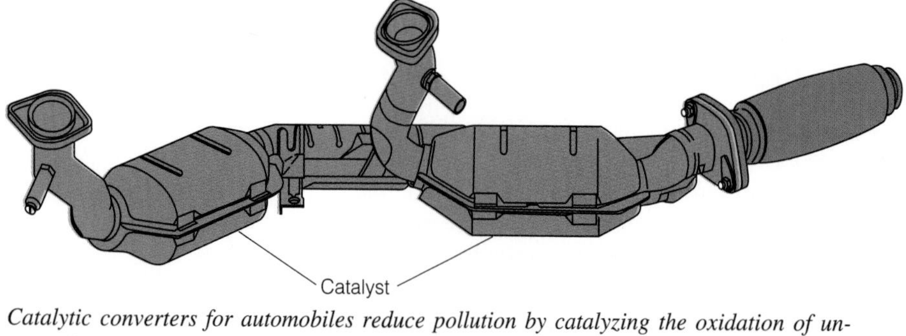

Catalytic converters for automobiles reduce pollution by catalyzing the oxidation of unwanted combustion by-products such as carbon monoxide.

(a) What does this information tell us about the activation energy for the oxidation of sucrose?
(b) How could it be possible for the oxidation of sugar to occur in the body at 37 °C but not when held in the hand at that temperature?
(c) Based on the information given, is the reaction endothermic or exothermic?

SOLUTION

(a) The activation energy is high. The oxidation does not occur readily when warmed in the hand or when heated by a match.
(b) The reaction occurs in the body at 37 °C in the presence of certain biological catalysts—enzymes—that lower the activation energy.
(c) The reaction is exothermic; energy is released during the oxidation.

See Problems 15.9–15.26.

EXERCISE 15.2
What do hibernating animals, food in refrigerators, and cold automobile batteries have in common?

15.3 Reversible Reactions and Equilibrium

Melting and freezing are reversible processes that were described in Chapter 13. The *rates* of these two opposing processes are equal at the melting/freezing point, and the two physical states—solid and liquid—are in *dynamic equilibrium.* At the boiling point, dynamic equilibrium is established when the rate of vaporization is equal to the rate of condensation. For a saturated solution, where a crystalline solute is dissolved in a liquid solvent (Chapter 14), dynamic equilibrium is established when the rate of dissolving equals the rate of precipitation.

The physical processes mentioned here are reversible, and they can involve dynamic equilibria, but they are not thought of as chemical reactions. A **reversible chemical reaction** is one that can proceed in either direction. The general equation for the reaction of A with B to produce C and D—the forward reaction—can be written

$$A + B \longrightarrow C + D$$

During the reverse reaction, substances C and D react to form A and B.

$$A + B \longleftarrow C + D$$

These two equations can be written as a single equation with arrows pointing in both directions ($\rightleftharpoons$) to indicate that the reaction is reversible.

$$A + B \rightleftharpoons C + D$$

Initially, as A reacts with B, the concentrations of A and B decrease while the concentrations of C and D increase. These initial changes in concentrations are represented graphically in Figure 15.10.

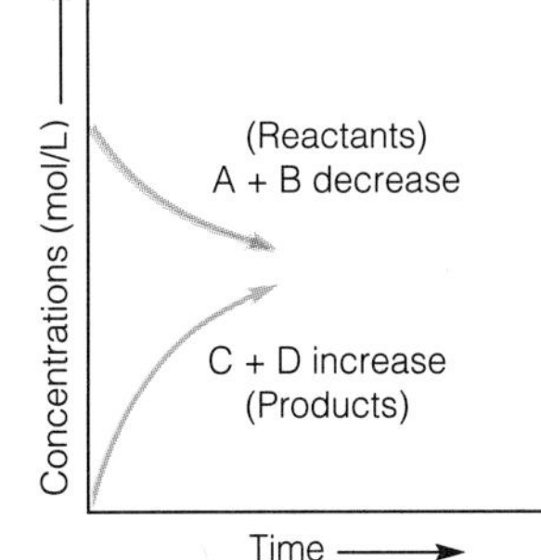

Figure 15.10 *As A reacts with B, the concentrations of these chemicals decrease while the concentrations of the products, C and D, increase. The graph traces the change in concentrations for the first few seconds of the reaction.*

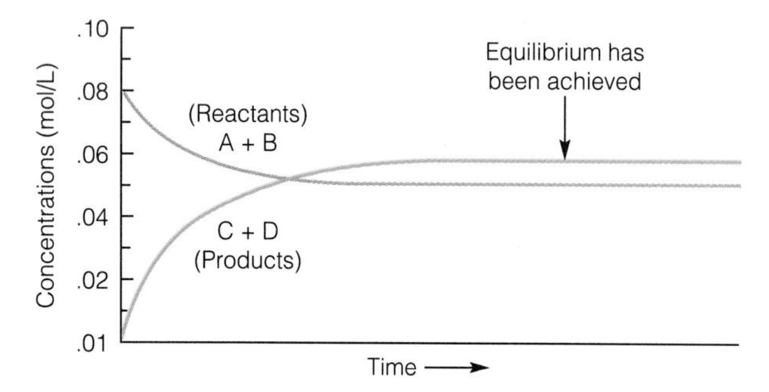

Figure 15.11 *After a period of time the concentrations of A and B become stabilized, and the concentrations of C and D become stabilized. At this point the rate of the forward reaction equals the rate of the reverse reaction; chemical equilibrium is established. Once equilibrium is achieved, there is no further change in concentrations. For each mass of reactants converted to products, an equal mass of products is converted to reactants.*

As the reaction proceeds, there is a point at which no further net change in concentrations can be detected; the concentrations of A and B as well as C and D become stabilized at specific values. At this point, dynamic **chemical equilibrium** is established. The rate of the forward reaction equals the rate of the reverse reaction (Figure 15.11).

Any chemical reaction will reach equilibrium when it is carried out at constant temperature and pressure in a closed container so no substances can enter or leave. For some reactions at equilibrium, the concentrations of products are much greater than the concentrations of reactants; we say the equilibrium lies *far to the right.* When the equilibrium concentrations of reactants are much greater than the concentrations of products, we say the equilibrium lies *far to the left.*

To get a mental picture of what is happening at the molecular level at the beginning of the reaction and when equilibrium is achieved, consider the following nuts and bolts illustration. If you put one nut on one bolt to make one assembled unit, the reversible ''reaction'' can be represented as an equation.

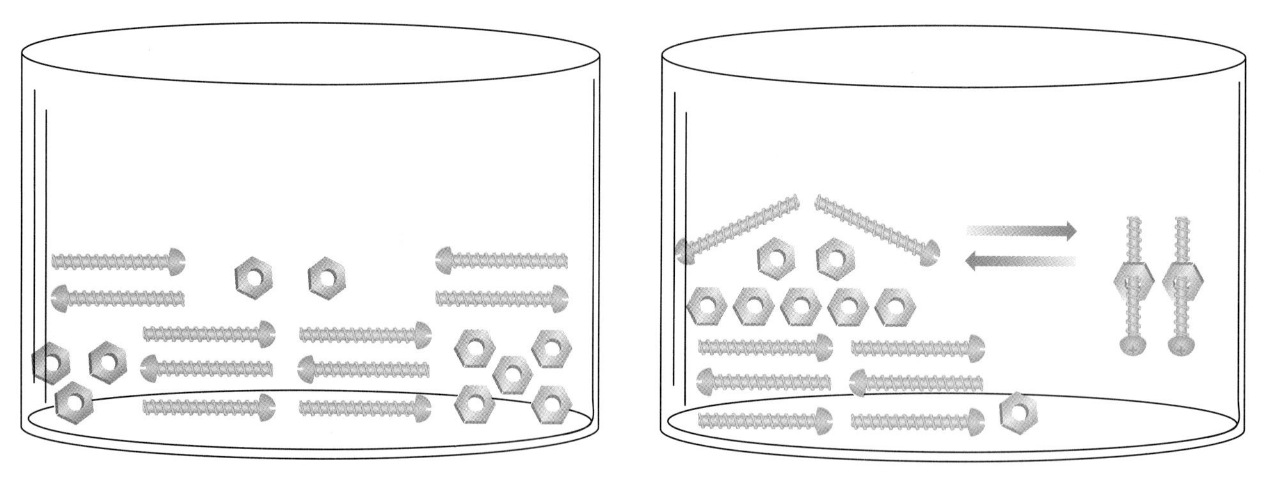

Figure 15.12 *A ''nuts and bolts'' representation of chemical equilibrium. (a) Initially there are 100 nuts and 100 bolts, which are represented in the figure by 10 of each. The concentration is 100 bolts per beaker for both nuts and bolts. (b) At equilibrium the concentrations are all constant, at 80 nuts per beaker, 80 bolts per beaker, and 20 assembled units per beaker. The rate of the forward reaction (assembling nuts and bolts) is equal to the rate of the reverse reaction (taking the units apart).*

$$1 \text{ Nut} + 1 \text{ Bolt} \rightleftharpoons 1 \text{ Assembled unit}$$

Initially, you pour 100 nuts and bolts (the reactants) into a beaker (Figure 15.12a), but there are no assembled units (no products). As you assemble nuts and bolts in a 1 : 1 ratio—and put them back in the beaker—the number of assembled units present in the beaker increases. This is the forward reaction. But since we are describing a reversible reaction, some of the assembled units must get taken apart, perhaps by your friend. After several minutes, the quantity of separate nuts and bolts has been reduced to 80 of each, but there are now 20 assembled units. By this time your friend has learned how to take the units apart as quickly as you put them together; for every unit you assemble, another unit gets taken apart. The rate of the forward reaction equals the rate of the reverse reaction; dynamic equilibrium is established (Figure 15.12b).

Once equilibrium is established, the quantity of reactants (the nuts and bolts) remains constant at 80 of each, and the quantity of assembled units remains constant at 20 units. (No nuts or bolts are created or destroyed.) Also, the *concentrations* of reactants remain constant at 80 nuts/beaker and 80 bolts/beaker, while the concentration of products is stabilized at 20 assembled units/beaker. Although a simple 1 : 1 nut-to-bolt ratio was used in this analogy, we could use a different ratio. Regardless of the ratio involved, dynamic equilibrium is established when the rate of the forward reaction equals the rate of the reverse reaction. We can make the following summarization.

At Equilibrium:
The rates of the forward and reverse reactions are equal.
The concentrations of reactants and products remain constant.

Figure 15.13 gives a model of an equilibrium being established in a closed container. At the beginning, we have 20 units of reactants (represented as colored dots). As time passes, the concentrations change until equilibrium is established with 6 units of reactants and 14 units of products. At this point—at equilibrium—there are no further changes in concentrations of reactants or of products because the rates of the forward and reverse reactions are equal.

The decomposition of orange mercury(II) oxide powder, HgO, to produce silvery Hg liquid and colorless O_2 gas, is reversible. Equilibrium can be established when the reaction is carried out at a fixed temperature in a closed vessel.

$$\underset{\text{Orange solid}}{2\,HgO(s)} \rightleftharpoons \underset{\text{Silvery liquid}}{2\,Hg(l)} + \underset{\text{Colorless gas}}{O_2(g)}$$

During the forward reaction, the heating of mercury(II) oxide produces mercury and oxygen. During the reverse reaction, tiny droplets of warm mercury on the surface of unreacted HgO react with oxygen gas to produce mercury(II) oxide.

When brown nitrogen dioxide gas, NO_2, is cooled, colorless dinitrogen tetroxide gas, N_2O_4, is formed, but warming $N_2O_4(g)$ produces $NO_2(g)$. Changing the temperature gives different equilibrium concentrations for the two gases when they are at equilibrium in a closed vessel.

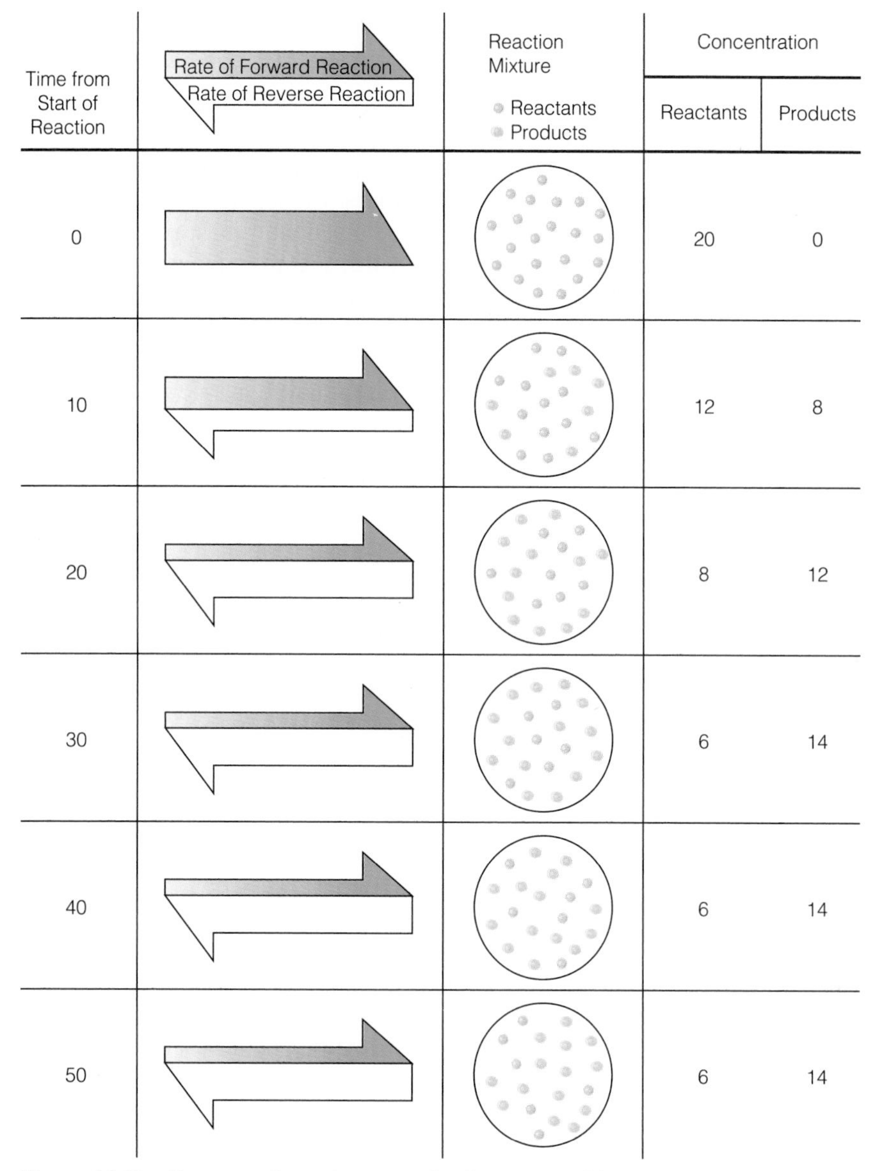

Figure 15.13 *Progress of reaction toward achieving equilibrium. How many minutes does it take to establish equilibrium?*

$$\underset{\text{Brown}}{2\,NO_2(g)} \underset{\text{heating}}{\overset{\text{cooling}}{\rightleftharpoons}} \underset{\text{Colorless}}{N_2O_4(g)}$$

You should be able to identify processes that are reversible and those that involve physical and chemical equilibria.

EXAMPLE 15.3

Which of the following illustrates chemical equilibrium?

(a) A saturated solution of salt in water.
(b) The melting of ice.

(c) The conversion of NO(g) to NO_2(g) in a closed vessel at a fixed temperature and pressure.

$$2\,NO(g) + O_2(g) \rightleftharpoons 2\,NO_2(g)$$

(d) The number of people who enter and leave a room (each hour) are equal.

SOLUTION

Only (c) is a reversible chemical reaction; it can be at chemical equilibrium. Choice (a) involves solution equilibria, but this is not chemical equilibrium. Choice (b) involves melting, an equilibrium between two states of matter. Choice (d) involves dynamic equilibrium, not chemical equilibrium.

See Problems 15.27–15.38

15.4 *Le Châtelier's Principle*

Chemical equilibrium can be established for a reversible reaction in a closed vessel, where no substances are added or taken away, and where conditions such as temperature and pressure do not change. But what happens if those conditions do change? For the answer to that question, we shall use as an example the reaction of nitrogen and hydrogen to produce ammonia.

$$N_2(g) + 3\,H_2(g) \rightleftharpoons 2\,NH_3(g) + \text{Energy}$$

All compounds involved in this reversible reaction are gases. From the work of Gay-Lussac and Avogadro, we know that the coefficients given in the equation correspond to the combining volumes of the gases. The equation shows a total of 4 moles of reactants to 2 moles of products—all are gases. From our study of gases (Chapter 12), we know that 2 moles of gas will exert less pressure than will 4 moles of gas. Therefore, as the reaction proceeds, there would be a *decrease* in pressure if volume and temperature are held constant. The specific equilibrium concentrations of reactants and products depend on the specific equilibrium conditions.

If we put the equilibrium mixture of gases under stress by increasing the pressure (by decreasing the volume of the vessel), the equilibrium is disrupted. Some of the N_2 and H_2 react to form more ammonia, and a new equilibrium is established (Figrue 15.14). Let us look at what has happened. When the system at equilibrium was put under the stress of increased pressure, the equilibrium shifted to relieve the pressure. The new equilibrium mixture contains a greater percentage of ammonia because the equilibrium has been shifted to the right. The higher the pressure, the farther the equilibrium is shifted to the right, and the more the production of ammonia is favored.

Many observations of shifts in equilibrium due to changes in concentration, pressure, and temperature were made in the nineteenth century. In 1888 Henri Louis Le Châtelier (1850–1936) explained these observed shifts in equilibrium with a simple statement known as **Le Châtelier's principle.**

> If a stress (such as a change in concentration, pressure, or temperature) is applied to a system at equilibrium, the equilibrium shifts in the direction that will partially relieve the stress.

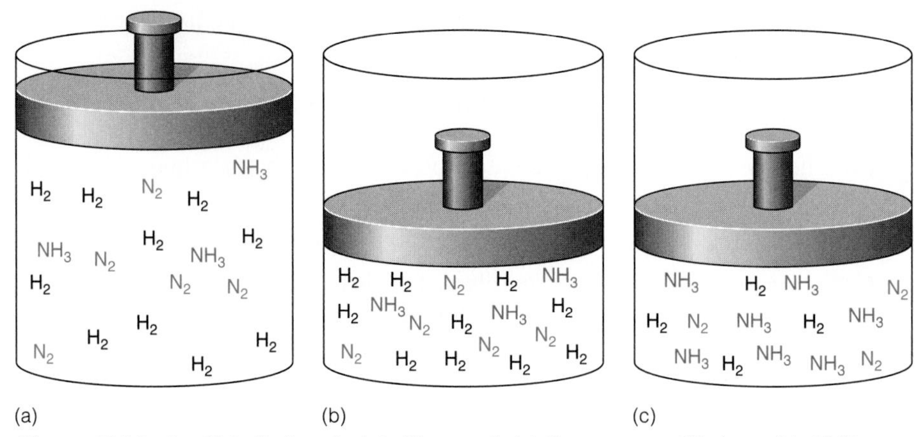

Figure 15.14 *Le Châtelier's principle illustrated. (a) System at equilibrium with 10 H_2, 5 N_2, and 3 NH_3, for a total of 18 molecules. (b) Same molecules forced into a smaller volume, creating a stress on the system. (c) Six H_2 and 2 N_2 have been converted to 4 NH_3. A new equilibrium has been established with 4 H_2, 3 N_2, and 7 NH_3, a total of 14 molecules. The stress is partially relieved by the reduction in the total number of molecules.*

The German chemist Fritz Haber became quite famous for his application of Le Châtelier's principle through his work in developing a method for the industrial synthesis of ammonia. The process he developed, known as the **Haber process,** involves the reaction of nitrogen gas and hydrogen gas at a moderately high temperature (about 450 °C) and a very high pressure (about 300 atm) in the presence of a special catalyst. As a result of his work, Haber was awarded the Nobel prize for chemistry in 1918. See the box "Ammonia Synthesis" for more details related to this important industrial process.

We shall describe three factors that affect the equilibrium of a system as described by Le Châtelier's principle.

Changes in Concentration

A chemical reaction at equilibrium can be represented by the general equation

$$A + B \rightleftharpoons C + D$$

If stress is applied by increasing the concentration of A, some of it will react with B, so the concentration of B decreases as the concentrations of C and D increase. When the new equilibrium is established, the rates of the forward and reverse reactions again become equal, but with greater concentrations of products. We say the equilibrium has shifted toward the products, or toward the right.

For the Haber process, increasing the concentration of either nitrogen or hydrogen causes a decrease in the concentration of the other reactant, but this leads to an increase in the concentration of ammonia. When the concentration of N_2 is increased, the equilibrium shift can be summarized as follows.

If the concentration of N_2 is increased:	$N_2(g)$	+	$3 H_2(g)$	$\rightleftharpoons$	$2 NH_3(g)$
	Increases		Decreases		Increases

Equilibrium shifts to the right to favor the products.

Chemistry at Work

Ammonia Synthesis: The Haber Process

Nitrogen compounds have long been valued chemicals with a wide range of uses, especially in the fertilizer and explosives industries. Before World War I, the main source of nitrogen compounds was from nitrate deposits, such as those mined in Chile. Then, commercial sources of nitrogen-containing compounds took a new turn when the noted German chemist Fritz Haber applied Le Châtelier's principle to the design of a viable industrial process for the production of ammonia from nitrogen gas and hydrogen gas. (His initial yield was only 0.26%.)

By 1914, a chemical plant was producing ammonia by the Haber process. The raw materials for the process are natural gas, water, and air. The hydrogen gas for the process is obtained by reacting natural gas with steam to produce carbon monoxide and hydrogen gas.

$$CH_4 + H_2O \longrightarrow CO + 3 H_2$$

The carbon monoxide produced is reacted with more steam to convert it to carbon dioxide and more hydrogen.

$$CO + H_2O \longrightarrow CO_2 + H_2$$

The carbon dioxide produced here can be used in a reaction with ammonia to produce urea, $CO(NH_2)_2$, for fertilizers.

The nitrogen gas for the Haber process is obtained by separating liquid nitrogen from liquid air. The exothermic reaction of nitrogen gas with hydrogen gas is then carried out in a closed vessel at a temperature of about 450 °C in the presence of a catalyst (iron oxide plus other metal oxides) so the rate is fast enough to make the reaction practical.

$$N_2(g) + 3 H_2(g) \xrightleftharpoons{\text{catalyst}} 2 NH_3(g) + 92 \text{ kJ}$$

Ammonia is one of the top ten chemicals produced in the United States. About 75% of the ammonia produced is applied directly to the soil to fertilize corn and other crops that deplete the soil of nitrogen.

The number of moles of gas decreases (from 4 to 2) as the reaction proceeds, so the reaction is carried out at a very high pressure (about 300–400 atm) to shift the equilibrium toward the products. The ammonia is removed from the system as a liquid while unreacted nitrogen and hydrogen are recycled through the process.

Approximately three-fourths of the 17 million tons of ammonia produced in the United States each year is used in fertilizers. Some of the ammonia is applied directly to the soil as anhydrous ammonia, NH_3. The ammonia is also converted to several ammonium salts that are applied to the soil in granular form such as ammonium nitrate, ammonium sulfate, and ammonium phosphate. Large quantities of ammonia are also used in the production of nitric acid, urea (used in fertilizers and in cattle feed), and numerous other compounds of nitrogen.

Removing ammonia from the system also shifts the equilibrium toward the right. However, removing either of the two reactants or increasing the concentration of ammonia will shift the equilibrium toward the left, favoring the reverse reaction.

Changes in Temperature

For *endothermic* reactions, heat can be treated as a *reactant* when we use Le Châtelier's principle to predict shifts in equilibrium. For *exothermic* reactions, heat can be treated as one of the *products.* That is, when heat is applied to a chemical reaction, the equilibrium shifts to use the additional heat energy. For the endothermic reaction involving the decomposition of $KClO_3$, adding heat energy shifts the equilibrium to the right to favor the forward reaction involving the production of more oxygen.

When temperature increases: $2\ KClO_3 + \text{Heat} \rightleftharpoons 2\ KCl + 3\ O_2(g)$ *endothermic*

Equilibrium shifts to favor production of more products.

For the exothermic Haber process, the higher the temperature, the more the equilibrium is shifted to the left to favor the reverse reaction.

When temperature increases: $N_2(g) + 3\ H_2(g) \rightleftharpoons 2\ NH_3(g) + \text{Heat}$ *exothermic*

Equilibrium shifts to favor production of more reactants.

See the box "Ammonia Synthesis" for more details.

Thus, to produce more ammonia, a lower temperature should be used, but there is a trade-off. As the temperature is lowered, the reaction rate is decreased. If there is not enough energy present for molecules to reach the activation energy, no reaction can occur. A temperature of about 450 °C is considered optimal for the Haber process.

Changes in Pressure

In order for changes in pressure to have any significant effect on a chemical reaction at equilibrium, one or more of the reactants or products must be a gas. In the reaction of H_2 and I_2 to produce HI at 700 °C, all chemicals involved are gases, and there are equal moles and volumes of reactants and products.

	$H_2(g) + I_2(g)$	$\rightleftharpoons$	$2\ HI(g)$
Moles:	2 mol		2 mol
Volumes:	2 vol		2 vol
Pressure increase:	No equilibrium shift.		

An increase in pressure brings the molecules closer together—and increases the concentration—so there is an increase in both the forward and the reverse reaction rates. However, the increased pressure does not affect the equilibrium because there are equal moles of gaseous reactants and products.

In the Haber ammonia synthesis, the number of moles of gaseous reactants and products are *not* equal. Four moles of reactants produce 2 moles of products (all gases), so an increase in pressure shifts the equilibrium toward the right.

$$N_2(g) + 3\,H_2(g) \rightleftharpoons 2\,NH_3(g) + \text{Heat}$$

Moles:	1 mol	3 mol	2 mol
Volumes:	4 vol		2 vol

Pressure increase: Equilibrium shifts to favor a greater concentration of ammonia.

A pressure of 1000 atm would be most effective in pushing the equilibrium toward the right, but equipment to handle pressure of this magnitude would be very massive and expensive. There is a capital investment and operating cost trade-off for each industrial operation. Because of such cost–benefit considerations, pressures of 300 to 400 atm are used in the Haber process.

EXAMPLE 15.4

What effect, if any, will each of the following changes have on the following reaction at equilibrium?

$$2\,CO(g) + O_2(g) \longrightarrow 2\,CO_2(g) + \text{Heat}$$

(a) Addition of CO
(b) Removal of O_2
(c) Cooling the reaction system
(d) Increasing the pressure

SOLUTION

(a) The equilibrium shifts to the right to use up the added CO.
(b) The equilibrium shifts to the left to replace the O_2 that is removed.
(c) The equilibrium shifts to the right to replace the lost heat.
(d) The equilibrium shifts to the right to relieve the pressure by converting 3 mol of gaseous reactants to 2 mol of gaseous products.

See Problems 15.39–15.46.

EXERCISE 15.3

What effect will removing NH_3, increasing the pressure, and increasing the temperature have on the Haber ammonia synthesis?

$$N_2(g) + 3\,H_2(g) \rightleftharpoons 2\,NH_3(g) + \text{Heat}$$

15.5 *The Effect of a Catalyst on a System*

A catalyst increases the rate of a reaction by lowering the activation energy (Section 15.2). For a catalyzed reaction, the rate of the reverse reaction is increased along with the rate of the forward reaction. In other words, when a catalyst is present, the system can come to equilibrium in a closed vessel more rapidly, but the concentrations of reactants and products at equilibrium will not be affected by the catalyst. The catalyst does not change the *position* of the equilibrium—the proportions of reactants and products—to favor either reactants or products.

Now let us look at the effect of a catalyst on a reaction that is not in a closed vessel. If a reaction product is removed as soon as it is produced, the reverse reaction never gets established at a rate equal to the forward reaction. The system never reaches equilibrium, so the forward reaction is the one that is catalyzed. For example, in the decomposition of potassium chlorate, $KClO_3$, described earlier, the catalyst increases the rate of production of oxygen—the forward reaction—but if the oxygen is removed as it is produced, equilibrium is never established. Catalysts are used extensively to increase the rate of production of all kinds of desired products.

EXAMPLE 15.5

Again consider the oxidation of carbon monoxide.

$$2\,CO(g) + O_2(g) \rightleftharpoons 2\,CO_2(g) + \text{Heat}$$

(a) If the reaction is carried out in a closed vessel, what effect will a catalyst have on the equilibrium?
(b) If the reaction takes place in an open vessel, such as in an automobile catalytic converter, what will be the effect of using a catalyst?

SOLUTION

(a) A catalyst has no effect on the position of equilibrium.
(b) In an open container, the CO_2 can leave as it is formed, so only the rate of the forward reaction is increased by the catalyst. The fraction of CO molecules that react is also increased because of the lowered activation energy.

See Problems 15.47–15.50.

15.6 The Equilibrium Constant Expression

We have described systems at equilibrium and various factors that control shifts in the position of the equilibrium, but there is more to this story, as you might expect. If we know the concentrations of all but one of the substances involved in the equilibrium, we can calculate that unknown concentration if we also know what is called the *equilibrium constant.*

We will first write a general equation to represent a reaction at equilibrium, where A and B are reactants, C and D are products, and a, b, c, and d are the coefficients.

$$a\,A + b\,B \rightleftharpoons c\,C + d\,D$$

We can then write an **equilibrium constant expression**

$$K_{eq} = \frac{[C]^c[D]^d}{[A]^a[B]^b}$$

where the **equilibrium constant** (K_{eq}) remains constant at a given temperature. If one concentration is changed, all other concentrations shift in a manner to maintain the same K_{eq} at that temperature. There is a new K_{eq} at each different temperature. The letters in square brackets, [ ], are used to represent concentrations in moles per liter, M. The superscript letters (the exponents) a, b, c, and d are the coefficients of the substances represented in the balanced equation. Notice that the concentration of each substance involved in the equilibrium is raised to a power that is equal to the numerical coefficient in the balanced equation. Here is a more formal definition.

The equilibrium constant, K_{eq}, is equal to the product of the equilibrium concentrations of reaction products divided by the product of the concentrations of reactants, each raised to a power corresponding to its coefficient in the balanced equation.

For the Haber ammonia synthesis

$$N_2(g) + 3\,H_2(g) \longrightarrow 2\,NH_3(g) + \text{Heat}$$

the equilibrium constant expression is written

$$K_{eq} = \frac{[NH_3]^2}{[N_2][H_2]^3}$$

In words, the equilibrium constant is equal to the square of the concentration of ammonia divided by the concentration of nitrogen and by the cube of the concentration of hydrogen.

Study Hint
Notice that the cube applies only to the concentration of hydrogen, not to the concentration of nitrogen.

EXAMPLE 15.6

Write the equilibrium constant expressions for the reactions

(a) $H_2(g) + Cl_2(g) \rightleftharpoons 2\,HCl(g)$
(b) $2\,CO(g) + O_2(g) \rightleftharpoons 2\,CO_2(g)$

SOLUTION

(a) $K_{eq} = \frac{[HCl]^2}{[H_2][Cl_2]}$ (b) $K_{eq} = \frac{[CO_2]^2}{[CO]^2[O_2]}$

Reactions that tend to go nearly to completion (to the right) have large equilibrium constants. However, when the equilibrium constant is very small, the equilibrium lies far to the left, toward the reactants. Notice that the concentrations of products are placed in the numerator of the equilibrium expression and that the concentrations of reactants are placed in the denominator. An equilibrium constant with a large positive value indicates that the equilibrium concentrations of products are greater than concentrations of reactants. When the equilibrium constant is quite small—between 0 and 1—equilibrium concentrations of reactants are greater than those of products. The following example shows how to determine the equilibrium concentration of a substance when all other concentrations are known, and when the equilibrium constant is known.

EXAMPLE 15.7

The equilibrium constant for the following reaction at a certain temperature is found to be 62. When the concentration of H_2 is 0.20 M and the concentration of I_2 is 0.25 M, what is the equilibrium concentration of HI(g)?

$$H_2(g) + I_2(g) \rightleftharpoons 2\,HI(g)$$

SOLUTION

The equilibrium concentration expression is

$$K_{eq} = \frac{[HI]^2}{[H_2][I_2]}$$

Substitute the equilibrium constant and concentrations in the equation and solve for [HI].

$$62 = \frac{[HI]^2}{[0.20\ M][0.25\ M]}$$

Note: All concentrations must be in moles per liter, M.

$$[HI]^2 = (62) \times (0.20\ M) \times (0.25\ M) = 3.10\ mol^2/L^2$$

To obtain the [HI], we take the square root of both sides of the equation.

See Problems 15.51–15.60.

$$[HI] = 1.76\ mol/L \text{ or } 1.76\ M$$

For the reaction described in Example 15.7, compare the concentration of HI(g) with the concentrations of $H_2(g)$ and $I_2(g)$ at equilibrium. Because the concentration of HI (the product) is greater than the concentrations of the reactants, the equilibrium lies to the right, and the equilibrium constant has a value that is greater than 1.

You may be wondering how K_{eq} values are obtained for a reaction. Determinations of this type can be made in a laboratory when a reaction is at equilibrium, and where concentrations of reactants and products are all measured at a constant temperature. After all equilibrium concentrations, in moles per liter, M, are obtained, the values are substituted into the equilibrium constant expression, and the missing K_{eq} can then be calculated. In the next chapter, we will look more closely at equilibrium expressions for acids and bases. The concept is quite important to our understanding of acidity.

Chapter Summary

Reversible chemical reactions can proceed in either the forward or reverse direction, depending on the conditions. When a reversible reaction takes place in a closed container at constant pressure and temperature, a balance is established between the forward and the reverse reactions. Chemical equilibrium is established when the rate of the forward reaction is equal to the rate of the reverse reaction. The equilibrium is said to lie far to the right if concentrations of products are quite large when compared with concentrations of reactants at equilibrium. The equilibrium is said to lie far to the left if concentrations of products are much smaller than concentrations of reactants at equilibrium.

The rate of a chemical reaction is controlled by three factors: (1) the frequency of collision, (2) the fraction of particles with the proper orientation at the time of the collision, and (3) the fraction of particles that have sufficient activation energy at the time of collision. The frequency of collision is controlled by concentration, temperature, and surface area of the reactants.

The series of individual steps that occur during a particular reaction are, collectively, referred to as the mechanism of the reaction. A catalyst is a substance that increases the rate of reaction without itself being consumed by the reaction. Catalysts lower the activation energy required for a reaction by changing the path—the mechanism—of the reaction, that is, by changing the small intermediate steps that occur during a reaction. When a catalyst is used, for a reaction at equilibrium, the activation energy is lowered for both the forward and the reverse reactions. Thus, a catalyst does not change the position of equilibrium nor the equilibrium constant. However, in an open container, where products are removed as they are formed, a catalyst can allow the forward reaction to occur at an increased rate.

As described by Le Châtelier's principle, if a stress due to a change in concentration, pressure, or temperature is applied to a system at equilibrium, the equilibrium will shift so as to relieve the stress. The equilibrium constant, K_{eq}, for a reaction is equal to the product of the concentrations of reaction products divided by the product of the concentrations of reactants, each raised to a power that equals the numerical coefficient in the balanced equation. There is a different equilibrium constant for each new temperature. Using the equilibrium constant expression, the concentration of one chemical can be calculated when the other values are known.

Assess Your Understanding

1. Use collision theory to discuss factors that determine reaction rate. [15.1]
2. Explain what activation energy is, and why it is important. [15.1]
3. Describe the effects of temperature, concentration, and surface area on reaction rate. [15.2]
4. Identify examples of dynamic equilibrium. [15.3]
5. Describe conditions required for chemical equilibrium. [15.3]
6. Discuss forward and reverse reaction rates at equilibrium. [15.3]
7. Use Le Châtelier's principle to predict shifts in equilibrium. [15.4]
8. Describe the effects of catalysts on reaction rate and equilibrium. [15.5]
9. Describe and use the equilibrium constant expression to determine equilibrium concentrations. [15.6]

Key Terms

activation energy [15.1]
catalyst [15.2]
chemical equilibrium [15.3]
chemical kinetics [15.1]
collision frequency [15.1]
enzyme [15.2]
equilibrium constant [15.6]
equilibrium constant expression [15.6]
Haber process [15.4]
Le Châtelier's principle [15.4]
rate of reaction [15.1]
reaction mechanism [15.2]
reversible chemical reaction [15.3]

Problems

REACTION RATES: COLLISION THEORY

15.1 What is meant by *chemical kinetics?*

15.2 List the three major factors that determine the rate of reaction.

15.3 What is meant by *collision frequency?*

15.4 How does collision frequency affect reaction rate?

15.5 What is meant by *activation energy?*

15.6 Use the concept of activation energy to explain why all exothermic reactions do not occur instantaneously.

15.7 Explain how the orientation of reactants affects reaction rate.

15.8 Make a sketch, using Figure 15.1, to give an example of one orientation that leads to an effective collision and another sketch of an orientation that leads to an ineffective collision.

FACTORS THAT CONTROL REACTION RATES

15.9 If all other factors are kept constant, what effect would decreasing the temperature of a reaction have on the rate of the reaction? Explain.

15.10 Using kinetic molecular theory, give two reasons why reactions occur more rapidly at higher temperatures.

15.11 Use temperature effects to explain how hibernating animals can go for long periods of time without food.

15.12 Explain how temperature effects can be used to an advantage in heart surgery.

15.13 If all other factors are kept constant, what effect would decreasing the concentration of reactants have on the rate of the reaction? Explain.

15.14 What effect does increasing surface area have on reaction rate? Explain.

15.15 What is a catalyst? Name a catalyst for the decomposition of H_2O_2.

15.16 What is an enzyme?

15.17 Describe the action of a catalyst using the "route from Browning to Kalispell" analogy.

15.18 Describe how a catalyst affects activation energy and reaction rate.

15.19 Describe why sugar can be metabolized in living cells even though it does not react with oxygen at room temperature to give CO_2 and H_2O.

15.20 Explain the function of a catalyst in the automobile catalytic converter.

15.21 How does manganese(IV) oxide, MnO_2, affect the decomposition rate of H_2O_2?

15.22 How does manganese(IV) oxide, MnO_2, affect the decomposition rate of $KClO_3$?

15.23 In what way does MnO_2 affect the activation energy for the decomposition of H_2O_2?

15.24 In what way does MnO_2 affect the activation energy for the decomposition of $KClO_3$?

15.25 Ozone in the outer atmosphere shields us from excessive ultraviolet radiation. However, the ozone layer can be destroyed by chlorine atoms released from chlorofluorocarbon (CFC) compounds in Freon gas. A two-step mechanism for the conversion of ozone, O_3, to O_2 is shown here.

$$\text{Cl(atomic)} + O_3(g) \longrightarrow \text{ClO}(g) + O_2(g) \quad (1)$$

$$\text{ClO}(g) + \text{O(atomic)} \longrightarrow \text{Cl(atomic)} + O_2(g) \quad (2)$$

a. Look at the reaction in step 1 and describe what is happening here.
b. Look at the reaction in step 2 and describe what is happening here.
c. Explain how the same chlorine atom can be involved in the decomposition of many ozone molecules.

15.26 The overall reaction of $H_2(g)$ with $I_2(g)$ to produce HI(g)

$$H_2(g) + I_2(g) \longrightarrow 2\ \text{HI}(g)$$

does not occur in one step. A two-step mechanism is involved.

$$I_2(g) \longrightarrow 2\ \text{I(atomic)} \quad \text{(fast)} \quad (1)$$

$$2\ \text{I(atomic)} + H_2(g) \longrightarrow 2\ \text{HI}(g) \quad \text{(slow)} \quad (2)$$

a. Look at the reaction in step 1 and describe what is happening here.
b. Look at the reaction in step 2 and describe what is happening here.
c. What does the mechanism show that the overall reaction does not show?

REVERSIBLE REACTIONS AND EQUILIBRIUM

15.27 Which of the following systems can involve dynamic physical equilibria and which can involve

chemical equilibria? At what point would each of these processes be at equilibrium?
a. vaporization and condensation of water
b. the conversion of oxygen gas to ozone
c. dissolving and crystallizing sugar

15.28 Which of the following systems can involve dynamic physical equilibria and which can involve chemical equilibria? At what point would each of these processes be at equilibrium?
a. melting and freezing water
b. dissolving and crystallizing sodium chloride
c. the conversion of $O_2(g)$ and $N_2(g)$ to oxides of nitrogen in an automobile engine and the decomposition of nitrogen oxides by a catalytic converter

15.29 What is *dynamic,* in terms of physical equilibria, at the melting point?

15.30 What is *dynamic* about a chemical reaction at equilibrium?

15.31 With the aid of Figure 15.10 describe what is happening to the concentrations of reactants and of products before equilibrium is established.

15.32 With the aid of Figure 15.11 describe what is happening to the concentrations of reactants and of products when chemical equilibrium is established.

15.33 Use the Figure 15.12 ''nuts and bolts'' analogy to compare concentrations of nuts, bolts, and assembled units *when* equilibrium is established.

15.34 Use the Figure 15.12 ''nuts and bolts'' analogy to compare concentrations of nuts, bolts, and assembled units *before* equilibrium is established.

15.35 Use the ''nuts and bolts'' analogy to compare the forward and reverse reaction rates at equilibrium. Explain.

15.36 Use the drawings in Figure 15.13 to compare reaction rates and concentrations before equilibrium is established and after equilibrium is established.

15.37 For the reaction

$$2\,CO(g) + O_2(g) \rightleftharpoons 2\,CO_2(g)$$

describe what happens to the concentrations of CO, O_2, and CO_2 before equilibrium is established and after equilibrium is established.

15.38 For the reaction

$$N_2(g) + 2\,O_2(g) \rightleftharpoons 2\,NO_2(g)$$

describe what happens to the concentrations of N_2, O_2, and NO_2 before equilibrium is established and after equilibrium is established.

LE CHÂTELIER'S PRINCIPLE

15.39 According to Le Châtelier's principle, what effect will increasing the temperature have on the following equilibria?
a. $H_2(g) + Cl_2(g) \rightleftharpoons 2\,HCl(g) + \text{Energy}$
b. $2\,CO_2(g) + \text{Energy} \rightleftharpoons 2\,CO(g) + O_2$
c. $3\,O_2 + \text{Energy} \rightleftharpoons 2\,O_3$

15.40 According to Le Châtelier's principle, what effect will increasing the temperature have on the following equilibria?
a. $CO(g) + 3\,H_2(g) \rightleftharpoons CH_4(g) + H_2O(g) + \text{Energy}$
b. $4\,NH_3(g) + 5\,O_2(g) \rightleftharpoons 4\,NO(g) + 6\,H_2O(g) + \text{Energy}$
c. $H_2(g) + I_2(g) + \text{Energy} \rightleftharpoons 2\,HI(g)$

15.41 Use Le Châtelier's principle to predict what effect increasing the total pressure (by decreasing the volume) will have on the equilibria listed in Problem 15.39.

15.42 Use Le Châtelier's principle to predict what effect increasing the total pressure (by decreasing the volume) will have on the equilibria listed in Problem 15.40.

15.43 The Haber process for producing ammonia is given in the equation

$$N_2(g) + 3\,H_2(g) \rightleftharpoons 2\,NH_3(g) + \text{Heat}$$

What would be the effect of each of the following on this equilibrium?
a. the removal of ammonia
b. the addition of nitrogen
c. the removal of heat
d. the addition of a catalyst

15.44 The hydrogen gas used for the Haber process can be obtained by the reaction of methane, CH_4, with steam to produce carbon monoxide, CO, and H_2. Additional H_2 is then produced by the following equilibrium reaction of carbon monoxide with steam.

$$CO(g) + H_2O(g) \rightleftharpoons CO_2(g) + H_2(g) + \text{Heat}$$

What would be the effect of each of the following on this equilibrium?

a. the removal of hydrogen gas
b. a decrease in total volume
c. the removal of heat
d. the addition of a catalyst

15.45 For any chemical equilibrium, what is the equilibrium shift caused by
a. increasing the concentration of a reactant?
b. increasing the concentration of a product?

15.46 For any chemical equilibrium, what is the equilibrium shift caused by
a. decreasing the concentration of a reactant?
b. decreasing the concentration of a product?

THE EFFECT OF A CATALYST ON A SYSTEM

15.47 How does a catalyst increase the rate of a reaction? Explain

15.48 What effect does a catalyst have on a system at equilibrium? Explain.

15.49 Automobiles are required by law to have catalytic converters to reduce pollution by converting carbon monoxide, CO, and unburned gasoline present in exhaust to CO_2. Describe why a catalyst is needed in the converter.

15.50 A certain biochemical reaction occurs in the human body at a rate that is about 10^4 times faster than it occurs in a laboratory at the same temperature without being catalyzed by an enzyme. How does the enzyme affect the activation energy and the rate of the reaction?

EQUILIBRIUM CONSTANTS

15.51 Write the equilibrium constant expression for each of the following reactions.
a. $H_2(g) + Cl_2(g) \rightleftharpoons 2\,HCl(g)$
b. $2\,CO_2(g) \rightleftharpoons 2\,CO(g) + O_2(g)$
c. $3\,O_2 \rightleftharpoons 2\,O_3(g)$

15.52 Write the equilibrium constant expression for each of the following reactions.
a. $CO(g) + 3\,H_2(g) \rightleftharpoons CH_4(g) + H_2O(g)$
b. $4\,NH_3(g) + 5\,O_2(g) \rightleftharpoons 4\,NO(g) + 6\,H_2O(g)$
c. $H_2(g) + I_2(g) \rightleftharpoons 2\,HI(g)$

15.53 What does an equilibrium constant with a value much greater than 1.0 tell us about the position of the equilibrium?

15.54 What does an equilibrium constant with a value much less than 1.0 tell us about the position of the equilibrium?

15.55 Where is the position of equilibrium for reactions with the following equilibrium constants?
a. 500 at 80 °C
b. 1.2×10^{-2} at 20 °C
c. 6.5×10^3 at 25 °C
d. 0.247 at 100 °C

15.56 Where is the position of equilibrium for reactions with the following equilibrium constants?
a. 0.038 at 20 °C b. 7.8×10^4 at 25 °C
c. 1.08×10^{-2} at 400 °C d. 36 at 0 °C

15.57 The equilibrium constant for the following reaction is 50 at 400 °C. When [HI] is 1.50 M and $[I_2]$ is 0.200 M, what is the equilibrium $[H_2]$?

$$H_2(g) + I_2(g) \rightleftharpoons 2\,HI(g)$$

15.58 The equilibrium constant for the following reaction is 0.212 at 100 °C. When $[NO_2] = 0.20$ M, what is the equilibrium $[N_2O_4]$?

$$N_2O_4(g) \rightleftharpoons 2\,NO_2(g)$$

Colorless Brown

15.59 The equilibrium constant for the reaction described in problem 15.57 is 69 at 340 °C. Based on the changes in equilibrium constants with temperature, are reaction products favored more by the higher or lower temperature?

15.60 For the reaction described in problem 15.58, how is K_{eq} affected by increasing the concentration of N_2O_4? How is K_{eq} affected by increasing the concentration of NO_2?

Additional Problems

15.61 Compare the activation energy for the oxidation of glucose (Figure 15.5) with the activation energy for the photosynthesis of glucose.

15.62 Chlorine gas can be prepared industrially by the following reaction.

$$4\,HCl(g) + O_2(g) \rightleftharpoons 2\,Cl_2 + 2\,H_2O(g) + \text{Heat}$$

Compare the activation energy for the forward reaction to produce $Cl_2(g)$ with the activation energy for the reverse reaction to produce HCl(g).

15.63 Give one reason insects move more slowly in the fall of the year. Explain.

15.64 Explain why meat keeps longer in the freezer than in the refrigerator.

15.65 Why do NO and NO_2 end up in automobile exhaust? How are these compounds affected by selected catalysts used in automobile catalytic converters?

15.66 What happens when carbon monoxide and oxygen come in contact with the platinum-metal-coated surfaces of a catalytic converter? Explain.

15.67 What would be the effect of each of the following on the equilibrium involving the synthesis of methanol, CH_3OH (also called methyl alcohol)?

$$CO(g) + 2\,H_2(g) \rightleftharpoons CH_3OH(g)$$

a. the removal of CH_3OH
b. an increase in pressure
c. lowering the concentration of H_2
d. the addition of a catalyst

15.68 What would be the effect of each of the following on the equilibrium involving the reaction of coke, C(s), with steam to give CO and H_2?

$$C(s) + H_2O(g) \rightleftharpoons CO(g) + H_2(g)$$

a. the addition of steam
b. an increase in pressure
c. the removal of H_2 as it is produced
d. the addition of a catalyst

15.69 A small percentage of nitrogen gas and oxygen gas in air combine at the high engine temperatures to produce NO(g), which is a pollutant.

$$N_2(g) + O_2(g) + \text{Heat} \rightleftharpoons 2\,NO(g)$$

a. Write the equilibrium constant expression for this reaction.
b. High engine temperatures are used to minimize carbon monoxide, CO, production. Do higher engine temperatures tend to increase or decrease NO(g) pollution problems? Explain.

15.70 Colorless NO(g)—see problem 15.69—can be further oxidized to nitrogen dioxide, $NO_2(g)$, which has an amber-brown color. Both gases contribute to pollution.

$$\underset{\text{Colorless}}{2\,NO(g)} + O_2(g) \rightleftharpoons \underset{\text{Brown}}{2\,NO_2(g)} + \text{Heat}$$

a. Write the equilibrium constant expression for this reaction.
b. Would increasing engine temperature tend to favor a higher equilibrium concentration of NO or of NO_2?

Millions of dollars are spent on antacids each year. An antacid containing carbonate or bicarbonate ions reacts with the hydrogen ions in an acid to produce carbon dioxide gas and water.

16

Acids and Bases

CONTENTS

Since ancient times vinegar has been obtained from apple cider and wine. The Latin word for vinegar, *acetum*, is closely related to another Latin word, *acidus* (sour), from which the word *acid* is derived. Sour-tasting liquids have long been known to contain acids. Chemicals that are said to be basic or alkaline have a bitter taste. Common examples of items containing acids and bases are pictured in Figures 16.1 and 16.2. Cleaning chemicals often contain acids or bases. We eat foods and drink liquids that contain acids and bases. Our own bodies even produce them. In fact, maintaining a delicate balance between the acids and bases in our bodies is—quite literally—a matter of life or death.

Antoine Lavoisier (Section 2.6) proposed in 1787 that all acids contain oxygen. In fact, the name *oxygen* in Greek means ''acid former.'' He was wrong. Sir Humphrey Davy (1778–1829) showed in 1811 that hydrochloric acid, HCl(aq),

All of us—even famous people—make mistakes. The important thing is to **learn from our mistakes**.

Figure 16.1 *Some common acids. From cider vinegar to carbonated beverages, from fruits and fruit juices to rust removers, acids are in our food and household chemicals.*

Figure 16.2 *Some common bases. From the antacids we take internally to the chemicals we use to remove grease and wax, we depend on these chemicals we call alkalies or bases.*

contains no oxygen. Davy concluded that the common component of acids is the presence of hydrogen, not oxygen.

Names and formulas of several common acids and their salts were given in Section 8.8. Some reactions that produce acids and bases were described in Chapter 10. Now, we are ready to look at important characteristics of acids and bases and to compare their properties.

16.1 Acids and Bases: The Arrhenius Theory

Many chemists tried to answer the "basic" question, "What is an acid?" A good answer was not available until about 100 years ago. In 1884, Svante Arrhenius (1859–1927), a Swedish chemist still in graduate school, proposed definitions of acids and bases that are still in common use today. **Acids** were generally recognized as substances that, in aqueous solution, would

Concentrated or dilute acids used in the laboratory are quite corrosive; they should never be tasted.

1. Taste sour when diluted enough to be tasted.
2. Cause litmus to change from blue to red (Figure 16.3).
3. React with active metals such as magnesium, zinc, and iron to produce hydrogen gas, $H_2(g)$.

   For example, hydrochloric acid, HCl(aq), reacts with magnesium metal to produce hydrogen gas and magnesium chloride.

Litmus is a pigment of plant origin isolated from certain types of lichen—a fungus in union with an alga—that grows on rocks and trees.

$$2\,HCl(aq) + Mg \longrightarrow H_2(g) + MgCl_2(aq)$$

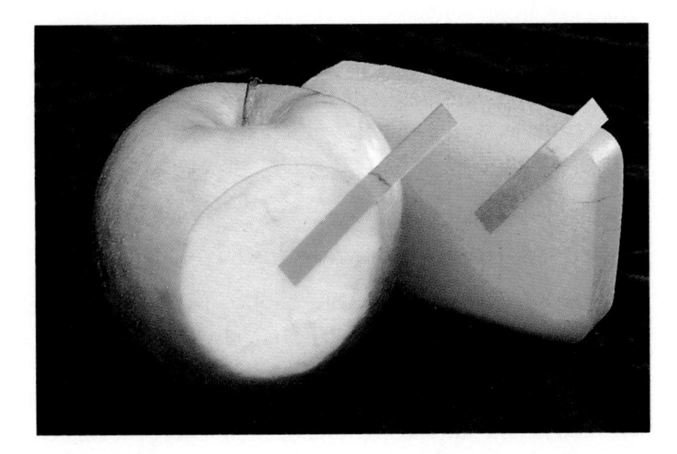

Figure 16.3 *Apples and all acid-containing products cause blue litmus to turn red (pink). A bar of soap and other products that contain bases cause red litmus to turn blue, but blue litmus remains blue in the presence of a base, as shown here.*

4. React with compounds called bases (they contain hydroxide ions, OH^-) to form water and compounds called salts. The **salt** that is formed is made up of the metal ion from the base and the nonmetal ion from the acid. Most salts are crystalline compounds that have high melting and boiling points.
   For example, hydrochloric acid, HCl(aq), reacts with sodium hydroxide, a base, to form water and sodium chloride, a salt.

$$\underset{\text{Acid}}{HCl(aq)} + \underset{\text{Base}}{NaOH(aq)} \longrightarrow \underset{\text{Water}}{H_2O} + \underset{\text{Salt}}{NaCl(aq)}$$

The reaction of an acid with a base (Sections 10.12 and 16.4) is called **neutralization**. If the correct amounts of acids and bases are mixed together, the original properties of the acids and bases are lost. The reaction product has a taste that is neither sour nor bitter, but salty. A salt and water are produced when an acid *neutralizes* a base.

Caution: Never taste *anything* in the lab!

Arrhenius proposed that the characteristic properties of acids are actually properties of the hydrogen ion, H^+, and that acids are compounds that release hydrogen ions in aqueous solutions.

Arrhenius was awarded the third Nobel prize in chemistry in 1903. Earlier, he almost failed his doctoral exam when he proposed the revolutionary theory that salts exist as positive and negative ions in aqueous solution.

Bases (also called **alkalies**) were generally recognized by Arrhenius and others as those substances that, in aqueous solution, would

1. Taste bitter. (Bases used in the laboratory should never be tasted.)
2. Feel slippery or soapy on the skin.
3. Cause litmus to change from red to blue.
4. React with acids to form water and salts.

Concentrated bases are quite *caustic*; they destroy body tissues.

Arrhenius explained that these properties of bases (alkalies) were actually properties of the hydroxide ion, OH^-. Bases, he proposed, are compounds that release hydroxide ions in aqueous solution. Arrhenius definitions are useful today—as long as we deal with aqueous solutions.

Study Hint
It may help to remember the letters "BB"—BASES cause litmus to turn BLUE.

Arrhenius Acids and Bases:
Acids release hydrogen ions in water.
Bases release hydroxide ions in water.

If you want to know whether a given compound is an Arrhenius acid, just dissolve some of it in pure water and put a drop or two of the solution on blue litmus paper. If the litmus turns red, the compound is an acid. However, if the solution turns red litmus to blue, the presence of a base is indicated.

Many foods are acidic. Cider vinegar contains acetic acid. Lemons and other citrus fruits as well as many fruit-flavored drinks contain citric acid. If a food tastes sour, it contains at least one acidic substance.

Hydrogen Ions or Hydronium Ions

The Arrhenius theory has been modified somewhat through the years. We know, for example, that simple hydrogen ions do not exist in water solutions. When a hydrogen atom is stripped of its only electron, all that is left is the bare nucleus containing a single proton, a H^+ ion. All other positive ions have nuclei that are surrounded and shielded by one or more energy levels completely filled with electrons. The hydrogen ion—with its single, unshielded proton—is simply too reactive to exist as a stable ion in solution. We know today that acidic solutions do not contain simple H^+ ions. Instead, the acidity is due to the presence of H_3O^+ ions—hydrated protons. Each H_3O^+ ion, called a **hydronium ion**, can be pictured as a hydrogen ion, H^+, bonded to and riding "piggyback" on a water molecule by sharing a pair of electrons with the oxygen of the water molecule. The electron-dot structure of a hydronium is

$$\left[\begin{array}{c} \mathrm{H:\ddot{\underset{\cdot\cdot}{O}}:H} \\ \mathrm{H} \end{array}\right]^+$$

Now, we can update the Arrhenius definition of an acid as follows.

> An acid is a substance that produces *hydronium ions*, H_3O^+, when dissolved in water.

Even this is an oversimplification. The hydronium ion is also hydrated by other water molecules, but we use the hydronium ion, H_3O^+, and ignore any further hydration. A hydronium ion is quite reactive; it can readily transfer a hydrogen ion, H^+ (which is simply a proton), to other molecules and ions. We may talk about the presence of hydrogen ions in a solution, but we are actually dealing with hydronium ions; the two are used interchangeably.

Because a hydrogen atom consists of one proton and one electron, the loss of that electron leaves only a proton as the hydrogen ion.

Monoprotic, Diprotic, and Triprotic Acids

Some acids such as hydrochloric acid, HCl(aq), and nitric acid, HNO_3(aq), give up one hydrogen ion—one proton—per formula unit. Thus, 1 mole of hydrochloric acid or nitric acid releases 1 mole of protons. These acids are called **monoprotic acids**.

Certain acids give up more than one proton (or hydrogen ion) per formula unit. For example, sulfuric acid, H_2SO_4, is a **diprotic acid**; it is capable of giving up two protons. Similarly, phosphoric acid, H_3PO_4, is a **triprotic acid**.

You should not assume that all of the hydrogen atoms in a compound are acidic. (An *acidic* hydrogen atom is one that is readily released in water solution.) None of the hydrogen atoms bonded to carbon atoms, such as in methane, CH_4, is given up in aqueous solution. Also, only one of the hydrogen atoms in acetic acid is acidic. For this reason, the formula for acetic acid is frequently written CH_3COOH (or

$HC_2H_3O_2$) to emphasize that only one proton—the proton at the end of the **carboxyl group** (the —COOH group)—is released. The following structure for acetic acid shows that the hydrogen of the carboxyl group is different; it is bonded to an oxygen rather than to carbon.

$$CH_3\overset{\overset{\displaystyle O}{\|}}{C}—OH$$

Only this proton of the carboxyl group (—COOH) is released in aqueous solution.

As shown for both methane and acetic acid, hydrogen atoms covalently bonded to carbon atoms do not tend to ionize; they are not acidic.

EXAMPLE 16.1

Using Arrhenius definitions, indicate whether each of the following is an acid, base, salt, or none of these.

(a) HNO_3 (b) $Ba(OH)_2$ (c) $CaCl_2$ (d) $CH_3CH_2CH_3$ (propane)

SOLUTION

(a) HNO_3 is an acid; it releases hydrogen ions, H^+, in solution.
(b) $Ba(OH)_2$ is a base; it releases hydroxide ions, OH^-, in solution.
(c) $CaCl_2$ is a salt formed during the neutralization reaction of hydrochloric acid with calcium hydroxide, $Ca(OH)_2(aq)$, a base.
(d) Propane is not an acid, base, or salt. It has hydrogen atoms, but they are covalently bonded to carbon atoms so they are not acidic.

See Problems 16.1–16.10.

16.2 Strong and Weak Acids

Strong acids are those that ionize completely—or nearly completely—in water. A strong acid molecule in dilute solution donates its acidic proton(s) to water to make hydronium ions. You should know names and formulas for these strong acids.

HCl, hydrochloric acid	H_2SO_4, sulfuric acid
HBr, hydrobromic acid	HNO_3 nitric acid
HI, hydroiodic acid	$HClO_4$, perchloric acid

The following equation for the reaction of hydrogen chloride gas with water shows that hydronium ions are produced in aqueous solution. Because the equilibrium lies so far to the right (the HCl is virtually all ionized), the equation is often written with a single arrow pointed toward the ionized products.

Stomach acid is equivalent to a dilute solution of hydrochloric acid. Although acid aids in the digestion of foods, especially proteins, it can cause problems—as anyone with an ulcer can attest.

$$HCl(g) + H_2O(l) \longrightarrow H_3O^+(aq) + Cl^-(aq)$$

Strong acids, if concentrated, can cause serious damage to skin or flesh. They produce chemical burns that are similar to burns from heat. They are often treated the same. Strong, corrosive acids also eat holes in clothing made of natural fibers

Figure 16.4 *Commercial vinegar is typically 5% acetic acid by mass.*

such as cotton, silk, or wool. They destroy most synthetics such as nylon, polyester, and acrylic fibers. Care should always be taken to prevent spills on either skin or clothing.

Weak acids are those that ionize only slightly in dilute solution. Acetic acid, CH_3COOH, is a typical weak acid. The equilibrium equation for the ionization of this acid is written with a longer arrow pointed toward the reactants to indicate that a large percentage of acetic acid molecules remain un-ionized.

$$CH_3COOH(aq) + H_2O(l) \rightleftharpoons CH_3COO^-(aq) + H_3O^+(aq)$$

Commercial vinegar solutions are about 5% acetic acid by mass (Figure 16.4).

Another common weak acid is carbonic acid, H_2CO_3, which is formed when CO_2 dissolves in water. Carbonated beverages all contain H_2CO_3.

Some Important Acids

Some important strong and weak acids are given in Table 16.1; you should learn names and formulas for those in red. Many of these will be encountered quite often as we continue our study. The table includes several organic acids; they contain carbon and one or more carboxyl groups. Each of the organic acids is a weak acid. Acids that are not organic acids are classified as inorganic acids.

More sulfuric acid is produced industrially than any other chemical. See the special topic regarding the production and uses of this acid. Hydrochloric acid, HCl(aq), can be used to clean the white hard-water deposits (calcium carbonate) from swimming pools and toilet bowls. It can be purchased from hardware stores under the name *muriatic acid*. The acid is also used extensively to clean mortar from bricks. Concentrated solutions of the acid (about 38% HCl) cause severe burns. Even dilute solutions, however, can cause skin irritation and inflammation. Ingestion of a small amount of any of the strong acids may be fatal.

EXAMPLE 16.2

For each of the following acids, give its formula and classify it as a strong or weak acid and as an inorganic or organic acid.

(a) acetic acid (b) boric acid (c) nitric acid

Table 16.1 **Some Important Acids***

Strong Acids		*Weak Acids*	
Name	*Formula*	*Name*	*Formula*
Hydrochloric acid	HCl	Phosphoric acid	H_3PO_4
Hydrobromic acid	HBr	Acetic acid	CH_3COOH
Hydroiodic acid	HI	Carbonic acid	H_2CO_3
Sulfuric acid	H_2SO_4	Citric acid	$C_3H_5(COOH)_3$
Nitric acid	HNO_3	Lactic acid	$CH_3CHOHCOOH$
Perchloric acid	$HClO_4$	Boric acid	H_3BO_3
		Hydrocyanic acid	HCN

*You should learn names and formulas for the strong acids and the first three weak acids.

CHEMISTRY AT WORK

Sulfuric Acid: Number One

Sulfuric acid, H_2SO_4, is by far the leading industrial chemical product of the United States. About 40 billion kg are produced annually. Sulfur can be burned in oxygen to produce $SO_2(g)$, which, in turn, is oxidized in the presence of a catalyst to form $SO_3(g)$, and then dissolved in water to produce H_2SO_4. Greater amounts of sulfur are now being recovered as SO_2 from smokestacks of power plants and metal ore smelters. The SO_2 can be used to make H_2SO_4.

Much of the sulfuric acid (around 70%) is used to convert phosphate rock into fertilizers. Another 10% is used in steelmaking and industrial metals processing. The acid is also used in petroleum refining, in the manufacture of numerous other chemicals, in automobile batteries, and in some drain cleaners. Concentrated sulfuric acid is a powerful dehydrating agent. Paper (cellulose) turns dark, becomes brittle, and crumbles after being dehydrated by the concentrated acid.

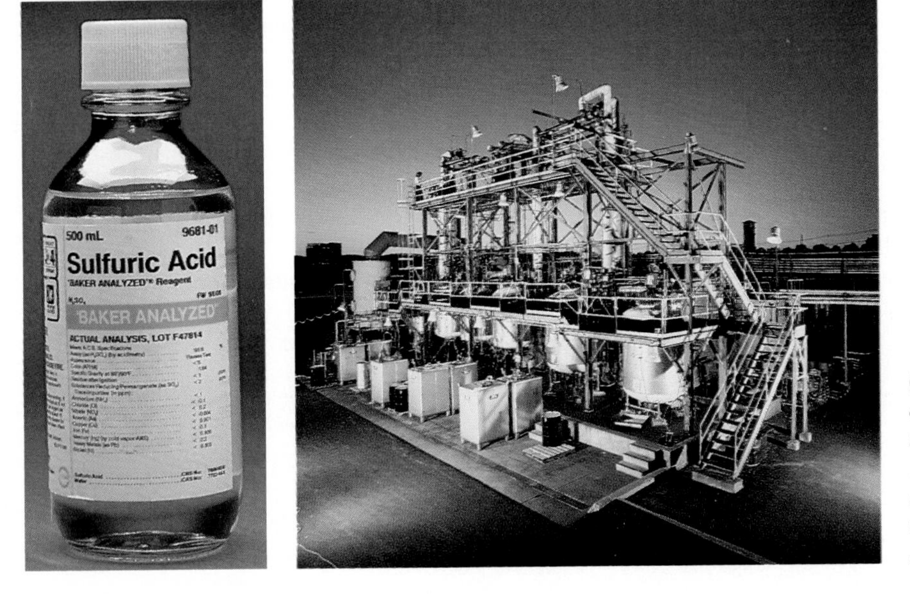

(left) *Concentrated sulfuric acid used in laboratories is supplied commercially as 18 M H_2SO_4. Battery acid is about 2 M H_2SO_4.* (right) *An industrial sulfuric acid production facility.*

SOLUTION

	Formula	*Relative Acid Strength*	*Organic or Inorganic*
(a) acetic acid	CH_3COOH	weak	organic
(b) boric acid	H_3BO_3	weak	inorganic
(c) nitric acid	HNO_3	strong	inorganic

EXERCISE 16.1
Give the names and formulas for the six strong acids.

16.3 Strong and Weak Bases

Figure 16.5 *Drano and certain other chemicals used to open clogged drains contain sodium hydroxide and other strong alkalies that dissolve grease and hair.*

Strong bases ionize completely—or nearly so—in water; **weak bases** do not. Sodium hydroxide, NaOH (also called lye), is perhaps the most familiar strong base. Even as a solid, sodium hydroxide is completely ionic; it exists as sodium ions and hydroxide ions. In solution, the hydroxide ions enter into the characteristic reactions of *basic* or *alkaline* solutions. Common products containing sodium hydroxide include drain cleaners and oven cleaners. The alkali destroys tissue rapidly, causing severe chemical burns.

The strong bases include all alkali metal hydroxides along with three Group II metal hydroxides: calcium hydroxide, $Ca(OH)_2$, strontium hydroxide, $Sr(OH)_2$, and barium hydroxide, $Ba(OH)_2$. They are all completely ionic. You should know their names and formulas. The alkali metal hydroxides are all quite soluble in water. Calcium hydroxide is only slightly soluble in water.

Magnesium hydroxide, $Mg(OH)_2$, is a weak base; it is so nearly insoluble in water that it can be safely taken internally as an antacid called milk of magnesia. Transition metal hydroxides are all weak bases; they have low water solubilities. Some important bases are listed in Table 16.2.

Ammonia, NH_3, another important weak base, is a gas at room temperature. It dissolves readily in water to give an alkaline solution. Ammonia reacts with water to a slight extent to produce ammonium ions and hydroxide ions.

$$NH_3(aq) + H_2O \rightleftharpoons NH_4^+(aq) + OH^-(aq)$$

Ammonia is classified as a weak base because an aqueous solution of ammonia

Table 16.2 **Some Important Bases**

Name	*Formula*	*Comments*
STRONG BASES		
Sodium hydroxide	NaOH	Called lye or caustic soda; over 11 million tons produced annually in U.S.
Potassium hydroxide	KOH	Used in industry and in certain oven cleaners
Calcium hydroxide	$Ca(OH)_2$	Slightly soluble in H_2O; saturated aqueous solution is called limewater; over 15 million tons of lime produced annually in U.S.
WEAK BASES		
Magnesium hydroxide	$Mg(OH)_2$	Slightly soluble in H_2O; suspension of solid $Mg(OH)_2$ in H_2O is called milk of magnesia
Aqueous ammonia	$NH_3(aq)$	Ammonia quite soluble in water; aqueous NH_3 solution used in household cleaners
Transition metal hydroxides		Very low solubilities in water

contains a relatively low concentration of hydroxide ions. In solution, only about 1% of the ammonia becomes ionized. Although the solution formed by dissolving ammonia in water is sometimes called ammonium hydroxide, **aqueous ammonia**, $NH_3(aq)$, is the correct name; most of the ammonia remains in the un-ionized form. Glass cleaners and numerous other familiar household cleaners contain ammonia; it can be readily detected by its characteristic odor.

EXAMPLE 16.3

Explain why saturated $Mg(OH)_2$—a suspension of $Mg(OH)_2$ in water—can be taken internally as the antacid called milk of magnesia, but NaOH is never used as an antacid and causes severe burns and damage to skin tissue.

SOLUTION

Magnesium hydroxide has a very low solubility in water so a saturated aqueous solution of the base has a *low* concentration of OH^- ions. However, NaOH is quite soluble in water; its solutions have *high* OH^- concentrations.

See Problems 16.11–16.22.

EXERCISE 16.2

Would an oven cleaner containing NaOH or one containing ammonia be more dangerous to use? Explain.

16.4 Reactions of Acids

Some typical reactions of acids were described in Chapter 10. In order for you to review these reactions and add to your understanding, representative reactions of acids are summarized here.

Reactions of Acids with Bases: Neutralization

The reaction of an acid (it contains H^+ ions) with a base (it contains OH^- ions) to produce water and a salt is called *neutralization.*

$$\text{Acid} + \text{Base} \longrightarrow \text{Water} + \text{Salt}$$

Examples:

$$HCl(aq) + NaOH(aq) \longrightarrow H_2O + NaCl(aq)$$

$$H_3PO_4(aq) + 3\ KOH(aq) \longrightarrow 3\ H_2O + K_3PO_4(aq)$$

Leaving out the spectator ions gives the same net ionic equation.

$$H^+(aq) + OH^-(aq) \longrightarrow H_2O(l)$$

Since hydrogen ions, H^+, exist in solution as hydronium ions, H_3O^+, we can also write the net ionic equation in the following way.

$$H_3O^+(aq) + OH^-(aq) \longrightarrow 2\ H_2O(l)$$

Reactions of Acids with Active Metals

Acids react with *active metals*, that is, metals placed above hydrogen in the activity series of metals (Section 10.8). The reaction produces hydrogen gas and a salt made up of the cation from the metal and the anion from the acid (Figure 16.6).

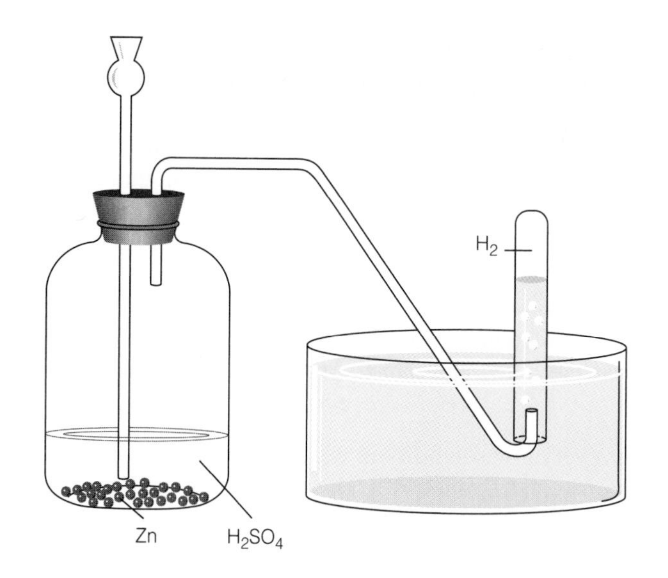

Figure 16.6 *Zinc reacts with sulfuric acid to produce hydrogen gas. The hydrogen can be collected by displacement of water from a test tube inverted in a pan of water.*

$$\text{Acid + Active metal} \longrightarrow \text{Hydrogen gas + Salt}$$

Example: $$H_2SO_4(aq) + Zn \longrightarrow H_2(g) + ZnSO_4(aq)$$

Reactions of Acids with Metal Oxides

Aqueous solutions of acids react with metal oxides to give water and a salt.

$$\text{Acid + Metal oxide} \longrightarrow \text{Water + Salt}$$

Example: $$2\,HCl(aq) + CaO(s) \longrightarrow H_2O + CaCl_2(aq)$$

Reactions of Acids with Carbonates and Bicarbonates

Carbon dioxide gas and water are produced when acids react with carbonates and bicarbonates. Carbonic acid, H_2CO_3, is not the reaction product because it is unstable and decomposes to give carbon dioxide and water.

$$\text{Acid + Carbonate} \longrightarrow \text{Salt + Carbon dioxide + Water}$$

Example: $$2\,HCl(aq) + CaCO_3(s) \longrightarrow CaCl_2(aq) + CO_2(g) + H_2O(l)$$

We can also write the following equation for the reaction of any acid with calcium carbonate.

$$CaCO_3 + 2\,H_3O^+(aq) \longrightarrow Ca^{2+}(aq) + CO_2(g) + 3\,H_2O$$

Figure 16.7 *Acid rain, resulting from nonmetal oxides in our atmosphere, is slowly dissolving sculptures like this gargoyle at Notre Dame Cathedral in Paris.*

Limestone and marble—common building materials—mainly consist of calcium carbonate, $CaCO_3$. Marble is also used in statues, monuments, and sculptures. The calcium carbonate in these materials is readily attacked by acids present in the atmosphere and in rain (Figure 16.7).

In recent years, the atmosphere has been made increasingly acidic, especially by the burning of sulfur-containing coal, which produces sulfur dioxide gas, SO_2. Dissolving SO_2 in rain gives sulfurous acid, H_2SO_3. Any SO_2 that is further oxidized to SO_3 dissolves to form sulfuric acid, H_2SO_4. All nonmetal oxides in the air (SO_x, NO_x, and CO_x) dissolve in rain to form what is called **acid rain**.

CHEMISTRY IN OUR WORLD

Acid Rain

Forests in the United States and Europe show the effects of acid rain. In the New England states, about 54% of the total lake area (1.5 million acres) is considered to be in danger, about 28% of the region is in an ''acid rain crisis,'' and over 10% (300,000 acres) is endangered. Acidic mist and acidic rainwater also attack metal and stone. Damage to automobiles, machinery, buildings, and other structures amounts to billions of dollars per year.

The region of greatest acid rain in the United States reaches from the Adirondack Mountains area in New York to the western borders of Indiana and Kentucky. The general region of moderately acidic rain reaches from the Great Lakes to the Mississippi River and into Georgia and South Carolina. A region of slightly acidic rainfall stretches from Minnesota into Texas. Although there are numerous contributing factors, sulfur dioxide emissions from coal-burning power plants contribute heavily to the problem. Much of the coal being used is about 2% sulfur by mass (2 tons sulfur/100 tons coal). Burning this coal can lead to the production of large amounts of sulfur dioxide, $SO_2(g)$.

A power plant may require two trainloads of coal each day. Those two trainloads of coal—each with 100 cars carrying 100 tons of coal per car—contain about 400 tons of sulfur.

$$\frac{2\text{ trains}}{\text{day}} \times \frac{100\text{ carloads}}{\text{train}} \times \frac{100\text{ tons coal}}{\text{carload}} \times \frac{2\text{ tons sulfur}}{100\text{ tons coal}} = \frac{400\text{ tons sulfur}}{\text{day}}$$

The mass of 1 mol of sulfur is 32.0 g and the mass of 1 mol of SO_2 is 64.0 g, so, by proportion, every ton of sulfur that is burned produces 2 tons of sulfur dioxide. Thus, the burning of two trainloads of coal—containing 400 tons of sulfur—can lead to the release of 800 tons of sulfur dioxide each day.

There is some good news. According to the Environmental Protection Agency (EPA), over 70% of river miles and 82% of the lakes tested can now support fishing and swimming. That substantial improvement over water conditions of the early 1970s is due to voluntary industrial controls as well as EPA regulations of power production, transportation, and other businesses.

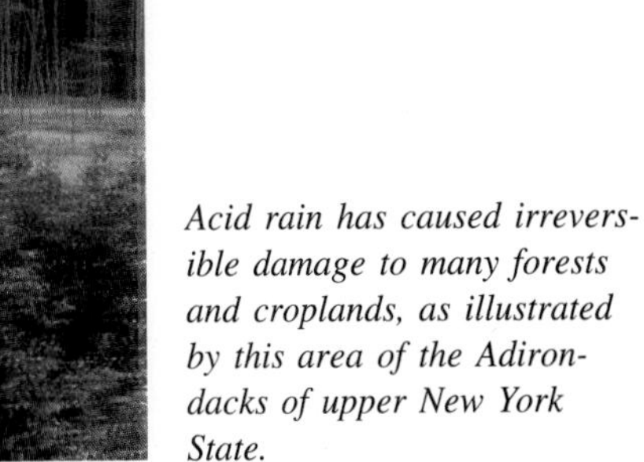

Acid rain has caused irreversible damage to many forests and croplands, as illustrated by this area of the Adirondacks of upper New York State.

CHEMISTRY IN OUR WORLD

Is There Acid in Your Dough?

Carbon dioxide gas is produced when sodium hydrogen carbonate (called sodium bicarbonate or baking soda) reacts with an acid. In the making of quick breads such as muffins or biscuits, sodium hydrogen carbonate is mixed with a dry acid such as potassium hydrogen tartrate, $KHC_4H_4O_6$ (called cream of tartar), which has one acidic hydrogen. No reaction occurs until water is added. The carbon dioxide gas produced by the reaction is trapped in the dough and expands when heated to give the characteristic light texture of muffins, biscuits, and pancakes.

$$NaHCO_3(aq) + KHC_4H_4O_6(aq) \rightarrow$$

Sodium bicarbonate; Potassium hydrogen tartrate

$$H_2O + CO_2(g) + KNaC_4H_4O_6(aq)$$

Potassium sodium tartrate

Chemistry is responsible for the "light" texture of muffins and other quick breads. When sodium bicarbonate and a dry acid such as cream of tartar are mixed with water, an acid–base reaction occurs that produces carbon dioxide gas, which is trapped in the dough.

Baking powder contains a mixture of the two dry solids described, but there are other ways to produce the carbon dioxide needed to make the dough rise. The essential ingredients are sodium hydrogen carbonate and an acid. Thus, it is possible to make biscuits, for example, by using sour milk, buttermilk, or even a little vinegar as the acid that is mixed with the baking soda.

Acid + Hydrogen carbonate → Water + Carbon dioxide

$$H_3O^+(aq) + HCO_3^+(aq) \rightarrow 2\,H_2O + CO_2(g)$$

Reactions of Acids with Metal Sulfides

Hydrogen sulfide gas is more toxic than hydrogen cyanide, HCN.

Hydrogen sulfide gas—with the odor of rotten eggs—is produced by the reaction of an acid with a metal sulfide.

Acid + Metal sulfide ⟶ Salt + Hydrogen sulfide gas

$$H_2SO_4(aq) + FeS(s) \longrightarrow FeSO_4(aq) + H_2S(g)$$

EXAMPLE 16.4

Write a balanced chemical equation for the

(a) reaction of sodium hydrogen carbonate with acetic acid, CH_3COOH (in vinegar)
(b) neutralization of hydrochloric acid by calcium hydroxide (in lime)

SOLUTION

(a) $NaHCO_3 + CH_3COOH(aq) \rightarrow NaCH_3COO(aq) + H_2O + CO_2(g)$
(b) $2\ HCl(aq) + Ca(OH)_2(s) \rightarrow CaCl_2(aq) + 2\ H_2O$

See Problems 16.23–16.26.

16.5 Reactions of Bases

The characteristic reactions of aqueous solutions of bases are due to the presence of hydroxide ions, OH^-. Reactions of metal oxides to produce bases were described in Section 10.8. Acid–base neutralization reactions were discussed in the last section. Other representative reactions of bases are included here.

Reactions of Bases with Salts of Transition Metals

Bases react with soluble salts of transition metals to give insoluble—or very slightly soluble—transition metal hydroxides and a soluble salt.

$$\text{Base} + \begin{matrix}\text{Soluble}\\\text{transition}\\\text{metal salt}\end{matrix} \longrightarrow \begin{matrix}\text{Insoluble}\\\text{transition metal}\\\text{hydroxide}\end{matrix} + \text{Salt}$$

$$3\ Ca(OH)_2(s) + 2\ CrCl_3(aq) \longrightarrow 2\ Cr(OH)_3(s) + 3\ CaCl_2(aq)$$

$$2\ KOH(aq) + Ni(NO_3)_2(aq) \longrightarrow Ni(OH)_2(s) + 2\ KNO_3(aq)$$

The driving force behind these reactions is the formation of a precipitate.

Reactions of Amphoteric Hydroxides with Acids and Bases

Hydroxides of certain metals, such as aluminum, chromium, and zinc, will react with either strong acids or strong bases; they are said to be **amphoteric**. Aluminum hydroxide is a typical amphoteric compound. Acting as an acid, aluminum hydroxide dissolves in a solution of a strong base.

$$\underset{\text{As an acid}}{Al(OH)_3(s)} + \underset{\text{Base}}{NaOH(aq)} \longrightarrow NaAl(OH)_4(aq)$$

Acting as a base, aluminum hydroxide dissolves in a strong acid to give an aluminum salt and water.

$$\underset{\text{As a base}}{Al(OH)_3(s)} + \underset{\text{Acid}}{3\ HCl(aq)} \longrightarrow AlCl_3(aq) + 3\ H_2O$$

EXAMPLE 16.5

What do these chemical equations tell us about zinc hydroxide?

$$Zn(OH)_2(s) + 2\ HCl(aq) \longrightarrow ZnCl_2(aq) + 2\ H_2O$$

$$Zn(OH)_2(s) + 2\ KOH(aq) \longrightarrow K_2Zn(OH)_4(aq)$$

SOLUTION

Zinc hydroxide is amphoteric. It reacts with both acids and bases.

See Problems 16.27–16.32.

CHEMISTRY IN OUR WORLD

Antacid Chemistry

Sometimes overindulgence or emotional stress leads to a condition called *hyperacidity* (too much acid). Numerous types of antacids*—many of which are aggressively advertised—are available to treat this condition. Sales of antacids in the United States are estimated to be about $650 million each year. From the standpoint of acid–base chemistry, antacids are alkaline compounds that react with acids.

One of the oldest and most familiar antacids is sodium hydrogen carbonate, $NaHCO_3$, called baking soda. It is thought to be safe and effective for occasional use by most people. Overuse will make the blood too alkaline, a condition called *alkalosis*. Sodium hydrogen carbonate is not recommended for those with hypertension (high blood pressure) because high concentrations of sodium ion tend to aggravate the condition. The antacid in Alka-Seltzer is sodium hydrogen carbonate. This popular remedy also contains citric acid and aspirin. When Alka-Seltzer is placed in water, the reaction of hydrogen carbonate ions with hydronium ions from the acid produces the familiar fizz.

$$HCO_3^- + H_3O^+ \rightarrow CO_2(g) + 2\,H_2O$$

Another common antacid ingredient is calcium carbonate, $CaCO_3$, also called precipitated chalk. It is fast acting and safe in small amounts, but regular use can cause constipation. It also appears that calcium carbonate can cause *increased* acid secretion after a few hours. Temporary relief may be achieved with the risk of a greater problem later. Tums is essentially flavored calcium carbonate. Alka-2 and Di-Gel liquid suspension also contain calcium carbonate as the antacid ingredient. The CO_3^{2-} ion neutralizes acid and gives $CO_2(g)$.

$$CO_3^{2-} + 2\,H_3O^+ \rightarrow CO_2(g) + 3\,H_2O$$

Aluminum hydroxide, $Al(OH)_3$, is another common antacid ingredient. Like calcium carbonate, it can cause constipation in large doses. The hydroxide

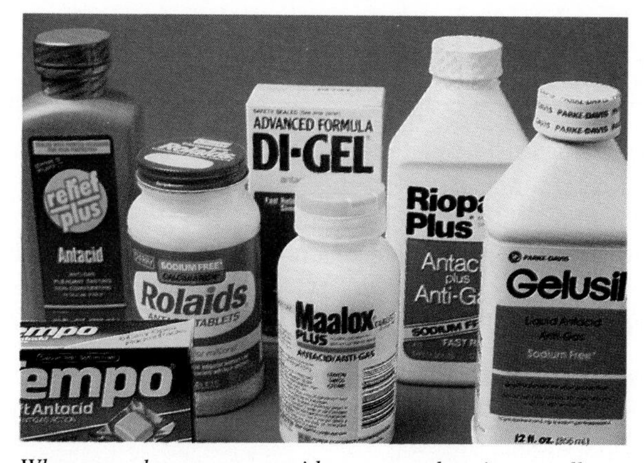

When you choose an antacid, you are choosing an alkaline compound—a base.

ions react with acids as shown by the neutralization reaction.

$$OH^- + H_3O^+ \rightarrow 2\,H_2O$$

There is concern that antacids containing aluminum ions deplete the body of essential phosphate ions. The aluminum phosphate formed is insoluble and is eliminated from the body.

$$Al^{3+} + PO_4^{3-} \rightarrow AlPO_4(s)$$

Aluminum hydroxide is the only antacid in Amphojel, but it is used in combination with other antacid ingredients in many popular products.

Certain magnesium compounds are also used as antacids. These include magnesium carbonate, $MgCO_3$, and magnesium hydroxide, $Mg(OH)_2$. Milk of magnesia is a suspension of magnesium hydroxide in water. It is sold under a variety of brand names, but the best known is probably Phillips. In small doses, magnesium compounds act as antacids. In large doses, they act as laxatives. Magnesium ions are poorly absorbed in the digestive tract. Instead,

*An *antacid* (pronounced ANT-acid) is not an acid extracted from ants! Antacids are bases; they work in opposition to acids, so they are actually *anti*-acids.

these small dipositive ions attract water into the colon (large intestine) causing the laxative effect.

Several popular antacids contain both aluminum hydroxide—which tends to cause constipation—and a magnesium compound—which acts as a laxative. These tend to counteract one another. Maalox and Mylanta are familiar brands. Another popular antacid, Rolaids, contains aluminum sodium dihydroxy carbonate, $AlNa(OH)_2CO_3$. Both the hydroxide ion and the carbonate ion consume acid.

Antacids interact with other medications. Anyone taking any type of medicine should consult a physician before taking antacids. Anyone with severe or repeated attacks of indigestion should consult a physician; self-medication in such cases can be dangerous. Generally, antacids are safe and effective for occasional use in small amounts. All antacids are *basic* compounds. If you are otherwise in good health, you can choose a base on the basis of price.

Reactions of Strong Bases with Amphoteric Metals

Amphoteric metals (such as aluminum and zinc) are those that react directly with acids to produce hydrogen gas and react directly with strong bases and water to produce hydrogen gas. The action of Drano in opening a clogged drain is a practical application of the latter type of reaction. Drano (pictured in Figure 16.5) contains small pieces of aluminum metal mixed with lye (sodium hydroxide). Hydrogen gas is produced when Drano is added to water.

$$2\ Al(s) + 2\ NaOH(s) + 6\ H_2O \longrightarrow 2\ NaAl(OH)_4(aq) + 3\ H_2(g)$$

16.6 *Brønsted–Lowry Definitions of Acids and Bases*

The Arrhenius definitions of acids and bases are quite useful for aqueous solutions, but by the 1920s chemists were working with solvents other than water. Compounds were found that acted like bases yet did not have OH in their formulas. A new theory was needed. More general definitions of acids and bases were suggested independently in 1923 by J. N. Brønsted (1897–1947), a Danish chemist (and colleague of Neils Bohr), and T. M. Lowry (1847–1936), an English chemist. The Brønsted–Lowry definitions are

In the process of science, we sometimes find that a theory, which had provided an acceptable explanation at one time, must be updated or rejected and replaced by one that can account for new findings or explain data more clearly.

A Brønsted–Lowry acid is a proton donor; it donates a hydrogen ion, H^+.

A Brønsted–Lowry base is a proton acceptor; it accepts a hydrogen ion, H^+.

All Arrhenius acids are also acids by the Brønsted–Lowry definition. For example, in the reaction of hydrogen chloride gas, HCl(g), with water to produce hydrochloric acid, the HCl(g) is the proton donor. All Arrhenius bases are also bases by the Brønsted definition, but there are other bases. In the hydrogen chloride and water reaction, the proton acceptor (the base) is water.

$$\underset{\text{Proton donor}}{HCl(g)} + \underset{\text{Proton acceptor}}{H_2O(l)} \longrightarrow H_3O^+(aq) + Cl^-(aq)$$

According to the Brønsted–Lowry definitions, this is an acid–base reaction; water acts as a base in this Brønsted–Lowry system. The driving force for the reaction is the formation of a weaker acid and a weaker base. A stronger acid reacts with a stronger base to produce a weaker base and a weaker acid.

$$\underset{\text{Stronger acid}}{HCl(g)} + \underset{\text{Stronger base}}{H_2O(l)} \longrightarrow \underset{\text{Weaker acid}}{H_3O^+(aq)} + \underset{\text{Weaker base}}{Cl^-(aq)}$$

It may not be obvious that HCl is a stronger acid than the hydronium ion, but keep in mind that a stronger acid is one that succeeds in giving away its proton. It may also seem odd to think of water as a base, but remember the definition: A base is a proton acceptor. A neutralization reaction can be thought of as a game of ''give-away.'' The species that succeeds in giving away the proton is the acid. The species that accepts the proton is the base. In the ''give-away'' game described here, water gets the proton nearly every time.

During the reaction, HCl(g)—the Brønsted–Lowry acid—donates a proton to H_2O, leaving a chloride ion, Cl^-, which is classified as the **conjugate base**. (The conjugate base is simply the acid minus its proton.) Together, HCl and Cl^- make up a **conjugate acid–base pair**. Water and the hydronium ion make up another conjugate pair. Here, water is a Brønsted–Lowry base; the hydronium ion is its conjugate acid. The conjugate acid–base pairs are shown here.

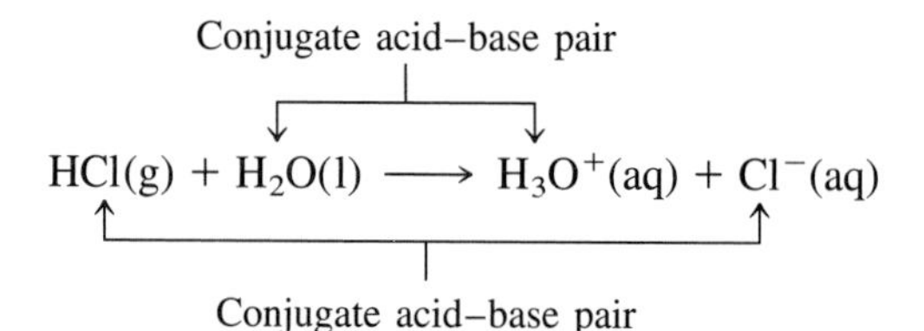

The definition of a base, according to the Brønsted–Lowry model, includes any species that accepts a proton. Thus, H_2O, Cl^-, NH_3, and many other species that are not hydroxides can act as Brønsted bases. When ammonia gas, $NH_3(g)$, reacts with hydrogen chloride gas, HCl(g), no hydroxide is involved, but $NH_3(g)$ accepts a proton from HCl(g) so it acts as a Brønsted–Lowry base.

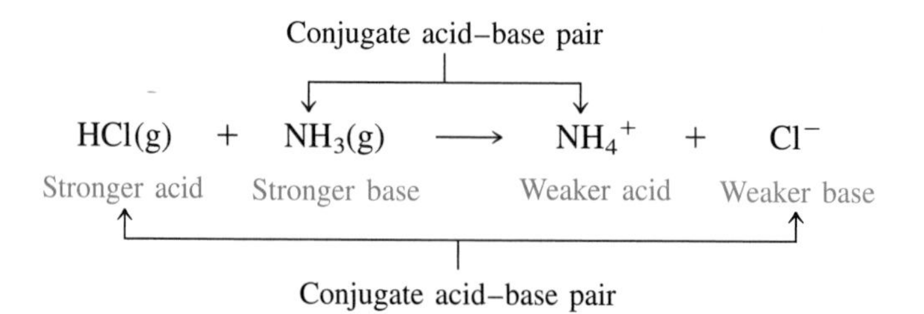

When ammonia gas dissolves in water, it reacts with water to a slight extent to form ammonium ions and hydroxide ions. In this system, ammonia acts as a proton acceptor (a base) and water acts as a proton donor (an acid).

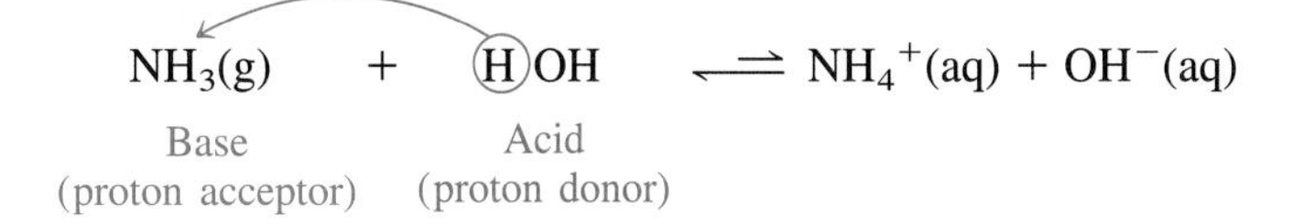

Ammonia and hydroxide ions compete for protons. Which is the stronger base? A dynamic equilibrium is established among all four species present, but the reverse reaction is dominant over the forward reaction. The hydroxide ion gets the proton most of the time. At any instant, the number of ammonia and water molecules is enormous; the number of ammonium and hydroxide ions is extremely small. Ammonia and the ammonium ion make up one conjugate pair, while water and the hydroxide ion make up another conjugate pair.

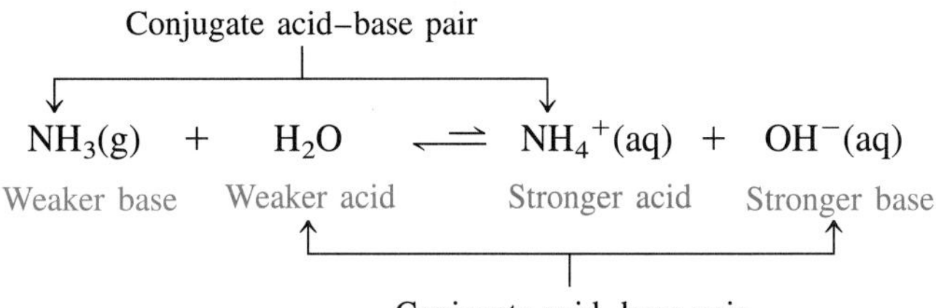

Since the driving force is toward the weaker base and the weaker acid (the reactants), it is reasonable that there would be more ammonia than ammonium ions and more water than hydroxide ions. As shown here, conjugate pairs always consist of a stronger and a weaker partner. This is certainly not a coincidence. What makes an acid stronger is its greater tendency to release—to give up—its proton. Its conjugate base must, therefore, be weaker in comparison.

In the reaction of water with ammonia, notice that the water acts as a Brønsted acid, but in the earlier reaction with hydrogen chloride, water acts as a Brønsted base. Water is said to be **amphiprotic**; it can lose or gain a proton to act either as an acid or as a base.

Table 16.3 lists a number of acids and their conjugate bases in order of their relative strengths. Notice that the strongest acid has the weakest conjugate base and the strongest base has the weakest conjugate acid.

EXAMPLE 16.6

For each acid or base, give the conjugate base or acid requested.

(a) H_2O can act as an acid. What is its conjugate base?
(b) H_2O can also act as a base. What is its conjugate acid?
(c) HCO_3^- can act as an acid. What is its conjugate base?
(d) HCO_3^- can also act as a base. What is its conjugate acid?
(e) From this information, what do we now know about H_2O and HCO_3^-?

SOLUTION

(a) The conjugate base is OH^-. The H_2O loses a proton to form OH^-.
(b) The conjugate acid is H_3O^+. The H_2O accepts a proton to form H_3O^+.
(c) The conjugate base is CO_3^{2-}. The HCO_3^- loses a proton to form CO_3^{2-}.
(d) The conjugate acid is H_2CO_3. The HCO_3^- accepts a proton to form H_2CO_3.
(e) Both H_2O and HCO_3^- are amphiprotic; they can donate or accept a proton.

Table 16.3 **Relative Strengths of Some Brønsted–Lowry Acids and Their Conjugate Bases**

	Acid		Base		
	Name	*Formula*	*Name*	*Formula*	
↑ Stronger acids	Perchloric acid	$HClO_4$	Perchlorate ion	ClO_4^-	Weaker bases ↑
	Sulfuric acid	H_2SO_4	Hydrogen sulfate ion	HSO_4^-	
	Hydrogen chloride	HCl	Chloride ion	Cl^-	
	Nitric acid	HNO_3	Nitrate ion	NO_3^-	
	Hydronium ion	H_3O^+	Water	H_2O	
	Sulfurous acid	H_2SO_3	Hydrogen sulfite ion	HSO_3^-	
	Phosphoric acid	H_3PO_4	Dihydrogen phosphate ion	$H_2PO_4^-$	
	Acetic acid	CH_3COOH	Acetate ion	CH_3COO^-	
	Carbonic acid	H_2CO_3	Hydrogen carbonate ion	HCO_3^-	
	Ammonium ion	NH_4^+	Ammonia	NH_3	
	Water	H_2O	Hydroxide ion	OH^-	
↓ Weaker acids	Ammonia	NH_3	Amide ion	NH_2^-	Stronger bases ↓

16.7 *Lewis Definitions of Acids and Bases*

The story of developments in acid–base theory would not be complete without at least a brief look at the Lewis model of acids and bases. The most general concept of acids and bases was proposed in 1923 by Gilbert N. Lewis (Section 5.7) who also introduced the use of electron-dot formulas. In fact, his use of electron pairs in writing chemical formulas is also the basis of the Lewis acid–base model. Lewis definitions of acids and bases are stated here.

A Lewis acid is a substance that can accept (and share) an electron pair.

A Lewis base is a substance that can donate (and share) an electron pair.

All chemicals that are acids according to the Arrhenius and Brønsted–Lowry theories are also acids by the Lewis theory. All chemicals that are bases according to the Arrhenius and Brønsted–Lowry theories are also bases by the Lewis theory. According to the Lewis theory, a hydrogen ion, H^+, is still an acid, and a hydroxide ion, OH^-, is still a base, but the Lewis definitions further expand the acid–base model beyond the Brønsted–Lowry and Arrhenius models.

The reaction of boron trifluoride with ammonia is a classic example of the Lewis model. There is no hydroxide ion (required by the Arrhenius definition), and there is no proton transfer (required by the Brønsted–Lowry definition).

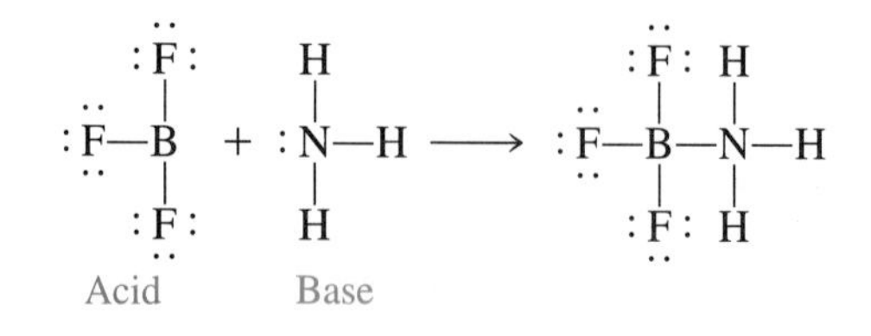

The boron of BF_3 has only six electrons involved in bonding, two fewer than the stable octet, so it needs an electron pair. Acting as a Lewis acid, BF_3 accepts—and then shares—an electron pair with nitrogen of NH_3. Ammonia is the Lewis base; it donates—and then shares—an electron pair.

There are three theories to explain acid–base reactions, but they do not contradict one another. Instead, each theory expands the previous model and takes a broader view (Figure 16.8). For example, hydroxide—recognized by Arrhenius as a base—is also a base according to the Brønsted–Lowry definition because it is a proton acceptor. Furthermore, the ion is a Lewis base; it is an electron pair donor. Lewis definitions of acids and bases are especially important in organic chemistry, but for explaining reactions in aqueous solutions, Arrhenius or Brønsted–Lowry definitions are usually adequate.

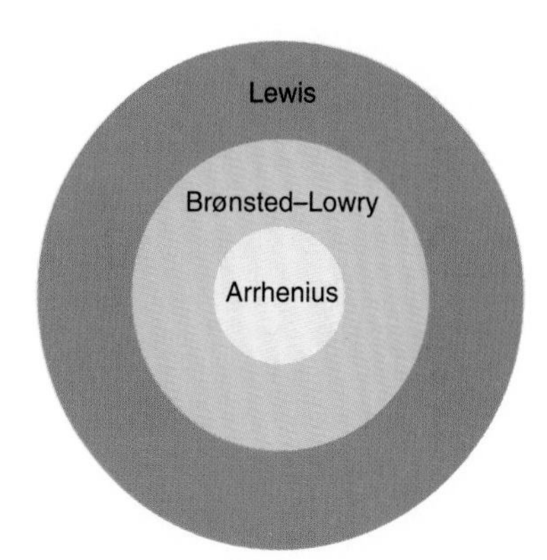

Figure 16.8 *The Arrhenius model of acids and bases was broadened in scope by the Brønsted–Lowry model. The Lewis acid–base model is the most general in scope. It extends the definition of an acid to include any substance that is an electron pair acceptor; a Lewis base is any substance that can act as an electron pair donor.*

EXAMPLE 16.7

Which theory is required to classify each of the following as an acid–base reaction? Identify which reactants are acids and which are bases.

(a) $AlCl_3 + :\ddot{\underset{..}{Cl}}:^- \rightarrow AlCl_4^-$

(b) $HCN(aq) + H_2O \rightleftharpoons H_3O^+(aq) + CN^-(aq)$

(c) $H_2SO_4(aq) + Ca(OH)_2(s) \rightarrow CaSO_4(s) + 2\,H_2O$

SOLUTION

(a) Only the Lewis theory applies here. The Cl^- ion is the base; it donates an electron pair. $AlCl_3$ is the acid; it accepts an electron pair.

(b) Both Lewis and Brønsted–Lowry theories apply. The HCN is the acid—the proton donor (Brønsted) and the electron pair acceptor (Lewis). The H_2O is the base—the proton acceptor (Brønsted) and the electron pair donor (Lewis).

(c) All three theories apply. H_2SO_4 is the acid, and $Ca(OH)_2$ is the base.

See Problems 16.33–16.46.

16.8 The Self-ionization of Water

We have compared reactions involving acids and bases and have compared three definitions of acids and bases. Now, before we describe quantitative methods of representing the acidity of a solution, we need, once again, to examine the bonding in water molecules.

When we think of water, we think of H_2O molecules. But even the most pure water isn't all H_2O. About 1 molecule in 500 million transfers a proton to another, to produce a hydronium ion and a hydroxide ion.

$$\underset{\text{Acid}}{H_2O} + \underset{\text{Base}}{H_2O} \rightleftharpoons \underset{\text{Acid}}{H_3O^+} + \underset{\text{Base}}{OH^-}$$

This equation represents the Brønsted–Lowry concept of what is happening: the water molecule, which acts as the acid, donates a proton to the other water molecule, which acts as a base. As represented by the equation, water is in equilibrium with hydronium ions and hydroxide ions, but the equilibrium lies far to the left. Few water molecules become involved in a proton transfer as they jostle against one another. The concentration of hydronium ions in pure water at 25 °C is 0.00000010, or 1.0×10^{-7} M. The concentration of hydroxide ions in water at

25 °C is also 1.0×10^{-7} M. Brackets, [ ], are used to represent concentrations in moles per liter, M. For water, we have

$$[H_3O^+] = 1.0 \times 10^{-7}$$
$$[OH^-] = 1.0 \times 10^{-7}$$

In pure water, the concentration of hydronium ions equals the concentration of hydroxide ions. The product of the hydronium ion concentration and the hydroxide ion concentration at 25 °C is

$$(1.0 \times 10^{-7})(1.0 \times 10^{-7}) = 1 \times 10^{-14}$$

This product, called the *ion product* of water, is always equal to a *constant*. This constant, called the **ion product constant of water**, K_w, is 1.0×10^{-14} at 25 °C. For simplicity, we will begin using $[H^+]$ instead of $[H_3O^+]$ with the understanding that H^+ ions are always hydrated in solution. Using this simplified notation, the expression for K_w is written

$$K_w = [H^+][OH^-] = 1.0 \times 10^{-14} \quad \text{at } 25\ °\text{C}$$

How does adding an acid to water affect the equilibrium concentrations of hydrogen ions and hydroxide ions? We can return to the chemical equation for the ionization of water and Le Châtelier's principle for the answer.

$$H_2O \rightleftharpoons H^+ + OH^-$$

As hydrogen ions (from the acid) are added, they will tend to react with—and decrease the concentration of—hydroxide ions. Thus, the $[OH^-]$ decreases when $[H^+]$ increases. This is in agreement with the expression for K_w.

$$K_w = [H^+][OH^-]$$
$$1.0 \times 10^{-14} = [H^+][OH^-]$$

If the $[H^+]$ is increased, the $[OH^-]$ will decrease until the product of the two concentrations equals 1.0×10^{-14}, and vice versa.

If we know either the hydrogen ion concentration, $[H^+]$, or the hydroxide ion concentration, $[OH^-]$, the concentration of the other ion can be calculated, as shown in these examples.

EXAMPLE 16.8

Lemon juice has a $[H^+]$ of 0.010 M. What is the $[OH^-]$?

SOLUTION

Given: $[H^+] = 1.0 \times 10^{-2}$ M in exponential form.

From K_w we have $\quad [H^+][OH^-] = 1.0 \times 10^{-14}$

Substitution for $[H^+]$ gives $\quad [1.0 \times 10^{-2}][OH^-] = 1.0 \times 10^{-14}$

Dividing both sides by 1.0×10^{-2} $\quad [OH^-] = \dfrac{1.0 \times 10^{-14}}{1.0 \times 10^{-2}}$

Hint: To divide exponential numbers, subtract the exponent in the denominator from the exponent in the numerator: $(-14) - (-2) = -14 + 2 = -12$.

$$[OH^-] = 1.0 \times 10^{-12}\ M$$

EXAMPLE 16.9

A sample of bile has a $[OH^-]$ of 1.0×10^{-6} M. What is the $[H^+]$?

SOLUTION

Given: $[OH^-] = 1.0 \times 10^{-6}$ M

From K_w we have $$[H^+][OH^-] = 1.0 \times 10^{-14}$$

Substitution for $[OH^-]$ gives $$[H^+][1.0 \times 10^{-6}] = 1.0 \times 10^{-14}$$

Dividing both sides by 1.0×10^{-6} $$[H^+] = \frac{1.0 \times 10^{-14}}{1.0 \times 10^{-6}}$$

$$[H^+] = 1.0 \times 10^{-8}\ M$$

The calculations for lemon juice (an acid) and the bile sample (which is alkaline or basic) illustrate the following.

For acidic solutions, the $[H^+]$ is greater than 1.0×10^{-7}.

For alkaline solutions, the $[H^+]$ is less than 1.0×10^{-7}.

For neutral solutions, the $[H^+]$ is equal to 1.0×10^{-7}.

The sample of bile described in Example 16.9 is alkaline (basic); it has a hydrogen ion concentration that is less than 1.0×10^{-7}.

16.9 The pH Scale

Instead of expressing all hydrogen ion concentrations in exponential form, as shown in Example 16.9, a more convenient method was introduced in 1909 by S. P. L. Sörensen, a Danish chemist. He proposed that the number in the exponent be used to express acidity. Sörensen's acidity scale came to be known as the pH scale, from the French *pouvoir hydrogene* ("power of hydrogen"). The **pH** of a solution is defined as the negative of the logarithm of the hydrogen ion concentration, $[H^+]$. The logarithm (log) of a number is the exponent—the power—to which 10 must be raised to give the specified number. Mathematically, pH is defined as follows.

Some Logarithms

$\log 10^2 = 2$
$\log 10^1 = 1$
$\log 10^0 = 0$
$\log 10^{-2} = -2$
$\log 10^{-1} = -1$

$$pH = -\log [H^+]$$

When first encountering this definition of pH, one might wonder if this is indeed "more convenient" than using the exponential notation for expressing the $[H^+]$, but with a little practice, you should find that it really is. For any solution where the hydrogen ion concentration is expressed as 1×10^{-n}, the pH is equal to the numerical value of n. Consider, for example, the following hydrogen ion concentrations in moles per liter, M, and their corresponding pH values.

Concentrations in brackets are always in moles per liter, M.

$[H^+]$	pH
1×10^{-1}	1
1×10^{-3}	3
1×10^{-9}	9
1×10^{-11}	11

Later in the chapter we will work with numbers with coefficients that are *not* exactly 1.

Whenever the $[H^+] = 1.00 \times 10^{-n}$ the pH = n.

If this number is *exactly* 1 — then the pH is this number.

We could say that a solution has a hydrogen ion concentration of 1×10^{-6} mol/L, or we can say it has a pH of 6. They have the same meaning, but it is simply easier to say that its pH is 6. This is what makes the pH scale quite convenient and reasonable for users. It has been universally adopted.

Pure water has a hydrogen ion concentration of 1×10^{-7} mol/L and a pH of 7. As represented on the pH scale shown here, any *neutral* solution has a pH of 7.

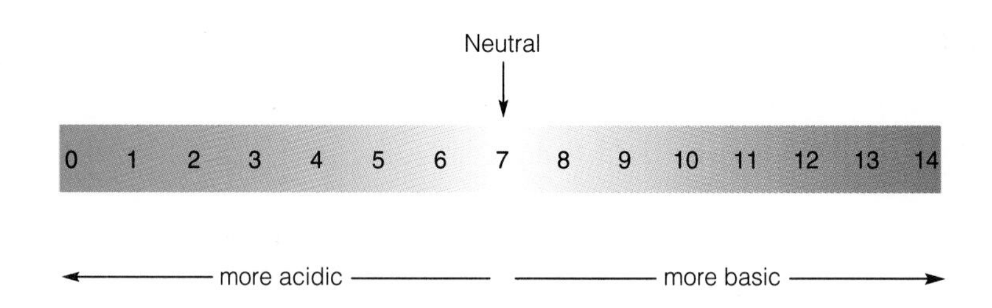

***Table 16.4* Approximate pH of Some Solutions**

Solution	*pH*	
0.10 M HCl	1.0	↑
Gastric juices	1.6–1.8	
Lemon juice	2.3	
Vinegar	2.4–3.4	
Soft drinks	2.0–4.0	Acidic
Milk	6.3–6.6	
Urine	5.5–7.5	
Rainwater (unpolluted)	5.6	
Saliva	6.2–7.4	↓
Pure water	7.0	Neutral
Blood	7.35–7.45	↑
Egg white (fresh)	7.6–8.0	
Bile	7.8–8.6	Basic
Milk of magnesia	10.5	
Household ammonia	11	
0.1 M NaOH (lye)	13	↓

An acidic solution has a pH that is less than 7. A basic (alkaline) solution has a pH that is greater than 7. The *lower* the pH, the more acidic the solution; the *higher* the pH, the more basic the solution. Approximate pH values of some familiar solutions are listed in Table 16.4.

A change in acidity of one full pH unit corresponds to a 10-fold change in the concentration of hydrogen ions.

For example, a solution with a pH of 3 is *10 times* as acidic as a solution with a pH of 4, *100 times* as acidic as a solution with a pH of 5, and so on. This will be more obvious if you keep in mind the corresponding hydrogen ion concentrations. A pH of 3 corresponds to a $[H^+]$ of 1×10^{-3} M or 0.001 M. A pH of 4 means a $[H^+]$ of 1×10^{-4} M or 0.0001 M. Notice that 0.001 is 10 times greater than 0.0001. Details involved in converting hydronium ion concentrations to pH are described in the following examples.

EXAMPLE 16.10

What is the pH of the bile sample in Example 16.9 that has a $[H^+]$ of 1.0×10^{-8} M? Is the solution acidic or basic?

SOLUTION

Substitute the known $[H^+]$ in the mathematical expression for pH.

$$\begin{aligned} pH &= -\log [H^+] \\ &= -\log(1.0 \times 10^{-8}) \end{aligned}$$

Warning!
Be sure to use the value for $[H^+]$, not $[OH^-]$.

When the coefficient of 10^{-n} for $[H^+]$ is *exactly* 1, as it is here, we do *not* need a calculator. The pH is equal to the number, n, in the exponent.

$$pH = 8.0$$

The pH of the bile sample is 8.0. Since the pH is greater than 7, the solution is basic.

See Problems 16.47 and 16.48.

EXERCISE 16.3

What is the pH of a vinegar solution where the $[H^+]$ is 1×10^{-3} M? Is the solution acidic or basic?

In the preceding example and exercise involving conversion from $[H^+]$ to pH, each hydrogen ion concentration had a ''nice'' number, fitting the form (1×10^{-n}). We could obtain the pH without using a calculator. However, in many cases, hydrogen ion concentrations are not ''nice'' numbers (1×10^{-n}). For numbers such as 4.0×10^{-3}, 4.73×10^{-3}, and others that fit the form $(m \times 10^{-n})$, where m is not 1, we can use a calculator (or table of logarithms) to determine the logarithm needed. Carefully study the next example before you do the problems at the end of this chapter.

EXAMPLE 16.11

Determine the pH of a solution with a $[H^+]$ of 4.5×10^{-3} M.

SOLUTION

$$\begin{aligned} pH &= -\log [H^+] \\ &= -\log(4.5 \times 10^{-3}) \end{aligned}$$

From the exponent, −3, we can make an approximation. The pH will be between 2 and 3.

To find the logarithm of 4.5×10^{-3}, use a calculator with a LOG key.

	Press	*Display*	
To enter the coefficient	4.5	4.5	Steps used to enter the $[H^+]$.
To access the exponent mode	EE or EXP	4.5 00	
To enter the exponent	3	4.5 03	
To change exponent sign	+/−	4.5 −03	
To obtain the log	LOG	−2.35	Rounded to hundredths.
To obtain the negative log	+/−	2.35	This is the pH.

$$pH = 2.35$$

The pH is between 2 and 3, as expected.

EXERCISE 16.4

What is the pH of a solution with $[H^+]$ of 5.7×10^{-4} M?

The pOH is defined in a manner similar to pH, except that in pOH we are dealing with the $[OH^-]$ rather than the $[H^+]$.

$$pOH = -\log [OH^-]$$

Table 16.5 **The Relationship Between pH and $[H^+]$ and Between pOH and $[OH^-]$ (at 20 °C)**

$[H^+]$	*pH*	*$[OH^-]$*	*pOH*	
1×10^{0}	0	1×10^{-14}	14	↑
1×10^{-1}	1	1×10^{-13}	13	
1×10^{-2}	2	1×10^{-12}	12	
1×10^{-3}	3	1×10^{-11}	11	Acidic solutions
1×10^{-4}	4	1×10^{-10}	10	
1×10^{-5}	5	1×10^{-9}	9	
1×10^{-6}	6	1×10^{-8}	8	↓
1×10^{-7}	7	1×10^{-7}	7	Neutral solution
1×10^{-8}	8	1×10^{-6}	6	↑
1×10^{-9}	9	1×10^{-5}	5	
1×10^{-10}	10	1×10^{-4}	4	
1×10^{-11}	11	1×10^{-3}	3	Basic solutions
1×10^{-12}	12	1×10^{-2}	2	
1×10^{-13}	13	1×10^{-1}	1	
1×10^{-14}	14	1×10^{0}	0	↓

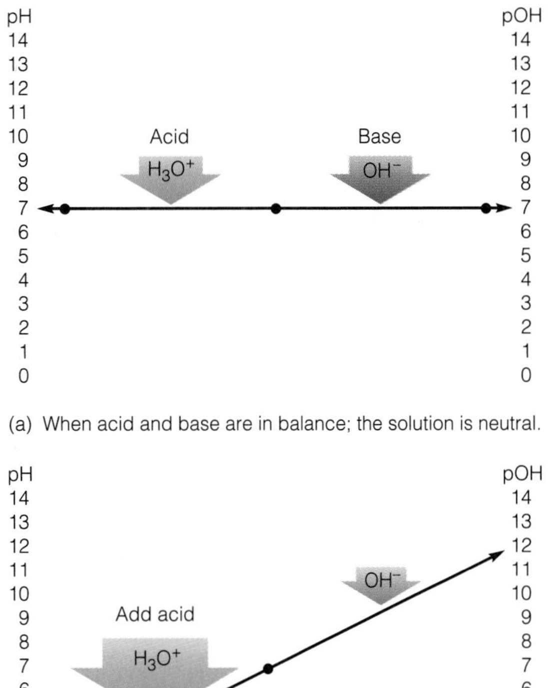

(a) When acid and base are in balance; the solution is neutral.

(b) When excess acid is added, pH decreases, and pOH increases.

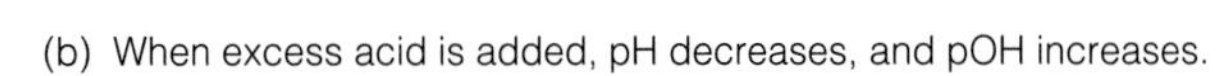

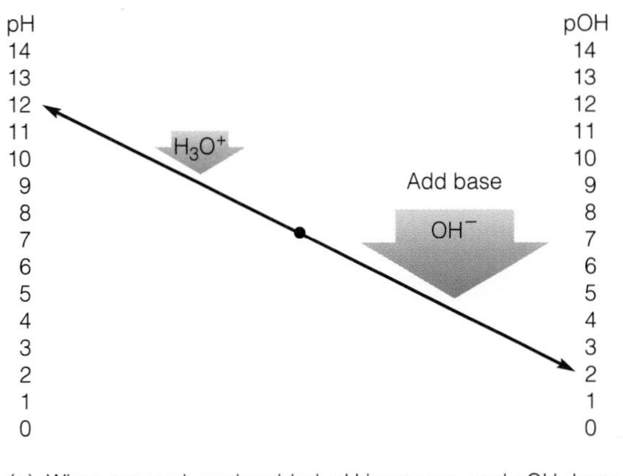

(c) When excess base is added, pH increases, and pOH decreases.

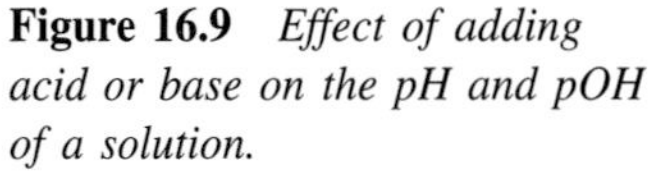

Figure 16.9 *Effect of adding acid or base on the pH and pOH of a solution.*

The relationship between pH and pOH is shown in Table 16.5. Notice that

$$\text{pH} + \text{pOH} = 14$$

The following statements and Figure 16.9 summarize the effect of adding acid or base on pH and pOH.

In a neutral solution, both pH and pOH are equal to 7.
The addition of an acid lowers the pH but increases the pOH.
The addition of a base lowers the pOH but increases the pH.
The sum of pH and pOH always equals 14.

EXAMPLE 16.12

What is the pOH of a solution that has a pH of 8.23?

SOLUTION

	pH + pOH = 14	By definition.
Thus,	pOH = 14 − pH	Subtract "pH" from both sides.
	pOH = 14 − 8.23	Substitute 8.23 for pH.
	pOH = 5.77	Answer.

See Problems 16.49–16.54.

EXERCISE 16.5

What is the pOH of an acidic solution with pH 4.83?

We have shown how pH, pOH, $[H^+]$, and $[OH^-]$ values are all interrelated. We have shown that the $[H^+]$ can be determined when the $[OH^-]$ is known, and vice versa. We have also shown conversions of hydrogen ion concentrations to pH values. In the following example, we show how to convert pH to $[H^+]$.

EXAMPLE 16.13

What is the $[H^+]$ for lemon juice that has a pH of 2.35?

SOLUTION

When the pH is known, the corresponding $[H^+]$ can be obtained by substitution in the pH expression. Follow the steps shown here.

$$pH = -\log [H^+]$$

$$2.35 = -\log [H^+]$$

Multiply both sides by −1.

$$-2.35 = \log [H^+]$$

Switch sides to give

$$\log [H^+] = -2.35$$

Take the inverse log (the antilog) of both sides to find $[H^+]$.

$$[H^+] = \text{antilog } -2.35 \qquad \text{which is} \qquad 10^{-2.35}$$

If your calculator has a 10^x key, press 2.35 to enter the number, press the +/− key to reverse signs, then press the 10^x key. (You have obtained the value of $10^{-2.35}$.) The display shows 4.4668 −03, so

Calculators are not all alike. For instructions on determining the inverse log (antilog) you may need to check your calculator user's manual.

$[H^+] = 4.47 \times 10^{-3}$ rounded to three significant figures

If your calculator does *not* have a 10^x key but has the INV and LOG keys, follow this sequence.

	Press	*Display*
To enter the number 2.35	2.35	2.35
To change the sign	+/−	−2.35
To access the exponential mode	EE or EXP	−2.35 00
To obtain the inverse log	INV then LOG	4.4668 −03

$[H^+] = 4.47 \times 10^{-3}$ rounded to three significant figures

See Problems 16.55 and 16.56.

EXERCISE 16.6
What is the $[H^+]$ in a can of carbonated beverage with pH 3.60?

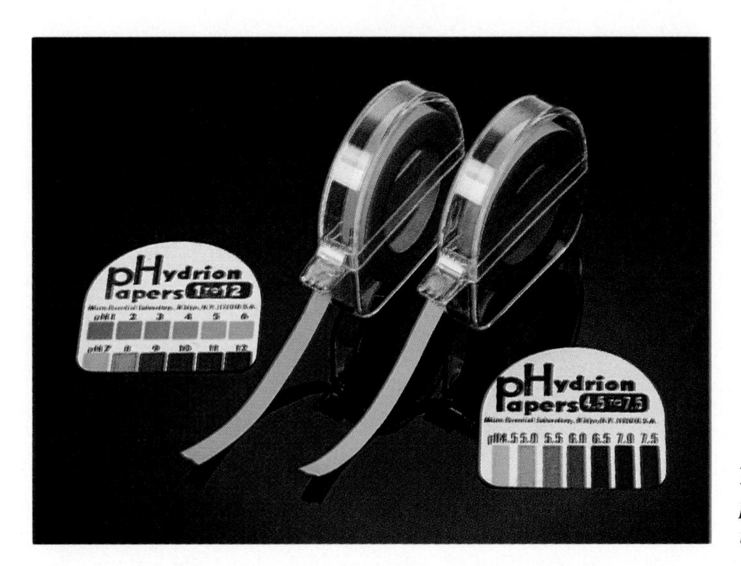

Figure 16.10 *Two common pH papers, wide range* (left) *and narrow range* (right).

Measurement of pH

One method of determining pH involves the use of certain chemicals called **acid–base indicators**. These indicators are dyes—some synthetic and some from plant and animal sources—that change color at specific pH values. Each indicator dye has one color in an acidic solution and a different color in a more basic solution (it changes from an acid to its conjugate base). For example, litmus is red in acidic solutions and blue in basic solutions. Combinations of certain indicator dyes exhibit a whole range of colors as the pH changes from strongly basic to strongly acidic (Figure 16.10). By selecting the proper acid–base indicator (see Table 16.6), one can determine the pH of almost any colorless aqueous solution.

More accurate pH measurements can be made electrically with pH meters. Generally, these instruments can measure pH to a precision of about 0.01 pH unit. A pH meter can be used for pH determinations of blood, urine, and other colored or complex mixtures. A typical pH meter is shown in Figure 16.11.

Table 16.6 **Some Acid–Base Indicators**

Indicator	*Approximate pH Range Where Color Changes*	*Color in the More Acidic Range*	*Color in the More Basic Range*
Methyl violet	0–2	Yellow	Violet
Thymol blue	1.2–2.8	Pink	Yellow
Methyl orange	3.2–4.4	Red	Yellow
Methyl red	4.2–6.2	Red	Yellow
Litmus	4.7–8.2	Red (pink)	Blue
Bromthymol blue	6.0–7.8	Yellow	Blue
Thymol blue	8.0–9.4	Yellow	Blue
Phenolphthalein	8.3–10.0	Colorless	Red
Alizarin yellow R	10.2–12.1	Yellow	Red

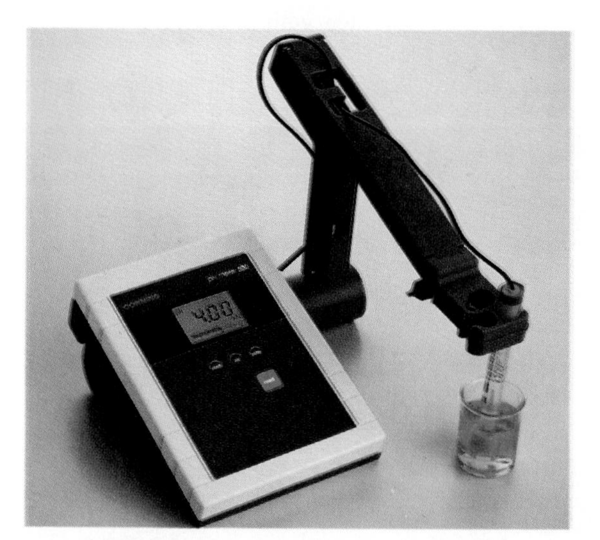

Figure 16.11 *A pH meter.*

EXAMPLE 16.14

Use the table of acid–base indicators to help you determine the color of the indicator specified in solutions with these pH values.

(a) bromthymol blue indicator in sauerkraut at a pH of 3.5
(b) phenolphthalein indicator in pure water at a pH of 7.0
(c) phenolphthalein indicator in milk of magnesia at a pH of 10.2

SOLUTION

(a) Bromthymol blue is yellow at this pH.
(b) Phenolphthalein is colorless at this pH.
(c) Phenolphthalein is red at this pH.

16.10 *Hydrolysis: Salts in Water*

When an acid reacts with a proportional amount of base, the products are water and a salt. The process is called neutralization, but is the solution neutral? What if we simply take a salt and dissolve it in water? Would the solution be acidic, basic, or neutral? It depends on the salt. The ions in many salts react to some extent with water to upset the 1 : 1 ratio of hydronium and hydroxide ions. The reaction of ions of salts with water is called **hydrolysis** from *hydro* (water) and *lysis* (to break loose or decompose).

How do we tell if a salt is acidic, basic, or neutral? Experimentally, it's simple. Just dissolve some of the salt in water and check the pH with an acid–base indicator or a pH meter. However, we really don't have to go into the lab and test the pH of each salt solution. We can *predict* whether a salt solution will be acidic, basic, or neutral by comparing the relative strengths of the acid and base from which the salt could be made. The rules needed are listed here.

Hydrolysis of Salts

1. The salt of a strong acid and a strong base gives a neutral solution.
2. The salt of a strong acid and a weak base gives an acidic solution.
3. The salt of a weak acid and a strong base gives a basic solution.
4. The salt of a weak acid and a weak base gives a slightly acidic, slightly basic, or neutral solution.

To apply these rules you must know the strong acids, strong bases, weak acids, and weak bases and the ions they release (Section 16.2).

Will a Salt Solution be Acidic, Basic, or Neutral?

Thinking It Through

To predict whether a salt is acidic, basic, or neutral, look at its formula and ask these questions: What ions would be released in solution? What base gives the positive ion released as the salt hydrolyzes? What acid gives the negative ion released as the salt hydrolyzes? Then, apply the four rules listed before this paragraph. For example, sodium carbonate, Na_2CO_3, gives Na^+ ions and CO_3^{2-} ions. The positive ion, Na^+, is released by the *strong* base, NaOH. The negative ion, CO_3^{2-}, is released by the *weak* acid, H_2CO_3. Sodium carbonate, then, is the salt of a weak acid and a strong base; hydrolysis gives a solution that is basic (rule 3).

EXAMPLE 16.15

Is a solution of KCl acidic, basic, or neutral?

SOLUTION

The positive ion in KCl is K^+, present in the strong base KOH. The negative ion in KCl is Cl^-, present in the strong acid HCl. This is a salt of a strong acid and a strong base, so the solution is neutral (rule 1).

See Problems 16.57–16.60.

EXAMPLE 16.16

Is a solution of $(NH_4)_2SO_4$ acidic, basic, or neutral?

SOLUTION

The solution is acidic (rule 2). The NH_4^+ ion is formed when NH_3, a weak base, dissolves in water. The SO_4^{2-} ion is present in H_2SO_4, a strong acid.

EXERCISE 16.7

Is a solution of sodium acetate acidic, basic, or netural?

16.11 Buffers: Controlling the pH

In chemistry, a **buffer** is a pair of chemicals that, if present in a given solution, can keep the pH almost constant when either an acid or a base is added. A buffer can be pictured as being something like a shock absorber; it tends to reduce the shock of drastic changes in H^+ and OH^- concentrations. Buffers are important in many manufacturing processes and are vital to life. If the buffer systems in your body would fail to function, so would you.

A buffer is prepared by using a weak acid and a salt of that acid (or a weak base and a salt of that base), usually in approximately equal concentrations. To understand how a buffer works, we will apply Le Châtelier's principle (Section 15.4). The **buffer solution** contains species that will react with—and tie up—added hydrogen ions or hydroxide ions.

We'll describe a buffer solution of acetic acid and sodium acetate. If a strong acid is added to this solution, the additional hydrogen ions react with acetate ions in solution to produce acetic acid, a weak acid. The equilibrium is shifted to the right as protons are taken up by the acetate buffer to form a weak acid.

Added acid

$$H^+(aq) + CH_3COO^-(aq) \rightleftharpoons CH_3COOH(aq)$$

The reaction is reversible to a slight extent, but the forward reaction that ties up the protons is, by far, the predominant reaction. The change in pH is extremely small until the **buffer capacity** is reached. That is, if a very large quantity of acid is added, there is a point at which the acetate ions of the buffer can no longer tie up protons. The buffer capacity has been reached.

When a strong base is added, the additional hydroxide ions are neutralized by hydrogen ions already in the buffer because of the presence of acetic acid.

Added base

$$OH^-(aq) + H^+(aq) \longrightarrow H_2O$$

As hydrogen ions are removed from the buffer solution, they are immediately replaced by further ionization of the acetic acid in the buffer.

$$CH_3COOH(aq) \rightleftharpoons CH_3COO^-(aq) + H^+(aq)$$

Table 16.7 **Some Important Buffers**

Buffer Components	*Buffer System Names*	*pH**
CH_3COOH/CH_3COO^-	Acetic acid/acetate ion	4.76
H_2CO_3/HCO_3^-	(Carbon dioxide) carbonic acid/ hydrogen carbonate ion	6.46†
$H_2PO_4^-/HPO_4^{2-}$	Dihydrogen phosphate ion/ monohydrogen phosphate ion	7.20
NH_4^+/NH_3	Ammonium ion/ammonia	9.25

*The values listed are for solutions that are 0.1 M in each compound at 25 °C.

†This value includes dissolved CO_2 molecules as undissociated H_2CO_3. The value for H_2CO_3 alone is about 3.8.

Once again, the concentration of hydrogen ions returns to approximately the original value, and the pH is only slightly changed.

There are many important buffer solutions. Most biochemical reactions, whether they occur in a laboratory or in our bodies, are carried out in buffered solutions. Some important buffers are listed in Table 16.7.

CHEMISTRY IN OUR WORLD

Buffers in the Blood

The pH of a person's blood plasma is held remarkably constant. If the pH drops below 7.35, the condition is called *acidosis*. If it rises above 7.45, the condition is called *alkalosis*. Should the pH rise above 7.8 or fall below 6.8—because of faulty respiration, starvation, kidney failure, or disease—the person may suffer irreversible damage to the brain or even die. Fortunately, human blood has not one, but at least three, buffering systems. Of these, the hydrogen carbonate/carbonic acid, HCO_3^-/H_2CO_3, buffering system is the most important.

If *acids* (hydrogen ions) enter the blood, they are taken up by the hydrogen carbonate (bicarbonate) ions to form carbonic acid, H_2CO_3 (a weak acid).

$$HCO_3^-(aq) + H^+(aq) \rightarrow H_2CO_3(aq)$$

As long as there is sufficient hydrogen carbonate to take up any added acid, the pH will undergo little change. The other member of this buffer system is H_2CO_3, which is slightly ionized to give hydrogen ions and hydrogen carbonate ions in solution.

$$H_2CO_3(aq) \rightleftharpoons H^+(aq) + HCO_3^-(aq)$$

Any *bases* that enter the bloodstream remove these hydrogen ions to form water, but more carbonic acid molecules ionize to replace the hydrogen ions that were taken out of solution. As the carbonic acid molecules are used up, more carbonic acid can be formed from the large quantity of dissolved CO_2 in the blood.

$$CO_2(g) + H_2O \rightleftharpoons H_2CO_3(aq)$$

Along with the hydrogen carbonate/carbonic acid buffer system in blood, a second buffer system is the dihydrogen phosphate/monohydrogen phosphate, $H_2PO_4^-/HPO_4^{2-}$, system. Any excess acid reacts with monohydrogen phosphate, HPO_4^{2-}, to form the weak conjugate acid, dihydrogen phosphate, $H_2PO_4^-$. Any excess base in the bloodstream would neutralize hydrogen ions released by dihydrogen phosphate.

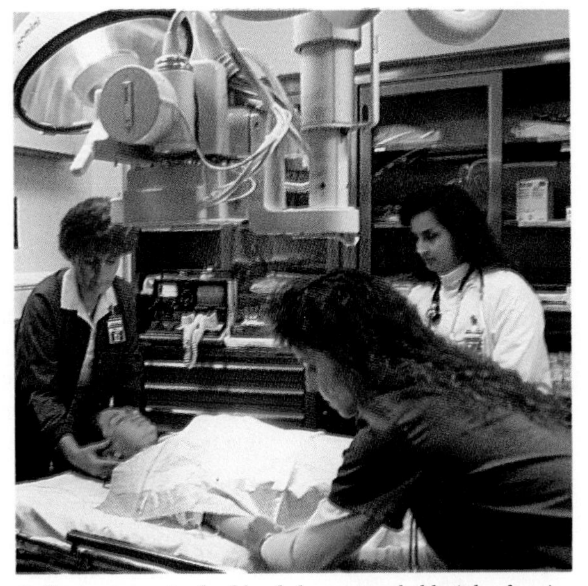

Buffers present in the blood do a remarkable job of maintaining a pH of 7.4, but an emergency situation arises if a person stops breathing and the level of carbon dioxide in the blood rises, causing a condition called acidosis.

Proteins act as a third type of blood buffer. These complex molecules contain carboxylate ions, $—COO^-$, which can act as proton acceptors. Proteins also contain $—NH_3^+$ ions, which can donate protons to neutralize excess base.

These three buffers act to keep the pH of the blood constant, but there is a limit to the buffering capacity of blood. Critical situations can occur if large quantities of acid or base enter the blood or when metabolism is greatly upset.

EXAMPLE 16.17

Explain, and show by equation, what happens when an acid is added to a sodium acetate/acetic acid buffer.

SOLUTION

As excess acid reacts with acetate ions in the buffer, the equilibrium shifts to the right and the $[H^+]$ decreases as the weak acid is formed.

See Problems 16.61–16.66.

$$H^+(aq) + CH_3COO^-(aq) \rightleftharpoons CH_3COOH(aq)$$

16.12 Acid–Base Titrations

The concentration of a particular solution of acid or base can be determined by a process called **titration**. When determining the concentration of an acid, a specific volume of the solution is carefully measured from a buret into a flask (Figure 16.12). A few drops of an acid–base indicator are also added to the acid in the flask. Then, a base of known concentration, called a **standard base**, is added from another buret, slowly and carefully, until finally just one additional drop of base changes the color of the indicator dye. This is the **end point** of the titration. The point at which stoichiometrically *equivalent* amounts of acid and base have been reacted is called the **equivalence point**. (The *end point* and the *equivalence point* are not necessarily the same.) An appropriate indicator must be selected for the titration so that the end point is as close as possible to the equivalence point. After performing the titration, the volumes of acid and base needed for the neutralization, along with the known concentration of the standard base, are used to calculate the unknown concentration of the acid.

To determine the concentration of a basic solution, the procedure is reversed. A quantity of the basic solution is measured into a flask. Then, acid of known concentration—**standard acid**—is added from another buret until the end point is reached. Using the concentration of the standard acid and the volumes of the acid and base, the unknown concentration of the base can be calculated. In carrying out these calculations, we will rely on the concepts presented in Chapter 11.

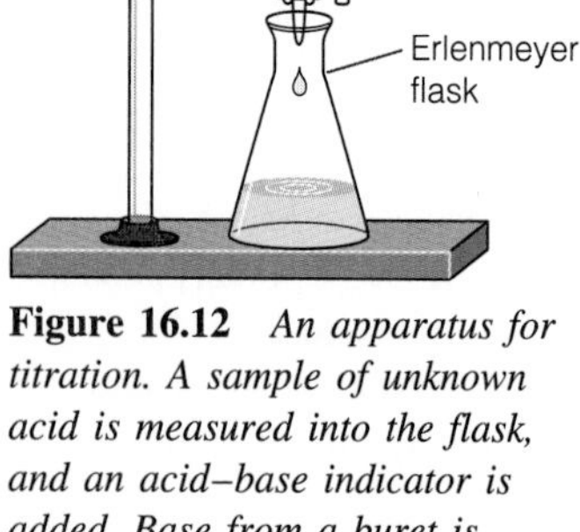

Figure 16.12 *An apparatus for titration. A sample of unknown acid is measured into the flask, and an acid–base indicator is added. Base from a buret is added dropwise until the indicator changes color.*

EXAMPLE 16.18

What is the molarity of an automobile battery sulfuric acid, H_2SO_4, solution if 22.53 mL of the acid neutralizes 42.11 mL of 1.923 M NaOH?

SOLUTION

STEP 1 Write a balanced equation for the neutralization reaction.

$$H_2SO_4 + 2\,NaOH \longrightarrow Na_2SO_4 + 2\,H_2O$$

We will need to use the mole ratio

$$\frac{1\text{ mol acid}}{2\text{ mol base}}$$

STEP 2 List the volume and molarity for the acid and the base.

	Acid	*Base*
Volume	22.53 mL	42.11 mL
Molarity	? M	1.923 M

STEP 3 Write a plan, using conversions to determine moles of acid.

$$\text{Plan: } \begin{matrix}\text{Volume of}\\ \text{base in L}\end{matrix} \xrightarrow{\text{M of base}} \begin{matrix}\text{Moles}\\ \text{of base}\end{matrix} \xrightarrow{\text{mole ratio}} \begin{matrix}\text{Moles}\\ \text{of acid}\end{matrix}$$

STEP 4 Carry out the conversions.

$$0.04211 \text{ L base} \times \frac{1.923 \text{ mol base}}{\text{L base}} \times \frac{1 \text{ mol acid}}{2 \text{ mol base}} = 0.04049 \text{ mol acid}$$

STEP 5 Divide moles of acid by liters of acid to obtain the molarity, M.

$$\frac{0.04049 \text{ mol acid}}{0.02253 \text{ L}} = 1.797 \text{ M } H_2SO_4$$

See Problems 16.67–16.74.

Chapter Summary

There are three important theories of acid–base reactions. According to the traditional Arrhenius definition, an acid is a substance that releases hydrogen ions, H^+, when dissolved in water. A base releases hydroxide ions, OH^-, in water. The neutralization reaction produces a salt and water.

According to the more general Brønsted–Lowry definitions of acids and bases, an acid is a proton, H^+, donor and a base is a proton acceptor. In aqueous solutions, individual protons, H^+, cannot be found; they are bonded to water to form hydronium ions, H_3O^+. When an acid loses a proton, the conjugate base is formed. When a base accepts a proton, the conjugate acid is formed. Strong acids produce weak conjugate bases, and strong bases produce weak conjugate acids. Strong acids are those that ionize completely in water; weak acids ionize only slightly in water.

The Lewis theory of acids and bases is the most inclusive. A Lewis acid is defined as any substance that can accept (and share) an electron pair. A Lewis base is any substance that can donate (and share) an electron pair. All chemicals that are acids according to the Arrhenius and Brønsted–Lowry theories are acids by the Lewis theory. The same holds true for bases. A hydrogen ion (a hydronium ion) is an acid by all three theories, and a hydroxide ion is a base by all three theories.

Water undergoes self-ionization to produce extremely small—but equal—concentrations of hydrogen and hydroxide ions. The product of these concentrations, K_w, is 1×10^{-14}. The addition of an acid or base to water affects the $[H^+]$

and $[OH^-]$, but the ion product remains equal to 1×10^{-14}. The hydrogen ion concentration, $[H^+]$, of a solution is often expressed as a pH.

$$pH = -\log [H^+]$$

A change in acidity of one full pH unit corresponds to a 10-fold change in $[H^+]$.

The reaction of a salt with water is called hydrolysis. Salts hydrolyze in water to produce acidic, basic, or neutral solutions, depending on the nature of the salt. A buffer can be prepared by mixing a weak acid with a salt of that acid. A buffer solution can keep the pH of a solution almost constant when a small amount of acid or base is added to the solution.

An unknown acid concentration can be determined by titrating the acid with a standard base when an appropriate indicator is used. An unknown base concentration can be determined by titrating the base with a standard acid.

Assess Your Understanding

1. Identify and list four principal properties of acids and bases. [16.1]
2. Identify and describe properties of strong and weak acids. [16.2]
3. Identify and describe properties of strong and weak bases. [16.3]
4. Complete and balance chemical equations for reactions of acids and bases. [16.4, 16.5]
5. Give examples of antacid ingredients and their reactions. [16.5]
6. Identify conjugate acid–base pairs. [16.6]
7. Define and compare acids and bases in terms of Arrhenius, Brønsted–Lowry, and Lewis theories. [16.7]
8. Use K_w to determine hydronium ion and hyroxide ion concentrations in acidic, basic, and neutral solutions. [16.8]
9. Define pH and compare changes in pH with changes in hydronium ion concentrations. [16.9]
10. Calculate the pH of a solution using the hydronium ion or the hydroxide ion concentration. [16.9]
11. Identify salts that give acidic, basic, and neutral solutions upon hydrolysis. [16.10]
12. Describe the composition of buffers and how they control pH. [16.11]
13. Make calculations that relate to acid–base titrations. [16.12]

Key Terms

acid [16.1]
acid–base indicator [16.9]
alkali [16.1]
amphiprotic [16.6]
amphoteric [16.5]
aqueous ammonia [16.3]
base [16.1]
buffer [16.11]
buffer capacity [16.11]
buffer solution [16.11]
carboxyl group [16.1]
conjugate acid–base pair [16.6]
diprotic acid [16.1]
end point [16.12]
equivalence point [16.12]
hydrolysis [16.10]
hydronium ion [16.1]
ion product constant for water [16.8]
monoprotic acid [16.1]
neutralization [16.1]

pH [16.9]
salt [16.1]
standard acid [16.12]
standard base [16.12]
strong acid [16.2]
strong base [16.3]
titration [16.12]
triprotic acid [16.1]
weak acid [16.2]
weak base [16.3]

Problems

ACIDS AND BASES: THE ARRHENIUS THEORY

16.1 List four general properties of acidic solutions.

16.2 List four general properties of basic solutions.

16.3 To what ion did Arrhenius attribute the properties of acidic solutions?

16.4 To what ion did Arrhenius attribute the properties of basic solutions?

16.5 What is a hydronium ion and how does it differ from a hydrogen ion? Why are the terms often used interchangeably?

16.6 What occurs during neutralization? What products are formed?

16.7 Indicate whether each of the following is an acid, a base, a salt, or none of these.
a. H_2SO_4 b. $Mg(OH)_2$ c. $MgSO_4$
d. HCN e. NaCN
f. $CH_3CH_2CH_2CH_3$ (butane)

16.8 Indicate whether each of the following is an acid, a base, a salt, or none of these.
a. HNO_3 b. KOH
c. KNO_3 d. CH_3COOH
e. CH_4 (methane) f. $CaCl_2$

16.9 Give the chemical formula for each of the following acids and indicate whether each is monoprotic, diprotic, or triprotic.
a. phosphoric acid b. nitric acid
c. carbonic acid d. lactic acid

16.10 Give the chemical formula for each of the following acids and indicate whether each is monoprotic, diprotic, or triprotic.
a. acetic acid b. boric acid
c. hydrochloric acid d. sulfuric acid

STRONG AND WEAK ACIDS AND BASES

16.11 Give formulas and names of six strong acids and any three weak acids.

16.12 Give formulas and names of any two strong bases and one common weak base.

16.13 What is the difference between a strong acid and a weak acid?

16.14 What is the difference between a strong base and a weak base?

16.15 Distinguish between a weak acid and a dilute acid.

16.16 Distinguish between a strong acid and a concentrated acid.

16.17 What is aqueous ammonia? Why is it sometimes called ammonium hydroxide?

16.18 Solutions of aqueous ammonia and magnesium hydroxide both have small percentages of OH^- ions in solution, but for different reasons. Explain.

16.19 Indicate whether each of the following is a strong or weak acid or base.
a. acetic acid
b. boric acid
c. nitric acid
d. hydrochloric acid
e. aqueous ammonia
f. magnesium hydroxide

16.20 Indicate whether each of the following is a strong or weak acid or base.
a. phosphoric acid
b. sulfuric acid
c. carbonic acid
d. lactic acid
e. calcium hydroxide
f. potassium hydroxide

16.21 Should you wear eye protection in the laboratory when working with acids and bases? Explain.

16.22 Should you wear eye protection and heavy, loose-fitting rubber gloves when working at home with oven cleaners and drain cleaners? How about when using muriatic acid to clean bricks? How about when using vinegar? Explain.

REACTIONS OF ACIDS AND BASES

16.23 Write a balanced equation for the complete neutralization of sulfuric acid by potassium hydroxide. Also write a net ionic equation for the reaction.

16.24 Write a balanced equation for the complete neutralization of carbonic acid by sodium hydroxide. Also write a net ionic equation for the reaction.

16.25 Write balanced equations for the following reactions.
a. hydrochloric acid with magnesium metal
b. hydrochloric acid with magnesium oxide
c. hydrochloric acid with magnesium carbonate

16.26 Write balanced equations for the following reactions.
a. sulfuric acid with zinc oxide
b. sulfuric acid with zinc metal
c. sulfuric acid with calcium carbonate

16.27 What is an amphoteric metal hydroxide? Give two examples.

16.28 What is an amphoteric metal? Give two examples.

16.29 Balance the following equations and explain what these reactions tell us about aluminum hydroxide.
a. $Al(OH)_3(s) + KOH(aq) \rightarrow KAl(OH)_4(aq)$
b. $Al(OH)_3(s) + HCl(aq) \rightarrow AlCl_3(aq) + H_2O$

16.30 Balance the following equations and explain what these reactions tell us about aluminum.
a. $Al(s) + KOH(aq) + H_2O \rightarrow KAl(OH)_4(aq) + H_2(g)$
b. $Al(s) + HCl(aq) \rightarrow ?$

16.31 Assuming that stomach acid is HCl(aq), write an equation for the reaction of this acid with sodium hydrogen carbonate (baking soda). Also give the net ionic equation.

16.32 Write an equation for the reaction of hydrochloric acid with milk of magnesia, a suspension of $Mg(OH)_2$. Also give the net ionic equation.

BRØNSTED–LOWRY AND LEWIS ACID–BASE THEORY

16.33 Give the Brønsted–Lowry and Lewis definitions for an acid.

16.34 Give the Brønsted–Lowry and Lewis definitions for a base.

16.35 Write an equation that shows water can be a Brønsted–Lowry base.

16.36 Write an equation that shows water can be a Brønsted–Lowry acid.

16.37 Give a balanced equation for the reaction of hydrogen chloride gas with ammonia gas. This is viewed as an acid–base reaction according to which definition(s)? It is not viewed as an acid–base reaction according to which definition(s)?

16.38 Give a balanced equation—showing electron dots—for the reaction of $FeBr_3$ with a bromide ion to give $FeBr_4^-$. This is viewed as an acid–base reaction according to which definition(s)? It is not viewed as an acid–base reaction according to which definition(s)?

16.39 Identify the first compound in each equation as a Brønsted–Lowry acid or base. Also identify the conjugate acid or base.
a. $C_5H_5N + H_2O \rightleftharpoons C_5H_5NH^+ + OH^-$
b. $C_6H_5OH + H_2O \rightleftharpoons C_6H_5O^- + H_3O^+$
c. $CH_3CHOHCOOH + H_2O \rightleftharpoons CH_3CHOHCOO^- + H_3O^+$

16.40 Identify the first compound in each equation as a Brønsted–Lowry acid or base. Also identify the conjugate acid or base.
a. $C_6H_5SH + H_2O \rightleftharpoons C_6H_5S^- + H_3O^+$
b. $CH_3NH_2 + H_2O \rightleftharpoons CH_3NH_3^+ + OH^-$
c. $NH_3 + H_2O \rightleftharpoons NH_4^+ + OH^-$

16.41 The following equations represent equilibria between two acids. In each case, identify the acids and indicate which is stronger. (See Table 16.3.)
a. $HBr + F^- \rightleftharpoons HF + Br^-$
b. $HCN + F^- \rightleftharpoons HF + CN^-$

16.42 The following equations represent equilibria between two acids. In each case, identify the acids and indicate which is stronger. (See Table 16.3.)
a. $H_3PO_4 + HSO_4^- \rightleftharpoons H_2PO_4^- + H_2SO_4$
b. $HClO_4 + NO_3^- \rightleftharpoons HNO_3 + ClO_4^-$

16.43 Each equilibrium in Problem 16.41 also includes a pair of bases. Identify the bases in each equation and indicate which is stronger.

16.44 Each equilibrium in Problem 16.42 also includes a pair of bases. Identify the bases in each equation and indicate which is stronger.

16.45 Is phenol, C_6H_5OH, acting as an acid or base in the following reaction?

$$C_6H_5OH + H_2O \rightleftharpoons C_6H_5O^- + H_3O^+$$

16.46 Is aniline, $C_6H_5NH_2$, acting as an acid or base in the following reaction?

$$C_6H_5NH_2 + H_2O \rightleftharpoons C_6H_5NH_3^+ + OH^-$$

HYDRONIUM ION CONCENTRATIONS, HYDROXIDE ION CONCENTRATIONS, AND PH

16.47 For each *hydronium* ion concentration given, determine the hydroxide ion concentration, and vice versa. Also give the pH and indicate which samples are acidic, basic, and neutral.
a. $[H_3O^+] = 1 \times 10^{-3}$ M
b. $[H_3O^+] = 1 \times 10^{-9}$ M

c. $[OH^-] = 1 \times 10^{-5}$ M
d. $[OH^-] = 1 \times 10^{-10}$ M

16.48 For each hydrogen ion concentration given, determine the hydroxide ion concentration, and vice versa. Also give the pH and indicate which samples are acidic, basic, and neutral.
a. $[H^+] = 1 \times 10^{-10}$ M
b. $[H^+] = 1 \times 10^{-2}$ M
c. $[OH^-] = 1 \times 10^{-8}$ M
d. $[OH^-] = 1 \times 10^{-4}$ M

16.49 Determine the pH where pOH is given, and vice versa.
a. pOH = 8.00 b. pOH = 4.7
c. pH = 3.00 d. pH = 9.4

16.50 Determine the pH where pOH is given, and vice versa.
a. pOH = 10.00 b. pOH = 8.6
c. pH = 5.00 d. pH = 10.3

16.51 Determine the $[H^+]$ and the pH for each of the following solutions. (Assume that each solution is completely ionized.)
a. 1.00×10^{-3} M HCl
b. 1.00×10^{-5} M NaOH
c. 1.00×10^{-4} M HNO_3
d. 1.00×10^{-3} M KOH

16.52 Determine the $[H^+]$ and the pH for each of the following solutions. (Assume that each solution is completely ionized.)
a. 1.00×10^{-6} M HBr
b. 1.00×10^{-5} M LiOH
c. 1.00×10^{-2} M HCl
d. 1.00×10^{-2} M NaOH

16.53 Determine the pH of each of the following solutions. (You will need a calculator with LOG key or a table of logarithms.)
a. $[H^+] = 3.4 \times 10^{-5}$ M
b. $[H^+] = 7.2 \times 10^{-12}$ M
c. $[OH^-] = 8.5 \times 10^{-7}$ M
d. $[OH^-] = 6.3 \times 10^{-6}$ M

16.54 Determine the pH of each of the following solutions. (You will need a calculator with LOG key or a table of logarithms.)
a. $[H_3O^+] = 5.6 \times 10^{-6}$ M
b. $[H_3O^+] = 9.2 \times 10^{-9}$ M
c. $[OH^-] = 1.8 \times 10^{-8}$ M
d. $[OH^-] = 7.8 \times 10^{-10}$ M

16.55 Determine the $[H^+]$ to two significant figures for solutions with the following pH and pOH values. (The method shown in the text requires a calculator with INV and LOG keys or a 10^x key.)
a. pH = 8.35 b. pH = 2.73
c. pOH = 9.10 d. pOH = 6.08

16.56 Determine the $[H^+]$ to two significant figures for solutions with the following pH and pOH values. (The method shown in the text requires a calculator with INV and LOG keys or a 10^x key.)
a. pH = 4.09 b. pH = 9.74
c. pOH = 10.22 d. pOH = 3.75

HYDROLYSIS: SALTS IN WATER

16.57 Indicate whether hydrolysis of the following salts would give solutions that are acidic, basic, or neutral.
a. K_2SO_4 b. $Ca(CH_3COO)_2$
c. NH_4Cl d. NH_4NO_3
e. $NaHCO_3$

16.58 Indicate whether hydrolysis of the following salts would give solutions that are acidic, basic, or neutral.
a. $MgBr_2$ b. $(NH_4)_3PO_4$
c. Na_2CO_3 d. $NaCH_3COO$
e. NaCN

16.59 Would an NH_4NO_3 fertilizer tend to make soil more acidic or alkaline?

16.60 Would crushed limestone (calcium carbonate) applied directly to the soil tend to make soil more acidic or alkaline?

BUFFERS

16.61 What types of substances are used in preparing a buffer?

16.62 Explain how a buffer works.

16.63 What substance could be used with sodium acetate to make a buffer?

16.64 What substance could be used with ammonium chloride to make a buffer?

16.65 Using chemical equations, explain how the hydrogen carbonate/carbonic acid buffer in blood works when an acid is added.

16.66 Using chemical equations, explain how the hydrogen carbonate/carbonic acid buffer in blood works when a base is added.

ACID–BASE TITRATIONS

16.67 What is an acid–base titration?

16.68 In an ideal titration, the end point occurs at the equivalence point. What is the difference between an end point and an equivalence point?

16.69 A weak acid is titrated with a strong base. Would

the solution at the equivalence point be acidic, basic, or neutral? Explain.

16.70 A weak base is titrated with a strong acid. Would the solution at the equivalence point be acidic, basic, or neutral? Explain.

16.71 Calculate the molarity of an HCl solution if 20.00 mL of it requires 40.00 mL of 0.2500 M NaOH for neutralization.

16.72 Calculate the molarity of an HCl solution if 20.00 mL of it requires 10.00 mL of 0.5000 M KOH for neutralization.

16.73 Calculate the molarity of a NaOH solution if a 31.22-mL sample of it requires 12.53 mL of 0.1000 M H_2SO_4 to neutralize the base.

16.74 Calculate the molarity of a NaOH solution if a 22.68-mL sample of it requires 18.77 mL of 0.1000 M H_2SO_4 to neutralize the base.

Additional Problems

16.75 Indicate whether each of these acids is monoprotic, diprotic, or triprotic.
a. H_3PO_4
b. $HClO_3$
c. $HOOC—CH_2CH_2—COOH$
d. H_2SO_3

16.76 Indicate whether each of these acids is monoprotic, diprotic, or triprotic.
a. H_2CO_3
b. CH_3COOH
c. HBr
d. H_3AsO_4

16.77 Ingesting a 5% solution of sulfuric acid would be quite injurious, but ingesting vinegar (5% acetic acid) is not. Explain.

16.78 Magnesium hydroxide is completely ionic, even in the solid state, yet it can be taken internally as an antacid. Why does it not cause injury as sodium hydroxide would?

16.79 Write an equation for the reaction of limestone (essentially $CaCO_3$) with sulfuric acid in acid rain.

16.80 Write an equation for the reaction of limestone with nitric acid present in acid rain.

16.81 Write an equation for the reaction of baking soda, $NaHCO_3$, with vinegar, CH_3COOH.

16.82 Write an equation for the reaction of baking powder in water. The baking powder contains $NaHCO_3$ (baking soda) and potassium hydrogen tartrate, $KHC_4H_4O_6$ (called cream of tartar).

16.83 Give names of three common active ingredients of antacids.

16.84 A person with hypertension (high blood pressure) should avoid high concentrations of sodium ion. Would there be any reason that a person with this condition should avoid heavy use of either baking soda or Alka-Seltzer as an antacid? Explain.

16.85 For the following reaction, label the "stronger base," the "weaker base," the "stronger acid," and the "weaker acid." Also use a line to connect the conjugate acid–base pairs.

$$HBr + H_2O \rightleftharpoons H_3O^+ + Br^-$$

16.86 For the following reaction, label the "stronger base," the "weaker base," the "stronger acid," and the "weaker acid." Also use a line to connect the conjugate acid–base pairs.

$$HF + H_2O \rightleftharpoons H_3O^+ + F^-$$

16.87 The box "Acid Rain" showed that a coal-fired electric power plant can release 400 tons of sulfur or 800 tons of SO_2 per day. How many tons of H_2SO_4 could be produced from 800 tons of SO_2?

16.88 Slaked lime, $Ca(OH)_2$, is sometimes used to neutralize acidic water in lakes resulting from acid precipitation. Assume that the acid is sulfuric acid and write an equation for the neutralization. How many metric tons of sulfuric acid can be neutralized by 10 metric tons of lime?

16.89 Examine the labels of at least four antacid preparations. Make a list of the active ingredients in each. What kind of chemical compound is each?

16.90 Examine the labels of at least five kitchen, bathroom, or automotive products that contain acids, bases, or salts. Make a list of the active ingredients in each, and indicate which ones are acids, bases, and salts.

16.91 Each of the following is classified as an acid–base reaction according to which definition(s)?

a. $FeCl_3 + Cl^- \rightarrow FeCl_4^-$
b. $H_2SO_4 + H_2O \rightarrow HSO_4^- + H_3O^+$

16.92 Each of the following is classified as an acid–base reaction according to which definition(s)?
a. $N_2H_4 + H_2O \rightarrow N_2H_5^+ + OH^-$
b. $SO_3 + CaO \rightarrow CaSO_4$

16.93 For each hydrogen ion concentration given, determine the hydroxide ion concentration, and vice versa. Also give the pH (to two significant figures) and indicate which samples are acidic, basic, and neutral.
a. $[H^+] = 1.00 \times 10^{-4}$ M
b. $[H^+] = 8.78 \times 10^{-8}$ M
c. $[OH^-] = 1.00 \times 10^{-6}$ M
d. $[OH^-] = 9.68 \times 10^{-9}$ M

16.94 For each hydrogen ion concentration given, determine the hydroxide ion concentration, and vice versa. Also give the pH (to two significant figures) and indicate which samples are acidic, basic, and neutral.
a. $[H^+] = 1.00 \times 10^{-9}$ M
b. $[H^+] = 6.15 \times 10^{-3}$ M
c. $[OH^-] = 1.00 \times 10^{-10}$ M
d. $[OH^-] = 1.77 \times 10^{-2}$ M

The reaction of copper with nitric acid is one of the oxidation–reduction reactions discussed in this chapter. The brownish gas you see in this photograph is nitrogen dioxide, which is produced by the vigorous reaction.

17

Oxidation and Reduction

CONTENTS

Foods and fossil fuels are high in energy. Their energy content is released through oxidation–reduction (called redox) reactions. Any time oxidation takes place, reduction is also taking place, and vice versa. You can't have one without the other. For convenience, however, we may choose to talk about one part of the process: the oxidation part or the reduction part.

Gasoline, coal, and sugars are examples of *reduced* forms of matter, which are high in energy (Figure 17.1). When glucose, $C_6H_{12}O_6$ (a simple sugar), reacts with oxygen during either combustion or metabolism, it gets *oxidized* as carbon dioxide and water are formed. Either reaction can be represented by the equation

$$C_6H_{12}O_6 + 6\,O_2 \longrightarrow 6\,CO_2 + 6\,H_2O + \text{Energy}$$

The CO_2 and H_2O are *oxidized* forms of matter, which are low in energy.

Green plants are able to use energy from sunlight to produce food by the reduction of carbon dioxide. This reaction—called photosynthesis—requires energy from the sun to drive the reaction. For the formation of glucose, the overall photosynthesis reaction is essentially the reverse of that for the oxidation of glucose, as shown here.

$$6\,CO_2 + 6\,H_2O + \text{Energy} \longrightarrow C_6H_{12}O_6 + 6\,O_2$$

The first reaction—metabolism—represents the oxidation of glucose by oxygen, which is reduced. The second reaction—photosynthesis—represents the reduction of carbon dioxide by water, which is oxidized (Figure 17.2).

Reduction processes are also required to obtain iron and other metals from their ores, but many of these metals are eventually lost to corrosion by slow oxidation. We maintain our technological civilization by oxidizing fossil fuels (coal, natural gas, and petroluem) to obtain the chemical energy that was stored in these materials eons ago by green plants.

All forms of life depend on redox processes. The metabolic processes that occur in every living cell involve oxidation–reduction reactions. In order for you to walk, talk, think, digest and metabolize food, or even read this page, oxidation–reduction reactions must take place. Many of these reactions are beyond the scope of this book because they are quite complex. Nevertheless, they serve to illustrate the importance of oxidation and reduction in our own lives.

Several reactions involving oxidation and reduction were presented in Sections 10.8 and 10.9. Furthermore, the use of oxidation numbers was first introduced in Section 8.7. We shall now review and extend our explanation of redox reactions. This time we will describe methods of balancing redox equations and deal with electrochemical processes that occur during the operation of batteries and during electrolysis.

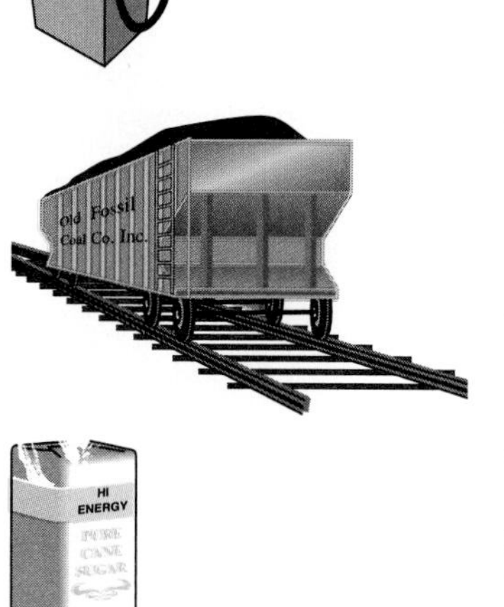

Figure 17.1 *Reduced forms of matter such as foods and fossil fuels are high in energy. The energy content is released through oxidation–reduction reactions.*

Figure 17.2 *The food we eat is oxidized to provide energy for our activities. The energy originates in the sun and is trapped by plants through photosynthetic reactions that reduce carbon dioxide to carbohydrates.*

17.1 Oxidation Numbers: A Review

Before dealing with processes involving oxidation and reduction, one must be comfortable with oxidation numbers of atoms and ions. The **oxidation number** (also called the **oxidation state**) is a number that is assigned to each kind of atom in a compound or ion, or to an element, using an arbitrary but consistent set of rules. The oxidation number represents the number of electrons that have been gained, lost, or shared by the species. Rules for assigning oxidation numbers were presented in Section 8.7 (page 237). If necessary, review them now. Test yourself by doing the following review examples and exercises before proceeding on to the next section.

EXAMPLE 17.1

What is the oxidation number of sulfur in H_2SO_4?

SOLUTION

The oxidation number of hydrogen is +1, and there are two hydrogen atoms. The oxidation number of oxygen is −2, and there are four oxygen atoms. The sum must be zero. Let the oxidation number of sulfur be x.

$$2(+1) + 4(-2) + x = 0$$

$$-6 + x = 0$$

$$x = +6$$

The sulfur has an oxidation number of +6.

EXERCISE 17.1

What is the oxidation number of sulfur in SO_2?

EXAMPLE 17.2

What is the oxidation number of chromium in $Cr_2O_7^{2-}$?

SOLUTION

The oxidation number of oxygen is −2, and there are seven oxygen atoms. The sum must be −2. Let the oxidation number of chromium be x.

$$2x + 7(-2) = -2$$

$$2x = 12$$

$$x = +6$$

The chromium has an oxidation number of +6.

EXERCISE 17.2

What is the oxidation number of carbon in CO_3^{2-}?

See Problems 17.1–17.4.

17.2 *Chemical Properties of Oxygen: Oxidation*

Oxidation occurs each time a substance combines with oxygen (Figure 17.3). The substances that combine with oxygen are said to be **oxidized**. The metabolism of food involves a complex series of slow oxidation reactions that produce carbon dioxide and water. **Combustion** (burning) involves rapid oxidation. **Spontaneous combustion** occurs when a substance begins to burn on its own, without being ignited by a spark or flame. Many costly fires have resulted from the spontaneous combustion of oily rags stored where there is poor air circulation and from freshly mown hay placed in unvented barns. When enough heat is released by slow oxidation, and the **kindling temperature** is reached, spontaneous combustion can occur.

The combustion of fuels involves oxidation and reduction. Natural gas, gasoline, wood, and coal need oxygen in order to burn and release their stored energy. Combustion of fossil fuels currently supplies over 90% of the energy used by our civilization.

Not all reactions that involve oxygen from the air are immediately desirable. Oxygen causes iron to rust and causes many metals to corrode. It also promotes the decay of wood and other organic materials. All these chemical processes—and many others—are called *oxidative reactions.*

When metals combine with oxygen, *metal oxides* are formed. For example, iron combines with oxygen of the atmosphere to form iron(III) oxide, Fe_2O_3, a reddish brown powder we call rust.

$$4\,Fe + 3\,O_2(g) \longrightarrow 2\,Fe_2O_3$$

(Iron is the only metal that forms an oxide called rust.) Many other metals also react with oxygen to form metal oxides, but metals with low reactivities such as copper, mercury, and silver are quite slow to react with oxygen.

Figure 17.3 *Cooking, breathing, and burning fuel all involve oxidation.*

CHEMISTRY IN OUR WORLD

Oxygen: An Essential Element

Oxygen is the most abundant element on this planet. It accounts for about one-half of the mass of Earth's crust, including the atmosphere, the hydrosphere, and the lithosphere. In the atmosphere (the gaseous layer surrounding Earth), oxygen occurs as "free" or "uncombined" diatomic molecules of O_2 gas. In the hydrosphere (the oceans, seas, lakes, and rivers of Earth), oxygen occurs combined with hydrogen in water, H_2O. Oxygen accounts for 89% of the mass of water. In the lithosphere (the solid portion of Earth's crust), oxygen occurs combined with silicon—in sand and clay—and with other minerals. Sand is largely SiO_2.

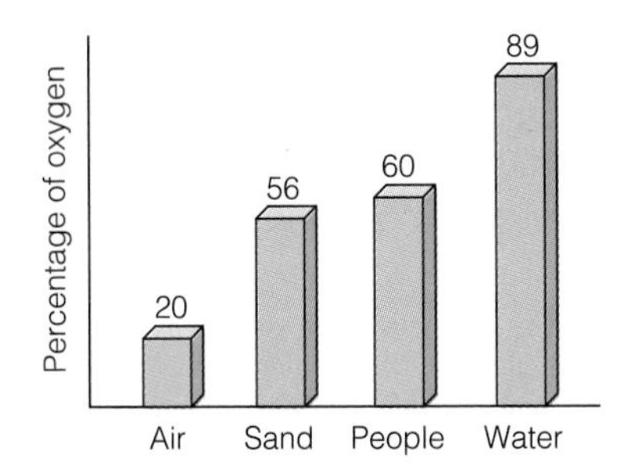

Huge quantities of liquid oxygen are used to burn the fuels that blast rockets into orbit.

Pure oxygen is obtained by liquefying air and then letting the nitrogen and argon boil off. (Nitrogen boils at −196 °C, argon at −186 °C, and oxygen at −183 °C.) About 15 billion kg of oxygen of 99.5% purity is produced annually in the United States. Most of this oxygen is used directly by industry, much of it in the metals industry. About 1% is compressed into steel tanks for use in welding and medicine and for many other purposes.

Many compounds in our bodies contain oxygen. About 60% of the weight of each one of us is oxygen. As we breathe, oxygen (in air) is taken into the lungs. From there it passes into the bloodstream, which delivers the oxygen to each cell for use in the metabolism of molecules of food we eat. Oxygen is required for the chemical processes that occur during metabolism. Heat energy released during metabolism is used to maintain body temperature. Metabolism also gives us the energy we need for mental and physical activity. Our very existence is dependent upon oxygen and oxidation.

Most nonmetallic elements react with oxygen to form *nonmetal oxides*. For example, carbon—present in coal, coke, and charcoal—reacts with oxygen during combustion to form carbon dioxide.

$$C + O_2(g) \longrightarrow CO_2(g)$$

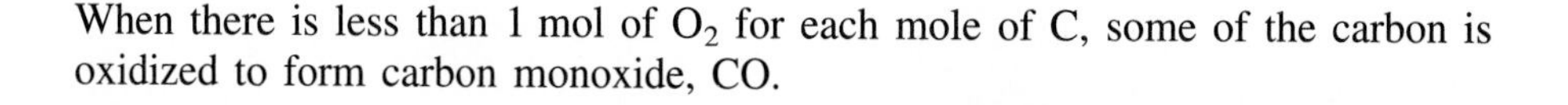

When there is less than 1 mol of O_2 for each mole of C, some of the carbon is oxidized to form carbon monoxide, CO.

$$2\,C + O_2(g) \longrightarrow 2\,CO(g)$$

Sulfur dioxide is formed when sulfur burns in oxygen. In the presence of an appropriate catalyst (Pt or V_2O_5), sulfur dioxide can be further oxidized rather rapidly to form sulfur trioxide.

$$2\ SO_2(g) + O_2(g) \xrightarrow{\text{catalyst}} 2\ SO_3(g)$$

At high temperatures, such as those that occur in automobile engines, oxygen combines with some of the ordinarily quite unreactive nitrogen gas from the air to form various oxides of nitrogen. For example,

$$N_2(g) + O_2(g) \longrightarrow 2\,NO(g)$$

The product of the reaction shown here is nitrogen monoxide, NO (also called nitric oxide).

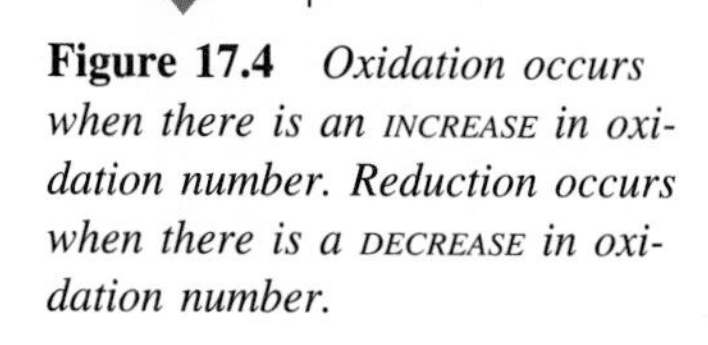

Figure 17.4 *Oxidation occurs when there is an* INCREASE *in oxidation number. Reduction occurs when there is a* DECREASE *in oxidation number.*

Besides reacting with metallic and nonmetallic elements, oxygen also reacts with many compounds. Methane, CH_4, the main component of natural gas, burns in air to give carbon dioxide and water. Stored energy is released as heat.

$$CH_4(g) + 2\,O_2(g) \longrightarrow CO_2(g) + 2\,H_2O(g) + \text{Heat}$$

Hydrogen sulfide, H_2S, a gaseous compound with a rotten egg odor, burns in oxygen to produce water and sulfur dioxide.

$$2\,H_2S(g) + 3\,O_2(g) \longrightarrow 2\,H_2O(g) + 2\,SO_2(g)$$

Notice that as these compounds are oxidized, oxygen combines with each kind of atom within the compound.

Originally, the term *oxidation* was restricted to reactions involving combination with oxygen, but chemists learned that combination with chlorine (or bromine or other reactive nonmetals) was not all that different from reaction with oxygen. In each case there is a change in the oxidation number. Thus, the concept of oxidation was broadened. We can define oxidation and reduction in terms of oxidation number (Figure 17.4) as follows.

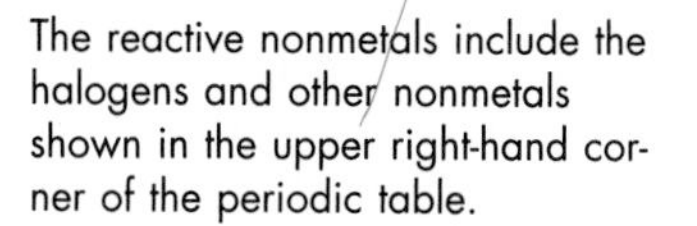

The reactive nonmetals include the halogens and other nonmetals shown in the upper right-hand corner of the periodic table.

Oxidation occurs when there is an *increase* in oxidation number.
Reduction occurs when there is a *decrease* in oxidation number.

Besides looking at the change in oxidation number, there are several other ways to determine quickly where oxidation—or reduction—processes occur.

More Oxidation Definitions

1. An *element* or *compound* is oxidized if it gains oxygen atoms. (In reactions described in this section, iron, carbon, nitrogen, sulfur dioxide, methane, and hydrogen sulfide gain oxygen atoms while being oxidized.)

2. A *compound* is oxidized if it loses hydrogen atoms. When methanol, CH_3OH (also called methyl alcohol), is passed over hot copper gauze, formaldehyde and hydrogen gas are produced.

$$\underset{\text{Methanol}}{CH_3OH} \xrightarrow{Cu} \underset{\text{Formaldehyde}}{CH_2O} + H_2$$

Here methanol loses hydrogen atoms; it is oxidized.

3. An *atom* or *ion* of an element is oxidized if it loses electrons. When magnesium metal (oxidation number of 0) reacts with chlorine gas, magnesium ions and chloride ions are formed.

$$Mg + Cl_2 \longrightarrow Mg^{2+} + 2\,Cl^-$$

Why so many definitions of oxidation? Simply for convenience. Which do we use? Whichever is most convenient. For the conversion of methanol into formaldehyde and hydrogen, we could assign oxidation numbers and determine that carbon is oxidized (it goes from -2 to 0, an *increase* of 2), but it is easier to see that the carbon loses hydrogen atoms; the carbon is oxidized. The four systems described are not in conflict. We simply use the system that is the most convenient in determining where oxidation is taking place.

We will now use the rules for assigning oxidation numbers (Section 17.1) to determine what is oxidized and reduced in the following example.

EXAMPLE 17.3

With the addition of a few drops of water as a catalyst, aluminum dust reacts spontaneously with pulverized iodine crystals to produce AlI_3.

(a) Write a balanced equation for the reaction.
(b) List the oxidation numbers of aluminum and iodine as reactants and as products.
(c) What is oxidized? What is reduced?

SOLUTION

(a) The balanced equation is

$$2\ Al + 3\ I_2 \xrightarrow{H_2O} 2\ AlI_3$$

(b) Oxidation numbers of reactants are

$$Al = 0 \qquad \text{and} \qquad I = 0 \qquad \text{(Both are uncombined.)}$$

Oxidation numbers in product are

$$Al = +3 \qquad \text{and} \qquad I = -1$$

(c) Aluminum is oxidized. (It undergoes an increase in oxidation number.) Iodine is reduced. (It undergoes a decrease in oxidation number.)

17.3 Chemical Properties of Hydrogen: Reduction

Hydrogen gas quietly burns in air with an almost colorless flame. If hydrogen gas and oxygen gas are mixed at room temperature, there is no perceptible reaction, but if this mixture is ignited by a spark, a tremendous explosion occurs. The product of the reaction, in both cases, is water.

$$2\ H_2 + O_2 \xrightarrow{\text{spark}} 2\ H_2O$$

If a piece of platinum metal is inserted into a container of hydrogen and oxygen, the two gases react at room temperature, even without a spark. The platinum acts as a catalyst: it lowers the activation energy (Section 15.5) for the reaction. The platinum glows from the heat evolved in the initial reaction and then ignites the mixture, causing an explosion.

Hydrogen gas reacts with a variety of metal oxides to remove oxygen and give the free metal. For example, when hydrogen gas is passed over heated copper(II) oxide, metallic copper and water are formed.

$$CuO + H_2(g) \longrightarrow Cu + H_2O$$

With lead(II) oxide, the products are metallic lead and water.

$$PbO + H_2(g) \longrightarrow Pb + H_2O$$

In both reactions, a metal oxide is acted upon to release a free, uncombined metal. The metal oxide is said to be **reduced**; the process is called **reduction**.

Oxidation once simply meant the combination of an element with oxygen, and reduction meant that a compound is changed, or reduced, to an element, but both terms now have broader meanings. Reduction and oxidation always occur simultaneously, but the following four changes are characteristic of reduction.

Reduction Definitions

1. Reduction occurs when there is a decrease in oxidation number. (This is the opposite of oxidation, as described in the previous section.) In the chemical equations shown in the last paragraph
   a. Copper has a +2 oxidation number in copper(II) oxide, CuO, but Cu metal has an oxidation number of 0. Here, copper(II) is *reduced.*
   b. Lead has a +2 oxidation number in lead(II) oxide, PbO, but Pb metal has an oxidation number of 0. Here lead(II) is *reduced.*
2. A *compound* is reduced if it loses oxygen atoms. In the last two reactions for the reduction of the metal oxides—copper(II) oxide and lead(II) oxide—oxygen atoms were lost by the metal oxides. As another example, chlorates release oxygen when heated.

$$2\ KClO_3 \xrightarrow{\text{heat}} 2\ KCl + 3\ O_2$$

The potassium chlorate loses oxygen atoms; it is reduced.

A CLOSER LOOK

Hydrogen: Abundance, Preparation, and Properties

Hydrogen is by far the most abundant element in the universe, but it ranks much lower in abundance on Earth. By mass, hydrogen makes up only about 0.9% of Earth's crust, placing it far down on the list of abundant elements. In terms of number of atoms, however, hydrogen is quite abundant. In a random sample of atoms from Earth's crust there are

5330 oxygen atoms in every 10,000 atoms
1590 silicon atoms in every 10,000 atoms
1510 hydrogen atoms in every 10,000 atoms

Most of the hydrogen on Earth is found combined with oxygen, as water. Nearly all compounds derived from living organisms contain hydrogen. Fats, starches, sugars, and proteins all contain combined hydrogen. Petroleum and natural gas also contain mixtures of hydrocarbons (compounds of hydrogen and carbon).

Since "free," uncombined, elemental hydrogen does not occur to any extent in nature on this planet, we must obtain it from hydrogen-containing compounds. In the laboratory, hydrogen gas can be prepared by the reaction of an acid with a moderately reactive metal such as zinc.

$$Zn(s) + H_2SO_4 \rightarrow ZnSO_4 + H_2(g)$$

The hydrogen is collected by displacement of water; that is, the gas is bubbled into a bottle of water turned upside down in a larger container of water. The gas is allowed to flow into the bottle until all the water is displaced.

Commercially, most hydrogen is obtained as a by-product of other processes. Much of it comes from petroleum refineries.

Hydrogen is a colorless and odorless gas that is essentially insoluble in water. Hydrogen is the least dense of all substances: its density is only about 7% that of air under comparable conditions. For this reason it was at one time used in dirigibles to give them the necessary buoyancy in air. However, since a spark or flame can trigger a hydrogen explosion or fire, the use of this gas represented a considerable danger. When a disastrous fire occurred aboard a luxury airship called the *Hindenburg* in 1937, the use of hydrogen was discontinued, as was the use of lighter-than-air ships. Modern blimps (smaller nonrigid dirigibles) use nonreactive helium gas, which is twice as dense as hydrogen but much less dense than air.

Hydrogen is the most effective buoyant gas, but it is highly flammable. The disastrous fire in the Hindenburg, *a hydrogen-filled dirigible, in 1937 led to the replacement of hydrogen by nonflammable helium.*

Certain metals, such as platinum, Pt, palladium, Pd, and nickel, Ni, can collect hydrogen on their surfaces. This *adsorbed* (not ab̲sorbed) hydrogen is more reactive than ordinary molecular hydrogen, so these metals are widely used as catalysts for reactions that require hydrogen gas as one of the reactants.

3. A *compound* is reduced if it gains hydrogen atoms. For example, methyl alcohol, CH_3OH, can be produced by the reaction of carbon monoxide with hydrogen gas in the presence of an appropriate catalyst.

$$CO + 2\ H_2 \xrightarrow{\text{catalyst}} CH_3OH$$

The carbon monoxide gains hydrogen atoms; it is reduced.

4. An *atom* or *ion* of an element is reduced if it gains electrons. When an electric current is passed through a solution containing copper(II) ions, Cu^{2+}, copper metal is plated out at the electrode called the cathode.

The letters in **OIL RIG** may help you remember that, for electrons,
Oxidation **I**s a **L**oss
Reduction **I**s a **G**ain

$$Cu^{2+} + 2\ e^- \longrightarrow Cu$$

As the copper ions gain electrons, they are reduced to copper metal.

These definitions of oxidation and reduction are summarized in Figure 17.5.

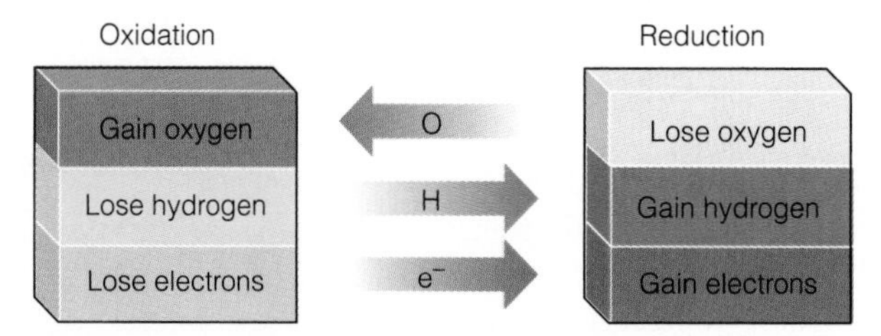

Figure 17.5 *Three different definitions of oxidation and reduction.*

Oxidation and reduction go hand in hand. You can't have one without the other. When one substance is oxidized, another is reduced. For the reaction

$$CuO + H_2 \longrightarrow Cu + H_2O$$

the oxidation number of copper changes from +2 to 0; copper in CuO is reduced. Also, the oxidation number of hydrogen changes from 0 to +1; hydrogen is oxidized. Furthermore, if one substance is being oxidized, the other must be causing it to be oxidized. In the example, hydrogen is oxidized by the copper(II) oxide. The copper(II) oxide is called the **oxidizing agent**. Conversely, copper(II) oxide is reduced by the hydrogen gas, so the hydrogen gas is the **reducing agent**. Among the reactants of each oxidation–reduction (redox) reaction are both an oxidizing agent and a reducing agent.

The *oxidizing agent* is the substance that is *reduced.*
The *reducing agent* is the substance that is *oxidized.*

The oxidizing and reducing agents along with the substances being oxidized and reduced are identified for the following equation.

Reduction: Copper in CuO is reduced to Cu metal.
CuO is the oxidizing agent.

+2 decreases to 0

$$CuO + H_2 \longrightarrow Cu + H_2O$$

0 increases to +1

Oxidation: Each H in H_2 is oxidized to form H_2O.
H_2 is the reducing agent.

Follow through with the following example.

EXAMPLE 17.4

Identify the oxidizing agent as OA and the reducing agent as RA in the following reactions.

(a) $C + O_2 \rightarrow CO_2$ (b) $N_2 + 3\,H_2 \rightarrow 2\,NH_3$
(c) $NO + O_3 \rightarrow NO_2 + O_2$ (d) $Mg + Cl_2 \rightarrow Mg^{2+} + 2\,Cl^-$

SOLUTION

To determine the oxidizing agent (OA) and reducing agent (RA) use the definitions of oxidation (Section 17.2) and reduction (Section 17.3).

(a) Carbon gains oxygen and is oxidized (oxidation definition 1, Section 17.2), so it must be the reducing agent, RA. O_2 is the oxidizing agent, OA.
(b) N_2 gains hydrogen and is reduced (reduction definition 3, Section 17.3), so it must be the oxidizing agent, OA. H_2 is the reducing agent, RA.
(c) NO gains oxygen and is oxidized (oxidation definition 1, Section 17.2), so it must be the reducing agent, RA. O_3 is the oxidizing agent, OA.
(d) Mg loses electrons and is oxidized (oxidation definition 3, Section 17.2), so it is the reducing agent, RA. Cl_2 gains electrons and is reduced (reduction definition 4, Section 17.3), so it is the oxidizing agent, OA.

See Problems 17.5–17.24.

EXERCISE 17.3

Identify the oxidizing agent as OA and the reducing agent as RA for the reaction

$$2\,Al + 3\,Br_2 \longrightarrow 2\,AlBr_3$$

17.4 *Some Important Oxidizing Agents*

Oxygen is undoubtedly the most common oxidizing agent. It quickly oxidizes the wood burned in our campfires and the gasoline in our automobile engines. During rusting and corrosion, oxygen slowly combines with metals that were obtained by reducing ores. Oxygen is also involved in the rotting of wood and food products. Even when the cells of our bodies "burn" or metabolize the foods we eat, oxygen

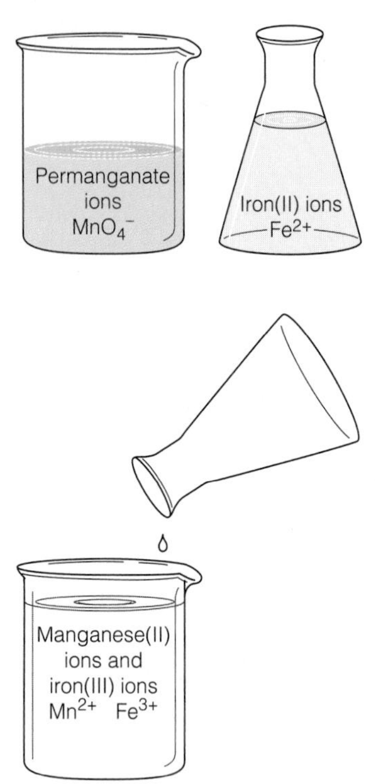

Figure 17.6 *Purple permanganate ions, MnO_4^-, are reduced to Mn^{2+} by iron(II) ions, Fe^{2+}, in acid solutions. The purple color disappears as the reaction proceeds.*

is consumed. We live in an oxidizing atmosphere. Cut an apple and watch it slowly turn brown. Some of the reactions involved are due to air oxidation. Fortunately, oxygen gas is a mild oxidizing agent. A piece of wood or even a container of gasoline can be in open air without catching fire until ignited.

Oxygen is often used industrially and in the laboratory as an oxidizing agent. For example, acetylene (used for cutting and welding torches) can be made by the *partial* oxidation of methane.

$$\underset{\text{Methane}}{4\,CH_4} + 3\,O_2 \longrightarrow \underset{\text{Acetylene}}{2\,C_2H_2} + 6\,H_2O$$

If the methane were completely oxidized (burned), the reaction would produce carbon dioxide and water. For many laboratory and industrial processes it is often more convenient to use other oxidizing agents such as potassium permanganate, $KMnO_4$, and sodium dichromate, $Na_2Cr_2O_7$.

Potassium permanganate is a black, shiny, crystalline solid. It dissolves in water to give deep purple solutions. This purple color disappears as the permanganate is reduced. (Remember, if permanganate is an oxidizing agent, it must be reduced.) Permanganate is often used to determine the quantity of oxidizable substances in a sample. For example, potassium permanganate can be used to determine the amount of iron(II) ion in a sample. When a purple permanganate solution is slowly added to a sample containing iron(II) ions, the purple color disappears as the permanganate is reduced (Figure 17.6). Once all the iron(II) has been oxidized, the excess purple permanganate becomes instantly visible.

During the determination of the quantity of Fe^{2+} present in a sample, a permanganate solution of known concentration is added dropwise to the solution containing iron(II) ions until the purple permanganate color ceases to disappear. At this point—the *equivalence* point—the quantities of Fe^{2+} ions and MnO_4^- ions are present in the proportions shown by the net ionic equation.

$$\underset{\substack{\text{Permanganate ion}\\\text{(purple)}}}{MnO_4^-} + 5\,Fe^{2+} + 8\,H^+ \longrightarrow \underset{\substack{\text{Manganese(II) ion}\\\text{(colorless)}}}{Mn^{2+}} + 5\,Fe^{3+} + 4\,H_2O$$

Permanganate solutions can also be used to oxidize oxalic acid (a poisonous compound found in rhubarb), sulfur dioxide, SO_2, and many other compounds.

Sodium dichromate, a strong oxidizing agent, can be used to oxidize alcohols to compounds called aldehydes and ketones. For the oxidation of ethyl alcohol (present in alcoholic beverages) to acetaldehyde, the equation is

$$8\,H^+ + \underset{\substack{\text{Dichromate ion}\\\text{(orange-red)}}}{Cr_2O_7^{2-}} + 3\,C_2H_5OH \longrightarrow 2\,Cr^{3+} + \underset{\substack{\text{Chromium(III) ion}\\\text{(green)}}}{3\,C_2H_4O} + 7\,H_2O$$

Some tests for intoxication—such as the Breathalyzer, used to check for drunken drivers—depend on a color change that occurs during oxidation–reduction.

CHEMISTRY AT WORK

Hydrogen Peroxide: A Household and Industrial Oxidizing Agent

Hydrogen peroxide, H_2O_2, is an important oxidizing agent used in the home and in industry. During reactions, it is reduced to products that are environmentally safe. Aqueous solutions of 3% hydrogen peroxide are often used as a topical antiseptic for minor cuts and wounds, and as a bleach. Commercially prepared hair bleaches also contain hydrogen peroxide.

Hydrogen peroxide decomposes to give water and oxygen gas. The decomposition reaction can be catalyzed by platinum metal, by manganese dioxide, MnO_2, by iodide ions, I^-, and even by substances in blood.

$$2\,H_2O_2(aq) \xrightarrow{\text{catalyst}} 2\,H_2O + O_2(g)$$

If you use hydrogen peroxide to clean a wound, you will see it foam as it decomposes. This is because hemoglobin in blood catalyzes the vigorous decomposition.

Hundreds of thousands of tons of hydrogen peroxide are produced industrially each year. It is used to bleach paper pulp, textiles, flour, leather, and hair. It is also used in the manufacture of other chemicals used in polymers, pharmaceuticals, and numerous other products.

Pure hydrogen peroxide is an unstable liquid that has a density of 1.47 g/cm^3 at 0 °C. The decomposition of hydrogen peroxide is used to supply oxygen for certain aircraft, for spacecraft attitude control, and for rockets. The stability of hydrogen peroxide solutions varies with concentration. The table below lists concentrations of hydrogen peroxide used for various purposes.

The F-104 plane is fueled by kerosene and hydrogen peroxide. The upper stage of the Saturn rocket uses H_2O_2. World War II German V-1 rockets, called buzz bombs, used H_2O_2 to propel the release of bombs.

The upper stage of a Saturn rocket used to launch this Apollo 17 uses hydrogen peroxide.

Uses of Hydrogen Peroxide Solutions of Various Concentrations

H_2O_2 Concentration	*Uses*
3%	Antiseptic
6%	Hair bleach
30%	Laboratory and industrial oxidizing agent
85% or greater	Strong oxidizing agent; rocket fuel oxidizers

EXAMPLE 17.5

Determine the oxidation number of manganese in the MnO_4^- ion and of chromium in the $Cr_2O_7^{2-}$ ion. For the two preceding reactions, what is oxidized? What is reduced? What are the oxidizing agents? What are the reducing agents?

SOLUTION

STEP 1 Determine the oxidation number of each element that changes charge from reactant to product. Show the oxidation numbers on lines drawn to connect species being oxidized and reduced.

STEP 2 Determine which elements are oxidized (they gain in oxidation number) and which are reduced (they lose in oxidation number).

STEP 3 Identify the oxidizing agents and the reducing agents.

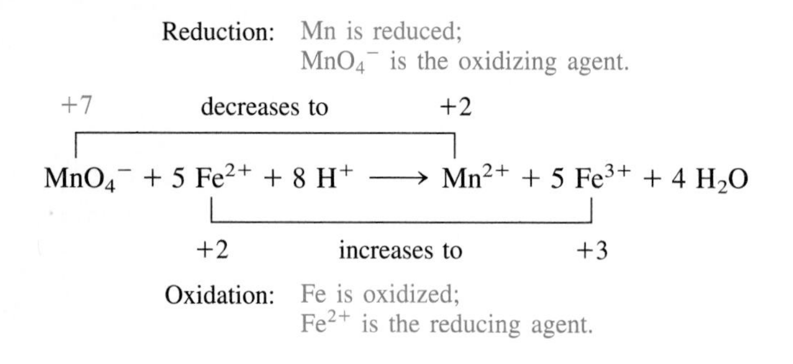

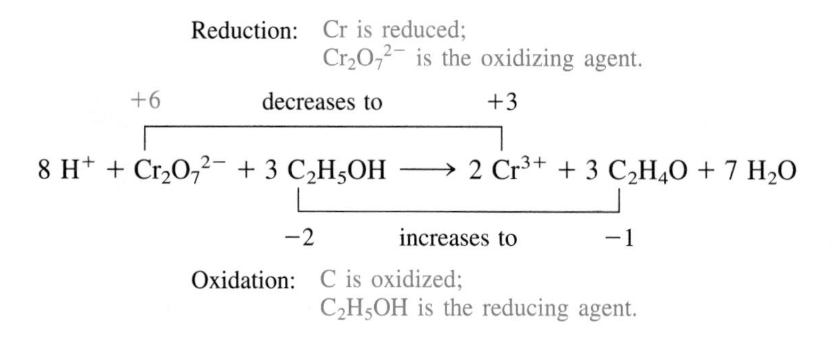

The halogens are important oxidizing agents. For example, chlorine, Cl_2, oxidizes magnesium metal to give magnesium ions.

$$Mg + Cl_2 \longrightarrow Mg^{2+} + 2\ Cl^-$$

In the process, chlorine is reduced to chloride ions, Cl^-, and the oxidation number of chlorine is reduced from 0 (in Cl_2) to −1 (in Cl^-).

Nitric acid, HNO_3, is another important oxidizing agent. Copper is below hydrogen in the activity series (Section 10.8) and does not replace hydrogen from acids, but metallic copper will dissolve in nitric acid because it is oxidized to copper(II) ions, Cu^{2+}. The reaction with concentrated HNO_3 is

$$Cu + 4\,HNO_3 \longrightarrow Cu(NO_3)_2 + 2\,NO_2(g) + 2\,H_2O$$

As copper is oxidized, half of the nitrate ions are reduced to NO_2 gas. The other half of the nitrate ions remain unchanged. (See the photograph on page 510.)

CHEMISTRY IN OUR WORLD

Antiseptics and Disinfectants: Oxidizing Agents and Health

Many common antiseptics (compounds that are applied to living tissue to kill or prevent the growth of microorganisms) are mild oxidizing agents. A 3% aqueous solution of hydrogen peroxide can be used as a topical antiseptic to treat minor cuts and abrasions.

Benzoyl peroxide, $(C_6H_5CO)_2O_2$, is a powerful oxidizing agent used at 5% and 10% concentrations in ointments for treating acne. When used on areas exposed to sunlight, benzoyl peroxide is thought to promote skin cancer, so its use is discouraged.

Solutions of iodine or iodine-releasing compounds are frequently used as antiseptics. In surgery, the area around the incision is usually disinfected with an iodine-containing solution. Iodine solutions discolor the skin, so it is easy to see if any of the critical area has been missed.

Sodium hypochlorite, NaClO, available in aqueous solutions as laundry bleach (Purex, Clorox, and the like), is also used as a disinfectant and deodorizer. Calcium hypochlorite, $Ca(ClO)_2$, is used in bleaching powder and in hospitals to disinfect clothing and bedding. Elemental chlorine, Cl_2, can be used to kill pathogenic (disease-causing) microorganisms in drinking water and to treat wastewater before it is returned to a stream or lake.

Chlorine treatment processes have been quite effective in preventing the spread of infectious diseases such as typhoid fever. Use of chlorine in this manner came under criticism in 1974, when it was shown that chlorine can react with organic compounds (presumably from industrial wastes) to form toxic chlorinated compounds that remain in the water. Among the alternatives is chlorine dioxide, ClO_2, which is gaining in support.

Ozone, O_3, has also been used to disinfect drinking water. Many European cities, including Paris and Moscow, use ozone to treat their drinking water. Ozone is more expensive than chlorine, but less of it is needed. An added advantage is that ozone kills viruses on which chlorine has little, if any, effect. Russian tests have shown ozone to be a hundred times more effective for killing polio virus. Ozone imparts no "chemical taste" to the water. Ozone acts by oxidizing the contaminant. However, on the negative side, that action quickly diminishes as it readily decomposes while being carried by water pipes.

Iodine-containing Betadine is a topical antiseptic and disinfectant used in hospitals to scrub skin tissue prior to surgery.

Chemistry in Our World

Bleaching and Stain Removal: Oxidation at Work

Bleaches are compounds that are used to remove unwanted color from white fabrics. Nearly any oxidizing agent will do the job. However, some also harm the fabric, some are unsafe, some produce undesirable by-products, and some are simply too expensive.

The most familiar laundry bleaches are aqueous solutions of sodium hypochlorite, NaClO. These formulations generally have 5.25% NaOCl, yet they vary widely in price, partially due to advertising costs for familiar brands.

Bleaching powder, containing calcium hypochlorite, $Ca(ClO)_2$, is generally preferred for large-scale bleaching operations. The paper industry uses it to make white paper, and the textile industry uses it to make whiter fabrics.

The active agent in common bleaches and many bleaching powders is the hypochlorite ion, OCl^-.

In aqueous bleaches and in bleaching powder, the active agent is the hypochlorite ion, ClO^-. Materials appear colored because loosely bound electrons are boosted to higher energy levels by absorption of visible light (Section 5.3). Bleaching agents do their work by removing or tying down these mobile electrons. For example, aqueous hypochlorite ions take up electrons from the substance being bleached to form chloride ions and hydroxide ions.

$$ClO^- + H_2O + 2\,e^- \rightarrow Cl^- + 2\,OH^-$$

Hypochlorite bleaches are safe and effective for cotton and linen fabrics, but they should not be used for wool, silk, or nylon.

Other bleaching agents include hydrogen peroxide, sodium perborate, $NaBO_2 \cdot H_2O_2$, and certain organic compounds that release chlorine molecules in water. When hydrogen peroxide is used to bleach hair, it acts as an oxidizing agent. In the process, the dark-colored pigment (called melanin) is oxidized to colorless products.

Stain removal is not nearly so simple a process as bleaching. A few stain removers are oxidizing agents or reducing agents; others operate by different processes. Nearly all stains require rather specific stain removers. Hydrogen peroxide in cold water can be used to remove bloodstains from cotton and linen. Oxalic acid can be used to remove rust spots, but not by redox; it bonds to and ties up the otherwise insoluble rust to form what is called a *complex*. This complex is carried in solution and washed away.

Many stain removers are simply absorbents (e.g., cornstarch, which removes grease spots), solvents (e.g., amyl acetate, which can remove ballpoint pen ink), or detergents (which can remove grease stains).

17.5 Some Important Reducing Agents

There is no single reducing agent that stands out as oxygen does among oxidizing agents. Hydrogen gas will reduce many compounds, but it is relatively expensive. Elemental carbon in the form of coke (obtained by driving off the volatile matter

CHEMISTRY AT WORK

Photography: The Reduction of Silver

Reducing agents are essential to photographic film development processes. The photographic film has a light-sensitive coating—called a photographic emulsion—that contains a silver halide such as silver bromide, AgBr, suspended in gelatin and applied to a plastic backing. When exposed to light, a small fraction of the AgBr becomes activated, depending on the intensity of the light.

During the film *developing* process, the activated silver bromide reacts (in an oxidation–reduction reaction) with the *developer*, a reducing agent such as the organic compound hydroquinone, to form black, metallic silver.

$$\underset{\text{Hydroquinone}}{C_6H_4(OH)_2} + 2\,Ag^+ \rightarrow C_6H_4O_2 + \underset{\text{Silver metal}}{2\,Ag(s)} + 2\,H^+$$

Only those silver ions that were exposed to light are reduced by the developer.

In the next step, called *fixing*, the film is treated with ''photographer's hypo,'' a solution of sodium thiosulfate, $Na_2S_2O_3$. The hypo solution is used to dissolve and wash away unexposed silver bromide. In this step, the hypo forms a soluble complex with the unexposed and insoluble silver bromide. If unexposed AgBr crystals were not removed by this process, they would slowly be reduced by light, and the *negative* would turn black.

After washing, the negative is left with the greatest amount of metallic silver in the areas that were exposed to the greatest intensity of light. Regions of the negative that were not exposed to light are left transparent after washing. The *positive* print is then made from the negative by shining light through the negative onto light-sensitive photographic paper.

(a)

(b)

A photographic negative (a) and a positive print (b).

from coal) is much cheaper and is frequently used as a reducing agent to release metals from their ores. For example, tin can be obtained from tin(IV) oxide by reduction, with carbon used as the reducing agent.

$$SnO_2 + C \longrightarrow Sn + CO_2$$

Hydrogen may be used for the production of expensive metals such as tungsten, W. The ore is first converted to an oxide, WO_3, and then is reduced in a stream of hydrogen gas at 1200 °C.

$$WO_3 + 3\,H_2 \longrightarrow W + 3\,H_2O$$

Carbon monoxide is one of the reducing agents used by the metallurgical industry to reduce metal oxide ores to free metals such as iron and copper.

$$FeO + CO(g) \longrightarrow Fe + CO_2(g)$$

$$CuO + CO(g) \longrightarrow Cu + CO_2(g)$$

These reactions must be carried out at high temperatures.

EXAMPLE 17.6

For the reduction of FeO to iron by carbon monoxide,

$$FeO + CO(g) \longrightarrow Fe + CO_2(g)$$

give the oxidation states of iron in FeO and Fe metal and of carbon in CO and in CO_2. Identify the substances that are oxidized and reduced. Also give the oxidizing agent and the reducing agent.

SOLUTION

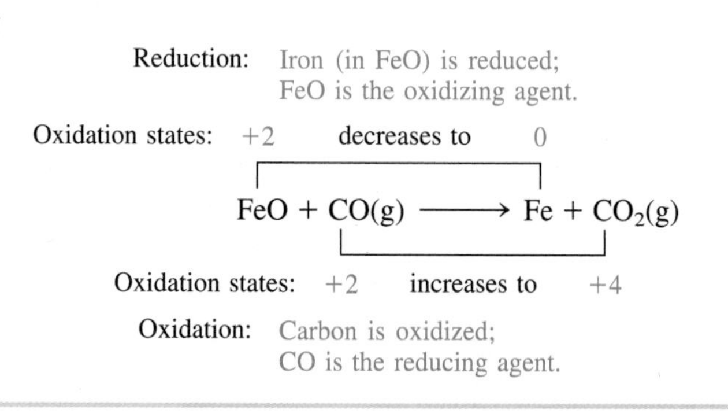

See Problems 17.25 and 17.26.

17.6 Oxidation and Reduction Half-reactions

Additional steps needed to balance redox equations for basic (alkaline) solutions are not included in this textbook.

In each redox reaction, one species is reduced. We can write what is called a **reduction half-reaction** equation to represent this part of the reaction. Similarly, we can write a separate **oxidation half-reaction** equation for the species that undergoes oxidation. One excellent method of balancing equations for redox reactions is based on the use of these half-reaction equations. The following steps should be used in balancing redox equations that occur in acidic solutions.

Balancing Redox Equations

Thinking It Through

Let us use half-reaction equations to balance the equation for the redox reaction of MnO_4^- with Fe^{2+} to produce Mn^{2+} and Fe^{3+}.

1. Write two separate equations, one using only the substances that are involved in the reduction, and another using only the substances involved in oxidation. (When balanced, these are called *half-reaction equations*.)

$$MnO_4^- \longrightarrow Mn^{2+}$$
$$Fe^{2+} \longrightarrow Fe^{3+}$$

2. Balance each kind of atom other than H and O by using coefficients. (They are already balanced in these equations.)
3. Balance O atoms by using H_2O.

$$MnO_4^- \longrightarrow Mn^{2+} + 4\,H_2O$$
$$Fe^{2+} \longrightarrow Fe^{3+}$$

4. Balance H atoms by using H^+ ions.

$$8\,H^+ + MnO_4^- \longrightarrow Mn^{2+} + 4\,H_2O$$
$$Fe^{2+} \longrightarrow Fe^{3+}$$

5. Use electrons as needed to obtain a charge that is balanced (not necessarily neutral). The reduction half-reaction shows a gain of electrons. The oxidation half-reaction shows a loss of electrons.

$$5\,e^- + 8\,H^+ + MnO_4^- \longrightarrow Mn^{2+} + 4\,H_2O$$
$$Fe^{2+} \longrightarrow Fe^{3+} + e^-$$

6. Multiply the half-reactions by the simplest set of whole numbers (1 and 5 here) so that electrons gained equal electrons lost. Then add the half-reactions together.

$$5\,e^- + 8\,H^+ + MnO_4^- \longrightarrow Mn^{2+} + 4\,H_2O$$
$$5\,Fe^{2+} \longrightarrow 5\,Fe^{3+} + 5\,e^-$$
$$\overline{5\,e^- + MnO_4^- + 5\,Fe^{2+} + 8\,H^+ \longrightarrow Mn^{2+} + 5\,Fe^{3+} + 4\,H_2O + 5\,e^-}$$

7. Cancel electrons and equal amounts of any substance that appears on both sides of the equation. The balanced equation is

$$MnO_4^- + 5\,Fe^{2+} + 8\,H^+ \longrightarrow Mn^{2+} + 5\,Fe^{3+} + 4\,H_2O$$

Check to make sure all atoms and charges are balanced, and to see that all substances are in simplest integer (whole-number) ratios.

The half-reaction equation method of balancing redox equations is used in the following example.

EXAMPLE 17.7

When a solution containing iron(II) ions is added to an orange dichromate solution, a green color appears, indicating that chromium(III) ions have formed. Iron(III) ions were also detected in the final solution. Write a balanced equation to represent this reaction.

$$Cr_2O_7^{2-} + Fe^{2+} \longrightarrow Cr^{3+} + Fe^{3+} \qquad \textit{(unbalanced)}$$

SOLUTION

STEP 1 Write half-reactions.

$$Cr_2O_7^{2-} \longrightarrow Cr^{3+}$$
$$Fe^{2+} \longrightarrow Fe^{3+}$$

STEP 2 Balance Cr atoms.

$$Cr_2O_7^{2-} \longrightarrow 2\,Cr^{3+}$$
$$Fe^{2+} \longrightarrow Fe^{3+}$$

STEP 3 Balance O atoms by using 7 H_2O.

$$Cr_2O_7^{2-} \longrightarrow 2\,Cr^{3+} + 7\,H_2O$$
$$Fe^{2+} \longrightarrow Fe^{3+}$$

STEP 4 Balance H atoms by using 14 H^+.

$$14\,H^+ + Cr_2O_7^{2-} \longrightarrow 2\,Cr^{3+} + 7\,H_2O$$
$$Fe^{2+} \longrightarrow Fe^{3+}$$

STEP 5 Balance electrical charges by using 6 e^- and 1 e^-.

$$6\,e^- + 14\,H^+ + Cr_2O_7^{2-} \longrightarrow 2\,Cr^{3+} + 7\,H_2O$$
$$Fe^{2+} \longrightarrow Fe^{3+} + e^-$$

STEP 6 Multiply the second equation by 6 and add the half-reactions together.

$$6\,e^- + 14\,H^+ + Cr_2O_7^{2-} \longrightarrow 2\,Cr^{3+} + 7\,H_2O$$
$$6\,Fe^{2+} \longrightarrow 6\,Fe^{3+} + 6\,e^-$$
$$6\,e^- + 14\,H^+ + Cr_2O_7^{2-} + 6\,Fe^{2+} \longrightarrow 2\,Cr^{3+} + 6\,Fe^{3+} + 7\,H_2O + 6\,e^-$$

STEP 7 Cancel electrons to obtain the simplest balanced equation.

$$14\,H^+ + Cr_2O_7^{2-} + 6\,Fe^{2+} \longrightarrow 2\,Cr^{3+} + 6\,Fe^{3+} + 7\,H_2O$$

Check to make sure all atoms and charges are balanced and to see that all substances are in simplest integer ratios.

See Problems 17.27–17.32.

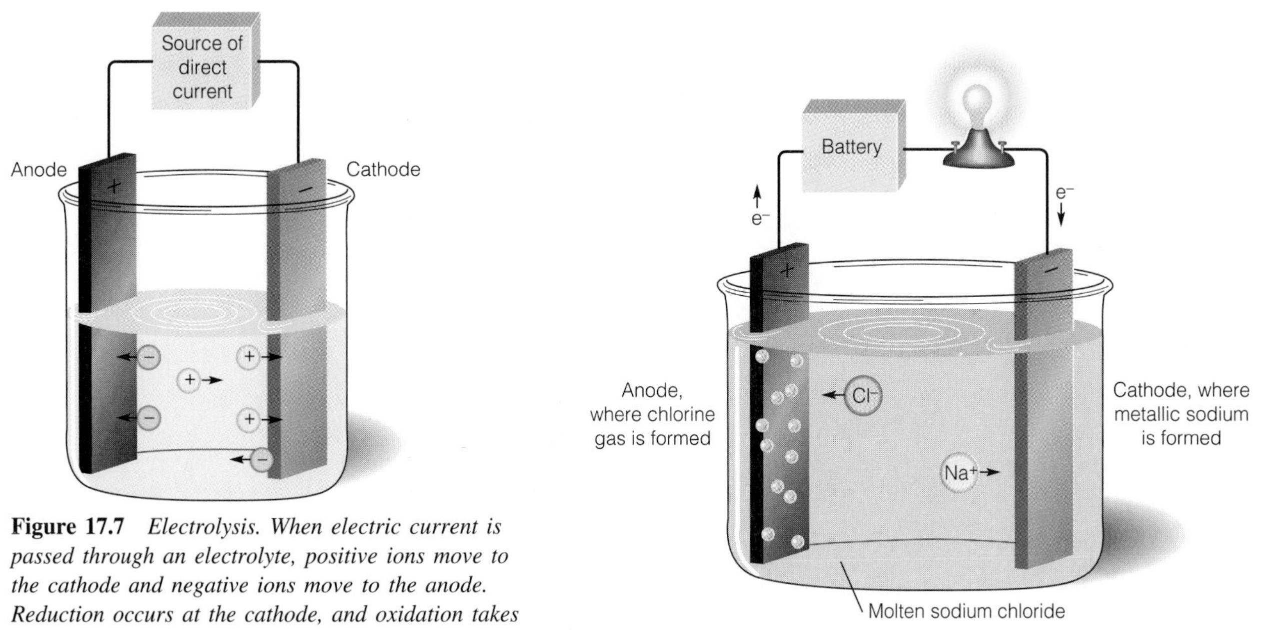

Figure 17.7 *Electrolysis. When electric current is passed through an electrolyte, positive ions move to the cathode and negative ions move to the anode. Reduction occurs at the cathode, and oxidation takes place at the anode.*

Figure 17.8 *The electrolysis of molten sodium chloride.*

17.7 Electrolytic Cells

When sodium chloride is melted, and when a direct current—not an alternating house current—is passed through the molten salt, there is an observable chemical reaction. Yellow-green chlorine gas forms at one of the electrodes, and silvery, metallic sodium is formed at the other electrode. The sodium that is formed is rapidly vaporized by the hot melt. During this reaction, molten sodium chloride is decomposed by the direct current into elemental sodium and chlorine.

$$2\,NaCl + Energy \longrightarrow 2\,Na + Cl_2(g)$$

The decomposition reaction that occurs when a direct current is passed through a compound is called an **electrolysis** reaction (Figure 17.7).

Electrolysis is a word of Greek origin that literally means "loosening by electricity." Michael Faraday was the first to use the term.

In crystalline form, sodium chloride does not conduct electricity. The ions occupy relatively fixed positions in the lattice and do not move very much, even when an electrical potential (a voltage) is applied. However, when the salt is melted, its ions move around freely. When a direct current from a battery or single cell is applied to the melt through a pair of electrodes, the sodium ions are attracted to the electron-rich cathode (Figure 17.8). Each sodium ion that picks up an electron is reduced to a sodium atom.

Battery: A grouping of two or more cells that are connected together to supply an electric current.

$$Na^+ + e^- \longrightarrow Na$$

Reduction

Recall that reduction occurs whenever there is a *gain* in electrons. By definition,

Reduction takes place at the electrode called the **cathode**.

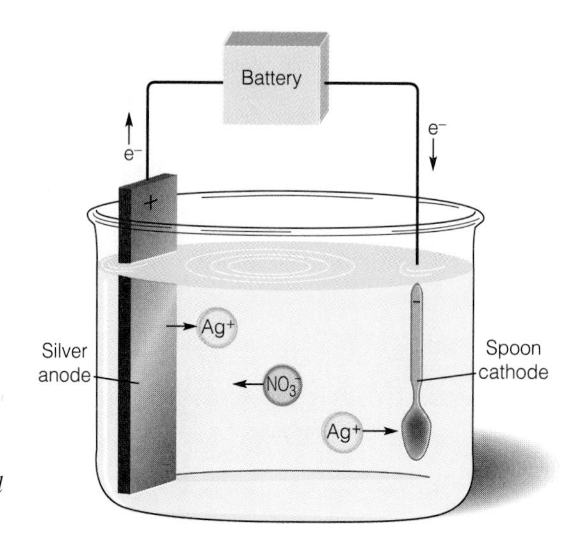

Figure 17.9 *An electrolytic call for the plating of silver.*

The chloride ions migrate to the electron-poor anode where they give up electrons and are oxidized to elemental chlorine, which forms diatomic molecules.

Oxidation

$$2\,Cl^- \longrightarrow Cl_2(g) + 2\,e^-$$

Recall that oxidation occurs whenever there is a *loss* in electrons. By definition,

Oxidation takes place at the electrode called the **anode**.

Electrons picked up at the anode move under the influence of the battery to the cathode, where they are transferred to sodium ions. The sodium ions are reduced to sodium metal.

All electrolytic processes involve redox reactions. In the NaCl electrolysis described here, the oxidation number of chlorine goes from -1 to 0 at the anode. At the cathode, the oxidation number of sodium is reduced from $+1$ to 0.

Electrolytic processes are used in the production of Al, Li, K, Na, and Mg and in the refining of Cu.

Electrolytic processes are used in the production and purification (refining) of several metals. Electrolysis is also used for coating one metal with another, an operation called **electroplating**. Usually the object to be electroplated, such as a fork or spoon, is cast of a cheaper metal. It is then coated with a thin layer of a more attractive and more corrosion-resistant metal, such as gold or silver. The cost of the finished product is far less than a product made entirely of silver or gold. A cell for the electroplating of silver is shown in Figure 17.9. A piece of silver metal is used as the anode, and the spoon or other item to be electroplated is used as the cathode. A solution of silver nitrate is used as the electrolyte. When the direct current is turned on, the silver ions, Ag^+, are attracted to the cathode (the spoon), where they pick up electrons and are deposited as silver atoms.

Cathode reaction: $Ag^+ + e^- \longrightarrow Ag$ (Silver is plated out.)

At the anode, electrons are removed from the silver metal bar. Some of the silver atoms lose electrons to become oxidized to silver ions.

Anode reaction: $Ag \longrightarrow Ag^+ + e^-$ (Silver ions are formed.)

The net process is one in which the silver from the bar is transferred to the spoon. The thickness of the deposit can be controlled by accurate measurement of the amount of current flow and the time needed for the electroplating process.

EXAMPLE 17.8

To protect steel sheet metal, nails, and other objects from rusting, they are often galvanized; that is, they are electroplated with a zinc coating in a cell similar to the one shown in Figure 17.9. Now, consider a metal bucket that is to be galvanized.

(a) Would a zinc metal rod be used as the anode or the cathode?
(b) Would the bucket be placed at the anode or the cathode?
(c) Write an equation for the anode reaction. What happens at the anode?
(d) Write an equation for the cathode reaction. What happens at the cathode?

SOLUTION

(a) The zinc rod is used as the anode. Zinc is oxidized to form Zn^{2+} ions.
(b) The bucket is placed at the cathode where Zn^{2+} ions are reduced to Zn metal, which is deposited as a layer on the bucket.
(c) The anode (oxidation) reaction is $Zn \rightarrow Zn^{2+} + 2\,e^-$. At the anode, some of the zinc atoms lose electrons to become zinc ions.
(d) The cathode (reduction) reaction is $Zn^{2+} + 2\,e^- \rightarrow Zn$. At the cathode, the metal bucket becomes coated with a layer of zinc.

EXERCISE 17.4

If a bracelet is to be gold-plated, at which electrode would it be attached? When gold atoms are deposited on the bracelet, is it oxidation or reduction?

17.8 Voltaic Cells

Electricity can cause chemical change (Section 17.7), but a chemical change can also produce electricity. That's what batteries and single electric cells are all about. Any system that either generates an electric current from a chemical reaction or uses an electric current to produce a chemical reaction is called an **electrochemical cell**. An electrochemical cell that uses electricity to drive a chemical reaction is called an **electrolytic cell**. An electrochemical cell that uses a spontaneous chemical reaction to generate an electric current is called either a **voltaic cell** (in honor of Alessandro Volta) or a **galvanic cell** (in honor of Luigi Galvani).

Alessandro Volta (1745–1827) invented the battery.

Luigi Galvani (1737–1798) discovered that electricity can cause muscles to contract.

If a strip of Zn metal is placed in a copper(II) sulfate solution, the Zn metal becomes coated with Cu metal. This reaction would be expected to occur spontaneously because zinc is more reactive than copper. (Zinc is above copper in the activity series of metals discussed in Section 10.8.) This is a spontaneous redox reaction; Zn metal is oxidized to Zn^{2+} ions by copper(II) ions, Cu^{2+}

$$\overbrace{Zn(s)}^{0 \text{ increases to } +2} + Cu^{2+}(aq) \longrightarrow Zn^{2+}(aq) + Cu(s)$$

(Zn: 0 increases to +2; Cu: +2 decreases to 0)

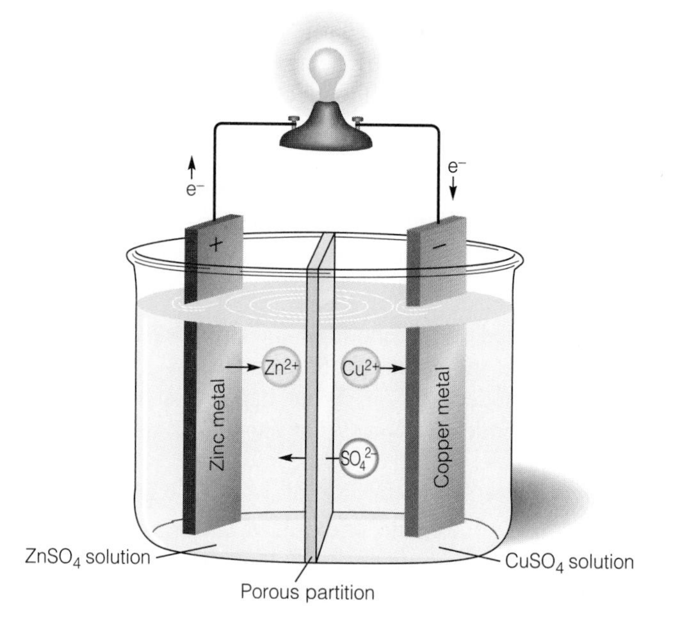

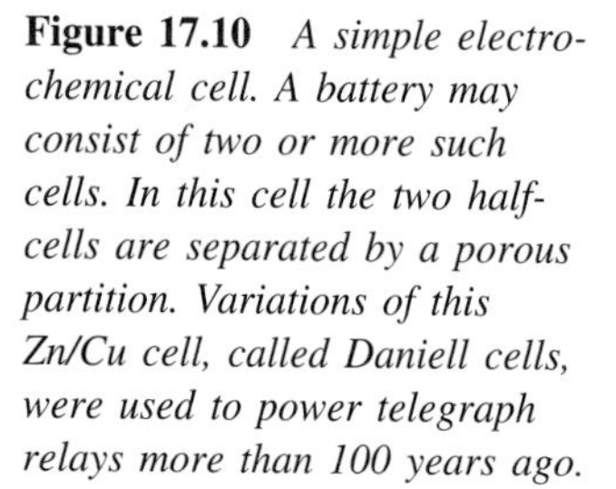

Figure 17.10 *A simple electrochemical cell. A battery may consist of two or more such cells. In this cell the two half-cells are separated by a porous partition. Variations of this Zn/Cu cell, called Daniell cells, were used to power telegraph relays more than 100 years ago.*

Although the reaction is spontaneous, simply placing a strip of zinc metal in a copper(II) sulfate solution does not produce useful electric current. However, if we carry out the reaction in an *electrochemical cell* like the one shown in Figure 17.10 (or 17.11), an electric current flows through the wire that connects the two metal strips. This gives us a voltaic cell.

A voltaic cell is constructed from two **half-cells**. Oxidation takes place in one half-cell; reduction takes place in the other half-cell. For this voltaic cell, one half-cell has a zinc metal strip placed in a solution of zinc sulfate. The other half-cell has a copper metal strip placed in a solution of copper(II) sulfate. The pair of half-cells must be connected in two ways: (1) The two metal strips are connected by a conductor that allows current to flow through a light bulb, voltmeter, or other electrical device. (2) The two solutions are connected in such a way as to retard the mixing of the metal ions in solution while permitting passage of the anions (the sulfate ions). The solutions can be connected by using a porous plate (a porous

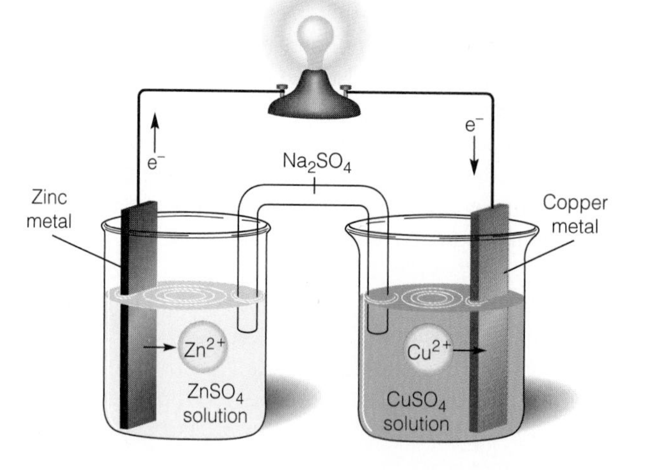

Figure 17.11 *In this electrochemical cell, solutions in the two half-cells are connected by a salt bridge. In the left compartment, zinc atoms give up electrons that flow through the wire to the copper metal in the right compartment. Copper(II) ions pick up the electrons and are reduced to copper atoms that are plated out as copper metal.*

partition) placed between the two solutions (Figure 17.10) or by a *salt bridge* (Figure 17.11).

A **salt bridge** (see Figure 17.11) is a tube filled with an aqueous solution or gel containing a strong electrolyte. The tube, fitted with porous plugs at each end, retards mixing of solutions in the half-cells. The anion (typically NO_3^-, Cl^-, or SO_4^{2-}) that is present in both half-cells is also used in the salt bridge.

In the half-cell with the Zn metal and $ZnSO_4$ solution, atoms of Zn lose electrons and are oxidized to zinc ions, Zn^{2+}. The anode (oxidation) reaction is

$$Zn(s) \longrightarrow Zn^{2+} + 2\,e^-$$

The two electrons are left behind on the zinc strip, but they flow through the conducting wire to the copper metal strip where Cu^{2+} ions are reduced to Cu metal atoms in the second half-cell. The cathode (reduction) reaction is

$$Cu^{2+} + 2\,e^- \longrightarrow Cu(s)$$

Adding the two half-cell reactions gives the net reaction for the voltaic cell.

$$Zn(s) + Cu^{2+}(aq) \longrightarrow Zn^{2+}(aq) + Cu(s)$$

The chemical reaction produces electricity—a flow of electrons—as the electrons pass through the conductor connecting the electrodes (the two metal strips). If 1.0 M solutions of $ZnSO_4$ and $CuSO_4$ are used, the cell will produce about 1.1 volts at 25 °C. The **volt**, V, is a measure of electrical potential, which is the tendency of electrons in a system to flow.

The cell is slightly more complicated than indicated by Figure 17.11. In addition to the metal cations, solutions in both half-cells and in the salt bridge contain sulfate ions, SO_4^{2-}. These sulfate ions don't enter into any chemical reaction, but, to maintain electrical balance, they tend to move in the opposing direction to the movement of metal ions. As positively charged metal cations move toward the cathode of copper metal, negatively charged sulfate anions move in the opposite direction across the salt bridge toward the anode. The cell continues to operate until the electron flow or ion flow is interrupted, or until the cell finally reaches equilibrium.

The tendency to reach equilibrium is what drives an electrochemical cell.

Why bother with the salt bridge or the porous partition between the half-cells? Why not just put both solutions and both electrodes in a single compartment so the ions don't have to move through the porous partition? If we were to do that, the solutions would quickly become mixed and the zinc metal would react directly with Cu^{2+} ions. The electrons lost by zinc would be transferred to Cu^{2+} ions without flowing through the wire; equilibrium would be established quickly.

Cells with the Zn/Cu couple (called Daniell cells) were commonly used some 100 years ago to supply power for telegraph relays. Each Daniell cell produces only 1.1 volts, but batteries with greater voltage were made by connecting together several Daniell cells in series (i.e., one after another, positive terminal to negative terminal). Although the zinc–copper Daniell cell is no longer used commercially, single "wet" cells of this type were the immediate ancestors of many types of

A CLOSER LOOK

Corrosion

Oxidation–reduction reactions involving the corrosion of metals are of considerable economic importance. It is estimated that in the United States alone, corrosion costs over \$70 billion a year. Perhaps as much as 20% of all the iron and steel production in the United States each year goes to replace corroded items. Let's look first at the corrosion of iron.

In moist air, especially at a nick or scratch in a painted steel surface, the iron is oxidized.

$$Fe \rightarrow Fe^{2+} + 2\,e^-$$

As iron is oxidized, oxygen is reduced.

$$O_2 + 2\,H_2O + 4\,e^- \rightarrow 4\,OH^-$$

The net result, initially, is the formation of insoluble iron(II) hydroxide.

$$2\,Fe + O_2 + 2\,H_2O \rightarrow 2\,Fe(OH)_2$$

This product is usually further oxidized to ion(III) hydroxide.

$$4\,Fe(OH)_2 + O_2 + 2\,H_2O \rightarrow 4\,Fe(OH)_3$$

Partial dehydration of $Fe(OH)_3$ gives a hydrated oxide solid that we call rust.

$$2\,Fe(OH)_3 \rightarrow \underset{\text{Rust}}{Fe_2O_3 \cdot H_2O} + 2\,H_2O$$

During the corrosion of iron, oxidation and reduction often occur at separate points on the metal surface. Electrons are transferred through the iron metal. The circuit is completed by an electrolyte in aqueous solution, such as the slush from road salt and melting snow. The metal is pitted in an *anodic* area, when iron is oxidized to Fe^{2+}. These ions migrate to the *cathodic* area, where they react with the hydroxide ions formed by reduction of oxygen.

$$Fe^{2+} + 2\,OH^- \rightarrow Fe(OH)_2$$

Further oxidation of iron(II) hydroxide gives $Fe(OH)_3$, or rust. During the rusting process the anodic area is protected from oxygen by a layer of water while the cathodic area is exposed to air.

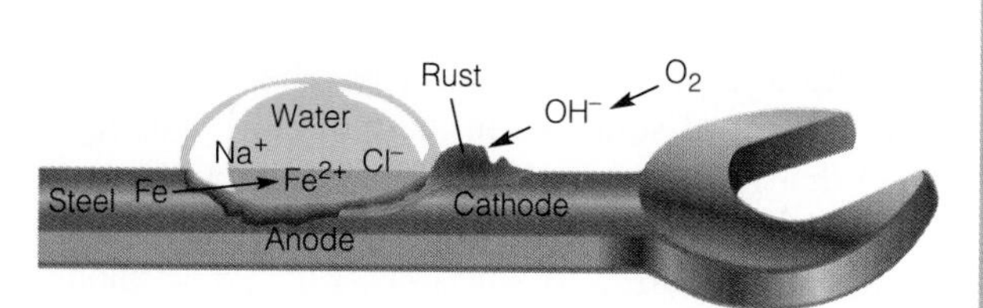

The corrosion of iron requires water, oxygen, and an electrolyte.

Cathodic protection is often used to minimize the corrosion of a buried steel pipeline or tank. A rod made of magnesium or another active metal is buried in the soil and is attached to the steel pipeline. With damp soil as the electrolyte, a voltaic cell is formed. The active metal acts as the anode, and oxygen reacts with water to produce hydroxide ions at the cathode.

Anode reaction: $Mg \rightarrow Mg^{2+}(aq) + 2\,e^-$ *(oxidation)*

Cathode reaction: $O_2 + 2\,H_2O + 4\,e^- \rightarrow 4\,OH^-$ *(reduction)*

Magnesium is oxidized in place of iron, so the steel pipe, at the cathode, is protected from oxidation. Eventually, the sacrificial Mg anode is consumed and must be replaced, but this is cheaper than replacing the pipe or the tank.

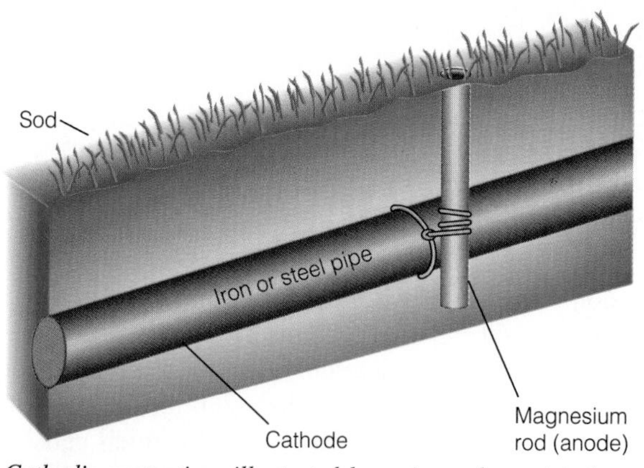

Cathodic protection, illustrated here, is used to minimize the corrosion of steel pipelines and tanks buried underground. The sacrificial anode, magnesium, is oxidized to Mg^{2+} ions. At the cathode, O_2 and water take up electrons and produce hydroxide ions, a reduction process.

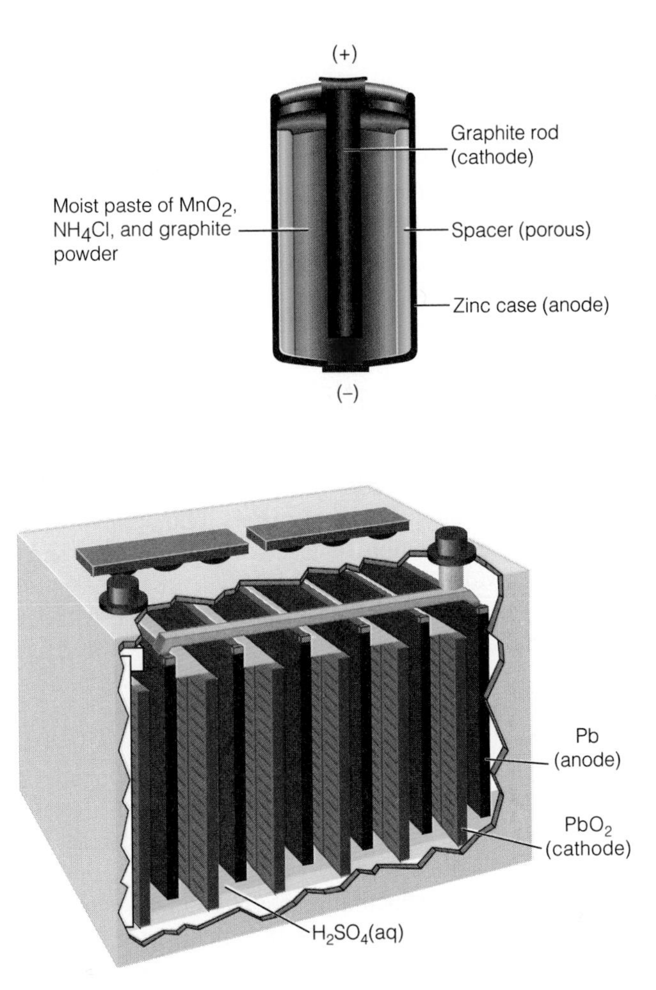

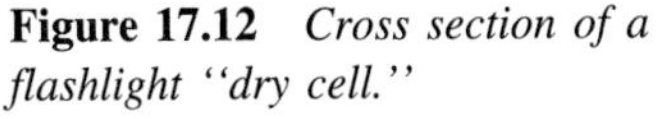

Figure 17.12 *Cross section of a flashlight "dry cell."*

Figure 17.13 *Cutaway diagram of one cell of a lead–acid storage battery. Six of these cells are used in a 12-volt automobile battery.*

"dry" cells (Figure 17.12) used today to operate battery-powered radios, television sets, cassette recorders, movie cameras, and so on.

The familiar 12-volt automobile storage battery consists of six cells wired in series, with each cell producing 2.0 volts (Figure 17.13). Instead of using the zinc–copper couple, cells in the automobile storage battery depend on a system involving lead, lead dioxide, and sulfuric acid. A distinctive feature of this battery is its capacity for being recharged. It acts as a voltaic cell and will discharge (supply electricity) when the car is being started or when the motor is off and the lights are on. The battery can be recharged (it acts as an electrolytic cell) when the engine is running and an electric current is supplied *to* the battery. The reaction that occurs spontaneously during discharge is reversed for the charging process.

EXAMPLE 17.9

Answer the following questions for the zinc–copper cell.

(a) Which metal is oxidized, and why? What is reduced?
(b) Write the oxidation half-reaction. Write the reduction half-reaction.
(c) Which metal is the anode? Which metal is the cathode?
(d) Which electrode gains mass? Which electrode loses mass?
(e) What is the direction of the electron flow?

SOLUTION

(a) Zinc is oxidized. It is more reactive than copper. Cu^{2+} ions are reduced.
(b) The oxidation half-reaction is $Zn(s) \rightarrow Zn^{2+} + 2\ e^-$. The reduction half-reaction is $Cu^{2+} + 2\ e^- \rightarrow Cu(s)$.
(c) Zinc is the anode. Copper is the cathode.
(d) The copper cathode gains mass while the zinc anode loses mass.
(e) The electron flow is from Zn to Cu (anode to cathode) through the connecting wire.

See Problems 17.33–17.44.

Chapter Summary

Oxidation–reduction (redox) reactions occur during combustion, metabolism, rusting, corrosion, the refining of metals, photographic film development, bleaching processes, the operation of electrochemical cells, and numerous other types of reactions. Oxidation always involves a loss of electrons. Reduction always involves a gain of electrons. You can't have one without the other. The oxidizing agent gains electrons and is reduced. The reducing agent loses electrons and is oxidized. In each redox reaction, the reducing agent gets oxidized, and the oxidizing agent gets reduced.

Oxidation takes place whenever (1) there is a loss of electrons; (2) there is a gain in oxidation number; (3) an element, compound, or ion gains oxygen atoms; and (4) a compound or ion loses hydrogen atoms. Reduction takes place whenever (1) there is a gain in electrons, (2) there is a decrease in oxidation number, (3) a compound or ion loses oxygen atoms, and (4) a compound or ion gains hydrogen atoms. Oxidation–reduction reactions can be balanced by using half-reactions in a series of seven steps.

There are two types of electrochemical cells. Those that use electricity to drive chemical reactions are called electrolytic cells. Those that use a spontaneous chemical reaction to generate an electric current are called voltaic or galvanic cells. Electrolytic cells can be used to reduce metal ions to metals, and to electroplate metals. Oxidation takes place at the anode and reduction takes place at the cathode in all electrochemical cells, whether electric current is being used or generated.

Assess Your Understanding

1. Assign oxidation numbers to each kind of atom within a given compound or ion. [17.1]
2. Identify which element is oxidized and which element is reduced in an oxidation–reduction reaction. [17.2, 17.3]
3. Identify the oxidizing agent and the reducing agent in an oxidation–reduction reaction. [17.2, 17.3]
4. Identify important oxidizing agents and reducing agents. [17.4, 17.5]
5. Identify oxidizing agents and reducing agents used in antiseptics, disinfectants, and bleaches. [17.4]
6. Describe how photographic processes involve oxidation and reduction. [17.5]

7. Balance oxidation–reduction equations using half-reactions. [17.6]
8. Distinguish between electrolytic and voltaic cells. [17.7, 17.8]
9. For a given electrolytic or voltaic cell, identify the anode, the cathode, and the reactions that occur at each. [17.7, 17.8]

Key Terms

anode [17.7]
cathode [17.7]
combustion [17.2]
electrochemical cell [17.8]
electrolysis [17.7]
electrolytic cell [17.8]
electroplating [17.7]
galvanic cell [17.8]
half-cell [17.8]
kindling temperature [17.2]
oxidation [17.2]
oxidation half-reaction [17.6]
oxidation number [17.1]
oxidation state [17.1]
oxidized [17.2]
oxidizing agent [17.3]
reduced [17.3]
reducing agent [17.3]
reduction [17.3]
reduction half-reaction [17.6]
salt bridge [17.8]
spontaneous combustion [17.2]
volt [17.8]
voltaic cell [17.8]

Problems

OXIDATION NUMBERS

17.1 Determine the oxidation number of
a. S in H_2S b. Cr in $CrO_4{}^{2-}$
c. vanadium in V_2O_3 d. Cl in ClO^-

17.2 Determine the oxidation number of
a. B in B_2O_3 b C in $C_2O_4{}^{2-}$
c. sulfur in S_8 d. Cl in $ClO_4{}^-$

17.3 Determine the oxidation state of P in each of the following.
a. P_4 b. $HPO_2{}^{2-}$
c. Na_3PO_4 d. P_4O_6

17.4 Determine the oxidation state of P in each of the following.
a. Na_3P b. $P_2O_7{}^{4-}$
c. H_3PO_3 d. $K_2H_2P_2O_7$

OXIDATION AND REDUCTION

17.5 Define *oxidation* in terms of the following.
a. oxygen atoms gained or lost
b. hydrogen atoms gained or lost
c. electrons gained or lost
d. change in oxidation number

17.6 Define *reduction* in terms of the following.
a. oxygen atoms gained or lost
b. hydrogen atoms gained or lost
c. electrons gained or lost
d. change in oxidation number

17.7 The following "equations" represent only part of a chemical reaction. Indicate whether each reactant shown is being oxidized or reduced.
a. $Cl_2 \rightarrow 2\,Cl^-$ b. $WO_3 \rightarrow W$
c. $2\,H^+ \rightarrow H_2$ d. $CO \rightarrow CO_2$

17.8 The following "equations" represent only part of a chemical reaction. Indicate whether each reactant shown is being oxidized or reduced.
a. $H_2O \rightarrow H_2(g)$ b. $Br_2 \rightarrow 2\,Br^-$
c. $C_2H_4O \rightarrow C_2H_6O$ d. $C_2H_4O \rightarrow C_2H_4O_2$

17.9 Which of the following are redox reactions?
a. $CaCl_2 + 2\,KF \rightarrow CaF_2 + 2\,KCl$
b. $CaI_2 + Cl_2 \rightarrow CaCl_2 + I_2$
c. $PbO_2 + 4\,HCl \rightarrow PbCl_2 + Cl_2 + 2\,H_2O$
d. $CaCO_3 + 2\,HCl \rightarrow CaCl_2 + CO_2 + H_2O$

17.10 Which of the following are redox reactions?
a. $Mg + H_2SO_4 \rightarrow MgSO_4 + H_2(g)$
b. $Pb(NO_3)_2 + Na_2SO_4 \rightarrow PbSO_4(s) + 2\,NaNO_3$
c. $CuO + CO(g) \rightarrow Cu + CO_2(g)$
d. $HCl + KOH \rightarrow KCl + H_2O$

17.11 Green grapes are exceptionally sour because of a high concentration of tartaric acid. As the grapes ripen, this compound is converted to glucose.

$$C_4H_6O_2 \rightarrow C_6H_{12}O_6$$

Tartaric acid   Glucose

Is the tartaric acid being oxidized or reduced?

17.12 Unsaturated vegetable oils react with hydrogen to form saturated fats. A typical reaction is

$$C_{57}H_{104}O_6 + 3\ H_2 \rightarrow C_{57}H_{110}O_6$$

Is the unsaturated oil being oxidized or reduced?

17.13 Place the letters OA above the oxidizing agent and RA above the reducing agent in each of these reactions.

a. $4\ Al + 3\ O_2 \rightarrow 2\ Al_2O_3$
b. $C_2H_2 + H_2 \rightarrow C_2H_4$
c. $2\ AgNO_3 + Cu \rightarrow Cu(NO_3)_2 + 2\ Ag$
d. $CuO + CO(g) \rightarrow Cu + CO_2(g)$

17.14 Place the letters OA above the oxidizing agent and RA above the reducing agent in each of these reactions.

a. $2\ Fe + 3\ Cl_2 \rightarrow 2\ FeCl_3$
b. $Mg + Cu(NO_3)_2 \rightarrow Cu + Mg(NO_3)_2$
c. $2\ PbO + C \rightarrow 2\ Pb + CO_2$
d. $Cl_2 + 2\ NaBr \rightarrow Br_2 + 2\ NaCl$

17.15 Given the reaction

$$2\ HNO_3 + SO_2 \rightarrow H_2SO_4 + 2\ NO_2$$

a. What element is oxidized?
b. What element is reduced?
c. What is the oxidizing agent?
d. What is the reducing agent?

17.16 Given the reaction

$$2\ CrO_3 + 6\ HI \rightarrow Cr_2O_3 + 3\ I_2 + 3\ H_2O$$

a. What element is oxidized?
b. What element is reduced?
c. What is the oxidizing agent?
d. What is the reducing agent?

17.17 The purple color of a solution containing permanganate ions, MnO_4^-, disappears when iron(II) ions are added, according to the reaction

$$MnO_4^- + 5\ Fe^{2+} + 8\ H^+ \rightarrow Mn^{2+} + 5\ Fe^{3+} + 4\ H_2O$$

a. What element is oxidized?
b. What element is reduced?
c. What is the oxidizing agent?
d. What is the reducing agent?

17.18 When copper metal is placed in concentrated nitric acid, the liquid turns green and amber-colored nitrogen dioxide gas, $NO_2(g)$, is produced.

$$Cu(s) + 4\ HNO_3(aq) \rightarrow Cu(NO_3)_2(aq) + 2\ NO_2(g) + 2\ H_2O$$

a. What element is oxidized?
b. What element is reduced?
c. What is the oxidizing agent?
d. What is the reducing agent?

17.19 Molybdenum metal, used in special kinds of steel, can be manufactured by the reaction of its oxide with hydrogen.

$$MoO_3 + 3\ H_2 \rightarrow Mo + 3\ H_2O$$

a. What element is oxidized?
b. What element is reduced?
c. What is the oxidizing agent?
d. What is the reducing agent?

17.20 When the water pump failed in the nuclear reactor at Three Mile Island in 1979, zirconium metal reacted with the very hot water to produce hydrogen gas.

$$Zr + 2\ H_2O \rightarrow ZrO_2 + 2\ H_2$$

a. What element is oxidized?
b. What element is reduced?
c. What is the oxidizing agent?
d. What is the reducing agent?

17.21 Indigo dye (used to color blue jeans) is formed from indoxyl by exposure of indoxyl to the oxygen in air.

$$2\ C_8H_7ON + O_2 \rightarrow C_{16}H_{10}N_2O_2 + 2\ H_2O$$

a. What substance is oxidized? Explain your answer.
b. What is the oxidizing agent? Explain your answer.

17.22 Silverware used with eggs becomes tarnished. The decomposition of protein in eggs produces hydrogen sulfide, H_2S, which is thought to react with silver by the following reaction to produce brown silver sulfide, Ag_2S.

$$4\ Ag + 2\ H_2S + O_2 \rightarrow 2\ Ag_2S + 2\ H_2O$$

a. What substance is oxidized? Explain your answer.
b. What is the oxidizing agent? Explain your answer.

17.23 Give the oxidizing agent and the reducing agent for each reaction. Explain your answer.
a. $Cl_2 + 2\,Br^- \rightarrow 2\,Cl^- + Br_2$
b. $6\,Fe^{2+} + Cr_2O_7^{2-} + 14\,H^+ \rightarrow 6\,Fe^{3+} + 2\,Cr^{3+} + 7\,H_2O$

17.24 Give the oxidizing agent and the reducing agents for each reaction. Explain your answer.
a. $2\,I^- + Br_2 \rightarrow I_2 + 2\,Br^-$
b. $2\,Ag^+ + Mg \rightarrow 2\,Ag + Mg^{2+}$

17.25 List four common oxidizing agents.

17.26 List three common reducing agents.

OXIDATION AND REDUCTION HALF-REACTIONS

17.27 Indicate whether sulfur is oxidized or reduced in the following half-reaction. Also show the change in oxidation number for sulfur.

$$SO_3^{2-} + H_2O \rightarrow SO_4^{2-} + 2\,H^+ + 2\,e^-$$

17.28 Indicate whether nitrogen is oxidized or reduced in the following half-reaction. Also show the change in oxidation number for nitrogen.

$$NO + 2\,H_2O \rightarrow NO_3^- + 4\,H^+ + 3\,e^-$$

17.29 Use redox half-reactions to write a balanced net ionic equation for the following reaction in an acidic solution. Show your work.

$$Cu + NO_3^- \rightarrow Cu^{2+} + NO_2$$

17.30 Use redox half-reactions to write a balanced net ionic equation for the following reaction in an acidic solution. Show your work.

$$I^-(aq) + SO_4^{2-}(aq) \rightarrow I_2(s) + S(s)$$

17.31 Use redox half-reactions to write a balanced net ionic equation for the following reaction in an acidic solution. Show your work.

$$MnO_4^-(aq) + Cl^-(aq) \rightarrow Mn^{2+}(aq) + Cl_2(aq)$$

17.32 Use redox half-reactions to write a balanced net ionic equation for the reaction of Cu with nitric acid in an acidic solution to produce NO(g).

$$Cu(s) + HNO_3(aq) \rightarrow Cu^{2+}(aq) + NO(g)$$

ELECTROCHEMICAL CELLS

17.33 What are electrolytic cells?

17.34 What are voltaic (or galvanic) cells?

17.35 The following half-reaction takes place at the positive electrode of the automobile battery (a lead storage battery). Also see Problem 17.36.

$$PbO_2 + SO_4^{2-} + 4\,H^+ + 2\,e^- \rightarrow PbSO_4 + 2\,H_2O$$

a. State whether the equation represents oxidation, reduction, or both.
b. Name the element that is oxidized or reduced.
c. State whether this half-reaction occurs at the anode or the cathode.

17.36 The following half-reaction takes place at the negative electrode of the automobile battery (a lead storage battery). Also see Problem 17.35.

$$Pb + SO_4^{2-} \rightarrow PbSO_4 + 2\,e^-$$

a. State whether the equation represents oxidation, reduction, or both.
b. Name the element that is oxidized or reduced.
c. State whether this half-reaction occurs at the anode or the cathode.

17.37 The following half-reaction takes place at one of the electrodes in an alkaline cell used in flashlights, portable radios, etc. (see Problem 17.38).

$$Cd + 2\,OH^- \rightarrow Cd(OH)_2 + 2\,e^-$$

a. State whether the equation represents oxidation, reduction, or both.
b. Name the element that is oxidized or reduced.
c. State whether this half-reaction occurs at the anode or the cathode.

17.38 The following half-reaction takes place at one of the electrodes in an alkaline cell used in flashlights, portable radios, etc. (see Problem 17.37).

$$Ni_2O_3 + 3\,H_2O + 2\,e^- \rightarrow 2\,Ni(OH)_2 + 2\,OH^-$$

a. State whether the equation represents oxidation, reduction, or both.
b. Name the element that is oxidized or reduced.
c. State whether this half-reaction occurs at the anode or the cathode.

17.39 A fuel cell converts chemical energy directly into electrical energy with about 90% efficiency. One of the simplest fuel cells uses the reaction of

$H_2(g)$ with $O_2(g)$ in an alkaline solution to produce water.

a. At one electrode, the half reaction is

$$2\,H_2(g) + 4\,OH^-(aq) \rightarrow 4\,H_2O + 4\,e^-$$

Is this oxidation or reduction? Is this the anode or the cathode reaction?

b. At the other electrode, the half-reaction is

$$4\,e^- + O_2(g) + 2\,H_2O \rightarrow 4\,OH^-(aq)$$

Is this oxidation or reduction? Is this the anode or the cathode reaction?

c. Add the half-reactions and write the net reaction.

17.40 Rechargeable nickel–cadmium cells (called nicad batteries) are used in some kitchen appliances, power tools, and toys.

a. At one electrode, the half-reaction is

$$Cd(s) + 2\,OH^-(aq) \rightarrow Cd(OH)_2(s) + 2\,e^-$$

Is this oxidation or reduction? Is this the anode or the cathode reaction?

b. At the other electrode, the half-reaction is

$$2\,e^- + NiO_2(s) + 2\,H_2O \rightarrow Ni(OH)_2(s) + 2\,OH^-(aq)$$

Is this oxidation or reduction? Is this the anode or the cathode reaction?

c. Add the half-reactions and write the net reaction.

17.41 Magnesium metal can be prepared commercially by the electrolysis of molten $MgCl_2$.

a. Write the cathode half-reaction equation.

b. Write the anode half-reaction equation.

17.42 Calcium metal and bromine can be prepared by the electrolysis of molten $CaBr_2$.

a. Write the cathode half-reaction equation.

b. Write the anode half-reaction equation.

17.43 Describe how to assemble a voltaic cell made with Cu metal, Ni metal, $Cu(NO_3)_2$, and $Ni(NO_3)_2$. Nickel is more reactive than copper. Nickel is involved in oxidation, while copper is involved in reduction. Write anode and cathode half-reactions and show the net ionic equation.

17.44 Describe how to assemble a voltaic cell made with Zn metal, Ag metal, $Zn(NO_3)_2$, and $AgNO_3$. Zinc is more reactive than silver. Zinc is involved in oxidation, while silver is involved in reduction. Write anode and cathode half-reactions and show the net ionic equation.

Additional Problems

17.45 Determine the oxidation number of nitrogen in

a. N_2O b. N_2O_3 c. N_2O_5 d. N_2

17.46 Determine the oxidation number of nitrogen in

a. NO b. NO_2 c. NH_4Cl d. HNO_3

17.47 Titanium dioxide is a white pigment used in paints, white eye shadow, soaps, and other products. What is the oxidation state of Ti in TiO_2?

17.48 The compound NbO has a gray metallic color. What is the oxidation number of niobium in NbO?

17.49 Perchloric acid, $HClO_4$, is used in analytical laboratories as a powerful oxidizing agent. What is the oxidation state of Cl in $HClO_4$?

17.50 Common liquid bleaches contain solutions of sodium hypochlorite, NaClO. What is the oxidation state of Cl in NaClO?

17.51 Clorox and other household bleaches containing the hypochlorite ion, ClO^-, take up electrons from the substance being bleached to form chloride ions and hydroxide ions.

$$ClO^- + H_2O + 2\,e^- \rightarrow Cl^- + 2\,OH^-$$

Is the hypochlorite ion acting as an oxidizing agent, a reducing agent, or neither one? Explain your answer in terms of redox definitions.

17.52 One industrial method of obtaining copper metal involves the reaction of carbon monoxide with copper oxide ores.

$$CuO + CO(g) \rightarrow Cu + CO_2(g)$$

Is carbon monoxide acting as an oxidizing agent, a reducing agent, or neither one? Explain your answer in terms of redox definitions.

17.53 Give the oxidizing agent and reducing agent for the following reaction.

$$4\ H_2SO_4 + 8\ KI \rightarrow 3\ K_2SO_4 + K_2S + 4\ I_2 + 4\ H_2O$$

17.54 Give the oxidizing agent and reducing agent for the following reaction.

$$3\ H_2SO_3 + 2\ HNO_3 \rightarrow 3\ H_2SO_4 + 2\ NO + H_2O$$

17.55 Use redox half-reactions to write a balanced net ionic equation for the following reaction in an acidic solution. Show your work.

$$H_2O_2 + I^- \rightarrow I_2 + H_2O$$

17.56 Use redox half-reactions to write a balanced net ionic equation for the following reaction in an acidic solution. Show your work.

$$HNO_2 + I^- \rightarrow NO + I_2$$

17.57 The outer container for a common (Leclanché) flashlight type of cell is made of zinc and serves as one of the electrodes. Zinc atoms are converted to Zn^{2+} ions at this electrode, which is labeled as the negative electrode.

a. Write a half-reaction equation for this electrode.

b. Is zinc oxidized or reduced at this electrode?

c. Is this the anode or the cathode?

17.58 The reaction that occurs at one electrode of a Leclanché flashlight type of cell is given in the preceding problem. The positive electrode at the center of the cell is a carbon rod, which does not enter into the reaction. Here $MnO_2(s)$ and water (as a paste) react to produce $Mn_2O_3(s)$ and hydroxide ions.

a. Write a half-reaction equation for this electrode.

b. Is manganese oxidized or reduced at this electrode?

c. Is this the anode or the cathode?

17.59 During the rusting of iron (see the box on corrosion), iron metal is converted to Fe^{2+} ions.

a. Write a half-reaction equation for what occurs.

b. Does this reaction involve oxidation or reduction?

17.60 During the rusting of iron (see previous problem), oxygen from the air reacts with water to produce hydroxide ions.

a. Write a half-reaction equation for what occurs.

b. Does this reaction involve oxidation or reduction?

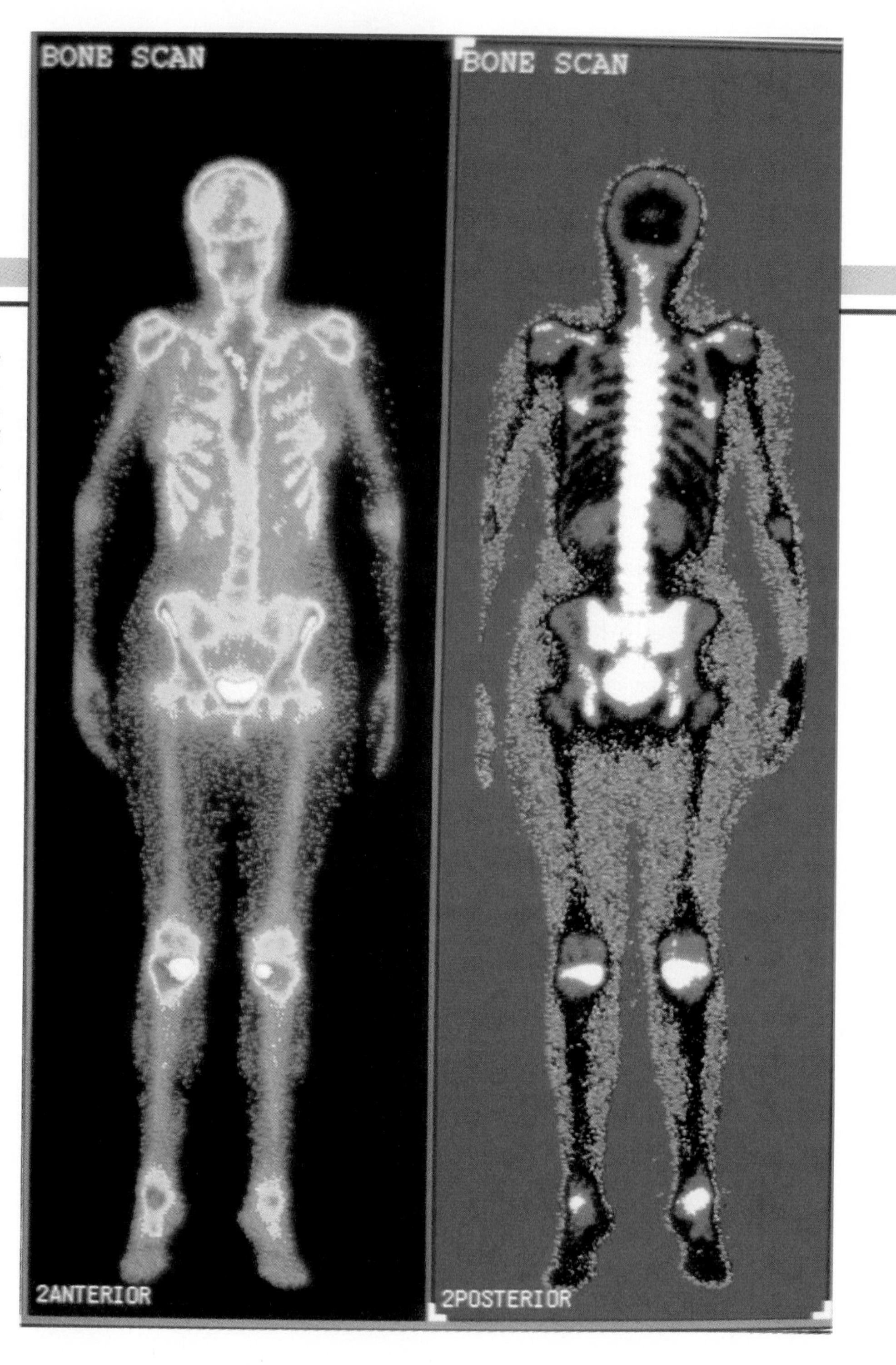

Nuclear medicine helps save lives. A wide variety of radioisotopes are used for radiation therapy and for diagnostic medical tests. For example, technetium-99^m is often used for anterior and posterior bone scans such as those shown in this photograph.

18

Fundamentals of Nuclear Chemistry

CONTENTS

Except for an introductory description of atoms and subatomic particles (Section 4.8), and a brief look at isotopes (Section 4.9), we have primarily been concerned with the atom's outermost electrons. The interactions of these electrons account for the ionic and covalent bonds present in every compound in our world. For this chapter, however, let us consider that tiny, central portion of each atom called the nucleus. Atomic nuclei are about 100,000 times smaller (in radius) than whole atoms, yet it is the nucleus that holds the power that has become the symbol of our age.

The nuclear age confronts us with a paradox. Although nuclear power can be used to destroy cities and perhaps civilizations, controlled nuclear power can provide the energy necessary to run our cities and maintain our civilization. Even here, there is another paradox. The controlled, peaceful use of nuclear power is not without its own potential dangers.

As citizens of the nuclear age, we have difficult decisions to make. We hear about problems at nuclear power plants, but coal-fired power plants have a different set of problems. For example, burning coal produces sulfur oxides and carbon dioxide, both of which produce acid rain. The carbon dioxide gas also traps heat and thus contributes to a problem known as the "greenhouse effect." The use of nuclear power, then, is another instance where there are "trade-offs." Tough decisions must be made, but tough decisions are even tougher to make when all the information is not known.

The concentration of carbon dioxide gas is now increasing at a rate of about 1 ppm per year.

Nuclear *chemistry*–other than nuclear power—does have many positive effects. Nuclear medicine saves lives. Diseases once regarded as incurable can be diagnosed and treated effectively with radioactive isotopes. Applications of nuclear chemistry to biology, industry, and agriculture have improved the human condition significantly. The use of radioisotopes in biological and agricultural research has led to increased crop production, which provides more food for a hungry world. Our knowledge of nuclear science gives us both power and responsibility. How we exercise that responsibility will determine how future generations remember us—even whether there will *be* future generations to remember us.

18.1 Natural Radioactivity

Some nuclei are unstable as they occur in nature. This is due to differences in attractions and repulsions within the nucleus. Those naturally occurring isotopes that spontaneously emit alpha or beta particles, or high-energy gamma rays, are said to possess **natural radioactivity**. Out of about 350 naturally occurring isotopes, approximately 80 are radioactive. Characteristics of the various forms of radiation—and the accompanying nuclear changes involved—are summarized in Table 18.1.

Alpha Particles

Let us first consider certain nuclei that emit alpha particles. Radium atoms with a mass number of 226 break down spontaneously, giving off alpha particles. The process is known as **alpha decay**. The **alpha particles** are identical to helium nuclei and are symbolized as $^{4}_{2}He$ (preferred) or with the Greek letter alpha, α. The equation for the alpha decay of radium is written

$$^{226}_{88}Ra \longrightarrow ^{4}_{2}He + ^{222}_{86}Rn$$

Table 18.1 **Radioactive Decay and Nuclear Change**

Type of Radiation	*Greek Letter*	*Mass Number*	*Charge*	*Change in Mass Number*	*Change in Atomic Number*
Alpha	α	4	2+	Decreases by 4	Decreases by 2
Beta	β	0	1−	No change	Increases by 1
Gamma	γ	0	0	No change	No change

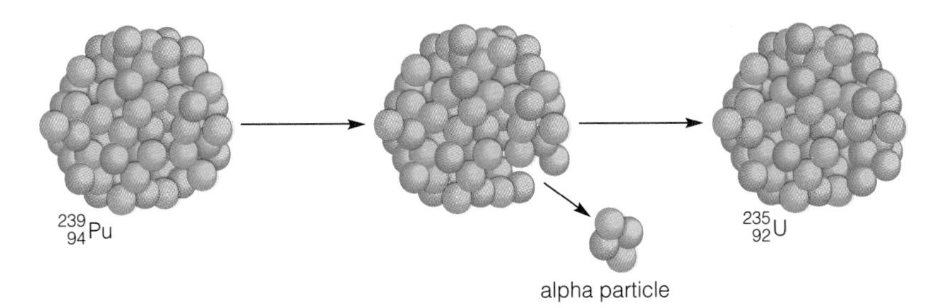

Figure 18.1 *Alpha decay. Nuclear changes that accompany the emission of an alpha particle.*

As shown here, each radium atom with a mass number of 226 (the superscript) ejects an alpha particle and is transformed into a new kind of atom with two fewer protons. The new atom, atomic number 86 (the subscript), is radon, Rn.

During any alpha particle emission, a **transmutation** occurs; that is, one kind of element is changed to another. Unlike chemical reactions, all nuclear reactions produce different isotopes. Each **nuclear equation** must be *balanced.*

The sums of mass numbers (the *superscripts*) on each side of the equation must be equal.

The sums of atomic numbers or nuclear charges (the *subscripts*) on each side of the equation must be equal.

A second example of alpha decay is shown in Figure 18.1. In that equation, plutonium with a mass number of 239, written either as $^{239}_{94}$Pu or as plutonium-239, is shown to decay to uranium-235. Notice that the equation is balanced.

Beta Particles

Tritium nuclei are also unstable. Tritium is one of the heavy isotopes of hydrogen (first mentioned in Chapter 4). Like all hydrogen nuclei, the tritium nucleus contains one proton. However, unlike the most common isotope of hydrogen, which has no neutrons, the tritium nucleus contains two neutrons and has a mass number of 3. Tritium, $^{3}_{1}$H, decomposes by a process that is called **beta decay** and produces a beta particle. A **beta particle** is a fast-moving electron and is represented as $^{0}_{-1}$e (preferred) or with the Greek letter beta, β. The equation for the beta decay of tritium is written

$$\underset{\text{Tritium}}{^{3}_{1}\text{H}} \longrightarrow \underset{\text{Beta}}{^{0}_{-1}\text{e}} + {}^{3}_{2}\text{He}$$

The isotope produced by this beta decay is identified by its atomic number as helium, He. How can a tritium nucleus, which contains only a proton and two neutrons, emit an electron? To answer this question, we can envision one of the neutrons in the original nucleus being changed into a proton as it ejects a beta particle (an electron).

$$^{1}_{0}\text{n} \longrightarrow {}^{1}_{1}\text{p} + {}^{0}_{-1}\text{e}$$

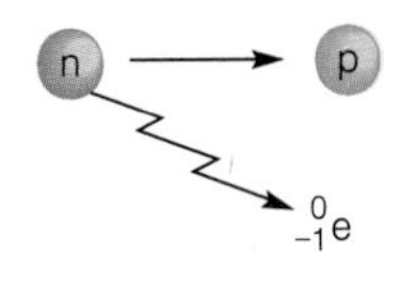

When the tritium nucleus ejects a beta particle, one neutron is transformed into one proton. The new proton is retained by the nucleus, so the atomic number is in-

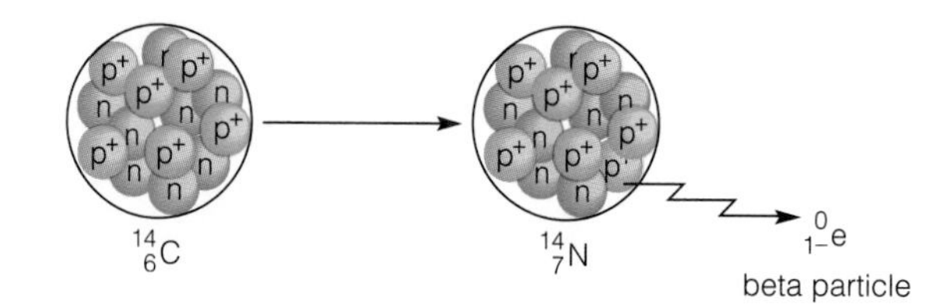

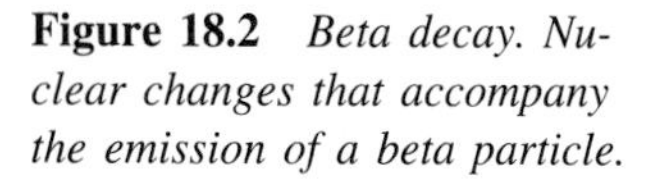

Figure 18.2 *Beta decay. Nuclear changes that accompany the emission of a beta particle.*

creased by 1 during this transmutation. After the almost massless electron or beta particle is ejected from the nucleus, the new nucleus has the same mass number as the original. A second example of beta decay is pictured in Figure 18.2.

Gamma Rays and X-Rays

Another type of radioactivity, called **gamma decay**, occurs when high-energy radiation is emitted by certain unstable isotopes. **Gamma rays**, symbolized by the Greek letter gamma, γ, have neither mass nor charge. This type of emission often accompanies alpha or beta radiation. The emission of an alpha or beta particle can leave the nucleus in a high-energy state. When this occurs, transitions between energy levels within the nucleus result in the subsequent emission of gamma rays as the nucleus returns to a more stable state.

Both gamma rays and X-rays are types of high-energy radiation, but **X-rays** have less energy than gamma rays. X-rays can be produced by bombarding a metal target with a high-energy electron beam. When the beam knocks electrons out of energy levels near the nucleus, other electrons at higher energy levels drop in to fill these "holes." As electron transitions occur within the higher energy levels, energy with frequencies in the X-ray region of the electromagnetic spectrum is emitted. While X-rays are produced during transitions of *electrons* between energy levels, gamma rays are emitted by the nucleus during transitions between *nuclear* energy levels.

X-rays were discovered in 1895 by Wilhelm Roentgen who, while working in a darkroom with a cathode ray tube, noticed that certain compounds glowed when exposed to the cathode rays. Barium sulfate, like bone, is opaque to X-rays. This compound can be used to yield X-ray photographs of some internal organs (Figure 18.3).

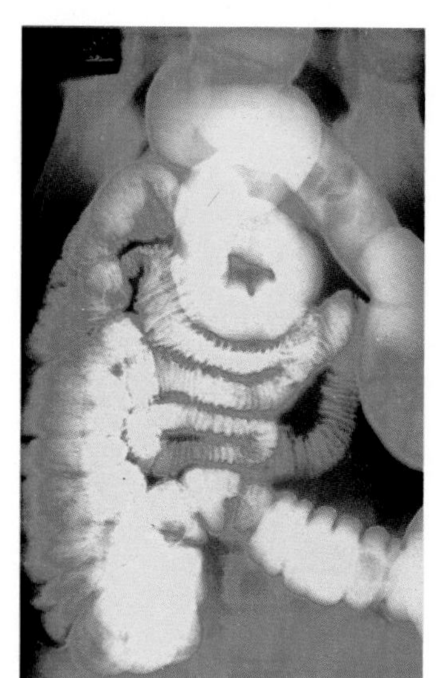

Figure 18.3 *Barium sulfate, $BaSO_4$, is insoluble in water and opaque to X-rays. This salt can be used to enhance the image of the stomach or of the intestine, as shown here, in X-ray photographs.*

Penetrating Power

The extent to which a material is penetrated by radiation depends on both the kind of radiation and the kind of material being bombarded. For materials with a low density, a greater thickness is required to reduce radiation intensity. Wood, water, and lead, for instance, will all act as shielding materials, but of the three, lead is best and water is the poorest.

All other things being equal, the more massive the particle, the less its *penetrating power.* When we compare alpha, beta, and gamma radiation (Figure 18.4), the alpha particles are the least penetrating; they do not travel through a sheet of paper. Alpha particles are helium nuclei; each particle has a mass number of 4. Beta particles, with a mass number of 0, are more penetrating than alpha particles, but are stopped by a thin sheet of metal such as aluminum. Beta particles are not really massless, but they are very much lighter than alpha particles. Gamma rays, being a form of high-energy radiation, are truly massless. They are the most penetrating of

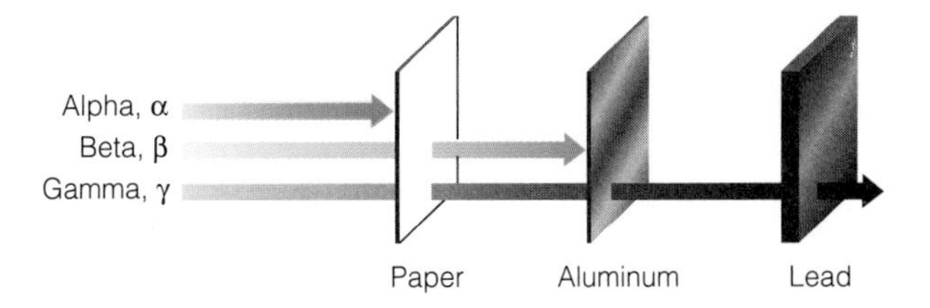

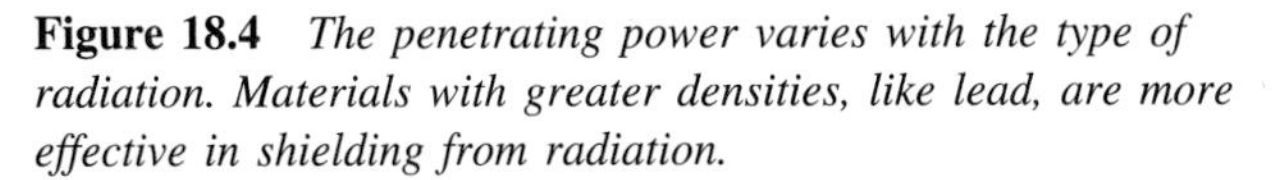
Figure 18.4 *The penetrating power varies with the type of radiation. Materials with greater densities, like lead, are more effective in shielding from radiation.*

the three. A body tissue thickness of about 50 cm is required to reduce the intensity of gamma rays by 10%. (The greater the thickness of the shielding material, the greater is the reduction in radiation intensity.)

It may seem contrary to common sense that the biggest particles have the least penetrating power. We can define penetrating power as a measure of the ability of the radiation to make its way through a sample of matter. It is as if you were trying to roll some rocks through a field of boulders (Figure 18.5). The alpha particle acts as if it were a boulder itself. Because of its size, it can't get very far before it bumps into and is stopped by the other boulders. The beta particle acts like a small stone. It can move between boulders and perhaps ricochet off one or another until it has made its way farther into the field. The gamma ray can be compared to a tiny insect that can get through the smallest openings, and although it may brush against some of the boulders, it can, in general, make its way through most of the field without being stopped.

At the beginning of this discussion of penetrating power we said that, all other things being equal, this is how things work. But all other things are not equal. The faster a particle moves or the more energetic the radiation is, the more penetrating power it has.

If a radioactive substance is *outside* the body, alpha particles are the least dangerous ones. These large particles damage the surface tissue, but they are stopped by the outer layer of skin. Gamma rays, however, are much more penetrating. They readily pass through body tissues, so an external gamma source can be quite dangerous internally.

If the radioactive substance is *inside* the body, the situation is reversed. Here, alpha particles can inflict great damage. When alpha particles—which cannot pass through body tissue—are trapped within the body, they release all their energy into a small amount of tissue. Beta particles are much smaller and more penetrating so they distribute their energy more widely and cause less damage to tissue.

There are essentially two ways to protect oneself from ionizing (damaging) radiation. One way is distance; the intensity of radiation decreases as the distance

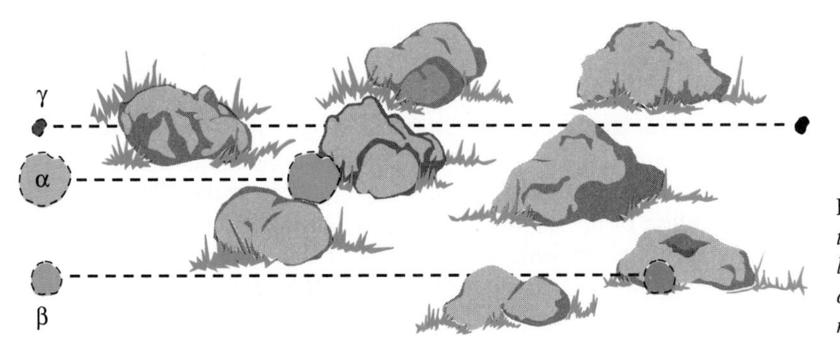

Figure 18.5 *Shooting radioactive particles through matter is like rolling rocks through a field of boulders—the larger rocks are more quickly stopped.*

from the source increases. The second way is protection by shielding. As stated earlier, the more dense a material is, the smaller the thickness needed to provide shielding against penetrating radiation.

EXAMPLE 18.1

Plutonium-239 emits an alpha particle when it decays What new element is formed?

$$^{239}_{94}\text{Pu} \longrightarrow {}^{4}_{2}\text{He} + ?$$

SOLUTION

Mass and charge are conserved. The new element must have a mass of 239 − 4 = 235 and a nuclear charge of 94 − 2 = 92. The nuclear charge (atomic number) of 92 identifies the element as uranium, U. The balanced nuclear equation is

See Problems 18.1–18.22.

$$^{239}_{94}\text{Pu} \longrightarrow {}^{4}_{2}\text{He} + {}^{235}_{92}\text{U}$$

EXERCISE 18.1

Radioactive phosphorus-32 decays by beta emission. What element is formed? Give the balanced nuclear equation.

18.2 *Half-life*

Thus far we have discussed radioactivity as applied to single atoms. If we could see the nucleus of an individual atom and observe its composition, we could tell *whether* or not it would undergo radioactive decay. This is because atoms with certain combinations of protons and neutrons are unstable. We cannot determine *when* a certain atom will undergo a change. But in the laboratory we generally deal with great numbers of atoms—numbers far larger than the number of people on Earth. With such large numbers of atoms, we can make accurate predictions of the fraction that will decay in a given period of time. Spontaneous decay is a random process; it is independent of outside influences.

Each large sample of atoms of a particular isotope decays or disintegrates at a specific and constant rate. The period of time required for one-half of a specific sample of radioactive atoms of a given isotope to decay is its **half-life.** Half-life is a characteristic property of each kind of **radioisotope** (radioactive isotope). For some radioisotopes the half-life is very long, but for others, it is very short. For example, uranium-238 has a half-life of 4.5 billion years, but boron-9 has a half-life of 8×10^{-19} second.

To get a mental picture of the meaning of half-life, consider the following example. Suppose that we had 16 billion atoms of tritium, the radioactive isotope of hydrogen. The half-life of this isotope is 12.3 years. This means that in 12.3 years, 8 billion of the atoms (one-half) would have undergone radioactive decay. In another 12.3 years, only 25% of the original atoms would remain unchanged. Thus, after two half-lives, the original sample is 75% ($\frac{1}{2} + \frac{1}{4}$) gone, not *all* gone!

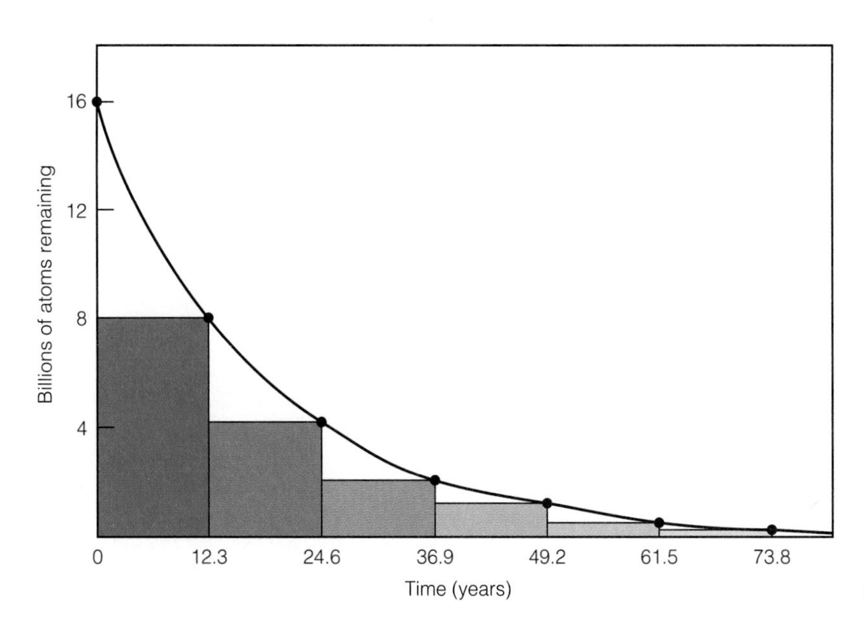

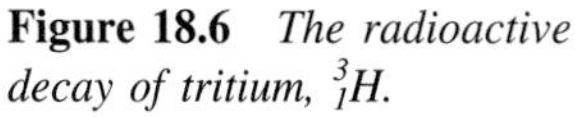
Figure 18.6 *The radioactive decay of tritium, ^3_1H.*

During each subsequent half-life, one-half of the remaining sample decays. A graph showing the number of atoms that remain for this sample after each of six subsequent half-lives is shown in Figure 18.6. For practical purposes, we can say that only a small fraction would remain after 10 half-lives. For the tritium sample, 10 half-lives would be 123 years.

The fraction of a radioactive sample that remains after each subsequent half-life can be determined as follows.

How Much Remains?

1. After one half-life, $\frac{1}{2}$ of the original atoms remain.
2. After two half-lives, $\frac{1}{2} \times \frac{1}{2} = 1/(2^2) = \frac{1}{4}$ of the original atoms remain.
3. After three half-lives, $\frac{1}{2} \times \frac{1}{2} \times \frac{1}{2} = 1/(2^3) = \frac{1}{8}$ of the original atoms remain.
4. After four half-lives, $\frac{1}{2} \times \frac{1}{2} \times \frac{1}{2} \times \frac{1}{2} = 1/(2^4) = \frac{1}{16}$ of the original atoms remain.
5. After five half-lives, $\frac{1}{2} \times \frac{1}{2} \times \frac{1}{2} \times \frac{1}{2} \times \frac{1}{2} = 1/(2^5) = \frac{1}{32}$ of the original atoms remain.

We can write a mathematical expression that will summarize this information. To determine what fraction of the original sample remains after a certain number (n) of half-lives, calculate the fraction as follows.

$$\text{Fraction remaining} = \frac{1}{2^n}$$

EXAMPLE 18.2

What fraction of radioactive atoms remains after six half-lives?

SOLUTION

We can use the expression $1/(2^n)$, where $n = 6$. The fraction remaining would be $1/(2^6) = \frac{1}{64}$ of the sample.

EXAMPLE 18.3

If you start with 4.00 g of a radioactive isotope, how many grams of the isotope would remain after four half-lives?

SOLUTION

Here $n = 4$. The fraction remaining would be $1/(2^4) = \frac{1}{16}$ of the sample. The mass of the remaining sample $= \frac{1}{16} \times 4.00$ g $= 0.250$ g.

See Problems 18.23–18.28.

EXERCISE 18.2

If you had 6.00 g of a radioactive substance, how many grams would remain after three half-lives?

18.3 Measuring Radiation: The Units

There are many ways of measuring radiation. The *rate* at which nuclear disintegrations occur in a particular sample—the nuclear activity—is measured in **disintegrations per second**. One unit of activity is the **curie**, Ci, named in honor of Marie Curie, the discoverer of radium. One curie is equal to the number of disintegrations per second (dis/s) produced by 1 g of radium.

$$1 \text{ Ci} = 3.7 \times 10^{10} \text{ dis/s}$$

A source used in the externally applied cobalt radiation treatment of cancer might be rated at 3000 Ci. For less active samples, metric prefixes are used.

A dose of radioactive isotope whose activity is 10 mCi might be taken internally by an adult for certain types of diagnostic scanning procedures.

$$1 \text{ millicurie (mCi)} = 0.001 \text{ Ci} = 3.7 \times 10^{7} \text{ dis/s}$$

$$1 \text{ microcurie } (\mu\text{Ci}) = 0.000001 \text{ Ci} = 3.7 \times 10^{4} \text{ dis/s}$$

A child would be subjected to a dose of perhaps 10–50 μCi for some kinds of diagnostic scanning.

We are perhaps more likely to be interested in our *exposure* to radiation than in simply counting disintegrations in some sample of radioactive material. The unit of exposure to gamma rays or X-rays is the **roentgen**, R. Both x-rays and gamma rays are forms of high-energy electromagnetic radiation, but the energy of gamma rays is higher than that of X-rays. The roentgen is a measure of the energy of the beam of rays that indicates how badly an air sample would be disrupted (ionized) by a particular source of X-rays or gamma rays. This is not quite the unit that relates directly to us; ionization in tissue is not the same as it is in air.

A roentgen, named after Wilhelm Roentgen who discovered X-rays in 1895, is defined as the amount of gamma or X-rays required to produce ions carrying a total of 2.1 billion units of electrical charge in 1 cm^3 of dry air at 0 °C and normal atmospheric pressure.

What most of us are interested in is damage to tissue. The SI unit for absorbed ionizing radiation is the Gray, named for Harold Gray, a British scientist. In the United States, a smaller unit, the rad, is frequently used.

$$1 \text{ Gray (gy)} = 100 \text{ rads}$$

Rad is the acronym for radiation absorbed dose. An exposure of 1 rad means that each gram of absorbing tissue has absorbed 100 ergs of radiation energy.

1 erg = 10^{-7} joule
= 2.39×10^{-8} cal

$$1 \text{ rad} = 100 \text{ ergs absorbed} = 2.39 \times 10^{-6} \text{ cal absorbed}$$

In terms of the amount of heat energy absorbed, the rad is an extremely small unit (1,000,000 rads = 2.39 cal). A ''diet'' cola provides perhaps a thousand times that much energy to the body. But it isn't the heat that is dangerous. It is the formation of highly reactive ions (hence the terms **ionizing radiation**) and other molecular fragments within the cell that makes radiation so hazardous to us and other living things. In fact, it is estimated that about 500 rads would kill most of us. A single dose of 1000 rads would kill nearly any mammal. For comparison, the average yearly radiation dose absorbed by an individual from medical and dental X-rays is about 1 rad.

Finally, there is the **rem**, which is an acronym for roentgen equivalency in man. It is a measure of the biological damage produced by a particular dose of radiation. The rem is a unit that takes into account not only the amount of radiation absorbed but also the kind of radiation. There is a precise definition for the rem, but a more general definition will be sufficient here. If a dose of radiation is rated at 4.5 rems, that dose can cause biological damage equivalent to 4.5 rads of radiation delivered by the standard sample.

It is not critical for you to know details such as how many disintegrations occur in 30 min in a 2.7-Ci sample of radium. What you should know is that

- The number of curies indicates how active the sample is.
- The number of roentgens indicates how much energy a particular dose transfers to the air.
- The number of rads indicates how much energy a dose transfers to tissue.
- The number of rems indicates how much biological damage the dose can deliver.

In all cases, the larger the number, the more potential danger is represented. For an approximation, a roentgen and a rad are about equal. When absorbed in muscle tissue, 1 R will generate about 1 rad of energy.

EXAMPLE 18.4

About how much radiation would a person be expected to receive yearly from medical and dental X-rays? About how many rads would kill us?

SOLUTION

We receive, on the average, about 1 rad/yr from medical and dental X-rays. About 500 rads would kill most of us.

18.4 Radiation Detectors

Geiger counters are instruments used to detect high-energy (ionizing) radiation. A Geiger counter has a metal tube with a thin glass or plastic window through which radiation can enter (Figure 18.7). The tube is filled with a gas at low pressure. A wire extending down the center of the tube is the anode; it has a high positive charge. The metal case for the tube is the cathode. Radiation entering the tube will produce ions by knocking electrons off the gas molecules—that's why it is called *ionizing radiation.* These electrons will be attracted to the positively charged wire in the center of the tube. Positive ions will move to the metal wall. A small electric

A Geiger counter is also known as a Geiger–Müller (G–M) counter.

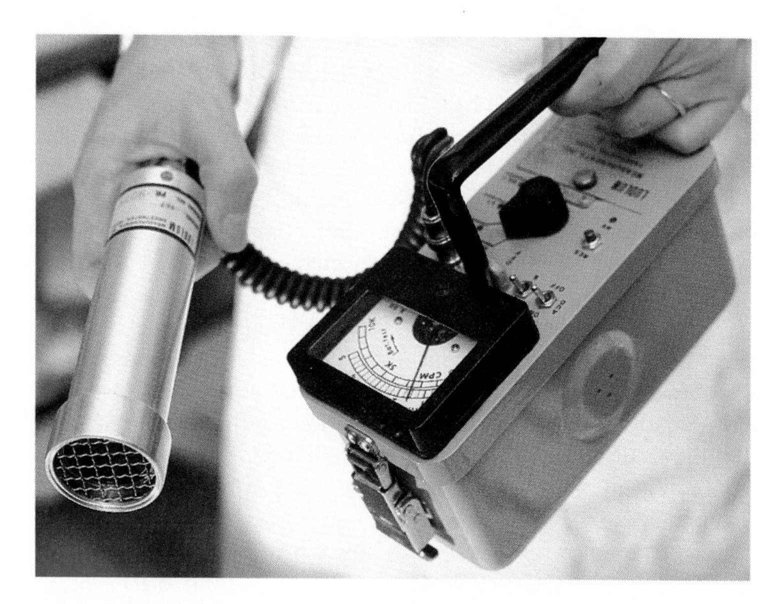

Figure 18.7 *A Geiger counter.*

current, which can be amplified and used to produce a meter reading or clicking sound or to cause a light to flash, will flow between the two electrodes. The meter can be calibrated in counts per minute or in other convenient rate units.

Numerous types of radiation detectors are available for special purposes. For example, individuals who work with radioactive materials use personal detectors. Generally, these are simple film badges (Figure 18.8) worn on the pocket or at the waist, or even as rings, that react to radiation as film does to light. The film badge records the total amount of radiation received by the wearer over a period of time. These badges are worn by workers in hospital radiology laboratories, nuclear power plants, and other industries that use radioactive materials.

EXAMPLE 18.5

Which of the devices described in this section

(a) can be used to monitor the total radiation received by a person?
(b) detects the *rate* of nuclear decay?

SOLUTION

See Problems 18.29–18.32.

(a) the film badge (b) the Geiger counter

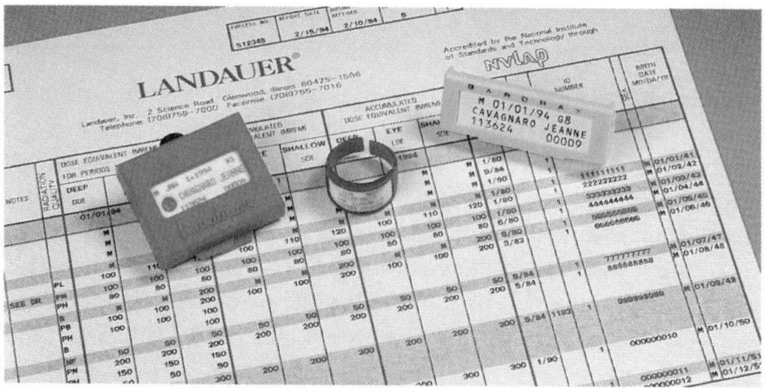

Figure 18.8 *A film badge or ring is worn by any person working near radioactive materials. Total radiation exposure is closely monitored.*

CHEMISTRY IN OUR WORLD

Radon-222 in Our Lives

Radon is a naturally occurring radioactive element that is chemically unreactive—it is a noble gas. The element is produced by the radioactive decay of uranium-238 and radium-226. There are three radon isotopes. Two of these, radon-219 and radon-220, decay very quickly: they have half-lives measured in seconds. The third isotope, radon-222, has a half-life of 3.82 days and decays by the emission of alpha particles and gamma radiation. Isotopes produced by the decay of radon—called radon daughters—include polonium-218 (half-life of 3.11 minutes) and polonium-214 (half-life of 0.0002 second). These radon daughters also decay by emission of alpha particles.

Once formed, the chemically unreactive radon-222 moves up through the ground. It can enter homes through cracks in basement floors. If the gas is inhaled, most of it is immediately exhaled, but because of its short half-life, some of the radon decays in the lungs before being exhaled. It is the two solid radon daughters—the alpha-emitting polonium radioisotopes—that are the greatest health concern. When trapped in the lungs, the radioactive radon daughters can damage nearby cells and cause cancer. The U.S. government estimates that high indoor radon concentrations account for 5000 to 20,000 cancer deaths annually.

The U.S. Environmental Protection Agency (EPA) has recommended an upper limit of 4 picocuries (pCi) of radiation per liter of air in homes. Outdoor air contains just 0.2 pCi of radon per liter. A 1988 EPA study estimates that 8 million U.S. homes have potentially hazardous levels of radon. Some 200,000 homes nationwide are estimated to have radon levels above 20 pCi per liter. Home radon detection kits are easy to use; they are available at local hardware stores. If high radon levels are found to be present, the homeowner or contractor will need to carry out corrective procedures such as sealing cracks in floors and installing pipes to vent the gas that can accumulate below a basement floor.

In terms of lung cancer risk, breathing radon-polluted air for 18 hr/day is estimated to be equivalent to smoking one to two packs of cigarettes per day. Although cigarette smoking is clearly the leading cause of lung cancer in the United States, radon gas is now believed to rank second among the causes of lung cancer. For smokers who live in an environment with a high concentration of radon, the risk of lung cancer is more than doubled—the combination of cigarettes and radon is especially carcinogenic.

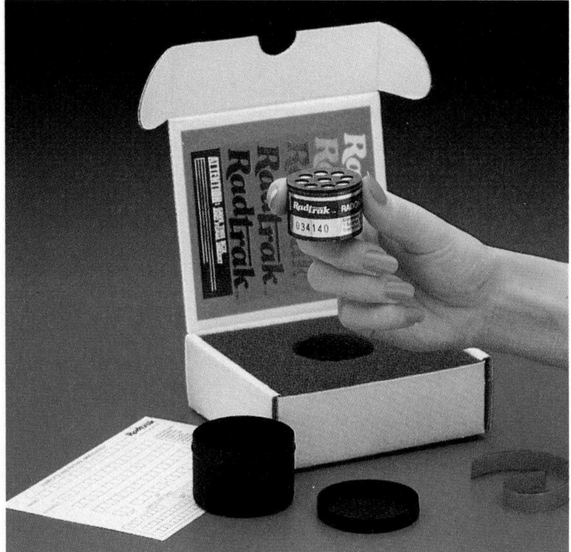

Home radiation detectors like those shown here can be purchased from a local hardware store. The detector is mailed to a laboratory for analysis after it has been used.

18.5 Background Radiation

A person receives about 0.6 mrem/hr while flying at 40,000 ft over the northern United States. In Denver, at ground level, one receives about 0.12 mrem/hr. **Those who smoke** one pack of cigarettes per day add a further 40 mrem or more of radiation, daily, to their lungs.

Radiation from natural sources is striking each of us at all times; no one can escape it. This ever-present **background radiation** is made up of cosmic rays (from the sun and outer space), radiation from naturally occurring radioactive isotopes in air, water, soil, and rocks, and radiation from artificial sources. Each person in the United States receives about 100 mrem of exposure each year, but the specific amount of background radiation varies from place to place. People who live at high altitudes receive more *cosmic* radiation than do those who live at low altitudes. This is because air absorbs some of the rays. People who live near certain mineral deposits also receive more background radiation.

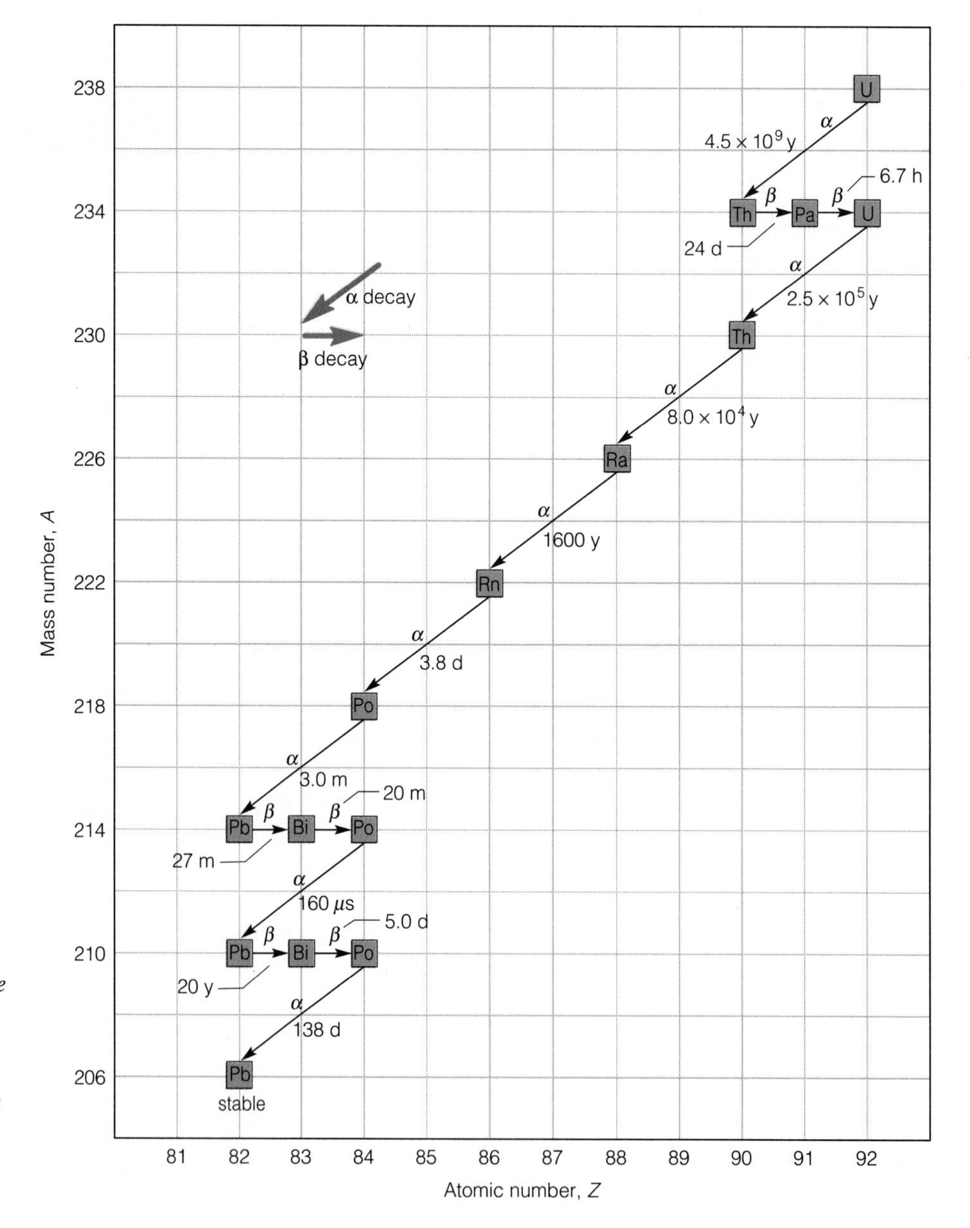

Figure 18.9 *The uranium-238 decay series. Each alpha particle emission reduces the mass number of an atom by 4 and the atomic number by 2. The emission of a beta particle raises the atomic number of an atom by 1 but does not change the mass number.*

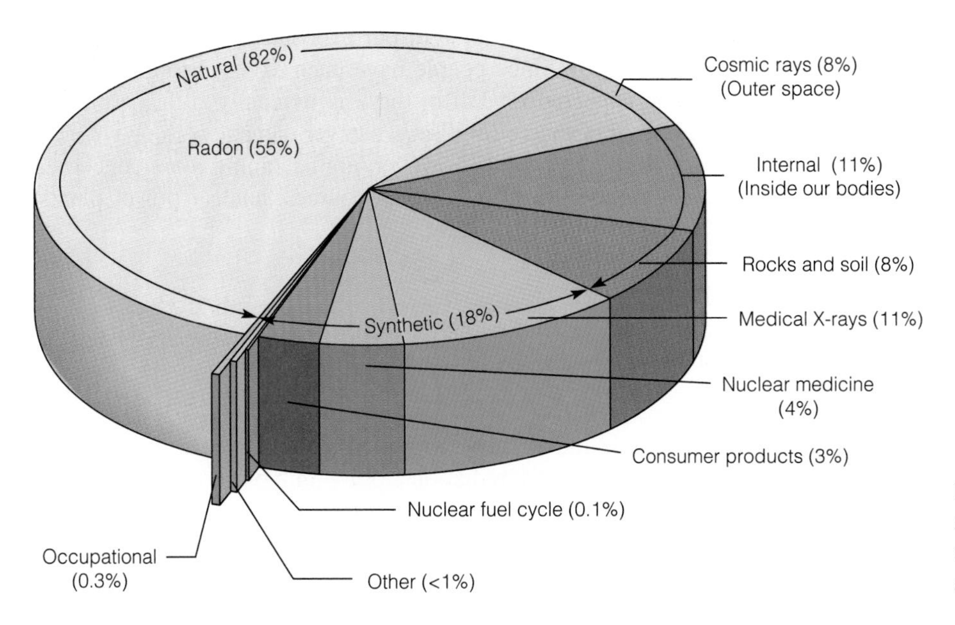

Figure 18.10 *Most of our exposure to radiation comes from natural sources. The largest single contribution is from radon.*

About 80 of the 350 or so natural isotopes are radioactive. Potassium-40, thorium-232, radon-222 (see the box "Radon-222 in Our Lives"), and the isotopes of uranium are important examples. Uranium-238, for example, eventually decays to become lead-206 in a series of 14 nuclear reactions (Figure 18.9).

Radioactive potassium-40 is present among all naturally occurring potassium atoms to an extent of 0.0012% by mass.

Radiation from uranium and thorium penetrates air and skin so little that it presents no external threat. Uranium and thorium present serious problems, however, when swallowed or inhaled: they break down to other radioactive elements, including radium. Radium resembles calcium in its chemical properties (both are Group IIA elements in the periodic table), and it readily replaces calcium in teeth and bones. The same lack of penetrating power that makes elements relatively innocuous on the outside makes them more damaging on the inside. If radiation is not penetrating, then all of it is trapped within the body, where it destroys vital components of our cells.

Our Exposure to Artificial Radiation Sources

About 82% of our exposure to radiation comes from natural sources (including 55% from radon), but about 18% comes from synthetic sources. For example, X-rays and other medical uses of radiation are responsible for about 15% of the total, as shown in Figure 18.10.

In 1979, radioactive material was released during an emergency shutdown of the Three Mile Island nuclear power plant near Harrisburg, Pennsylvania. Even though no lives were lost, the danger associated with this release was disputed by two different groups. On the one hand, it was argued that the people in neighboring communities were exposed to very little radiation—not much more than the normal background radiation. On the opposite side of the argument was a group that maintained that the released radioactive isotopes would ultimately enter the body through food or inhaled air. The radioisotopes would then concentrate within the body, these people argued, and remain to do long-term damage to various organs. Only the future can tell us which group is correct in its assessment.

Not only are we being constantly bombarded by radiation from external sources, but we ourselves are radioactive. Some 11% of our exposure comes from internal radiation sources (Figure 18.10). Since people have been exposed to background radiation ever since they appeared on Earth, there is evidently little permanent damage to our bodies from this source (or else, whatever damage is done we have come to accept as normal). Some people, however, are becoming more concerned about additional radiation exposure from medical sources, nuclear power plants, and radon gas in our homes.

EXAMPLE 18.6

About what percent of background radiation is from natural sources? Give four natural sources of background radiation.

SOLUTION

About 82% of background radiation is from natural sources. The four natural sources are radon gas, cosmic rays, radiation from within our bodies, and radiation from rocks and soil.

See Problems 18.33–18.38.

18.6 *Artificial Transmutation*

The forms of radioactivity described thus far occur in nature. Other nuclear reactions can be brought about by bombardment of stable nuclei with alpha particles and various other positive ions. When given a sufficient amount of energy, these charged particles can penetrate and become captured by a target nucleus being bombarded. The target nucleus undergoes transmutation. That is, one kind of atom is changed into another. Because the change would not have occurred naturally, the process is called **artificial transmutation**.

Ernest Rutherford studied the bombardment of several light elements with alpha particles. During his bombardment of nitrogen, protons were produced.

These investigations were conducted by Rutherford in 1919, a few years after his famous gold foil experiment (Chapter 5).

$$^{14}_{7}N + ^{4}_{2}He \longrightarrow ^{17}_{8}O + ^{1}_{1}H$$

(Because the hydrogen nucleus is a proton, the symbol $^{1}_{1}H$ is used to represent 1 proton.) Note that the sum of the mass numbers on the left equals the sum of the mass numbers on the right. The subscripts (particle charges) are also balanced.

By 1914 Rutherford had predicted that protons are present in all nuclei. In 1919 he gave the first experimental verification of the existence of protons as fundamental particles. Rutherford's experiment showed that protons can be obtained from the nucleus of an atom other than hydrogen, so protons are fundamental particles. By *fundamental particles* we mean basic units from which more complicated structures (such as the nitrogen nucleus) can be fashioned. Rutherford's experiment—described here—was the first induced nuclear reaction.

Eugen Goldstein had earlier produced protons in his discharge tube experiments (Chapter 5). He obtained these particles from hydrogen gas in the tube by knocking electrons off of hydrogen atoms.

Many transmutations were carried out during the 1920s. Then, in 1932, one such reaction led to the discovery of another fundamental particle. When James Chadwick, an English scientist, bombarded beryllium with alpha particles, he identified neutrons among the reaction products.

$$^{9}_{4}Be + ^{4}_{2}He \longrightarrow ^{12}_{6}C + ^{1}_{0}n$$

Chadwick had made an important discovery; he was awarded the 1935 Nobel prize in physics.

EXAMPLE 18.7

When potassium-39 is bombarded with neutrons, chlorine-36 is produced. What other particle is emitted?

$$^{39}_{19}K + ^{1}_{0}n \longrightarrow ^{36}_{17}Cl + ?$$

SOLUTION

To balance the equation, a particle with a mass of 4 and an atomic number of 2 is required. That is an alpha particle.

$$^{39}_{19}K + ^{1}_{0}n \longrightarrow ^{36}_{17}Cl + ^{4}_{2}He$$

18.7 *Induced Radioactivity*

The first artificial nuclear reactions produced stable isotopes already known to occur in nature. It was inevitable, however, that an unstable nucleus would be produced sooner or later. Irène Curie (daughter of Marie and Pierre Curie, the 1903 Nobel prize winners) and her husband, Frédéric Joliot (Figure 18.11), were studying the bombardment of aluminum with alpha particles. During the bombardment, neutrons were produced, leaving behind an isotope of phosphorus.

The work of the Curies is described in Section 5.1.

$$^{27}_{13}Al + ^{4}_{2}He \longrightarrow ^{30}_{15}P + ^{1}_{0}n$$

Much to their surprise, the target continued to emit particles even after the bombardment had been ended. This is because phosphorus-30 is also an unstable isotope. It emits particles called **positrons**, $^{0}_{+1}e$, that have a mass equal to that of an electron but with a positive charge. The equation for the reaction they were observing is written

$$^{30}_{15}P \longrightarrow ^{0}_{+1}e + ^{30}_{14}Si$$

Figure 18.11 *Frédéric and Irène Joliot-Curie discovered artificially induced radioactivity in 1934.*

The Joliot-Curies adopted the combined surname to perpetuate the Curie name. Irène Curie's parents, Marie and Pierre Curie, had two daughters but no son.

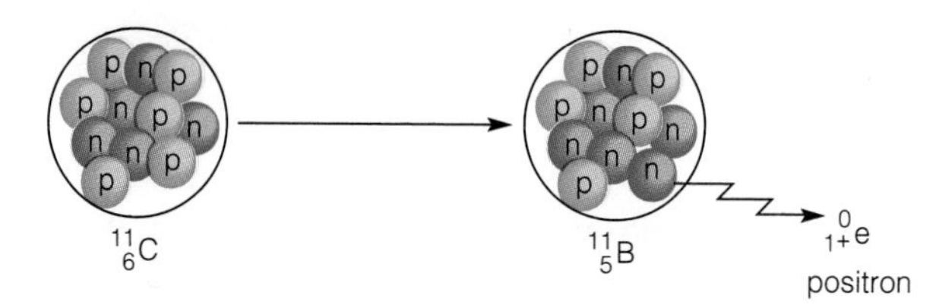

Figure 18.12 *Nuclear change accompanying positron emission.*

Once again the question arises: If the nucleus contains only protons and neutrons, where does this particle come from? To answer this question, try to visualize a proton (a hydrogen nucleus) being changed into a neutron and a positron that is ejected from the nucleus.

$$^{1}_{1}H \longrightarrow {}^{0}_{+1}e + {}^{1}_{0}n$$

When the positron is emitted, the original nucleus suddenly has one less proton, but it has one more neutron than before. Therefore, the product nucleus has the same mass number, but its atomic number is decreased by 1. Another example involving positron emission is presented in Figure 18.12.

As a result of their 1934 discovery of artificially induced radioactivity, the Joliot-Curies were awarded the Nobel prize in chemistry in 1935—the same year that Chadwick received the physics prize for discovery of the neutron.

Uranium, U, with an atomic number of 92 ($Z = 92$) was the heaviest element known until 1940 when neptunium ($Z = 93$) was synthesized at the University of California, Berkeley, by bombarding uranium with neutrons. Since then, other transuranium elements (94 through 109) have been prepared by Glenn T. Seaborg and others using the cyclotron and modern particle accelerators (Table 18.2).

To honor Glenn T. Seaburg's work, it was proposed, at a national American Chemical Society meeting in 1994, that element 106 be named seaborgium, Sg.

Table 18.2 **Preparation of the Transuranium Elements**

Atomic Number	*Name*	*Symbol*	*Year Discovered*	*Reaction*
93	Neptunium	Np	1940	$^{238}_{92}U + ^{1}_{0}n \rightarrow ^{239}_{93}Np + _{-1}^{0}e$
94	Plutonium	Pu	1940	$^{238}_{92}U + ^{2}_{1}H \rightarrow ^{238}_{93}Np + 2\,^{1}_{0}n$
				$^{238}_{93}Np \rightarrow ^{238}_{94}Pu + _{-1}^{0}e$
95	Americium	Am	1944	$^{239}_{94}Pu + ^{1}_{0}n \rightarrow ^{240}_{95}Am + _{-1}^{0}e$
96	Curium	Cm	1945	$^{239}_{94}Pu + ^{4}_{2}He \rightarrow ^{242}_{96}Cm + ^{1}_{0}n$
97	Berkelium	Bk	1949	$^{241}_{95}Am + ^{4}_{2}He \rightarrow ^{243}_{97}Bk + 2\,^{1}_{0}n$
98	Calfornium	Cf	1950	$^{242}_{96}Cm + ^{4}_{2}He \rightarrow ^{245}_{98}Cf + ^{1}_{0}n$
99	Einsteinium	Es	1952	$^{238}_{92}U + 15\,^{1}_{0}n \rightarrow ^{253}_{99}Es + 7\,_{-1}^{0}e$
100	Fermium	Fm	1952	$^{238}_{92}U + ^{16}_{8}O \rightarrow ^{250}_{100}Fm + 4\,^{1}_{0}n$
101	Mendelevium	Md	1955	$^{253}_{99}Es + ^{4}_{2}He \rightarrow ^{256}_{101}Md + ^{1}_{0}n$
102	Nobelium	No	1958	$^{246}_{96}Cm + ^{12}_{6}C \rightarrow ^{254}_{102}No + 4\,^{1}_{0}n$
103	Lawrencium	Lr	1961	$^{250}_{98}Cf + ^{11}_{5}B \rightarrow ^{257}_{103}Lr + 4\,^{1}_{0}n$
104	Unnilquadium	Unq	1964	$^{249}_{98}Cf + ^{12}_{6}C \rightarrow ^{257}_{104}Unq + 4\,^{1}_{0}n$
105	Unnilpentium	Unp	1970	$^{249}_{98}Cf + ^{15}_{7}N \rightarrow ^{260}_{105}Unp + 4\,^{1}_{0}n$
106	Unnilhexium	Unh	1974	$^{249}_{98}Cf + ^{18}_{8}O \rightarrow ^{263}_{106}Unh + 4\,^{1}_{0}n$
107	Unnilseptium	Uns	1981	$^{209}_{83}Bi + ^{54}_{24}O \rightarrow ^{262}_{107}Uns + ^{1}_{0}n$
108	Unniloctium	Uno	1984	$^{208}_{82}Pb + ^{58}_{26}Fe \rightarrow ^{265}_{108}Uno + ^{1}_{0}n$
109	Unnilennium	Une	1988	$^{209}_{83}Bi + ^{58}_{26}Fe \rightarrow ^{266}_{109}Une + ^{1}_{0}n$

CHEMISTRY AT WORK

Positron Emission Tomography Scans

The positron is more than a subatomic particle of interest only to scientists; it has been put to work in medicine, both in diagnosis and in medical research. Positron emission tomography (PET) is a technique that makes use of positron-emitting radioisotopes and modern computer technology to scan and obtain images of internal organs.

PET can be used to measure dynamic processes occurring in the body, such as blood flow or the rate at which oxygen or glucose is being metabolized. PET scans are being used to pinpoint the area of brain damage that triggers severe epileptic seizures. Prior to a scan, compounds containing positron-emitting isotopes, such as carbon-11, are inhaled or injected. The emitted positron cannot travel very far in the body. It soon encounters and collides with an electron, producing two gamma rays.

$$^{11}_{6}C \longrightarrow ^{11}_{5}B + ^{0}_{+1}e$$

$$^{0}_{+1}e + ^{0}_{-1}e \longrightarrow 2\,\gamma$$

These gamma rays simultaneously exit the body in exactly opposite directions Detectors positioned on opposite sides of the patient record these rays while other gamma rays from background radiation are ignored. A computer is used to calculate the point within the body at which the annihilation of the positron and electron occurred and to produce an image of that area.

The PET technique is quite important in medical research. For example, PET has been used recently to trace the site of anxiety and panic to the temporal lobes (situated at the base of the brain, behind the eyes). If researchers could determine the chemical processes that trigger anxiety, new drugs could be developed to prevent or arrest panic attacks.

PET equipment costs millions of dollars, and a single scan is quite expensive, but the technique can provide information that could otherwise be obtained only by subjecting the individual to the risks of surgery. Furthermore, certain information provided by this technique is not available by any other means.

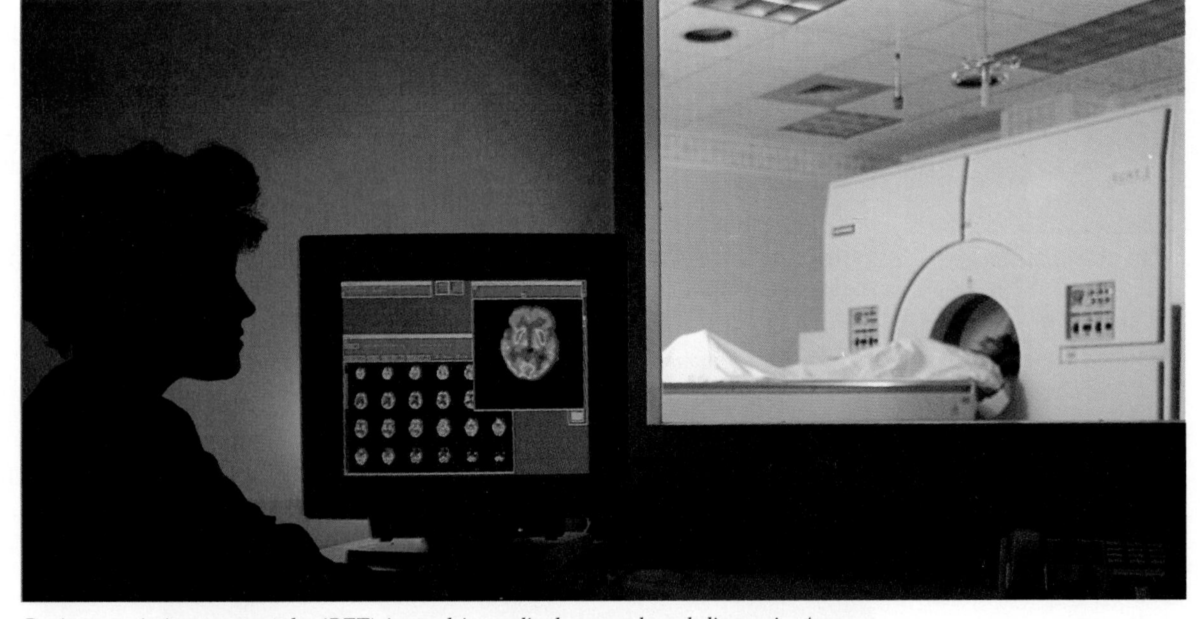

Positron emission tomography (PET) is used in medical research and diagnosis. A patient in position for PET is shown at the right, and the image appears on the computer screen at the left.

EXAMPLE 18.8

Carbon-10 is a radioactive isotope that emits a positron when it decays. Write an equation for this process.

$$^{10}_{6}C \longrightarrow {}^{0}_{+1}e + ?$$

SOLUTION

To balance the equation, a particle with a mass of 10 and an atomic number of 5 (boron) is required.

See Problems 18.39–18.44.

$$^{10}_{6}C \longrightarrow {}^{0}_{+1}e + {}^{10}_{5}B$$

EXERCISE 18.3

Neptunium-238 (element 93) can be synthesized by bombarding uranium-238 with hydrogen-2. The neptunium-238 is unstable and emits a beta particle when it decays. Write a balanced equation showing this decay.

18.8 Practical Uses of Some Radioisotopes

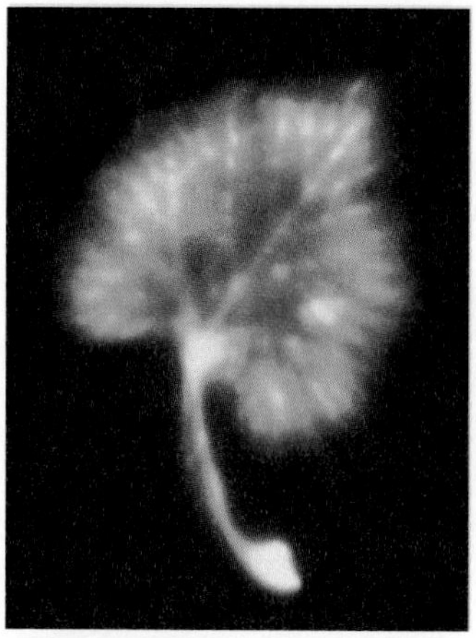

Figure 18.13 *Autoradiograph showing the uptake of phosphorus in a green plant.*

Scientists in a wide variety of fields make use of radioactive isotopes as **tracers** in physical, chemical, and biological systems. Isotopes of a given element—whether radioactive or not—behave nearly identically both chemically and physically, so radioactive isotopes, which are readily detected, can be used to study complicated systems.

For example, a plant can be fed fertilizer containing radioactive phosphorus, and the uptake of phosphorus can be measured. A simple method of detection involves placing the plant on a photographic film where the radiation exposes the film, much as light does. This type of exposure, called an *autoradiograph,* shows the distribution of phosphorus in the plant (Figure 18.13). Radioactive tracers are used widely in agricultural research to study the effectiveness of fertilizers and weed killers, to compare the nutritional value of various kinds of feed, and to determine the best methods for controlling insects.

One method used to help preserve foodstuffs is to **irradiate** them, that is, to expose them to sources of gamma radiation (Figure 18.14). The radiation destroys microorganisms that cause food spoilage. Irradiated food shows little change in

Figure 18.14 *Gamma radiation delays the decay of mushrooms. The mushrooms pictured on the right were irradiated; those on the left were not.*

taste or appearance. Some people are concerned about possible harmful effects of substances produced by the radiation, but there is no good evidence of harm to laboratory animals fed irradiated food, nor are there any known adverse effects in humans in countries where irradiation has been used for several years. No residual radiation can be detected in the food after irradiation.

Even the radioisotopes of transuranium elements (the synthetic elements with atomic numbers greater than 92) have found useful places in today's world. For example, the energy produced by the decay of artificially produced plutonium-238 is used to power heart pacemakers. This isotope has a relatively long half-life so the device can be used for about 10 years before being replaced. Another artificially produced radioactive isotope, americium-241, is used in home smoke detectors (ionization type).

Many kinds of radioisotopes have been used in basic scientific research. The mechanism of photosynthesis was worked out in large part by using carbon-14 as a tracer. Metabolic pathways in plants, animals, and humans are being studied by radioactive tracers. The potential for the use of this knowledge for human good is as enormous as the potential for the use of nuclear bombs for evil.

Nuclear Medicine

Nuclear medicine involves two distinct uses of radioisotopes: therapeutic and diagnostic. In *radiation therapy,* an attempt is made to treat or cure disease with radiation. The *diagnostic* use of radioisotopes is aimed at obtaining information about the type or extent of illness. Table 18.3 lists a variety of radioisotopes that are used in medicine.

Cancer is not one disease, but many. Some forms are particularly susceptible to radiation therapy. Radiation is carefully aimed at the cancerous tissue, and exposure of normal cells is minimized. If the cancer cells are killed by the high energy

Table 18.3 **Some Radioisotopes and Their Application in Medicine**

Isotope	*Name*	*Radiation*	*Half-life**	*Uses*
^{51}Cr	Chromium-51	γ	27.8 d	Determination of volume of red blood cells and total blood volume
^{57}Co	Cobalt-57	γ	270 d	Determination of uptake of vitamin B_{12}
^{60}Co	Cobalt-60	β, γ	5.3 y	Radiation treatment of cancer
^{153}Gd	Gadolinium-153	γ	242 d	Determination of bone density
^{131}I	Iodine-131	β, γ	8.0 d	Detection of thyroid malfunction; treatment of thyroid cancer; measurement of liver activity
^{59}Fe	Iron-59	β, γ	45 d	Measurement of rate of formation and lifetime of red blood cells
^{32}P	Phosphorus-32	β	14.3 d	Detection of skin cancer
^{226}Ra	Radium-226	α, γ	1590 y	Radiation therapy for cancer
^{24}Na	Sodium-24	β, γ	15.0 h	Detection of constrictions and obstructions in the circulatory system
^{99m}Tc	Technetium-99^{m}	γ	6.0 h	Imaging of brain, thyroid, liver, kidney, lung, and cardiovascular system
^{3}H	Tritium	β	12.3 y	Determination of total body water

*h = hours, d = days, y = years.

Nausea and vomiting are the usual symptoms of radiation sickness.

of the radiation, the malignancy is halted. But persons undergoing radiation therapy often get sick from the treatment or experience exhaustion. The aim of radiation therapy is to destroy the cancerous cells before too much damage is done to healthy tissue.

For many years, compounds of radium were used for the radiation treatment of cancer. Radium-226 is an alpha and gamma emitter.

$$^{226}_{88}\text{Ra} \longrightarrow {}^{222}_{86}\text{Rn} + {}^{4}_{2}\text{He} + \gamma$$

The radon, Rn, product is a radioactive gas. To prevent its escape, the radium was sealed in tiny hollow needles made of gold or platinum. These could be inserted directly into a tumor to irradiate the tumor tissue until the desired dosage had been administered. Unfortunately, the needles were so small that they were sometimes lost; frantic efforts to find them were not always successful.

During recent years, cobalt-60 has come into widespread use for cancer therapy. It is easily made by neutron bombardment of ordinary cobalt-59 in certain nuclear reactors that specialize in the synthesis of radioisotopes.

$$^{59}_{27}\text{Co} + {}^{1}_{0}\text{n} \longrightarrow {}^{60}_{27}\text{Co}$$

The half-life of cobalt-60 is 5.3 years.

The cobalt-60 emits beta particles and strong gamma rays, but in medical practice the beta particles are screened out. Cobalt-60 teletherapy units with intensities of over 1000 Ci are now quite common (Figure 18.15). Eventually the radioactive source becomes too weak and must be replaced.

Radioactive iodine-131 is used for both therapeutic and diagnostic purposes. It is used to determine the size, shape, and activity of the thyroid gland as well as to control a hyperactive thyroid and to treat cancer of the thyroid. In either case, the patient drinks a solution of potassium iodide containing iodine-131. The body concentrates iodine in the thyroid. Large doses are used for treatment of thyroid cancer; the radiation from the isotope concentrates in the thyroid cancer cells even if the cancer has spread to other parts of the body. For diagnostic purposes, only a

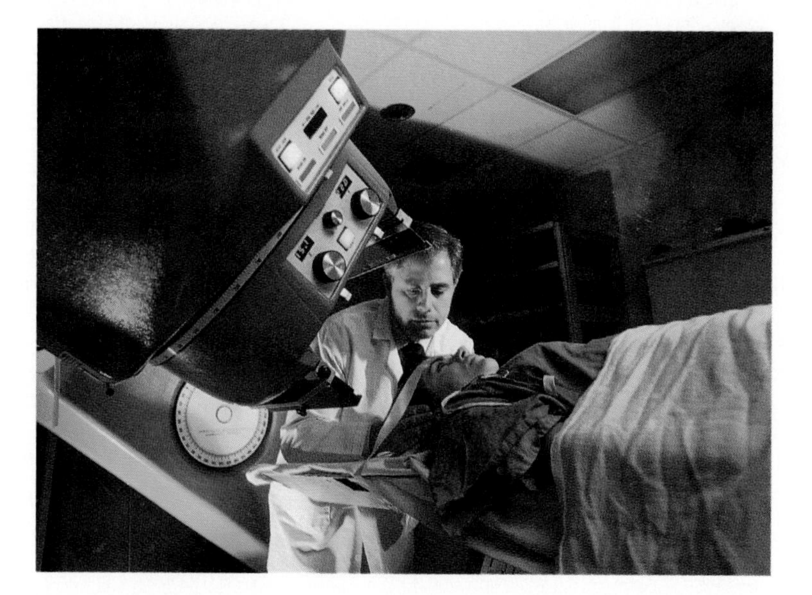

Figure 18.15 *A cobalt-60 unit for radiation therapy.*

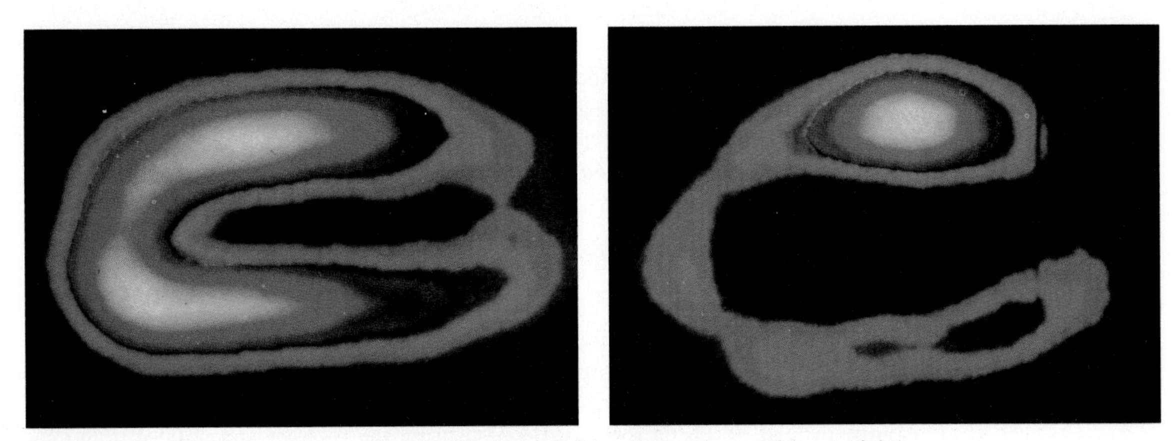

Figure 18.16 *Blood flow patterns in a healthy heart (*left*) and in a damaged heart (*right*). The highlighted images from a technetium-99^m compound indicate regions receiving adequate blood flow.*

small amount is needed. A detector is set up so that readings are translated into a permanent visual record showing the differential uptake of the isotope. The "picture" that results is referred to as a *photoscan* (Figure 18.16).

The radioisotope most widely used in medicine is gadolinium-153. It is used to determine bone mineralization, especially for persons who suffer from *osteoporosis* (reduction in the quantity of bone) as they grow older. Gadolinium-153 gives off two characteristic radiations, a gamma ray and an X-ray. A scanning device compares these radiations after they pass through bone. Bone densities are then determined by differences in absorption of the rays.

Technetium-99^{m} is used in a variety of diagnostic tests. The "m" stands for *metastable,* which means that this isotope will give up some energy to become a more stable form of the same isotope (same atomic number, same atomic mass). The energy it gives up is the gamma ray needed to detect the isotope.

$$^{99\text{m}}_{43}\text{Tc} \longrightarrow {}^{99}_{43}\text{Tc} + \gamma$$

The decay of technetium-99^{m} produces no alpha or beta particles, which could cause unnecessary damage to the body. Technetium-99^{m} has a short half-life (about 6 hours), which means that the radioactivity does not linger in the body after the scan has been completed. With this short half-life, use of the isotope must be carefully planned. In fact, the isotope itself is not what is purchased. Technetium-99^{m} is formed by the decay of molybdenum-99.

$$^{99}_{42}\text{Mo} \longrightarrow {}^{99\text{m}}_{43}\text{Tc} + {}^{0}_{-1}\text{e} + \gamma$$

A container of this molybdenum isotope is obtained, and the decay product, technetium-99^{m}, is "milked" (this is the terminology actually used) from the container as needed.

Hospital patients who have ingested therapeutic doses of radioisotopes must themselves be regarded as sources of radiation while the radioisotope maintains significant activity. Health personnel who are exposed to many such patients during their careers must exercise caution in order to avoid exposing themselves to a damaging dose of radiation over a long period of time.

EXAMPLE 18.9

List some practical uses of the radioisotopes (a) cobalt-60, (b) gadolinium-153, (c) iodine-131, and (d) technetium-99^{m}.

SOLUTION

(a) Cobalt-60 is used in the radiation treatment of cancer.
(b) Gadolinium-153 is used in the determination of bone density.
(c) Iodine-131 is used in the detection and treatment of thyroid cancer.
(d) Technetium-99^m is used in diagnostic testing of the brain, liver, and lungs.

Radioisotopic Dating

The half-lives of certain isotopes can be used to estimate the age of rocks and archaeological artifacts. Uranium-238 decays with a half-life of 4.5 billion years. The products of this decay are also radioactive, and breakdown continues until an isotope of lead, ^{206}Pb, is formed. By measuring the relative amounts of uranium-238 and lead-206, chemists can estimate the age of a rock. Some of the rocks on Earth have been found to be from 3.0 to 3.5 billion years old. Moon rocks and meteorites have been dated at a maximum age of about 4.5 billion years. Thus, the age of Earth is generally estimated to be about 4.5 to 5.0 billion years.

Carbon-14 dating, as outlined here, assumes that the formation of the isotope was constant over the years. This is not quite the case. However, for the most recent 7000 years or so, carbon-14 dates are quite reliable; they have been correlated with the annual growth rings of trees. Generally, carbon-14 is reasonably accurate for dating objects up to about 50,000 years old. Objects older than that have too little of the isotope left for accurate measurement.

The dating of artifacts usually involves a radioactive isotope of carbon. Carbon-14 is formed in the upper atmosphere by the bombardment of ordinary nitrogen by neutrons from cosmic rays.

$$^{14}_{7}N + ^{1}_{0}n \longrightarrow ^{14}_{6}C + ^{1}_{1}H$$

This process leads to a steady-state concentration of carbon-14 on Earth. Living plants and animals incorporate this isotope as carbon dioxide. When they die, however, the incorporation of carbon-14 ceases, and the carbon-14 in the plants and organisms decays—with a half-life of 5730 years—to nitrogen-14.

$$^{14}_{6}C \longrightarrow ^{14}_{7}N + ^{0}_{-1}e$$

Thus, we merely need to measure the carbon-14 activity remaining in an artifact of plant or animal origin to determine its age. For example, a sample that has half the ^{14}C activity of new plant material is 5730 years old; it has been dead for one half-life. Similarly, an artifact with 25% of the ^{14}C activity of new plant material is 11,460 years old; it has been dead for two half-lives.

Charcoal from the fires of an ancient people, dated by determining the ^{14}C activity, is used to estimate the age of artifacts found at the same archaeological

Table 18.4 **Several Isotopes Useful in Radioactive Dating**

Isotope	*Half-life (years)*	*Useful Range*	*Dating Applications*
Carbon-14	5730	500 to 50,000 years	Charcoal, organic material
Tritium ($^{3}_{1}H$)	12.3	1 to 100 years	Aged wines
Potassium-40	1.3×10^9	10,000 years to the oldest Earth samples	Rocks, the Earth's crust, the moon's crust
Rhenium-187	4.3×10^{10}	4×10^7 years to the oldest samples in universe	Meteorites
Uranium-238	4.5×10^9	10^7 years to the oldest Earth samples	Rocks, the Earth's crust

CHEMISTRY AT WORK

Dating the Shroud of Turin

Carbon-14 dating techniques were used to determine the age of a piece of linen that is known as the Shroud of Turin. This large, very old piece of linen bears a yellowish image of a man. The primary question was whether this cloth could have been used as a burial shroud for Christ.

To determine the carbon-14 content and, ultimately, the age of the cloth, researchers at radiocarbon dating labs at Zurich, Switzerland, Oxford University in England, and the University of Arizona were given small, 50-mg samples of the shroud along with three ancient samples of cloth of known age. Independent tests carried out by these laboratories were reported in 1989. The independent tests were in agreement; they placed the age of the shroud at 1260 to 1390 A.D. According to these results, the linen dates from the Middle Ages and could not have been the burial shroud of Christ.

A photographic negative of the image of a person on a 4-m-long piece of linen cloth, known as the Shroud of Turin. It was alleged to have been the burial shroud of Jesus Christ.

site. Carbon-14 dating also has been used to detect forgeries of supposedly ancient artifacts (see the box "Dating the Shroud of Turin"). Thus, dating procedures based on the structure and stability of atomic nuclei have become the routine, dependable tools for establishing the age of artifacts.

Tritium, the radioactive isotope of hydrogen, also is useful for dating. Its half-life of 12.3 years makes it useful for dating items up to about 100 years old. An interesting application is the dating of brandies aged for 10 to 50 years. Tritium dating can be used to check the truthfulness of advertising claims about the aging process of the most expensive brands. Many other isotopes are also useful for estimating the ages of objects and materials. Several of the more important ones are listed in Table 18.4.

EXAMPLE 18.10

How old is a piece of fossilized wood that has a carbon-14 activity that is $\frac{1}{8}$ that of new wood? (The half-life of carbon-14 is 5730 years.)

SOLUTION

Because the fraction remaining is $\frac{1}{8}$, the carbon-14 has gone through three half-lives.

$$\tfrac{1}{2} \times \tfrac{1}{2} \times \tfrac{1}{2} = (\tfrac{1}{2})^3 = \tfrac{1}{8}$$

The wood is therefore about 3 × 5730 years = 17,200 years old

See Problems 18.45–18.52.

18.9 Nuclear Fission: Splitting Atoms

Certain chemical reactions—the explosion of nitroglycerine or TNT, for example—can release considerable amounts of energy through the breaking and making of chemical bonds. However, during nuclear fission and fusion reactions, far greater amounts of energy can be released. In **nuclear fission,** the nucleus of a heavy atom absorbs a neutron and then splits apart to give lighter atoms plus two or more neutrons and great quantities of energy. Nuclear energy can be used peacefully to power nuclear reactors that produce electricity, but the energy unleashed during fission was first used to build atomic bombs. Many books have been written about the people and the events that led to the discovery and impending use of nuclear fission; only parts of the story can be included here.

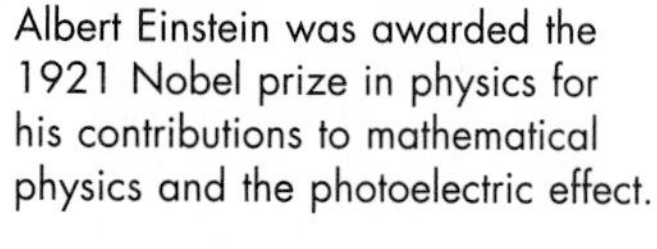
Albert Einstein was awarded the 1921 Nobel prize in physics for his contributions to mathematical physics and the photoelectric effect.

Let's pick up the story with a look at the work of Albert Einstein (1879–1955). Albert Einstein may well be the best known scientist of all time, yet his achievements were those of the mind, not the laboratory. By 1905, Einstein had worked out his special theory of relativity. In doing this, he derived a relationship between matter and energy. The now-famous equation is

$$E = mc^2$$

The energy equivalent to 1 g of matter is 9×10^{13} J. That is enough to heat the average home for a thousand years.

where E represents energy, m represents mass, and c is the speed of light. According to Einstein, energy and mass are interrelated; they are different aspects of the same thing. Einstein's reasoning was clearly verified 40 years later. This verification shook the world.

In 1934 the Italian physicists Enrico Fermi (1901–1954) and Emilio Segrè (1905–1989) bombarded uranium atoms with neutrons. Fermi (Figure 18.17) and Segrè were trying to make elements higher in atomic number than uranium, which had the highest known atomic number at that time. To their surprise, they found *four* radioactive species among the products. They presumed that one of these was element 93, but the scientists were unable to explain the remaining radioactivity. Actually, they had failed to interpret their work properly.

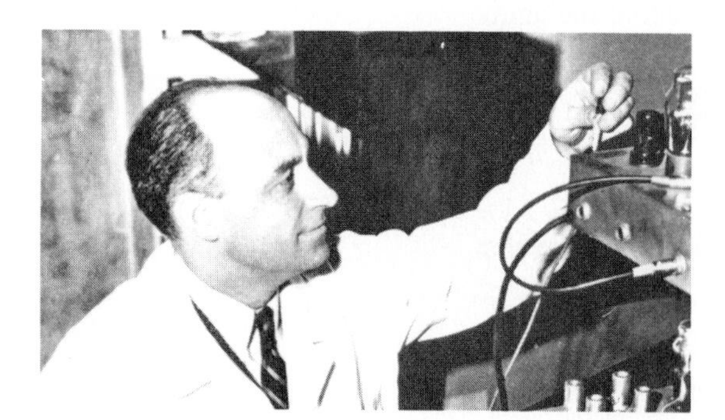

Figure 18.17 *Enrico Fermi (1901–1954) was born in Italy. He received the Nobel prize in physics in 1938. He and his wife, who was Jewish, left fascist Italy and came to the United States in 1939. Fermi worked on the Manhattan project, which developed the first atomic bombs. Element 100, fermium, was later named in his honor.*

Figure 18.18 *Lise Meitner and Otto Hahn in Hahn's laboratory.*

Otto Hahn was awarded the 1944 Nobel prize in chemistry for his work with atomic fission in heavy nuclei.

Two German chemists, Otto Hahn (1879–1968) and Fritz Strassmann (1902–1981) repeated the Fermi–Segrè experiment in 1938. Among the reaction products, they identified barium, Ba, lanthanum, La, and cerium, Ce—elements with atomic masses that are a little more than half that of uranium. They correctly concluded that the uranium atom had been split! In this reaction, it wasn't a matter of a small piece (an alpha or beta prticle, for example) being chipped from the original nucleus. Here, the nucleus was cleaved; it had been split into two major fragments—a process called *nuclear fission.*

Hahn was perplexed by these discoveries and relayed them to Lise Meitner (1878–1968), a Jewish Austrian, who had once worked with him in Berlin (Figure 18.18). She fled to Sweden when Germany annexed Austria in 1938. That is where she was when she received the news from Hahn about the splitting of uranium atoms. She and her nephew, Otto Frisch, an undergraduate student at the University of Copenhagen who was visiting her during the Christmas season, calculated the energy associated with the fission of uranium. They found the energy to be several times greater than that of any previously known nuclear reaction. In addition, the splitting resulted in the release of more neutrons. These could split other uranium atoms, producing enormous amounts of energy (Figure 18.19).

The news of these momentous discoveries was carried to the United States by Niels Bohr (1885–1962), the Danish physicist known for his quantum theory of the electron structure of atoms. Fermi and his wife, who was Jewish, were already in the United States. They had taken advantage of his trip to Stockholm, Sweden, to receive the 1938 Nobel prize in physics to flee fascist Italy and take refuge in the United States.

Scientists quickly realized that massive amounts of energy could be obtained from the fission of uranium. Leo Szilard (1898–1964)—a brilliant Jewish Hungar-

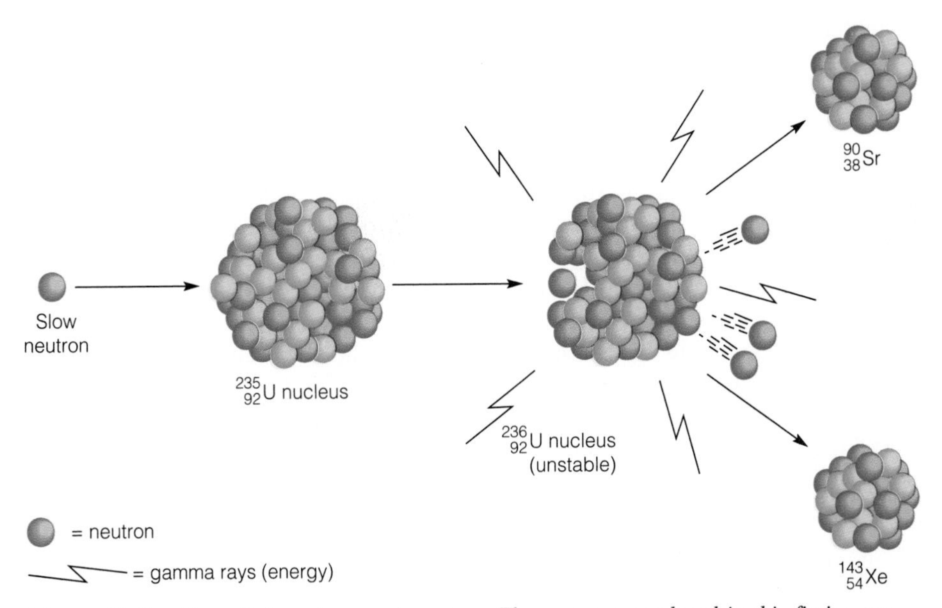

Figure 18.19 *The splitting of a uranium atom. The neutrons produced in this fission can split other uranium atoms, thus sustaining a chain reaction. The splitting of one uranium-235 atom yields 8.90 × 10^{-15} kWh (3.20 × 10^{-14} kJ) of energy. Fission of a mole of uranium-235 (6.02 × 10^{23} atoms) produces 5,300,000 kWh (1.91 × 10^{10} kJ) of energy.*

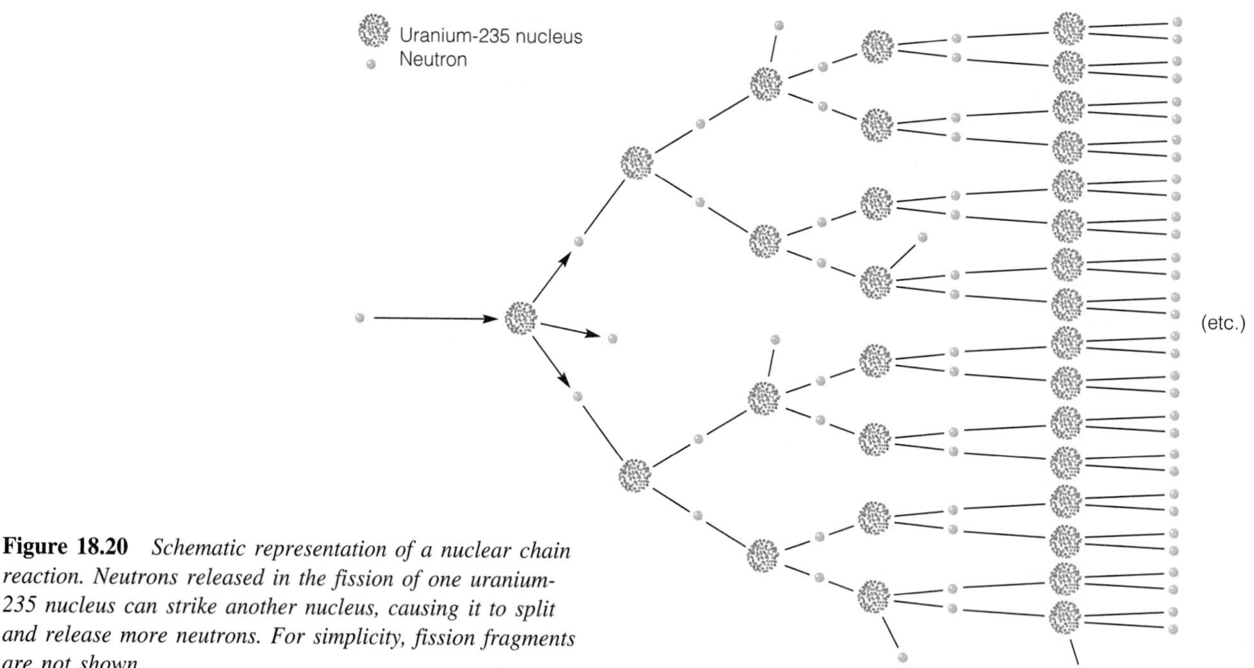

Figure 18.20 *Schematic representation of a nuclear chain reaction. Neutrons released in the fission of one uranium-235 nucleus can strike another nucleus, causing it to split and release more neutrons. For simplicity, fission fragments are not shown.*

ian physicist who left Germany on one of the last trains before the country was sealed off—had envisioned the concept that neutrons released by the fission of one atom could trigger the fission of other uranium atoms and set off a **chain reaction** (Figure 18.20). If rapid enough, this process might produce a bomb with tremendous explosive force. Although Lord Rutherford—known for his discovery of the nucleus—had told Szilard that a nuclear chain reaction was not possible, Szilard visited and discussed his ideas with Fermi.

Aware of the destructive forces that could be produced, and concerned that Germany might develop such a bomb, Fermi and Szilard prevailed on Einstein to write a letter to President Franklin D. Roosevelt indicating the importance of the discovery. After considerable delay, the U.S. government finally launched a massive research project for the study of atomic energy; its code name was the Manhattan Project. Uranium for the project had to be obtained and purified. Very little—less than 2 ounces—of the metal was available in the United States. The Mallinckrodt Company in St. Louis was chosen to extract the uranium from ores obtained from Canada and Belgian Congo. But pure uranium wasn't good enough; the isotopes also had to be separated because only the relatively rare uranium-235 isotope is fissionable.

Having enough uranium-235 was only part of the problem. The neutrons released by the uranium-235 had to be slowed down by a **moderator** to increase the probability that they would be captured by other uranium nuclei prior to fission. Heavy water would make a good moderator—and scientists in Germany had access to heavy water separated from ocean water at a plant in Norway—but none was available in the United States.

As a substitute, Fermi and his group of elite scientists who were gathered at the University of Chicago chose to use long bars of graphite as the moderator to slow down the neutrons. Working in the squash courts under the bleachers of an abandoned football field (Stagg Field) at the University of Chicago, they set out to produce a controlled nuclear reaction. After drilling holes in these graphite bars, and filling them with chunks—some say lumps—of uranium the size of eggs, Fermi's team stacked them tightly in a pile. Over and over they assembled new, better piles with the best uranium available placed nearest the center of the pile. Running through the pile were **control rods** made of cadmium metal, which absorbs neutrons. These rods could be pulled out or pushed back into the pile to control the neutrons and the reaction rate. Finally, on December 2, 1942, under Fermi's direction, the first sustained nuclear reaction was achieved.

To obtain enough of the fissionable uranium-235 isotope to make a bomb is not easy. Every 5000 tons of uranium ore yields only about 10 tons of pure uranium metal, and only about 0.7% of this metal is uranium-235. To make a bomb, ordinary uranium has to be *enriched* to roughly 90% uranium-235. This enrichment proceeded slowly at an isolated, top-secret installation at Oak Ridge, Tennessee.

General Leslie R. Groves acquired Appalachian semiwilderness hill country near Knoxville, Tennessee, and the Great Smoky Mountain National Park for use by the Manhattan Project. Facilities to separate the uranium isotopes were built here, along with a new town that was named Oak Ridge after the rugged local terrain.

Separation of the isotopes could not be accomplished by chemical reactions because the isotopes are nearly identical, chemically. Rather, separation was accomplished by conversion of the uranium to volatile uranium hexafluoride, UF_6. Gaseous molecules of UF_6 containing the uranium-235 isotope are slightly lighter and move a little more rapidly than molecules containing the uranium-238 isotope. Based on these differences, Harold C. Urey of Columbia University designed an elaborate gaseous diffusion process that allowed the gases to pass through thousands of consecutive pinholes. The gaseous molecules that contained uranium-235 gradually outdistanced the others. The scientists eventually obtained 15 kg of the separated uranium-235 isotope—enough to make a small explosive device.

Gaseous diffusion is the procedure used in the production of most of the enriched uranium in the world today. Nuclear power plants require enriched uranium that is about 3% uranium-235. H. C. Urey, the American chemist who designed the gaseous diffusion process, had received the 1934 Nobel prize in chemistry for his discovery of the isotope called deuterium.

While the tedious work of separating uranium isotopes was under way at Oak Ridge, other workers, led by Glenn T. Seaborg at the University of California, Berkeley, approached the problem of obtaining fissionable material by another route. They had been bombarding uranium-238 with neutrons to obtain samples of a new element (element 93) named neptunium, Np, which has a short half-life. Seaborg predicted that traces of element 94 might be found in the decay products of neptunium, and after much work, they were able to obtain and identify a minute sample of element 94, which was later named plutonium, Pu, by Seaborg. (Half-lives are shown beneath the arrows in the equations.)

$$^{238}_{92}U + ^{1}_{0}n \longrightarrow ^{239}_{92}U$$

$$^{239}_{92}U \xrightarrow[23.5\text{ min}]{} ^{239}_{93}Np + ^{0}_{-1}e$$

$$^{239}_{93}Np \xrightarrow[2.35\text{ days}]{} ^{239}_{94}Pu + ^{0}_{-1}e$$

The plutonium-239 isotope was found to be fissionable and, therefore, could be used in making an atomic bomb. To produce this plutonium, several large reactors were built near Hanford, Washington.

Before a fissionable material can sustain a chain reaction, a certain minimum amount, called the **critical mass**, must be brought together. There must be enough fissionable nuclei that the neutrons released in one fission process will have a good chance of being captured by another fissionable nucleus before escaping from the mass. For uranium-235, this critical mass is about 4 kg, a mass about the size of a baseball. To construct a bomb, separate smaller masses are used. These subcritical masses are then brought together forcefully to trigger the runaway chain reaction of a nuclear explosion.

The task of constructing the first uranium and plutonium bombs was put under the direction of J. Robert Oppenheimer, a brilliant physicist with outstanding leadership ability from the University of California, Berkeley, who would head the project at a new city to be built at Los Alamos, New Mexico. By July 1945, enough plutonium had been made for a bomb to be assembled. The first atomic bomb was tested in the desert near Alamogordo, New Mexico, on July 16, 1945. The heat from the explosion vaporized the 30-m steel tower on which it was placed and melted the sand in an area of several thousand square meters around the site. The light released was the brightest anyone had ever seen.

Several of the scientists were so awed by the force of the blast that they argued against its use on Japan. Leo Szilard was perhaps the most outspoken in his opposition to the use of nuclear weapons. He circulated a petition among the Manhattan Project scientists protesting the bomb's impending use. He also hand-carried a document to the White House that argued against use of atomic bombs, saying that even a demonstration of their use would ''precipitate a race in the production of these devices between the United States and Russia.''

The campaign of the concerned scientists was not successful. Fear of a well-publicized ''dud'' and the desire to avoid millions of additional casualties led President Harry S Truman to order the use of the new weapon on, according to his diary, ''military objectives . . . and not women and children.'' However, after the Japanese refused to answer to the unconditional surrender demanded by the Potsdam Declaration, a 4-ton uranium bomb called ''Little Boy'' was loaded onto a B-29 to be piloted by Colonel Paul Tibbets. The 65-ton plane was more than 10% overweight as it thundered down the runway, finally becoming airborne at the last possible takeoff point. It was on its way for a 12-hour round-trip flight. The bomb that was exploded over Hiroshima that morning of August 6, 1945, had a yield of 12,500 tons of TNT (Figure 18.21). It resulted in over 100,000 casualties. Three days later, a plutonium bomb called ''Fat Man'' was dropped on Nagasaki with comparable results. World War II ended with the surrender of Japan on August 14, 1945.

The B-29 that carried that atomic bomb is being restored by the Smithsonian Institution amidst controversy and mixed emotions. Colonel Tibbets had his mother's name, Enola Gay, painted on the side of the plane a few hours before the historic flight to Hiroshima.

The earth-shaking scientific and political events described here point to a fundamental principle: science, politics, and civilization are interwoven. The thread that binds together all of civilization is the recognition of our ultimate vulnerability.

Figure 18.21 *The now familiar mushroom cloud that follows a nuclear explosion.*

Although scientists are sometimes blamed for the nuclear threat, placing such blame on scientists is like blaming the messenger for the message. Otto Hahn and Fritz Strassman did not *invent* nuclear fission—they discovered it. Edwin L. Drake did not invent oil, but his 1859 oil well near Titusville, Pennsylvania, brought a message of its availability. Our future existence is dependent upon our listening to the messengers and the messages of the past and the present and making responsible decisions that impact the quality of life in the future.

EXAMPLE 18.11

From the discussion provided in this section, make one or more relevant statements about each of the following.

(a) uranium-235
(b) heavy water and carbon rods
(c) cadmium metal
(d) enriched uranium
(e) critical mass

SOLUTION

(a) Uranium-235 is a fissionable isotope.
(b) Heavy water and carbon rods are moderators used to slow neutrons.
(c) Cadmium metal is used in control rods that absorb neutrons.
(d) Enriched uranium contains more ^{235}U than typical uranium.
(e) Critical mass is the amount of a fissionable isotope required to sustain a chain reaction.

EXERCISE 18.4

Describe the work of these three pairs: Fermi and Segrè, Hahn and Strassmann, and Meitner and Frisch.

18.10 *Nuclear Power Plants*

Nuclear fission reactors have been used in the United States to produce electrical power since 1957. By 1994 there were 110 licensed nuclear power plants. These reactors permit use of controlled chain reactions to produce great quantities of heat for generating electricity. Although nuclear power plants use the same fission reactions as nuclear bombs, there are some important differences.

Nuclear power supplied 19% of the electricity used in the United States in 1990. Only coal (55%) supplied more electrical energy.

1. The reaction is controlled by the insertion and removal of boron steel or cadmium control rods. Both boron and cadmium absorb neutrons readily and can be used to control the number of neutrons participating in the chain reaction. Pulling the rods out part way starts the chain reaction; pushing the rods back in stops the reaction.
2. Ordinary uranium cannot be used to fuel nuclear power reactors; the uranium must first be *enriched.* That is, the concentration of uranium-235 must be increased from about 0.7% in natural uranium to a concentration of about 3% before it can be used in nuclear power reactors. Because this uranium-235 concentration is quite low when compared with the 85 or 90% concentration required in nuclear fission bombs, a nuclear explosion cannot occur in an electric power reactor.

A CLOSER LOOK

Nuclear Power: No Easy Answers

Nuclear power plants have one great advantage over coal- and oil-burning plants—they do not pollute the air with soot, fly ash, sulfur dioxide, and other chemicals that contribute to acid rain. However, nuclear plants offer some disadvantages as well.

1. The reactor requires heavy shielding to protect operating personnel.
2. Fissionable fuel is rare and expensive. The supply of high-grade uranium ore is quite limited.
3. Fuel rods must eventually be replaced as the fuel becomes spent and as neutron-absorbing fission products accumulate in these rods. The original intent was that spent fuel rods would be sent to reprocessing plants where the remaining fuel could be separated from radioactive wastes, but there has been considerable public opposition to the construction of these plants. Spent fuel rods are currently being stored on reactor sites, but this cannot continue indefinitely.
4. The radioactive fission products present a serious disposal problem. Putting them in deep wells or mines or burying them at sea is like sweeping them under the rug. Do we have the right to leave our descendants with a problem that they will have to contend with for 10,000 years?
5. The waste heat from nuclear power plants heats up the environment. This effect is known as thermal pollution. The problem, however, is not unique to nuclear plants; all fossil-fueled plants also produce thermal pollution.
6. Nuclear power plants release some radioactivity into the environment, no matter how carefully they are constructed. Although proponents of nuclear power say that the amount is negligible, others say that *any* exposure is dangerous.
7. There is the possibility of a major accident—not a nuclear explosion—at a nuclear power plant. Such an accident could release large amounts of radioactivity to surrounding areas as occurred in 1986 at Chernobyl, Ukraine.

There is considerable controversy over most of these points, and scientists stand on each side. While they may agree on the results of laboratory investigations, scientists obviously do not agree on what is best for society.

Public fear of nuclear power was increased by the meltdown of the reactor core of the Soviet reactor at Chernobyl, Ukraine. That reactor did not have a reinforced concrete containment building as is required of all power reactors in the United States.

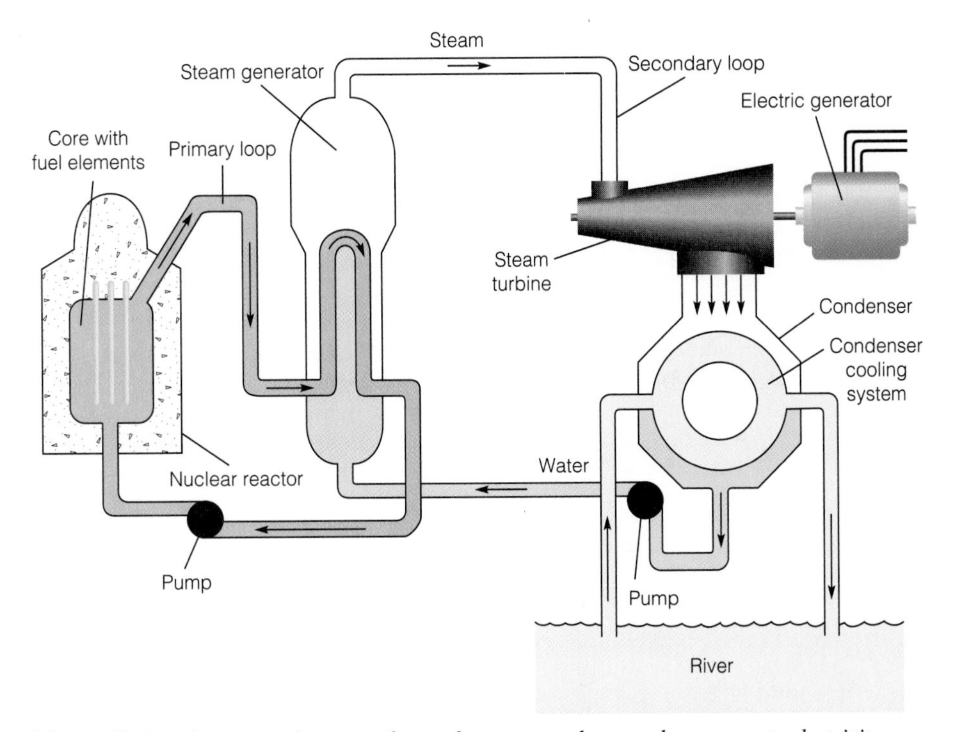

Figure 18.22 *Schematic diagram of a nuclear power plant used to generate electricity.*

3. In light-water reactors commonly used in the United States, ordinary water acts both as a moderator and as a coolant. In order to allow the water to circulate at a temperature of about 350 °C without boiling, a pressure of about 150 atm is maintained in the reactor vessel.

Graphite moderators are used in only a few commercial reactors in the United States, but many Russian reactors still use graphite.

The superheated water in the light-water reactor (Figure 18.22) is not used directly to drive a steam turbine. Instead, the hot water is circulated in a primary loop (a closed pipe system) through a heat exchanger—the steam generator. Water in the reactor vessel does not come in contact with—or mix with—water that is converted into steam. Steam produced in this manner is delivered by way of the secondary loop to the turbine that drives an electric generator. As the cycle continues, great quantities of river water are required to cool the water coming off of the steam turbine before it is circulated back through the steam generator. Characteristics of several types of nuclear reactors are shown in Table 18.5.

In spite of engineering attempts to build safe reactors, there have been some problems. In 1979, a loss-of-coolant accident at the Three Mile Island nuclear power plant near Harrisburg, Pennsylvania, released small amounts of radioactivity into the environment. Although no one was killed or seriously injured, this accident whetted public fear of nuclear power. The 1986 accident at Chernobyl, Ukraine, was even more frightening. The reactor core melt-down killed several people outright. Others died during the following weeks and months from radiation sickness. Thousands were evacuated. A large area will remain contaminated for decades. Radioactive fallout spread across much of Europe. Thousands, particularly those close to the accident, are at greatly increased risk of cancer from exposure to the radiation. At Three Mile Island, a containment building kept most of the radioactive material inside. The Chernobyl plant had no such protective structure.

Table 18.5 **Characteristics of Several Types of Nuclear Reactors**

Reactor	*Abbreviation*	*Fuel*	*Moderator*	*Primary Coolant*	*Fluid in Secondary Loop*
Light water	LWR				
Boiling water	BWR	^{235}U (enriched to 3%)	Water	Water	(No secondary loop)
Pressurized water	PWR	^{235}U (enriched to 3%)	Water	Water	Water
Canadian uranium–deuterium	CANDU	^{235}U (natural abundance)	Heavy water*	Heavy water	Water
High-temperature gas-cooled	HTGR	^{235}U (enriched) to 3%)	Graphite	Helium gas	Water
Liquid-metal fast breeder	LMFBR	^{235}U (at start), then ^{238}U converted to ^{239}Pu	—	Molten sodium	Molten sodium†

*Heavy water is water enriched with the heavy isotope of hydrogen called deuterium (^2_1H). Heavy water often is indicated by the formula D_2O. Other references to water are to ordinary, or light, water.

†A third loop in the LMFBR contains water, which is converted to steam to power the turbine.

With nuclear power, as with most other forms of technology, there are trade-offs. Today's industrialized society depends on electricity to power our businesses and our homes, but satisfying this need requires making some difficult choices.

EXAMPLE 18.12

Describe why a nuclear explosion cannot take place within an electric power reactor.

SOLUTION

See the second point listed in Section 18.10.

See Problems 18.53–18.66.

18.11 *Nuclear Fusion*

We have seen that very large atoms can be split, but it is also possible for nuclei of very small atoms to be united to make larger nuclei—a process known as **nuclear fusion**. During fusion, even more energy can be released than during fission, but fusion cannot be accomplished easily. Exceedingly high temperatures are required if nuclei are to be united. The mass lost during the fusion process is released as energy.

Nuclear fusion is not a new phenomenon. Nearly all of our energy on Earth is derived directly or indirectly from thermonuclear fusion reactions taking place in the sun (Figure 18.23). At the intense temperature in the center of the sun, nuclei fuse and release tremendous amounts of energy. The principal net reaction is believed to be the fusion of four hydrogen nuclei to produce one helium nucleus.

$$4\,^1_1H \longrightarrow \,^4_2He + 2\,_{+1}^{\;\;0}e + \text{Energy}$$

Figure 18.23 *The sun is the ultimate source of nearly all our energy.*

To get an idea of just how much energy is released, consider this comparison: Upon fusing, 1 gram of hydrogen releases an amount of energy equivalent to that produced by burning 17,000 kg (nearly 20 tons) of coal.

The hydrogen bomb—a fusion bomb—makes use of a uranium or plutonium (fission) bomb to provide the tremendous heat necessary to start the fusion reaction. The fusion of ordinary hydrogen, ${}^{1}_{1}H$, occurs much too slowly, so the heavier isotopes deuterium, ${}^{2}_{1}H$, and tritium, ${}^{3}_{1}H$, are used. The intense heat of the fission explosion starts the fusion of hydrogen nuclei.

$${}^{2}_{1}H + {}^{3}_{1}H \longrightarrow {}^{4}_{2}He + {}^{1}_{0}n$$

The neutron released splits lithium atoms, forming more tritium.

$${}^{6}_{3}Li + {}^{1}_{0}n \longrightarrow {}^{4}_{2}He + {}^{3}_{1}H$$

To date, the fusion reactions are useful only for making bombs, but research toward the development of controlled nuclear fusion is progressing. Controlled fusion would have several advantages over nuclear fission reactors. The principal fuel, deuterium, is plentiful and is obtained easily by the fractional electrolysis of water. Although only 1 atom in 5000 hydrogen atoms is a deuterium atom, we have oceans full of water to work with. The problem with radioactive wastes would be minimized. The end product—helium—is stable and biologically inert. Escape of tritium might be a problem because this radioactive hydrogen isotope would be readily incorporated into organisms. Tritium, ${}^{3}_{1}H$, which undergoes beta decay, has a half-life of 12.3 years. Another problem—associated with any production and use of energy—is the unavoidable loss of part of the energy as heat. We would still have to be concerned with thermal pollution.

Great technical difficulties would have to be overcome before a controlled fusion reaction could be used to produce useful energy. It is generally assumed that

1. Temperatures of 50 to 100 million °C would have to be attained, and no material on Earth could withstand more than a few thousand degrees. At such temperatures, all atoms would be stripped of their electrons, and the nuclei and free electrons would form a mixture called a **plasma**, which would need a magnetic field for confinement.
2. An extremely high density of the plasma must be attained.
3. The plasma must be confined long enough at a high temperature and at a high density for the fusion reaction to occur and become sustained.

A CLOSER LOOK

The Nuclear Age: Challenges and Choices

These are exciting times. The goal of the alchemists, to change one element into another, has been achieved through applications of scientific principles. Today, plutonium (atomic number, Z, of 94) is synthesized by the ton. Neptunium ($Z = 93$), americium ($Z = 95$), and curium ($Z = 96$) are produced by the kilogram. Berkelium ($Z = 97$) and einsteinium ($Z = 99$) are produced by the milligram. These new elements have been used in medicine, in home smoke detectors, to power spacecraft, and to build bombs.

Fantastic forces of nature have been unleashed. Nuclear bombs have been used for destruction. Nuclear power reactors are being used to power our cities. Science and scientists have been deeply involved in it all. Still, it is hard to believe that the world would be a better place if we had not discovered the secrets of the atomic nucleus. For one thing, more lives have been saved through nuclear medicine than have been destroyed by nuclear bombs. And no nuclear bombs have been used in warfare since 1945. Perhaps the terror of nuclear holocaust, more than anything else, has prevented World War III.

Nuclear power plants have not been the ultimate answer to all of our energy problems, as was once predicted. Yet, nuclear fission remains—despite many problems—one of our best alternative energy sources. It can supply us with energy until well into the twenty-first century. Most of all, nuclear power gives us more choices. But those choices bring with them even greater responsibilities. Can we handle them?

About 19% of the electrical power produced in the United States is from the 110 operating nuclear power plants that produce about 620 billion kWh of electricity.

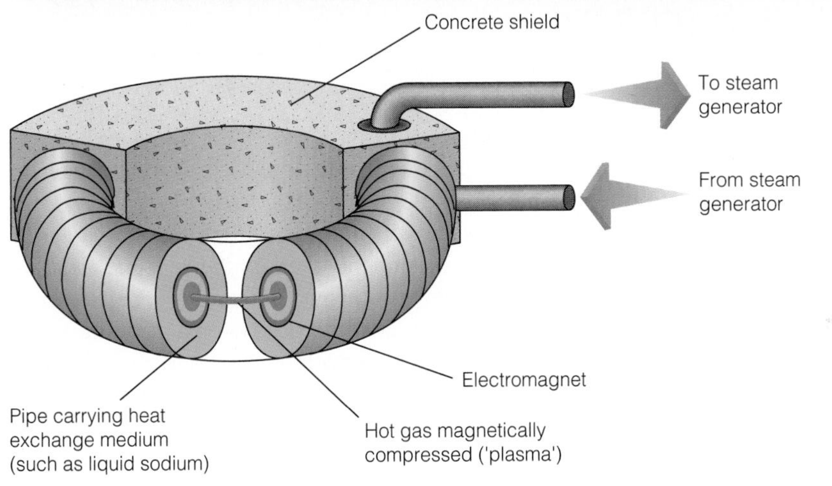

Figure 18.24 *The giant donut-shaped Tokamak Fusion Test Reactor at Princeton University is a 50-foot-high, 40-foot-diameter facility that uses a magnetic field to confine a hot ionized gas (a plasma). Using a 50-50 mixture of deuterium and tritium, the reactor achieved a temperature of 410 million °C and generated a record 9 MW (9 million watts) in 1994. It set the previous power record of 6 MW in 1993. These records are impressive, but the reactor delivered less energy than it consumed, and it can operate only a few seconds at a time. Commercial fusion power is thought to be some 40 years away.*

Research to develop nuclear fusion for power plants is progressing slowly. Two approaches that have some promise are being investigated: (1) *laser fusion* reactors and (2) *magnetic confinement.* The laser approach is being developed at the Los Alamos National Laboratory in New Mexico and at the Lawrence Livermore National Laboratory in California. These reactors use several lasers placed in a spherical arrangement and aimed at a central point where pellets containing deuterium and tritium are dropped into place. The Livermore laboratory is working with Shiva, the world's most powerful laser, but a new laser is under construction. Livermore workers hope to obtain a ''break-even'' point (to produce as much energy as the fusion reactor uses) early in the next century.

The magnetic confinement approach is being developed at the Massachusetts Institute of Technology and at Princeton University in New Jersey. A Tokamak Fusion Test Reactor (TFTR) has been constructed at each institution. These reactors use a doughnut-shaped magnetic field to confine the plasma so it does not come in contact with any material (Figure 18.24). A temperature of 410 million °C has been attained at the Princeton reactor.

Cold Fusion

In 1989 two scientists announced that they had produced ''cold fusion'' (at room temperature), but their claims have not been upheld.

Nuclear fusion may be our best hope for relatively clean, abundant energy in the future, but much work remains to be done. If controlled fusion is achieved in the laboratory, a prototype power plant will have to be built and tested before commercial plants could be constructed. It is unlikely that we will get any significant amount of energy from this source until well into the next century.

EXAMPLE 18.13

Give (a) two examples of *uncontrolled* nuclear fusion reactions and (b) two approaches being used to develop *controlled* nuclear fusion.

SOLUTION

(a) Uncontrolled fusion occurs in the sun and in hydrogen bombs.
(b) Controlled fusion approaches include laser fusion and magnetic confinement.

Chapter Summary

Approximately 80 of some 350 naturally occurring isotopes are radioactive. Each radioactive isotope is unstable and spontaneously emits radiation from its nucleus. The types of radiation emitted include (1) alpha particles, ${}^4_2He$, where the atomic number of the original nucleus is decreased by 2 and the mass number is decreased by 4; (2) beta particles, ${}^{0}_{-1}e$, where the atomic number of the original nucleus increases by 1 and the mass number remains unchanged; and (3) high-energy gamma radiation, γ, which has neither mass nor charge. Gamma rays are often emitted along with alpha and beta particles when a nucleus returns to a more stable state, in contrast to X-rays, which are produced during certain transitions of electrons between energy levels.

The period of time required for one-half of a specific sample of a radioactive isotope to decay is its *half-life*. The fraction of the original sample remaining after a certain number of half-lives, n, is equal to $1/(2^n)$.

When an isotope of one element decays, an isotope of a different element is produced; that is, one type of atom is changed into another. A nuclear equation represents the changes involved. Each particle in a nuclear equation is symbolized in the form ${}^A_ZX$, to indicate the atomic number (or nuclear charge), Z, and the mass number, A, of the particle. When balancing a nuclear equation, the sums of mass numbers of each particle (the superscripts) on each side of the equation must be equal, and the sums of nuclear charges (the subscripts) on each side of the equation must be equal.

By bombarding the nuclei of certain elements with various kinds of particles, different elements can be produced—one kind of atom is changed into another. The process is called artificial transmutation because the change would not have occurred naturally. Work with nuclear reactions of this type led to the discovery of several subatomic particles and to the synthesis of several transuranium elements. Artificial transmutation is also used to prepare radioisotopes for industrial and medical applications.

Radioisotopes are used frequently as tracers in research, in industry, and in medicine. Nuclear radiation is used in medicine to assist in the diagnosis and treatment of disease. Examples include the use of iodine-131 in the diagnosis and treatent of thyroid disorders, cobalt-60 in cancer therapy, and technetium-99^m and gadolinium-153 for various diagnostic tests.

When bombarded by neutrons of an appropriate energy, certain unstable nuclei (uranium-235 or plutonium-239) undergo fission. After absorbing a neutron, the fissionable nucleus splits to give two medium-sized nuclei and two or more neutrons. The process also releases tremendous quantities of energy. If a certain minimum amount—a critical mass—of a fissionable isotope is brought together, a self-sustaining chain reaction can take place. In a fission bomb, the chain reaction is uncontrolled. In nuclear power plants, the reaction is controlled by neutron-absorbing control rods. The concentration of uranium-235 is low (enriched to about 3%) for use in power plants, but is quite high (enriched to about 85 to 90%) in fission bombs.

During nuclear fusion, small nuclei—such as deuterium and tritium—are united. Reactions of this type release even more energy than is released during fission. Uncontrolled nuclear fusion occurs in the sun and in hydrogen bombs. Controlled nuclear fusion has not been achieved, but research continues.

Nuclear chemistry has become extremely important in medicine; it has made substantial contributions to industry and agriculture. Hardly a single facet of our lives has been left untouched by developments in nuclear science. The historical events described in this chapter also point to the fact that knowledge is not limited to any nation or group. When people from diverse backgrounds work together, much can be achieved. Also, science does not stand alone: solving the major problems of our age depends on our ability to blend business, science, politics, and human values.

Assess Your Understanding

1. Balance nuclear equations. [18.1]
2. Give characteristics of alpha, beta, and gamma radiation. [18.1]
3. Make calculations involving half-life and the sample fraction remaining. [18.2]
4. Distinguish differences in meanings of curies, roentgens, rads, and rems. [18.3]
5. Describe ionizing radiation and how it can be detected. [18.4]
6. Describe background radiation and contributing sources. [18.5]
7. Describe transmutation and give examples. [18.6, 18.7]
8. Give examples of medical and other practical uses of radioisotopes. [18.8]
9. Describe major events surrounding the discovery of nuclear fission. [18.9]
10. Explain why nuclear power plants cannot explode like nuclear bombs. [18.10]
11. Compare nuclear fission and nuclear fusion. [18.10, 18.11]

Key Terms

alpha decay [18.1]
alpha particle [18.1]
artificial transmutation [18.6]
background radiation [18.5]
beta decay [18.1]
beta particle [18.1]
chain reaction [18.9]
control rods [18.9]
critical mass [18.9]
curie [18.3]
disintegrations per second [18.3]
gamma decay [18.1]
gamma rays [18.1]
Geiger counter [18.4]
half-life [18.2]
ionizing radiation [18.3]
irradiate [18.8]
moderator [18.9]
natural radioactivity [18.1]
nuclear equation [18.1]
nuclear fission [18.9]
nuclear fusion [18.11]
plasma [18.11]
positrons [18.7]
rad [18.3]
radioisotope [18.2]
rem [18.3]
roentgen [18.3]
tracers [18.8]
transmutation [18.1]
X-rays [18.1]

Problems

NATURAL RADIOACTIVITY

18.1 Approximately how many isotopes occur naturally? How can this number be greater than the number of elements shown on the periodic table?

18.2 Approximately how many naturally occurring isotopes are radioactive?

18.3 Using the form $^{A}_{Z}X$, give the symbol for an alpha particle. Show the Greek letter that is also used for alpha radiation.

18.4 Using the form $^{A}_{Z}X$, give the symbol for a beta particle. Show the Greek letter that is also used for beta radiation.

18.5 For alpha decay, explain what happens to the atomic number and the mass number of the original nucleus. Refer to Figure 18.1.

18.6 For beta decay, explain what happens to the atomic number and the mass number of the original nucleus. Refer to Figure 18.2.

18.7 Describe gamma rays, their mass, charge, and source.

18.8 How are gamma rays and X-rays similar? How are they different?

18.9 Using the form $^{A}_{Z}X$, where Z = atomic number, A = mass number, and X = particle symbol, write nuclear symbols for
a. an isotope with $Z = 27$ and $A = 60$
b. protium (hydrogen-1)
c. an alpha particle
d. radium-226

18.10 Using the form $^{A}_{Z}X$ where Z = atomic number, A = mass number, and X = particle symbol, write nuclear symbols for
a. an isotope with $Z = 53$ and $A = 131$
b. tritium (hydrogen-3)
c. a neutron
d. radon-222

18.11 Using the form $^{A}_{Z}X$, write nuclear symbols for
a. deuterium (hydrogen-2)
b. potassium-39 (not radioactive)
c. potassium-40 (radioactive)
d. phosphorus-32

18.12 Using the form $^{A}_{Z}X$, write nuclear symbols for
a. a beta particle
b. iron-56 (not radioactive)
c. iron-59 (radioactive)
d. carbon-14

18.13 Which of the following pairs represent isotopes of the same element?
a. $^{70}_{34}X$ and $^{70}_{33}X$
b. $^{186}_{74}X$ and $^{184}_{74}X$
c. $^{22}_{11}X$ and $^{44}_{22}X$
d. $^{35}_{17}X$ and $^{37}_{17}X$

18.14 Which of the following pairs represent isotopes of the same element?
a. $^{57}_{28}X$ and $^{66}_{28}X$
b. $^{8}_{2}X$ and $^{6}_{4}X$
c. $^{123}_{51}X$ and $^{123}_{52}X$
d. $^{55}_{26}X$ and $^{56}_{26}X$

18.15 Polonium, a naturally occurring radioactive element that emits alpha particles, was discovered in 1898 by Marie Curie. She named the element after her native country of Poland. Write a balanced nuclear equation for the alpha decay of $^{214}_{84}Po$. What element is produced by this decay?

18.16 Radium is a naturally occurring radioactive element that was discovered by Marie and Pierre Curie. Write a balanced nuclear equation for the alpha decay of $^{226}_{88}Ra$. How can radon-222 be present in homes?

18.17 Complete each of the following nuclear equations.
a. $^{238}_{92}U \rightarrow ^{234}_{90}Th + ?$ b. $^{234}_{90}Th \rightarrow ^{234}_{91}Pa + ?$
c. $^{234}_{91}Pa \rightarrow ^{234}_{92}U + ?$ d. $^{234}_{92}U \rightarrow ^{230}_{90}Th + ?$

18.18 Complete each of the following nuclear equations.
a. $^{214}_{82}Pb \rightarrow ^{214}_{83}Bi + ?$ b. $^{214}_{83}Bi \rightarrow ^{214}_{84}Po + ?$
c. $^{214}_{84}Po \rightarrow ^{210}_{82}Pb + ?$ d. $^{210}_{82}Pb \rightarrow ^{210}_{83}Bi + ?$

18.19 Considering alpha, beta, and gamma radiation, which form(s) would travel through a sheet of paper?

18.20 Considering alpha, beta, and gamma radiation, which form(s) would travel through a sheet of aluminum?

18.21 A pair of rubber gloves would be sufficient to shield the hands from which type of radiation: the heavy alpha particles or the massless gamma rays?

18.22 Heavy lead shielding is necessary as protection from which type of radiation: alpha, beta, or gamma?

HALF-LIFE

18.23 C. E. Bemis and colleagues at Oak Ridge National Laboratory confirmed the synthesis of element 104, which has a half-life of 4.5 s. Only 3000 atoms of the element were created in the tests. Give the number of atoms left after 4.5 s and after 9.0 s.

18.24 Iron-59 can be used to study the intestinal absorption of iron. If 6.00 g of the isotope is present when purchased, how many grams of the isotope would be present 90 days later? How many grams remain after 135 days? The half-life of iron-59 is 45 days.

18.25 How old is a bottle of wine if the tritium activity is 25% that of new wine? The half-life of tritium is 12.3 years.

18.26 Living matter has a carbon-14 content that gives 16 counts per minute per gram of carbon. What is the age of an artifact for which the carbon-14 gives 4 counts per minute per gram of carbon? The half-life of $^{14}_{6}C$ is 5730 years.

18.27 Krypton-81^{m} is used for lung ventilation studies. Its half-life is 13 s. How long will it take the activity of this isotope to reach one-fourth of its original value?

18.28 Iodine-131 is used in the diagnosis and treatment of thyroid cancer. Its half-life is 8.0 days. How long will it take the activity of this isotope to reach one-eighth of its original value?

RADIATION UNITS AND DETECTORS

18.29 What makes ionizing radiation hazardous to living things?

18.30 How many rads would kill most people? What is the average number of rads received by an individual from medical and dental X-rays each year?

18.31 Name one type of instrument used to detect ionizing radiation.

18.32 Name one type of detector used to record the total amount of radiation received by a person over a period of time (such as one month).

BACKGROUND RADIATION

18.33 Of some 350 naturally occurring isotopes, about how many are radioactive?

18.34 Name four elements that have naturally occurring isotopes.

18.35 The box "Radon-222 in Our Lives" indicates that radon-222 causes greater concern than the other two isotopes of radon. Explain.

18.36 Describe the connection between cigarette smoking, radon-222, and lung cancer.

18.37 List four sources of natural background radiation (see Figure 18.10).

18.38 Figure 18.10 shows that 18% of background radiation is from synthetic sources. What percentages are attributable to medical X-rays, nuclear medicine, and consumer products?

TRANSMUTATION AND INDUCED RADIOACTIVITY

18.39 Write a balanced nuclear equation for the emission of a positron by sulfur-31.

18.40 An individual receives a small amount of a positron-emitting isotope prior to a PET scan. Write a balanced nuclear equation for positron emission by carbon-11.

18.41 Several kinds of particles can be used to bombard atoms while carrying out transmutation reactions. Complete the following equations involving transmutations.
a. $^{10}_{5}B + ^{1}_{0}n \rightarrow ^{4}_{2}He + ?$
b. $^{12}_{6}C + ^{2}_{1}H \rightarrow ^{13}_{6}C + ?$
c. $^{121}_{51}Sb + ? \rightarrow ^{121}_{52}Te + ^{1}_{0}n$

18.42 Several kinds of particles can be used to bombard atoms while carrying out transmutation reactions. Complete the following equations involving transmutations.
a. $^{154}_{62}Sm + ^{1}_{0}n \rightarrow 2\ ^{1}_{0}n + ?$
b. $^{7}_{3}Li + ^{1}_{1}H \rightarrow ^{7}_{4}Be + ?$
c. $^{106}_{46}Pd + ^{4}_{2}He \rightarrow ? + ^{1}_{1}H$

18.43 Complete the following equations to show which transuranium element can be produced during bombardment of the target nucleus.

a. ${}^{239}_{94}Pu + {}^{1}_{0}n \rightarrow {}^{0}_{-1}e + ?$

b. ${}^{239}_{94}Pu + {}^{4}_{2}He \rightarrow {}^{1}_{0}n + ?$

c. ${}^{241}_{95}Am + {}^{4}_{2}He \rightarrow 2\ {}^{1}_{0}n + ?$

d. ${}^{242}_{96}Cm + {}^{4}_{2}He \rightarrow {}^{1}_{0}n + ?$

18.44 Complete the following equations to show which transuranium element can be produced during bombardment of the target nucleus.

a. ${}^{238}_{92}U + {}^{14}_{7}N \rightarrow 6\ {}^{1}_{0}n + ?$

b. ${}^{238}_{92}U + {}^{16}_{8}O \rightarrow 4\ {}^{1}_{0}n + ?$

c. ${}^{253}_{99}Es + {}^{4}_{2}He \rightarrow {}^{1}_{0}n + ?$

d. ${}^{246}_{96}Cm + {}^{12}_{6}C \rightarrow 4\ {}^{1}_{0}n + ?$

PRACTICAL USES OF SOME RADIOISOTOPES

18.45 Radioactive iodine-131 is used in therapy for the treatment of cancer of the thyroid. What happens to the cancer cells of the thyroid?

18.46 How is radioactive iodine-131 used for diagnostic purposes in obtaining a photoscan of the thyroid?

18.47 Give two examples of how radioisotopes other than iodine-131 can be used in the treatment of disease (therapy).

18.48 Give two examples of how radioisotopes other than iodine-131 can be used as a diagnostic tool in medicine.

18.49 Does exposing food to radiation (irradiation) make it radioactive?

18.50 When foodstuffs are irradiated, how does this help in their preservation?

18.51 Give one way plutonium-238, a radioisotope of a transuranium element, is used to help people.

18.52 Give one practical use of americium-241.

NUCLEAR FISSION AND FUSION

18.53 What is meant by a *critical mass*?

18.54 Uranium has a density of 19 g/cm^3. What volume is occupied by a critical mass of 8.0 kg of uranium?

18.55 Which subatomic particles are responsible for carrying on the chain of reactions that are characteristic of nuclear fission?

18.56 In order for a chain reaction to occur, how does the number of neutrons captured by a fissionable atom compare with the number of neutrons released during fission?

18.57 What is the purpose of using a *moderator* in a fission reaction? Give two examples of materials that can be used as moderators.

18.58 What is the function of control rods in nuclear reactors? How are the control rods manipulated in a nuclear reactor?

18.59 What is the only naturally occurring isotope that can undergo fission?

18.60 It was discovered that one of the first synthetic elements also has a fissile isotope. What is the name of this element?

18.61 When Fermi and Segrè bombarded uranium with neutrons, what did they think had happened?

18.62 When two German chemists, Hahn and Strassmann, repeated the Fermi–Segrè experiment, they identified barium, lanthanum, and cerium among the reaction products. What was their (correct) conclusion?

18.63 Give three important differences between a fission reaction in a nuclear bomb and fission in a nuclear power plant.

18.64 Give three problems with nuclear power plants.

18.65 What must occur during nuclear fusion? Give two cases where fusion takes place.

18.66 Give a specific equation that illustrates nuclear fusion.

Additional Problems

18.67 What are some practical medical uses of technetium-99^{m}?

18.68 Describe some of the characteristics that make technetium-99^{m} such a useful radioisotope for diagnostic purposes.

18.69 Give the number of protons, number of neutrons, mass number, and symbol for the following.

a. ${}^{62}_{30}X$ b. ${}^{241}_{94}X$ c. ${}^{32}_{15}X$

18.70 Give the number of protons, number of neutrons, mass number, and symbol for the following.

a. ${}^{40}_{19}X$ b. ${}^{59}_{27}X$ c. ${}^{51}_{24}X$

18.71 Complete the following equations to determine which particle would be used in a particle accelerator to bombard the given target nucleus in preparing transuranium elements with atomic numbers 103 and 104.

a. ${}^{250}_{98}Cf + ? \rightarrow {}^{257}_{103}Lr + 4\ {}^{1}_{0}n$
b. ${}^{249}_{98}Cf + ? \rightarrow {}^{257}_{104}Unq + 4\ {}^{1}_{0}n$

18.72 Complete the following equations to determine which particle would be used in a particle accelerator to bombard the given target nucleus in preparing transuranium elements with atomic numbers 105 and 106.
a. ${}^{249}_{98}Cf + ? \rightarrow {}^{260}_{105}Unp + 4\ {}^{1}_{0}n$
b. ${}^{249}_{98}Cf + ? \rightarrow {}^{263}_{106}Unh + 4\ {}^{1}_{0}n$

18.73 Determine the number of neutrons produced by the fission reaction

$${}^{235}_{92}U + {}^{1}_{0}n \longrightarrow {}^{139}_{56}Ba + {}^{94}_{36}Kr + ?\ {}^{1}_{0}n$$

How does the number of neutrons released compare to the number captured?

18.74 Determine the number of neutrons produced by the fission reaction

$${}^{235}_{92}U + {}^{1}_{0}n \longrightarrow {}^{144}_{54}Xe + {}^{90}_{38}Sr + ?\ {}^{1}_{0}n$$

How does the number of neutrons released compare to the number captured?

18.75 Did President Harry S Truman make the right decision when he decided to drop the nuclear bombs on Japanese cities? Would your answer be the same if you were living in 1945 and had relatives among the U.S. troops preparing for the invasion of Japan? If you were an inhabitant of one of the bombed cities?

18.76 Describe the scientific events that led up to the race toward a fission bomb in the United States. What international political events were involved?

18.77 One atom of element 109 with a mass number of 266 was produced in 1982 by bombarding a target of bismuth-209 with iron-58 nuclei for 1 week. How many neutrons were released in the process?

$${}^{209}_{83}Bi + {}^{58}_{26}Fe \longrightarrow {}^{266}_{109}Une + ?\ {}^{1}_{0}n$$

18.78 Element 109 undergoes alpha emission to form element 107, which in turn also emits an alpha particle. Write balanced nuclear equations for the two reactions.

18.79 What does PET in PET scan stand for? What are positrons? What happens to them?

18.80 Give five uses for PET scans.

From disposable plastic products for parties and packaging, to polymer fibers for swimsuits, stylish clothes, and carpets, organic chemicals are all around us.

19

Organic Chemistry

CONTENTS

More than 9 million known compounds contain carbon. Organic chemistry is defined simply as the chemistry of these carbon-containing compounds. By contrast, inorganic chemistry refers to the chemistry of all other elements and approximately 500,000 inorganic compounds.

Organic chemistry originally referred to the study of carbon-containing compounds that are present in—or are produced by—living organisms, including substances of either plant or animal origin. The definition of *organic* had to be expanded when chemists discovered that certain organic compounds can be produced from inorganic compounds. Other organic compounds can be synthesized that are totally unlike any compounds present in living organisms.

A small number of carbon-containing compounds are classified as inorganic compounds. These include compounds that contain the carbonate ion, CO_3^{2-}, the

Figure 19.1 *The word* organic *has several different meanings. Organic fertilizer is organic in the original sense that it is derived from a living organism. Organic foods are those produced without pesticides, hormones, or synthetic additives. Organic chemistry is the chemistry of carbon compounds.*

bicarbonate ion, HCO_3^-, and the cyanide ion, CN^-, because compounds containing these ions have properties that are similar to those of other inorganic compounds; they are not at all like typical organic compounds. Most organic chemicals we use are derived from compounds present in oil, natural gas, and coal. They include fuels, synthetic fibers, plastics, resins, medicines, pesticides, and herbicides. Organic compounds are also present in living organisms. These include carbohydrates, lipids (fats and related compounds), amino acids, proteins, vitamins, hormones, and enzymes.

Considerable confusion is associated with everyday use of the word *organic.* Organic fertilizer is organic in the original sense that it is derived from living organisms. There is no federal legal definition of organic foods, but the U.S. Department of Agriculture is developing a set of standards to assure that foods labeled organic are grown and handled without pesticides, hormones, or synthetic additives (Figure 19.1).

Because there are so many organic compounds, they are grouped into classes or families with similar structural and bonding characteristics. In this chapter we will consider a wide variety of important organic compounds.

19.1 General Properties of Organic and Inorganic Chemicals

Some specific properties of sodium chloride, an inorganic compound, and benzene, an organic compound, are listed in Table 19.1. Consider the following comparisons of typical organic and inorganic chemicals.

Melting Point. Typical organic compounds have relatively low melting points. Many, like benzene and ethanol (ethyl alcohol), are liquids at room temperature. Most inorganic salts, by contrast, have high melting points.

Solubility and Density. Most organic compounds are insoluble in water but are soluble in organic liquids. (Polar organic compounds with low molar masses and those that can form strong hydrogen bonds are water-soluble.) Most organic liquids are less dense than water, and, like oil, they float on top of water if we attempt to dissolve them.

Flammability. Typical organic compounds are flammable—some are highly flammable. Some, like gasoline, form explosive mixtures with air and must not be used near an open flame. Typical inorganic compounds are nonflammable. Some, such as water and sodium bicarbonate, are even used in fighting fires.

Table 19.1 **Comparison of an Organic and an Inorganic Compound**

	Benzene	*Sodium Chloride*
Formula	C_6H_6	NaCl
Solubility in H_2O	Insoluble	Soluble
Solubility in gasoline	Soluble	Insoluble
Flammable?	Yes	No
Melting point	5.5 °C	801 °C
Boiling point	80 °C	1413 °C
Density	0.88 g/cm^3	2.7 g/cm^3 (crystal)
Bonding	Covalent	Ionic

Bonding. Typical properties of organic compounds are related to the fact that these compounds are composed of molecules with covalent bonds. The typical inorganic compound is ionic. Recall that water solutions of ionic compounds conduct an electric current, but solutions of molecular substances—those with covalent bonds—are nonconductors.

EXAMPLE 19.1

Upon investigation of an automobile gas line antifreeze, it was found that the liquid was flammable and had a low boiling point of 64 °C. The liquid was water-soluble, but it did not conduct an electric current. Classify the chemical as being either organic or inorganic.

SOLUTION

These properties are consistent with those of many low molecular weight organic compounds. (Further analysis showed that the chemical in question was actually methanol, CH_3OH.)

See Problems 19.1–19.6.

19.2 Alkanes: The Saturated Hydrocarbons

Before one can even begin to understand the large, complex molecules on which life is based, it is necessary to learn something about simpler organic molecules. We will start with organic molecules containing only two elements, carbon and hydrogen. These compounds are called **hydrocarbons**. In Section 7.12 we described the covalent bonding of the simplest hydrocarbon called methane, CH_4. Recall from Chapter 7 that methane has a tetrahedral shape (Figure 19.2).

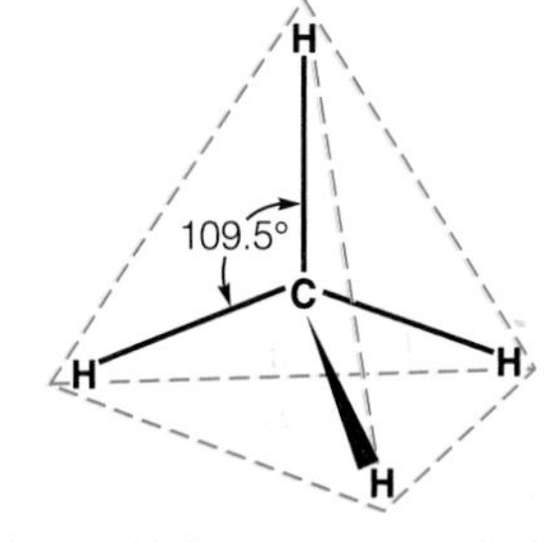

Figure 19.2 *The tetrahedral methane, CH_4, molecule. The solid lines represent covalent bonds, the dashed lines outline the tetrahedron.*

Methane is the first member of a series of related compounds called the **alkanes**, which are **saturated hydrocarbons**. Saturated, in this case, means that each carbon atom in the compound is bonded to four other atoms by single bonds. (Hydrocarbons with double or triple bonds are members of other families and are discussed in subsequent sections.)

Structural formulas of the simplest two alkanes, methane and ethane, with molecular formulas of CH_4 and C_2H_6, respectively, are shown here.

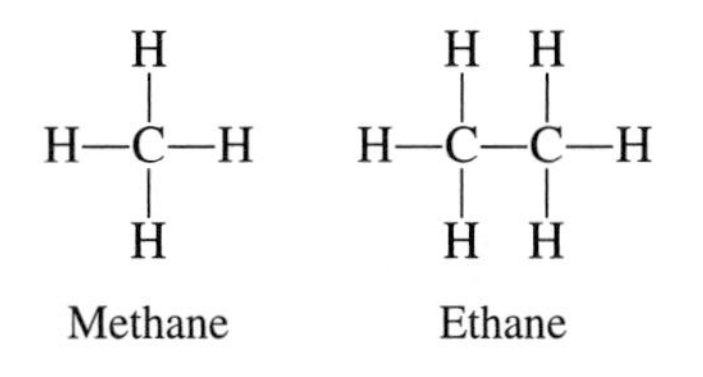

Structural formulas do not accurately show the three-dimensional geometry of molecules. The bond angles in these alkanes—and all others—are 109.5°. However, unlike molecular formulas, structural formulas show the order in which atoms are attached. Ball and stick and space-filling models of methane and ethane are shown in Figure 19.3.

The three-carbon alkane, C_3H_8, is called propane. Models of propane are shown in Figure 19.4. The structural formula of propane is shown at the right.

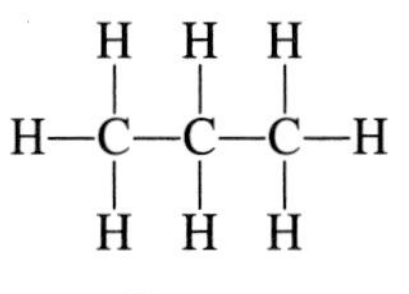

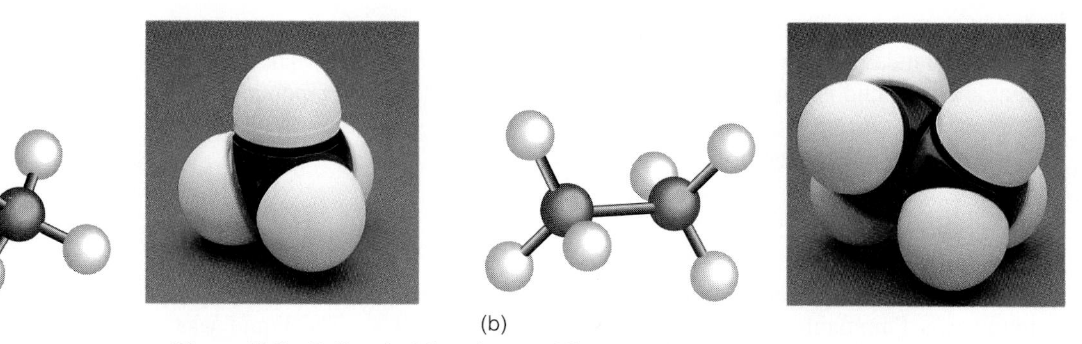

Figure 19.3 *Ball-and-stick and space-filling models of methane (a) and ethane (b).*

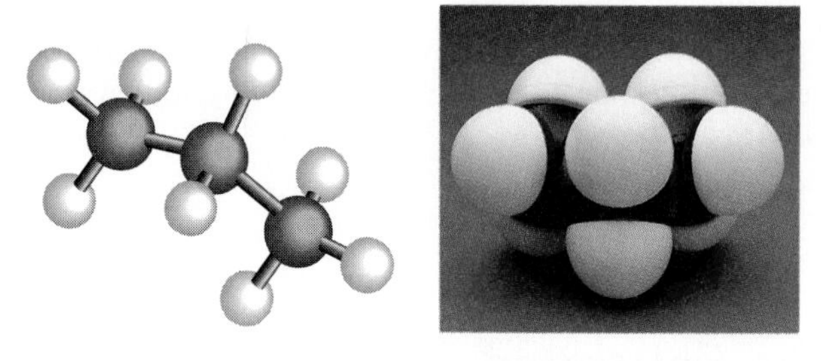

Figure 19.4 *Ball-and-stick and space-filling models of propane.*

Table 19.2 **Prefixes That Indicate the Number of Carbon Atoms in Organic Molecules**

Prefix	*Number*
Meth-	1
Eth-	2
Prop-	3
But-	4
Pent-	5
Hex-	6
Hept-	7
Oct-	8
Non-	9
Dec-	10

A pattern is now becoming apparent. We can build alkanes of any length simply by putting carbon atoms together in long chains and adding sufficient hydrogen atoms to give each of the carbon atoms a total of four covalent bonds. Even the naming of these compounds follows a pattern. For compounds of five carbon atoms or more, each stem or root is derived from the Greek or Latin name for the number of carbon atoms in the molecule, as shown in Table 19.2. The suffix, *-ane*, indicates that the compound is an alk*ane*. Table 19.3 gives structural formulas and names for the first 10 continuous-chain alkanes; that is, alkanes where each carbon is bonded to no more than two other carbon atoms. We do not need to stop at 10 carbon atoms; alkanes can contain 100 or 1000 or more than 1 million carbon atoms. Although we can make a tremendous number of alkanes simply by lengthening the chain, this is not the only option. With four carbon atoms or more, chain branching is also possible. For example, butane, C_4H_{10}, boils at −0.5 °C and has the structural formula

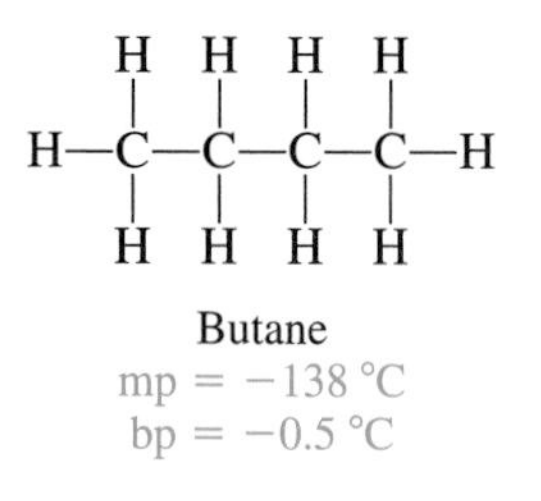

Butane
mp = −138 °C
bp = −0.5 °C

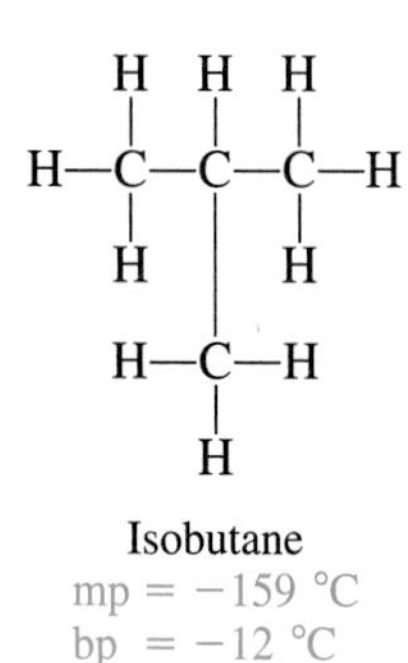

Isobutane
mp = −159 °C
bp = −12 °C

A second compound has the same molecular formula, C_4H_{10}, but it boils at −12 °C. The structural formula of this second compound is different from that of butane. Instead of having four carbon atoms connected in a continuous chain, this new compound has a continuous chain of only three carbon atoms. The fourth carbon is branched off the middle carbon of the three-carbon chain.

Table 19.3 **The First 10 Continuous-Chain Alkanes**

Name	*Molecular Formula*	*Structural Formula*	*Number of Possible Isomers*
Methane	CH_4	$\begin{array}{ccccc} & & H & & \\ & & \vert & & \\ H & - & C & - & H \\ & & \vert & & \\ & & H & & \end{array}$	1
Ethane	C_2H_6	$\begin{array}{ccccccc} & & H & & H & & \\ & & \vert & & \vert & & \\ H & - & C & - & C & - & H \\ & & \vert & & \vert & & \\ & & H & & H & & \end{array}$	1
Propane	C_3H_8	$\begin{array}{ccccccccc} & & H & & H & & H & & \\ & & \vert & & \vert & & \vert & & \\ H & - & C & - & C & - & C & - & H \\ & & \vert & & \vert & & \vert & & \\ & & H & & H & & H & & \end{array}$	1
Butane	C_4H_{10}	$\begin{array}{ccccccccccc} & & H & & H & & H & & H & & \\ & & \vert & & \vert & & \vert & & \vert & & \\ H & - & C & - & C & - & C & - & C & - & H \\ & & \vert & & \vert & & \vert & & \vert & & \\ & & H & & H & & H & & H & & \end{array}$	2
Pentane	C_5H_{12}	$\begin{array}{ccccccccccccc} & & H & & H & & H & & H & & H & & \\ & & \vert & & \vert & & \vert & & \vert & & \vert & & \\ H & - & C & - & C & - & C & - & C & - & C & - & H \\ & & \vert & & \vert & & \vert & & \vert & & \vert & & \\ & & H & & H & & H & & H & & H & & \end{array}$	3
Hexane	C_6H_{14}	$\begin{array}{ccccccccccccccc} & & H & & H & & H & & H & & H & & H & & \\ & & \vert & & \vert & & \vert & & \vert & & \vert & & \vert & & \\ H & - & C & - & C & - & C & - & C & - & C & - & C & - & H \\ & & \vert & & \vert & & \vert & & \vert & & \vert & & \vert & & \\ & & H & & H & & H & & H & & H & & H & & \end{array}$	5
Heptane	C_7H_{16}	$\begin{array}{ccccccccccccccccc} & & H & & H & & H & & H & & H & & H & & H & & \\ & & \vert & & \vert & & \vert & & \vert & & \vert & & \vert & & \vert & & \\ H & - & C & - & C & - & C & - & C & - & C & - & C & - & C & - & H \\ & & \vert & & \vert & & \vert & & \vert & & \vert & & \vert & & \vert & & \\ & & H & & H & & H & & H & & H & & H & & H & & \end{array}$	9
Octane	C_8H_{18}	$\begin{array}{ccccccccccccccccccc} & & H & & H & & H & & H & & H & & H & & H & & H & & \\ & & \vert & & \vert & & \vert & & \vert & & \vert & & \vert & & \vert & & \vert & & \\ H & - & C & - & C & - & C & - & C & - & C & - & C & - & C & - & C & - & H \\ & & \vert & & \vert & & \vert & & \vert & & \vert & & \vert & & \vert & & \vert & & \\ & & H & & H & & H & & H & & H & & H & & H & & H & & \end{array}$	18
Nonane	C_9H_{20}	$\begin{array}{ccccccccccccccccccccc} & & H & & H & & H & & H & & H & & H & & H & & H & & H & & \\ & & \vert & & \vert & & \vert & & \vert & & \vert & & \vert & & \vert & & \vert & & \vert & & \\ H & - & C & - & C & - & C & - & C & - & C & - & C & - & C & - & C & - & C & - & H \\ & & \vert & & \vert & & \vert & & \vert & & \vert & & \vert & & \vert & & \vert & & \vert & & \\ & & H & & H & & H & & H & & H & & H & & H & & H & & H & & \end{array}$	35
Decane	$C_{10}H_{22}$	$\begin{array}{ccccccccccccccccccccccc} & & H & & H & & H & & H & & H & & H & & H & & H & & H & & H & & \\ & & \vert & & \vert & & \vert & & \vert & & \vert & & \vert & & \vert & & \vert & & \vert & & \vert & & \\ H & - & C & - & C & - & C & - & C & - & C & - & C & - & C & - & C & - & C & - & C & - & H \\ & & \vert & & \vert & & \vert & & \vert & & \vert & & \vert & & \vert & & \vert & & \vert & & \vert & & \\ & & H & & H & & H & & H & & H & & H & & H & & H & & H & & H & & \end{array}$	75

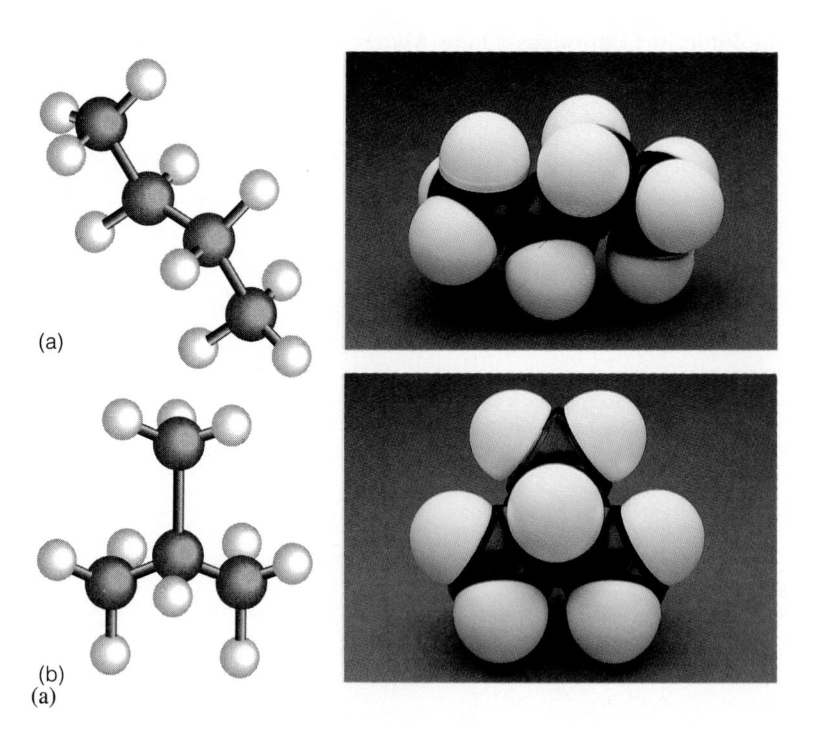

Figure 19.5 *Ball-and-stick models of butane (a) and isobutane (b).*

Compounds that have the same molecular formula but different structural formulas are called **isomers**. Because this branched four-carbon alkane is an isomer of butane, it is called isobutane. Figure 19.5 shows models of the two isomers—butane and isobutane.

The number of isomers increases rapidly with the number of carbon atoms, as shown in Table 19.3. There are three pentanes, five hexanes, nine heptanes, and so

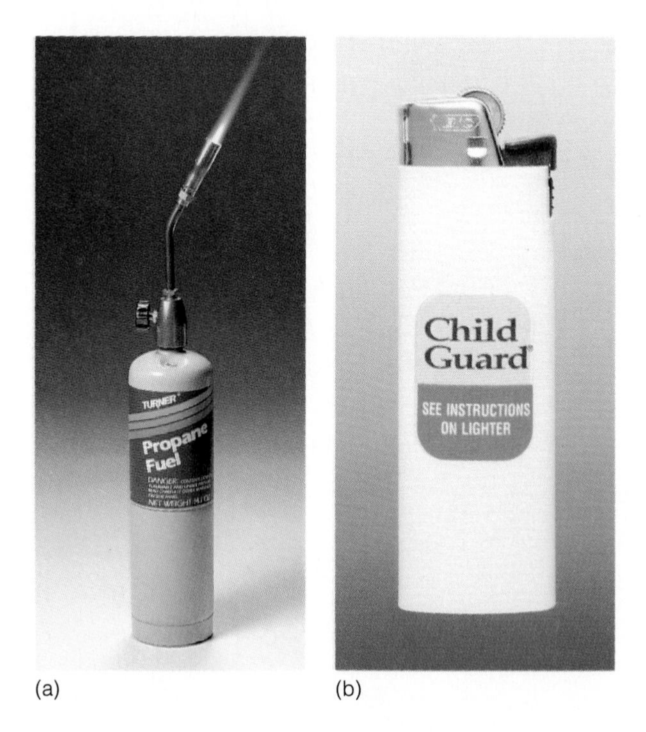

Figure 19.6 *Propane and butane are common portable fuels. (a) A propane torch; (b) a butane lighter.*

on. Isomerism is common in the compounds of carbon; it is another reason there are millions of organic compounds.

Propane and the butanes are familiar fuels (Figure 19.6). They usually are supplied under pressure in tanks. Although they are gases at ordinary temperatures and under normal atmospheric pressure, they liquefy under pressure and are sold as liquefied petroleum gas (called LP gas or LPG). Gasoline is a mixture of hydrocarbons, most of which are alkanes with 5 to 12 carbon atoms.

Let's return to Table 19.3 for a moment. Notice that the molecular formula of each alkane in the table differs from the one preceding it by precisely one carbon atom and two hydrogen atoms, that is, by a CH_2 group. Such a series of compounds has properties that vary in a regular and predictable manner; they are called **homologs**. The alkanes are homologs with the general formula C_nH_{2n+2}. Instead of studying the chemistry of a bewildering array of individual carbon compounds, organic chemists study a few members of a *homologous series* from which they can deduce the properties of other compounds in the series.

Consider one other point before leaving Table 19.3: Not all the possible isomers of the larger molecules have been isolated. Indeed, the task rapidly becomes even more prohibitive as you proceed through the series. There are, for example, over 4 billion possible isomers with the molecular formula $C_{30}H_{62}$.

Physical Properties of the Alkanes

Some physical properties of the first 20 alkanes are given in Table 19.4. Notice the fairly regular increase in the melting point, boiling point, and density as the number of carbon atoms—and the molar mass—increases. As shown in the table, at

Table 19.4 **Physical Properties of Selected Alkanes**

Name	*Molecular Formula*	*Melting Point (°C)*	*Boiling Point (°C)*	*Density at 20 °C (g/mL)*
Methane	CH_4	−183	−164	(Gas)
Ethane	C_2H_6	−183	−89	(Gas)
Propane	C_3H_8	−190	−42	(Gas)
Butane	C_4H_{10}	−138	0	(Gas)
Pentane	C_5H_{12}	−130	36	0.626
Hexane	C_6H_{14}	−95	69	0.659
Heptane	C_7H_{16}	−91	98	0.684
Octane	C_8H_{18}	−57	126	0.703
Nonane	C_9H_{20}	−54	151	0.718
Decane	$C_{10}H_{22}$	−30	174	0.730
Undecane	$C_{11}H_{24}$	−26	196	0.740
Dodecane	$C_{12}H_{26}$	−10	216	0.749
Tridecane	$C_{13}H_{28}$	−6	235	0.757
Tetradecane	$C_{14}H_{30}$	6	254	0.763
Pentadecane	$C_{15}H_{32}$	10	271	0.769
Hexadecane	$C_{16}H_{34}$	18	280	0.775
Heptadecane	$C_{17}H_{36}$	22	302	(Solid)
Octadecane	$C_{18}H_{38}$	28	316	(Solid)
Nonadecane	$C_{19}H_{40}$	32	330	(Solid)
Eicosane	$C_{20}H_{42}$	37	343	(Solid)

room temperature alkanes having 1 to 4 carbon atoms per molecule are gases, alkanes having 5 to 16 carbon atoms per molecule are liquids, and those having more than 16 carbon atoms per molecule are solids.

Densities of the liquid alkanes are less than the density of water (1.0 g/mL). Since these compounds are nonpolar and essentially insoluble in water, they float on top of water. Alkanes are often used as solvents to dissolve substances of low polarity such as fats, oils, and waxes.

Chemical Properties of the Alkanes

Chemically, the alkanes are the least reactive of all organic compounds. They are generally unreactive toward strong acids (such as sulfuric acid), strong bases (such as sodium hydroxide), most oxidizing agents (such as potassium dichromate), and most reducing agents (such as sodium metal). The alkanes do undergo a few very important reactions, including combustion. Natural gas (chiefly methane) is one of our most important fossil fuels. The equation for the reaction involving the combustion of methane is given here.

$$CH_4(g) + 2\ O_2(g) \longrightarrow CO_2(g) + 2\ H_2O(g) + \text{Heat}$$

Bottled gas—mainly propane—is used as a fuel for some homes, for portable barbecue grills, and for soldering water pipes. Butane is the fuel used in small pocket-size lighters. Gasoline, kerosene, and aviation fuel are mixtures of hydrocarbons obtained from petroleum. Besides burning in oxygen, the alkanes react with chlorine and bromine to give **alkyl halides** such as methyl bromide, CH_3Br. Alkyl halides are used to synthesize numerous other compounds.

EXAMPLE 19.2

Give names of alkanes with three, four, five, six, seven, and eight carbon atoms.

SOLUTION

Number of Carbon Atoms	*Name of Alkane*	*Number of Carbon Atoms*	*Name of Alkane*
3	Propane	6	Hexane
4	Butane	7	Heptane
5	Pentane	8	Octane

EXERCISE 19.1

Give names of four alkanes that are gases at room temperature.

19.3 *Drawing Structural Formulas*

Structural formulas that show bond lines for all covalent bonds present in a molecule were used in the previous section. A structural formula gives more information than is provided by a molecular formula. For example, the molecular formula C_4H_{10} doesn't tell us whether we are dealing with butane or isobutane, but the structural formulas—shown in Section 19.2—allow us to distinguish between these two compounds by showing the order of attachment of the various atoms.

Unfortunately, it takes a considerable amount of time and space to type structural formulas on a printed page. Chemists often use **condensed structural formulas** to alleviate these problems. The condensed structures show the hydrogen atoms right next to the carbon atoms to which they are attached. Condensed structural formulas of the two butanes are shown here.

$$CH_3{-}CH_2{-}CH_2{-}CH_3 \qquad \text{and} \qquad \begin{array}{c} CH_3{-}\underset{\displaystyle CH_3}{\underset{|}{CH}}{-}CH_3 \end{array}$$

These structures can be simplified further by omitting horizontal bond lines.

$$CH_3CH_2CH_2CH_3 \qquad \text{and} \qquad CH_3\underset{\displaystyle CH_3}{\underset{|}{C}}HCH_3$$

EXAMPLE 19.3

Draw the structural formula and the condensed structural formula for heptane.

SOLUTION

The name heptane tells us that the compound is an alkane (-ane) with seven (hept-) carbon atoms. Begin by showing a string of seven carbon atoms linked by single bonds.

$$C{-}C{-}C{-}C{-}C{-}C{-}C$$

Each carbon must have four covalent bonds, so attach enough hydrogen atoms to the string of carbon atoms to give four bonds to each carbon atom.

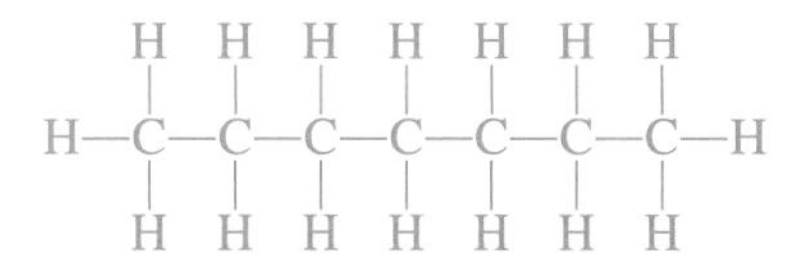

For the condensed form, use a subscript to show the number of hydrogen atoms attached to each carbon. Note that the carbon on each end has three hydrogen atoms, while each carbon in between has only two hydrogen atoms.

$$CH_3CH_2CH_2CH_2CH_2CH_2CH_3$$

There are five consecutive CH_2 groups, so the structure can also be written

$$CH_3(CH_2)_5CH_3$$

See Problems 19.7 and 19.8.

19.4 The Universal Language: IUPAC Nomenclature

To bring order to the chaotic naming of newly discovered compounds, the *International Union of Pure and Applied Chemistry* or **IUPAC** (usually pronounced ''eye you pack'') held what was to be the first of several meetings on nomenclature—that is, a system of naming—in 1892. This conference established formal rules for naming compounds. Some fundamental rules for naming alkanes are summarized here.

1. The names of individual members end in **-ane**, indicating that they are alkanes. The names of the continuous-chain members having up to 10 carbon atoms are given in Table 19.3.
2. The names of branched-chain alkanes are made up of two parts. The end of the name is taken from the *longest* continuous chain of carbons present in the compound. This is sometimes referred to as the parent compound. For example, the compound with the structure

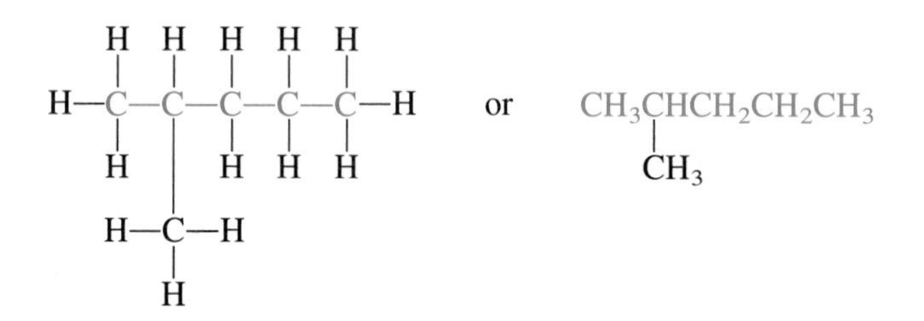

   would be named as a derivative of pentane because there are five carbon atoms in the longest continuous chain. The second part of the name would be *pentane*.
3. The first part of the name consists of prefixes that indicate the groups attached to the parent chain. If the attached group contains only carbon and hydrogen atoms joined by single bonds, it is called an **alkyl group**. The **alk-** indicates that these groups are similar to alkanes; the **-yl** indicates that a group of atoms is attached to some parent chain. The name of a specific alkyl group is derived from the alkane with the same number of carbons. For example, the one-carbon group derived from methane is

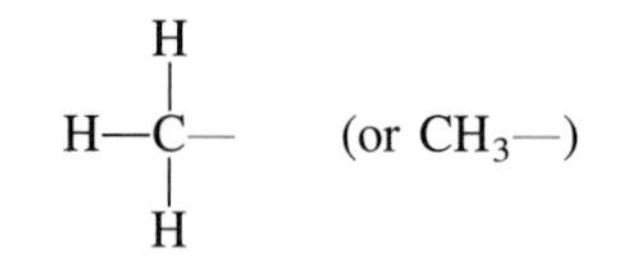

   It is called a **methyl** group. The alkyl group derived from ethane is

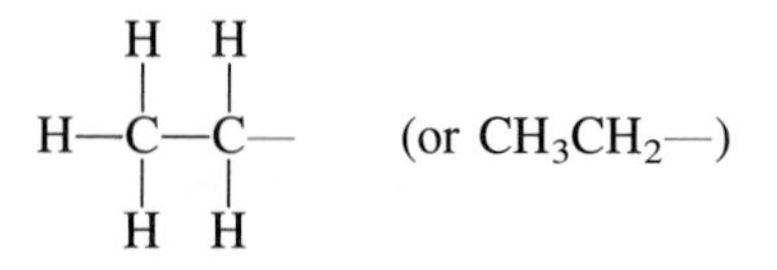

It is called an **ethyl** group. *Two* alkyl groups can be derived from propane. One of these is the **propyl** group. Notice that the attachment is through the end carbon of the three-carbon chain.

```
    H  H  H
    |  |  |
 H—C—C—C—        (or CH3CH2CH2—)
    |  |  |
    H  H  H
```

The other alkyl group derived from propane is the **isopropyl** group. It is attached through the middle carbon and can be represented as

```
    H  H  H                                     CH3
    |  |  |                                      |
 H—C—C—C—H       (or CH3CHCH3   or   CH3CH—)
    |  |  |                  |
    H     H
```

There are many other alkyl groups. The ones that we are most likely to encounter are listed in Table 19.5. Notice that each alkyl group has one less hydrogen than the corresponding alkane. The removal of a hydrogen gives a site where the alkyl group can bond to a parent chain or to another group.

4. The locations of substituent groups (alkyl groups or other groups) on the longest chain are designated by Arabic numerals. For example, in naming the compound represented by the structure

```
    H  H  H  H  H  H
    |  |  |  |  |  |
 H—C—C—C—C—C—C—H      or      CH3CHCH2CH2CH2CH3
    |  |  |  |  |  |                   |
    H  |  H  H  H  H                   CH3
       |
    H—C—H
       |
       H
```

we first identify the longest continuous chain.

```
    H  H  H  H  H  H
    |  |  |  |  |  |
 H—C—C—C—C—C—C—H      or      CH3CHCH2CH2CH2CH3
    |  |  |  |  |  |                   |
    H  |  H  H  H  H                   CH3
       |
    H—C—H
       |
       H
```

There are six carbon atoms in this chain, so the compound is considered to be a derivative of hexane. The group attached to the chain is a methyl group, CH_3—. It is on the second carbon atom of the parent chain. The name of this compound is 2-methylhexane. Always number the parent chain by starting at the end nearest the substituent. If there are more than

Table 19.5 **Common Alkyl Groups**

Name	Structural Formula	Condensed Structural Formula
Methyl	$\begin{array}{ccc} & H & \\ & \vert & \\ H- & C & - \\ & \vert & \\ & H & \end{array}$	CH_3-
Ethyl	$\begin{array}{cccccc} & & H & & H & \\ & & \vert & & \vert & \\ H & - & C & - & C & - \\ & & \vert & & \vert & \\ & & H & & H & \end{array}$	CH_3CH_2-
DERIVED FROM PROPANE		
Propyl	$\begin{array}{cccccccc} & & H & & H & & H & \\ & & \vert & & \vert & & \vert & \\ H & - & C & - & C & - & C & - \\ & & \vert & & \vert & & \vert & \\ & & H & & H & & H & \end{array}$	$CH_3CH_2CH_2-$
Isopropyl	$\begin{array}{ccccccccc} & & H & & H & & H & & \\ & & \vert & & \vert & & \vert & & \\ H & - & C & - & C & - & C & - & H \\ & & \vert & & \vert & & \vert & & \\ & & H & & & & H & & \end{array}$	$CH_3\underset{\vert}{C}HCH_3$
DERIVED FROM BUTANE		
Butyl	$\begin{array}{cccccccccc} & & H & & H & & H & & H & \\ & & \vert & & \vert & & \vert & & \vert & \\ H & - & C & - & C & - & C & - & C & - \\ & & \vert & & \vert & & \vert & & \vert & \\ & & H & & H & & H & & H & \end{array}$	$CH_3CH_2CH_2CH_2-$
Secondary butyl (*sec*-Butyl)	$\begin{array}{ccccccccccc} & & H & & H & & H & & H & & \\ & & \vert & & \vert & & \vert & & \vert & & \\ H & - & C & - & C & - & C & - & C & - & H \\ & & \vert & & \vert & & \vert & & \vert & & \\ & & H & & & & H & & H & & \end{array}$	$CH_3\underset{\vert}{C}HCH_2CH_3$
DERIVED FROM ISOBUTANE		
Isobutyl	$\begin{array}{cccccccc} & & & & H & & & \\ & & & & \vert & & & \\ & & & H- & C & -H & & \\ & & & & \vert & & & \\ & & H & & \vert & & H & \\ & & \vert & & \vert & & \vert & \\ H & - & C & - & C & - & C & - \\ & & \vert & & \vert & & \vert & \\ & & H & & H & & H & \end{array}$	$CH_3\overset{CH_3}{\overset{\vert}{C}}HCH_2-$
Tertiary butyl (*tert*-Butyl)	$\begin{array}{ccccccccc} & & & & H & & & & \\ & & & & \vert & & & & \\ & & & H- & C & -H & & & \\ & & & & \vert & & & & \\ & & H & & \vert & & H & & \\ & & \vert & & \vert & & \vert & & \\ H & - & C & - & C & - & C & - & H \\ & & \vert & & \vert & & \vert & & \\ & & H & & & & H & & \end{array}$	$CH_3\underset{\vert}{\overset{CH_3}{\overset{\vert}{C}}}CH_3$

two substituents, and the substituents nearest each end of the main chain are equidistant from the ends, number from the end nearest the first point of difference.

5. When two or more identical groups are attached to the main chain, a number is required to specify the location of each substituent. Further, we must indicate whether there are two, three, or four identical groups attached to the parent chain. We do this by using prefixes: **di-** for two, **tri-** for three, and **tetra-** for four. Even if two identical groups are attached to the same carbon, the number must be repeated for each group.

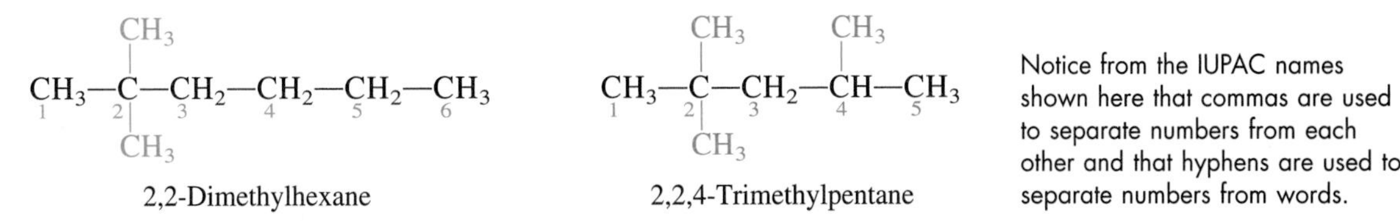

2,2-Dimethylhexane 2,2,4-Trimethylpentane

Notice from the IUPAC names shown here that commas are used to separate numbers from each other and that hyphens are used to separate numbers from words.

Groups should be listed in alphabetical order—ethyl before methyl and so on—according to IUPAC rules, as shown here.

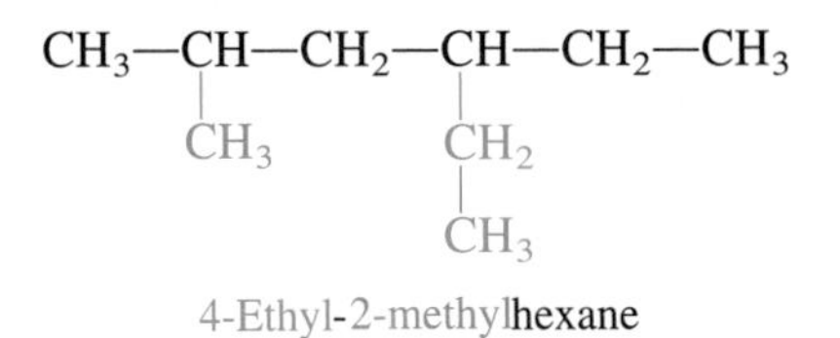

4-Ethyl-2-methylhexane

The best way to learn how to name organic structures is by working out examples, not just by memorizing rules. It's easier than it sounds. Try these.

EXAMPLE 19.4

Give the IUPAC name for the compound represented by the formula.

```
                        CH3
                        |
CH3—CH2—CH—CH—CH3
                |
                CH3
```

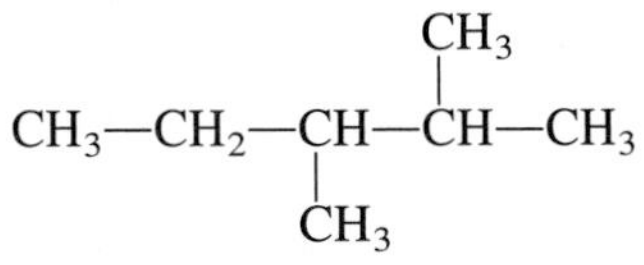

SOLUTION

The longest continuous chain has five carbon atoms. There are two methyl groups attached to the second and third carbon atoms (not the third and fourth). Give these methyl groups the lowest numbers possible, counting from one end. The correct name is 2,3-dimethylpentane.

EXAMPLE 19.5

Give the IUPAC name for the compound represented by the formula

```
CH3—CH—CH2—CH—CH3
        |               |
        CH2           CH3
        |
        CH3
```

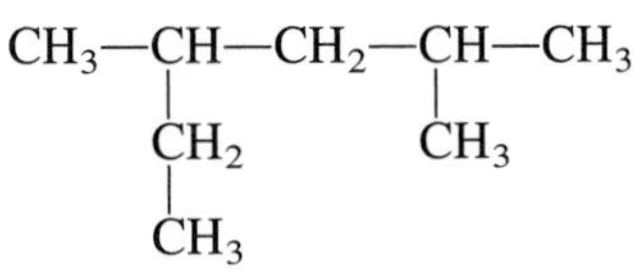

SOLUTION

The IUPAC name is 2,4-dimethylhexane, not 2-ethyl-4-methylpentane. This one could be tricky! The parent compound is the longest continuous chain, not necessarily the chain drawn straight across the page. For this structure, the longest chain contains six—not five—carbon atoms.

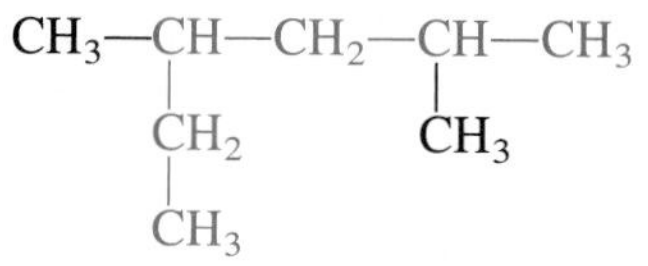

```
CH3—CH—CH2—CH—CH3
    |       |
    CH2     CH3
    |
    CH3
```

EXAMPLE 19.6

Give the IUPAC name for the compound represented by the formula

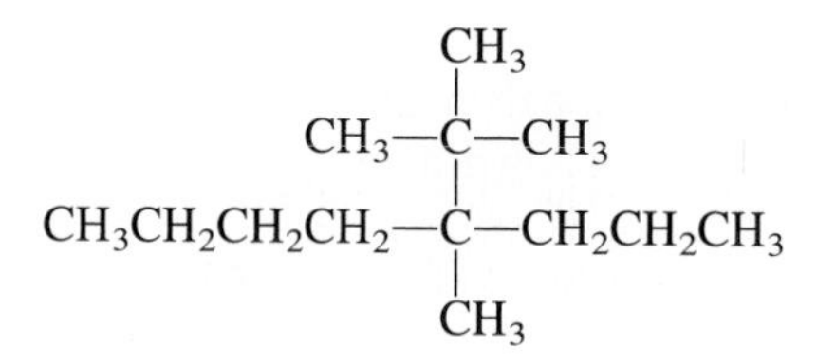

SOLUTION

There is a *tert*-butyl group (see Table 19.5) shown above the longest chain. It is named, alphabetically, before the methyl group shown below the parent chain. The correct name is 4-*tert*-butyl-4-methyloctane.

EXAMPLE 19.7

Draw the structure for 4-isopropyl-2-methylheptane.

SOLUTION

When drawing structures, start with the parent chain—heptane in this case.

C—C—C—C—C—C—C

Then add the groups at the proper carbon numbers. You can number the parent chain from either direction as long as you follow through consistently (don't change directions in the middle of a problem).

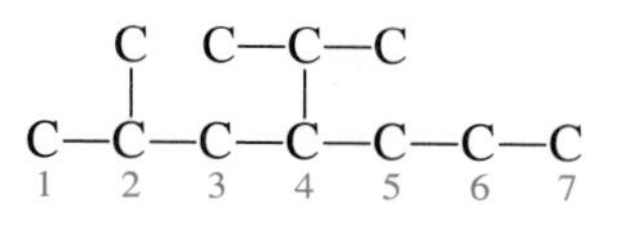

Complete the condensed formula by including the appropriate number of hydrogens.

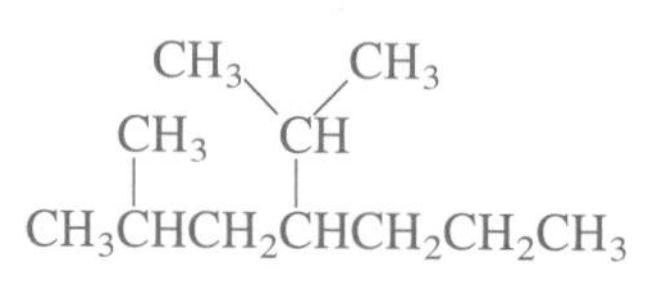

19.5 *Alkenes and Alkynes: Unsaturated Hydrocarbons*

Two carbon atoms can share more than one pair of electrons. In ethene (the common name is ethylene), C_2H_4, the two carbon atoms share *two* pairs of electrons to form a *double bond*, as shown here and in Figure 19.7.

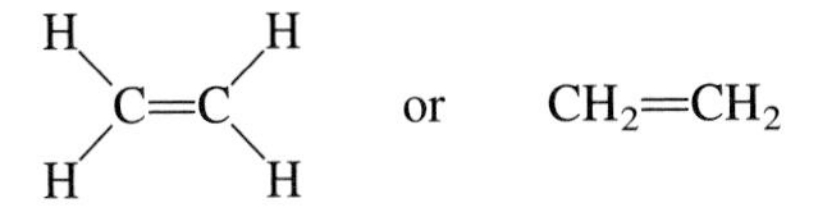

Figure 19.7 *Ball-and-stick and space-filling models of ethylene.*

Ethene is the first member of a family of hydrocarbons called **alkenes** (notice the **-ene** ending). Each alkene contains a carbon-to-carbon double bond. Names, structures, and physical properties of several alkenes are given in Table 19.6.

Table 19.6 **Physical Properties of Some Selected Alkenes**

IUPAC Name	*Molecular Formula*	*Condensed Structure*	*Melting Point (°C)*	*Boiling Point (°C)*
Ethene	C_2H_4	$CH_2{=}CH_2$	−169	−104
Propene	C_3H_6	$CH_3CH{=}CH_2$	−185	−47
1-Butene	C_4H_8	$CH_3CH_2CH{=}CH_2$	−185	−6
1-Pentene	C_5H_{10}	$CH_3CH_2CH_2CH{=}CH_2$	−138	30
1-Hexene	C_6H_{12}	$CH_3(CH_2)_3CH{=}CH_2$	−140	63
1-Heptene	C_7H_{14}	$CH_3(CH_2)_4CH{=}CH_2$	−119	94
1-Octene	C_8H_{16}	$CH_3(CH_2)_5CH{=}CH_2$	−102	121

Figure 19.8 *A variety of consumer products are sold in containers made of polyethylene.*

Ethene (ethylene) is the most important commercial organic chemical. Annual U.S. production is about 16 billion kg. About 45% of this ethylene goes into the manufacture of polyethylene, one of the most familiar plastics (Figure 19.8). Another 15% or so is converted to ethylene glycol, the major component of most brands of antifreeze for automobile radiators.

Propene (the common name is propylene) is the second alkene listed in the table of alkenes. It is used in the production of polypropylene and other plastics, isopropyl alcohol (rubbing alcohol), and a variety of other organic chemicals.

Two carbon atoms can also share *three* pairs of electrons to form a triple bond between two carbon atoms, as in ethyne, C_2H_2, which is usually called acetylene. Its structure is shown here; a ball-and-stick model is shown in Figure 19.9.

$$H{-}C{\equiv}C{-}H$$

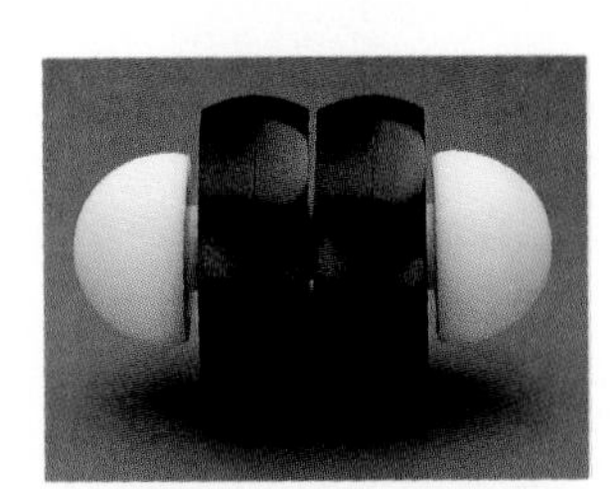

Figure 19.9 *Ball-and-stick and space-filling models of acetylene.*

Ethyne is the first member of a family of hydrocarbons called **alkynes**. Each alkyne contains a carbon-to-carbon triple bond. About 10% of the acetylene produced in the United States is used in oxyacetylene torches for cutting and welding metals. Most acetylene, however, is converted to chemical intermediates—that is, chemicals that are used to make a variety of other products.

There are specific IUPAC rules for naming alkenes and alkynes. Some of the most fundamental IUPAC rules for naming alkenes are listed here.

1. Names of alkenes end with **-ene**. The **-diene** ending is used for alkenes that contain two double bonds.
2. The longest chain of carbon atoms *containing the double bond* is the parent compound. The base name for this string is the same as that of the alkane, but the ending is *-ene*. For example, the base name for a string of four carbons containing a double bond is *butene*.
3. The carbon atoms in the parent compound are numbered from the end of the string nearer the double bond. We designate the location of the double bond by using the number of the first carbon atom of the double bond as a prefix. The compound $CH_3CH{=}CHCH_2CH_3$, for example, has the double bond between the second and third carbon atoms. Its name is 2-pentene.
4. Substituent groups are named by methods described for alkanes. For example, the compound

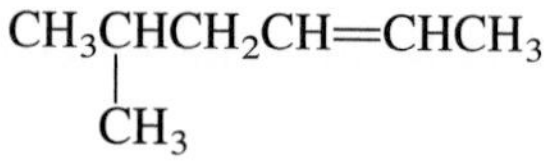

$$\begin{array}{l}CH_3\underset{\displaystyle CH_3}{\underset{|}{C}}HCH_2CH{=}CHCH_3\end{array}$$

is 5-methyl-2-hexene. Notice that the double bond gets the lowest number possible, even if that forces a substituent group to have a higher number. We say the double bond has *priority* in numbering.

The rules are more easily learned through examples.

EXAMPLE 19.8

Name the compound with the structure

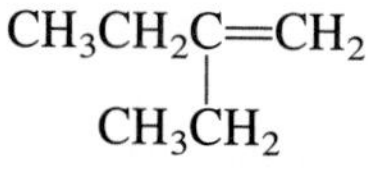

$$CH_3CH_2\underset{\displaystyle CH_3CH_2}{\underset{|}{C}}{=}CH_2$$

SOLUTION

The longest continuous chain in the structure contains five carbon atoms, but the longest continuous chain *containing the double bond* has only four carbon atoms. This four-carbon chain serves as the parent compound; it is numbered from right to left to give the double bond the lowest possible number.

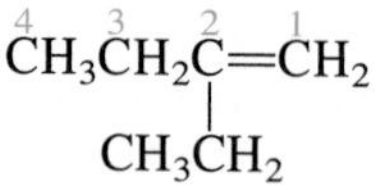

$$\overset{4}{C}H_3\overset{3}{C}H_2\underset{\displaystyle CH_3CH_2}{\underset{|}{\overset{2}{C}}}{=}\overset{1}{C}H_2$$

An ethyl group is attached to carbon number 2. The name of this compound is 2-ethyl-1-butene.

EXERCISE 19.2

Draw the structure for 5-bromo-4-methyl-2-hexene.

Cis-Trans Isomerism

Atoms and groups of atoms are free to rotate about single bonds, as the models of alkanes would indicate, but this is not the case with double bonds. As indicated by the models pictured in Figure 19.10, the second and third carbon atoms of 2-butene are prevented from free rotation—with respect to each other—at the double bond. (The double bond in the figure is symbolized by the two curved connectors between carbon atoms.) Thus, the two arrangements of 2-butene shown in Figure 19.10, actually represent two *different* compounds; they have different geometries. One compound cannot be converted to the other unless the double bond is broken first. The *cis-* or *trans-* prefix is used with each name to distinguish between the two isomers.

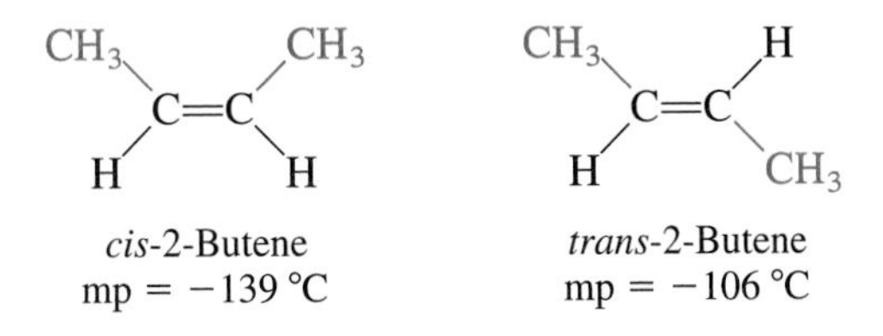

This **cis** isomer is the one with both identical groups on the same side (top or bottom) of the double bond. In the **trans** isomer, the methyl groups are on opposite sides of the double bond. The melting points, boiling points, and other physical properties of these isomers are different—they are different chemicals.

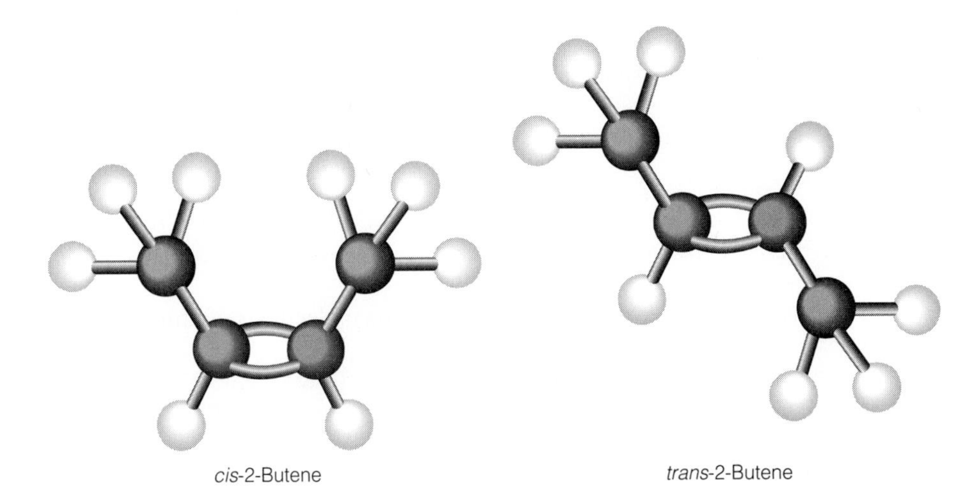

Figure 19.10 *Ball-and-stick models of the 2-butenes. While the carbon atoms joined by single bonds are free to spin about these bonds, the doubly bonded carbon atoms are restricted in this regard.*

The mere presence of a double bond—with its restricted rotation—does not always mean there will be cis and trans isomers. Isomers of this type must also have two *nonidentical* groups on *each* carbon involved in the double bond. Although 2-butene has cis and trans isomers, there are no such isomers for 1-butene, $CH_2{=}CHCH_2CH_3$, or propene, $CH_2{=}CHCH_3$, because carbon number 1 in these two compounds contains two identical groups—two hydrogen atoms in both cases.

EXAMPLE 19.9

Draw structures and give IUPAC names for all compounds with the formula $C_2H_2Cl_2$.

SOLUTION

Because there are two carbon atoms and a total of only four substituents (2 H and 2 Cl atoms), there must be a double bond between the carbon atoms. First, draw the carbon-to-carbon skeleton with a double bond.

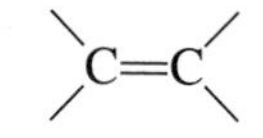

Then place the 2 H and 2 Cl atoms in all geometrically different arrangements, and name the structures. There are three different isomers.

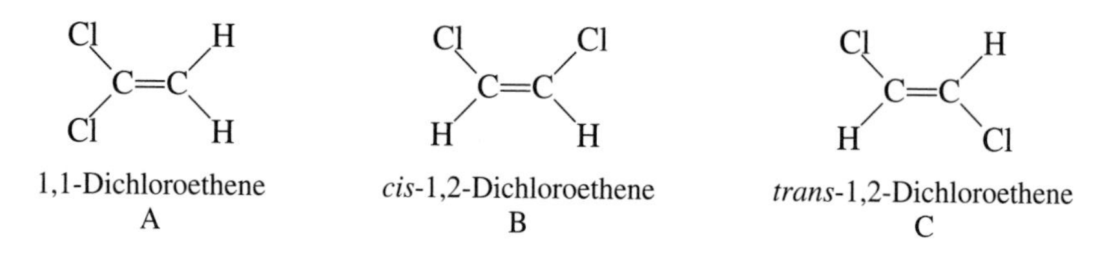

1,1-Dichloroethene A | *cis*-1,2-Dichloroethene B | *trans*-1,2-Dichloroethene C

There is only one 1,1-dichloro- isomer, but there are two 1,2-dichloro- isomers (compounds B and C), which we distinguish by using the cis or trans prefix.

While cis and trans isomerism may initially seem to be of little practical importance to us, all things are not what they appear to be. For example, the female housefly secretes *cis*-9-tricosene as an attractant for the male. The male has little, if any, affinity for the trans isomer.

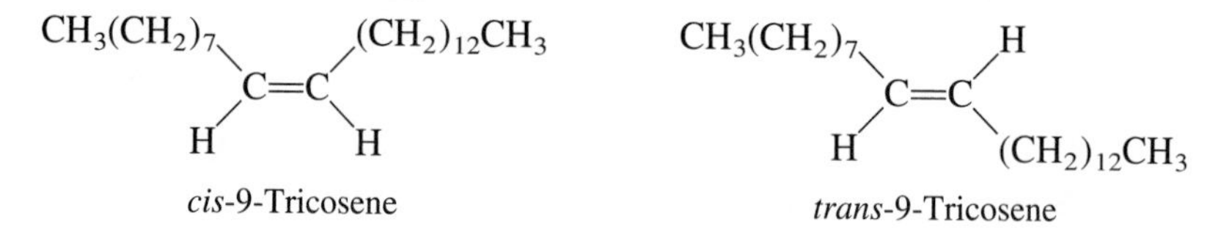

cis-9-Tricosene | *trans*-9-Tricosene

Indeed, in most biological systems, the geometry of a molecule is of utmost importance.

See Problems 19.9–19.22.

Chemical Properties of the Alkenes

Like the alkanes—and all other hydrocarbons—the alkenes burn, but these compounds are not used as fuels. The typical reactions of the alkenes are classified as **addition reactions**. During an addition reaction one of the bonds in the double bond is broken, permitting each of these carbon atoms to bond to an additional atom or group. For example, ethene (ethylene) undergoes addition of hydrogen in the presence of a nickel, Ni, platinum, Pt, or palladium, Pd, catalyst to form ethane.

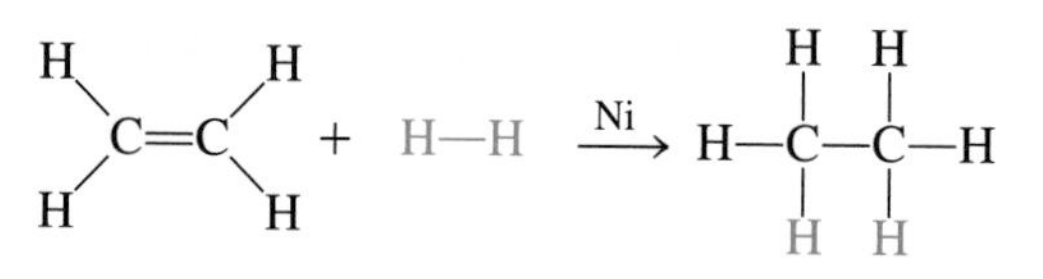

Ethene (Ethylene) | Hydrogen | Ethane

Compounds containing carbon-to-carbon double bonds are said to be **unsaturated**; they can take up—and react with—hydrogen in the presence of an appropriate catalyst to form an alkane. This process, called **hydrogenation**, is used industrially in the conversion of unsaturated vegetable oils (liquids) into the saturated fats used in making Crisco and other semisolid vegetable shortening products. The difference between the liquid oil and the solid fat is due to a difference in the number of double bonds present; there are more double bonds in the unsaturated oils and fewer in the saturated fats. Hydrogenation is also used to produce margarine with properties that resemble those of butter.

Alkenes readily add halogen molecules. When bromine reacts with an alkene, a distinct color change occurs, making it possible to use this reaction as a test for unsaturation. Specifically, when an alkene is mixed with a solvent containing a small amount of dark red bromine, the color disappears as the alkene reacts with the bromine.

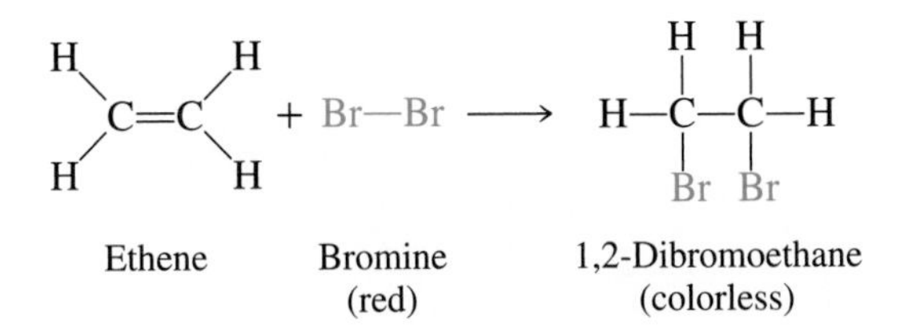

The reaction of water with an alkene to form an alcohol is another important type of addition reaction. This reaction, called *hydration*, requires a mineral acid, such as sulfuric acid, H_2SO_4, as a catalyst.

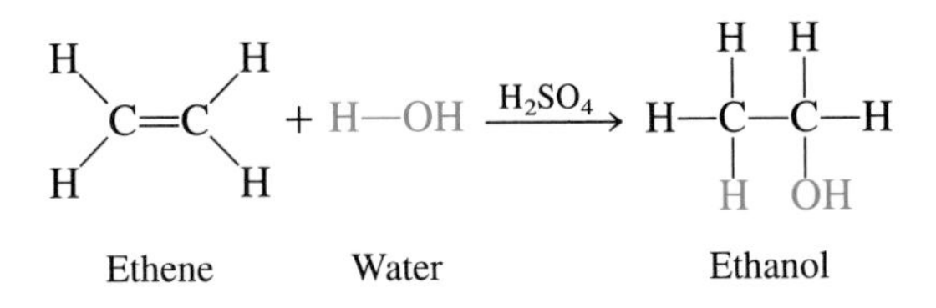

Vast quantities of ethanol (commonly called ethyl alcohol) for use as an industrial solvent are made from ethylene.

See Problems 19.9–19.22.

Polymerization

Perhaps the most important alkene reactions of all are those that involve polymerization. **Polymers** (from the Greek *poly*, ‘‘many’’ + *meros*, ‘‘parts’’) are giant molecules assembled from many small molecules. The polymerization of ethylene to form polyethylene can be represented in the following manner.

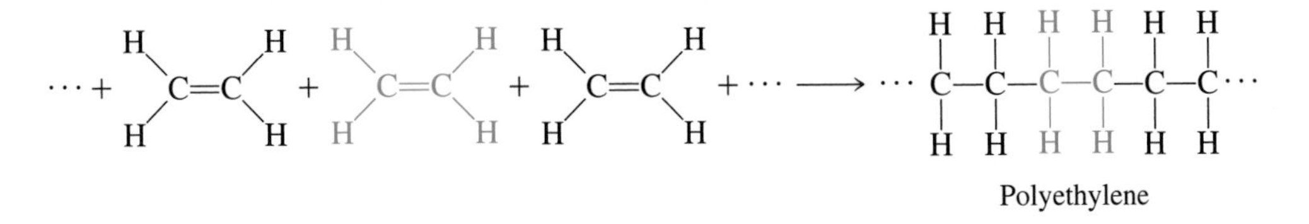

The dotted lines in the formula of the product indicate that the structure is extended for many units in each direction. Notice that the two carbon atmos and four hydro-

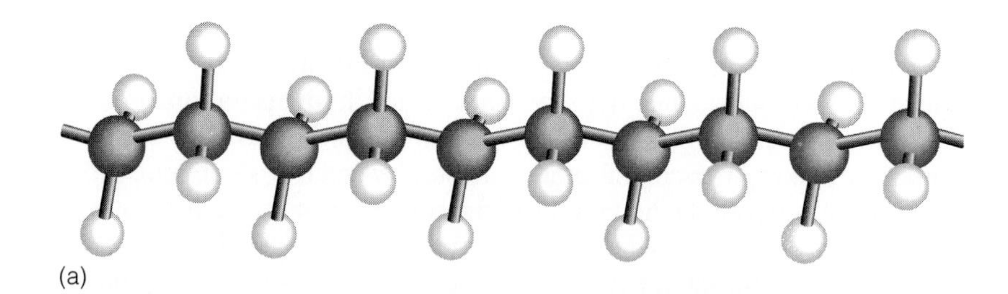
(a)

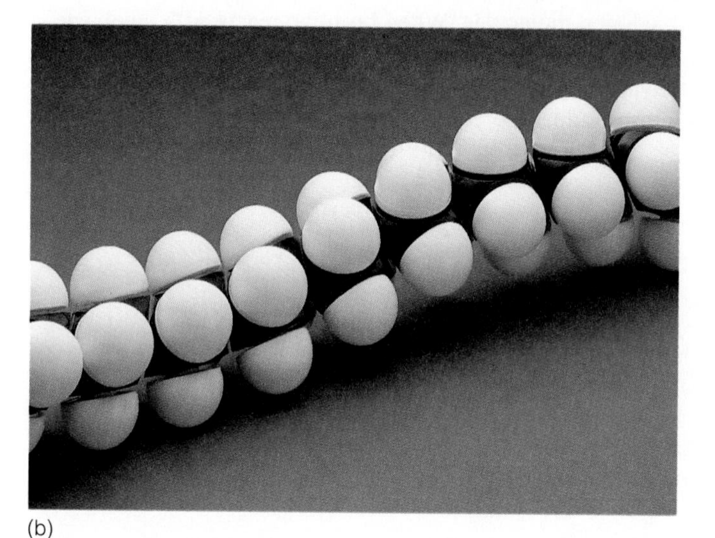
(b)

Figure 19.11 *Ball-and-stick (a) and space-filling (b) models of a segment of a polyethylene molecule.*

gen atoms of each monomer molecule are incorporated into the polymer structure. Models of polyethylene are pictured in Figure 19.11, but each represents only a small portion of an actual molecule. Real polyethylene molecules have varying numbers of carbon atoms; the average is 6000 carbon atoms.

By substitution of various groups for one or more hydrogens of the simple ethylene molecule, a fantastic array of synthetic polymers can be obtained. Some of these are listed in Table 19.7. In this table, the repeating polymer unit is placed within brackets with bonds extended to both sides. The subscript n indicates that this molecular fragment is repeated a very large number of times in the full polymer structure.

Occurrence of Alkenes

One kg of tomatoes can be ripened by exposure to as little as 0.1 mg of ethylene for 24 hr.

Alkenes occur widely in nature. Ripening fruits and vegetables give off ethylene, which triggers further ripening. Fruit suppliers artificially introduce ethylene to hasten the ripening process. Lemon oil contains 1-octene; octadecene, $C_{18}H_{36}$, is found in fish liver. Butadiene, $H_2C{=}CH{-}CH{=}CH_2$, is found in coffee. Lycopene and the carotenes are isomeric polyenes, $C_{40}H_{56}$, that give the red, orange, and yellow colors to watermelons, tomatoes, carrots, and other vegetables and fruits. Vitamin A, essential to good vision, is derived from carotene. Vitamin A is converted in the body to *trans*-retinene, which absorbs visible light and is converted to *cis*-retinene. It is this process—occurring on the retina of the eye—that is responsible in part for vision. The world would be a much darker place without the chemistry of the alkenes.

Table 19.7 A Selection of Addition Polymers

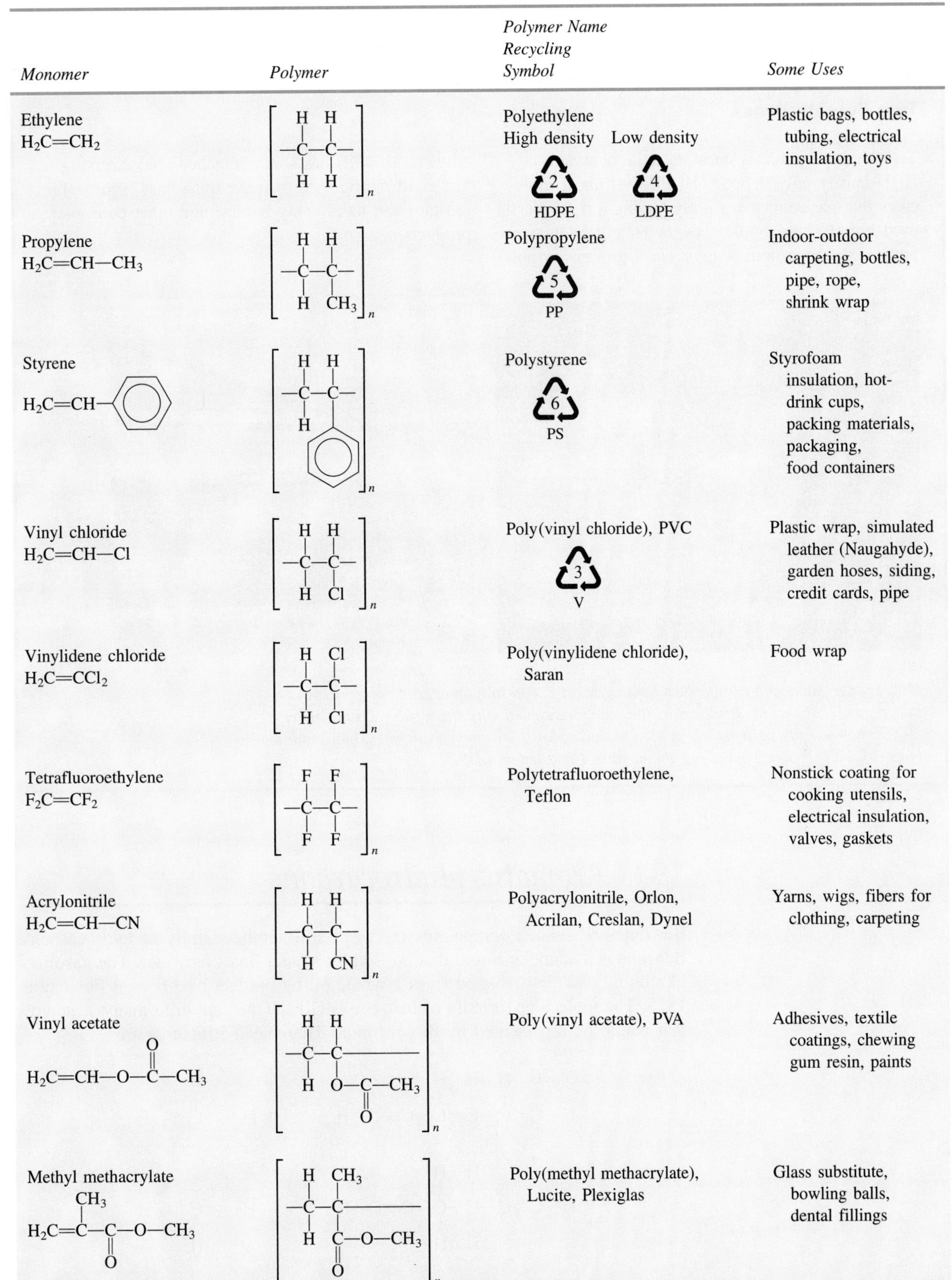

Monomer	*Polymer*	*Polymer Name Recycling Symbol*	*Some Uses*
Ethylene $H_2C{=}CH_2$	$\left[-CH_2-CH_2- \right]_n$	Polyethylene; High density: 2 HDPE; Low density: 4 LDPE	Plastic bags, bottles, tubing, electrical insulation, toys
Propylene $H_2C{=}CH-CH_3$	$\left[-CH_2-CH(CH_3)- \right]_n$	Polypropylene; 5 PP	Indoor-outdoor carpeting, bottles, pipe, rope, shrink wrap
Styrene $H_2C{=}CH-C_6H_5$	$\left[-CH_2-CH(C_6H_5)- \right]_n$	Polystyrene; 6 PS	Styrofoam insulation, hot-drink cups, packing materials, packaging, food containers
Vinyl chloride $H_2C{=}CH-Cl$	$\left[-CH_2-CH(Cl)- \right]_n$	Poly(vinyl chloride), PVC; 3 V	Plastic wrap, simulated leather (Naugahyde), garden hoses, siding, credit cards, pipe
Vinylidene chloride $H_2C{=}CCl_2$	$\left[-CH_2-CCl_2- \right]_n$	Poly(vinylidene chloride), Saran	Food wrap
Tetrafluoroethylene $F_2C{=}CF_2$	$\left[-CF_2-CF_2- \right]_n$	Polytetrafluoroethylene, Teflon	Nonstick coating for cooking utensils, electrical insulation, valves, gaskets
Acrylonitrile $H_2C{=}CH-CN$	$\left[-CH_2-CH(CN)- \right]_n$	Polyacrylonitrile, Orlon, Acrilan, Creslan, Dynel	Yarns, wigs, fibers for clothing, carpeting
Vinyl acetate $H_2C{=}CH-O-C({=}O)-CH_3$	$\left[-CH_2-CH(O-C({=}O)-CH_3)- \right]_n$	Poly(vinyl acetate), PVA	Adhesives, textile coatings, chewing gum resin, paints
Methyl methacrylate $H_2C{=}C(CH_3)-C({=}O)-O-CH_3$	$\left[-CH_2-C(CH_3)(C({=}O)-O-CH_3)- \right]_n$	Poly(methyl methacrylate), Lucite, Plexiglas	Glass substitute, bowling balls, dental fillings

CHEMISTRY IN OUR WORLD

Two Polyethylenes

Two kinds of polyethylenes are quite common. High-density polyethylene (HDPE) has linear molecules that pack together tightly, giving a rigid plastic used in milk jugs, bottle caps, butter tubs, detergent bottles, shampoo bottles, toys, etc. Low-density polyethylene (LDPE) has some branched chains, giving a less rigid plastic used to make trash bags, squeezable mustard and nasal spray bottles, and other consumer products.

(a) (b)

Milk, orange juice, and certain liquid detergents are sold in high-density polyethylene containers (a). They can be identified by the recycling symbol with the number 2 and letters HDPE. Low-density polyethylene is used to make plastic sheeting, trash bags, and sandwich bags (b). Its recycling symbol has the number 4 and letters LDPE.

19.6 Aromatic Hydrocarbons

Benzene and related compounds represent still another family of hydrocarbons. Benzene is a nonpolar liquid that boils at 80 °C and floats on water. The gasoline-like liquid was first isolated from a whale oil by-product by Michael Faraday in 1825. The molecular formula of benzene is C_6H_6. One can write many structures that would be represented by this formula. Three such structures are

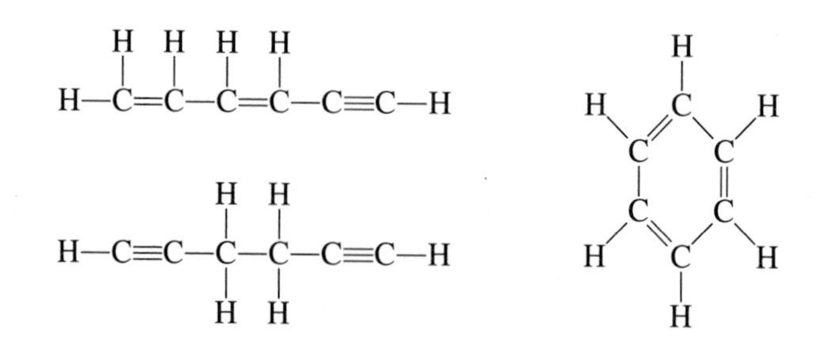

The real substance, benzene, does not have the properties that one would predict from these structures. For example, if benzene really contained double or triple bonds, it would be expected to readily undergo addition reactions. It does not.

Friedrich August Kekulé, a German chemist, proposed the six-membered ring structure pictured here as one of the possible structures. Determining an appropriate structure that would account for the unusual properties of benzene had presented quite a challenge for chemists at that time. Kekulé's structure—with a ring of carbon atoms joined by alternate single and double bonds and with one hydrogen atom bonded to each carbon atom—is by far the best representation that was proposed at that time. It is accurate in some respects. Benzene does have a six-membered ring structure, and all hydrogen atoms are equivalent, but in order for benzene to behave as it does, there must be no ordinary double bonds. Instead, all carbon-to-carbon bonds are actually equivalent and stable; they resist being disrupted. To emphasize this special nature of the bonding in the benzene ring, a hexagon with a circle—either with or without the hydrogen atoms—is generally used to represent benzene.

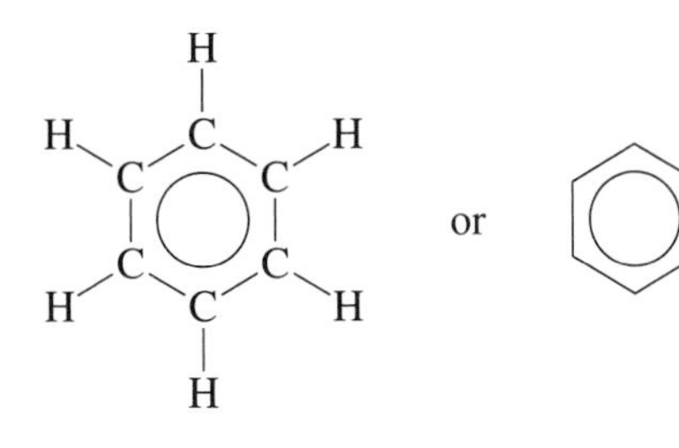

There is, once again, a trade-off. One cannot count four covalent bonds attached to each carbon atom when the circle inside a hexagon structure is used. Many chemists use the circle inside a hexagon symbol, but others still use the simplified Kekulé structure shown at the right.

Annual U.S. production of benzene is about 4.5 billion kg. It is used widely as a solvent and in the production of related compounds. Benzene is thought to cause leukemia in workers receiving long-term exposure; its use has been restricted.

Many compounds—both natural and synthetic—contain one or more benzene rings, but the chemical nature of these compounds varies widely. Various explosives, amino acids, and medications contain one or more benzene rings. Compounds that contain one or more benzene rings are classified as **aromatic** compounds. The ring-containing portion of a large molecule is called an **aryl group** (symbolized Ar). Hydrocarbons containing no benzene rings are classified as **aliphatic** compounds.

A number of aromatic compounds can be derived from benzene by substitution of a variety of groups for one or more of the hydrogen atoms of benzene. For example, substitution of a methyl group for one hydrogen gives methylbenzene, better known as **toluene** (pronounced to rhyme with "doll, you mean"). Toluene is an important solvent and a starting material for the synthesis of other aromatic compounds such as the explosive called TNT, which is short for 2,4,6-trinitrotoluene (see Figure 19.12).

—CH_3

Toluene

Substitution of an ethyl group for a hydrogen atom of benzene gives ethylbenzene. This compound is used in the synthesis of styrene, from which the common plastic polystyrene is made.

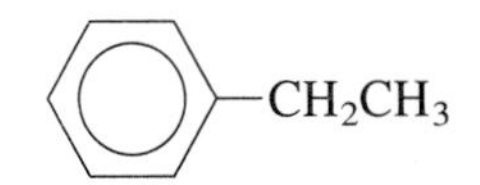

Ethylbenzene

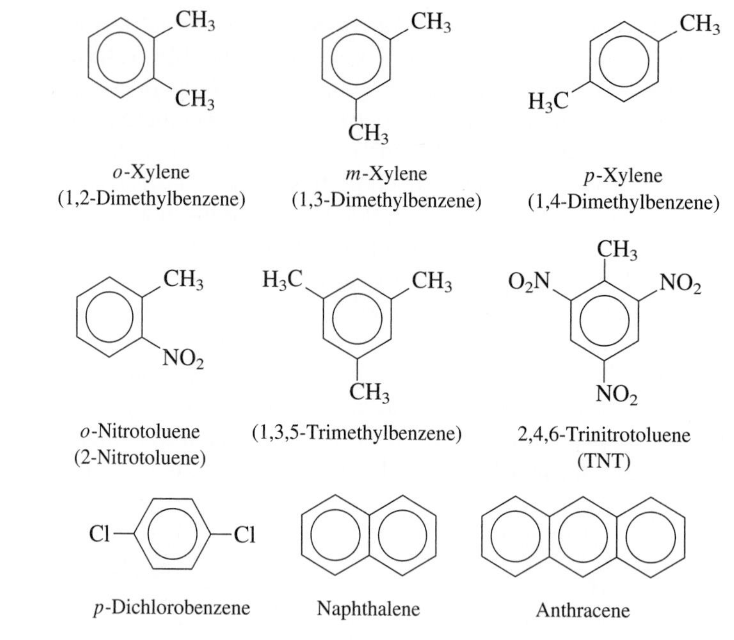

Figure 19.12 *Some aromatic hydrocarbons and derivatives.*

EXAMPLE 19.10

Name the compound whose molecular structure is

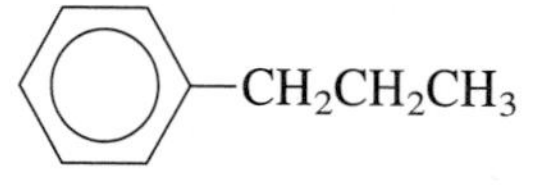

SOLUTION

Think of this compound as an alkyl group (a propyl group) on benzene. The compound is propylbenzene.

When a benzene ring is joined to a large organic group, it is the benzene ring that is named as the substituent group. The ring is called a **phenyl group**, symbolized

or $C_6H_5—$

Thus,

$CH_3CHCH_2CH_2CH_2CH_2CH_2CH_3$

is named 2-phenyloctane.

When two substituents are attached to a benzene ring, we must use some way of indicating their relative positions. There are two methods of doing this. One is the familiar method of using numbers. The other method uses the prefixes *ortho-*, *meta-*, and *para-* to indicate the relative positions. The prefix *ortho-* is abbreviated *o-* and indicates that substituents are on adjacent carbon atoms. An ortho-substituted compound is a 1,2-disubstituted benzene. The prefix *meta-*, or *m-*, is used for 1,3-disubstituted benzenes, and the prefix *para-*, or *p-*, is used with 1,4-disubstituted benzenes. The following four structures represent the same compound, 1,3-dinitrobenzene or *m*-dinitrobenzene. The NO_2— group is called the *nitro* group.

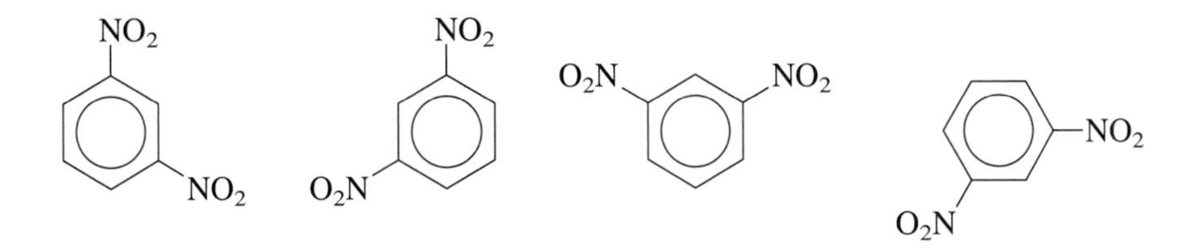

When three or more substituents are attached to a ring, the numbering system must be used to indicate the positions of these groups. Structures and names of several aromatic compounds are shown in Figure 19.12.

See Problems 19.23–19.30.

19.7 Alcohols

An **alcohol** is an organic compound that contains a hydroxyl group, —OH, covalently bonded to an alkyl group. Thus, a methyl group, CH_3—, bonded to a hydroxyl group to give CH_3OH is called methyl alcohol. This is the common name. Its IUPAC name is methanol. The —OH group—present in all alcohols—is called the *functional group*. We shall later describe a wide variety of compounds with different functional groups.

To give the IUPAC name of an alcohol, simply replace the final *-e* of the parent alkane with the letters *-ol*. Thus, CH_3OH is methanol, and CH_3CH_2OH is ethanol. There are two isomers of propanol, one with the —OH group positioned on the first carbon, 1-propanol, and the other with the —OH group on the number 2 carbon, 2-propanol. The 2-propanol isomer is likely to be more recognizable when we use its common name, isopropyl alcohol. In 70% solutions it is used as rubbing alcohol. Structures of these isomers are

$CH_3CH_2CH_2$—OH	$CH_3CH(OH)CH_3$
1-Propanol	2-Propanol (Isopropyl alcohol)

There are four butyl alcohols. They are formed when an —OH group bonds to one of the four butyl groups (Table 19.5).

All alcohols can be placed into one of three classes known as the **primary** (1°). **secondary** (2°), and **tertiary** (3°) **alcohols**. These classes are based on the number of carbon atoms attached to the carbon atom that bears the hydroxyl group. An alcohol with the structure

```
    |
   —C—
    |
  H—C—H
    |
    OH
```

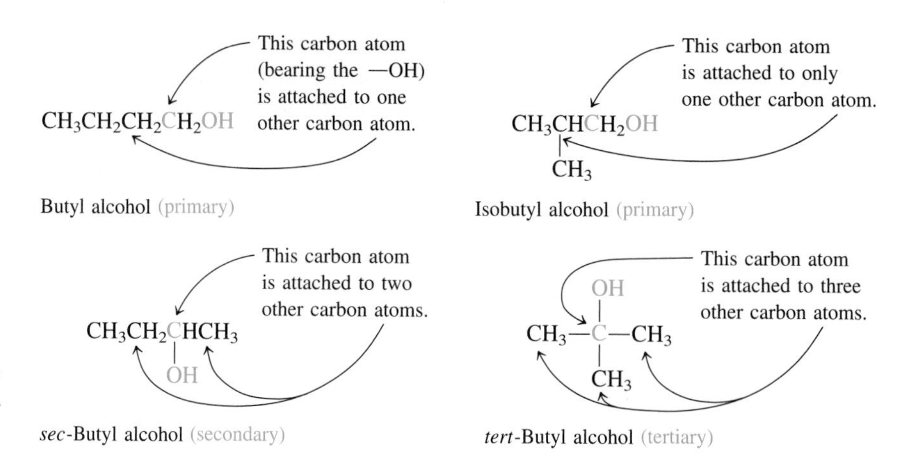

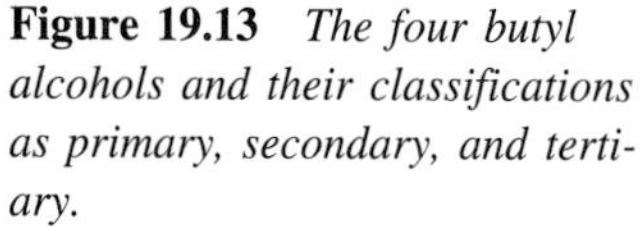

Figure 19.13 *The four butyl alcohols and their classifications as primary, secondary, and tertiary.*

is classified as a primary alcohol because the —OH group is attached to a primary carbon atom, that is, a carbon that is attached to only one alkyl or aryl group (and two hydrogen atoms). The following alcohol structure

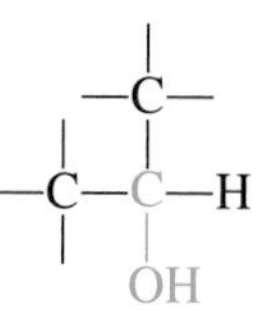

is secondary because the —OH group is attached to a secondary carbon, that is, one that is bonded to *two* other alkyl or aryl groups and one hydrogen atom. Finally, if the hydroxyl group is attached to a carbon bonded to three alkyl or aryl groups, the alcohol is tertiary.

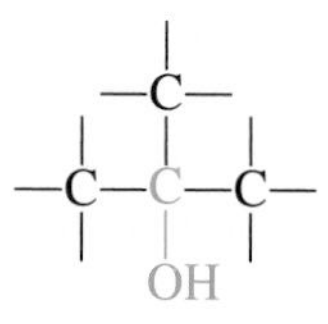

Figure 19.13 shows that each of the four butyl alcohols can be classified as a primary, secondary, or tertiary alcohol. Differences in reactivities of the alcohols are related to the primary, secondary, or tertiary structure at the site of the —OH group.

Methanol, CH_3OH, the simplest alcohol, is a liquid that boils at 34 °C and is quite toxic.* Ingestion of as little as 30 mL (1 oz) can cause permanent blindness or death. Methanol (methyl alcohol) is sometimes called wood alcohol because it can be made by heating wood in the absence of air until the wood breaks down. The vapor produced during the reaction contains methanol—along with other sub-

*The common term *poison* is often applied to a variety of substances without any indication of the degree of toxicity. To avoid this uncertainty, biochemists use the abbreviation $\mathbf{LD_{50}}$ to indicate the dosage—in grams per kilogram of body weight—that would be lethal to 50% of a population of test animals. Toxicities of some alcohols are listed in Table 19.8.

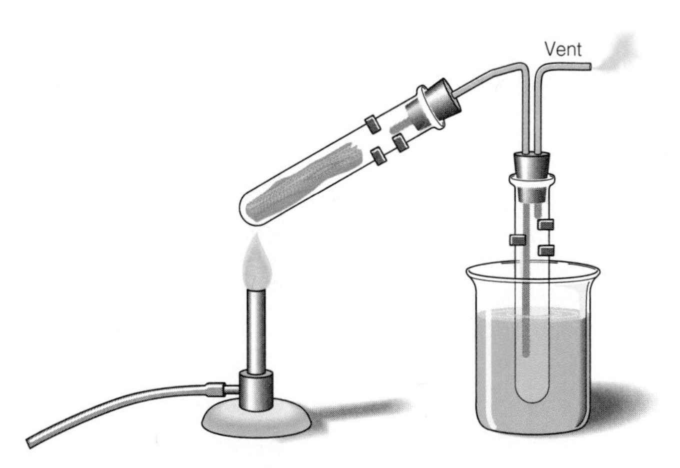

Figure 19.14 *An apparatus for the destructive distillation of wood. The wood is heated in an enclosed tube, and alcohol is condensed in the second tube by the cold water in the beaker. Gases formed in the process can be burned as they exit through the vent tube.*

stances—which is collected as it forms. This process is called **destructive distillation** (Figure 19.14). Commercial methanol production involves a reaction of carbon monoxide with hydrogen.

$$CO + 2\,H_2 \longrightarrow CH_3OH$$

The catalyzed reaction is carried out at high temperature and pressure.

Approximately 4 billion kg of methanol is produced annually in the United States. It is used as a solvent, as a fuel, and as a starting material for the production of fuel additives, adhesives, fibers, and plastics.

Ethanol, CH_3CH_2OH, with a boiling point of 78 °C, is the second member of the homologous series of alcohols. It is perhaps better known by its other names: ethyl alcohol and grain alcohol. For beverage use, the alcohol is produced by the fermentation of grain. Like methanol, and other low molecular weight alcohols, ethanol is toxic. One pint (about 500 mL) of pure ethanol, rapidly ingested, will kill most people.

Table 19.8 **Lethal Oral Doses (in Rats) for Some Alcohols**

Alcohol	*Structure*	*Boiling Point (°C)*	*LD_{50} (g/kg of body weight)*	*Uses*
Methyl alcohol	CH_3OH	64	*	Solvent, fuel additive
Ethyl alcohol	CH_3CH_2OH	78	7.06	Solvent, beverages
Propyl alcohol	$CH_3CH_2CH_2OH$	97	1.87	Solvent
Isopropyl alcohol	$CH_3CHOHCH_3$	82	5.8	Solvent, body rubs
Butyl alcohol	$CH_3CH_2CH_2CH_2OH$	118	4.36	Solvent
Hexyl alcohol	$CH_3(CH_2)_4CH_2OH$	156	4.59	—
Ethylene glycol	$HOCH_2CH_2OH$	198	8.54	Antifreeze
Glycerol	$HOCH_2CHOHCH_2OH$	290 (dec)†	> 25	Moisturizer

*No LD_{50} is given for methanol. Its acute (short-term) toxicity is not terribly high. However, it is metabolized to formaldehyde (HCHO) in the body, so that the chronic (long-term) toxicity is quite high. The LD_{50} for formaldehyde administered orally to rats is 0.070 g/kg of body weight. The LD_{50} for acetaldehyde, the metabolite of ethanol, is 1.9 g/kg.

†Glycerol decomposes.

Source: Susan Budani (Ed.), *The Merck Index,* 11th ed. Rathway, NJ: Merck and Co., 1989.

CHEMISTRY IN OUR WORLD

Ethyl Alcohol: The Beverage

Ethanol, CH_3CH_2OH, is the alcohol that is produced for beverages by the fermentation of grain or other materials that contain sugar or starch. An enzyme catalyzes the hydrolysis of starch to sugar units. If the sugar is glucose—a six-carbon simple sugar—the reaction is

$$C_6H_{12}O_6 \xrightarrow{\text{yeast}} 2\,CH_3CH_2OH + 2\,CO_2$$

Fermentation of the sugar present in grapes produces wine. Beer is made from malted barley and other starchy cereal grains that are fermented after being flavored with hops. Whiskey can be obtained by distilling fermented corn, barley, or wheat. Vodka can be prepared by distilling spirits of mash from potatoes. Rum is obtained by distilling fermented molasses from sugar cane. The oriental drink called saki is prepared from rice. In each case, the fermentation process gives ethanol. Alcoholic beverages are seldom more than 45% ethanol, or 90 proof.

The **proof** is merely twice the percentage of alcohol by volume. The term has its origin in a seventeenth-century English method for testing whiskey for alcohol content. A test for the whiskey was to pour some of it on gunpowder and ignite it. If the gunpowder ignited after the alcohol had burned away, that was considered ''proof'' that the whiskey did not contain too much water.

Approximate Relationship Between Drinks Consumed, Blood-Alcohol Level, and Behavior for a 70-kg (154-lb) Moderate Drinker

*Number of Drinks**	*Blood-Alcohol Level (percent by volume)*	*Behavior†*
2	0.05	Mild sedation; tranquility
4	0.10	Lack of coordination
6	0.15	Obvious intoxication
10	0.30	Unconsciousness
20	0.50	Possible death

*A *drink* is a 30-mL (1-oz) ''shot'' of 90-proof whiskey, 360-mL (12-oz) bottle of beer, or 150-mL (5-oz) glass of wine.

†An inexperienced drinker would be affected more strongly, or more quickly, than one who is ordinarily a moderate drinker. Conversely, an experienced *heavy* drinker would be affected less.

Ethanol is the key chemical present in beer, wine, tequila, mixed drinks, and all alcoholic beverages.

More than two-thirds of the adult population in the United States drinks alcoholic beverages at least occasionally. Although the majority do so responsibly, misuse—and abuse—is common. There are approximately 10 million alcoholics in the United States who are so severely addicted that they are unable to hold a steady job or maintain stable family relationships. Furthermore, over half of the fatal automobile accidents involve at least one drinking driver.

Generally alcohol acts as a mild depressant; it slows down both physical and mental activity. Large amounts can produce unconsciousness or even death. Listed here are typical effects of various blood-alcohol levels, but the specific effects for each person vary with body weight, the amount of food in the stomach, the experience of the drinker, and other factors.

Taxes on alcoholic beverages raise about $6 billion a year in the United States. An estimated 100,000 to 200,000 deaths are attributed annually to alcohol. The total cost to society—from medical treatment, accidents, lost work time, and other factors—is estimated to be $120 billion a year.

In addition to its use as a beverage, about 230,000 kg (a half million lb) of ethanol is produced synthetically in the United States each year. This industrial ethanol is used as a fuel, as a solvent, and as a starting material for many other chemical products. Most ethanol produced industrially is prepared by the acid-catalyzed reaction of ethylene, an alkene (Section 19.5), with water.

$$CH_2{=}CH_2 + H_2O \xrightarrow{H_2SO_4} CH_3CH_2OH$$

Synthetic ethanol has exactly the same properties as ethanol produced by fermentation. Synthetic ethanol is generally cheaper, and, unlike alcoholic beverages, it does not carry an excise tax. To prevent the consumption of synthetic ethanol as a beverage, poisonous or noxious substances such as gasoline and methanol are added. The resulting **denatured alcohol** is unfit for drinking.

Polyhydroxy Alcohols

Each of the simple alcohols described so far contains only one hydroxyl group per molecule. Alcohols that contain more than one —OH group are called **polyhydroxy alcohols**. The most common **diol** is 1,2-ethanediol, which is generally known as ethylene glycol. It has the structure

$$\begin{array}{l} H_2C{-}OH \\ \;\;\;| \\ H_2C{-}OH \end{array}$$

With an annual production of over 2 billion kg (about 5 billion lb), it ranks among the top 30 chemicals in the United States. This syrupy liquid made from ethylene is the common ingredient in antifreeze and is used extensively in the manufacture of polyester fibers and films (Figure 19.15). The hydrogen bonding between —OH groups of different molecules gives this chemical a relatively high boiling point of 197 °C compared to other chemicals with a similar molar mass.

Glycerol, also known as glycerine, is an important **triol** with the structure

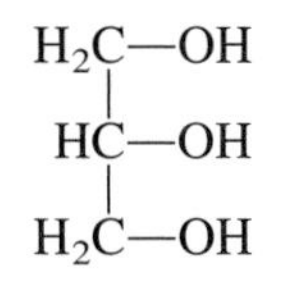

Glycerol is a by-product of soap manufacturing. It is used in hand lotions and cosmetics as a moistening agent and is also an important biochemical. Typical fat (lipid) molecules are derivatives of glycerol; that is, they contain various groups attached to glycerol. Physicians routinely test blood samples for the level of triglycerides. These include simple fats and other molecules that have three groups attached to a glycerol structure.

There are many alcohols; only a few have been presented here. All alcohols are characterized by the presence of a hydroxyl group (—OH group), which is responsible for most of the characteristic reactions of alcohols. The —OH group is the **functional group** in all alcohols. The general formula for an alcohol can be represented as R—OH, where R represents the alkyl group. We can categorize organic compounds and typical characteristic reactions by the specific functional groups involved. Table 19.9 lists some of the more important organic functional groups.

Figure 19.15 *Ethylene glycol is the principal active ingredient in permanent antifreeze solutions.*

Table 19.9 **Selected Organic Functional Groups**

Name of Class	*Functional Group*	*General Formula of Class*
Alkane	None	R—H
Alkene	$>C=C<$	$R(R')C=C(R'')R'''$
Alkyne	—C≡C—	R—C≡C—R′
Alcohol	$-\overset{\vert}{\underset{\vert}{C}}-O-H$	R—O—H
Ether	$-\overset{\vert}{\underset{\vert}{C}}-O-\overset{\vert}{\underset{\vert}{C}}-$	R—O—R′
Aldehyde	$-\overset{O}{\overset{\Vert}{C}}-H$	$R-\overset{O}{\overset{\Vert}{C}}-H$
Ketone	$-\overset{O}{\overset{\Vert}{C}}-$	$R-\overset{O}{\overset{\Vert}{C}}-R'$
Amine	$-\overset{\vert}{\underset{\vert}{C}}-\overset{\vert}{N}-$	$R-\overset{H}{\overset{\vert}{N}}-H$ $R-\overset{H}{\overset{\vert}{N}}-R'$ $R-\overset{R'}{\overset{\vert}{N}}-R''$
Carboxylic acid	$-\overset{O}{\overset{\Vert}{C}}-O-H$	$R-\overset{O}{\overset{\Vert}{C}}-O-H$
Ester	$-\overset{O}{\overset{\Vert}{C}}-O-\overset{\vert}{\underset{\vert}{C}}-$	$R-\overset{O}{\overset{\Vert}{C}}-O-R'$
Amide	$-\overset{O}{\overset{\Vert}{C}}-\underset{\vert}{N}-$	$R-\overset{O}{\overset{\Vert}{C}}-\underset{H}{\underset{\vert}{N}}-H$ $R-\overset{O}{\overset{\Vert}{C}}-\underset{H}{\underset{\vert}{N}}-R'$ $R-\overset{O}{\overset{\Vert}{C}}-\underset{R''}{\underset{\vert}{N}}-R'$

EXAMPLE 19.11

Give a name for each of the following structures and indicate which, if any, are structural isomers.

(a) $CH_3\underset{OH}{\underset{\vert}{C}}HCH_3$

(b) $CH_3\underset{CH_3}{\underset{\vert}{C}}HCH_2OH$

(c) $CH_3CH_2\underset{OH}{\underset{\vert}{C}}HCH_3$

(d) $CH_3CH_2\underset{OH}{\underset{\vert}{C}}HCH_2CH_3$

SOLUTION

(a) isopropyl alcohol or 2-propanol
(b) isobutyl alcohol or 2-methylpropanol
(c) *sec*-butyl alcohol or 2-butanol
(d) 3-pentanol

Structures (b) and (c) are structural isomers; both are alcohols with four carbon atoms.

See Problems 19.31–19.42.

EXERCISE 19.3
Draw structures for *tert*-butyl alcohol and 2-methyl-2-butanol. What do these alcohols have in common?

19.8 Phenols

Compounds with a hydroxyl group attached to a benzene ring are called **phenols**. The general formula for a phenolic compound can be written Ar—OH, where Ar symbolizes a benzene ring or another aromatic (substituted benzene) group. The parent compound, C_6H_5OH, is itself called phenol.

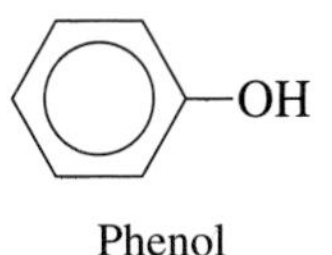

Phenol

The phenols have found wide use as antiseptics (substances that kill microorganisms on living tissue) and disinfectants (substances intended to kill microorganisms on furniture, floors, and around the house in general).

Phenol—once commonly known as carbolic acid—was the first widely used antiseptic. In 1867 Joseph Lister used it for the first antiseptic surgery. Phenol has also been used as the active ingredient in various first-aid salves and ointments. Unfortunately, phenol doesn't kill just the undesirable microorganisms. It kills all types of cells. When applied to the skin, it can cause severe burns. For the most part, it has been replaced by safer phenolic antiseptics such as 4-hexylresorcinol, which is used in some mouthwashes.

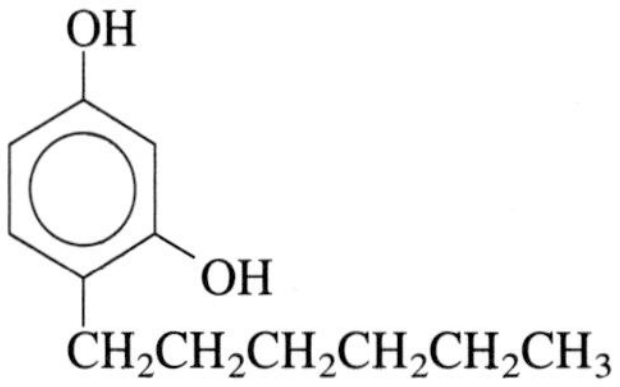

OH
OH
$CH_2CH_2CH_2CH_2CH_2CH_3$

4-Hexylresorcinol

See Problems 19.43 and 19.44.

19.9 Ethers

Alcohols and phenols may both be considered derivatives of water in which one of the hydrogen atoms of water has been replaced by an alkyl (—R) or an aryl (—Ar) group. **Ethers** may be thought of as compounds in which both hydrogen atoms of water have been replaced by alkyl or aryl groups. General structural formulas can be written

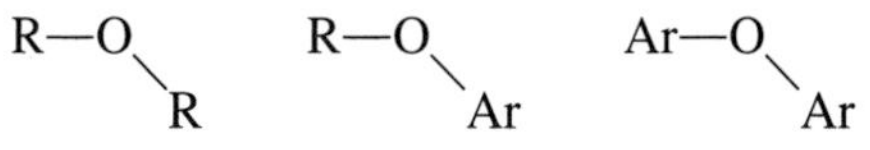

R—O—R R—O—Ar Ar—O—Ar

Figure 19.16 *Diethyl ether—once a well-known anesthetic—is used in automotive starting fluids and as a solvent for fats, oils, and other organic substances.*

Simple ethers are named simply. Just name the groups attached to oxygen and then add the generic name *ether*. For example, methyl *tert*-butyl ether (often abbreviated MTBE) is widely used as an octane booster in gasoline. (Check the gasoline pump next time you fill up!) Its structural formula is

$$CH_3—O—\underset{\displaystyle CH_3}{\overset{\displaystyle CH_3}{\overset{|}{\underset{|}{C}}}}—CH_3$$

Diethyl ether, $CH_3CH_2—O—CH_2CH_3$, was once widely used as an anesthetic. Its main use now is as a solvent. It dissolves many organic substances such as fats and oils. Diethyl ether boils at 36 °C, so it is readily evaporated away, enabling one to recover the dissolved substance. Often simply called ''ether,'' diethyl ether is highly flammable. It is used in aerosol cans of automotive quick-starting fluid (Figure 19.16). Whenever this chemical is being used, great care must be taken to avoid sparks or flames.

In the IUPAC System, ethers are named as alkoxy derivatives of alkanes. First, select the name of the longer carbon chain joined to the oxygen and give it the alkane name. The remaining alkoxy group (the RO— group) is named as a functional group bonded to the alkane. Thus, $CH_3—O—CH_2CH_2CH_3$ is named methoxypropane.

EXAMPLE 19.12

Give the IUPAC name and structural formula for ethyl methyl ether.

SOLUTION

This compound has an ethyl group and a methyl group attached to oxygen. Its IUPAC name is methoxyethane. Its structural formula is

$$CH_3CH_2—O—CH_3$$

See Problems 19.45 and 19.46.

19.10 *Aldehydes and Ketones*

The aldehydes and ketones are two related families of organic compounds. Aldehydes and ketones both are characterized by the presence of a **carbonyl group**, that is, a carbon atom with a double bond to oxygen and single bonds to hydrogen atoms or alkyl (or aryl) groups.

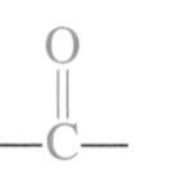

Aldehydes have at least one hydrogen atom attached to the carbonyl group. In **ketones**, only alkyl groups (R—) or aryl groups (Ar—) are bonded to the carbonyl group.

$$H—\overset{\displaystyle O}{\overset{\|}{C}}—H \qquad H—\overset{\displaystyle O}{\overset{\|}{C}}—(R \text{ or } Ar) \qquad (Ar \text{ or } R)—\overset{\displaystyle O}{\overset{\|}{C}}—(R \text{ or } Ar)$$

Aldehydes Ketones

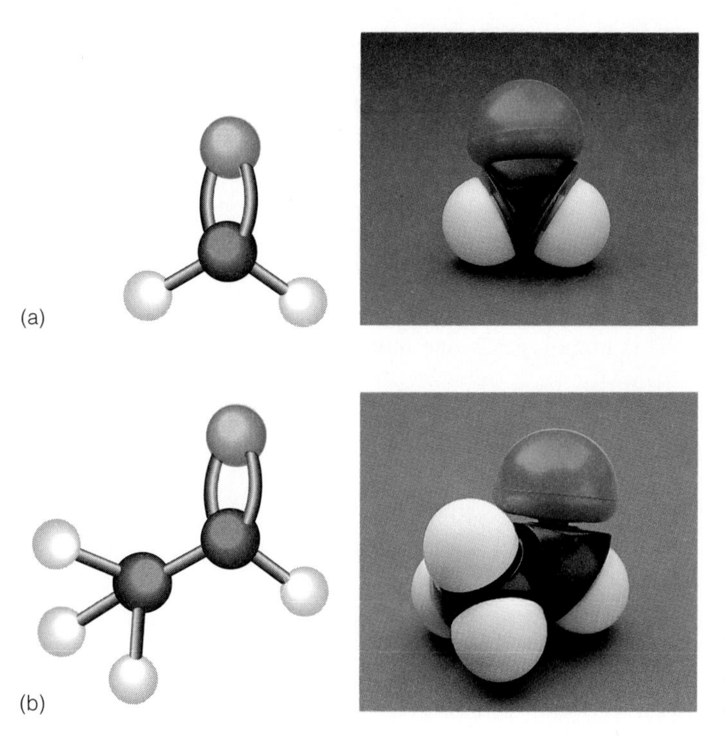

Figure 19.17 *Ball-and-stick and space-filling models of formaldehyde (a) and acetaldehyde (b).*

The simplest and most widely used aldehyde is formaldehyde (Figure 19.17a).

$$\mathrm{H{-}\overset{\overset{\displaystyle O}{\|}}{C}{-}H}$$

It is a gas at room temperature and is readily soluble in water. As a 40% solution called *formalin*, it is used in embalming fluid and sometimes used as a preservative for biological specimens, but the chemical is toxic. Formaldehyde is used extensively in the manufacture of adhesives and plastics. It is made by the oxidation of methanol and ranks among the top 25 chemicals produced annually in the United States. This is basically the same reaction that occurs in the human body when methanol is ingested.

$$\underset{\text{Methanol}}{\mathrm{CH_3OH}} \xrightarrow{\text{oxidation}} \underset{\text{Formaldehyde}}{\mathrm{H{-}\overset{\overset{\displaystyle O}{\|}}{C}{-}H}}$$

The equations shown here are not balanced. Only the major reactants and products are shown.

The next member of the homologous series of aldehydes is acetaldehyde, CH_3CHO (Figure 19.17b). It is formed by the oxidation of ethanol.

In linear form, the aldehyde group is written as CHO. Thus, CH_3CHO is equivalent to $\mathrm{CH_3{-}\overset{\overset{\displaystyle O}{\|}}{C}{-}H}$.

$$\underset{\text{Ethanol}}{\mathrm{CH_3CH_2OH}} \xrightarrow{\text{oxidation}} \underset{\text{Acetaldehyde}}{\mathrm{CH_3{-}\overset{\overset{\displaystyle O}{\|}}{C}{-}H}}$$

Notice that the oxidation of a primary alcohol gives an aldehyde.

Ethanol is also oxidized to acetaldehyde in the human body. A buildup of acetaldehyde is responsible in part for the hangover effect of heavy drinking.

To write the IUPAC name of an aldehyde, drop the final *-e* and add *-al* to the parent alkane (the longest carbon chain containing the carbonyl group). Thus, formaldehyde is named methanal, acetaldehyde—with two carbon atoms—is ethanal, and so on.

EXAMPLE 19.13

Give the structure and IUPAC name for the next homolog after ethanal in the homologous series of aldehydes.

SOLUTION

This homolog has one more CH_2 unit. It is named propanal.

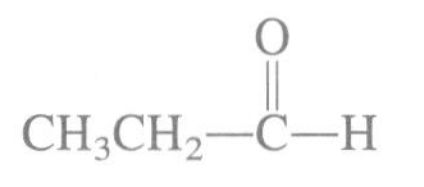

Benzaldehyde is an aromatic aldehyde; it has an aldehyde group attached to a benzene ring.

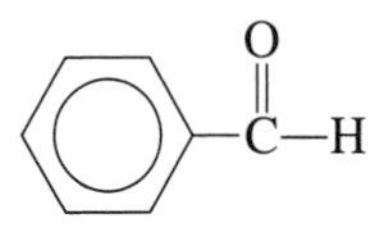

Also called (synthetic) oil of almond, benzaldehyde is used in flavorings and fragrances.

The simplest ketone is called *acetone* (Figure 19.18). It can be made by the oxidation of isopropyl alcohol, a secondary alcohol.

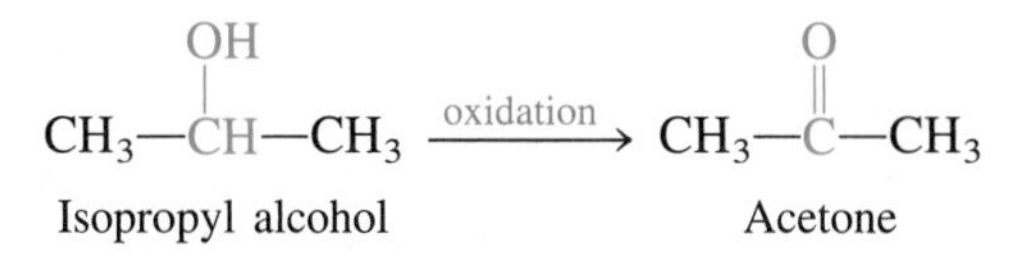

Notice that the oxidation of a secondary alcohol gives a ketone.

Acetone is a common solvent for such organic materials as fats, plastics, varnishes, and some rubber materials. It is the major—sometimes the only—ingredient in fingernail polish removers.

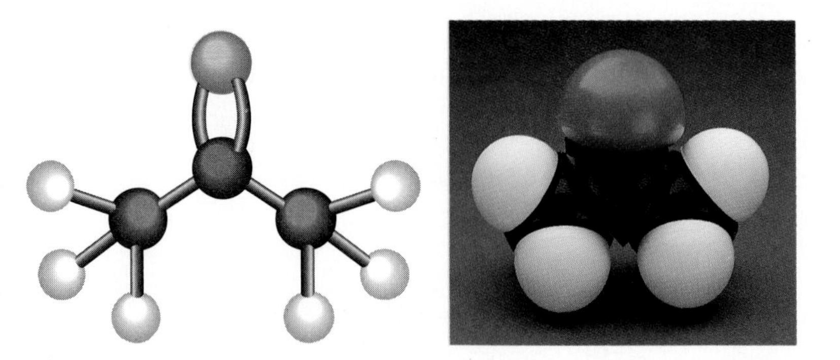

Figure 19.18 *Ball-and-stick and space-filling models of acetone.*

Acetone has a name that is unique; it does not correspond to the first in a series of common names of ketones. Generally, ketones are given common names consisting of the names of the groups attached to the carbonyl group—in alphabetical order—followed by the word ketone. (Note the similarity to the naming of ethers.) In the IUPAC system, the longest continuous chain containing the carbonyl group is selected as the parent chain. The final *-e* of the corresponding alkane name is dropped and replaced with *-one*. Using IUPAC rules, acetone is named propanone and ethyl methyl ketone is butanone. In higher ketones, a number placed in front of the name indicates the position of the carbonyl group. This number is arrived at by counting carbons, beginning at the end of the parent chain nearest the carbonyl group. Consider the following ketones. (The IUPAC name is printed above the common name.)

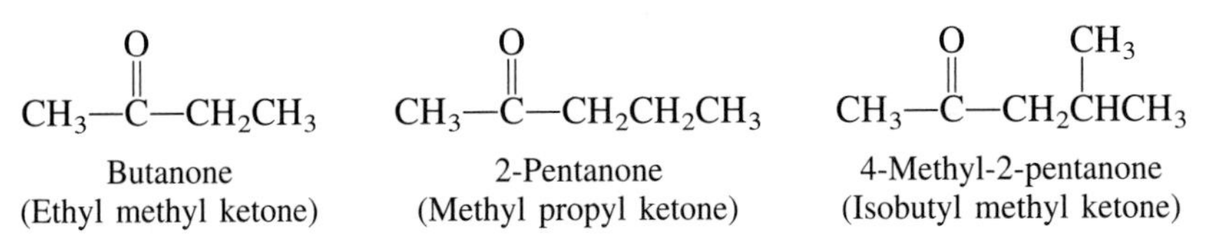

EXAMPLE 19.14

Give the structural formula for 4-methyl-3-hexanone.

SOLUTION

The longest chain has six carbon atoms with a doubly bonded oxygen located at the third carbon and a methyl group at the fourth carbon.

$$\overset{1}{C}H_3\overset{2}{C}H_2-\overset{3}{C}(=O)-\overset{4}{C}H(CH_3)\overset{5}{C}H_2\overset{6}{C}H_3$$

See Problems 19.47–19.56.

EXERCISE 19.4

Give the common name for the ketone in Example 19.14.

19.11 Carboxylic Acids and Esters

The familiar strong acids (Chapter 16) such as sulfuric, hydrochloric, and nitric acids are called *mineral acids* because they are derived from inorganic materials. Many of the weak acids contain carbon; these are the organic acids. The functional group of the organic acids is called the **carboxyl group**.

$$-C(=O)-OH \quad \text{or} \quad -COOH$$

Acids containing this group are called **carboxylic acids**; they can be represented by the general formula RCOOH. Several carboxylic acids are listed in Table 19.10. Notice that their names fit the general form -ic acid.

The simplest carboxylic acid is called formic acid. It was first obtained by the destructive distillation of ants (Latin *formica*, ''ant''). The bite of an ant smarts because the ant injects formic acid as it bites. The stings of wasps and bees also contain formic acid along with other poisonous materials.

$$H-C(=O)-OH$$

Methanoic acid
(Formic acid)

Table 19.10 Some Common Carboxylic Acids

Condensed Formula	*IUPAC Name*	*Common Name*	*Derivation of Common Name*
$HCOOH$	Methanoic acid	Formic acid	Latin *formica*, "ant"
CH_3COOH	Ethanoic acid	Acetic acid	Latin *acetum*, "vinegar"
CH_3CH_2COOH	Propanoic acid	Propionic acid	Greek *protos*, "first" + *pion*, "fat"
$CH_3CH_2CHCOOH$	Butanoic acid	Butyric acid	Latin *butyrum*, "butter"
$CH_3(CH_2)_3COOH$	Pentanoic acid	Valeric acid	Latin *valere*, "powerful"
$CH_3(CH_2)_4COOH$	Hexanoic acid	Caproic acid	
$CH_3(CH_2)_6COOH$	Octanoic acid	Caprylic acid	Latin *caper*, "goat"
$CH_3(CH_2)_8COOH$	Decanoic acid	Capric acid	
$CH_3(CH_2)_{10}COOH$	Dodecanoic acid	Lauric acid	Laurel tree
$CH_3(CH_2)_{12}COOH$	Tetradecanoic acid	Myristic acid	*Myristica fragans* (nutmeg)
$CH_3(CH_2)_{14}COOH$	Hexadecanoic acid	Palmitic acid	Palm tree
$CH_3(CH_2)_{16}COOH$	Octadecanoic acid	Stearic acid	Greek *stear*, "tallow"

$CH_3—\overset{\overset{O}{\|}}{C}—OH$

Ethanoic acid
(Acetic acid)

$CH_3CH_2—\overset{\overset{O}{\|}}{C}—OH$

Propanoic acid
(Propionic acid)

$CH_3CH_2CH_2—\overset{\overset{O}{\|}}{C}—OH$

Butanoic acid
(Butyric acid)

Acetic acid can be made by the aerobic (in the presence of oxygen) fermentation of a mixture of apple cider and honey. This produces a vinegar solution that contains about 4 to 10% acetic acid, plus a number of other compounds that give vinegar its flavor. Acetic acid is one of the most familiar *weak* acids used in educational and industrial chemistry laboratories (Figure 19.19).

The third member of the homologous series of acids is propionic acid. It is seldom encountered in everyday life. The fourth member is much more familiar, at least by its odor. If you've ever smelled rancid butter or buttermilk, you know what butyric acid smells like. It is one of the most foul-smelling substances imaginable. Small amounts of butyric acid are present in strong cheeses and can be isolated from butterfat or synthesized in the laboratory. It is also one of the ingredients of body odor. Extremely small amounts of this and other chemicals enable bloodhounds to track fugitives.

The acid with a carboxylic acid group attached directly to a benzene ring is called benzoic acid.

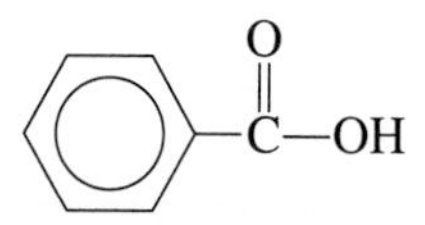

Figure 19.19 *Vinegars are 5% solutions of acetic acid.*

Like the mineral acids, the carboxylic acids form salts. Calcium propionate, sodium benzoate, and other carboxylate salts are widely used as food additives to prevent molds. Carboxylic acids are quite common in nature. Many of the higher homologs are obtained from fats.

Esters

Esters are derived from reactions between carboxylic acids and alcohols. The general acid-catalyzed reaction involves splitting out a molecule of water for each molecule of acid and alcohol that react.

$$\underset{\text{An acid}}{\text{R—}\overset{\overset{\text{O}}{\|}}{\text{C}}\text{—OH}} + \underset{\text{An alcohol}}{\text{R'OH}} \xrightarrow{H^+} \underset{\text{An ester}}{\text{R—}\overset{\overset{\text{O}}{\|}}{\text{C}}\text{—OR'}} + H_2O$$

Although some carboxylic acids have a disagreeable odor, esters generally have pleasant odors. They are often responsible for the characteristic fragrances of fruits and flowers. For example, foul-smelling butyric acid reacts with ethanol to produce ethyl butyrate, an ester that is present in pineapples and is used in artificial flavoring.

Once a fruit or flower has been chemically analyzed, flavor chemists can attempt to duplicate the natural odor or taste. Artificial fruit flavors are often mixtures of esters (Figure 19.20). Esters are also used as solvents for numerous organic substances. Several esters are shown in Figure 19.21. The name of an ester consists of the name of the alkyl group from the alcohol followed by the name of the acid with the *-ic* suffix changed to *-ate*.

Figure 19.20 *The flavor and aroma in fruit and fruit-flavored candy (shown here) come from a blend of esters.*

EXAMPLE 19.15

Draw the structure for ethyl butyrate. (Its IUPAC name is ethyl butanoate from the IUPAC name of the carboxylic acid.)

SOLUTION

It is easier to start with the acid portion, butyrate, with its four-carbon structure.

$$CH_3CH_2CH_2\overset{\overset{\text{O}}{\|}}{\text{C}}\text{—}$$

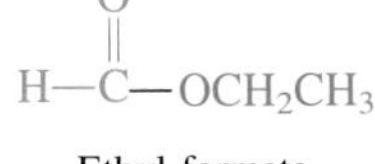

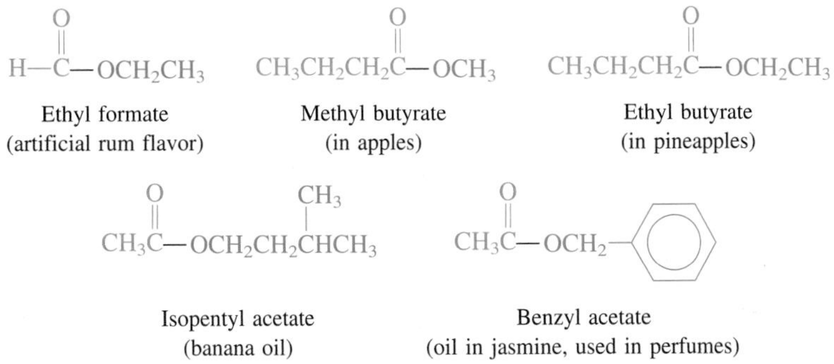

$\text{H—}\overset{\overset{\text{O}}{\|}}{\text{C}}\text{—}OCH_2CH_3$

Ethyl formate
(artificial rum flavor)

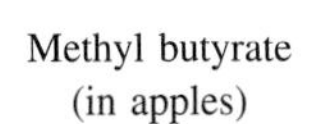

$CH_3CH_2CH_2\overset{\overset{\text{O}}{\|}}{\text{C}}\text{—}OCH_3$

Methyl butyrate
(in apples)

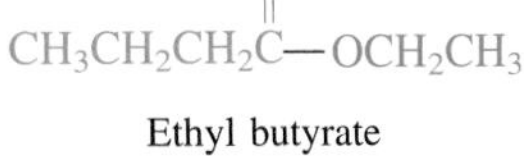

$CH_3CH_2CH_2\overset{\overset{\text{O}}{\|}}{\text{C}}\text{—}OCH_2CH_3$

Ethyl butyrate
(in pineapples)

$CH_3\overset{\overset{\text{O}}{\|}}{\text{C}}\text{—}OCH_2CH_2\overset{\overset{CH_3}{|}}{\text{C}}HCH_3$

Isopentyl acetate
(banana oil)

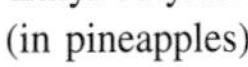

$CH_3\overset{\overset{\text{O}}{\|}}{\text{C}}\text{—}OCH_2\text{—}C_6H_5$

Benzyl acetate
(oil in jasmine, used in perfumes)

Figure 19.21 *Some esters of interest.*

Then attach the alcohol portion, the —OR′ group, to the R—C(=O)— from the acid structure.

See Problems 19.57–19.62.

$$CH_3CH_2CH_2C(=O)—O—CH_2CH_3$$

19.12 Amines and Amides

Many organic compounds contain nitrogen. The **amines** contain the elements carbon, hydrogen, and nitrogen. They are derived from ammonia by replacing one, two, or three of the hydrogen atoms by alkyl or aryl groups to give primary (1°) amines, secondary (2°) amines, and tertiary (3°) amines.

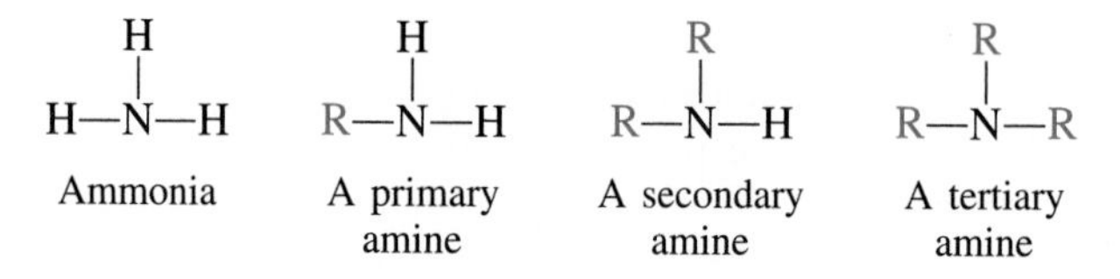

$CH_3—N(CH_3)—CH_3$

Trimethylamine

The simplest amine is methylamine, CH_3NH_2. The next higher homolog is ethylamine, $CH_3CH_2NH_2$. With two carbon atoms we can also have dimethylamine, CH_3NHCH_3. Notice that ethylamine and dimethylamine are isomers; both have the molecular formula C_2H_7N. With three atoms, there are several possibilities, including trimethylamine.

EXAMPLE 19.16

The structure of trimethylamine was given. Show structures and give names for the other amines that contain three carbon atoms.

SOLUTION

Two of these amines are derived from the two kinds of propyl groups.

$CH_3CH_2CH_2NH_2$ — Propylamine

$CH_3CH(NH_2)CH_3$ — Isopropylamine

The other three-carbon amine has one methyl and one ethyl group. (The alkyl groups are named in alphabetical order.)

$CH_3CH_2NHCH_3$

Ethylmethylamine

The amine with an NH_2 group attached directly to a benzene ring has the special name *aniline*.

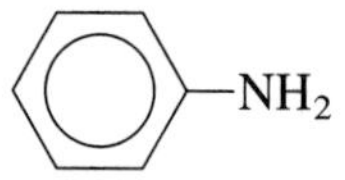

$C_6H_5—NH_2$

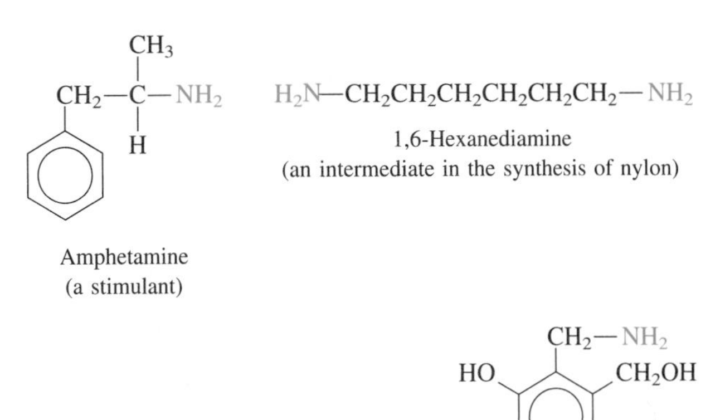

$H_2N—CH_2CH_2CH_2CH_2CH_2—NH_2$

Cadaverine
(odor of decaying flesh)

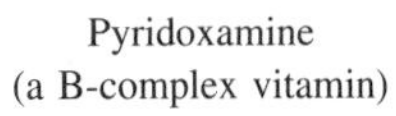

Pyridoxamine
(a B-complex vitamin)

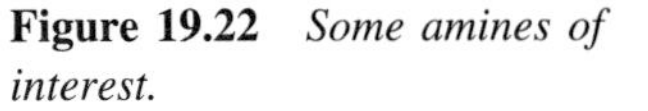

Figure 19.22 *Some amines of interest.*

The simplest amines are similar to ammonia in basicity and other properties. It is the higher amines that are especially interesting. Several of these are shown in Figure 19.22. Notice that each structure contains an $—NH_2$ group, called an **amino group**.

Another family of nitrogen-containing compounds is the **amides**. (What a difference a letter makes!) These compounds are nitrogen-containing derivatives of carboxylic acids that contain the amide group.

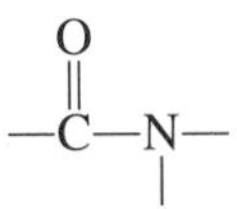

Amides can be made by reacting a carboxylic acid with ammonia or an amine to form a salt and then heating the salt to drive off water.

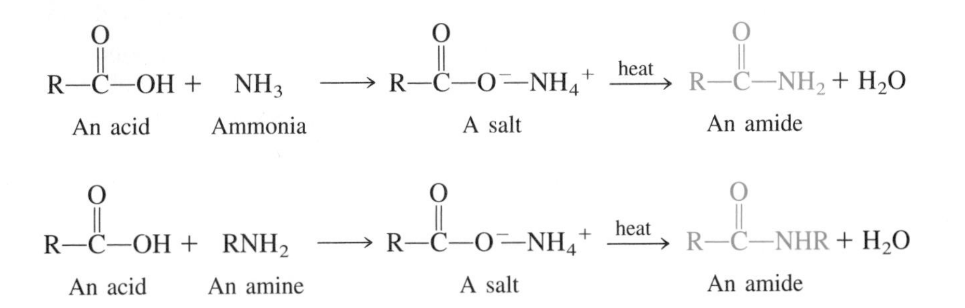

The simple amides are of little interest to us here, but the complex amides are of tremendous importance. For example, nylon is a polymer with amide linkages. The *polyamide* can be made by reacting 1,6-hexanediamine (see Figure 19.22) with 1,6-hexanedioic acid (a six-carbon acid with a carboxyl group at each end) to form amid linkages. The polymer consists of amide linkages that alternately join acid to amine to acid to amine and so on. Even more important—and fundamental to life itself—are the proteins (described in the next chapter). They contain many amino acid units joined together by amide linkages.

More than 1 million tons of nylon are produced each year in the United States for use in clothing, carpets, rope, and various other products.

Nylon 6 6 was the first polyamide to be produced commercially. Its name (pronounced nylon six-six) is said to be based on the fact that both reactants have six carbon atoms. "Nylons" went on sale on May 15, 1940, and 4 million pairs were sold in New York City.

EXAMPLE 19.18

The polyamide called nylon 6 6 was first synthesized by the Nobel prize–winning Wallace H. Carothers of the Du Pont laboratories. Draw a segment of this structure based on the information given in the preceding paragraph.

SOLUTION

Nylon, a polyamide, has alternate units of 1,6-hexanediamine and 1,6-hexanedioic acid (also called adipic acid).

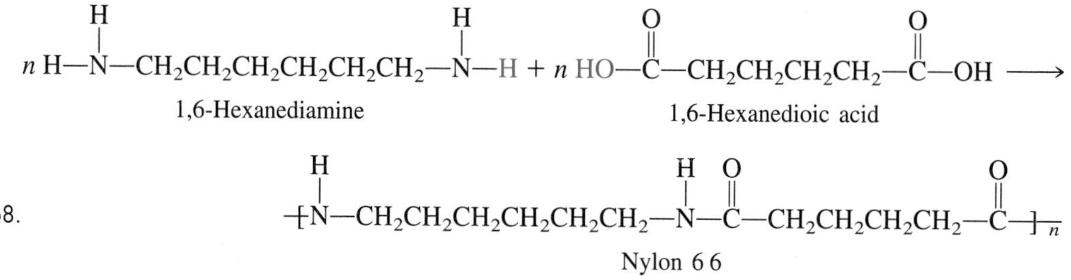

$$n\,\text{H}-\overset{\overset{\displaystyle\text{H}}{|}}{\text{N}}-\text{CH}_2\text{CH}_2\text{CH}_2\text{CH}_2\text{CH}_2\text{CH}_2-\overset{\overset{\displaystyle\text{H}}{|}}{\text{N}}-\text{H} + n\,\text{HO}-\overset{\overset{\displaystyle\text{O}}{\|}}{\text{C}}-\text{CH}_2\text{CH}_2\text{CH}_2\text{CH}_2-\overset{\overset{\displaystyle\text{O}}{\|}}{\text{C}}-\text{OH} \longrightarrow$$

1,6-Hexanediamine 1,6-Hexanedioic acid

See Problems 19.63–19.68.

$$\left[\overset{\overset{\displaystyle\text{H}}{|}}{\text{N}}-\text{CH}_2\text{CH}_2\text{CH}_2\text{CH}_2\text{CH}_2\text{CH}_2-\overset{\overset{\displaystyle\text{H}}{|}}{\text{N}}-\overset{\overset{\displaystyle\text{O}}{\|}}{\text{C}}-\text{CH}_2\text{CH}_2\text{CH}_2\text{CH}_2-\overset{\overset{\displaystyle\text{O}}{\|}}{\text{C}}\right]_n$$

Nylon 6 6

Chapter Summary

Organic chemistry is the chemistry of carbon-containing componds. Unlike typical inorganic compounds, most organic compounds are flammable and have relatively low melting points. Those that are nonpolar have low water solubilities. The alkanes are a homologous series of hydrocarbons that have only single covalent bonds between carbon atoms. They are said to be "saturated" because they contain the maximum number of hydrogen atoms. Structural formulas are used to show the order in which atoms are attached. Compounds that have the same molecular formula but different structural formulas are called isomers.

Alkenes contain a carbon-to-carbon double bond and alkynes contain a carbon-to-carbon triple bond. Both classes are said to be "unsaturated" because they can undergo addition reactions with hydrogen. Unlike single bonds, there is restricted motion about double bonds. When there are two nonidentical groups on each carbon involved in the double bond, cis and trans isomers are possible. Cis isomers have two identical groups positioned on the same side of the double bond. Trans isomers have two identical groups positioned on opposite sides of the double bond. Alkenes can react with one another to form giant molecules called polymers. With the IUPAC system, alkenes are named in a manner similar to alkanes, but they have the -ene suffix; alkynes have the -yne suffix. The position of the double or triple bond is designated by a number.

Compounds that contain one or more benzene rings are classified as aromatic compounds. Removal of a hydrogen atom from a benzene ring gives a structure called a phenyl group. Phenols are aromatic compounds that have an —OH group attached to a benzene ring. An alcohol molecule contains a hydroxyl group, —OH, covalently bonded to an alkyl group, —R. The general formula of an alcohol is R—OH. Alcohols are classified as primary (1°), secondary (2°), or tertiary (3°), depending on whether one, two, or three alkyl groups are attached to the carbon bonded to the —OH group.

A carbonyl group, $-\overset{\overset{\displaystyle\text{O}}{\|}}{\text{C}}-$, is the functional group present in both aldehydes and ketones. Aldehydes have at least one hydrogen atom attached to the carbonyl group. In ketones, only alkyl or aromatic groups are bonded to the carbonyl group. The oxidation of a primary alcohol gives the corresponding aldehyde. Oxidation of a secondary alcohol gives a ketone.

The carboxyl group, —COOH, is the functional group of all organic acids (also called carboxylic acids). Esters are derived from reactions between carboxylic acids and alcohols. Replacing one, two, or three hydrogen atoms of ammonia gives primary, secondary, and tertiary amines. Aniline is the simplest aromatic amine; it has an NH_2 group attached directly to a benzene ring. The amides are carboxylic acid derivatives obtained by replacing the hydroxyl group of the acid with a nitrogen-containing group. Nylon is a polyamide.

Assess Your Understanding

1. Give general characteristics of organic compounds that distinguish them from inorganic compounds. [19.1]
2. Identify and give examples of isomers and of a homologous series [19.2]
3. Classify organic compounds by type when structural formulas are given. [19.3–19.12]
4. Give the IUPAC name for an organic compound when the structural formula is provided. [19.4–19.12]
5. Identify saturated and unsaturated hydrocarbons and cis-trans isomers. [19.5]
6. Describe and give examples of polymers [19.5]
7. Compare properties of benzene with open-chain unsaturated compounds. [19.6]
8. Compare structures of alcohols, phenols, and ethers. [19.7–19.9]
9. Compare structures of simple aldehydes and ketones. [19.10]
10. Identify examples of structures representing carboxylic acids, esters, amines, and amides. [19.11, 19.12]

Key Terms

addition reaction [19.5]
alcohol [19.7]
aldehyde [19.10]
aliphatic [19.6]
alkanes [19.2]
alkenes [19.5]
alkyl group [19.4]
alkyl halides [19.2]
alkynes [19.5]
amide [19.12]
amine [19.12]
amino group [19.12]
aromatic [19.6]
aryl group [19.6]
benzene [19.6]
carbonyl group [19.10]
carboxyl group [19.11]
carboxylic acid [19.11]
cis [19.5]
condensed structural formulas [19.3]
denatured alcohol [19.7]
destructive distillation [19.7]
diol [19.7]
ethanol [19.7]
ethers [19.9]
ethyl [19.4]
functional group [19.7]
homologs [19.2]
hydrocarbons [19.2]
hydrogenation [19.5]
isomers [19.2]
isopropyl [19.4]
IUPAC [19.4]
ketone [19.10]
methanol [19.7]
methyl [19.4]
phenols [19.8]
phenyl group [19.6]
polyhydroxy alcohols [19.7]
polymers [19.5]
primary alcohol [19.7]
proof (alcohol) [19.7]
propyl [19.4]
saturated hydrocarbons [19.2]

secondary alcohol [19.7]
structural formulas [19.3]
tertiary alcohol [19.7]
toluene [19.6]
trans [19.5]
triol [19.7]
unsaturated [19.5]

Problems

ORGANIC VERSUS INORGANIC CHEMICALS

19.1 Compare the definition of organic chemistry used in earlier times with the definition used by modern organic chemists.

19.2 Classify the following compounds as organic or inorganic.
a. C_6H_{10}
b. $NaHCO_3$
c. $C_{12}H_{22}O_{11}$
d. CH_3NH_2
e. synthetic vitamin C, $C_6H_8O_6$
f. vitamin C, $C_6H_8O_6$, from oranges

19.3 Which member of each pair has a higher melting point? Explain.
a. CH_3OH and $NaOH$
b. CH_3Cl and KCl
c. C_6H_{12} and $C_{20}H_{42}$

19.4 Which member of each pair has a higher solubility in water? Explain.
a. octane, C_8H_{18}, and NaCl
b. ethanol, CH_3CH_2OH, and oil containing $C_{14}H_{30}$
c. octane and NH_4NO_3

19.5 You find an unlabeled jar containing a solid material. The substance melts at 48 °C. It ignites readily and burns cleanly. The substance is insoluble in water and floats on the surface of the water. Classify the substance as organic or inorganic.

19.6 You had an unlabeled can in the garage that contains a liquid. When a wood stick was dipped into the liquid and then ignited, the substance burned with a smoky flame. The liquid was immiscible with water. Classify the substance as organic or inorganic.

ALKANES, ALKENES, AND ALKYNES

19.7 How many carbon atoms are there in each of the following? Draw their condensed structural formulas.
a. ethane
b. butane
c. isobutane
d. 1-pentene
e. 2,2-dimethylhexane
f. acetylene

19.8 How many carbon atoms are there in each of the following? Draw their condensed structural formulas.
a. propane
b. hexane
c. ethyne
d. 2-pentene
e. 2,2,4-trimethylpentane (in gasoline)
f. 2-bromo-2-methylpropane

19.9 Name these compounds.
a. $CH_2{=}CH_2$
b. $CH_3(CH_2)_4CH_3$
c. CH_3CH_2Br
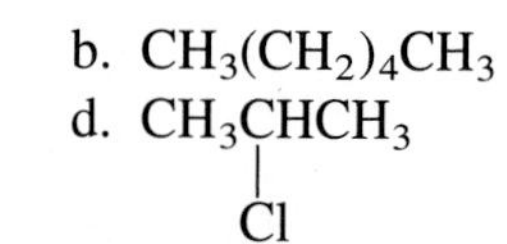

19.10 Name these compounds.
a. CH_3CH_3
b. $CH_3(CH_2)_2CH_3$
c. $CH_3CH{=}CH_2$
d. $CH_3CH{-}Br$ (with CH_3 on the CH)

19.11 Indicate whether the structures in each set represent the same compound or isomers.
a. CH_3CH_3 and $CH_3{-}CH_3$ (drawn vertically)
b. $CH_3CH(CH_3)CH_2CH_2CH_3$ and $CH_3CH_2CH(CH_3)CH_2CH_3$
c. $CH_3CH_2CH(CH_3)CH_2CH_3$ and $CH_3CH(CH_3)CH_2CH_2(CH_3)$

19.12 Indicate whether the structures in each set represent the same compound or isomers.
a. $CH_3CH_2(CH_3)$ and $CH_3CH_2CH_3$
b. $CH_3CH_2CH(CH_3)(CH_3)$ and $CH_3CH(CH_3)CH_2CH_3$
c. $CH_3CH(CH_3){-}Br$ and $CH_3CH(Br)CH_3$

19.13 Indicate whether each compound is saturated or unsaturated.
a. ethyne
b. butene

c. oleic acid,
$CH_3(CH_2)_7CH{=}CH(CH_2)_7COOH$

d. $CH_3(CH_2)_6COOH$

19.14 Indicate whether each compound is saturated or unsaturated.

a. pentene

b. propyne

c. cyclohexane

d. linoleic acid,
$CH_3(CH_2)_4CH{=}CHCH_2CH{=}CH(CH_2)_7COOH$

19.15 Complete these structures by adding the appropriate number of hydrogen atoms. Give an IUPAC name for each.

a. 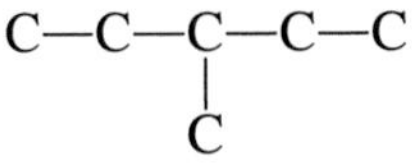

b. 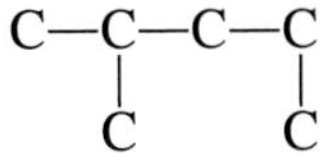

19.16 Complete these structures by adding the appropriate number of hydrogen atoms. Give an IUPAC name for each.

a. 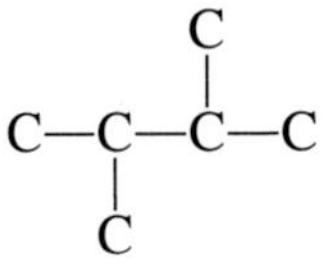

b. 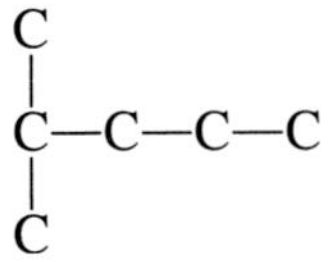

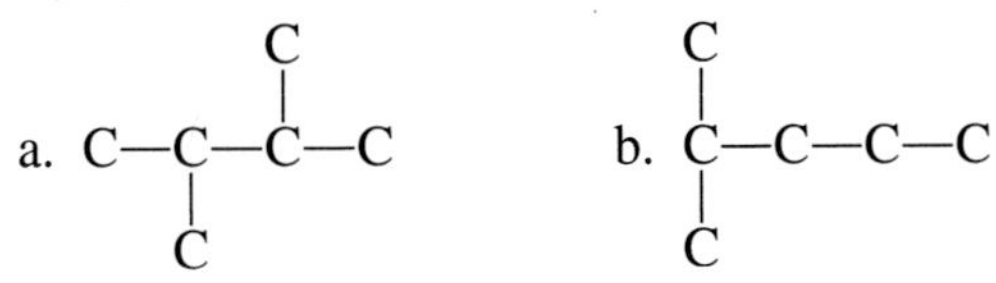

19.17 There are four isomers with the molecular formula C_4H_9OH. Draw all structures and classify each as a primary, secondary, or tertiary alcohol.

19.18 Indicate which of the following are isomers of $C_5H_{11}Br$ and classify each as a primary, secondary, or tertiary alkyl bromide. Identify any structures that represent the same compound.

a. 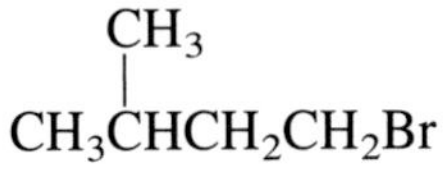

b. 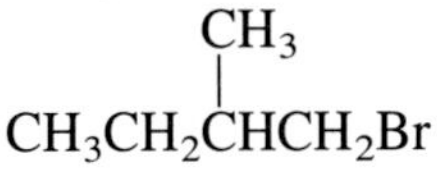

c. 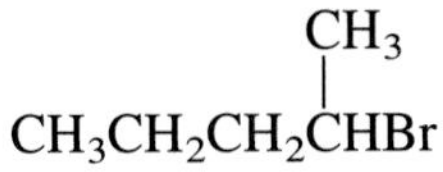

d. 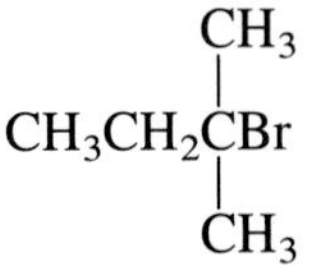

e. 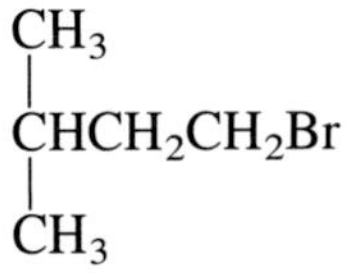

19.19 The compound $CH_3CH_2CH_2CH_2CH_2CH_3$ is a solvent used to extract oil from soybeans. What is the name of the compound?

19.20 The compound $CH_3CH_2CH_2CH_3$ is used as the fuel in many disposable cigarette and fireplace lighters. What is the name of the compound?

19.21 Give structural formulas for the following.

a. 3-isopropyl-1-hexene

b. *cis*-2,3-dichloro-2-butene

c. *trans*-2,3-dichloro-2-butene

19.22 Give structural formulas for the following.

a. 3-ethyl-2-pentene

b. *cis*-2-hexene

c. *trans*-2-hexene

AROMATIC COMPOUNDS

19.23 Indicate whether the compound represented is aromatic or aliphatic.

a. 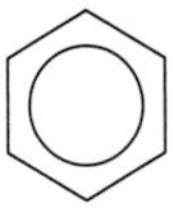

b. 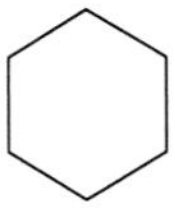

c. 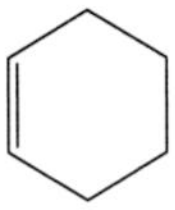

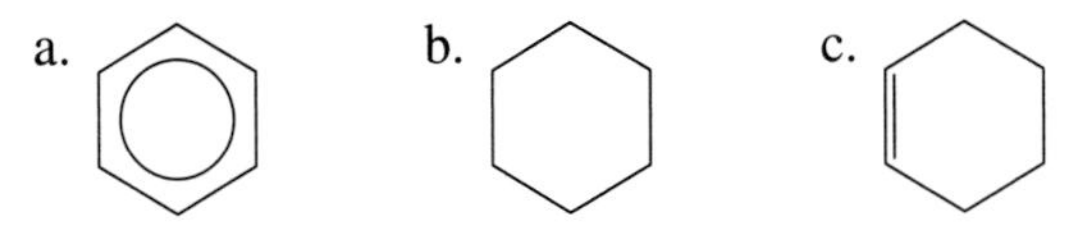

19.24 Indicate whether the compound represented is aromatic or aliphatic.

a. 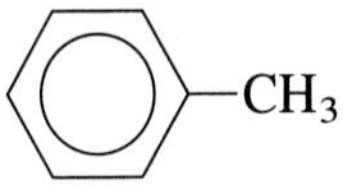

b. 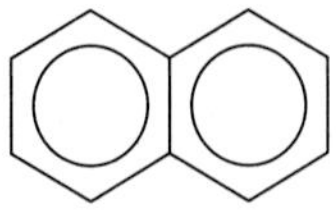

c. 

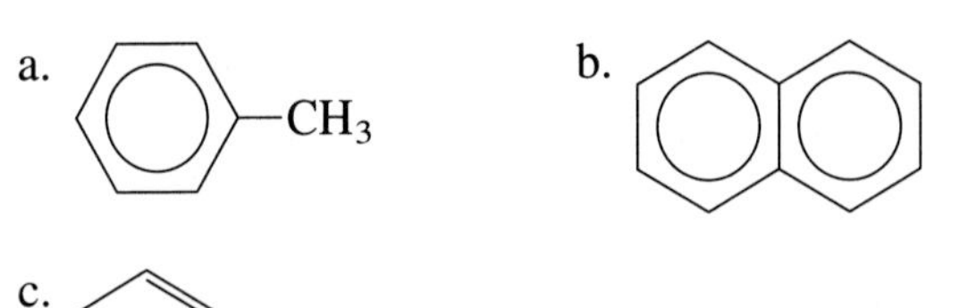

19.25 Draw two equivalent Kekulé resonance structures for benzene. Describe how the true benzene structure differs from the Kekulé resonance structures.

19.26 Draw a benzene structure represented by a simple hexagon containing a circle. Give an advantage and a disadvantage of replacing the Kekulé structure for benzene with this alternate structure.

19.27 Identify the substitution pattern as ortho, meta, or para.

a. 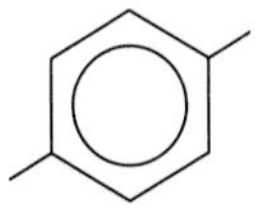

b. 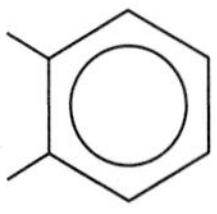

c. 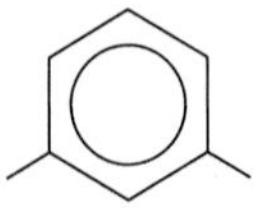

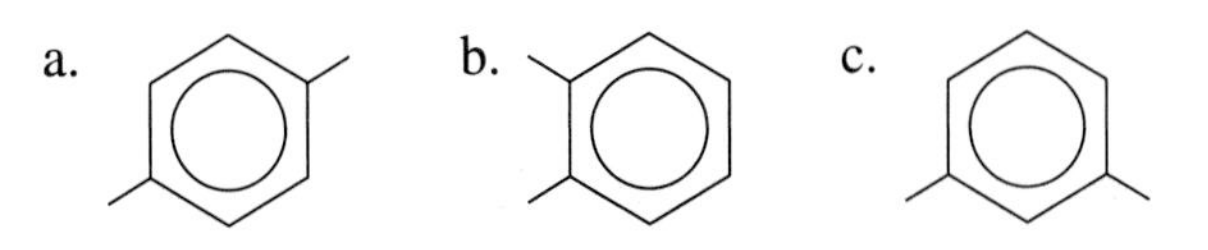

19.28 Identify the substitution pattern as ortho, meta, or para.

a. 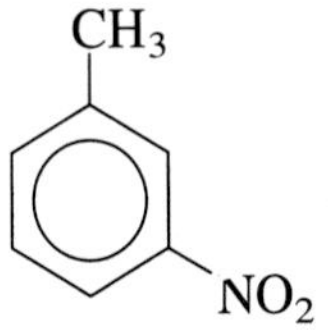

b. 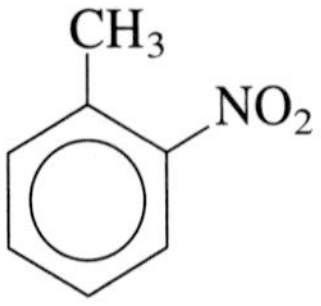

c. 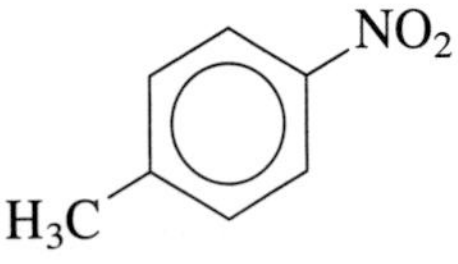

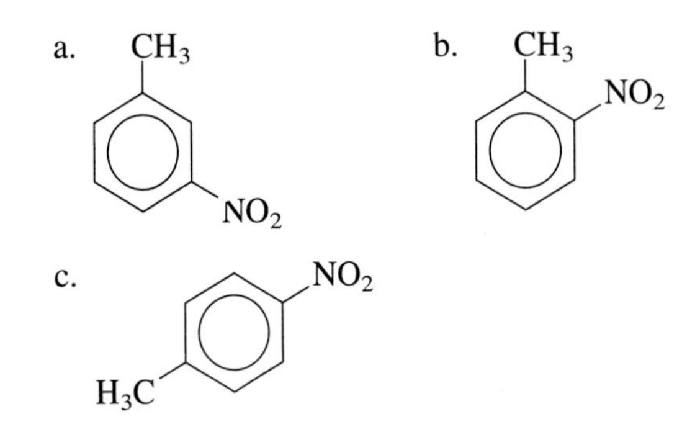

19.29 Give structural formulas for the following.
a. *p*-dichlorobenzene (used in moth flakes)
b. 1,4-dinitrobenzene
c. *o*-nitrochlorobenzene (used to prevent a fungus in ears of corn)
d. naphthalene (used in moth balls)

19.30 Give structural formulas for the following.
a. *o*-dibromobenzene
b. toluene (used as a solvent and in some automotive gasoline additives)
c. *m*-xylene
d. 2,4,6-trinitrotoluene (called TNT—you can get a ''bang'' out of it!)

ALCOHOLS, PHENOLS, ETHERS

19.31 Give structural formulas for the following compounds and indicate whether each is a primary, secondary, or tertiary alcohol.
a. methanol (also called methyl alcohol)
b. *tert*-butyl alcohol
c. 2-butanol
d. 2-methylpropanol

19.32 Give structural formulas for the following compounds and indicate whether each is a primary, secondary, or tertiary alcohol.
a. isopropyl alcohol (rubbing alcohol)
b. 2-methyl-2-propanol
c. isobutyl alcohol
d. 3-methylbutanol

19.33 What is destructive distillation? Give the name of an alcohol that can be produced by the destructive distillation of wood.

19.34 Give two other names for methanol. Give two or more uses for methanol.

19.35 What is denatured alcohol?

19.36 Give two other names for ethanol. Give two or more uses for ethanol.

19.37 What is the proof of a beverage that is 43% alcohol by volume?

19.38 What is the percent alcohol by volume in 80-proof vodka?

19.39 Describe one method—including the equation—for the commercial production of methanol.

19.40 Describe one method—including the equation—for the commercial production of ethanol.

19.41 Give the name of and uses for one diol.

19.42 Give the name of and uses for one triol.

19.43 What is a phenolic compound? Give the general formula for a phenolic compound.

19.44 Give the formula for phenol. Give two uses for phenol.

19.45 Give the general formula for an ether.

19.46 Give the structural formula and two uses for diethyl ether.

ALDEHYDES AND KETONES

19.47 Give the general structural formula for an aldehyde. Give the name and structure of the functional group present in all aldehydes.

19.48 Give the general structural formula for a ketone. Give the name and structure of the functional group present in all ketones.

19.49 Give the structure of formaldehyde. Give two uses. What is formalin?

19.50 Give the structure of acetone. Give two uses.

19.51 Give the structure for each of the following.
a. ethyl methyl ketone
b. acetaldehyde
c. isopropyl methyl ketone

19.52 Give the structure for each of the following.
a. butanal
b. benzaldehyde
c. methyl propyl ketone

19.53 What type of alcohol can be oxidized to produce a ketone?

19.54 What type of alcohol can be oxidized to produce an aldehyde?

19.55 Give the structure of the alcohol from which each of the following aldehydes and ketones can be made by oxidation.
a. 2-pentanone b. acetaldehyde

19.56 Give the structure of the alcohol from which each of the following aldehydes and ketones can be made by oxidation.
a. acetone b. butanal

CARBOXYLIC ACIDS AND ESTERS

19.57 Give the general structural formula for a carboxylic acid.

19.58 Give the structure of the carboxyl group.

19.59 Give the general structural formula for an ester.

19.60 Write an equation, using structural formulas, that shows the general reaction for the formation of an ester from an acid and an alcohol.

19.61 Write an equation for the formation of methyl butyrate. (It is present in apples.) Name each compound involved in the reaction.

19.62 Write an equation for the formation of isopentyl acetate. (It is present in banana oil, a flavoring.) Name each compound involved in the reaction.

AMINES AND AMIDES

19.63 Give the structure of an amino group. What is an amine?

19.64 What type of compounds contain the

$$-\overset{\overset{\displaystyle O}{\|}}{C}-\underset{|}{N}-$$ group?

19.65 Give structural formulas for the following compounds and classify each as a primary, secondary, or tertiary amine.

a. ethylamine b. dimethylamine c. trimethylamine

19.66 Give structural formulas for the following compounds and classify each as a primary, secondary, or tertiary amine.

a. methylamine b. ethylmethylamine c. isopropylamine

19.67 Give the structure of the product of the reaction of acetic acid with methyl amine. What type of compound is this product?

19.68 What two molecular units are present in nylon 6 6? What type of linkage is present in the polymer?

Additional Problems

19.69 Is sulfuric acid, H_2SO_4, or candle wax containing $C_{20}H_{42}$ more soluble in hexane? Explain.

19.70 Is charcoal lighter fluid—it contains decane—or water a better solvent for removing motor oil from the garage floor? Explain.

19.71 Why is ethanol, rather than methanol, used as a solvent for medicines taken internally. (Compare the toxicities of ethanol and methanol.)

19.72 Describe the meaning of the term LD_{50} as applied to alcohols.

19.73 Name these compounds.

a. $CH_3CH_2\underset{\displaystyle OH}{\underset{|}{C}}HCH_3$

b. $CH_3CH_2CH_2\underset{\displaystyle OH}{\underset{|}{C}}HCH_3$

c. $CH_3\underset{\displaystyle CH_3}{\underset{|}{C}}HCH_2CH_2OH$

d. $CH_3CH_2-O-CH_2CH_3$

19.74 Name these compounds.

a. $CH_3\underset{\displaystyle OH}{\underset{|}{C}}HCH_3$

b. $CH_3CH_2\underset{\displaystyle OH}{\underset{|}{C}}HCH_2CH_3$

c. $CH_3CH_2\underset{\displaystyle CH_3}{\underset{|}{C}}HCH_2OH$

d. $CH_3CH_2-O-\underset{\displaystyle CH_3}{\underset{|}{C}}HCH_3$

19.75 Name these compounds.

a. $CH_3-\underset{\displaystyle O}{\underset{\|}{C}}-CH_2CH_3$

b. $CH_3CH_2-\underset{\displaystyle O}{\underset{\|}{C}}-H$

c. $CH_3CH_2-\underset{\displaystyle O}{\underset{\|}{C}}-CH_2CH_2CH_3$

19.76 Name these compounds.

a. CH_3CH_2CHO

b. $CH_3CH_2-\underset{\displaystyle O}{\underset{\|}{C}}-CH_2CH_3$

c. $CH_3CH_2-\underset{\displaystyle O}{\underset{\|}{C}}-CH_2\underset{\displaystyle CH_3}{\underset{|}{C}}HCH_3$

19.77 Give the structure and the IUPAC name for each of the following.

a. butyric acid b. lauric acid c. ethyl butyrate

19.78 Give the structure and the IUPAC name for each of the following.

a. palmitic acid b. formic acid c. ethyl formate

19.79 Name each of the following compounds, and classify each as a primary, secondary, or tertiary amine.

a. $CH_3CH_2CH_2NH_2$ b. $CH_3NHCH_2CH_3$ c. $(CH_3)_3N$

19.80 Name each of the following compounds, and classify each as a primary, secondary, or tertiary amine.

a. $(CH_3CH_2)_2NH$

b. $CH_3CH_2CH_2CH_2NH_2$

c. $C_6H_5-NH_2$ (benzene ring with $-NH_2$)

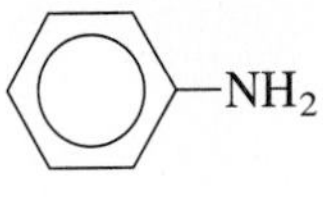

A model of a double-stranded deoxyribonucleic acid (DNA) molecule is shown. The nucleus of each cell in an organism contains DNA molecules that carry the genetic code—the information that gives the organism is unique characteristics and functions.

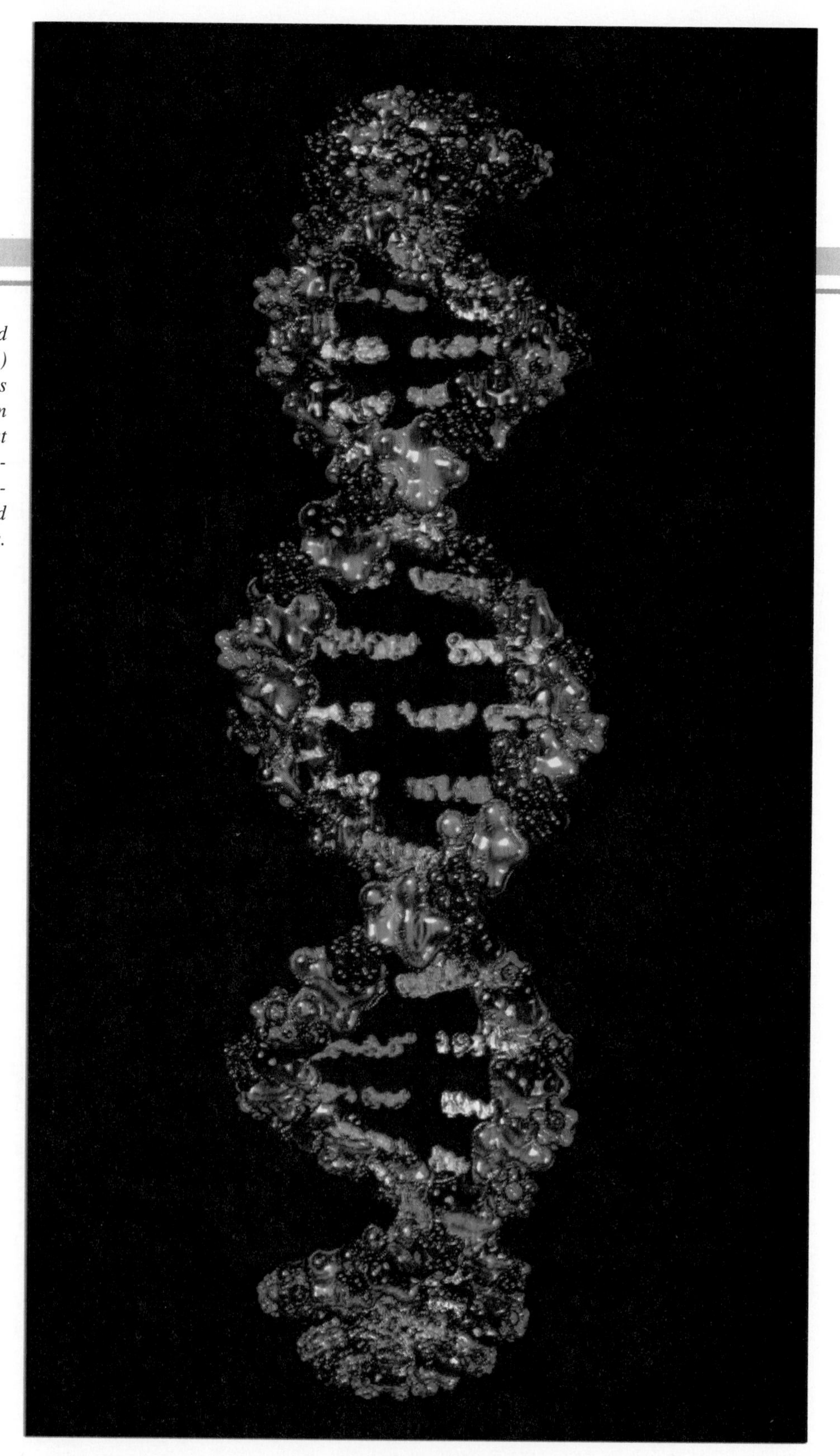

20

Biochemistry

CONTENTS

The study of chemistry that takes place in our bodies and in all living cells is called *biochemistry*. Each person is able to digest and metabolize many different substances present in the diet. These chemicals—obtained from food—are the building blocks of muscle, bone, fat, and other complex body tissues. The body is also able to manufacture a tremendous number of different molecules—some with molar masses in the millions—that control and regulate body functions and protect the body from disease.

Important developments in biochemical research are reported each month in thousands of journal articles. This type of information is essential to understanding about the control and prevention of disease and about life processes. Much biochemistry is known, but much more remains to be discovered.

In this chapter we can include only a limited number of fundamental biochemical topics. Those described here provide a general basis for further reading and study in this area of active research. Our knowledge of chemistry and biochemistry is rapidly expanding. It can benefit all of us. We hope that your study of chemistry has just begun.

Figure 20.1 *Carbohydrates occur in fruits, vegetables, bread, potatoes, pasta, and other foods.*

20.1 Carbohydrates

Almost everyone knows what carbohydrates are: They're what you eat or don't eat depending on whose diet book you follow (Figure 20.1). Any dietitian will tell you that carbohydrates must be included in any well-balanced diet. When we eat and metabolize these compounds, the atoms rearrange themselves to make simple, stable compounds and, in the process, release their stored energy for our use.

Chemically, **carbohydrates** are compounds of the elements carbon, hydrogen, and oxygen with the empirical formula $C_x(H_2O)_y$. The formula of glucose, which is $C_6H_{12}O_6$, could also be written $C_6(H_2O)_6$. When the white crystalline glucose is heated, it gives off water and leaves behind a black, charred form of carbon. Indeed, early scientists erroneously thought that sugars were hydrates of carbon—carbohydrates, they called them. However, one cannot add the charred product back to water and obtain glucose. Glucose is not a hydrate at all. The term carbohydrate is misleading. (Recall that when a *true* hydrate is heated, it gives an anhydrous salt that will recombine with water to form the original hydrate).

Monosaccharides

Simple carbohydrates are called the **monosaccharides**. They cannot be further hydrolyzed; that is, they cannot be broken into simpler sugar units. Carbohydrates that can be hydrolyzed to two monosaccharide units (two simple sugars) are called **disaccharides**, and carbohydrates that can be hydrolyzed to many monosaccharide units are called **polysaccharides**. Monosaccharides can be further classified by the number of carbon atoms per molecule.

Number of Carbon Atoms	*Class*
3	Triose
4	Tetrose
5	Pentose
6	Hexose

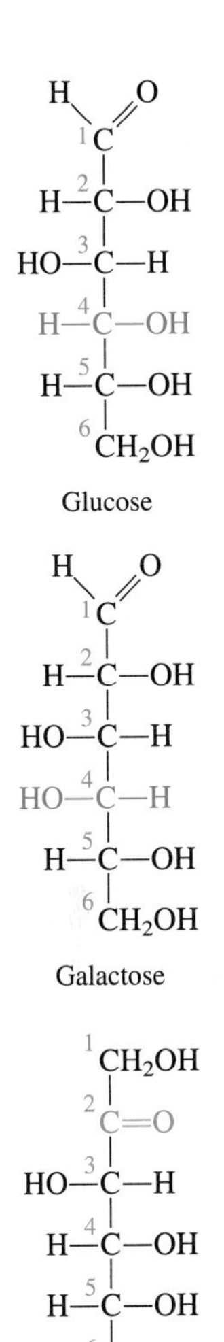

Figure 20.2 *Structures of three important hexoses.*

Each monosaccharide has the **-ose** ending. Still another system can also be used to classify monosaccharides. Those that have an aldehyde group are called aldoses, and those with a ketone group are called ketoses. The two systems are often combined. For example, *glucose* is an aldohexose; it is a monosaccharide with six carbon atoms and an aldehyde group. *Fructose,* another monosaccharide, is a ketohexose; it has six carbon atoms and a ketone group. Structures of three monosaccharides are shown in Figure 20.2.

There are 16 different aldohexoses. All of these have the formula $C_6H_{12}O_6$; they are isomers. Of these, glucose and galactose are the most familiar. As you look at Figure 20.2, you will notice that glucose and galactose appear to have precisely the same structures with one exception: the hydrogen and hydroxyl (—H and —OH) groups on the fourth carbon atom—counting from the aldehyde end—are reversed. This is a very important and fundamental difference, much like that between your right hand and your left hand: each hand has one thumb and four fingers, but a glove designed to fit your right hand does not fit your left hand. Any organic molecule that has a carbon atom with four different groups attached always has this right-handedness or left-handedness property (Figure 20.3). The carbon is said to be chiral (pronounced KY-ral). Each of the aldohexoses has four chiral carbon atoms (carbon atom numbers 2, 3, 4, and 5). An —OH group shown on the right-hand side at one of these positions designates a different structure than an —OH group on the left-hand side.

D and L Designations. Half of the aldohexose structures that can be drawn would have the —OH group on the right-hand side at the carbon number 5 posi-

Figure 20.3 *A chiral* carbon atom is one that is attached to four different kinds of atoms or four different groups of atoms.

tion. These monosaccharides carry the D designation, as used in D-glucose and D-galactose. Other aldohexoses could be drawn with the —OH group on the left-hand side at the carbon number 5 position. These are known as L sugars. Nearly all naturally occurring carbohydrates are members of the D series. This is not a point to be taken lightly; we cannot obtain energy from L carbohydrates.

Cyclic Structures. Monosaccharides do not remain in the stretched out or extended form like the structures pictured in Figure 20.2. As a molecule of glucose flops around and folds back on itself, the oxygen on carbon number 5 can easily be situated near carbon number 1 (the carbonyl carbon) as shown in Figure 20.4. When this carbonyl carbon is ''bumped'' by the oxygen of the hydroxyl group, the cyclic (ring) structure is formed as pictured in Figure 20.5. Notice that as the ring forms, the oxygen at carbon number 5 becomes bonded to carbon number 1 to complete the six-membered ring.

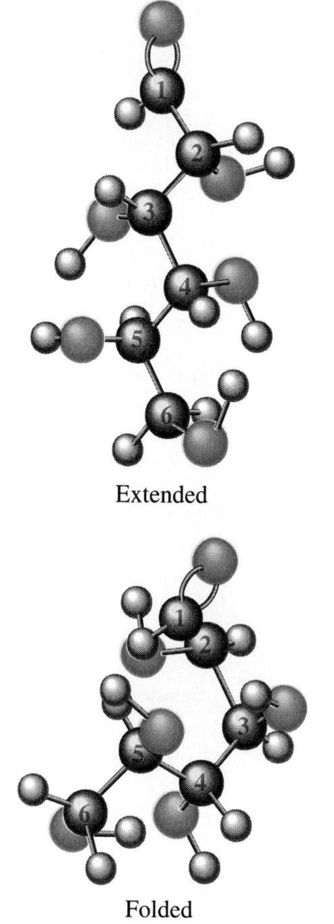

Figure 20.4 *Models of the free-aldehyde form of glucose. Note that in the folded model the oxygen atom on the fifth carbon is near the carbonyl carbon.*

During the ring-closing reaction, the originally doubly bonded oxygen of the carbonyl group—at carbon number 1—is converted to a hydroxyl group that can be in either of two positions to make two different closed-ring structures called the **alpha** form and the **beta** form. In the alpha form, the hydroxyl group at carbon number 1 is projected downward, and in the beta form, this hydroxyl is projected upward when the molecule is situated in the position shown in Figure 20.5.

Crystalline glucose can exist in either the alpha or beta form. In either case, there are four different groups —OH, —OR, —R, and —H) attached to carbon number 1. The presence of precisely these four groups produces a structure called a **hemiacetal**. A hemiacetal is quite unstable. One can start out with either the pure alpha or beta form, but as soon as it dissolves in water, the unstable hemiacetal group opens, giving the aldehyde structure, and then closes again to give either the alpha or beta form. Opening and reclosing occur in succession; the interconversion is referred to as **mutarotation**. At equilibrium, the mixture is about 36% alpha and 64% beta. There is less than 0.02% of the aldehyde (open) form. Nevertheless, that is enough to give most of the characteristic reactions of aldehydes. All aldoses readily react with mild oxidizing agents, so they are called **reducing sugars**.

The difference between the alpha and beta forms may seem trivial, but such differences are often crucial in biochemical reactions. For example, we shall soon be looking at the structure of starch, which is a polymer with glucose units in the alpha form, and comparing it with cellulose, which is a polymer with glucose units in the beta form. What a difference a chiral carbon can make!

Selected Monosaccharides. Let us look more closely at three important monosaccharides. D-Glucose, which is also known as *dextrose,* is the most impor-

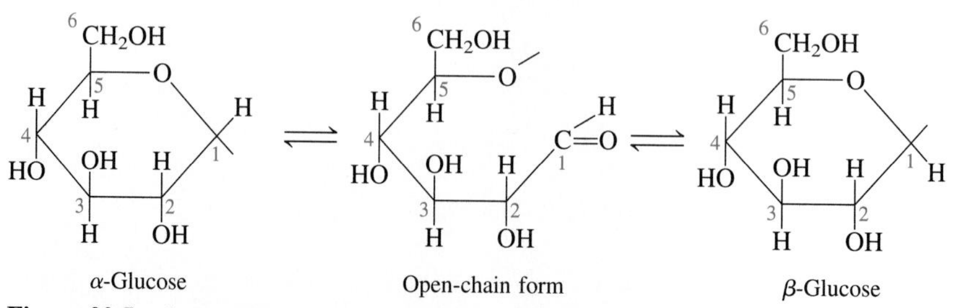

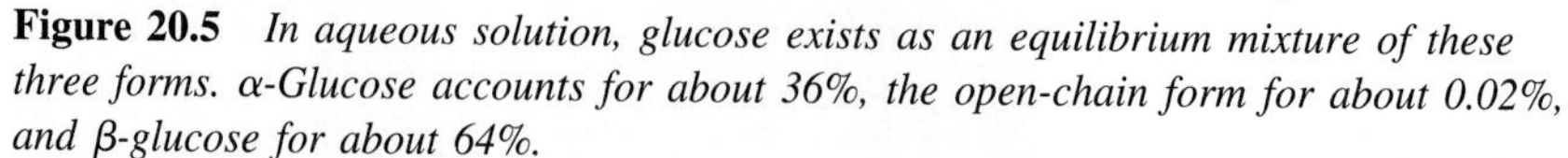

Figure 20.5 *In aqueous solution, glucose exists as an equilibrium mixture of these three forms. α-Glucose accounts for about 36%, the open-chain form for about 0.02%, and β-glucose for about 64%.*

tant hexose. It is the normal "blood sugar" that makes up about 0.065 to 0.11% of our blood. It is essential to life because it is the main sugar our cells use directly for the production of energy. It has been estimated that nearly half of the carbon atoms in the biosphere are tied up in glucose. Unfortunately, for the hungry people of the world, much of it is in the form of cellulose, which has little or no food value for humans.

D-Galactose—our second monosaccharide—is an aldohexose that, along with glucose, can be obtained upon hydrolysis of a disaccharide called lactose or milk sugar. It is present in certain compounds found in the brain and nerves.

The only ketohexose we shall consider is D-fructose. It has a carbonyl group at the carbon number 2 position, as shown in Figure 20.2. Fructose is the sweetest of the common sugars. It occurs, along with glucose and sucrose, in honey and fruit juices. Hydrolysis of the disaccharide called sucrose or table sugar yields both fructose and glucose.

Disaccharides

There are three common disaccharides: maltose, lactose, and sucrose. Hydrolysis of 1 mol of disaccharide gives 2 mol of monosaccharide. Using word equations, we can write

$$\text{Maltose} + H_2O \xrightarrow{H^+ \text{ or enzyme}} 2\ \text{Glucose}$$

$$\text{Lactose} + H_2O \xrightarrow{H^+ \text{ or enzyme}} \text{Glucose} + \text{Galactose}$$

$$\text{Sucrose} + H_2O \xrightarrow{H^+ \text{ or enzyme}} \text{Glucose} + \text{Frutose}$$

Maltose is present in sprouting grain, but its major source is the partial hydrolysis of starch. In maltose, two glucose units are joined by an alpha **acetal** linkage; that is, the hydroxyl group at carbon number 1 of the glucose shown on the left is in the downward position—the alpha position—when it bonds to the hydroxyl group at carbon number 4 of the glucose shown on the right.

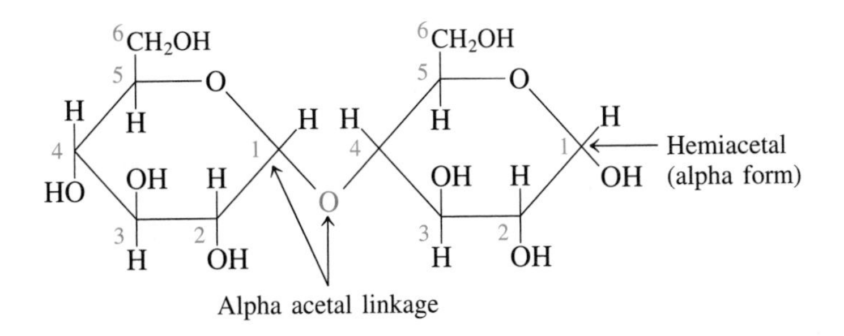

The glucose ring on the right is pictured with the hemiacetal in the alpha form, but this ring can be in either the alpha or the beta form since it can continually open and close.

Lactose occurs to the extent of 5 to 7% in human milk and 4 to 6% in cow's milk. In lactose, a galactose unit—at the left—is joined to a glucose unit by a beta acetal linkage (the oxygen at carbon number 1 of galactose that links the two rings is in the upward position).

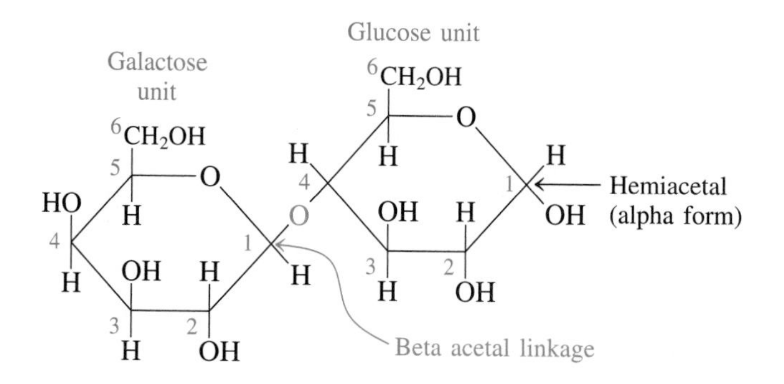

Sucrose, commonly called table sugar, is present in cane and beet sugar. It is composed of a glucose unit joined to a fructose unit by an alpha acetal linkage. (Notice that the fructose ring contains five members, not six.) This is the first nonreducing sugar we have encountered. Neither ring can open up because there is no hemiacetal.

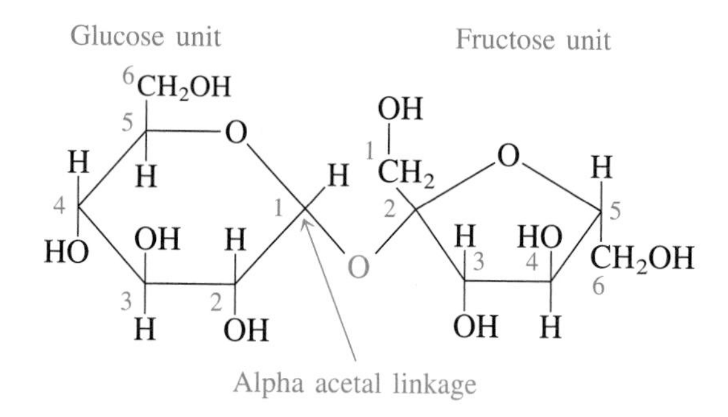

Polysaccharides

Our Diet
From 65 to 80% of our energy should come from carbohydrates, preferably the complex carbohydrates found in cereal grains rather than from simple sugars.

Both *starch* and *cellulose* are polymers of glucose. They are known as *complex carbohydrates.* Starch is an important source of energy in any balanced diet. Cellulose contributes fiber to our diet. Figure 20.6 shows short segments of the structures present in starch and cellulose. You will notice that the glucose units in starch are joined by alpha linkages, whereas in cellulose the glucose units are joined by beta linkages. This difference may appear to be trivial, but its significance is tremendous. Most animals can digest and metabolize starch, but humans and certain other mammals get no food value from cellulose. That is, we can eat and digest potatoes (starch), but we cannot digest grass. Certain bacteria are present in the digestive tracts of grazing animals and termites that make it possible for them to utilize cellulose. Humans have no such microorganisms.

The differences between alpha and beta linkages also result in different three-dimensional forms for cellulose and starch. For example, cellulose in the cell wall of plants is arranged in bundles of parallel *fibrils* to form fibers. In alternate layers, the cellulose fibers are perpendicular, giving greater strength (Figure 20.7).

Stored Glycogen
A person can store only about 500 g of glycogen, enough for about a day's reserve of energy. The glycogen is stored primarily in the liver and in muscle tissue.

There are two kinds of starch. One, called *amylose,* has glucose units joined in a continuous chain, like beads on a string. The other kind, called *amylopectin,* has branched chains of glucose units. Animal starch is called *glycogen.* Like amylopectin, it is composed of branched chains of glucose units, but there are many more branches in glycogen than in amylopectin. In amylose, there may be from 60 to

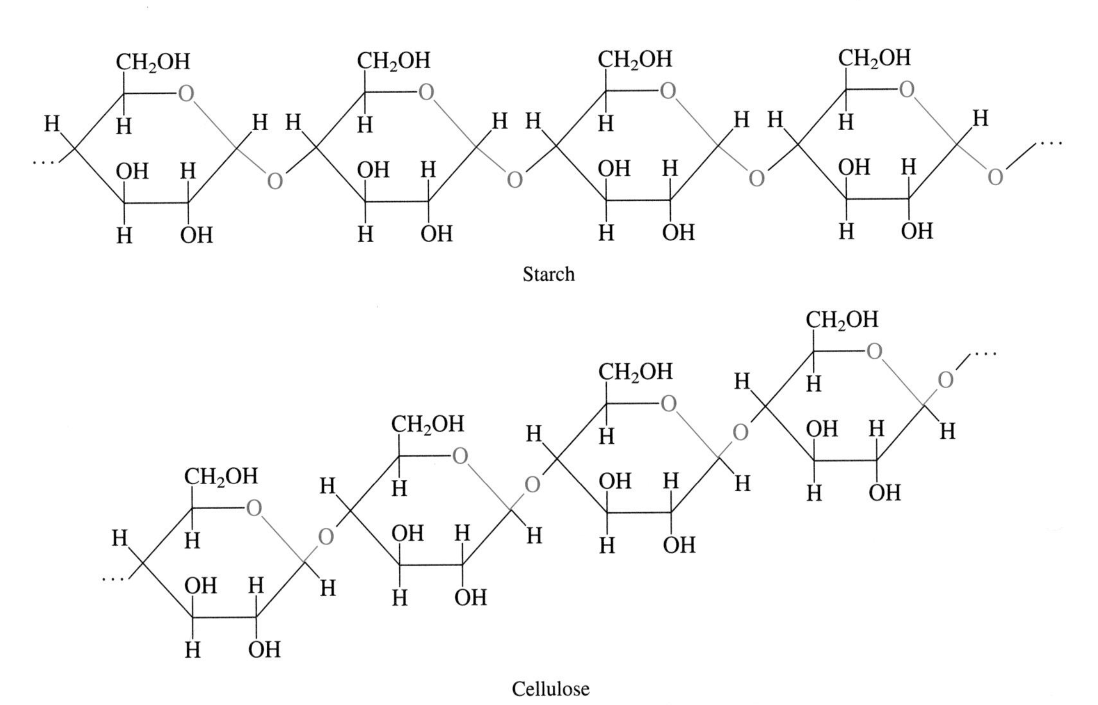

Figure 20.6 *Carbohydrates. Both starch and cellulose are polymers of glucose.*

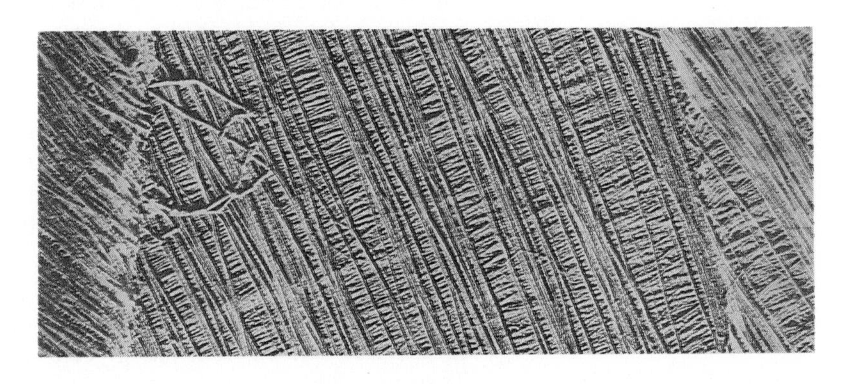

Figure 20.7 *Successive layers of cellulose fibers in parallel arrangement are shown in this electron micrograph of the cell wall of an alga.*

300 glucose units per molecule. The branched-chain structures of amylopectin may contain from 300 to 6000 glucose units per molecule.

Starch is broken down to glucose when it is digested. This simple sugar is readily absorbed through the intestinal walls and taken into the bloodstream. Glucose is broken down through a complex series of more than 50 chemical reactions to produce carbon dioxide and water with the release of energy.

$$C_6H_{12}O_6 + 6\ O_2 \xrightarrow{\text{enzymes}} 6\ CO_2 + 6\ H_2O + \text{Energy}$$

Enzymes are described at the end of Section 20.3.

These reactions are essentially the reverse of photosynthesis. In this way, animals are able to make use of the energy from the sun that was captured by plants in the process of photosynthesis.

Carbohydrates supply about 4 kcal of energy per gram. When we eat more than we can use, a small amount of carbohydrate can be converted to glycogen for storage in liver and muscle tissue. Large excesses, however, are converted to lipids (fat) for storage.

EXAMPLE 20.1

Give one or more examples of each of the following.

(a) a monosaccharide that is an aldohexose
(b) a disaccharide with a glucose unit joined to a fructose unit by an alpha acetal linkage
(c) a polysaccharide of glucose units joined by beta linkages

SOLUTION

(a) Glucose and galactose are two examples of aldohexoses. (There are others.)
(b) Sucrose is the only disaccharide that fits this description.
(c) Cellulose is the polysaccharide that fits this description.

See Problems 20.1–20.12.

EXERCISE 20.1

Compare structural similarities and differences for starch and cellulose and for glucose and galactose.

20.2 Lipids

The food that we eat is divided into three primary groups: the carbohydrates, the proteins, and the lipids. Although the carbohydrates have structural similarities, the **lipids** have both varying structures and functions, but their solubility characteristics are similar. Compounds isolated from body tissue are classified as lipids if they are more soluble in organic solvents than in water. Lipids are greasy, oily substances that can be simple lipids (esters of glycerol and fatty acids), phospholipids (common in nerve tissue), steroids (cholesterol and common male and female hormones), or the prostaglandins (a group of compounds with a wide range of effects on heart rate, blood pressure, fertility, and allergic responses).

Simple lipids are esters derived from glycerol and long-chain carboxylic acids, commonly called **fatty acids**. The simple lipids are also called **triglycerides**, each of which is derived from one glycerol molecule and three fatty acid molecules. Tristearin, shown here, is a typical simple lipid.

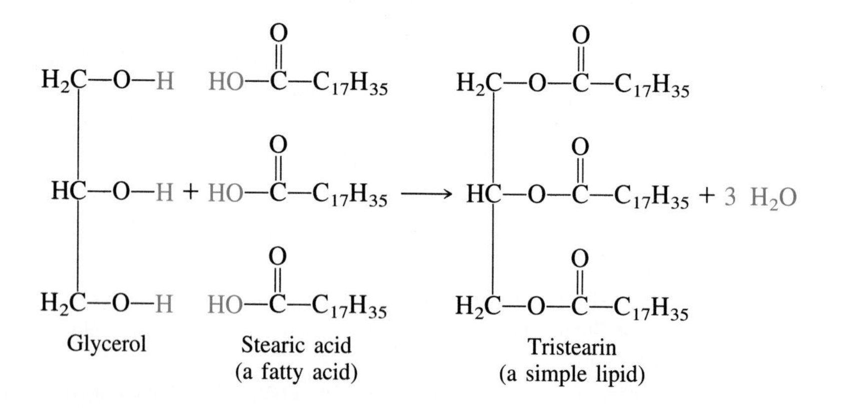

Naturally occurring fatty acids (Table 20.1) nearly always have an even number of carbon atoms like the stearic acid shown here. In general, fats are obtained from animal sources while oils are primarily from vegetable sources.

Table 20.1 Some Fatty Acids in Natural Fats

Number of Carbon Atoms	*Condensed Structure**	*Name*	*Common Source*
4	$CH_3CH_2CH_2COOH$	Butyric acid	Butter
6	$CH_3(CH_2)_4COOH$	Caproic acid	Butter
8	$CH_3(CH_2)_6COOH$	Caprylic acid	Coconut oil
10	$CH_3(CH_2)_8COOH$	Capric acid	Coconut oil
12	$CH_3(CH_2)_{10}COOH$	Lauric acid	Palm kernel oil
14	$CH_3(CH_2)_{12}COOH$	Myristic acid	Oil of nutmeg
16	$CH_3(CH_2)_{14}COOH$	Palmitic acid	Palm oil
18	$CH_3(CH_2)_{16}COOH$	Stearic acid	Beef tallow
18	$CH_3(CH_2)_7CH{=}CH(CH_2)_7COOH$	Oleic acid	Olive oil
18	$CH_3(CH_2)_4CH{=}CHCH_2CH{=}CH(CH_2)_7COOH$	Linoleic acid	Soybean oil
18	$CH_3CH_2(CH{=}CHCH_2)_3(CH_2)_6COOH$	Linolenic acid	Fish oils
20	$CH_3(CH_2)_4(CH{=}CHCH_2)_4CH_2CH_2COOH$	Arachidonic acid	Liver

*All double bonds are in the cis configuration.

Typical animal fat and vegetable oils contain both saturated and unsaturated portions (Figure 20.8). In animal fat, the saturated components predominate. In vegetable oils, the unsaturated components usually predominate, but there are exceptions. A *saturated fatty acid* is one that contains no double bonds. A *monounsaturated fatty acid* contains one double bond per molecule, and a *polyunsaturated fatty acid* contains two or more double bonds.

A **saturated fat** is one that contains a relatively high proportion of saturated fatty acids. For example, beef tallow is classified as a saturated fat. It is, on the average, 52% saturated, 44% monounsaturated, and 4% polyunsaturated. **Polyunsaturated fats** (oils) incorporate mainly unsaturated fatty acids. For example, corn

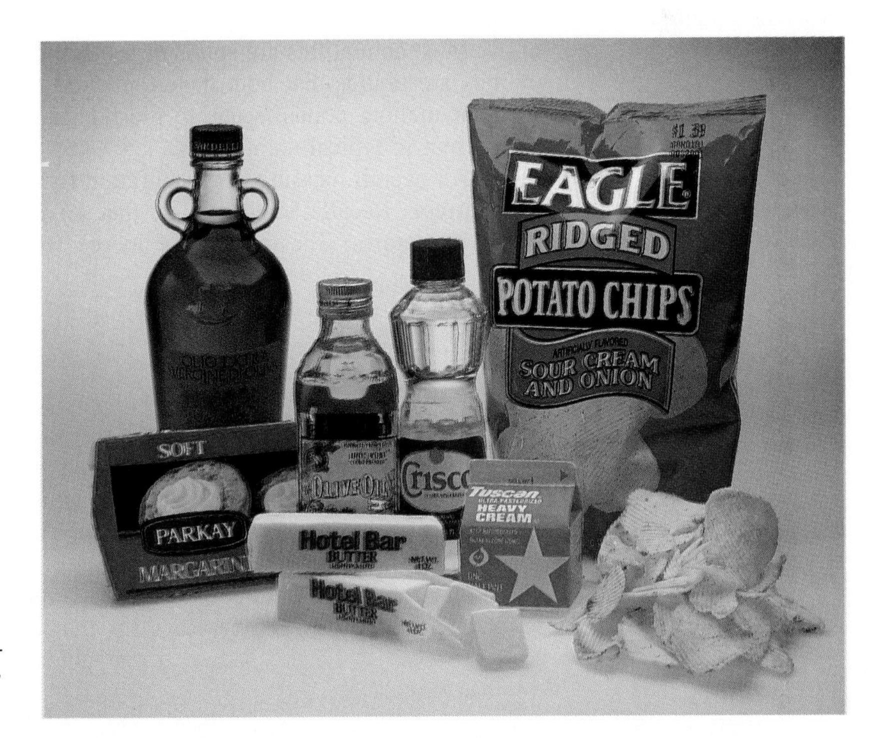

Figure 20.8 *Butter, margarine, cream, cooking oils, and food fried in fat are examples of substances that provide dietary lipids.*

CHEMISTRY IN OUR WORLD

Saturated and Unsaturated Fats in Our Diet

Saturated fats have been implicated, along with cholesterol, in one type of *arteriosclerosis* (hardening of the arteries). As this condition develops, deposits form on the walls of arteries. Eventually, these deposits—called plaque—become calcified (hardened), robbing the vessels of their elasticity. There is a strong correlation between diets rich in saturated fats and incidence of the disease. It is this correlation that has led to concern over the relative amounts of saturated and unsaturated fats in our diet.

Nutritionists recommend that dietary fat be restricted to supply no more than 30% of a person's calorie intake. Unsaturated fatty acids with their double bonds in the cis configuration are thought to be most desirable; they have a bent shape.

There is some statistical evidence that fish oils—which contain polyunsaturated fatty acid components—can prevent heart disease. Researchers at the University of Leiden in the Netherlands have found that Greenlanders who eat a lot of fish have a low risk of heart disease, despite a diet that is high in cholesterol and total fat. Furthermore, studies have shown that the use of fish oils in diets leads to lower cholesterol and triglyceride levels in the blood. Your body chemistry is affected by your diet. What you don't know *can* hurt you.

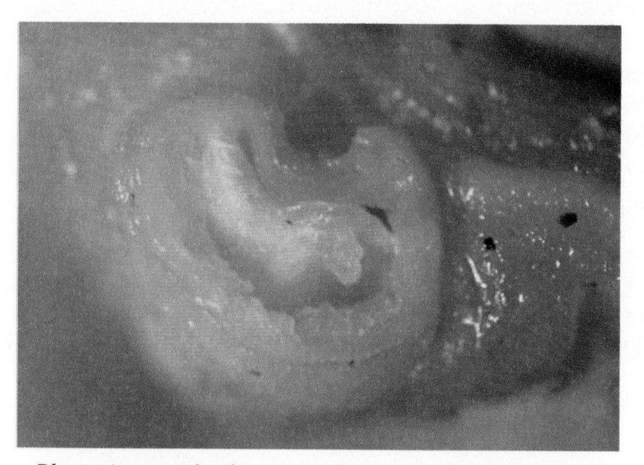

Photomicrograph of a cross section of a hardened artery with deposits of plaque. The deposits contain cholesterol.

Table 20.2 Iodine Numbers for Some Fats and Oils

Fat or Oil	*Iodine Number*
Coconut oil	8–10
Butter	25–40
Beef tallow*	30–45
Palm oil	37–54
Lard*	45–70
Olive oil	75–95
Peanut oil	85–100
Cottonseed oil	100–117
Corn oil	115–130
Fish oils*	120–180
Soybean oil	125–140
Safflower oil	130–140
Sunflower oil	130–145
Linseed oil	170–205

*From animals; the majority are from plants.

oil is, on the average, 13% saturated, 25% monounsaturated, and 62% polyunsaturated.

Most animal fats are solids or semisolids at room temperature, while vegetable oils are liquids, but both have the same triglyceride structure. The difference in their melting points is due to the higher proportion of saturated fatty acid units present in typical animal fats.

An overall measure of unsaturation for a fat or an oil is indicated by its **iodine number**. Recall (Section 19.5) that all halogens add readily to double-bonded carbon atoms. The reaction of iodine with a double bond is represented here.

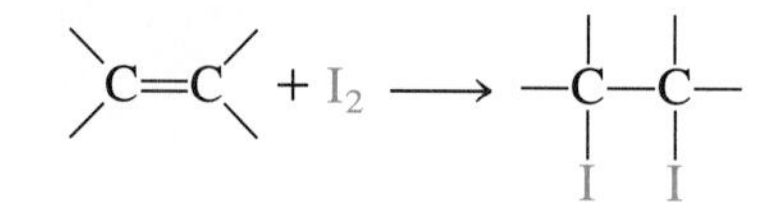

The iodine number is defined as the number of grams of iodine that will be consumed by 100 g of fat or oil. The more double bonds a fat contains, the more iodine it takes up during the addition reaction. Thus, a high iodine number means a high degree of unsaturation. Iodine numbers of several fats and oils are given in Table 20.2. Generally, the animal fats such as butter and tallow have lower values than do the vegetable oils. Two exceptions that can be noted in the table are coconut oil—which is a highly saturated vegetable oil—and fish oils—which are relatively unsaturated.

Fats are high-energy foods. They yield about 9 kcal of energy per gram. Fats eaten in excess are stored in the body, where they serve as energy reserves. This stored fat serves to insulate the body against loss of heat and to protect vital organs from injury by acting as extra padding. Stored fats presumably got our ancestors through lean times. Our ability to store large amounts of fats probably is genetic.

Fat in our diet comes from meat and dairy products, vegetable oils and shortenings, and some seeds and nuts. Generally not more than 30% of our energy should come from fat, and this should be divided evenly among three types: saturated, monounsaturated, and polyunsaturated fats. The average American diet is about 37% fat, a figure that is considerably higher than recommended by most medical authorities. The triglyceride level in your blood will reveal your eating habits.

EXAMPLE 20.2

From each group, identify the one that is the most saturated and the one that is the most unsaturated.

(a) oleic acid, linoleic acid, palmitic acid
(b) soybean oil, peanut oil, coconut oil, olive oil

SOLUTION

(a) See Table 20.1. Palmitic acid is saturated. Of the three, linoleic acid is the most unsaturated; it has two double bonds.
(b) See Table 20.2 for iodine numbers. Of the four given, coconut oil is the most saturated and soybean oil is the most unsaturated.

See Problems 20.13–20.22.

20.3 *Amino Acids and Proteins*

The third class of food, the proteins, is the vital component of all life. No living part of the human body—or of any living cell, for that matter—is without protein. There is protein in the blood, the muscles, the brain, and even in tooth enamel. The smallest cellular organisms—the bacteria—contain protein. Each type of cell makes its own kinds of proteins. The proteins serve as structural materials in muscle and skin tissue. For example, silk, wool, nails, claws, feathers, horns, and hooves are proteins. The structure of a short segment of a typical protein molecule is shown in Figure 20.9.

Proteins are copolymers of about 20 different **amino acids** joined by peptide bonds. A list of these amino acids is shown in Table 20.3. Each amino acid has two functional groups: a carboxyl group, —COOH, and an amino group, $—NH_2$, which is on the carbon next to the carboxyl group. This carbon is known as the alpha (α) carbon; thus, all naturally occurring amino acids are alpha amino acids.

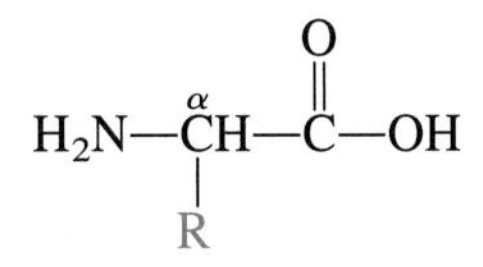

The identity of a specific amino acid is determined by the specific group, symbolized by —R, that is attached to the carbon that bears the amino group.

The amino acid general formula shown here is only partially correct. Although it

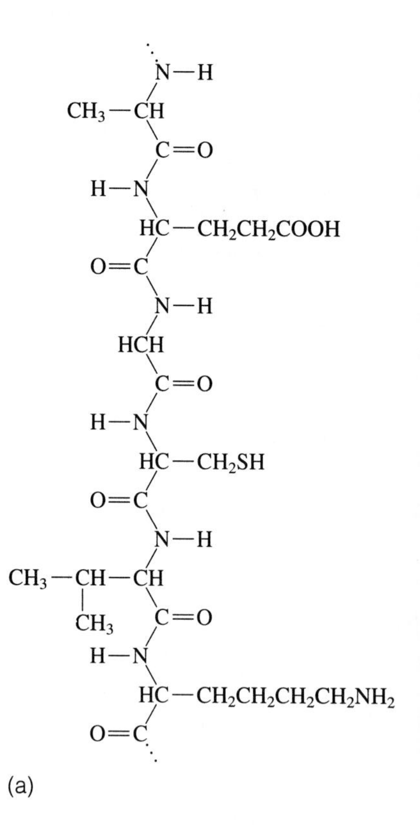

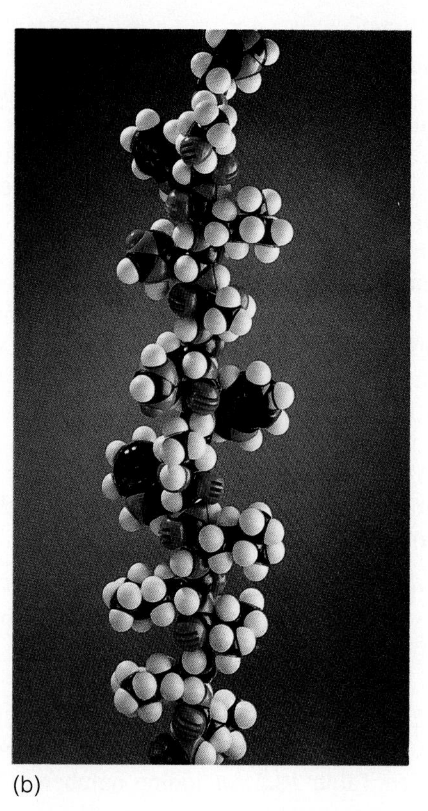

Figure 20.9 *(a) Structural formula of a segment of a protein molecule. (b) A space-filling model of a segment of a protein chain.*

Table 20.3 **Some Representative Amino Acids**

Name	*Abbreviation*	*Essential*	*Structure*
Glycine	Gly	No	$H{-}CH(^{+}NH_3){-}COO^-$
Alanine	Ala	No	$CH_3{-}CH(^{+}NH_3){-}COO^-$
Phenylalanine	Phe	Yes	$C_6H_5{-}CH_2{-}CH(^{+}NH_3){-}COO^-$
Valine	Val	Yes	$CH_3{-}CH(CH_3){-}CH(^{+}NH_3){-}COO^-$
Leucine	Leu	Yes	$CH_3CH(CH_3)CH_2{-}CH(^{+}NH_3){-}COO^-$
Isoleucine	Ile	Yes	$CH_3CH_2CH(CH_3){-}CH(^{+}NH_3){-}COO^-$
Proline	Pro	No	ring: $CH_2{-}CH_2{-}C(H)(COO^-){-}^{+}NH_2{-}CH_2$

***Table 20.3* (continued)**

Name	*Abbreviation*	*Essential*	*Structure*
Methionine	Met	Yes	$CH_3{-}S{-}CH_2CH_2{-}CH({}^{+}NH_3){-}COO^-$
Serine	Ser	No	$HO{-}CH_2{-}CH({}^{+}NH_3){-}COO^-$
Threonine	Thr	Yes	$CH_3CH(OH){-}CH({}^{+}NH_3){-}COO^-$
Asparagine	Asn	No	$H_2N{-}C({=}O){-}CH_2{-}CH({}^{+}NH_3){-}COO^-$
Glutamine	Gln	No	$H_2N{-}C({=}O){-}CH_2CH_2{-}CH({}^{+}NH_3){-}COO^-$
Cysteine	Cys	No	$HS{-}CH_2{-}CH({}^{+}NH_3){-}COO^-$
Tyrosine	Tyr	No	$HO{-}C_6H_4{-}CH_2{-}CH({}^{+}NH_3){-}COO^-$
Tryptophan	Trp	Yes	$C_8H_6N{-}CH_2{-}CH({}^{+}NH_3){-}COO^-$ (indole ring: C, CH, NH)
Lysine	Lys	Yes	$H_3\overset{+}{N}CH_2CH_2CH_2CH_2{-}CH(NH_2){-}COO^-$
Arginine	Arg	*	$H_2N{-}C({=}{}^{+}NH_2){-}NHCH_2CH_2CH_2{-}CH(NH_2){-}COO^-$
Histidine	His	†	imidazole (N, HN) $-CH_2{-}CH({}^{+}NH_3){-}COO^-$
Aspartic acid	Asp	No	$HOOC{-}CH_2{-}CH({}^{+}NH_3){-}COO^-$
Glutamic acid	Glu	No	$HOOC{-}CH_2CH_2{-}CH({}^{+}NH_3){-}COO^-$

*Essential to growing children but not to adult humans.
†Essential to human infants.

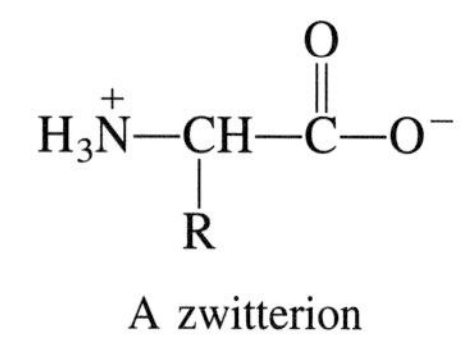
A zwitterion

indicates the proper carbon atom to which the functional groups are attached, amino groups and carboxyl groups do not coexist in the forms shown. The carboxyl group, which is acidic, reacts with the amino group, which is basic. When these two functional groups interact, the acid group transfers a proton to the base. The resulting product is an inner salt, called a **zwitterion**; that is, a compound in which the anion and the cation are parts of the same molecule.

Essential Amino Acids

The human body can synthesize all but eight of the amino acids needed for making proteins. Those eight—shown in Table 20.3—are called **essential amino acids**. They must be obtained through the diet. We eat proteins, break them down in our bodies to their constituent amino acids, and then use some of these amino acids to build other protein structures.

Corn lacks
lysine
tryptophan
Rice lacks
threonine
Wheat lacks
lysine

Most proteins from plant sources are deficient in one or more of the essential amino acids. For example, corn protein is lacking in lysine and tryptophan. People whose diet consists mainly of corn may suffer from malnutrition due to the shortage of these amino acids. Protein from rice is lacking in lysine and threonine. Wheat protein lacks lysine. Soy protein lacks methionine. By contrast, most proteins from animal sources contain adequate amounts of all essential amino acids. Our daily requirement for protein is about 0.8 g per kilogram of body weight. Protein deficiency leads to both physical and mental retardation.

Peptide Bonds

The human body contains about 30,000 different proteins. Each of us has his or her own tailor-made set. Proteins are polyamides. Amide linkages can join numerous types of structures, but when an amide linkage, —CONH—, joins two amino acid units, the linkage is called a **peptide bond**.

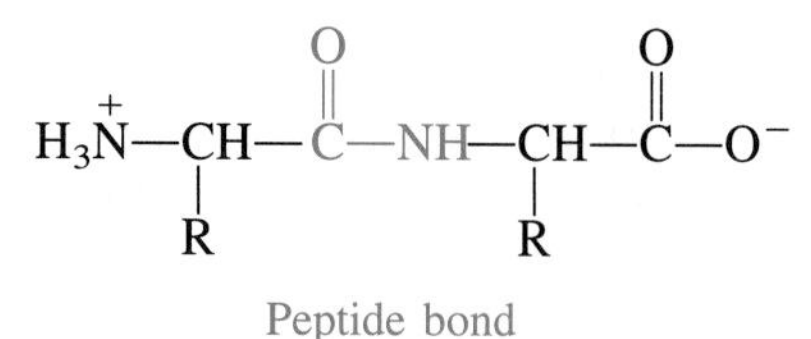
Peptide bond

The reactive amino group on the left end and the carboxyl group on the right can react with additional amino acids. When only two amino acids are joined, the product is called a *dipeptide*. When three amino acids are combined, the product is a *tripeptide*.

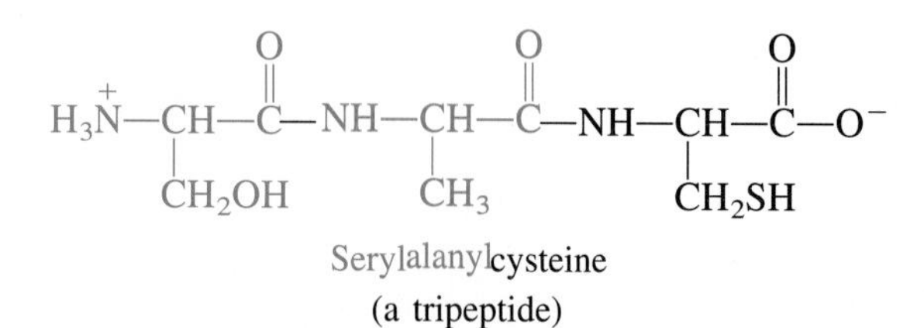
Serylalanylcysteine
(a tripeptide)

A string of about 10 or more amino acid units is called a **polypeptide.** When the molecular weight of the polypeptide exceeds 10,000, it is called a *protein*. Either a polypeptide or a protein can be represented by the structure shown here.

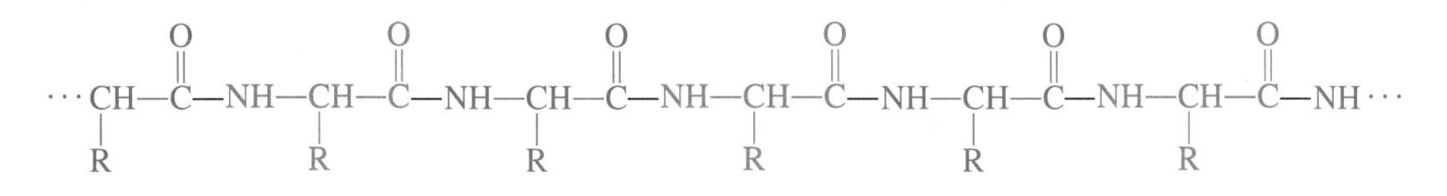

For peptides and proteins to be physiologically active, it is not enough that they incorporate a certain number of specific amino acids. The order or *sequence* in which the amino acids are connected is also critical.

When peptides and proteins are being described by scientists, they find it much simpler to use three-letter abbreviations for the amino acids (Table 20.3). For example, the sequence for serylalanylcysteine is written Ser-Ala-Cys. It is understood from the shorthand that the peptide is arranged with the free amino group to the left and free carboxyl group to the right.

As the length of a peptide chain increases, the possible sequential variations become almost infinite. And this potential for many different arrangements is exactly what is needed in a material that will make up such diverse things as hair, skin, eyeballs, toenails, certain hormones, and a thousand different enzymes.

As a comparison, consider the millions of different words we can make with our 26-letter English alphabet. We can also have millions of different proteins made with the 20 or so amino acids. Furthermore, just as one can write gibberish with the English alphabet, one can make nonfunctioning proteins by putting together the wrong sequence of amino acids. Sometimes a protein with one or two amino acids out of sequence will continue to function—just as you can continue to ''read'' the ''sentence'' in spite of a spelling error. In other cases, a seemingly minor change can have disastrous effect. Some people have hemoglobin with one particular incorrect amino acid unit in about 300. That ''minor'' error is responsible for sickle cell anemia, an inherited condition that usually is fatal.

What we have described thus far is the amino acid sequence called the **primary structure** of a protein. For a protein to function properly, simply having the correct primary structure is not enough; the overall shape of the protein must be correct. Strands of certain proteins tend to coil around like the cord on a telephone receiver. This shape, known as a helix, is the type of **secondary structure** present in wool. In silk, the secondary structure forms what is known as a pleated sheet arrangement. The overall general shape taken on by the protein as it folds around on itself, like a plate of spaghetti, is called its **tertiary structure**. We could say that with proteins we have a case where it's not only what you have that's important, it's how you arrange what you have that counts!

Enzymes

Certain proteins called **enzymes** act as biological catalysts. With molecular weights of about 10^4 to 10^6 amu, enzymes increase reaction rates by factors of 10^6 to 10^{12}, enabling reactions to occur at normal body temperature that otherwise would not be possible. The action of an enzyme is often described in terms of a lock-and-key model (Figure 20.10). The enzyme for a particular reaction must fit precisely with the molecule it is acting upon; for example, enzymes that break down starch to glucose don't fit the cellulose molecule.

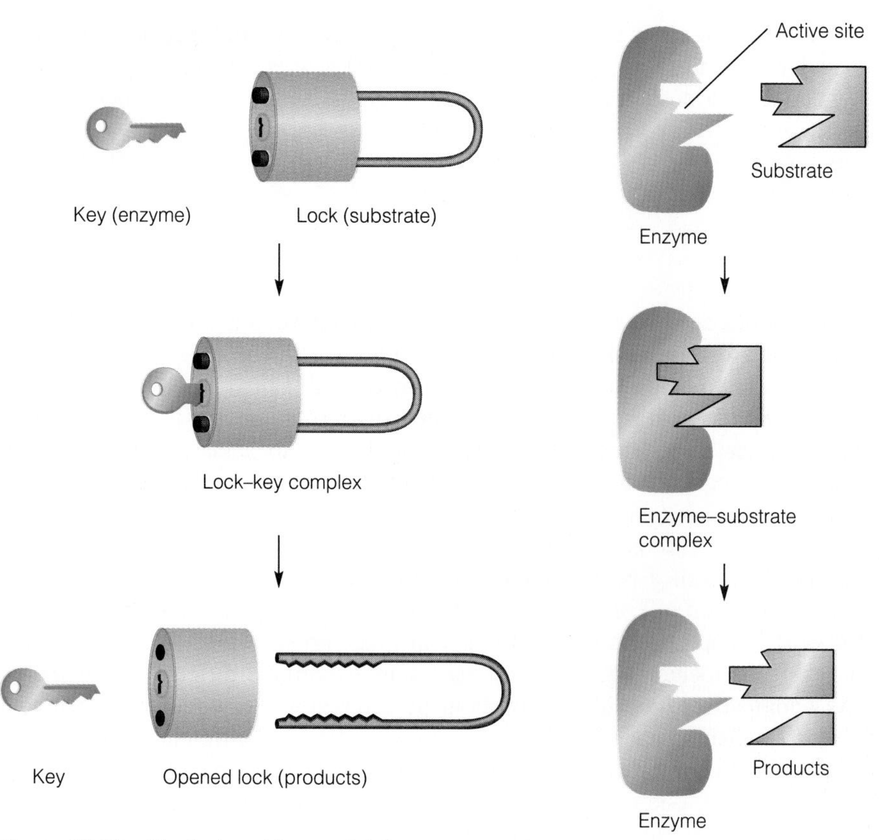

Figure 20.10 *The lock-and-key model for enzyme action.*

EXAMPLE 20.3

Classify each of the following: an enzyme composed of 129 amino acid units, silk, wool, Gly-Phe-Val, stearic acid, aspartic acid, and lysine.

SOLUTION

Silk, wool, and the enzyme are proteins. Gly-Phe-Val is a tripeptide. Stearic acid is a fatty acid. Both aspartic acid and lysine are amino acids.

See Problems 20.23–20.32.

EXERCISE 20.2

Give names of the eight essential amino acids.

20.4 Nucleic Acids

Complex compounds called **nucleic acids** are found in every living cell. They serve as the information and control centers of the cell. There are actually two kinds of nucleic acids: **deoxyribonucleic acid** (DNA) is found primarily in the cell nucleus; **ribonucleic acid** (RNA) is found in all parts of the cell. Both types of nucleic acids are long chains of repeating units called *nucleotides.*

A **nucleotide** consists of three parts: a sugar (ribose or deoxyribose), a heterocyclic amine base, and a phosphate unit. The sequence is

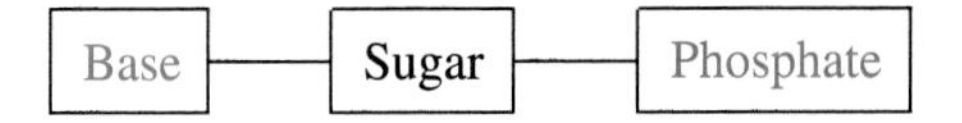

We'll describe each part. The sugar is either ribose, present in RNA, or deoxyribose, present in DNA. As shown in Figure 20.11, the only structural difference between the two pentoses is at carbon number 2 where deoxyribose lacks an oxygen atom present in ribose.

To visualize building a nucleic acid structure, imagine starting with ribose or deoxyribose as the sugar unit. Then, replace the hydroxyl group at carbon number 1 with one of five heterocyclic amine bases shown in Figure 20.12. Two of the five bases have two fused rings; these are purine bases. Three of the bases have single heterocyclic rings; these are pyrimidine bases.

The third part of a nucleotide is a phosphate ester group present at the fifth carbon of the sugar unit. Adenosine monophosphate (AMP), shown in Figure 20.13, is a representative nucleotide. In AMP, the base is adenine and the sugar is ribose.

Nucleotides are joined to one another through the phosphate group to form chains of nucleic acids. The process is repeated to build up a long nucleic acid

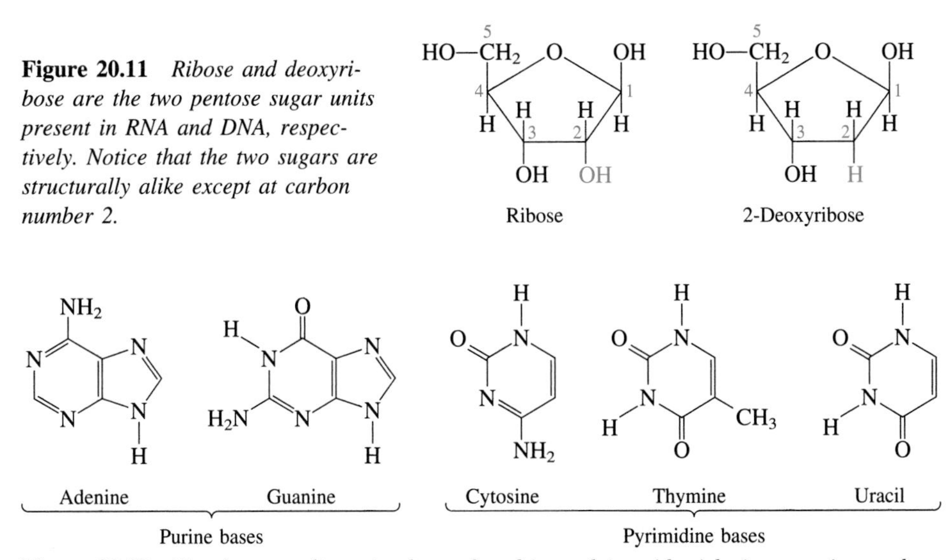

Figure 20.11 *Ribose and deoxyribose are the two pentose sugar units present in RNA and DNA, respectively. Notice that the two sugars are structurally alike except at carbon number 2.*

Figure 20.12 *Five heterocyclic amine bases found in nucleic acids. Adenine, guanine, and cytosine are found in both DNA and RNA. Thymine is present only in DNA, and uracil only in RNA.*

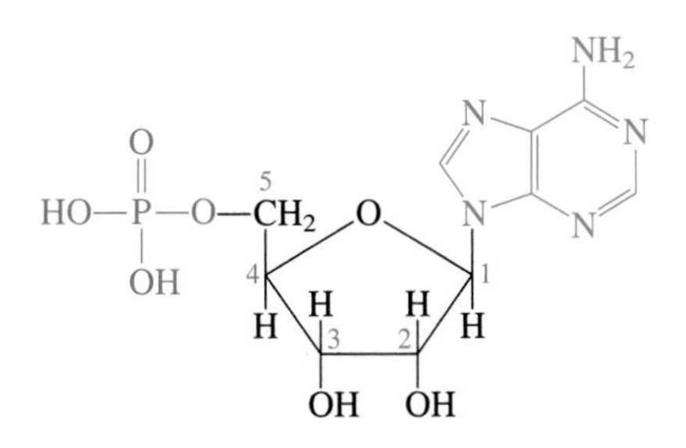

Adenosine monophosphate (a representative nucleotide)

Figure 20.13 *Adenosine monophosphate is a representative nucleotide. The base, adenine, is shown in blue. The phosphate group is red.*

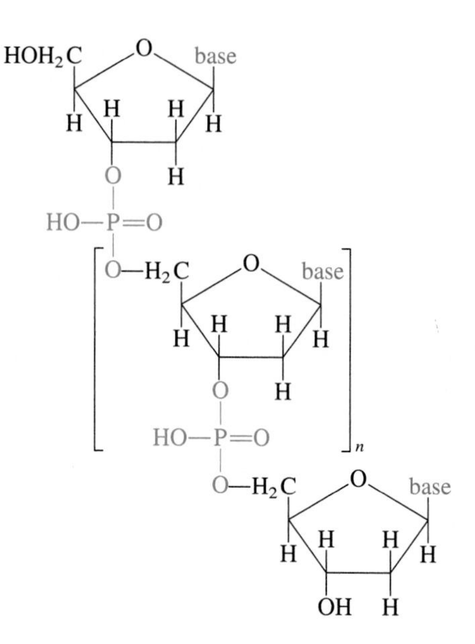

Figure 20.14 *The backbone of a deoxyribonucleic acid molecule. The* n *indicates that the unit is repeated many times.*

chain as represented in Figure 20.14. Notice that the backbone of the chain consists of alternating phosphate and sugar units. The heterocyclic bases are branched off of this backbone at each sugar unit.

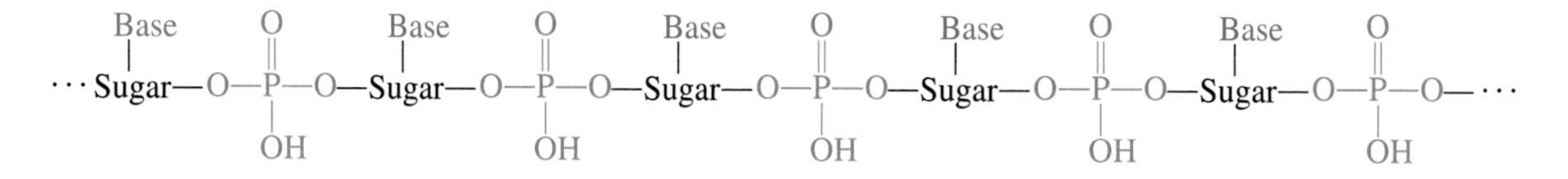

As we have seen, the sugar in DNA is deoxyribose, but in RNA the sugar is ribose. The bases in DNA are adenine, guanine, cytosine, and thymine. Those in RNA are adenine, guanine, cytosine, and uracil (Table 20.4).

Table 20.4 **Components of DNA and RNA**

	DNA	*RNA*
Purine bases	Adenine	Adenine
	Guanine	Guanine
Pyrimidine bases	Cytosine	Cytosine
	Thymine	Uracil
Pentose sugar	Deoxyribose	Ribose
Inorganic acid	Phosphoric acid	Phosphoric acid

A most important feature of a DNA or RNA molecule is the *sequence* of the four bases along the strand. The molecules are huge, with thousands of nucleotides and with molecular weights ranging into the billions for mammalian DNA. Along these great chains, the four bases may be arranged in essentially infinite variations. That is a crucial feature of DNA and RNA molecules, because it is the base sequence that is used to store the tremendous amount of information needed to build living organisms.

Figure 20.15 *James D. Watson and Francis Crick, discoverers of the double helix model of DNA.*

The Double Helix

Many experiments have been designed to probe the structure of DNA. By 1950 it was quite clear that this determination would bring with it a Nobel prize. Although many respected researchers worked on the problem, a team of two relatively unknown scientists made this discovery. In 1953, James D. Watson, an American biologist, and Francis Crick, a British physicist, shown in Figure 20.15, made their now famous discovery that DNA must be composed of a double-stranded helix structure with the two helixes wound about one another. The phosphate and sugar backbone of the polymer chains form the outside of the structure and give it a spiral staircase appearance. The heterocyclic amines are paired on the inside, with guanine always opposite cytosine and adenine always opposite thymine. In our staircase analogy, these base pairs are the steps (Figure 20.16).

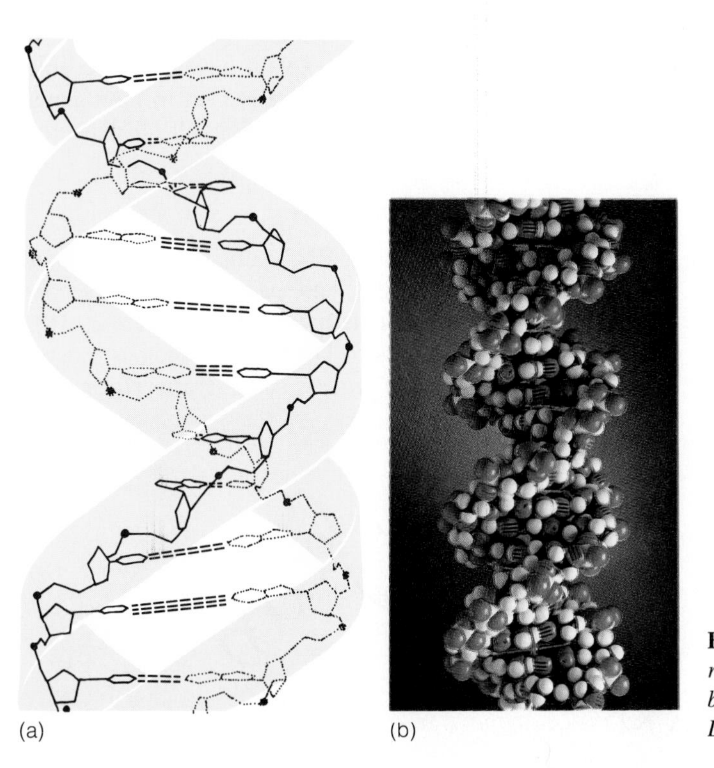

Figure 20.16 *(a) A schematic representation of the DNA double helix. (b) A model of the DNA molecule.*

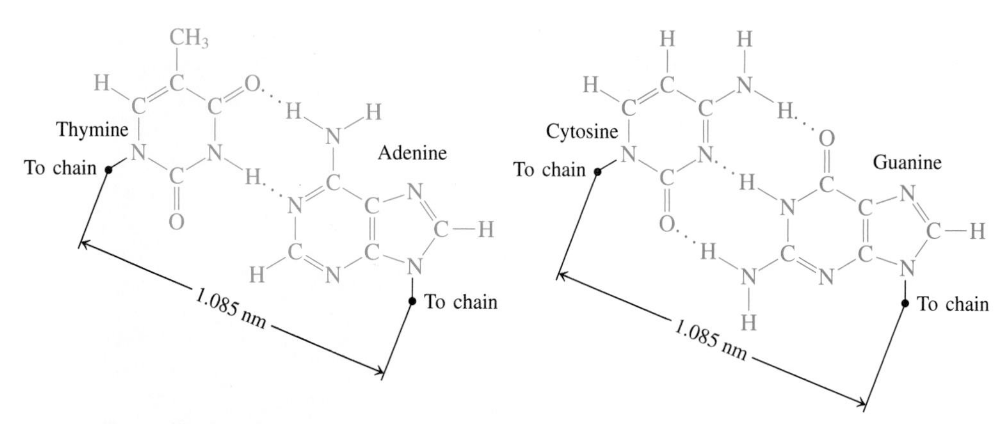

Figure 20.17 *The pairing of bases in the DNA double helix.*

Why do the bases pair in this precise pattern, always A (adenine) to T (thymine) and T to A, always G (guanine) to C (cytosine) and C to G? The answer is hydrogen bonding and a truly elegant molecular design. Figure 20.17 shows the two sets of base pairs. You should notice two things. First, a pyrimidine—it is the single heterocyclic ring structure—is paired with a purine—it has two heterocyclic rings—and the total lengths of both pairs are identical (1.085 nm).

The second thing you should notice in Figure 20.17 is the hydrogen bonding between the bases in each pair. When guanine is paired with cytosine, three hydrogen bonds can be drawn between the bases. No other pyrimidine to purine pairing will permit such extensive interaction. Indeed, in the combination shown in Figure 20.17 both pairs of bases fit like lock and key.

Watson and Crick received the Nobel prize in 1962 for discovering, as Crick put it, ''the secret of life.'' The structure these two scientists proposed was accepted almost immediately by other scientists around the world because it answered so many crucial questions. It can explain how cells are able to divide and go on functioning, how genetic data are passed on to new generations, and even how proteins are built to required specifications. It all depends on base pairing.

RNA and Protein Synthesis

Unlike DNA, the molecules of RNA consist of single strands of the nucleic acid. Some internal (intramolecular) base pairing occurs in sections where the molecule folds back on itself in a double helical form (Figure 20.18). The information contained in one of the strands of DNA in the nucleus must be relayed to the cytoplasm and acted upon. In a process called **transcription**, DNA transfers its infor-

Figure 20.18 *RNA occurs as single strands that can form double helical portions by internal base pairing.*

BIOCHEMISTRY IN OUR WORLD

Genetic Engineering and Recombinant DNA

Over 3000 human diseases have a genetic component. Researchers have linked specific genes to specific diseases. Now the ability to use this information to diagnose and cure genetic diseases appears to be within our grasp. By determining the location of a gene on the DNA molecule, scientists have been able to identify and isolate genes with a specific function. For example, in 1989 a U.S.–Canadian research team identified the gene that causes cystic fibrosis, the most common lethal genetic disease in North America. The next step is for chemists to identify the protein the gene is supposed to make.

There are about 10,000 genes on each human chromosome. Isolating the one gene that is defective—for a particular genetic disease—is quite difficult, but there are new detection approaches to give DNA segments that are easier to work with. If the DNA pattern of a person matches that of a relative with a genetic disease, the person may also develop the disease. Thus, it is possible to identify and even predict the occurrence of a genetic disease.

A future hope of genetic engineering is that the action of a defective gene can be corrected by introducing a functioning gene into a person's cells. It should also be possible to place a gene from one organism into the genetic material of another. Actually, recombinant DNA (rDNA) technology does just that. The gene must first be identified, isolated, and placed in a separate piece of DNA. The recombined DNA is then transferred into bacteria or another suitable organism. The final step, called cloning, yields many copies of the modified bacteria that can produce relatively large amounts of protein coded by the gene.

By working backward from the amino acid sequence of the protein, scientists can work out the base sequence of the gene that codes for the protein. Many valuable materials are now made using rDNA technology. People with diabetes formerly had to use insulin from pigs or cattle. Now human insulin, a protein coded by human DNA, is being produced by the cell machinery of bacteria. Newly diagnosed insulin-dependent diabetics in the United States are now treated with rDNA-produced human insulin.

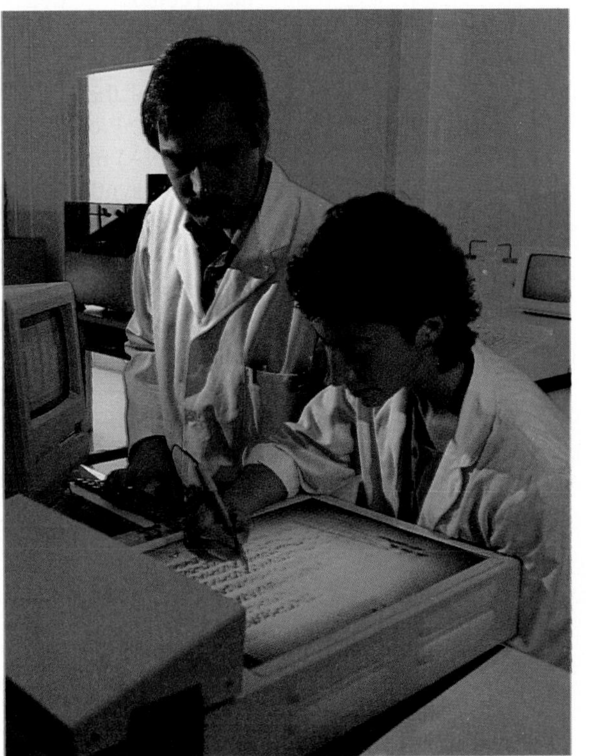

Researchers shown here are studying human genes by using DNA sequencing techniques.

Human growth hormone, used to treat children who fail to grow properly, is now readily available through rDNA technology. This technology has even allowed scientists to design bacteria to "eat" the oil released in an oil spill. However, there is public concern over the potential for disaster in research of this type. Strict guidelines for recombinant DNA research have been instituted.

The new field of molecular genetics has provided some impressive achievements. Future possibilities are mind boggling: elimination of genetic defects, a cure for cancer, enhanced intelligence, and so on. Knowledge gives power, but it does not necessarily give wisdom. Who will decide what sort of creature the human species should be? The greatest problem we may face in our use of bioengineering is that of choosing who is to play God with the new "secret of life."

mation to a special type of RNA called *messenger RNA* (mRNA). The base sequence of DNA specifies the base sequence of mRNA. For example, thymine in DNA calls for adenine in mRNA, cytosine specifies guanine, and so on. The DNA to RNA relationship is like that of a rubber mold used in making ceramic objects: the "contours" of the RNA that is formed complement those of DNA, the "mold."

The base sequence carried by the messenger RNA specifies the amino acid sequence of the protein. Each set of three consecutive bases specifies a particular amino acid. Floating around in the cytoplasm surrounding the mRNA are *transfer RNA* (tRNA) molecules, each carrying its own type of amino acid. The actual site of protein synthesis is at a ribosome that is composed of *ribosomal RNA* (rRNA) and protein. The process described here is simplified, but it shows that the function of RNA is to direct the building of a specific protein in accordance with information contained in DNA.

EXAMPLE 20.4

What is DNA? How is it classified? Give the three component parts of DNA. Describe the secondary structure of DNA; how is it held in place?

SOLUTION

DNA is deoxyribonucleic acid. It is classified as a nucleic acid—a polymer of nucleotides. Each DNA nucleotide consists of three parts (a heterocyclic amine base, deoxyribose, and a phosphate unit). The secondary structure of DNA is a double-stranded helix. Hydrogen bonding between heterocyclic amine bases (guanine with cytosine and adenine with thymine) keeps the helixes wound about one another.

See Problems 20.33–20.38.

20.5 *Some Vitamins and Hormones*

Vitamins are specific organic compounds that are required in the diet in small amounts for proper functioning; they cannot be synthesized in the body. Furthermore, the absence or shortage of a vitamin results in a vitamin-deficiency disease. Some of the vitamins, their structures, sources, and deficiency symptoms are shown in Table 20.5.

The role of vitamins in the prevention of deficiency diseases has been well established. In recent years, massive doses of certain vitamins have been recommended as preventives or cures for diseases as varied as the common cold and schizophrenia. This type of treatment is called megavitamin therapy.

As you can see from Table 20.5, the vitamins do not share a common chemical structure. They can, however, be divided into two broad categories: the **fat-soluble vitamins**—including A, D, E, and K—and the **water-soluble vitamins**—made up of the B complex and vitamin C. The fat-soluble vitamins are nonpolar molecules. In contrast, a water-soluble vitamin molecule contains a high proportion of the electronegative oxygen and nitrogen atoms. These atoms become involved in hydrogen bonding, which accounts for the water solubility of these vitamins.

Fat-soluble vitamins dissolve in the fatty tissue of the body; reserves of these vitamins can be stored there for future use. For example, while on an adequate diet, an adult can store several years' supply of vitamin A. If the diet becomes deficient in vitamin A, these reserves are mobilized for use. On the other hand, a small child who has not had the opportunity to build up a store of the vitamin soon exhibits

Table 20.5 Some of the Vitamins

Vitamin	*Structure and Name*	*Sources*	*Deficiency Symptoms*
	FAT-SOLUBLE VITAMINS		
A	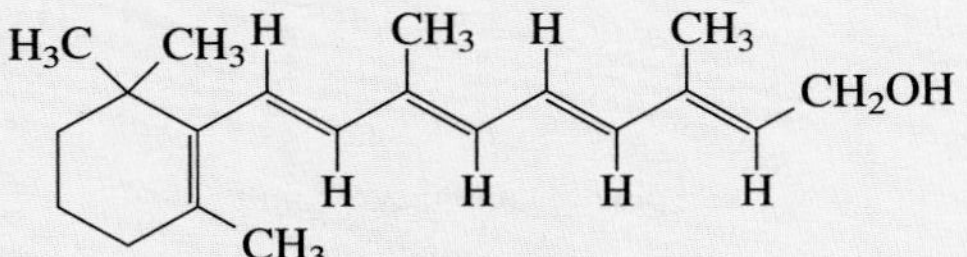 Retinol	Fish, liver, eggs, butter, cheese; also a vitamin precursor in carrots and other vegetables	Night blindness
D	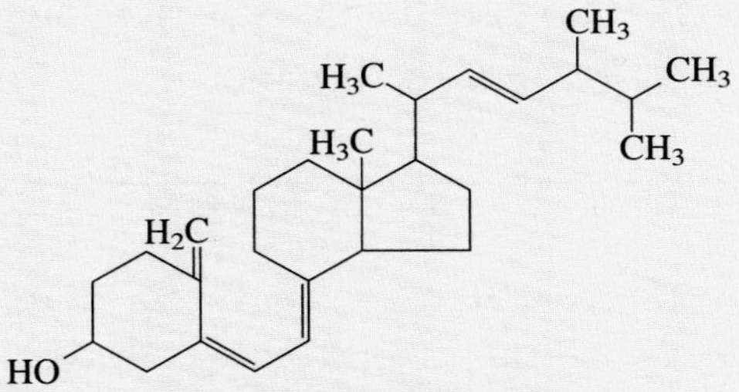 Vitamin D_2 (calciferol)	Cod liver oil, irradiated ergosterol (milk supplement)	Rickets
E	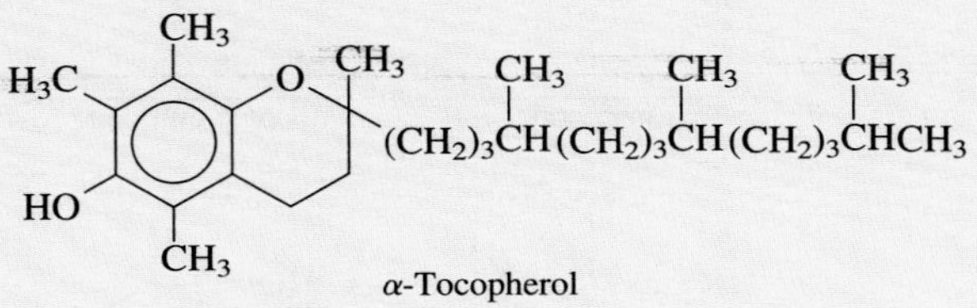 α-Tocopherol	Wheat germ oil, green vegetables, egg yolks, meat	Sterility, muscular dystrophy
K	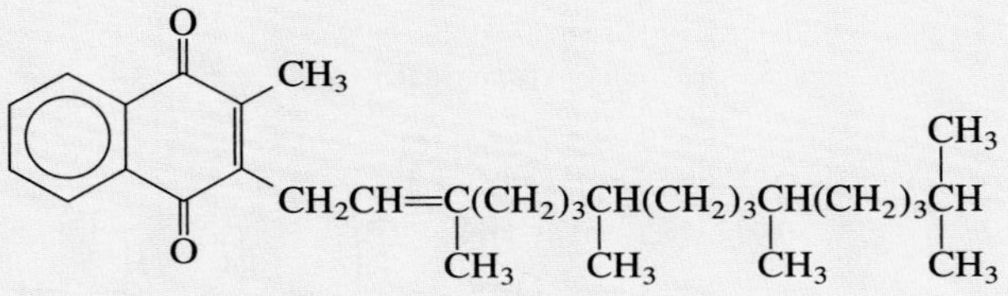 Vitamin K_1 (phylloquinone)	Spinach, other green leafy vegetables	Hemorrhage
	WATER-SOLUBLE VITAMINS		
The B Complex	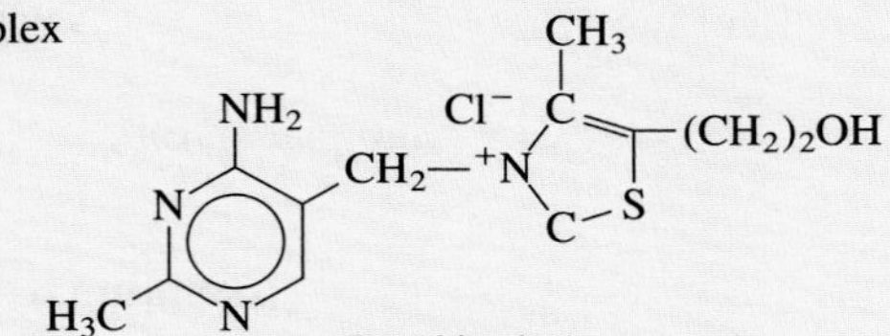 B_1 (thiamine)	Germ of cereal grains, legumes, nuts, milk, and brewers yeast	Beriberi—polyneuritis resulting in muscle paralysis, enlargement of heart, and ultimately heart failure

(continued)

Table 20.5 **(continued)**

Vitamin	*Structure and Name*	*Sources*	*Deficiency Symptoms*
The B Complex (continued)	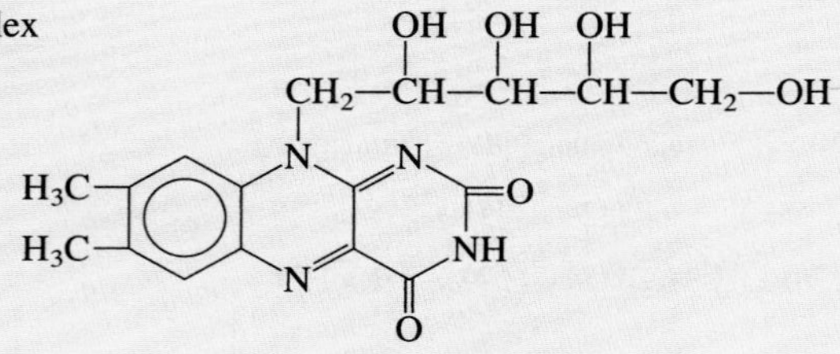 B_2 (riboflavin)	Milk, red meat, liver, egg white, green vegetables, whole wheat flour (or fortified white flour), and fish	Dermatitis, glossitis (tongue inflammation)
	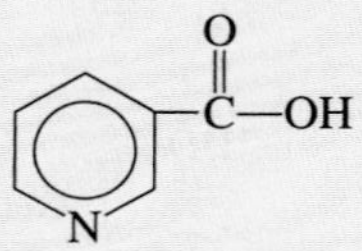 Nicotinic acid; 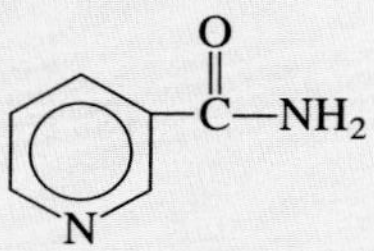 Nicotinamide — Niacin	Red meat, liver, collards, turnip greens, yeast, and tomato juice	Pellagra—skin lesions, swollen and discolored tongue, loss of appetite, diarrhea, various mental disorders
	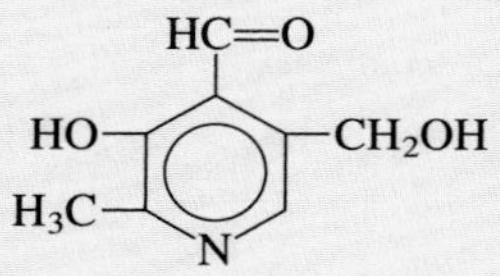 Pyridoxal; 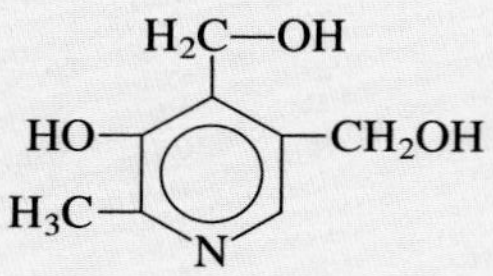 Pyridoxol; 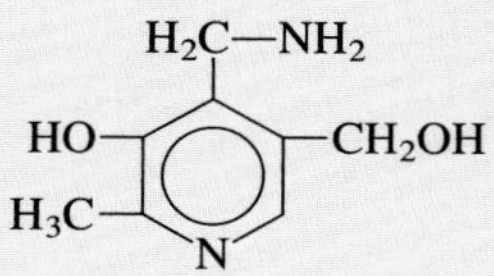 Pyridoxamine — B_6	Eggs, liver, yeast, peas, beans, and milk	Dermatitis, apathy, irritability, and increased susceptibility to infections; convulsions in infants
	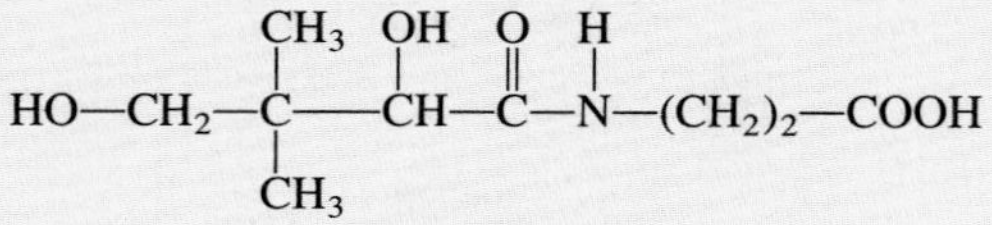 $HO{-}CH_2{-}C(CH_3)_2{-}CH(OH){-}C(=O){-}N(H){-}(CH_2)_2{-}COOH$ Pantothenic acid	Liver, eggs, yeast, and milk	(Possibly) emotional problems and gastrointestinal disturbances
	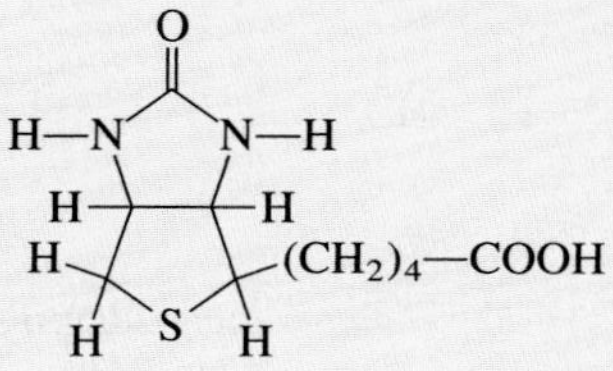 Biotin	Beef liver, yeast, peanuts, chocolate, and eggs (although this vitamin cannot be synthesized by humans, it is a product of their intestinal bacteria)	Dermatitis

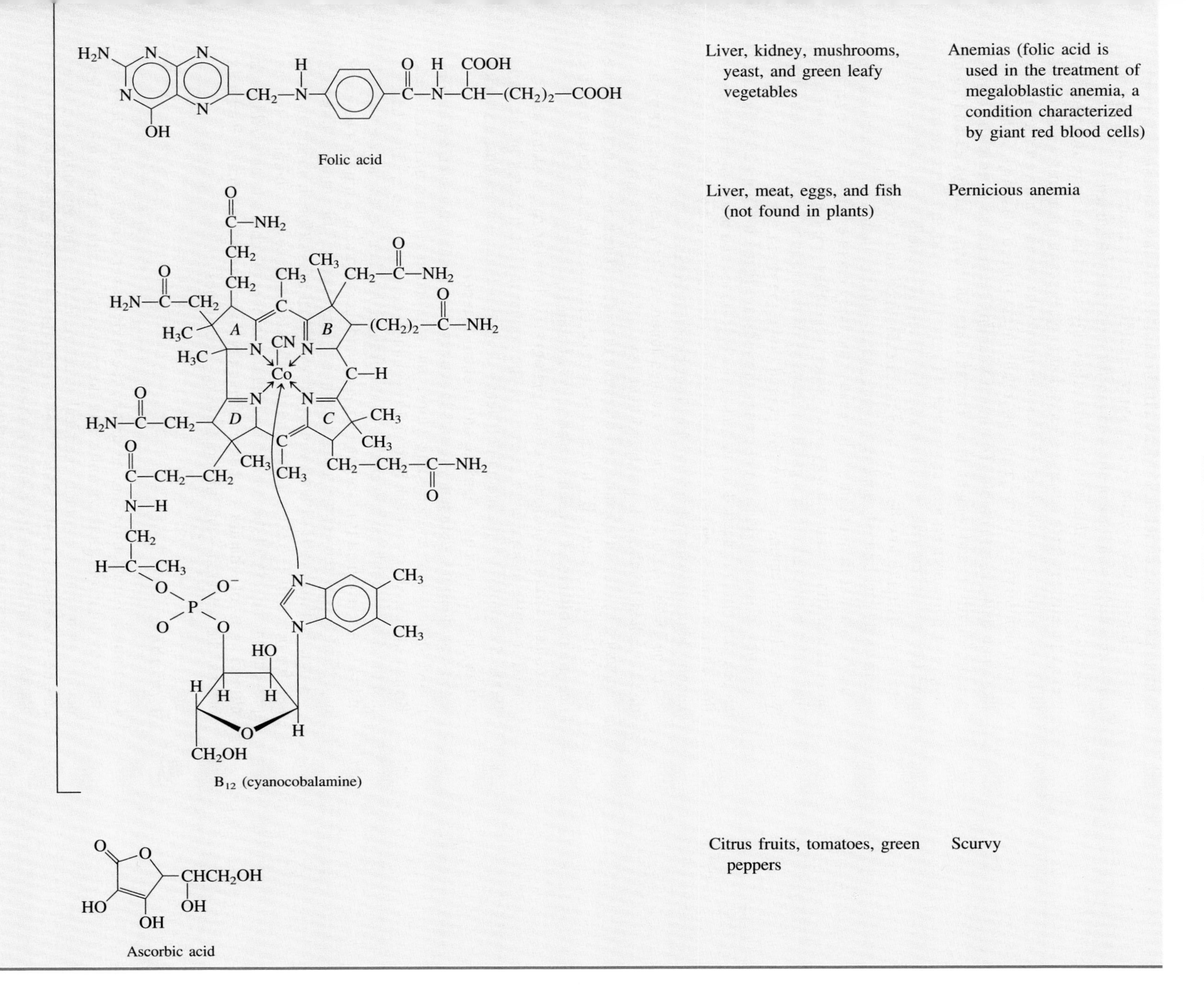

	Folic acid	Liver, kidney, mushrooms, yeast, and green leafy vegetables	Anemias (folic acid is used in the treatment of megaloblastic anemia, a condition characterized by giant red blood cells)
	B_{12} (cyanocobalamine)	Liver, meat, eggs, and fish (not found in plants)	Pernicious anemia
C	Ascorbic acid	Citrus fruits, tomatoes, green peppers	Scurvy

symptoms of the deficiency. Many children in developing countries are permanently blinded by a vitamin A deficiency. Health workers in these countries often carry injectable solutions of the vitamin for emergency treatment.

Because the fat-soluble vitamins are efficiently stored in the body, overdoses of these vitamins can have adverse effects. Large excesses of vitamin A cause irritability, dry skin, and a feeling of pressure inside the head. Massive doses of the vitamin administered to pregnant rats result in malformed offspring. Vitamin D, like vitamin A, is fat soluble. Too much vitamin D can cause pain in the bones, nausea, diarrhea, and weight loss. The amounts of both vitamins D and A in nonprescription preparations are regulated by the U.S. Food and Drug Administration. In contrast, the risk of storing vitamins E and K—another pair of fat-soluble vitamins—is quite low because they are metabolized and excreted.

The body has limited capacity to store water-soluble vitamins. These vitamins must be taken at frequent intervals because the body excretes any excess over the amount that can be immediately used. When vegetables are cooked in water, and that water is poured off, a significant portion of the vitamin content is lost. The water-soluble vitamins are literally poured down the drain.

Names of several vitamin-deficiency diseases are listed in Table 20.5. For example, vitamin D is needed for normal growth of bones and teeth. A deficiency in this vitamin causes **rickets**. The condition is characterized by bowed legs and knobby bone growths where the ribs join the breastbone.

In the 1870s, sailors of the newly created Japanese Navy developed a crippling disease called **beriberi**. It paralyzed their legs, affected their hearts, and was accompanied by loss of appetite and digestive disorders. Their diet of polished rice was deficient in vitamin B_1. It was later learned that the outer husk portion of the rice grain contains B vitamins, especially vitamin B_1.

Another debilitating disease called **pellagra** (Italian for rough skin) was observed in quite a number of persons in the southern United States in the early 1900s. Symptoms of this niacin-deficiency disease include skin lesions, a swollen tongue, loss of appetite, diarrhea, and mental disorders.

Severe vitamin C deficiency results in **scurvy**, a condition characterized by thin, porous bones, sore and bleeding gums, and a pronounced muscular weakness. British sailors, short on fresh fruit and vegetables, developed scurvy. They learned that the disease could be prevented by fresh fruit. Ships then began carrying barrels of limes, a convenient fruit for long voyages, and the sailors ate a lime or two every day. That is how they came to be known as "lime eaters," or simply "limeys."

Linus Pauling (1901–1994), winner of two Nobel prizes (for chemistry in 1954 and for peace in 1962), proposed the use of massive doses of vitamin C for the prevention and cure of the common cold and a number of other ailments. Although clinical tests of vitamin C therapy generally have not substantiated Pauling's claims, the research—and the controversy—continues.

Hormones

Hormones, like vitamins, are needed by the body in very small amounts for normal body function. Both vitamins and hormones play critical biochemical roles; neither group has a common chemical structure. Unlike the vitamins, hormones can be synthesized in the body. They are produced in the endocrine (ductless) glands, which include the pituitary, thyroid, parathyroid, adrenal, ovaries, testes, placenta, pancreas, and various portions of the gastrointestinal tract.

Hormones are discharged directly into the bloodstream and act as "chemical messengers" by signaling profound physiological changes in other parts of the

Table 20.6 Some Human Hormones and Their Physiological Effects

Name	*Gland and Tissue*	*Chemical Nature*	*Effect*
Various releasing and inhibitory factors	Hypothalamus	Peptide	Triggers or inhibits release of pituitary hormones
Human growth hormone (HGH)	Pituitary, anterior lobe	Protein	Controls the general body growth; controls bone growth
Thyroid-stimulating hormone (TSH)	Pituitary, anterior lobe	Protein	Stimulates growth of the thyroid gland and production of thyroxin
Adrenal cortex-stimulating hormone (ACTH)	Pituitary, anterior lobe	Protein	Stimulates growth of the adrenal cortex and production of cortical hormones
Follicle-stimulating hormone (FSH)	Pituitary, anterior lobe	Protein	Stimulates growth of follicles in ovaries of females, sperm cells in testes of males
Luteinizing hormone (LH)	Pituitary, anterior lobe	Protein	Controls production and release of estrogens and progesterone from ovaries, testosterone from testes
Prolactin	Pituitary, anterior lobe	Protein	Maintains the production of estrogens and progesterone, stimulates the formation of milk
Vasopressin	Pituitary, posterior lobe	Protein	Stimulates contractions of smooth muscle; regulates water uptake by the kidneys
Oxytocin	Pituitary, posterior lobe	Protein	Stimulates contraction of the smooth muscle of the uterus; stimulates secretion of milk
Parathyroid	Parathyroid	Protein	Controls the metabolism of phosphorus and calcium
Thyroxine	Thyroid	Amino acid derivative	Increases rate of cellular metabolism
Insulin	Pancreas, beta cells	Protein	Increases cell usage of glucose; increases glycogen storage
Glucagon	Pancreas, alpha cells	Protein	Stimulates conversion of liver glycogen to glucose
Cortisol	Adrenal gland, cortex	Steroid	Stimulates conversion of proteins to carbohydrates
Aldosterone	Adrenal gland, cortex	Steroid	Regulates salt metabolism; stimulates kidneys to retain Na^+ and excrete K^+
Epinephrine (adrenaline)	Adrenal gland, medulla	Amino acid derivative	Stimulates a variety of mechanisms to prepare the body for emergency action including the conversion of glycogen to glucose
Norepinephrine (noradrenaline)	Adrenal gland, medulla	Amino acid derivative	Stimulates sympathetic nervous system; constricts blood vessels, stimulates other glands
Estradiol	Ovary, follicle	Steroid	Stimulates female sex characteristics; regulates changes during menstrual cycle
Progesterone	Ovary, corpus luteum	Steroid	Regulates menstrual cycle; maintains pregnancy
Testosterone	Testis	Steroid	Stimulates and maintains male sex characteristics

body. By causing reactions to speed up or slow down, they control growth, metabolism, reproduction, and many other functions of the body and mind. A hormone that is produced by one animal is generally active in other species. For example, the insulin obtained from various species can be used to treat diabetes mellitus in humans.

Some of the more important human hormones and their physiological effects are listed in Table 20.6. The general ''chemical nature'' of hormones falls into several categories. Some, for example, have complicated protein structures. Others are, by comparison, rather simple. Those classified as **steroids** all have the same four-ring skeletal structure (Figure 20.19).

It should be understood that not all steroids have hormonal activity. For example, cholesterol is a steroid that is a common component of all animal tissues. The brain is about 10% cholesterol, but cholesterol's function there is not known. Cholesterol is a major component of gallstones. It is also found in deposits in hardened arteries. To avoid high cholesterol levels and heart disease, physicians advise patients to avoid heavy use of eggs, dairy products, and other cholesterol-rich foods.

EXAMPLE 20.5

List (a) the fat-soluble vitamins, (b) the water-soluble vitamins, (c) two steroid hormones, and (d) two nonsteroid hormones.

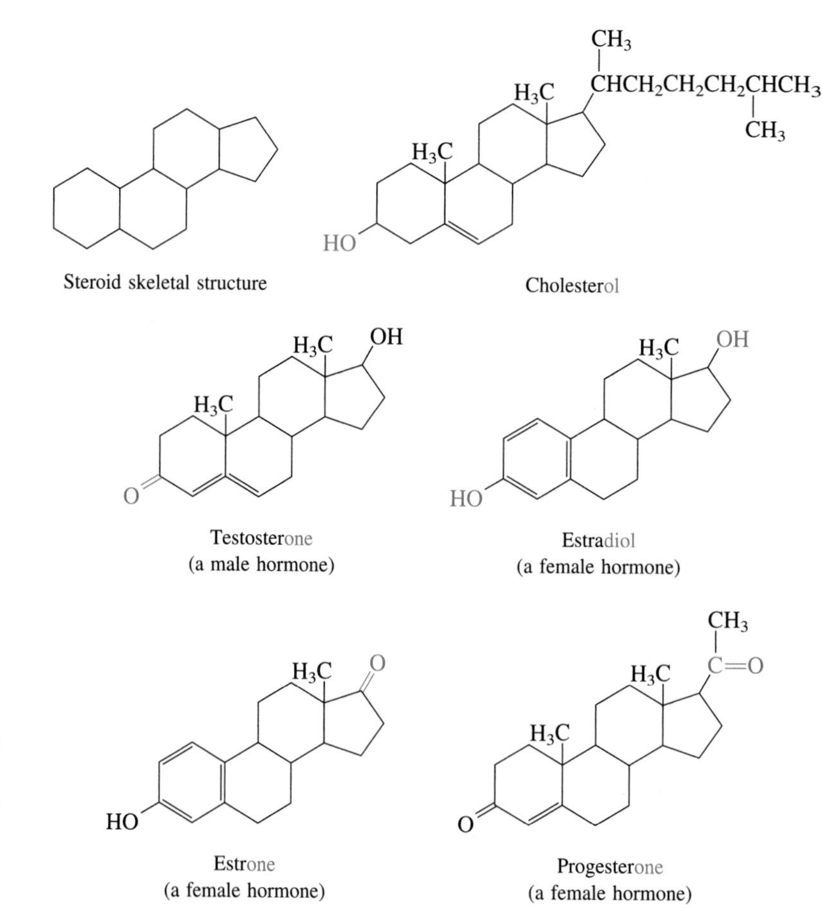

Figure 20.19 *Some steroids. The steroid skeletal structure is shown along with structures of four steroid hormones and cholesterol, which is present in all body tissues but is not a hormone.*

SOLUTION

(a) A, D, E, and K
(b) the B complex and vitamin C
(c) estradiol, progesterone
(d) insulin, thyroxine

See Problems 20.39–20.44.

From the information contained in this chapter, and within this book, it should be abundantly clear that all things—living and nonliving—are composed of chemicals. The chemical reactions that they undergo, and the rates at which they occur, account for the various changes in the makeup of all substances present on this planet. From baked bread to corrosion protection to computer parts to chemotherapy, controlling these chemical reactions and their rates is at the heart of business, industry, and the health professions. We are, in fact, dependent upon chemical changes. Indeed, without chemicals, life itself would be impossible.

Chemistry is everywhere.

EXERCISE 20.3
Describe at least three ways this course has affected your life and how chemistry affects everyone.

Chapter Summary

Biochemistry is the chemistry of all living organisms. Carbohydrates are classified as mono-, di-, and polysaccharides depending on whether hydrolysis gives one, two, or many simple sugar units. Each monosaccharide can be further classified—by the number of carbon atoms per molecule—as a hexose (six carbon atoms), a pentose (five carbon atoms), and so on. Monosaccharides form internal ring structures that continually open and close when dissolved in water. At equilibrium there is a mixture of the alpha and beta closed forms—depending on the orientation of the hydroxyl group at carbon number 1—along with a small percentage of the open form. Maltose, lactose, and sucrose are common disaccharides. Starch is a polysaccharide made up of a large number of glucose units joined by alpha linkages. In cellulose, the glucose units are joined by beta linkages. The fact that we can metabolize starch, but not cellulose, is based on this difference in linkages between glucose units.

Lipids are compounds classified by their similar solubilities. However, they have widely varying structures. The triglycerides are ''simple lipids'' that are esters of glycerol and fatty acids. The steroids, including various steroid hormones, comprise another subgroup of lipids.

Proteins are copolymers of amino acids joined by peptide bonds. For a protein to function properly, it must have the proper primary, secondary, and tertiary structure. The two types of nucleic acids—deoxyribonucleic acid (DNA) and ribonucleic acid (RNA)—are long chains of nucleotides. Each nucleotide consists of three parts: a sugar, a heterocyclic amine base, and a phosphate unit. The sequence of heterocyclic bases determines the information stored by genes on the DNA molecule. During transcription, this information is transferred to RNA, which directs the synthesis of proteins.

Vitamins are organic compounds that are required in the diet in small amounts; they cannot be synthesized in the body. The absence of a vitamin results in a specific deficiency disease. Hormones are also needed by the body in small amounts, but these compounds are produced by the endocrine glands. Hormones are messengers that signal physiological changes within the body to control growth, metabolism, reproduction, and other functions.

Assess Your Understanding

1. Identify structures of mono-, di-, and polysaccharides. [20.1]
2. Compare the component parts and linkages in starch and cellulose. [20.1]
3. Describe fats, oils, and the component parts of simple lipids. [20.2]
4. Describe the makeup and properties of saturated and unsaturated fats. [20.2]
5. Describe the makeup of amino acids, proteins, and peptide bonds. [20.3]
6. Describe structures, component parts, and functions of nucleic acids. [20.4]
7. Compare the functions of vitamins and hormones in the body. [20.5]

Key Terms

acetal [20.1]
alpha/beta forms [20.1]
amino acid [20.3]
carbohydrate [20.1]
deoxyribonucleic acid [20.4]
disaccharide [20.1]
enzyme [20.3]
essential amino acids [20.3]
fat-soluble vitamins [20.5]
fatty acids [20.2]
hemiacetal [20.1]
hormones [20.5]
iodine number [20.2]
lipids [20.2]
monosaccharide [20.1]
mutarotation [20.1]
nucleic acid [20.4]
nucleotide [20.4]
peptide bond [20.3]
polypeptide [20.3]
polysaccharide [20.1]
polyunsaturated fat [20.2]
primary protein structure [20.3]
protein [20.3]
reducing sugar [20.1]
ribonucleic acid [20.4]
saturated fat [20.2]
secondary protein structure [20.3]
steroids [20.5]
tertiary protein structure [20.3]
transcription [20.4]
triglycerides [20.2]
vitamins [20.5]
water-soluble vitamins [20.5]
zwitterion [20.3]

Problems

CARBOHYDRATES

20.1 Give two other names for D-glucose. Is the compound a monosaccharide, a disaccharide, or a polysaccharide? Explain.

20.2 Give the chemical name for common table sugar. Name the two simple sugars produced upon hydrolysis of this compound.

20.3 Give an example of each of the following.
a. an aldohexose
b. a disaccharide containing only glucose units

20.4 Give an example of each of the following.
a. a ketohexose
b. a disaccharide that contains fructose units

20.5 What structural similarities are present in amylose, amylopectin, and glycogen?

20.6 Describe differences between amylose, amylopectin, and glycogen.

20.7 Describe structural similarities and differences between starch and cellulose.

20.8 Describe differences between alpha and beta forms of D-glucose. What happens when either form is dissolved in water?

20.9 When an ordinary soda cracker—often served with soup and salad—is held in the mouth for several minutes, a sweet taste can be detected. Explain.

20.10 Sucrose (table sugar) produces energy when it is consumed by mouth, but this sugar should never be injected into the bloodstream. Explain.

20.11 Tartaric acid from grapes has the structure shown. Is its configuration D or L?

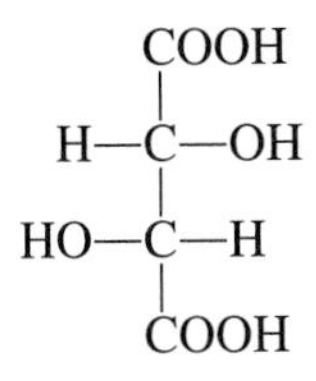

20.12 DNA molecules present in all cells contain ribose sugar units with the structure shown. Is its configuration classified as D or L?

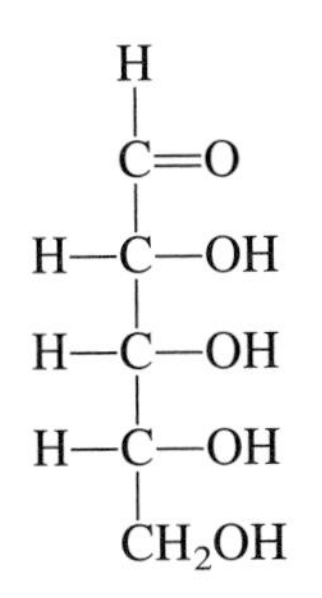

LIPIDS

20.13 What is a saturated fatty acid? Give an example.

20.14 What is a monounsaturated fatty acid? Give an example.

20.15 Describe the chemical structure of a simple lipid.

20.16 In what way are fats and oils similar? How do they differ?

20.17 Using information provided in a table in this chapter, classify each of the following as a saturated, monounsaturated, or polyunsaturated fatty acid. Also give a natural source for each fatty acid.
a. oleic acid b. palmitic acid
c. myristic acid d. arachidonic acid

20.18 Using information provided in a table in this chapter, classify each of the following as a saturated, monounsaturated, or polyunsaturated fatty acid. Also give a natural source for each fatty acid.
a. linoleic acid b. linolenic acid
c. stearic acid d. capric acid

20.19 What information is given by the iodine number of a fat or oil?

20.20 In determining the iodine number of a fat, what part of the fat molecule reacts with the iodine reagent?

20.21 Which would you expect to have the higher iodine number, corn oil or beef tallow? Explain your reasoning.

20.22 Which would you expect to have the higher iodine number, hard or liquid margarine? Explain your reasoning.

AMINO ACIDS, PROTEINS, AND NUCLEIC ACIDS

20.23 Where in the body are proteins found? What tissues are largely protein?

20.24 What are the simple ''building units'' in proteins? How are they joined?

20.25 What functional groups are found on amino acid molecules?

20.26 What is a zwitterion? Why are such ions important?

20.27 How many different amino acids are incorporated into proteins?

20.28 What are essential amino acids?

20.29 What is the difference between a polypeptide and a protein?

20.30 Is the dipeptide represented by Ser-Ala the same as the one represented by Ala-Ser? Explain.

20.31 When we give the sequence of amino acids in a protein, are we describing the primary structure, the secondary structure, or the tertiary structure?

20.32 Certain strands of amino acids coil around to form a helix. Is this a description of the primary structure, the secondary structure, or the tertiary structure?

NUCLEIC ACIDS

20.33 Give two kinds of nucleic acids. Where is each type found?

20.34 What are the three component parts of a nucleotide?

20.35 Give the different sugar units that can be present in a nucleotide.

20.36 Give the five base units that can be present in a nucleotide.

20.37 Which type of nucleic acid—DNA or RNA—exists as a double-stranded helix? How are the strands held together?

20.38 Which type of nucleic acid exists in single strands? What is meant by ''transcription''?

VITAMINS AND HORMONES

20.39 What is a vitamin?

20.40 What is a hormone?

20.41 List the fat-soluble vitamins. Are they polar or nonpolar molecules?

20.42 List the water-soluble vitamins. Are they polar or nonpolar molecules?

20.43 Indicate whether each of the following hormones is a protein or a steroid and give the physiological effect of each hormone.

a. prolactin b. insulin
c. aldosterone d. progesterone

20.44 Indicate whether each of the following hormones is a protein or a steroid and give the physiological effect of each hormone.

a. oxytocin b. vasopressin
c. estradiol d. testosterone

Additional Problems

20.45 What monosaccharide is obtained upon hydrolysis of maltose?

20.46 What monosaccharide is obtained upon hydrolysis of cellulose?

20.47 Give the proper name for the chemical present in each of the following.

a. milk sugar b. cane sugar

20.48 Give the proper name for the chemical present in each of the following.

a. table sugar b. dextrose

20.49 How can chemists and biochemists help in fighting genetic diseases?

20.50 List some products now available as a result of rDNA technology.

20.51 Give the vitamin (A, B_6, C, etc.) corresponding to these compounds: ascorbic acid, calciferol, retinol, and tocopherol.

20.52 Identify the vitamin deficiency associated with these diseases: scurvy, night blindness, rickets, and beriberi.

APPENDIX A

Metric and English Conversions and Some Physical Constants

Length

SI unit: meter (m)

1 meter	= 1.0936 yards
	= 39.37 inches
1 centimeter	= 0.3937 inch
1 inch	= 2.54 centimeters (exactly)
1 kilometer	= 0.62137 mile
1 mile	= 5280 feet
	= 1.609 kilometers

Mass

SI unit: kilogram (kg)

1 kilogram	= 1000 grams
	= 2.20 pounds
1 pound	= 453.59 grams
	= 0.45359 kilogram
	= 16 ounces
1 ton	= 2000 pounds
	= 907.185 kilograms
1 ounce (avoir.)	= 28.35 grams
1 ounce (troy)	= 31.10 grams

Volume

SI unit: cubic meter (m^3)

1 liter	= 10^{-3} m^3
	= 1 dm^3
	= 1.0567 quarts
1 gallon	= 4 quarts
	= 8 pints
	= 3.785 liters
1 quart	= 32 fluid ounces
	= 0.946 liter
1 fluid ounce	= 29.6 mL
1 $in.^3$	= 16.39 cm^3

Temperature

SI unit: kelvin (K)

0 K	= −273.15 °C
K	= °C + 273.15
°C	= $\frac{°F - 32}{1.8}$
°F	= 1.8(°C) + 32

Pressure

SI unit: pascal (Pa)

1 atmosphere = 101.325 kilopascals
= 760 torr (mm Hg)
= 14.70 pounds per square inch (psi)

Energy

SI unit: joule (J)

1 joule = 0.23901 calorie
1 calorie = 4.184 joules

Physical Constants

Avogadro's number:	$N = 6.022045 \times 10^{23}$
Speed of light:	$c = 2.9979 \times 10^{8}$ m/s
Gas constant:	$R = 0.08206 \dfrac{\text{liter-atm}}{\text{mol-K}}$
	$= 62.363 \dfrac{\text{liter-torr}}{\text{mol-K}}$

APPENDIX B

Exponential Notation

Scientists often use numbers that are so large—or so small—that they boggle the mind. For example, light travels at 300,000,000 m/s. There are 602,200,000,000,000,000,000,000 carbon atoms in 12 g of carbon. On the small side, the diameter of an atom is about 0.0000000001 m. The diameter of an atomic nucleus is about 0.000000000000001 m.

It is obviously difficult to keep track of the zeros in such quantities. Scientists fidn it convenient to express such numbers as *powers of ten*. Tables B.1 and B.2 contain partial lists of such numbers.

The speed of light is usually expressed as 3×10^8 (i.e., $3 \times 10 \times 10 \times 10 \times 10 \times 10 \times 10 \times 10 \times 10$) m/s. The mass of an atom of cesium (Cs) is expressed as 2.21×10^{-22} g, that is, as

$$2.21 \times \frac{1}{10{,}000{,}000{,}000{,}000{,}000{,}000{,}000} \text{ g}$$

Numbers such as 10^6 are called exponential numbers, where 10 is the *base* and 6 is the *exponent*. Numbers in the form 6.02×10^{23} are said to be written in *scientific notation*.

Table B.1 **Positive Powers of Ten**

$10^0 = 1$
$10^1 = 10$
$10^2 = 10 \times 10 = 100$
$10^3 = 10 \times 10 \times 10 = 1000$
$10^4 = 10 \times 10 \times 10 \times 10 = 10{,}000$
$10^5 = 10 \times 10 \times 10 \times 10 \times 10 = 100{,}000$
$10^6 = 10 \times 10 \times 10 \times 10 \times 10 \times 10 = 1{,}000{,}000$
$10^{23} = 100{,}000{,}000{,}000{,}000{,}000{,}000{,}000$

Table B.2 **Negative Powers of Ten**

$10^{-1} = 1/10 = 0.1$
$10^{-2} = 1/100 = 0.01$
$10^{-3} = 1/1000 = 0.001$
$10^{-4} = 1/10{,}000 = 0.0001$
$10^{-5} = 1/100{,}000 = 0.00001$
$10^{-6} = 1/1{,}000{,}000 = 0.000001$
⋮
$10^{-13} = 1/10{,}000{,}000{,}000{,}000 = 0.0000000000001$

Exponential numbers are often used in calculations. The most common operations are multiplication and division. Two rules must be followed: (1) to *multiply* exponentials, *add* the exponents, and (2) to *divide* exponentials, *subtract* the exponents. These rules can be stated algebraically as

$$(x^a)(x^b) = x^{a+b}$$

$$\frac{x^a}{x^b} = x^{a-b}$$

Some examples follow.

$$(10^6)(10^4) = 10^{6+4} = 10^{10}$$

$$(10^6)(10^{-4}) = 10^{6+(-4)} = 10^{6-4} = 10^2$$

$$(10^{-5})(10^2) = 10^{(-5)+2} = 10^{-5+2} = 10^{-3}$$

$$(10^{-7})(10^{-3}) = 10^{(-7)+(-3)} = 10^{-7-3} = 10^{-10}$$

$$\frac{10^{14}}{10^6} = 10^{14-6} = 10^8$$

$$\frac{10^6}{10^{23}} = 10^{6-23} = 10^{-17}$$

$$\frac{10^{-10}}{10^{-6}} = 10^{(-10)-(-6)} = 10^{-10+6} = 10^{-4}$$

$$\frac{10^3}{10^{-2}} = 10^{3-(-2)} = 10^{3+2} = 10^5$$

$$\frac{10^{-8}}{10^4} = 10^{(-8)-4} = 10^{-12}$$

$$\frac{10^7}{10^7} = 10^{7-7} = 10^0 = 1$$

Problems involving both a coefficient (a numerical part) and an exponential are solved by multiplying (or dividing) coefficients and exponentials separately.

EXAMPLE B.1

To what is the following expression equivalent?

$$(1.2 \times 10^5)(2.0 \times 10^9)$$

SOLUTION

First, multiply the coefficients.

$$1.2 \times 2.0 = 2.4$$

Then multiply the exponentials.

$$10^5 \times 10^9 = 10^{5+9} = 10^{14}$$

The complete answer is 2.4×10^{14}.

EXAMPLE B.2

To what is the following expression equivalent?

$$\frac{(8.0 \times 10^{11})}{(1.6 \times 10^{4})}$$

SOLUTION

First, divide the coefficients.

$$\frac{8.0}{1.6} = 5.0$$

Then divide the exponentials.

$$\frac{10^{11}}{10^{4}} = 10^{11-4} = 10^{7}$$

The answer is 5.0×10^{7}.

EXAMPLE B.3

Give an equivalent for the following expression.

$$\frac{(1.2 \times 10^{14})}{(4.0 \times 10^{6})}$$

SOLUTION

Before carrying out the division, it is convenient to rewrite the dividend (the numerator) so that the coefficient is larger than that of the divisor (the denominator).

$$1.2 \times 10^{14} = 12 \times 10^{13}$$

Note that the coefficient was made larger by a factor of 10 and the exponential was made smaller by a factor of 10. The quantity as a whole is unchanged. Now divide.

$$\frac{12 \times 10^{13}}{4.0 \times 10^{6}} = 3.0 \times 10^{7}$$

EXAMPLE B.4

Give an equivalent for the following expression.

$$\frac{(3 \times 10^{7})(8 \times 10^{-3})}{(6 \times 10^{2})(2 \times 10^{-1})}$$

SOLUTION

In problems such as this, you can carry out the multiplications specified in the numerator and in the denominator separately and then divide the resulting numbers.

$$(3 \times 10^7)(8 \times 10^{-3}) = 24 \times 10^4$$
$$(6 \times 10^2)(2 \times 10^{-1}) = 12 \times 10^1$$
$$\frac{24 \times 10^4}{12 \times 10^1} = 2 \times 10^3$$

The multiplications and divisions in problems like this can be carried out in any convenient order.

Only one other mathematical function involving exponentials is of importance to us. What happens when you raise an exponential to a power? You just multiply the exponent by the power. To illustrate,

$$(10^3)^3 = 10^9$$
$$(10^{-2})^4 = 10^{-8}$$
$$(10^{-5})^{-3} = 10^{15}$$

If the exponential is combined with a coefficient, the two parts of the number are dealt with separately, as in the following example.

$$(2 \times 10^3)^2 = 2^2 \times (10^3)^2 = 4 \times 10^6$$

For a further discussion of—and more practice with—exponential numbers, see Dorothy M. Goldish, *Basic Mathematics for Beginning Chemistry,* 4th ed., Macmillan Publishing Company, New York, 1990. Chapter 3 covers exponential notation.

Problems

1. Express each of the following in scientific notation.

a. 0.00001 b. 10,000,000
c. 0.0034 d. 0.0000107
e. 4,500,000,000 f. 406,000
g. 0.02 h. 124×10^3

2. Carry out the following operations. Express the numbers in scientific notation.

a. $(4.5 \times 10^{13})(1.9 \times 10^{-5})$
b. $(6.2 \times 10^{-5})(4.1 \times 10^{-12})$
c. $(2.1 \times 10^{-6})^2$
d. $\dfrac{(4.6 \times 10^{-12})}{(2.1 \times 10^3)}$
e. $\dfrac{(9.3 \times 10^9)}{(3.7 \times 10^{-7})}$
f. $\dfrac{(2.1 \times 10^5)}{(9.8 \times 10^7)}$
g. $\dfrac{(4.3 \times 10^{-7})}{(7.6 \times 10^{22})}$

Answers

1. a. 1×10^{-5} b. 1×10^7 c. 3.4×10^{-3}
d. 1.07×10^{-5} e. 4.5×10^9 f. 4.06×10^5
g. 2×10^{-2} h. 1.24×10^5

2. a. 8.6×10^8 b. 2.5×10^{-16} c. 4.4×10^{-12}
d. 2.2×10^{-15} e. 2.5×10^{16} f. 2.1×10^{-3}
g. 5.7×10^{-30}

APPENDIX C

Using Conversion Factors to Solve Problems

Problems in chemistry are often solved by a method that involves converting from one kind of unit to another. This approach is called **dimensional analysis** or the **factor-label method.** Whatever we call it, the method employs units—such as L, mi/hour, cm/ft, or g/cm^3—as aids in setting up and solving problems. The general approach is to multiply the known quantity (and its units!) by one or more conversion factors so that the answer is obtained in the desired units.

$$\text{Known quantity and unit} \times \text{Conversion factor} = \text{Answer (in desired unit)}$$

The method is best learned by practice. We urge you to learn it now to save yourself a lot of time and wasted effort later.

C.1 Conversions Within a System

Quantities can be expressed in a variety of units. For example, you can buy beverages by the 12-oz can or by the pint, quart, gallon, or liter. If you wish to compare prices, you must be able to convert from one unit to another. Such a conversion changes the numbers and units, but it does not change the quantity. Your actual weight, for example, remains unchanged whether it is expressed in pounds, ounces, or kilograms.

You know that multiplying a number by 1 doesn't change its value. Multiplying by a fraction equal to 1 also leaves the value unchanged. A fraction is equal to 1 when the numerator is equal to the denominator. For example, we know that

$$1 \text{ ft} = 12 \text{ in.}$$

Therefore,

$$\frac{1 \text{ ft}}{12 \text{ in.}} = 1$$

Similarly,

$$\frac{12 \text{ in.}}{1 \text{ ft}} = 1$$

Now, if you want to convert an answer from inches to feet, you can do so by choosing one of the above fractions as a **conversion factor.** Which one do you choose? The one that gives you an answer with the right units! Let's illustrate by an example.

EXAMPLE C.1

My bed is 72 in. long. What is its length in feet?

SOLUTION

You know the answer, of course, but let us proceed to show *how* the answer is obtained by using dimensional analysis. We need to multiply 72 in. by one of the above fractions. Which one? The known quantity and unit is 72 in.

$$72 \text{ in.} \times \text{conversion factor} = ?\text{ ft}$$

For the conversion factor, choose the fraction that, when inserted in the equation, cancels the unit *in.* and becomes the unit *ft.*

$$72 \text{ in.} \times \frac{1 \text{ ft}}{12 \text{ in.}} = 6.0 \text{ ft}$$

Just for kicks, let us try the other conversion factor.

$$72 \text{ in.} \times \frac{12 \text{ in.}}{1 \text{ ft}} = \frac{860 \text{ in.}^2}{\text{ft}}$$

Absurd! How can a bed be 860 in.^{2}/ft? You should have no difficulty in choosing between the two possible answers.

One of the advantages of the metric system is the ease of conversion between units. Let us try to demonstrate that with a few examples. Remember that a list of equivalent values is actually a list of conversion factors. Thus, the equality

$$1 \text{ kilogram} = 1000 \text{ grams}$$

can be rearranged into two useful conversion factors.

$$\frac{1 \text{ kilogram}}{1000 \text{ grams}} \quad \text{and} \quad \frac{1000 \text{ grams}}{1 \text{ kilogram}}$$

EXAMPLE C.2

Convert 0.371 kg to grams.

SOLUTION

$$0.371 \text{ kg} \times \frac{1000 \text{ g}}{1 \text{ kg}} = 371 \text{ g}$$

EXAMPLE C.3

Convert 0.371 lb to ounces.

SOLUTION

$$0.371 \text{ lb} \times \frac{16 \text{ oz}}{1 \text{ lb}} = 5.94 \text{ oz}$$

EXAMPLE C.4

Convert 2429 cm to meters.

SOLUTION

$$2429 \text{ cm} \times \frac{1 \text{ m}}{100 \text{ cm}} = 24.29 \text{ m}$$

EXAMPLE C.5

Convert 2429 in. to yards.

SOLUTION

$$2429 \text{ in.} \times \frac{1 \text{ yd}}{36 \text{ in.}} = 67.47 \text{ yd}$$

In conversions of customary units, you multiply and divide by factors such as 16 or 36. In metric conversions, you multiply and divide by 100 or 1000 and so on. You need only shift the decimal point when doing metric conversions.

Conversion factors are not usually given in a problem. They may be obtained from listings such as those in Appendix A. However, you would be wise to learn to convert within the metric system without the need of tables. Also, you should remember that this equality

$$1 \text{ centimeter} = 0.01 \text{ meter}$$

is equivalent to

$$100 \text{ centimeters} = 1 \text{ meter}$$

All these fractions

$$\frac{1 \text{ cm}}{0.01 \text{ m}} \qquad \frac{0.01 \text{ m}}{1 \text{ cm}} \qquad \frac{100 \text{ cm}}{1 \text{ m}} \qquad \frac{1 \text{ m}}{100 \text{ cm}}$$

are valid conversion factors.

EXAMPLE C.6

Knowing that 1 mL = 0.001 L, write four conversion factors relating milliliters and liters.

SOLUTION

The first two conversion factors can be formed by arranging the two sides of the equality in the form of a fraction.

$$\frac{1\ \text{mL}}{0.001\ \text{L}} \qquad \frac{0.001\ \text{L}}{1\ \text{mL}}$$

To derive the other two conversion factors, first multiply both sides of the equality by 1000 (in order to obtain the equality in terms of 1 L).

$$1000 \times 1\ \text{mL} = 1000 \times 0.001\ \text{L}$$

$$1000\ \text{mL} = 1\ \text{L}$$

Now just arrange this last equality in fractional form.

$$\frac{1000\ \text{mL}}{1\ \text{L}} \qquad \frac{1\ \text{L}}{1000\ \text{mL}}$$

The conversion factors 1 mL/0.001 L and 1000 mL/1 L would give exactly the same answer if used in a problem. The only difference is convenience. Some people would rather multiply by 1000 than divide by 0.001. In this age of the electronic calculator, perhaps even this difference is no longer significant.

Let us try some more conversions within the metric system.

EXAMPLE C.7

How many milliliters are there in a 2-L bottle of soda pop?

SOLUTION

From memory or from the tables in Appendix A you find that

$$1\ \text{L} = 1000\ \text{mL}$$

$$2\ \cancel{\text{L}} \times \frac{1000\ \text{mL}}{1\ \cancel{\text{L}}} = 2000\ \text{mL}$$

Notice that we picked the conversion factor that allowed us to cancel liters and obtain an answer in the desired units, milliliters.

Sometimes it is necessary to carry out more than one conversion in a problem.

EXAMPLE C.8

In the United States, the usual soda pop can holds 360 mL. How many such cans could be filled from one 2.0-L bottle?

SOLUTION

The problem tells us that

$$1 \text{ can} = 360 \text{ mL}$$

Using that equivalence, we can calculate the answer.

$$2.0 \text{ L} \times \frac{1000 \text{ mL}}{1 \text{ L}} \times \frac{1 \text{ can}}{360 \text{ mL}} = 5.6 \text{ cans}$$

EXAMPLE C.9

How many 325-mg aspirin tablets can be made from 275 g of aspirin?

SOLUTION

The problem asks us to convert the *given* value of 275 g to tablets. The problem also includes a necessary conversion factor.

$$1 \text{ tablet} = 325 \text{ mg}$$

From memory or the tables, we have another required conversion factor.

$$1 \text{ g} = 1000 \text{ mg}$$

By multiplying the given value by the appropriately arranged conversion factors, we arrive at the answer.

$$275 \text{ g} \times \frac{1000 \text{ mg}}{1 \text{ g}} \times \frac{1 \text{ tablet}}{325 \text{ mg}} = 846 \text{ tablets}$$

C.2 Conversions Between Systems

To convert from one system of measurement to another, you need a list of equivalents such as that in Appendix A. Let us plunge right in and work some examples.

EXAMPLE C.10

How many kilograms are there in 33 lb?

SOLUTION

$$33 \text{ lb} \times \frac{1.0 \text{ kg}}{2.2 \text{ lb}} = 15 \text{ kg}$$

EXAMPLE C.11

You know that your weight is 142 lb, but the job application form asks for your weight in kilograms. What is it?

SOLUTION

From the table we find

$$1.00\text{ lb} = 0.454\text{ kg}$$

The solution is simple.

$$142\text{ lb} \times \frac{0.454\text{ kg}}{1.00\text{ lb}} = 64.5\text{ kg}$$

EXAMPLE C.12

A recipe calls for 750 mL of milk, but your measuring cup is calibrated in fluid ounces. How many ounces of milk will you need?

SOLUTION

$$750\text{ mL} \times \frac{1.00\text{ fl oz}}{29.6\text{ mL}} = 25.3\text{ fl oz}$$

EXAMPLE C.13

How many meters are there in 764 ft (1.00 m = 39.4 in.)?

SOLUTION

$$764\text{ ft} \times \frac{12\text{ in.}}{1\text{ ft}} \times \frac{1.00\text{ m}}{39.4\text{ in.}} = 233\text{ m}$$

EXAMPLE C.14

How would you describe a young man who is 1.6 m tall and weighs 91 kg?

SOLUTION

$$1.6\text{ m} \times \frac{39\text{ in.}}{1.0\text{ m}} \times \frac{1\text{ ft}}{12\text{ in.}} = 5.2\text{ ft}$$

$$91\text{ kg} \times \frac{2.2\text{ lb}}{1.0\text{ kg}} = 200\text{ lb}$$

The young man is 5 ft 2 in. tall and weighs 200 lb. Let us be generous and say that he is well muscled.

It is possible (and frequently necessary) to manipulate units in the denominator as well as in the numerator of a problem. Just remember to use conversion factors in such a way that the unwanted units cancel.

EXAMPLE C.15

A sprinter runs the 100-m dash in 11 s. What is her speed in kilometers per hour?

SOLUTION

The given speed is 100 m per 11 s.

$$\frac{100\ \text{m}}{11\ \text{s}}$$

The conversion factors that we need are found in the tables or recalled from memory.

$$\frac{100\ \cancel{\text{m}}}{11\ \cancel{\text{s}}} \times \frac{1\ \text{km}}{1000\ \cancel{\text{m}}} \times \frac{60\ \cancel{\text{s}}}{1\ \cancel{\text{min}}} \times \frac{60\ \cancel{\text{min}}}{1\ \text{hr}} = 33\ \text{km/hr}$$

Note that the first conversion factor changes m to km. It takes two factors to change s to hr. Note also that we could have first applied the factors that convert s to hr and then converted m to km. The answer would have been the same.

EXAMPLE C.16

If your heart beats at a rate of 72 times per minute, and your lifetime will be 70 years, how many times will your heart beat during your lifetime?

SOLUTION

Two equivalences are given in the problem.

$$72\ \text{beats} = 1\ \text{min}$$
$$1\ \text{lifetime} = 70\ \text{yr}$$

Three others that you will need you can recall from memory.

$$1\ \text{yr} = 365\ \text{days}$$
$$1\ \text{day} = 24\ \text{hr}$$
$$1\ \text{hr} = 60\ \text{min}$$

Start now with the factor 72 beats/1 min (the known quantities and units) and apply the conversion factors as needed to get an answer in beats/lifetime (the desired units).

$$\frac{72\ \text{beats}}{1\ \cancel{\text{min}}} \times \frac{60\ \cancel{\text{min}}}{1\ \cancel{\text{hr}}} \times \frac{24\ \cancel{\text{hr}}}{1\ \cancel{\text{day}}} \times \frac{365\ \cancel{\text{days}}}{1\ \cancel{\text{yr}}} \times \frac{70\ \cancel{\text{yr}}}{1\ \text{lifetime}}$$
$$= 2{,}600{,}000{,}000\ \text{beats/lifetime}$$

APPENDIX D

Glossary*

absolute zero The lowest temperature possible: 0 K = −273.15 °C. [3.12]

accuracy How closely experimental measurements agree with the true value. [3.7]

acetal A carbon atom that is bonded to two —OR groups, an —R, and an —H. An acetal is present, among other places, in the linkage of disaccharides. [20.1]

acid A substance that produces hydrogen ions or, more precisely, hydronium ions, H_3O^+, in solution. [8.8, 10.12, 16.1]

acid–base indicator A dye—natural or synthetic—that changes from one color to another color at one or more specific pH values. [16.9]

acid salt A salt of a partially neutralized polyprotic acid containing hydrogen atoms that can be neutralized (replaced by other cations); for example, $NaHCO_3$ and KH_2PO_4. [8.8]

actinides The group of fourteen elements with numbers 90 through 103. [6.9]

activation energy The minimum kinetic energy that colliding molecules must possess for a reaction to occur. [15.1]

activity series A listing of metals in order of decreasing reactivity. [10.8]

actual yield The quantity of product (usually in grams) that is actually obtained from a particular reaction. [11.7]

addition reaction A reaction that involves the breaking of a double or triple bond and the attachment of an additional atom or group to each carbon formerly involved in the double or triple bond. [19.5]

alcohol An organic compound that contains a hydroxyl group, —OH, as a substituent replacing one or more hydrogen atoms of a hydrocarbon. [19.7]

aldehyde An organic compound containing a carbonyl group bonded to a hydrogen—on one side—and also to a hydrogen, an alkyl, or an aryl group. [19.10]

aliphatic Pertaining to hydrocarbon compounds having no benzene rings. [19.6]

alkali A substance that has the properties of a base. (*Also see* base.) [16.1]

alkali metal A Group I element (except hydrogen) in the periodic table. [4.4, 5.7]

alkaline earth metal An element in Group IIA of the periodic table. [5.7]

alkanes Compounds of hydrogen and carbon with single bonds; the saturated hydrocarbons. [19.2]

alkenes Hydrocarbon compounds that contain one or more double bonds. [19.5]

alkyl group A hydrocarbon group (symbolized —R) that is formed by removing a hydrogen atom from an alkane. [19.4]

alkyl halide A compound formed by substituting a halogen atom for a hydrogen atom in an alkane. [19.2]

alkynes Hydrocarbon compounds that contain one or more triple bonds. [19.5]

*For additional information, refer to the section(s) or page shown in brackets.

allotropes Two or more physical forms of an element in the same state, such as diamond and graphite, which are allotropes of carbon. [6.7, 13.5]

alpha decay Radioactive decay that produces alpha particles. [18.1]

alpha particle A particle that consists of two protons and two neutrons. It is identical to the nucleus of a helium atom and is symbolized $^{4}_{2}He^{2+}$. [18.1]

alpha/beta forms (of acetals and hemiacetals) The two relative positions of the —OH group of a hemiacetal or an —OR group of an acetal. [20.1]

alpha rays Fast-moving particles emitted by an atomic nucleus during radioactive decay. Each alpha particle is a helium nucleus (mass = 4, charge = +2). [5.1]

amide An organic compound having a functional group containing a carbon atom double-bonded to an oxygen atom and single-bonded to a nitrogen atom. [19.12]

amine An organic compound derived from ammonia by replacing one, two, or three hydrogens of NH_3 with the same number of alkyl or aryl groups. [19.12]

amino acid A substance that contains an amine group, $—NH_2$, and a carboxyl group, —COOH. Both groups are joined to the same carbon in α-amino acids, which are the building blocks of proteins. [20.3]

amino group The $—NH_2$ group. [19.12]

amorphous solid A noncrystalline solid. Examples: plastics and wax. [13.5]

amphiprotic The ability of a substance to accept or donate a proton. [16.6]

amphoteric The ability of a substance to react with either an acid or a base. Certain metals and certain metal hydroxides are amphoteric. [16.5]

analytical balance An instrument used to determine the mass of a sample to the nearest tenth of a milligram (to the nearest 0.0001 g). [3.5]

angular (bent) shape An unbalanced arrangement of atoms in a molecule where its atoms are not bonded together in a linear A—B—A arrangement. [7.9]

anhydrous salt A chemical compound (a salt) that has no water of hydration. [8.9]

anion (pronounced AN-ion) An ion with a negative charge. [6.3]

anode The electrode of any electrochemical cell at which oxidation (the loss of electrons) takes place. [5.1, 17.7]

aqueous ammonia A water solution of ammonia, NH_3. It is a weak base. [16.3]

aqueous solution A solution obtained by dissolving a solute (the general name for the chemical being dissolved) in water, the solvent. [8.8, 10.1, 14.1]

aromatic compound Any hydrocarbon compound containing a benzene ring. [19.6]

artificial transmutation The conversion of atoms of one kind of element into atoms of another element by bombarding the target nucleus with particles that are absorbed by the target nucleus, which becomes a different kind of atom. [18.6]

aryl group A benzene-ring-containing group (symbolized —Ar) on a large organic molecule. [19.6]

atmosphere (pressure unit) A pressure of 1 atmosphere will support a 760-mm column of mercury. 1 atm = 760 mm Hg = 760 torr = 14.7 psi = 101.325 kPa. [12.3]

atmospheric pressure The force per unit area exerted on objects on the Earth as a result of the attraction of the Earth for the blanket of air that surrounds our planet. [12.3]

atom The smallest particle that retains the property of the element. [2.3]

atomic mass (average) The weighted average of the atomic masses for the mixture of all naturally occurring isotopes of an element. This mass is the one given in the periodic table. [4.10]

atomic mass unit (amu) The unit for expressing the relative masses of atoms. An amu equals one-twelfth of the mass of a carbon-12 atom. [4.7]

atomic number The number of protons in the nucleus of an atom. [4.8]

atomic weight *See* atomic mass (average).

Avogadro hypothesis Equal volumes of gases at the same temperature and pressure contain equal numbers of molecules; gas volume is proportional to the number of moles of gas at a constant pressure and temperature. [11.5, 12.10]

Avogadro's number The number of particles—atoms, ions, or molecules—in a mole of the particles being counted, that is, 6.02×10^{23} particles. [4.11]

background radiation The ever-present radiation in the environment from cosmic rays and from radioactive materials in the air, water, soil, and rocks. [18.5]

barometer A device used to measure atmospheric pressure. [12.3]

base A substance that produces hydroxide ions in solution. Also defined as a substance that can accept a hydrogen ion (a proton). [10.12, 16.1]

benzene A hydrocarbon compound having the formula C_6H_6 with all six carbon atoms bonded to each other in a ring and with each carbon also bonded to one hydrogen atom. [19.6]

beta decay Radioactive decay that produces beta particles. [18.1]

beta rays (particles) Fast-moving electrons emitted by a radioactive nucleus during decay. A beta particle, symbolized $_{-1}^{0}e$, is emitted when a neutron is converted to a proton during radioactive decay. [5.1, 18.1]

boiling point The temperature at which the vapor pressure of a liquid is equal to the total pressure exerted on the liquid. [13.4]

Boyle's law The volume, V, occupied by a sample of gas is inversely proportional to pressure, P, at constant temperature. $P_1V_1 = P_2V_2$. [12.4]

buffer A pair of chemicals—a weak acid and its salt or a weak base and its salt—that, if present in a given solution, can keep the pH almost constant when small amounts of acid or base are added to the solution. [16.11]

buffer capacity The maximum amount of acid or base that can be added to a buffered solution before the pH begins to change significantly. [16.11]

calorie A metric unit of heat energy. One calorie will raise the temperature of 1 gram of water by 1 °C. 1 calorie = 4.184 joules. One Calorie (capital C) equals one kilocalorie. [3.13]

carbohydrate An organic substance composed of carbon, hydrogen, and oxygen that can be classified as a sugar or a compound made up of sugar units. [20.1]

carbonyl group The organic functional group having the structure $>C{=}O$. [19.10]

carboxyl group The —COOH group; it is present in all organic acids. [8.8, 16.1, 19.11]

carboxylic acid A compound that contains the carboxyl, —COOH, group. [8.8, 19.11]

catalyst A substance that speeds up a chemical reaction without itself undergoing chemical change. [10.1, 15.2]

cathode The electrode of any electrochemical cell at which reduction (the gain of electrons) takes place. [5.1, 17.7]

cathode ray The beam of electrons that is emitted by the cathode and moves toward the anode of a partially evacuated gas discharge tube. [5.1]

cation (pronounced CAT-ion) An ion with a positive charge. [6.3]

Celsius scale The temperature scale on which water freezes at 0 °C and boils at 100 °C at standard pressure (1 atm). [3.12]

chain reaction A self-sustaining change in which one or more products of a reaction or event cause one or more new events. [18.9]

characteristic properties The physical and chemical properties that can be used to identify a substance and distinguish that substance from other substances. (They do not depend on the quantity of the substance.) [2.5]

Charles's law The volume, V, occupied by a sample of gas is directly proportional to its Kelvin temperature, T, at constant pressure. $V_1/T_1 = V_2/T_2$. [12.5]

chemical bonds The forces of attraction that hold atoms or ions together in chemical compound. [7.1]

chemical change A chemical reaction; a change in the atomic makeup (composition) of a substance. One or more substances are used up as others are formed. [2.5]

chemical equation A symbolic representation of a chemical reaction showing chemical formulas of substances present before and after a chemical reaction and showing the mole ratios of reactants and products. [10.1]

chemical equilibrium The dynamic system in which concentrations of reactants and products remain constant and the rate of the forward reaction equals the rate of the reverse reaction. [15.3]

chemical family *See* group of elements.

chemical formula A symbolic way of representing the composition of a substance using symbols of the elements and subscripts to represent the appropriate number of atoms of each kind. [4.7]

chemical kinetics The study of the rates of chemical reactions and factors that affect reaction rates. [15.1]

chemical nomenclature *See* nomenclature (chemical).

chemical properties Characteristic properties of a substance that relate to how a substance changes in composition or how it interacts with other substances. Examples: tendency to explode, to burn, or to corrode. [2.5]

chemical reaction The process that involves a chemical change; that is, one or more substances are used up while one or more substances are formed. [10.1]

chemistry The branch of science that deals with the characteristics and composition of all materials and with the reactions they can undergo. [1.1, 2.1]

cis configuration The geometrical isomer in which two groups in reference are positioned on the same side of a double bond or ring arrangement. [19.5]
coefficient The number that is placed in front of a chemical formula in a chemical equation. [10.3]
colligative properties Properties of solutions that depend on the number of solute particles present rather than on the identities of the particles. Examples: boiling point elevation, freezing point depression, and osmotic pressure. [14.8]
collision frequency The number of particle collisions per unit time. [15.1]
colloidal dispersion A mixture in which particles of one component are intermediate in size (about 1 to 100 nm) between those in solutions and those in suspensions. [14.9]
combination reaction *See* synthesis reaction.
combined gas law The mathematical relationship involving pressures, volumes, and Kelvin temperatures for gases at two different sets of conditions. The equation takes the form $P_1V_1/T_1 = P_2V_2/T_2$. [12.8]
combustion reaction The vigorous chemical reaction of oxygen with a fuel that usually contains carbon and hydrogen. Combustion is exothermic, it is accompanied by the liberation of heat and/or light. [10.4, 10.5, 17.2]
compound A pure substance made up of two or more kinds of elements that are combined together chemically in fixed proportions. [2.3, 4.1]
concentrated solution A solution containing a relatively large quantity of the solute. [14.2]
concentration A measure of the quantity of solute dissolved in a particular volume of solution. [9.6, 14.7]
condensation The conversion of a vapor to a liquid by cooling the vapor. [13.4]
condense To pass from the gaseous state to the liquid state. [2.2]
condensed structural formula A simplified representation of a structural formula that shows hydrogen atoms clustered next to the carbon atom to which they are bonded; that is, single bonds to hydrogen atoms are omitted. [19.3]
conductivity The capacity for transmitting heat or electrical energy. [4.5]
conjugate acid–base pair A proton donor and the ion formed by the loss of the proton. The two species differ by only one proton. [16.6]
conjugate base The ion that remains after an acid releases one proton. [16.6]
continuous spectrum The ''rainbow of colors'' consisting of all wavelengths of the visible spectrum. [5.2]
control rods Neutron-absorbing metal rods (e.g., cadmium) used to control the rate of fission in a nuclear reactor. [18.9]
conversion factor A fraction with a set of units—such as 24 hr/day—used in problem solving to convert a quantity with the given units to a quantity with the desired units. [3.2]
coordinate covalent bond A chemical bond formed when one atom donates both of the electrons shared in a covalent bond. [7.11]
covalent bond A pair of electrons shared by two atoms in a molecule. [7.2]
covalent network (macromolecular) solids Solids that have atoms joined together by covalent bonds in a continuous manner throughout the material. [13.5]
critical mass The minimal mass of a fissionable material required for a self-sustaining chain reaction. [18.9]
crystal A solid having plane faces at definite angles and whose atoms, ions, or molecules have a regular three-dimensional arrangement. [7.1]
crystal lattice The three-dimensional, ordered, repeating pattern of atoms, ions, or molecules that occurs in a crystalline solid. [7.1, 13.5]
crystalline solid A solid chemical that has particles that are arranged in an orderly (systematic) pattern that is repeated throughout the solid. [13.5]
curie (Ci) The quantity of radioactive material that undergoes 3.70×10^{10} disintegrations per second (the number produced by 1 g of radium). [18.3]

Dalton's atomic theory Elements are made up of atoms that combine in fixed, small, whole-number ratios. A chemical reaction involves a change, not in the atoms themselves, but in the way atoms are combined to form compounds. [4.7]
Dalton's law of partial pressures The total pressure exerted by a mixture of gases is equal to the sum of the partial pressures exerted by the separate gases. [12.11]
decomposition reaction A chemical reaction in which a compound is broken down into two or more simpler substances. [10.4, 10.7]
denatured alcohol Ethanol (ethyl alcohol) that has been made unfit for consumption as a beverage by the addition of poisonous or noxious substances. [19.7]
density The ratio obtained by dividing an object's mass by its volume. [3.11]

deoxyribonucleic acid (DNA) The type of nucleic acid found primarily in the nuclei of cells. [20.4]

destructive distillation The process of heating a material in the absence of oxygen to give decomposition products. [19.7]

deuterium The isotope of hydrogen—sometimes called heavy hydrogen—with 1 proton, 1 neutron, and a mass number of 2. [4.9]

dialysis The selective passage of ions and small molecules—not large molecules or colloidal particles—along with the solvent through a semi-permeable membrane. [14.10]

diatomic elements The seven elements—H_2, N_2, O_2, F_2, Cl_2, Br_2, and I_2—that exist as diatomic molecules, that is, as molecules composed of two atoms. [4.4]

diffusion The spontaneous mixing of gases at constant pressure. [2.2, 12.2]

dilute solution A solution containing a relatively small quantity of the solute. [14.2]

dimensional analysis (also called factor-label method) An approach to problem solving where the given quantity and units are multiplied by one or more conversion factors to obtain an answer with the desired units. (*See* conversion factor.) [3.3, 3.6]

diol An alcohol that has two hydroxyl, —OH, groups. [19.7]

dipoles Molecules that have separate centers of partial negative and partial positive charges. For example, in HCl, the chlorine atom carries a partial negative charge and the hydrogen atom carries a partial positive charge. [13.2]

diprotic acid An acid that has two ionizable (acidic) hydrogen atoms per molecule. [16.1]

disaccharide A carbohydrate that can give two monosaccharides (simple sugars) when it is hydrolyzed. [20.1]

disintegrations per second The measure of nuclear activity that counts the total number of radioactive nuclei that break down each second as they undergo nuclear decay. [18.3]

dissociation The process whereby a chemical substance breaks up into simpler component parts (molecules or ions) when it melts or when it dissolves in a solvent. [7.1, 10.11]

dissolve To pass into solution; to mix uniformly and completely at the molecular level. [14.1]

distillate The condensed vapor sample obtained during distillation. [13.4]

distillation A method of separating volatile liquids by boiling the mixture and collecting the condensed vapor. The separation process is based on differences in the boiling points of the components of the mixture. [13.4]

double bond A covalent bond where two electron pairs are shared between two atoms. [7.2]

double-replacement reaction (also called a metathesis reaction) A chemical reaction between two salts, symbolized AB and CD, that switch cation partners to form two different compounds, symbolized AD and CB. [10.4, 10.10]

ductility A metal's ability to be drawn (stretched) into wire. [4.5]

Einstein equation $E = mc^2$, where E, m, and c represent energy, mass, and the speed of light, respectively. [2.9]

electrochemical cell Any system that either generates an electric current from a chemical reaction or uses an electric current to produce a chemical reaction. (*See* galvanic cell and electrolytic cell) [17.8]

electrode An electrical point of contact that carries a charge (+ or −). [5.1]

electrolysis A process where a direct electrical current (dc power) is used to decompose a compound. [4.6, 10.7, 11.8, 17.7]

electrolyte Any substance that dissolves in water to give a solution that contains ions and will, therefore, conduct electricity. [7.1]

electrolytic cell An electrochemical cell that uses electricity to drive a chemical reaction that would otherwise not occur. [17.8]

electromagnetic radiation The general term for the entire range of energy that travels through space as waves and at the speed of light. [5.2]

electron The negatively charged particle with a mass of 0 amu that occupies the space around the nucleus of an atom. [4.8]

electron configuration The symbolic representation of the electron arrangement in sublevels of an atom using the form $1s^22s^22p^63s^23p^6$ and so on. [5.10]

electronegative elements Elements—especially fluorine, oxygen, and nitrogen—that have a very strong attraction for electrons involved in a chemical bond. [7.3]

electronegativity The relative attraction of an atom in a molecule for the electrons in a covalent bond. [7.3]

electroplating The electrolysis process whereby a thin layer of a metal is deposited on the surface of another metal. [17.7]

element A material composed of only one kind of atom; a substance that cannot be decomposed into simpler substances by chemical or physical means. [2.3, 4.1]

empirical formula (simplest formula) The chemical formula that gives the simplest whole-number ratio of atoms of each element in a compound. [9.7]

emulsifying agent A substance that stabilizes emulsions. [14.9]

emulsion The type of colloid that is produced when a liquid is dispersed (not dissolved) in a liquid or solid. Examples: mayonnaise and butter. [14.9]

endergonic reaction A chemical reaction that proceeds only when it takes up or absorbs heat and/or other forms of energy (e.g., light or electricity). [2.7]

endothermic reaction A chemical reaction that proceeds only when it takes up or absorbs heat from the surroundings. [2.7, 11.8]

end point The point at which a titration is stopped because the indicator has changed color or another signal indicates the completion of the reaction. [16.12]

energy The capacity to do work or to transfer heat. [2.7]

energy levels of atoms The allowed specific energy values—or regions around the nucleus—that electrons can occupy in atoms. [5.4]

enthalpy change The change in heat energy (the energy released or absorbed) by a reaction carried out at constant pressure. [11.8]

enzyme A complex protein molecule that can act as a biological catalyst; each enzyme is highly specific for a particular biological reaction. [15.2, 20.3]

equilibrium (dynamic) The state of dynamic balance where the rates of forward and reverse processes are equal. [13.4, 15.3]

equilibrium constant The value, K_{eq} (at a particular temperature), obtained when equilibrium concentrations are substituted into the equilibrium constant expression. [15.6]

equilibrium constant expression A mathematical equation where the equilibrium constant, K_{eq}, equals the product of the equilibrium concentrations of reaction products divided by the product of the concentrations of reactants, each raised to a power equal to its coefficient in the balanced chemical equation. [15.6]

equivalence point The point of a titration at which stoichiometrically equivalent amounts of acid and base have reacted. [16.12]

essential amino acid One of eight amino acids that are not produced in the body and must be included in one's diet. [20.3]

ethanol CH_3CH_2OH; its common name is ethyl alcohol. [19.7]

ethers Organic compounds having the structure R—O—R. [19.9]

ethyl The alkyl group CH_3CH_2—, obtained by removing one hydrogen atom from ethane. [19.4]

evaporation The conversion of a volatile liquid to a gas (vapor). [13.4]

exact number One that has no uncertain digits, so it has an infinite number of significant figures. One obtained by direct count or by definition. [3.8]

excited state An atom with electron(s) that are in higher energy levels than their most stable state (called the ground state). [5.4]

exergonic reaction A chemical reaction that releases heat and/or other forms of energy (e.g., light, sound, electricity). [2.7]

exothermic reaction A chemical reaction that releases heat energy. [2.7, 11.8]

experiment Controlled investigation used to test or obtain facts, to test or establish a hypothesis, or to illustrate a known scientific law. [1.3]

extensive properties Properties that relate to the amount of a material present in a sample, including mass, volume, and length. [2.5]

factor-label method *See* dimensional analysis.

Fahrenheit scale The temperature scale on which water freezes at 32 °F and boils at 212 °F at standard pressure (1 atm). [3.12]

family of elements *See* group of elements.

fat-soluble vitamins Nonpolar vitamins—including A, D, E, and K—that dissolve in the fatty tissues of the body where they are stored for future use. [20.5]

fatty acid A carboxylic acid having a long hydrocarbon tail (usually with 10 to 20 carbon atoms). [20.2]

first ionization energy The amount of energy required to remove the outermost electron from a gaseous atom in its ground state. [6.4]

first law of thermodynamics *See* law of conservation of energy.

fluoresce To emit visible light when bombarded by UV radiation or other forms of high-energy radiation. [5.2]

fluorocarbons Compounds that contain both fluorine and carbon. [6.7]

foam The type of colloid that is produced when a gas is dispersed in a liquid or solid. Examples: shaving cream and marshmallows. [14.9]

formula unit The specific group of atoms or ions symbolized and expressed in the chemical formula. [4.11]

formula weight The sum of the atomic masses of all atoms in one formula unit of a substance (molecular or ionic), expressed in amu. [4.11, 9.1]

freezing The process that occurs when a liquid changes to a solid when cooled. [2.2, 13.6]

freezing point The temperature at which a substance changes from a liquid to a solid; the temperature at which the liquid and solid are in dynamic equilibrium. [13.6]

frequency The number of peaks of a wave that pass a point in one second. [5.2]

functional group A specific group of atoms—such as a carboxyl or hydroxyl group—that gives an organic compound certain characteristic properties. [19.7]

galvanic cell (also called a voltaic cell) An electrochemical cell that uses a spontaneous chemical reaction to generate an electric current. [17.8]

gamma rays High-energy radiation emitted by radioactive nuclei. [5.1, 8.1]

gas The state of matter that has neither definite shape nor definite volume. [2.2]

Gay-Lussac's law For a sample of gas with a constant volume, its pressure, P, is directly proportional to its Kelvin temperature, T. $P_1/T_1 = P_2/T_2$. [12.6]

Geiger counter A device used to detect and measure the rate of ionizing radiation. [18.4]

gram A metric unit of mass equal to 0.001 kg. The kilogram is the SI base unit of mass. (1 lb = 454 g.) [3.1, 3.5]

ground state The most stable electron arrangement in an atom. [5.4]

group of elements A vertical column of elements in the periodic table. [4.4, 6.2]

Haber process An industrial process for the chemical reaction of nitrogen gas with hydrogen gas in the presence of a catalyst to produce ammonia gas. [15.4]

half-cell One of two compartments that make up a galvanic (voltaic) cell. Each half-cell contains an electrode and an electrolyte. [17.8]

half-life The time required for one half of the atoms of an unstable isotope (nuclide) to undergo radioactive decay. [18.2]

halogen family Elements in Group VIIA of the periodic table. [4.4]

heat The form of energy that is transferred between samples of matter because of differences in their temperatures. [3.13]

heating curve The graphical plot of temperature (vertical axis) versus the heating time or energy (horizontal axis). The curve levels off at the melting point and at the boiling point. [13.7]

hemiacetal A carbon atom that is bonded to four different groups (—OH, —OR, —R, and —H); it is present in certain simple sugars called aldoses. [20.1]

heterogeneous mixture A mixture of substances that does not have uniform composition and properties throughout. [2.4]

homogeneous mixture A solution; a mixture that has the same composition and properties throughout. [2.4, 14.1]

homologs A series of compounds each of which differs from the next one in the series by some constant unit; e.g., —CH_2— for the alkanes. [19.2]

hormone A chemical messenger that is secreted into the blood by an endocrine (ductless) gland. Hormones regulate physiological processes such as metabolism, growth, and reproduction. [20.5]

Hund's rule Electrons within a sublevel of an atom do not pair up in an orbital until each orbital in that sublevel has one electron. Unpaired electrons in an orbital have parallel spins. [5.10]

hydrate A crystalline compound that contains a definite number of water molecules in each formula unit. [8.9]

hydration The process in which water molecules surround the solute particles being dissolved. [14.3]

hydrocarbons Organic compounds composed of only hydrogen and carbon. [19.2]

hydrogen bonding The strong attraction between molecules that have hydrogen atoms covalently bonded to atoms of fluorine, oxygen, or nitrogen. [7.14, 13.2]

hydrogenation The addition of H_2 to unsaturated organic compounds (those containing double or triple bonds). [19.5]

hydrolysis The reaction of a compound (especially a salt) with water to give a solution that is acidic, basic, or neutral. [16.10]

hydrometer A device that is calibrated so the specific gravity of a liquid can be determined directly as it floats in the liquid by reading the number on the hydrometer stem at the surface of the liquid. [3.11]

hydronium ion The H_3O^+ ion, which is a hydrogen ion (a proton) bonded to a water molecule. [8.8, 16.1]

hypothesis A tentative, reasonable explanation of facts or of a law. [1.3]

ice Water in the solid state (below its melting point). [2.2]

ideal gas A gas that obeys the gas laws exactly under all conditions. [12.10]

ideal gas law The equation, $PV = nRT$, used to determine any one of the variables, P, V, and T for a given sample of gas; R is the universal gas constant. [12.10]

immiscible Liquids that, when mixed, do not dissolve in each other to give a solution. [2.2, 14.2]

impermeable A material through which other materials cannot pass. [14.10]

infrared radiation Radiant heat. Electromagnetic radiation having wavelengths longer than visible light but shorter than microwaves that interact with matter to cause characteristic molecular vibrations. [5.2]

inner transition elements Elements in the two rows of elements represented at the bottom of the periodic table; they have partially filled f sublevels. [6.9]

inorganic chemical Any element or compound that is not classified as organic. (*See* organic chemical.) [p. 226]

insoluble Not soluble; a substance that does not dissolve to any easily detectable extent. [14.2]

intensive properties The characteristic physical and chemical properties (those not dependent on sample size) that are used to identify a substance. [2.5]

interionic forces The forces of attraction between ions in crystalline, ionic solids. [13.2]

intermolecular forces The forces of attraction between neighboring molecules. [13.1, 13.2]

intramolecular forces The forces that hold atoms together within a molecule due to chemical bonding. [13.2]

iodine number The number of grams of iodine that will react with 100 g of fat or oil; this gives a measure of the overall degree of unsaturation. [20.2]

ion The electrically charged particle formed when an atom, or group of atoms, either gains or loses electrons. [5.4, 6.3]

ion product of water The value obtained when the hydrogen ion concentration is multiplied by the hydroxide ion concentration in a solution. This value, K_w, is equal to 1.00×10^{-14} at 25 °C. [16.8]

ionic bond The attraction between ions having opposite charges. [7.1]

ionic compound A chemical substance that is made up of positive and negative ions. [7.1]

ionic equation A balanced chemical equation that shows all water-soluble ionic substances written in ionic form, while insoluble solids and covalently bonded substances are written in molecular form. [10.11]

ionic solid A crystalline substance composed of positive and negative ions. [13.5]

ionization energy The amount of energy required to remove an electron from a gaseous atom in its ground state. [5.4, 6.4]

ionizing radiation High-energy radiation capable of causing molecules to be ionized or torn apart into fragments. [18.3]

irradiation The exposure of a material to gamma radiation. [18.8]

isoelectronic Particles with the same total number of electrons, such as Na^+, Ne, and F^-. [6.3]

isomers Compounds that have the same molecular formula but different structural formulas; that is, the atoms are arranged and bonded differently. [19.2]

isopropyl The alkyl group $CH_3\overset{|}{C}HCH_3$, derived from propane by removing a hydrogen from its middle carbon, the site of attachment of a substituent group. [19.4]

isotonic solution A solution that exhibits the same osmotic pressure as that of the fluid within cells. [14.10]

isotopes Atoms of a particular kind of element that have different numbers of neutrons and, therefore, different mass numbers. [4.9]

IUPAC The abbreviation for the International Union of Pure and Applied Chemistry, which makes recommendations on chemical nomenclature. [19.4]

joule The SI base unit of heat energy. 1 calorie = 4.184 joules. [3.13]

Kelvin scale The temperature scale where temperatures are expressed in kelvins (K); the lowest temperature possible, absolute zero (−273.15 °C), is defined as 0 K. K = °C + 273.15. [3.12]

ketone An organic compound containing a carbonyl group bonded to two alkyl or aryl groups. Its general formula is $R{-}\overset{\overset{O}{\|}}{C}{-}R$. [19.10]

kindling temperature The temperature required for combustion to occur. [17.2]

kinetic energy (K.E.) The energy that objects or molecules possess because of their motion. K.E. = $\frac{1}{2}mv^2$. [2.7, 12.2]

kinetic molecular theory (KMT) A model that describes the behavior of ideal gases in terms of tiny particles that are in constant, random motion. [12.2]

lanthanides The group of fourteen elements with numbers 58 through 71. [6.9]

law (natural or scientific) A statement that summarizes experimental facts about nature where behavior is consistent and has no known exceptions. [1.3, 2.6]

law of combining volumes *See* Avogadro hypothesis.

law of conservation of energy (also called the first law of thermodynamics) Energy is neither created nor destroyed during chemical reactions. [2.8]

law of conservation of mass Mass is neither created nor destroyed (neither gained nor lost) during a chemical reaction. [2.6]

law of definite composition *See* law of definite proportions.

law of definite proportions A given compound always has a specific atom ratio and a specific mass ratio (a specific percentage by mass) of each element in the compound. [2.3, 4.6]

law of multiple proportions If two elements form more than one compound, the masses of one element that combine with a fixed mass of the second element are in a simple, whole-number ratio. [4.7]

Le Châtelier's principle If a stress (such as a change in concentration, pressure, or temperature) is applied to a system at equilibrium, the equilibrium shifts in the direction that will partially relieve the stress. [15.4]

Lewis electron-dot symbol The symbol of an element surrounded by one to eight dots to represent the valence electrons of an atom or ion of the element. [5.7]

limiting reagent The reactant that is totally consumed during a chemical reaction. It limits the quantities of products that can be formed. [11.6]

line spectrum A spectrum that consists of the discrete lines that result from wavelengths produced by atoms whose electrons are excited. [5.3]

lipids A variety of naturally occurring greasy substances, with varying structures and functions, that are more soluble in organic solvents than in water. Examples: fats, oils, phospholipids, and steroids. [20.2]

liquid The state of matter that has a definite volume but takes the shape of its container (except for having a flat upper surface). [2.2]

liter (pronounced LEE-ter) The metric (SI) volume of 1000 cm^3, which is slightly larger than one quart (1 L = 1.057 qt). [3.4]

London (dispersion) forces The weak forces of attraction between molecules due to shifts of electrons within nonpolar molecules that result in temporary dipoles. [13.2]

malleable Capable of being rolled or hammered into shape. [4.5]

mass A measure of the quantity of matter in an object. [2.1]

mass number The total number of protons and neutrons in an atom. [4.8]

matter Anything that has mass and, therefore, must also take up space. [2.1]

melting The process that occurs when a solid is changed to a liquid. [13.6]

melting point The temperature at which a substance changes from a solid to a liquid; the temperature at which the solid and liquid are in equilibrium. [13.6]

meniscus The crescent-shaped liquid surface caused by the attraction of a liquid for glass. (For mercury, the meniscus is inverted.) [3.8]

metallic bonding The bonding of metal atoms in solids where positive metal ions are arranged in a regular (fixed) three-dimensional array and the loosely held valence electrons are able to move freely throughout the crystal. [7.5]

metallic solid A solid that is composed of metal atoms. (*Also see* metallic bonding.) [13.5]

metalloids The elements whose symbols are positioned adjacent to the heavy stepped diagonal line on the periodic table, between metals and nonmetals. Their properties also lie between those of metals and nonmetals. [4.4]

metals The group of elements that are to the left of the heavy stepped diagonal line on the periodic table. [4.4]

metathesis reaction *See* double-replacement reaction.

meter The metric (or SI) base unit of length (approximately 39.37 in.). [3.1]

methane The compound whose formula is CH_4. It is a gas at room temperature and is the major component of natural gas. [7.11]

methanol CH_3OH; its common name is methyl alcohol. [19.7]

methyl The alkyl group CH_3—, obtained by removing one hydrogen atom from methane. [19.4]

metric system The decimal system of weights and measures based on the meter, the liter, and the kilogram. [3.1]

miscible Liquids that can dissolve in one another to form a solution. [2.2, 14.2]

mixture A material that is made up of two or more substances that can be in variable proportions but are not combined together chemically. [2.4]

moderator A substance used to slow down the fission neutrons within a nuclear reactor. [18.9]

molar heat of fusion The energy (usually in joules or calories) required to melt 1 mol of a solid. [13.6]

molar heat of vaporization The energy (usually in joules or calories) required to vaporize 1 mol of a liquid. [13.4]

molar mass (M.M.) The mass in grams of a mole of any substance (atoms, molecules, or formula units), that is, the sum of the atomic masses of all the atoms represented in the formula and expressed in grams. [4.11, 9.2]

molar volume The volume of one mole of gas, which is 22.4 L/mol at STP. [12.9]

molarity A concentration unit that gives the number of moles of solute per liter of solution, expressed in mol/L or M. [9.6, 14.7]

mole (mol) The amount of a substance whose mass in grams is numerically equal to the formula weight of the substance (and contains as many formula units as there are atoms in exactly 12 grams of the carbon-12 isotope). One mole can represent 6.022×10^{23} atoms, molecules, formula units, or ions. [4.11, 9.2]

molecular formula The chemical formula that gives the actual number of atoms of each kind present in a molecule of the substance. [4.11, 9.7]

molecular solid A solid compound that has discrete covalent molecules at each lattice point of the crystal. [13.5]

molecular weight (M.W.) The sum of the atomic masses of all atoms in one molecule of a particular compound. [4.11, 9.1]

molecule An electrically neutral cluster of two or more atoms joined together chemically by covalent bonds so it behaves as a single particle with a neutral electrical charge. [7.2]

monoprotic acid An acid that has only one ionizable (acidic) hydrogen atom per molecule. [16.1]

monosaccharide A simple sugar molecule (with structural characteristics of a carbohydrate) that combines with other sugar units to form complex sugars. [20.1]

mutarotation The interconversion (Latin *mutare*, "to change") of pure alpha or beta cyclic forms of simple sugars in solution to give an equilibrium mixture of the two forms. [20.1]

natural radioactivity The spontaneous emission of alpha particles, beta particles, and gamma rays by the disintegration of unstable atomic nuclei. [18.1]

net ionic equation A chemical equation that is obtained by omitting spectator ions from the ionic equation. [10.11]

neutralization The reaction of an acid with a base to produce a salt and water. [10.12, 16.1]

neutron An electrically neutral particle with a mass of 1 amu that is found in the nucleus of atoms. [4.8]

noble gases Elements in the right-hand column of the periodic table. [4.5]

nomenclature (chemical) The system of names and formulas used to identify all chemicals. [p. 216]

nonbonding electron pair A pair of valence electrons that is not involved in covalent bonding with other atoms in the molecule. [7.9]

nonmetals The group of elements that are to the right of the heavy stepped diagonal line on the periodic table. [4.4]

nonpolar covalent bond A chemical bond in which one or more pairs of electrons are shared equally between two atoms of the same element. [7.2]

nuclear equation The symbolic representation that shows the target nucleus and bombarding particle on the left side of an equation with the product nucleus and ejected particle(s) on the right side. [18.1]

nuclear fission A radioactive decay process in which certain heavy nuclei can absorb a neutron and then split apart to give lighter nuclei along with the release of several neutrons and energy. [18.9]

nuclear fusion The joining together of two small nuclei to produce one larger nucleus with some of their mass being converted to energy. [18.11]

nucleic acid A polymer—present in every living cell—of repeating units call nucleotides. DNA and RNA are nucleic acids. (*Also see* nucleotide.) [20.4]

nucleotide The monomer unit of a nucleic acid. Each unit is composed of a five-carbon sugar (ribose or deoxyribose), a phosphate, and a heterocyclic amine base. [20.4]

nucleus The tiny dense center of an atom that contains an atom's protons and neutrons. The diameter of a nucleus is about 1×10^{-15} m. [4.8]

octet of electrons Used to describe the complete set of eight valence electrons that are present in the outermost energy level of noble gases that follow helium in the periodic table. [5.7]

octet rule The tendency for a nonmetal atom to gain or share electrons until it has eight valence electrons. [7.2]

orbital A region within an atom's sublevel that can be occupied by a maximum of two electrons that have opposite spin. There are *s*, *p*, *d*, and *f* orbitals. [5.5, 5.8]

organic chemical A covalently bonded compound that contains carbon. [p. 226]

osmosis The selective passage of a solvent through a semipermeable membrane. The direction of solvent passage is from the compartment with lower solute concentration to that of high solute concentration. [14.10]

osmotic pressure The amount of pressure needed to prevent osmosis. [14.10]

oxidation The process that occurs when a substance combines with oxygen, or any chemical process where there is a loss of electrons. [7.2, 10.8, 17.2]

oxidation half-reaction The balanced half-reaction that is written to show the loss of electrons that occurs during oxidation. [17.6]

oxidation number (also called oxidation state) The charge of a simple ion or the "apparent charge" assigned to an atom within a compound or polyatomic ion. [8.7, 17.1]

oxidation–reduction (redox) reaction A chemical reaction that involves the transfer of electrons from one substance to another. [10.8]

oxidizing agent The reactant that accepts electron(s) and becomes reduced when another substance is oxidized. [10.8, 17.3]

oxyacid An acid whose anion contains one or more oxygen atoms. [8.8]

partial pressure The pressure exerted independently by a particular gas present in a mixture of gases. [12.11]

parts per billion (ppb) The number of parts (regardless of the unit used) present in a billion parts (using the same unit); a concentration of 1 ppb = 1 microgram per liter. [14.7]

parts per million (ppm) The number of parts (regardless of the unit used) present in a million parts (using the same unit); a concentration of 1 ppm = 1 milligram per liter. [14.7]

parts per trillion (ppt) The number of parts (regardless of the unit used) present in a trillion parts (using the same unit); a concentration of 1 ppt = 1 microgram per milliliter. [14.7]

Pauli exclusion principle The key point of this rule is that there can be no more than two electrons in any orbital of an atom, and when two electrons occupy the same orbital, they must have *opposite* spins. [5.8]

peptide bond The amide linkage that joins amino acids in chains to build polypeptides and proteins. [20.3]

percent by mass The mass of solute divided by the total mass of solution times 100%. [14.7]

percent by volume The volume of solute divided by the total volume of solution times 100%. [14.7]

percent by weight The number of grams of a particular substance that can be found in 100. grams of the sample. [9.3]

percent composition A listing of the percents by mass (weight) of each element in a compound. [9.3]

percent yield The mass of the actual yield divided by the mass of the theoretical yield multiplied by 100%. [11.7]

period of elements A horizontal row of elements in the periodic table. [4.4, 6.2]

periodic law There is a periodic variation in the physical and chemical properties of elements when they are arranged in order of increasing atomic number. [6.1]

periodic table An arrangement of the elements according to increasing atomic numbers so that elements with similar chemical properties are in the same column. A periodic table is shown inside the front cover of this textbook. [6.1]

permeable Materials that have extremely small holes or pores that allow the passage of small particles (ions and small molecules). [14.10]

pH A method of expressing the hydrogen ion concentration (the acidity) of a solution where pH = $-\log [H^+]$. [16.9]

phenols Organic compounds that have a hydroxyl group, —OH, covalently bonded to a benzene ring. Phenols have the general formula, Ar—OH. [19.8]

phenyl group The aryl group C_6H_5—, obtained by removing one hydrogen atom from a benzene ring. [19.6]

photon A tiny packet of electromagnetic radiation. [5.3]

physical change A change in which the composition of a substance is not affected. [2.5]

physical properties Characteristic properties of a substance that identify the substance without causing a change in its composition. Physical properties (e.g., color and odor) do not depend on the quantity of the substance. [2.5]

physical states *See* states of matter.

plasma A high-energy state of matter similar to a gas but composed of isolated electrons and nuclei rather than discrete whole atoms or molecules. [18.11]

polar covalent bond A covalent bond in which electrons are not shared equally because of differences in electronegativities (unequal attractions) of the atoms joined by the bond. [7.4]

polar molecule A molecule in which polar covalent bonds are in an unsymmetrical three-dimensional arrangement about a central atom. [7.6–7.12]

polyatomic ion A cluster of two or more atoms that are covalently bonded together and carry an overall charge and act as a single particle. [8.2]

polyhydroxy alcohols Alcohols that contain more than one —OH group. [19.7]

polymers Giant molecules made by linking together many small molecules. [19.5]

polypeptide A polymer of amino acids linked together by peptide bonds; usually of lower molar mass than that of a protein. (*Also see* peptide.) [20.3]

polysaccharide A carbohydrate such as starch or cellulose that is composed of more than ten simple sugars. [20.1]

polyunsaturated fat An oily liquid—typically a vegetable oil—that contains a relatively high proportion of fatty acids with two or more double bonds. [20.2]

positron A particle with a positive charge and the mass of an electron. [18.7]

potential energy The stored energy an object possesses because of its position (e.g., on a hill) or its chemical composition (chemical bonding). [2.7]

precipitate A solid that forms and separates out of solution as the result of a chemical reaction. [10.10]

precision The degree to which repeated measurements of a quantity are in agreement. Measurements in close agreement are said to have good precision. [3.7]

primary alcohol An alcohol with its —OH group attached to a primary carbon (a carbon attached to only one alkyl or aryl group and two hydrogen atoms). [19.7]

primary protein structure The sequence of amino acids in proteins. [20.3]

principal energy level Any one of the main energy levels of an atom, designated by the letter n, and assigned a whole number 1, 2, and so on. [5.6]

products The substances formed in a chemical reaction; they appear on the right-hand side of a chemical equation. [10.1]

proof (alcohol) A number that is twice the percentage of the ethanol content by volume. [19.7]

propyl The alkyl group $CH_3CH_2CH_2$—, derived from propane by removing a hydrogen atom from the carbon at the end of the three-carbon chain. [19.4]

protein A polymer of amino acids joined by peptide bonds and having a specific biological function and a molar mass of 10,000 or more. [20.3]

protium The most prevalent isotope of hydrogen; it has only a proton (no neutrons) and a mass number of 1. [4.9]

proton The positively charged particle with a mass of 1 amu found in the nucleus of all atoms. The number of protons determines an element's identity. [4.8]

pure substance A single chemical—an element or a compound—composed of the same kind of matter with the same kind of particles throughout. [2.3]

pycnometer (also called a specific gravity bottle) A small bottle—with a narrow marked opening—used to repeatedly measure a fixed volume precisely for the determination of the density or specific gravity of a liquid. [3.11]

quantum(a) The smallest countable, discrete packet or increment of radiant energy that can be absorbed or emitted. [5.3]

quantum mechanics The theory of atomic structure that is based on the wave properties of matter and the probability of finding electrons in specific energy levels and sublevels. [5.5]

rad The rad—taken from the first letters of radiation absorbed dose—is the unit for the quantity of ionizing radiation that results in the absorption of 100 ergs of energy per gram of absorbing tissue. [18.3]

radioactivity The spontaneous emission of certain types of radiation (e.g., alpha, beta, and gamma radiation) by unstable atomic nuclei. [5.1]

radioisotope A radioactive isotope (nuclide). [18.2]

rate of reaction A measure of the change in concentration per unit time. (*Also see* chemical kinetics.) [15.1]

reactants The starting materials that enter into a chemical reaction; they appear on the left-hand side of a chemical equation. [10.1]

reaction mechanism The individual step-by-step small reactions that show how molecules interact to give an overall reaction. [15.2]

real gas A gas that actually exists. (*Also see* ideal gas.) [12.10]

reducing agent The reactant that donates the electron(s) and becomes oxidized when another substance is reduced. [10.8, 17.3]

reducing sugar A sugar that must have available an aldehyde group that can be oxidized to an acid by a mild oxidizing agent such as the Cu^{2+} ion. [20.1]

reduction The process that occurs when a substance releases oxygen, or any chemical process where there is a gain of electrons. [7.2, 10.8, 17.3]

reduction half-reaction The balanced half-reaction that is written to show the gain of electrons that occurs during reduction. [17.6]

rem The rem—taken from the first letters of roentgen equivalency in man—is the unit for the quantity of biological damage produced by a particular dose of radiation. The rem takes into account both the amount of radiation and the kind of radiation. A dose of 3.2 rems, for example, can cause biological damage equivalent to 3.2 rads of radiation delivered by a standard sample. [18.3]

representative elements Elements in the A groups (the first two and last six columns) of the periodic table. [5.9]

resonance structures Two or more Lewis structures having the same arrangements of atoms but different arrangements of electrons. [7.7]

reversible chemical reaction A chemical reaction that can proceed in either direction, depending on the reaction conditions. [15.3]

ribonucleic acid (RNA) The form of nucleic acid found mainly in the cytoplasm but also present in all parts of the cell; it is involved in protein synthesis. [20.4]

roentgen (R) A unit of exposure to gamma rays or X rays. (1 R will produce ions carrying a total of 2.1 billion units of electrical charge in 1 cm^3 of dry air at 0 °C and a pressure of 1 atm.) [18.3]

salt Any ionic compound that does not contain hydroxide, OH^-, or oxide, O^{2-}, ions. During acid–base neutralization, a salt is formed that contains the cation from the base and the anion from the acid. [8.8, 10.12, 16.1]

salt bridge A tube—filled with an aqueous solution or gel containing a strong electrolyte—that is used to connect two half-cells of a galvanic cell. [17.8]

saturated fat An ester of glycerol and fatty acids having relatively high proportions of saturated fatty acids. (*Also see* saturated hydrocarbon.) [20.2]

saturated hydrocarbon An organic compound made up of carbon and hydrogen atoms where all carbon-to-carbon bonds are single bonds. [19.2]

saturated solution A solution that contains as much dissolved solute as it can hold at a given temperature while in equilibrium with undissolved solute. [14.5]

scientific law *See* law.

scientific method The process of following specific procedures when solving problems or carrying out planned (research) investigations. It involves making observations and tests to develop laws, hypotheses, and theories. [1.3]

scientific notation The exponential notation form where a number is expressed as a decimal between 1 and 10 that is multiplied by 10 raised to the appropriate power, for example, 6.022×10^{23} particles per mole. [3.10]

scintillation counter An instrument that detects the tiny flashes of light produced when radiation strikes materials called phosphors. [18.4]

secondary alcohol An alcohol with its —OH group attached to a secondary carbon (a carbon attached to two alkyl or aryl groups and one hydrogen atom.) [19.7]

secondary protein structure The coiled or pleated-sheet arrangement formed by a polypeptide chain or protein. [20.3]

semipermeable Materials that have minute holes or pores large enough to allow the passage of small particles but not large particles. [14.10]

significant figures All certain digits of a measurement plus one additional rounded off or estimated digit (called an uncertain digit). [3.8]

single bond A covalent bond in which a single pair of electrons is shared between two atoms. [7.2]

single-replacement reaction An oxidation–reduction reaction in which one metal (or nonmetal) replaces the ion of another metal (or nonmetal). [10.4, 10.8]

solid The state of matter that has a definite shape and volume. [2.2]

solubility A measure of how much solute will dissolve in a given quantity of solvent. [14.2]

soluble (substance) A substance that dissolves to an appreciable extent. [14.2]

solute The substance being dissolved (the substance present in a solution in the smaller amount). [9.6, 14.1]

solution A homogeneous mixture of two or more substances. [2.4, 14.1]

solvation The process in which solvent molecules surround the solute particles being dissolved. [14.3]

solvent The component of a solution whose physical state is retained (also the substance present in the greater amount). [9.6, 14.1]

specific gravity The value (ratio) obtained by dividing the density of a substance by the density of water at the same conditions; it has no units. [3.11]

specific heat The amount of heat needed to raise the temperature of 1 gram of a substance by 1 °C. The units are in J/(g-°C) or cal/(g-°C). [3.13]

spectator ion An ion that is present during a chemical reaction but does not undergo change during the reaction. Spectator ions are not included when an ionic equation is simplified to give a net ionic equation. [10.11]

spectroscope An instrument used to separate visible light into discrete wavelengths having "the colors of the rainbow." [5.3]

spontaneous combustion The rapid oxidation that occurs when a substance begins to burn on its own without being ignited by a spark or flame. [17.2]

standard acid An acid with a precisely known concentration. A standard acid is used in a titration to determine the concentration of a base. [16.12]

standard base A base with a precisely known concentration. A standard base is used in a titration to determine the concentration of an acid. [16.12]

standard conditions *See* standard temperature and pressure.

standard pressure A pressure of 1 atmosphere, or 760 torr. [12.3, 12.7]

standard temperature and pressure (STP) A temperature of 273 K (0 °C) and a pressure of 1 atmosphere (760 torr). [12.7]

states of matter (also called physical states) The three physical states—solid, liquid, and gas—in which matter can exist, depending upon temperature. [2.2]

steam Invisible water vapor at a high temperature. [2.2]

steroids A class of lipids having a complex four-ring structure (three rings with six carbon atoms and one ring with five carbon atoms) and specific biological activity. [20.5]

stoichiometry Calculations that deal with the amounts of materials and the energy changes that are involved in chemical reactions. [p. 317]

strong acid An acid that is completely ionized in water. [16.2]

strong base A metal hydroxide that dissociates completely in water. [16.3]

structural formulas Chemical formulas that show which atoms are joined to each other by using a single bond line for a single bond between atoms, double lines for double bonds, and three parallel lines for triple bonds. [19.3]

subatomic particles Any one of over 100 particles that are smaller than an atom. The three main subatomic particles are electrons, protons, and neutrons. [4.8]

sublevels of atoms The *s*, *p*, *d*, and *f* sublevels of atoms with 1, 3, 5, and 7 orbitals, and up to 2, 6, 10, and 14 electrons, respectively. [5.8]

sublimation The process that occurs when a substance is transformed directly from a solid to a vapor (gas) without passing through the liquid state. [6.7]

superheated steam Water vapor that is heated above the boiling point. [13.7]

supersaturated solution A solution that contains more dissolved solute at a given temperature than could be present in a saturated solution in equilibrium with excess solute. [14.6]

surface tension The force of attraction that causes the surface of a liquid to contract and form a "bead" or spherical drop. [13.3]

surfactants Chemicals that reduce the surface tension of water and, thereby, increase its wetting action. [13.3]

suspension A mixture of substances that are temporarily dispersed one in the other, but when left alone will separate to give the component parts. [14.9]

synthesis reaction The reaction of elements to produce a compound, symbolized $A + B \rightarrow AB$ (also called a combination reaction), or the planned preparation of a specific compound. [10.4, 10.6]

temperature A measure of the hotness or coldness of matter, expressed usually in degrees Fahrenheit, degrees Celsius, and kelvins. [3.12]

tertiary alcohol An alcohol with its —OH group attached to a tertiary carbon (a carbon attached to three alkyl or aryl groups). [19.7]

tertiary protein structure The overall three-dimensional globular shape taken on by a protein as it folds around on itself. [20.3]

tetrahedron A three-dimensional geometric shape having four triangular faces. A *tetrahedrally shaped* molecule is one having four atoms positioned at the four points of a tetrahedron and bounded to a central atom at an angle of 109°. [7.8]

theoretical yield The maximum quantity of a substance that can be produced by the complete reaction of all of the limiting reagent in accordance with the chemical equation. [11.7]

theory A hypothesis that has withstood extensive testing. [1.3, 4.7]

titration The laboratory procedure used to determine the concentration of a solution by slowly adding one solution to another solution. When equivalent quantities have reacted, and when the volumes of both solutions and the concentration of one of the solutions is known, the concentration of the other solution can be calculated. [16.12]

toluene The organic compound having the formula $C_6H_5—CH_3$; it has a benzene ring with a methyl group replacing one hydrogen atom of the benzene. [19.6]

tracers Radioactive isotopes used to trace movement or to locate the site of radioactivity in either living or nonliving materials. [18.8]

trans configuration The geometrical isomer in which two groups in reference are positioned on opposite sides of a double bond or ring arrangement. [19.5]

transcription The process by which DNA directs the synthesis of an mRNA molecule during protein synthesis. [20.4]

transition elements Elements in the central region of the periodic table—between the two left-hand and six right-hand columns—with inner *d* orbitals that are partially filled. [5.9]

transmutation The process that occurs when atoms of one kind of an element are changed into atoms of a different element. [18.1]

transuranium elements All elements that follow uranium (atomic number 92) in the periodic table. These elements are all synthetic and radioactive. [6.9]

triglycerides The simple lipids (esters of glycerol with three long-chain fatty acids) and related compounds. [20.2]

triol A polyhydroxy alcohol whose structure has three —OH groups. [19.7]

triple bond A covalent bond where three pairs of electrons are shared between two atoms. [7.2]

triprotic acid An acid that has three ionizable (acidic) hydrogen atoms per molecule. [16.1]

tritium The radioactive isotope of hydrogen with 1 proton, 2 neutrons, and a mass number of 3. [4.9]

Tyndall effect The scattering of light to give a visible beam when viewed from the side. The effect is due to the presence of colloidal particles that scatter and reflect light off to the side. [14.9]

ultraviolet radiation Electromagnetic radiation—sometimes called *black light*—having high frequencies and wavelengths that lie between those of visible light and X-rays. [5.2]

uncertainty principle (Heisenberg's) We are limited, mathematically, in our ability to determine both the speed (or momentum) and position of an electron, so its path has a degree of uncertainty. [5.5]

universal gas constant (*R*) The constant, *R*, in the equation $PV = nRT$ has the value of 0.0821 L-atm/mol-K. [12.10]

unsaturated organic compound Any organic compound containing carbon-to-carbon double or triple bonds; addition of H_2 gives a *saturated* compound. [19.5]

unsaturated solution A solution that contains a smaller amount of solute than it can hold at a given temperature. [14.5]

valence electrons The electrons in the outermost occupied energy level of an atom. There can be from one to eight valence electrons. [5.4]

valence shell electron pair repulsion theory A theory used to predict the shape of a molecule by looking at the structure produced when all valence electron pairs surrounding a central atom in a molecule repel each other and stay as far apart as possible. [7.8]

van der Waals forces The intermolecular forces of attraction, that is, those that exist between neighboring molecules. [13.2]

vapor pressure The partial pressure exerted by a vapor above a liquid that is in equilibrium with the liquid. [12.11]

vaporization The process by which molecules of a volatile liquid break away or escape and go into the gas (vapor) phase. [13.4]

viscosity A measure of the resistance of a liquid to flow; the higher the viscosity, the slower is the rate of flow. [2.2, 13.3]

visible spectrum All wavelengths of the electromagnetic spectrum that are visible (red, orange, yellow, green, blue, indigo, and violet). [5.2]

vitamins Organic compounds required in the diet in small amounts for proper body functioning; they cannot be produced in the body in sufficient quantities. The absence of a vitamin results in a vitamin deficiency disease. [20.5]

volt The unit of electrical potential (tendency of electrons in a system of flow); a volt is one joule of work per coulomb of charge transferred. [17.8]

voltaic cell *See* galvanic cell.

water of hydration *See* hydrate.

water-soluble vitamins Polar vitamins—including C and the B complex—that are able to form hydrogen bonds with water and dissolve. [20.5]

wavelength The distance between peaks or valleys of consecutive waves. [5.2]

weak acid An acid that ionizes only slightly (less than 5%) in water. [16.2]

weak base A compound that reacts with water to produce hydroxide ions to only a slight extent (less than 5%). [16.3]

weight The force of gravity acting on the mass of a particular object. [2.1]

wetting action The ability of a liquid, especially water, to spread out evenly over a surface as a thin film. The wetting action is increased when the surface tension is decreased. [13.3]

work The movement of a mass through a distance. [2.7]

X-rays Penetrating electromagnetic radiation with a higher energy (higher frequency and shorter wavelength) than ultraviolet radiation; produced by bombarding a metal with high-energy electrons. [5.1, 18.1]

zwitterion An inner salt formed when an anion and a cation are parts of the same molecule. At body pH, amino acids exist as zwitterions. [20.3]

APPENDIX E

Solutions to Exercises and Answers to Odd-Numbered Problems

Chapter 1

1.1 All items.

1.3 There are many possible answers.

1.5 Chemistry deals with (a) the characteristics, (b) the composition, and (c) the chemical changes that occur in all materials.

1.7 The rusting of iron.

1.9 Benefit: Burning fossil fuels provides energy for homes and industry.
Problem: Gaseous products of burning cause acid rain, an environmental problem. Chemists are studying these effects, reporting the problems, and suggesting solutions to save the enivronment.

1.11 Chemists are developing chemicals that prevent insects, plant disease, and weeds from destroying crops.

1.13 Chemicals for high technology include materials for computer chips, digital displays, semiconductors, and superconductors, materials for audio tapes, laser disks and small movie cameras, and so on.

1.15 The chemical industry is the fifth largest in the United States and ranks first in worker safety.

1.17 See end of Section 1.2.

1.19 Researchers use controlled investigations (experiments) to test or obtain facts, to test a hypothesis, or even to illustrate a scientific law. A natural law summarizes experimental facts mathematically or in words.

1.21 A hypothesis is a tentative explanation of the facts; a theory is a hypothesis that has withstood extensive testing.

1.23 Chemists in applied research are involved Chemists in the development of products—and solving problems related to the development of products—for business, industry, and the rest of society. Chemists in basic research are not looking for new products but, instead, are searching for solutions to unanswered questions. Much basic research is used by applied researchers.

1.25 a. basic research b. applied research

1.27 Chemistry is a central science that relates to all other sciences. More important, learning how to solve problems in chemistry is a good background for approaching and solving many types of technical problems.

Chapter 2

Ex. 2.1 Because the gravity on Mars is 0.38 of the gravity of Earth, your weight on Mars is equal to 0.38 times your own weight. (Although your weight would be different, your mass would be the same on Mars as it is on Earth.)

Ex. 2.2 There is no overall mass loss during the burning of a log or in any other chemical change. The sum of the masses of the wood and oxygen used to burn the wood equals the sum of the masses of carbon dioxide, water vapor, and any unburned wood.

Ex. 2.3 When butane—or any fuel—burns, heat energy is released. The reaction is exothermic. During combustion, the high-energy fuel molecules release stored-up energy and are converted to low-energy molecules of carbon dioxide and water.

2.1 Earth's gravity is greater than the moon's, so the rock's weight is greater on Earth. The rock's mass is not affected by gravity. See Example 2.1.

2.3 a and c contain matter; b and d do not.

2.5 immiscible.

2.7 solid.

2.9 a. liquid b. liquid

2.11 See Table 2.3. Hydrogen and oxygen are colorless gases, but water is a liquid at room temperature.

2.13 See Table 2.3.

2.15 an atom

2.17 compounds

2.19 a compound

2.21 Bronze is homogeneous.

2.23 a homogeneous mixture—a solution

2.25 an element

2.27 a–c, e. physical properties d. chemical properties

2.29 a, d. chemical change b, c. physical change

2.31 a, d. chemical properties b, c. physical properties

2.33 Iron combines with oxygen in air to produce rust. The rust must have a mass equal to the sum of the mass of the iron plus the mass of the oxygen.

2.35 During heating the limestone breaks down (decomposes) to give a solid substance and a gas.

2.37 the diver on the 10-m platform

2.39 a, b, e. exothermic c, d. endothermic

2.41 A scientific law is a general statement that summarizes experimental data regarding natural phenomena that occur without variation.

2.43 The energy released is the difference between the chemical energy of the substances produced (the products) and the chemical energy of the substances that react (the wood and the oxygen).

2.45 The energy released is due to the difference in chemical energy of the substances present before and after the reaction.

2.47 A chemical change occurs when wood burns. Energy is neither created nor destroyed.

2.49 c

2.51 a, c. homogeneous b, d. heterogeneous

2.53 Liquid water has more energy than ice.

2.55 The kinetic energy of motion of the car is transferred into heat energy—as the brakes get hot—and into frictional energy that wears off the brake pads and tires. Energy is conserved as it is transformed from kinetic energy into heat and friction.

Chapter 3

Ex. 3.1 Plan: Days → hours → minutes

$$4.5 \text{ days} \times \frac{24 \text{ hr}}{1 \text{ day}} \times \frac{60 \text{ min}}{1 \text{ hr}} = 6480 \text{ min}$$

Ex. 3.2 Plan: km → m → cm

$$0.00451 \text{ km} \times \frac{1000 \text{ m}}{1 \text{ km}} \times \frac{100 \text{ cm}}{1 \text{ m}} = 451 \text{ cm}$$

Ex. 3.3 Answer (4). A quart is a little less than a liter, so 1 gal (4 qt) is a little less than 4 L. Answer (4), 3.8 L, is the approximate volume of a gallon.

Ex. 3.4 $0.075 \text{ L} \times \frac{1000 \text{ mL}}{1 \text{ L}} = 75 \text{ mL}$

Ex. 3.5 Answer (3). A petite cheerleader would be expected to have a mass that is about half as great as the 100-kg (220-lb) football player, or 50 kg.

Ex. 3.6 $0.497 \text{ g} \times \frac{1000 \text{ mg}}{1 \text{ g}} = 497 \text{ mg}$

Ex. 3.7 Plan: ft → in. → cm → m

$$\text{Height in ft} \times \frac{12 \text{ in.}}{1 \text{ ft}} \times \frac{2.54 \text{ cm}}{1 \text{ in.}} \times \frac{1 \text{ m}}{100 \text{ cm}} = \text{Height in meters}$$

Ex. 3.8 Plan: L → mL → g

$$2.5 \text{ L} \times \frac{1000 \text{ mL}}{1 \text{ L}} \times \frac{0.67 \text{ g}}{\text{mL}} = 1.7 \times 10^3 \text{ g or } 1.7 \text{ kg}$$

(The answer is in two significant figures.)

Ex. 3.9 °F = (1.8 × °C) + 32
°F = [1.8 × (−10)] + 32
°F = −18 + 32 = 14°F

3.1

Numerical	*Unit*	*Substance*
a. 10	lb	Sugar
b. 5	kg	Potatoes
c. 2	L	Cola

3.3 a. 0.001, 0.001 b. 0.000001, 0.000001 c. 100, 100
3.5 a. (2) b. (3)
3.7 a. 1.820 km b. 0.01400 km c. 1.700 m
3.9 a. 125 mm b. 3.45 m c. 3.45×10^4 μm d. 1.05 cm e. 4250 cm f. 92 mm
3.11 a. (2) b. (3) c. (2)
3.13 a. 0.025 L b. 5 μL c. 5.0×10^{-2} mL d. 750. cm^3
3.15 60. cm^3
3.17 2000 dm^3
3.19 a. $0.895/L in bottles b. $0.937 in cans c. the cola in 2-L bottles
3.21 a. (2) b. (3)
3.23 a. 1×10^{-4} g b. 250 mL c. 500 μg d. 0.524 g
3.25 a. 6.50 in. b. 1.27 qt c. 65.9 kg d. 45.7 cm e. 16.9 fl oz f. 0.606 lb
3.27 141 kg
3.29 2.99×10^8 m/s
3.31 24.6 m/s
3.33 2.31 days
3.35 a. 5 b. 3 c. 1 d. 2 e. uncertain f. 4
3.37 a. 86.0 b. 30.0 c. 6.13 d. 0.00823
3.39 a. 7.0×10^{-6} b. 2.53×10^5 c. 8.25000×10^5 d. 8.267×10^{-3}
3.41 a. 169.6 g b. 83 mm^3 c. 1.18×10^5
3.43 46 kg
3.45 630 mL
3.47 a. 8.95 g/mL b. copper c. Another metal or alloy might have a density close to that of copper.
3.49 38 mL
3.51 1.34
3.53 a. 20 °C b. 102 °F c. 312 K d. −23 °C e. 14 °F
3.55 −321 °F
3.57 100 °C (or 212 °F)
3.59 1500 cal
3.61 3.14×10^3 J
3.63 a. 8.45×10^5 J b. 1.2×10^4 g
3.65 a. 1.5×10^3 cal/hr-kg b. 6.3×10^3 J/hr-kg
3.67 232 $in.^3$
3.69 7.15×10^4 g

Chapter 4

Ex. 4.1 (a) potassium, (b) copper, (c) iron, (d) silver.
Ex. 4.2 Magnesium, a metal, and phosphorus, a nonmetal, are in the same period (period 3, the third row) of the periodic table.
Ex. 4.3 The only two elements that are liquids at room temperature are mercury (a metal) and bromine (a nonmetal).
Ex. 4.4 Berzelius and Proust both conducted experiments that showed that elements combine in definite proportions to form compounds.
Ex. 4.5 (a) Number of protons = mass number − number of neutrons = 35 − 18 = 17
(b) Number of electrons = number of protons = 17
(c) Atomic number = number of protons = 17
(d) The element with 17 protons (see periodic table) is chlorine.
Ex. 4.6 (a) The number of protons—the atomic number—is the lower number shown at the left side of the symbol, Am, which is 95.
(b) Number of neutrons = mass number − number of protons = 241 − 95 = 146
Ex. 4.7 1 × atomic mass of Ca = 40.1 amu
2 × atomic mass of O = 32.0 amu
2 × atomic mass of H = 2.0 amu
Formula weight = 74.1 amu
1 mol $Ca(OH)_2$ = 74.1 g

$$2.47 \text{ mol } Ca(OH)_2 \times \frac{74.1 \text{ g}}{1 \text{ mol } Ca(OH)_2} = 183 \text{ g}$$

4.1 Alchemists were the experimentalists of the Middle Ages.
4.3 An element is a substance that cannot be broken down into simpler substances.
4.5 Antoine Lavoisier
4.7 Technetium.
4.9 J. J. Berzelius
4.11 a. Na b. Mg c. Cr d. Fe e. Hg f. Ag
4.13 a. strontium b. bromine c. chlorine d. tin e. tungsten f. lead
4.15 See Table 4.1.
4.17 hydrogen

4.19 a, c. metal b, e. nonmetal d, f. metalloid
4.21 Ar_2, Cu_2
4.23 hydrogen, nitrogen, oxygen, fluorine, and chlorine; all nonmetals
4.25 tungsten
4.27 phosphorus
4.29 Matter is atomistic.
4.31 J. Priestley and W. Scheeler; Priestley is usually credited with the discovery.
4.33 During combustion, substances combine with oxygen of the air.
4.35 the law of definite proportions
4.37 A scientific model is a description—also a picture—or theory used to explain experimental observations.
4.39 a. the law of definite proportions
b. the law of multiple proportions and the law of definite proportions
c. The Cl mass ratio is 2:1.
4.41 24.0 g oxygen
4.43 2.4×10^{11} H atoms
4.45 A scientific law merely summarizes experimental fact, often in mathematical form.
4.47 A proton has a positive charge and a mass number of 1 amu. An electron has a negative charge and a mass number of 0 amu. A neutron has no charge but has a mass number of 1 amu.
4.49 protons
4.51 element, protons, electrons (in that order):
a. Na, 11, 11 b. Ra, 88, 88 c. N, 7, 7
d. F, 9, 9
4.53 a. no b. yes c. 21 amu d. 20 amu
4.55 ${}^{2}_{1}H$, atomic number = 1, mass number = 2, 1 proton, 1 neutron, and 1 electron.
4.57 a. 38 b. 52 c. 38
4.59 63.5451
4.61 12.0 g, 6.02×10^{23} atoms
4.63 a. 100.1 g/mole b. 80.0 g/mole
c. 164.0 g/mole
4.65 a. 122 g $CaCO_3$ b. 97.6 g NH_4NO_3
c. 200. g Na_3PO_4
4.67 a. Co is the symbol for cobalt; CO is the formula for carbon monoxide.
b. Pb is the symbol for lead; PB is the formula for a compound of P and B.
4.69 Only 11 g of carbon dioxide can be produced.
4.71 10^7 atoms
4.73 ${}^{99}_{43}Tc$ has 43 protons, 56 neutrons, and 43 electrons; atomic number = 43, mass number = 99.
4.75 6.02×10^{23} molecules

Chapter 5

Ex. 5.1 (a) infrared, (b) radar or microwaves, (c) visible and ultraviolet light. These forms of energy and the way they interact with matter are compared in Figure 5.15.
Ex. 5.2 A photon is a small packet or quantum of light energy. As shown in quantum theory, the frequency of light increases proportionally with an increase in the energy of the photons of light.
Ex. 5.3 Phosphorus, atomic number 15, has 15 electrons. Two of these electrons fill the first energy level, 8 go into the second energy level, and the remaining 5 go into the third energy level.

P) $2\,e^-$) $8\,e^-$) $5\,e^-$

The outer 5 electrons are valence electrons.
Ex. 5.4 The Lewis electron-dot symbols for phosphorus (5 valence electrons), potassium (1 valence electron), and barium (2 valence electrons) are as follows.

:P· (with dots above and below) K· ·Ba·

Ex. 5.5 Oxygen has 8 electrons. Two are in the 1*s* sublevel and 2 are in the 2*s* sublevel. The remaining 4 electrons go in the three orbitals of the 2*p* sublevel. In the orbital diagram, the 2*p* electrons are shown to be distributed according to Hund's rule.

	Electron Configuration	*Orbital Diagram*
		2*s* 2*p*
Oxygen	$1s^2 2s^2 2p^4$	[He] ⇅ ⇅ ↑ ↑

5.1 Berzelius, like Dulong and Petit, determined atomic masses of metals from their specific heats.
5.3 A beam of light, called a cathode ray, passed between the cathode and the anode in a straight line.
5.5 a modern fluorescent tube
5.7 J. J. Thomson; knowing this ratio was a step toward knowing the mass of the electron, which could be calculated once the charge of the electron was determined.
5.9 Goldstein used a metal disk filled with holes; he observed rays in the region behind the cathode that had a positive charge.
5.11 the charge of an electron
5.13 X-rays
5.15 See Figure 5.9 and Table 5.1.

5.17

Observation	*Conclusion*
(1) Most of the alpha particles went through the gold foil.	The mass is concentrated into a tiny nucleus; the atom is mostly empty space.
(2) Some of the alpha particles were deflected.	Alpha particles passing near the nucleus are deflected.
(3) A few alpha particles came back.	Alpha particles bounce back when they squarely hit the nucleus.

Model: The positive charge and nearly all the mass of an atom are contained in the tiny nucleus of the atom.

5.19 gamma rays (highest frequencies), X-rays, and ultraviolet light

5.21 Speed of light = frequency × wavelength

$$c = \nu \times \lambda$$

5.23 ultraviolet light

5.25 infrared (IR) radiation

5.27 red light

5.29 infrared

5.31 Each element has a specific set of frequencies that are emitted by atoms of an excited (energized) element.

5.33 strontium

5.35 $E = h\nu$; energy and frequency increase proportionally.

5.37 Bohr proposed that electrons in atoms must exist in specific energy levels.

5.39 An ion is a charged particle formed when an atom or group of atoms gains or loses electrons. The phenomenon is called ionization. Ionization energy is the energy required to remove an electron from a gaseous atom in its ground state.

5.41 Excited electrons within an atom fall back to lower energy states.

5.43 Listed here are element symbols followed by number of electrons in each energy level, beginning at the nucleus. Valence electrons are in the outermost energy level.
a. Mg: 2, 8, 2; two valence electrons
b. Ca: 2, 8, 8, 2; two valence electrons
c. N: 2, 5; five valence electrons
d. S: 2, 8, 6; six valence electrons
e. F: 2, 7; seven valence electrons.

5.45 Schrödinger and Heisenberg showed mathematically that electrons do not exist in simple spherical planetary orbits.

5.47 a. Li· b. ·Al· (three dots) c. :P· (five dots) d. :O· (six dots)
e. :Br· (seven dots)

5.49 Two electrons, and they must have opposite spins.

5.51 a *d* sublevel

5.53 a. Be: $1s^22s^2$ [He] ⇅
b. B: $1s^22s^22p^1$ [He] ⇅ ↑ __ __
c. N: $1s^22s^22p^3$ [He] ⇅ ↑ ↑ ↑
d. S: $1s^22s^22p^63s^23p^4$ [Ne] ⇅ ⇅ ↑ ↑
e. Cl: $1s^22s^22p^63s^23p^5$ [Ne] ⇅ ⇅ ⇅ ↑

5.55 The excited electrons in different atoms can fall back to several lower states before reaching the ground state.

5.57 These elements all have seven valence electrons; they are called the halogens.

5.59 a. Dalton did not know about any of the subatomic particles or isotopes.
b. Thomson did not know about the nucleus of an atom.
c. Bohr did not know about sublevels and orbitals.

5.61 three (each in one of the $2p$ orbitals)

5.63 They each have one valence electron. They are all in Group IA.

5.65 a. oxygen, six valence electrons (in the $2s$ and $2p$ sublevels)
b. phophorus, five valence electrons (in the $3s$ and $3p$ sublevels)

Chapter 6

Ex. 6.1 Elements in periods 4, 5, 6, and 7 contain transition metals. Elements in periods 1, 2, and 3 do not contain transition metals. (See Figure 6.4 as a reference.)

Ex. 6.2 A potassium atom is much larger than a potassium ion. When the metal atom loses its valence electrons—its outer energy level—the radius (size) decreases. A chloride ion is much larger than a chlorine atom. Adding electrons increases the radius (size).

Ex. 6.3 Ionization energies rise and fall in a *periodic* fashion. They increase for elements within each period—from the alkali metals to the noble gases—then drop sharply for the first element in each subsequent period. The pattern in repeated for each period.

6.1 When atomic masses for Ca and Ba are averaged, the result is very close to the atomic mass of Sr.

6.3 Mendeleev left gaps in his periodic table for unknown elements. He boldly predicted that elements would be discovered with various predicted properties.

6.5 Although the periodic table developed by Meyer was submitted for publication before Mendeleev's, it was not published until after Mendeleev's table was published. Further, Mendeleev gained notoriety by predicting properties of certain undiscovered elements.

6.7 When elements are arranged by atomic number rather than by atomic mass, Te, I, Co, Ni, Ar, and K fall in appropriate families in the periodic table.

6.9 A period of elements is a horizontal row of elements in the periodic table. There are seven periods of elements.

6.11 For the second and third periods of elements, the shiny reactive metals are at the left, followed by dull solids, reactive nonmetals, and a nonreactive noble gas. This accompanies the metallic to nonmetallic trend.

6.13 Each element in Group IIA (alkaline earth elements) has two valence electrons.

6.15 a. Li and K are in Group IA and have one valence electron.
b. Cl and I are in Group VIIA and have seven valence electrons.
c. N and P are in Group VA and have five valence electrons.
d. Al and B are in Group IIIA and have three valence electrons.

6.17 The alkali metals are Group IA, the alkaline earths are Group IIA, and the halogens are Group VIIA.

6.19 Period 1 has two elements. Period 2 has eight elements.

6.21 Group IA elements have the greatest atomic size (Figure 6.7).

6.23 Atomic size within a period decreases as atomic number increases. This is because each element within a period has one more proton than the previous element and the increase in positive charge draws the electron cloud closer to the nucleus.

6.25 a. A sodium ion is smaller than a sodium atom.
b. A chloride ion is larger than a chlorine atom.

6.27 A F^- ion is larger than a Na^+ ion. They are isoelectronic with 10 electrons.

6.29 Ionization energy—often meaning the first ionization energy—is the amount of energy required to remove an electron from a neutral gaseous atom in its ground state. Yes.

6.31 The first ionization energy increases for elements within each period of elements; the higher the atomic number, the higher the ionization energy. Elements in period 3 have lower ionization energies than corresponding elements in period 2.

6.33 Ionization energy within a group decreases as atomic number increases.

6.35 a. Cs has a smaller ionization energy than Na and a greater tendency to form a positive ion because the outermost electron of Cs is farther from the nucleus and is more easily removed.
b. Na has a smaller ionization energy than Si and a greater tendency to form a positive ion because the outermost electron of Si is held more tightly by a higher positive nuclear charge.
c. Si has a smaller first ionization energy than Cl and a greater tendency to form a positive ion than Cl, but neither kind of atom generally loses electrons in chemical reactions because both have relatively high ionization energies.

6.37 The alkali metals have the lowest first ionization energies.

6.39 For the alkali metals, melting points decrease as atomic number increases; for the halogens, the trend is the opposite. See Table 6.2.

6.41 Carbon. The attraction between carbon atoms must be quite high.

6.43 With few exceptions, there is an increase in density for elements in a group (the alkali metals, the halogens, etc.) as atomic number increases.

6.45 a. Al, closer to the center of the period, has a higher density than Mg.
b. Au, closer to the center of the period, has a higher density than Pb.
c. Pt, in the same group but with a higher atomic number, has a higher density than Ni.

6.47 Because hydrogen has one valence electron, it can be placed in Group IA. Because hydrogen, like the halogens, needs one more electron to fill an energy level, it is sometimes placed in Group VIIA. Because properties of hydrogen are not like those of the halogens or of the alkali metals, it is sometimes shown set apart by itself at the top of a periodic table.

6.49 Calcium; it is the fifth most abundant element in the Earth's crust. Calcium carbonate is the chemical present in chalk, limestone, marble, and calcite.

6.51 Boron. Borax is used in water softening and as a cleanser. Boron compounds are used in Pyrex glassware, abrasives, cutting tools, porcelain enamels, etc.

6.53 diamond, graphite, and buckyballs

6.55 Phosphorus is not reactive in water, but is quite reactive in oxygen.

6.57 A carbon compound that contains fluorine. Freon-12 (CCl_2F_2) is a common fluorocarbon used as a refrigerant in air conditioners.

6.59 When iodine sublimes, it goes directly from the solid to the gaseous state without passing through the liquid state.

6.61 Iron is an element. Steel is an iron alloy that contains small amounts of Cr, Mn, and various other metals for strength, hardness, and durability.

6.63 The lanthanides, elements 58 to 71, follow lanthanum in the periodic table. The actinides, elements 90 to 103, follow actinium in the periodic table.

6.65 An oxide ion, O^{2-}, is larger than an oxygen atom.

6.67 The second ionization energy of magnesium is higher than the first. It takes more energy to remove each subsequent electron.

6.69 a. Au, in the same group but with a higher atomic number, has a greater density than Cu.
b. Fe, closer to the center of the period, has a greater density than K.
c. W, in the same group but with a higher atomic number, has a greater density than Cr.

6.71 On the detrimental side, ozone is an air pollutant, especially in urban high-traffic areas, and is involved in deterioration of rubber. It is also injurious to some plants. On the beneficial side, the ozone in the outer atmosphere helps filter out damaging ultraviolet rays from the sun.

6.73 One allotropic form of phosphorus is a reddish, noncrystalline form. The other is a yellowish, crystalline, waxy form, with the formula P_4.

Chapter 7

Ex. 7.1 Calcium, shown in Group II on the periodic table, has two valence electrons. It is represented ·Ca· with two dots. Oxygen has six valence electrons. It is represented ·Ö: with six dots. The reaction is represented

$$\cdot\mathrm{Ca}\cdot + \cdot\ddot{\mathrm{O}}: \longrightarrow \mathrm{Ca}^{2+} + :\ddot{\mathrm{O}}:^{2-}$$

Ex. 7.2 Each iodine atom has seven valence electrons and needs one more electron. Two iodine atoms come together to form a covalent bond with one shared pair of electrons.

$$:\ddot{\mathrm{I}}\cdot + \cdot\ddot{\mathrm{I}}: \rightarrow :\ddot{\mathrm{I}}:\ddot{\mathrm{I}}:$$

Ex. 7.3 A sulfur-to-oxygen bond is polar covalent; the difference in electronegativities is less than 1.7. An oxygen-to-sodium bond is ionic; the difference in electronegativities is much greater than 1.7.

Ex. 7.4 Oxygen and sulfur atoms have six valence electrons and share electron pairs with two atoms of hydrogen. The two nonbonding electron pairs on oxygen and sulfur atoms give a bent shape to molecules of H_2O and H_2S, making them polar. Because both molecules are polar, H_2S gas should—and does—dissolve in water.

7.1 $\mathrm{K}\cdot \rightarrow \mathrm{K}^+ + e^-$; oxidation.

7.3 $:\ddot{\mathrm{Br}}\cdot + e^- \rightarrow :\ddot{\mathrm{Br}}:^-$; reduction.

7.5 $\mathrm{K}\cdot + :\ddot{\mathrm{Br}}\cdot \rightarrow \mathrm{K}^+ + :\ddot{\mathrm{Br}}:^-$

7.7 An electrolyte is a substance that dissolves in water to give a solution that conducts an electric current. Yes, common table salt, NaCl, is an electrolyte. When the salt dissolves, the dissociated sodium ions and chloride ions in solution allow the solution to conduct electricity.

7.9 An ionic bond is formed when electrons are transferred from one atom to another to form positive and negative ions. A covalent bond is formed when electrons are shared between two atoms.

7.11 In a coordinate covalent bond, one atom donates both of the electrons involved in the shared pair of electrons.

7.13 Sodium chloride has ionic bonds rather than the covalent bonds present in molecular substances.

7.15 a, c, f. nonpolar covalent b. ionic
d, e. polar covalent

7.17 The molecule is symmetrical with the carbon at the center and bromine atoms spaced equidistantly around the central carbon atom.

7.19 a. 0, nonpolar covalent b. 0.3, covalent
c. 1.9, ionic

7.21 Only $CaCl_2$.

7.23 Nonpolar covalent.

7.25 Metallic.

7.27 Nonpolar covalent.

7.29 a. $:\ddot{\mathrm{Cl}}\cdot$ b. $:\ddot{\mathrm{Cl}}:\ddot{\mathrm{Cl}}:$ c. $:\ddot{\mathrm{Cl}}:^-$

7.31 a.
```
   H
   ..
H: N :
   ..
   H
```
b.
```
 ┌   H   ┐+
 |   ..  |
 | H:N:H |
 |   ..  |
 └   H   ┘
```
c.
```
     :Cl:
      ..  ..
  H : C : Cl:
      ..  ..
     :Cl:
      ..
```
d.
```
    H
    ..  ..
 H: C : O :H
    ..  ..
    H
```

7.33
```
 ..          ..
:S=O:  <->  :S—O:
 |           ||  ..
:O:         :O
 ..          ..
```

7.35 Tetrahedral shape with 109.5° bond angles.

7.37 a. 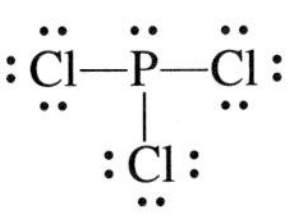 :Cl̤—P̤—Cl̤: (with :Cl̤: bonded below P) pyramidal shape

b. :Cl̤—B—Cl̤: (with :Cl̤: bonded below B) 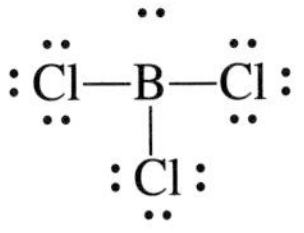 trigonal planar shape

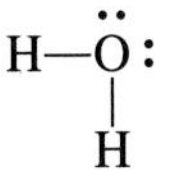

c. 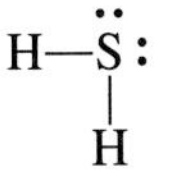 H—S̤: (with H bonded below S) bent shape

d. 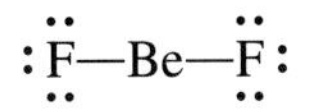 :F̤—Be—F̤: linear shape

7.39 H—Ö: (with H bonded below O) bent shape

7.41

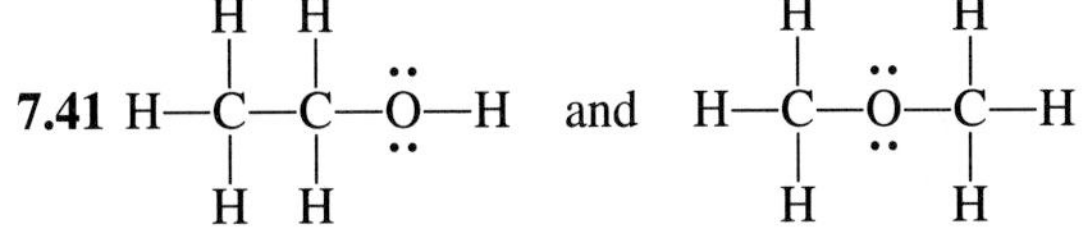

7.43 The electrons shared between hydrogen and oxygen atoms within water molecules form covalent bonds. Hydrogen bonding occurs between water molecules and is due to the attraction of hydrogen atoms in one molecule for the nonbonding electron pair held by oxygen in another water molecule.

7.45 These attractions cause water to expand as it freezes, and the expanding water causes surrounding solid materials to crack and break up.

7.47 a. X belongs to Group VIIA. Y belongs to Group VIA. Z belongs to Group VA.

b. 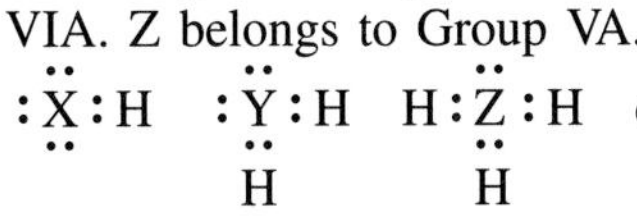 :X̤:H :Y̤:H (H below Y) H:Z̤:H (H below Z) c. 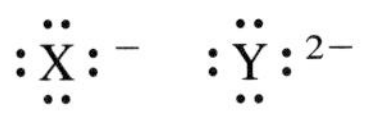 :X̤:⁻ :Y̤:²⁻

7.49 HBr is quite polar; Br_2 is nonpolar.

7.51 a. 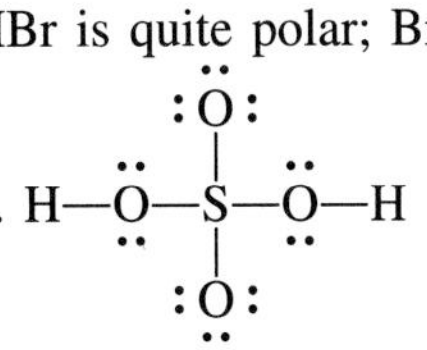 H—Ö—S—Ö—H (with :Ö: above and :Ö: below S) b. 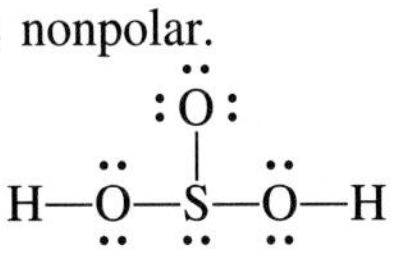 H—Ö—S̤—Ö—H (with :Ö: above S)

7.53 Gasoline molecules must be nonpolar.

7.55 $CaCl_2$, which contains ions; chloride ions; and CH_3CH_2OH, which can form hydrogen bonds.

7.57 CH_3OH and CH_3NH_2.

7.59 a. nonpolar covalent b. ionic c. polar covalent

Chapter 8

Ex. 8.1 (a) N_2 (b) N^{3-} (c) S^{2-} (d) Ni^{2+}

Ex. 8.2 (a) CrO_4^{2-} (b) ClO_4^- (c) ClO^-

Ex. 8.3 (a) calcium chloride (b) zinc nitrate (c) sodium sulfate

Ex. 8.4 Since the charge of the sulfate ion is 2− and there are three of these ions, the total negative charge is $3(-2) = -6$. The total positive charge must be +6 to balance the negative charge. For the two iron ions to give a +6 charge, each iron ion must be Fe^{3+}, iron (III). The name for the compound is iron (III) sulfate, or ferric sulfate.

Ex. 8.5 In $NaNO_3$, the oxidation number of sodium is +1. Oxygen is −2, so three oxygen atoms have a total value of −6. Let x = nitrogen, N, the unknown. Write an equation: $+1 + x + (-6) = 0$ and solve for x, which it is +5. In this compound, the oxidation number of nitrogen, N, is +5.

Ex. 8.6 (a) calcium sulfate dihydrate (b) cobalt (II) chloride hexahydrate

8.1 Oxygen.

8.3 a. HCO_3^-, 5 atoms b. CO_3^{2-}, 4 atoms c. $S_2O_3^{2-}$, 5 atoms d. CrO_4^{2-}, 5 atoms e. CH_3COO^-, 7 atoms f. PO_4^{3-}, 5 atoms g. PO_3^{3-}, 4 atoms h. P^{3-}, 1 atom

8.5 a. Cu^+, cuprous, and Cu^{2+}, cupric b. Fe^{2+}, ferrous, and Fe^{3+}, ferric c. Sn^{2+}, stannous, and Sn^{4+}, stannic d. Hg^+, mercurous (always in pairs), and Hg^{2+}, mercuric

In each case, the *-ic* ending designates the ion with the higher charge number.

8.7 a. chlorate b. perchlorate c. carbonate d. thiocyanate e. sulfate

8.9 a. Al^{3+}, Cl^-, $AlCl_3$ b. Al^{3+}, OH^-, $Al(OH)_3$
c. NH_4^+, SO_4^{2-}, $(NH_4)_2SO_4$
d. Fe^{2+}, PO_4^{3-}, $Fe_3(PO_4)_2$
e. Sb^{3+}, S^{2-}, Sb_2S_3 f. Ca^{2+}, OH^-, $Ca(OH)_2$
g. Sn^{2+}, F^-, SnF_2
h. NH_4^+, PO_4^{3-}, $(NH_4)_3PO_4$
i. Ca^{2+}, CO_3^{2-}, $CaCO_3$
j. Ca^{2+}, ClO^-, $Ca(ClO)_2$

8.11 a. tin(IV) chloride, stannic chloride
b. mercury(I) chloride, mercurous chloride
c. iron(II) oxide, ferrous oxide
d. manganese(II) chloride, manganous chloride
e. copper(II) sulfide, cupric sulfide
f. cobalt(III) nitrate, cobaltic nitrate

8.13 a. N_2O b. CCl_4 c. sulfur hexafluoride d. dinitrogen pentoxide e. nitrogen dioxide f. phosphorus trichloride

8.15 a. +5 b. +3 c. +7 d. +4 e. +5 f. +3

8.17 a. HBr b. HNO_3 c. HNO_2 d. H_2CO_3 e. AgBr f. $AgNO_3$ g. $NaNO_2$ h. K_2CO_3

8.19 a. phosphoric acid b. potassium phosphate
c. dipotassium hydrogen phosphate
d. potassium dihydrogen phosphate
e. sulfuric acid f. potassium hydrogen sulfate
g. nitric acid h. potassium nitrate

8.21

```
           ..                      ..      2-
          :O:                     :O:
       ..  ..  ..              ..  ..  ..
    H:O:S:O:H      and       [:O:S:O:]
       ..  ..  ..              ..  ..  ..
          :O:                     :O:
           ..                      ..
```

8.23 a. $CaSO_4 \cdot 2H_2O$, calcium sulfate dihydrate, dry wall sheets and figurines
b. $MgSO_4 \cdot 7H_2O$, magnesium sulfate heptahydrate, medicine and dyeing
c. $Na_2B_4O_7 \cdot 10H_2O$, sodium tetraborate decahydrate, water-softening agent
d. NaOH, sodium hydroxide, neutralizing acids and paper production
e. CaO, calcium oxide, cement and mortar
f. $CaCO_3$, calcium carbonate, antacids, medicine, and cement production
g. CH_3COOH, acetic acid, pickling and manufacture of other chemicals
h. $C_{12}H_{22}O_{11}$, sucrose, sweetener

8.25 a. H_2SO_4, oxidation of sulfur + water, 74 billion lb, manufacture of fertilizers and chemicals.
b. H_3PO_4, phosphate rock + acid, 19 billion lb, fertilizers and detergents
c. NaOH, electrolysis of NaCl, 22 billion lb, chemical and paper manufacturing
d. Na_2CO_3, minerals or brine, 17 billion lb, glass and chemicals

8.27 $4.95 billion/yr; plastics, antifreeze, and fibers.

8.29 cation

8.31 All acids contain ionizable hydrogens. In water solution, each ionizable hydrogen becomes a hydronium ion, H_3O^+.

8.33 CO is carbon monoxide; Co is the symbol for cobalt.

8.35 In peroxide, two oxygen atoms are covalently bonded; in a dioxide compound, two separate oxygen atoms are joined to a different central atom.

8.37 NO_2 is the formula for nitrogen dioxide; NO_2^- is the formula for a nitrite ion.

8.39 Parentheses are placed around a polyatomic ion in a formula when—and only when—two or more of the polyatomic ions are represented in the formula.

8.41 A hydrate is a compound that contains a definite, fixed number of water molecules.

8.43 a. 3 b. 3 c. 2

8.45 sodium chloride, NaCl; magnesium chloride, $MgCl_2$; aluminum chloride, $AlCl_3$; silicon tetrachloride, $SiCl_4$; phosphorus trichloride, PCl_3; sulfur dichloride, SCl_2; chlorine, Cl_2

8.47 a. HCN(g) b. $H_2C_2O_4$ c. Na_2O_2
d. $NaKSO_4$ e. Au_2O_3 f. ICl
g. $KAl(SO_4)_2 \cdot 10H_2O$ h. NaCl

8.49 a. sodium thiocyanate b. nitrogen dioxide
c. nitrite ion d. ammonium oxalate
e. magnesium bicarbonate f. phosphorous acid

8.51 a. binary covalent compound b. acid c. salt

8.53 a. nickel(II) chloride b. nickel(III) chloride
c. phosphorus trichloride

Chapter 9

Ex. 9.1 Multiply the number of moles of sodium carbonate given by the ratio of one mole of carbonate ions per mole of sodium carbonate.

$$3.84 \text{ mol } Na_2CO_3 \times \frac{1 \text{ mol } CO_3^{2-}}{1 \text{ mol } Na_2CO_3} = 3.84 \text{ mol } CO_3^{2-} \text{ ions}$$

Ex. 9.2 To obtain the desired mass, multiply the 2.63 mol of washing soda given by the molar mass, 286.0 g/mol (determined in Example 9.4). This gives 752 g, the desired mass of washing soda.

Ex. 9.3 $$1.64 \text{ mol } CO_2 \times \frac{44.0 \text{ g } CO_2}{1 \text{ mol } CO_2} = 72.2 \text{ g } CO_2$$

Ex. 9.4 $$91.6 \text{ g } CO_2 \times \frac{1 \text{ mol } CO_2}{44.0 \text{ g } CO_2} = 2.08 \text{ mol } CO_2$$

Ex. 9.5 Multiply the molar mass of CO_2 by Avogadro's number (inverted).

$$\frac{44.0 \text{ g } CO_2}{1 \text{ mol } CO_2} \times \frac{1 \text{ mol } CO_2}{6.02 \times 10^{23} \text{ molecules}} = 7.31 \times 10^{-23} \text{ g } CO_2/\text{molecule}$$

Ex. 9.6 Step 1: Start with the known quantity—the number of grams of NaCl.
Step 2: Convert grams to moles using the molar mass of NaCl.
Step 3: Convert moles to liters and milliliters of solution.

$$1.36 \text{ g NaCl} \times \frac{1 \text{ mol NaCl}}{58.5 \text{ g NaCl}} \times \frac{1 \text{ L}}{0.125 \text{ mol}} \times \frac{1000 \text{ mL}}{1 \text{ L}} = 186 \text{ mL}$$

Ex. 9.7 To find the empirical formula of hydrogen peroxide, divide the subscripts by 2 to obtain HO.

To find the empirical formula of butane, divide the subscripts by 2 to obtain C_2H_5.

9.1 The formula weight is an appropriate name for the sum of the atomic masses for any compound. When the compound is molecular—not ionic—the sum may be called the molecular weight.
9.3 a. 58.3 b. 149.0 c. 74.1 d. 60.0
9.5 a. 32.0 g/mol b. 2 atoms/molecule c. 1.20×10^{24} atoms/mol d. 80.0 g O_2 e. 2.19 mol O_2
9.7 a. 194.0 g/mol b. 24 atoms/molecule c. 1.44×10^{25} atoms/mol d. 24.2 g caffeine e. 0.258 mol caffeine
9.9 a. one Ca^{2+} ion/formula unit of $Ca(OH)_2$ b. two OH^- ions/formula unit of $Ca(OH)_2$ c. 2.50 mol Ca^{2+} ions d. 5.00 mol OH^- ions
9.11 282 g N
9.13 3.93 g Na
9.15 8.66 kg Pb
9.17 82.4% N, 17.6% H
9.19 21.2% N, 6.06% H, 24.2% S, 48.5% O
9.21 NH_3 with 82.4% N
9.23 a. 0.179 mol Fe b. 0.0625 mol Fe_2O_3 c. 2.00 mol C_2H_5OH d. 0.467 mol Au
9.25 a. 44.7 g Fe b. 128 g Fe_2O_3 c. 69.0 g C_2H_5OH d. 296 g Au
9.27 1.99×10^{-23} g C
9.29 7.31×10^{-23} g CO_2
9.31 1.67×10^{21} molecules H_2O
9.33 2.32×10^{19} OH^- ions
9.35 1.63×10^{22} Cl^- ions
9.37 6.75 $C_6H_{12}O_6$; dissolve the glucose in enough water to make a total volume of 250. mL.
9.39 74.1 mL
9.41 16.7 mL
9.43 a. C_4H_9 b. $C_{12}H_{22}O_{11}$ c. HgCl d. $CaCl_2$
9.45 a. C_2H_6 b. C_4H_8 c. C_5H_7N d. P_4O_{10}
9.47 P_3N_5
9.49 P_4S_7
9.51 a. NH_2 b. N_2H_4
9.53 a. CH_2Cl b. $C_2H_4Cl_2$
9.55 18.0 g/mol; 2.99×10^{-23} g H_2O/molecule
9.57 149 g/mol; 1.21×10^{21} NH_4^+ ions
9.59 50. mL
9.61 50.0 g $CuSO_4 \cdot 5H_2O$
9.63 30.0 mL
9.65 1.20×10^{20} Cl^- ions
9.67 9.33 g N
9.69 CCl_2F_2
9.71 $$15.0 \text{ mL} \times \frac{1 \text{ L}}{1000 \text{ mL}} \times \frac{0.200 \text{ mol NaHCO}_3}{\text{L}} \times \frac{84.0 \text{ g NaHCO}_3}{1 \text{ mol NaHCO}_3} = 0.252 \text{ g NaHCO}_3$$
9.73 $(V_1)(6.00 \text{ M}) = (500. \text{ mL})(0.100 \text{ M})$

$$V_1 = \frac{(500. \text{ mL})(0.100 \text{ M})}{6.00 \text{ M}} = 8.33 \text{ mL}$$

Chapter 10

Ex. 10.1 $3\ NaOH + FeCl_3 \longrightarrow 3\ NaCl + Fe(OH)_3(s)$
Ex. 10.2 The reaction in Example 10.1 is a *synthesis* reaction involving the synthesis of ammonia. The reaction in Example 10.3 is a *combustion* reaction involving the combustion of methane.
Ex. 10.3 Step 1: Write the unbalanced equation.

$$C_2H_5OH + O_2 \longrightarrow CO_2 + H_2O \quad \text{(unbalanced)}$$

Step 2: Balance the C, H, and then O, with the O_2 balanced last.

$$C_2H_5OH + 3\ O_2 \longrightarrow 2\ CO_2 + 3\ H_2O \quad \text{(balanced)}$$

Ex. 10.4 No. Mercury and gold to not react with acids to release hydrogen gas, as indicated by their placement below hydrogen in the activity series.
Ex. 10.5 Bromine is more reactive than iodide ions, so a reaction occurs.

$$Br_2 + NaI \longrightarrow NaBr + I_2$$

10.1 In a balanced chemical equation the total masses of reactants and products are equal. This is in agreement with the law of conservation of mass.
10.3 all the correct formulas
10.5 a. $4\ Al + 3\ O_2 \rightarrow 2\ Al_2O_3$
b. $N_2 + 2\ O_2 \rightarrow 2\ NO_2$
c. $2\ H_2O_2 \rightarrow 2\ H_2O + O_2(g)$
d. $2\ LiOH + CO_2 \rightarrow Li_2CO_3 + H_2O$
e. $Fe_2(SO_4)_3 + 6\ NaOH \rightarrow 2\ Fe(OH)_3 + 3\ Na_2SO_4$
10.7 a–c. true d, e. false.
10.9 a. $Mg + 2\ H_2O(g) \rightarrow Mg(OH)_2 + H_2(g)$
b. $2\ NaHCO_3 + H_3PO_4 \rightarrow Na_2HPO_4 + 2\ H_2O + 2\ CO_2$
c. $2\ Al + 3\ H_2SO_4(aq) \rightarrow Al_2(SO_4)_3 + 3\ H_2(g)$
d. $C_3H_8 + 5\ O_2 \rightarrow 3\ CO_2 + 4\ H_2O$
e. $CH_3OH + O_2 \rightarrow CO_2 + 2\ H_2O$
10.11 a. $CaCO_3 + 2\ HCl \rightarrow CaCl_2 + H_2O + CO_2(g)$
b. $PCl_5 + 4\ H_2O \rightarrow H_3PO_4 + 5\ HCl$

c. $2\ KClO_3 \rightarrow 2\ KCl + 3\ O_2(g)$
d. $3\ Ba(OH)_2 + 2\ H_3PO_4 \rightarrow Ba_3(PO_4)_2 + 6\ H_2O$
e. $2\ C_2H_6(g) + 7\ O_2(g) \rightarrow 4\ CO_2(g) + 6\ H_2O(g)$

10.13 a. decomposition b. single replacement
c. combustion d. double replacement
e. decomposition

10.15 $2\ C_4H_{10} + 13\ O_2 \rightarrow 8\ CO_2 + 10\ H_2O$

10.17 a. $H_2(g) + Cl_2(g) \rightarrow 2\ HCl(g)$
b. $P_4 + 6\ Br_2 \rightarrow 4\ PBr_3$
c. $2\ H_2(g) + O_2(g) \rightarrow 2\ H_2O(g)$
d. $2\ SO_2(g) + O_2(g) \rightarrow 2\ SO_3(g)$
e. $2\ Al + 3\ Cl_2 \rightarrow 2\ AlCl_3$

10.19 a. $2\ Al_2O_3 \xrightarrow{dc} 4\ Al + 3\ O_2(g)$
b. $2\ PbO_2 \rightarrow 2\ PbO + O_2(g)$
c. $2\ NaClO_3 \rightarrow 2\ NaCl + 3\ O_2(g)$
d. $2\ KNO_3 \rightarrow 2\ KNO_2 + O_2(g)$
e. $2\ H_2O_2 \xrightarrow{I^-} 2\ H_2O + O_2(g)$

10.21 a. $4\ Fe + 3\ O_2 \rightarrow 2\ Fe_2O_3$
b. $2\ Sb + 3\ Cl_2 \rightarrow 2\ SbCl_3$
c. $2\ Ca + O_2(g) \rightarrow 2\ CaO$

10.23 a. $Li_2O + H_2O \rightarrow 2\ LiOH$
b. $2\ Na + 2\ H_2O \rightarrow 2\ NaOH + H_2(g)$
c. $Mg + 2\ H_2O(g) \rightarrow Mg(OH)_2 + H_2(g)$
d. $Ag + H_2O \rightarrow$ no reaction
e. $SrO + H_2O \rightarrow Sr(OH)_2$

10.25 a. $Zn + 2\ HCl(aq) \rightarrow ZnCl_2(aq) + H_2(g)$
b. $Cu + HCl(aq) \rightarrow$ no reaction
c. $3\ Mg + 2\ Fe(NO_3)_3(aq) \rightarrow 3\ Mg(NO_3)_2(aq) + 2\ Fe$
d. $3\ AgNO_3(aq) + Al \rightarrow Al(NO_3)_3(aq) + 3\ Ag$
e. $Fe + MgCl_2(aq) \rightarrow$ no reaction

10.27 a. $Ca + 2\ HCl(aq) \rightarrow CaCl_2(aq) + H_2(g)$
b. $Au + H_2SO_4(aq) \rightarrow$ no reaction
c. $CuSO_4(aq) + Zn \rightarrow ZnSO_4(aq) + Cu$
d. $Cu + NaCl(aq) \rightarrow$ no reaction
e. $2\ Al + 6\ HCl(aq) \rightarrow 2\ AlCl_3(aq) + 3\ H_2(g)$

10.29 a. $S_8 + 8\ O_2(g) \rightarrow 8\ SO_2(g)$
b. $SO_2 + H_2O \rightarrow H_2SO_3$
c. $N_2O_5 + H_2O \rightarrow 2\ HNO_3$
d. $2\ KBr + Cl_2 \rightarrow 2\ KCl + Br_2$
e. $KCl + I_2 \rightarrow$ no reaction

10.31 The chlorine would react with NaI to produce elemental iodine, I_2.

$$Cl_2 + 2\ NaI \rightarrow 2\ NaCl + I_2$$

10.33 a. $AgNO_3(aq) + KCl(aq) \rightarrow \underline{AgCl(s)} + KNO_3(aq)$
b. $FeCl_3(aq) + 3\ NaOH(aq) \rightarrow \underline{Fe(OH)_3(s)} + 3\ NaCl(aq)$
c. $Al_2(SO_4)_3(aq) + 3\ Ba(NO_3)_2(aq) \rightarrow 2\ Al(NO_3)_3(aq) + 3\ \underline{BaSO_4(s)}$
d. $Pb(NO_3)_2(aq) + K_2Cr_2O_7(aq) \rightarrow \underline{PbCr_2O_7(s)} + 2\ KNO_3(aq)$
e. $2\ AgNO_3(aq) + K_2CrO_4(aq) \rightarrow \underline{Ag_2CrO_4(s)} + 2\ KNO_3(aq)$

10.35
a. $Ag^+(aq) + NO_3^-(aq) + K^+(aq) + Cl^-(aq) \rightarrow \underline{AgCl(s)} + K^+ + NO_3^-(aq)$
b. $Fe^{3+}(aq) + 3\ Cl^-(aq) + 3\ Na^+(aq) + 3\ OH^-(aq) \rightarrow \underline{Fe(OH)_3} + 3\ Na^+(aq) + 3\ Cl^-(aq)$
c. $2\ Al^{3+}(aq) + 3\ SO_4^{2-}(aq) + 3\ Ba^{2+}(aq) + 6\ NO_3^-(aq) \rightarrow 2\ Al^{3+}(aq) + 6\ NO_3^-(aq) + 3\ \underline{BaSO_4(s)}$
d. $Pb^{2+}(aq) + 2\ NO_3^-(aq) + 2\ K^+(aq) + CrO_4^{2-}(aq) \rightarrow \underline{PbCrO_4(s)} + 2\ K^+(aq) + 2\ NO3^-(aq)$
e. $2\ Ag^+(aq) + 2\ NO_3^-(aq) + 2\ K^+(aq) + CrO_4^{2-}(aq) \rightarrow \underline{Ag_2CrO_4(s)} + 2\ K^+(aq) + 2\ NO_3^-(aq)$

10.37 a. $Ag^+(aq) + Cl^-(aq) \rightarrow \underline{AgCl(s)}$
b. $Fe^{3+}(aq) + 3\ OH^-(aq) \rightarrow \underline{Fe(OH)_3(s)}$
c. $Ba^{2+}(aq) + SO_4^{2-}(aq) \rightarrow \underline{BaSO_4(s)}$
d. $Pb^{2+}(aq) + Cr_2O_7^{2-}(aq) \rightarrow \underline{PbCr_2O_7(s)}$
e. $2\ Ag^+(aq) + CrO_4^{2-}(aq) \rightarrow \underline{Ag_2CrO_4(s)}$

10.39
a. $H_2SO_4(aq) + 2\ KOH(aq) \rightarrow K_2SO_4(aq) + 2\ H_2O$
b. acid base salt water
c. $H^+(aq) + OH^-(aq) \rightarrow H_2O$
d. the hydrogen ion, H^+
e. $2\ HCl(aq) + Mg(OH)_2(aq) \rightarrow MgCl_2(aq) + 2\ H_2O$

10.41 Each H^+ ion of the acid combines with a OH^- ion of the base to give water. The cation of the base combines with the anion of the acid to give the salt.

10.43 $Cu + 2\ H_2SO_4 \rightarrow SO_2 + CuSO_4 + 2\ H_2O$

10.45 double replacement

10.47 synthesis

10.49 $Fe_2O_3 + 3\ Co(g) \rightarrow 2\ Fe + 3\ CO_2(g)$

Chapter 11

Ex. 11.1 $1.20\text{ mol } C_2H_5OH \times \dfrac{2\text{ mol } CO_2}{1\text{ mol } C_2H_5OH} = 2.40\text{ mol } CO_2$

Ex. 11.2 $10.0\text{ g } C_2H_5OH \times \dfrac{1\text{ mol } C_2H_5OH}{46.0\text{ g } C_2H_5OH} \times \dfrac{2\text{ mol } CO_2}{1\text{ mol } C_2H_5OH} \times \dfrac{44.0\text{ g } CO_2}{1\text{ mol } CO_2} = 19.1\text{ g } CO_2$

Ex. 11.3 $0.243\text{ g } CaCO_3 \times \dfrac{1\text{ mol } CaCO_3}{100.0\text{ g } CaCO_3} \times \dfrac{2\text{ mol HCl}}{1\text{ mol } CaCO_3} \times \dfrac{1\text{ L solution}}{0.100\text{ M HCl}} = 0.0486\text{ L}$ or 48.6 mL HCl(aq)

Ex. 11.4 $3000.\ \text{L CH}_4 \times \dfrac{2\ \text{mol H}_2}{1\ \text{mol CH}_4} = 6000.\ \text{L H}_2\ \text{gas}$

Ex. 11.5 $\dfrac{13.3\ \text{g (actual yield)}}{13.9\ \text{g (theoretical yield)}} \times 100\% = 95.7\%$

11.1 a. $\dfrac{3\ \text{mol BaSO}_4}{1\ \text{mol Al}_2(\text{SO}_4)_3}$ b. $\dfrac{1\ \text{mol Al}_2(\text{SO}_4)_3}{3\ \text{mol BaSO}_4}$

c. $\dfrac{2\ \text{mol Al(NO}_3)_3}{3\ \text{mol Ba(NO}_3)_2}$ d. $\dfrac{1\ \text{mol Al}_2(\text{SO}_4)_3}{3\ \text{mol Bs(NO}_3)_2}$

11.3 a. 0.675 mol $BaSO_4$ b. 1.10 mol $Al_2(SO_4)_3$ c. 0.500 mol $Al(NO_3)_3$ d. 3.11 mol $Al_2(SO_4)_3$

11.5 a. 25.3 g $Ca(OH)_2$ b. 12.3 g H_2O c. 38.0 g $CaCl_2$ d. 25.0 g HCl e. There are 50.3 g of reactants and 50.3 g of products for the reaction; mass is conserved.

11.7 a. 2.34 mol O_2 b. 75.0 g O_2 c. 1.25 mol CH_3OH d. 40.0 g CH_3OH

11.9 a. $Mg + 2\ HCl(aq) \rightarrow MgCl_2 + H_2(g)$ b. 0.823 mol H_2 c. 60.1 g HCl

11.11 274 mL

11.13 10.6 mL

11.15 a. 2.50 tanks O_2 b. The volume ratio of oxygen to acetylene is 5:2 when pressures are the same for both gases, so the tank of oxygen needs to be 2.5 times as large as the acetylene tank if the tanks are to become empty at the same time.

11.17 a. Al b. 39.5 g $AlBr_3$ c. 81.5%

11.19 a. H_2 b. 43.5 kg CH_3OH c. 90.8%

11.21 During an exothermic reaction, heat is released; during an endothermic reaction, heat is absorbed.

11.23 −501 kJ; the enthalpy change is negative because the reaction is exothermic. The amount of energy released by burning methane is 3.5 times larger than that released by burning an equal number of grams of glucose.

11.25 $CaCO_3(s) \rightarrow CaO(s) + CO_2(g)$ $\Delta H = 178$ kJ, +3178 kJ

11.27 232 g

11.29 a. $2\ C_4H_{10} + 13\ O_2(g) \rightarrow 8\ CO_2 + 10\ H_2O$ b. 0.560 mol O_2 c. 17.9 g O_2

11.31 42.3 g

11.33 1.68 g

11.35 a. 200. tons/day b. 400. tons

11.37 a. 22.2 g b. 76.6%

11.39 1500. L

11.41 44.4 kg

11.43 18.5 kg

11.45 a. 0.500 mol b. 35.5 g c. The reaction of bleach with acid produces Cl_2 gas, which can be quite harmful if inhaled. The gas should not be released in a closed room.

11.47 1.55×10^4 kJ

11.49 100. mL

11.51 $8.64\ \text{kg CaCO}_3 \times \dfrac{1000\ \text{g CaCO}_3}{1\ \text{kg CaCO}_3} \times \dfrac{1\ \text{mol CaCO}_3}{100\ \text{g CaCO}_3} \times \dfrac{1\ \text{mol CaO}}{1\ \text{mol CaCO}_3} \times \dfrac{56.0\ \text{g CaO}}{1\ \text{mol CaO}} \times \dfrac{1\ \text{kg CaO}}{1000\ \text{g CaO}} = 4.84\ \text{kg CaO}$

Chapter 12

Ex. 12.1 Given: $P_1 = 740.$ torr, $V_1 = 10.0\ \text{m}^3$, $P_2 = 370.$ torr; V_2 is unknown.

$$V_2 = 10.0\ \text{m}^3 \times \frac{740.\ \text{torr}}{370.\ \text{torr}} = 20.0\ \text{m}^3$$

Ex. 12.2 Given: $V_1 = 2.00$ L, $T_1 = 27$ °C (300. K), $T_2 = 47$ °C (320 K); V_2 is unknown.

$$V_2 = 2.00\ \text{L} \times \frac{320.\ \text{K}}{300.\ \text{K}} = 2.13\ \text{L}$$

Ex. 12.3 Given: $P_1 = 32.0 + 14.7$ or 46.7 psi, $T_1 = 20. + 273 = 293$ K, $P_2 = 34.0 + 14.7 = 48.7$ psi; T_2 is unknown.

$$T_2 = 293\ \text{K} \times \frac{48.7\ \text{psi}}{46.7\ \text{psi}} = 306\ \text{K}$$ or (by subtracting 273) = 33 °C

Ex. 12.4 $V_2 = 4.5\ \text{L} \times \dfrac{744\ \text{torr}}{760\ \text{torr}} \times \dfrac{273\ \text{K}}{294\ \text{K}} = 4.09\ \text{L}$

Ex. 12.5 Start with the molar mass of Cl_2, 71.0 g/mol, and multiply by the molar volume (inverted so units cancel) to find density in g/L.

$$\frac{71.0\ \text{g}}{\text{mol}} \times \frac{1\ \text{mol}}{22.4\ \text{L}} = 3.17\ \text{g/L at STP}$$

Ex. 12.6 Solving the ideal gas law for P gives $P = \dfrac{gRT}{M_m V}$ or we can break it down:

$$P = \frac{g}{M_m} \times R \times \frac{T}{V}.$$

$$P = \frac{10.0\ \text{g}}{28.0\ \text{g/mol}} \times \frac{0.0821\ \text{L-atm}}{\text{mol-K}} \times \frac{295\ \text{K}}{2.00\ \text{L}} = 4.32\ \text{atm}$$

Ex. 12.7 To obtain the partial pressure of nitrogen in air at 738 torr, multiply the percentage of nitrogen in air by the air pressure.

$$P_{\text{nitrogen}} = 0.780 \times 738 \text{ torr} = 576 \text{ torr}$$

12.1 Earth's atmosphere is the layer of gases that surrounds our planet.

12.3 Pressure is the force exerted per unit area; atmospheric pressure is the total force exerted by the air on each unit of area.

12.5 The mercury in a mercury barometer flows in or out to maintain equal pressures on the inside and outside of the tube. A measurement of the height of the mercury column is reported as the barometric pressure.

12.7 a. 760 torr b. 1.20 atm c. 380. mm Hg d. 81.6 atm e. 202 kPa f. 0.10 atm

12.9 Gas particles are moving rapidly, randomly, and continuously in all directions with negligible attraction between them.

12.11 If the volume of the container is decreased, the particles will travel shorter distances before they strike the walls so they strike the walls more frequently, thus increasing the pressure—the force per unit area.

12.13 When the speed of the particles decreases, the temperature also decreases.

12.15 The two gases have the same kinetic energy when they are at the same temperature, but the He atoms move faster. The tank of He is lighter, but the number of particles in both tanks is the same because equal volumes contain equal numbers of particles at the same conditions.

12.17 Oxygen sample A, which is at a higher pressure, has a higher density than oxygen sample B because there are more molecules per unit volume for sample A.

12.19 Because gas molecules move continuously, rapidly, and randomly in all directions, the aroma moves throughout all available space.

12.21 Additional air can be added to a pressurized tire because the particles of air are extremely tiny and the distance between them is large.

12.23 Pressure times volume equals a constant for a sample of gas at a fixed temperature.

12.25 7.45×10^4 for each set of data.

12.27 1200 mL

12.29 1.47×10^3 L

12.31 At constant pressure, volume increases as temperature increases. Volume is proportional to temperature. The graph of V vs. T is a straight line.

12.33 73 mL/73 K = 1,173 mL/173 K = 1, etc. Each V/T ratio equals 1.00 for these data when temperatures are in kelvins. V/T ratios are constant only when temperatures are in kelvins.

12.35 3.00 L

12.37 69 °C; the temperature change is 49 °C.

12.39 Models of Gay-Lussac's law include a pressurized can (hair spray, etc.) or a sealed can of food being heated. As T increases, P increases when V remains constant.

12.41 2.51×10^3 torr

12.43 39.6 psi (actual pressure) gives 24.9 psi (gauge pressure).

12.45 129 mL

12.47 1.24 atm

12.49 2.22 L

12.51 5.71×10^3 mg

12.53 0.268 mol

12.55 1.96 g/L

12.57 An ideal gas is one that perfectly obeys the gas laws. There is no ideal gas; only "real" gases actually exist.

12.59 $\dfrac{\text{liters} \times \text{atmosphere}}{\text{moles} \times \text{kelvins}}$

12.61 1.84 atm

12.63 7.77 L

12.65 0.95 atm

12.67 95.0%

12.69 716 torr, 3.66×10^{-3} mol

12.71 963 mL

12.73 1.98 L at STP, 2.24 L at 735 torr and 25 °C

12.75 a bicycle tire pump

12.77 0.759 g/L

12.79 19.1 L

12.81 944 L

12.83 1 mol SO_2 = 64.0 g SO_2; 1 mol SO_2 = 22.4 L SO_2 (at STP)

$$d = \frac{64.0 \text{ g}}{22.4 \text{ L}} = 2.86 \frac{\text{g}}{\text{L}} \text{(at STP)}$$

Chapter 13

Ex. 13.1 Iodine, I_2, is nonpolar so it has only London dispersion forces. Hydrogen iodide, HI, is polar so it has dipole forces along with London dispersion forces. Neither I_2 nor HI has hydrogen bonding—the only other type of intermolecular force described.

Ex. 13.2 Start with the quantity in grams and use the

heat of vaporization as a conversiion factor to determine the energy change.

$$400.\ \text{g} \times \frac{2.26\ \text{kJ}}{\text{g}} = 904\ \text{kJ}$$

Ex. 13.3 To convert the heat of fusion from kJ/mol to kJ/g, start with the value given in kJ/mol and use the molar mass of naphthalene (128 g/mol) as a conversion factor to find the equivalent value in kJ/g.

$$\frac{19.3\ \text{kJ}}{\text{mol}} \times \frac{1\ \text{mol}}{128\ \text{g}} = 0.151\ \text{kJ/g}$$

13.1 In liquids and solids, the individual particles are in contact with one another, so very little compression is possible.

13.3 8.03×10^{26} molecules

13.5 Substances with large molar masses are more likely to be in the solid state than in the liquid or gaseous state.

13.7 a. nonpolar covalent b. ionic

13.9 *Inter*molecular forces are the attractions of molecules for one another. *Intra*molecular forces are those that exist within a molecule due to bonding.

13.11 Intermolecular forces, as a group, are often called van der Waals forces.

13.13 a. London forces b, e. hydrogen bonding, dipole, and London forces
c, d. dipole and London forces

13.15 N_2 (London forces) < HCl (London and dipole forces) < H_2O (hydrogen bonding + London and dipole forces) < NaCl (interionic forces).

13.17 London forces are greater in Br_2 (larger molecules) than in F_2 molecules.

13.19 Butane and hexane are similar nonpolar molecules, but hexane—with the greater molar mass and greater London forces—has a higher boiling point.

13.21 only CH_3OH

13.23 Liquids whose molecules have strong intermolecular forces also have strong intermolecular attractions, so they have a higher surface tension.

13.25 Water beads up on a freshly waxed surface, which is made up of nonpolar molecules, because the attraction of water molecules for one another is greater than is their attraction for the waxed surface.

13.27 Motor oil is more viscous than gasoline. Viscosity decreases as temperature increases.

13.29 Small, symmetrical molecules generally have low viscosities.

13.31 When water in a closed bottle is at equilibrium, liquid is evaporating at the same rate vapor is condensing; the mass of liquid remains constant.

13.33 The boiling point of a liquid is the temperature at which its vapor pressure becomes equal to the atmospheric pressure.

13.35 Foods cook in water in a pressure cooker more quickly because water boils at a higher temperature in the pressure cooker. As pressure increases, the boiling point increases, so the cooking temperature can be above 100 °C.

13.37 b.p. low to high: O_2 (nonpolar) < CO (polar) < H_2O (hydrogen bonding)

13.39 ethyl alcohol

13.41 Because of the lower atmospheric pressure, the boiling point is lower at the higher altitude and more time is needed to cook the egg.

13.43 Explain that the thermometer is not broken. The thermometer levels off at both 79 °C and 99 °C because these are the boiling points of the two fractions being separated. Each fraction is distilled off before the temperature climbs further.

13.45 30.1 kJ/mol

13.47 0.935 kcal

13.49 Carbon atoms in diamond are bonded to four other carbon atoms in three-dimensional tetrahedral arrangements. In graphite, hexagonal rings of carbon atoms join to form stacked planes (or layers) of carbon atoms. Both allotropes are crystalline, with carbon atoms joined covalently.

13.51 In both S_8 and graphite, atoms form covalently bonded rings.

13.53 The properties of aspirin are like those of molecular crystals.

13.55 Energy is absorbed by a substance during melting. Energy is released by a substance during freezing. These are reverse processes. The amount of energy absorbed during melting is the same as that released during freezing.

13.57 1200 kcal

13.59 228 kJ

13.61 0.600 Calorie

13.63 Unlike nearly every other substance, water is less dense as the solid (ice) than is liquid water. That is why ice floats in water.

13.65 The specific heat of water, 1 cal/g-°C, is higher than that of any other substance listed in Table 13.6. Thus, great amounts of heat are needed to warm a given quantity of water. However, the water has a great capacity to hold the heat, so vast amounts of water on the Earth alternately store and release heat. This helps moderate daily temperature fluctuations.

13.67 As an ionic compound dissolves, ions are attracted away from the crystal lattice. Water molecules surround positive ions in solution so that the negative dipole end of water is near a positive ion. Water molecules surround negative ions so the positive dipole end is near the anion.

13.69 a, c. dipole and London forces
b. hydrogen bonding, dipole, and London forces
d. London forces

13.71 Molecules in liquid O_2 are held by London forces of attraction.

13.73 Ethylene glycol has two —OH groups that can enter into hydrogen bonding, so it would be expected to have a higher boiling point than ethanol, which has only one —OH group. Ethylene glycol could also be expected to evaporate more slowly than ethanol for the same reason.

13.75 Because water evaporates more slowly than gasoline, water can be expected to have greater intermolecular forces and greater surface tension than gasoline. Thus, water is expected to be less volatile than gasoline.

13.77 Xenon can be expected to have a higher boiling point than neon because xenon has a greater molar mass and greater London forces than neon.

13.79 CO_2 has covalent bonding and is nonpolar. It cannot have hydrogen bonding, so its intermolecular forces are London forces.

13.81 145 kJ

13.83

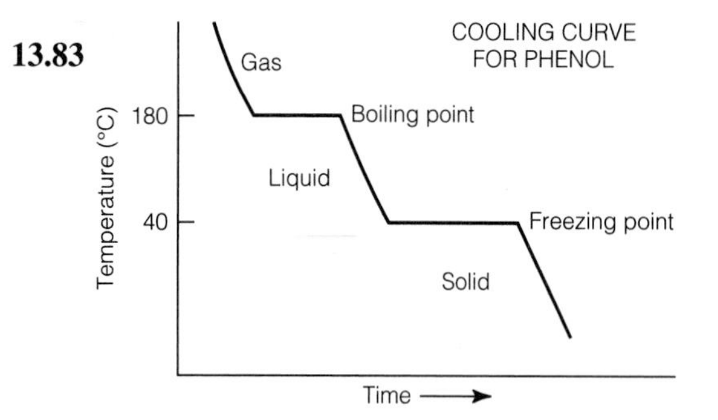

13.85 The heat of vaporization of water is 40.7 kJ/mol, which is the amount of heat energy difference between a mole of liquid water and a mole of steam at 100 °C. Burns caused by steam are more severe for this reason.

13.87 68.5 kJ

Chapter 14

Ex. 14.1 When an ionic solid such as NaCl dissolves in water, three things must happen: (1) The attractive forces holding the positive and negative ions together must be overcome. (2) The attractive forces holding at least some of the water molecules must be overcome. (3) The solute (NaCl) and solvent molecules (water) must interact and attract one another. The energy released by the interaction must be greater than the sum of the energy needed to overcome the forces holding the ions together and the energy needed to separate the solvent molecules.

Ex. 14.2 $375 \text{ g water} \times \dfrac{23.8 \text{ g KCl}}{100. \text{ g water}} = 89.3 \text{ g KCl}$

Ex. 14.3 $15.0 \text{ mL} \times \dfrac{1 \text{ L}}{1000 \text{ mL}} \times \dfrac{12.0 \text{ mol}}{1 \text{ L}} \times \dfrac{36.5 \text{ g HCl}}{1 \text{ mol HCl}} = 6.57 \text{ g HCl}$

Ex. 14.4 To prepare 250. g of 6.0% table sugar—or any other solute:
Solute: 6.0% × 250. g = 250. g × 0.060 = 15 g sugar (the solute)
Solvent: 250. g total − 15 g solute = 235 g water (the solvent)
Dissolve 15 g of sugar in 235 g of water.

14.1 The *solute* is the substance being dissolved. The *solvent* dissolves the solute and is the substance present in the greater amount. A *solution* is a homogeneous mixture of solute and solvent.

14.3

	Solute	*Solvent*	*Solution State*
a.	chlorine	water	liquid
b.	acetic acid	water	liquid
c.	zinc	copper	solid

14.5 A small amount of table sugar or table salt is soluble in a glass of water, but large amounts of sugar or salt are not soluble in the water.

14.7 completely miscible

14.9 Nonpolar solutes dissolve best in nonpolar solvents, and polar solutes dissolve best in polar solvents.

14.11 Glucose has several —OH groups, which can be involved in hydrogen bonding with water, so the solubility of glucose is much greater than would be expected on the basis of molar mass alone.

14.13 a. soluble; a polar compound
b. less soluble than (a) but perhaps slightly soluble; a polar compound
c. soluble; a soluble ionic compound
d. insoluble; by the solubility rules (Chapter 10)
e. soluble; —OH groups can form hydrogen bonds
f. soluble; a nitrate compound (all nitrates are water soluble)

14.15 Ammonia is quite polar so it does not dissolve in hexane (which is nonpolar), but it readily dissolves in water because of hydrogen bonding.

14.17 When ionic solids dissolve in water, the energy released by the interaction of the solute with water must be greater than the sum of the energy needed to overcome the forces holding the ions together in the crystal and the energy needed to separate the water molecules. For insoluble compounds, the energy released during hydration is less than the energy absorbed.

14.19 *Solvation* describes the process of dissolving a solute in a solvent. When water is the solvent, the process is called *hydration.*

14.21 a, c. soluble
b, d. insoluble (although dolomite is insoluble in water, it reacts with acids to give CO_2 and soluble HCO_3^- ions)

14.23 dissolving and crystallizing

14.25 a. none b. 40. g

14.27 Most soluble solids become more soluble with increasing temperature because the motion of particles increases at higher temperature, allowing more particles to be broken loose from the crystal. Once free, the particles in solution move too rapidly to be recaptured by the solid. Gases become less soluble with increasing temperature because the rapidly moving gaseous molecules escape from the liquid surface in an open container.

14.29 Once the bottle is opened, the pressure in the bottle drops back to atmospheric pressure and gas solubility decreases. As gas bubbles escape from the liquid, some of the foaming liquid may overflow the container.

14.31 yes

14.33 saturated

14.35 17 g; dissolve the 17 g of sucrose in enough water to make a total volume of 500. mL of solution.

14.37 400. mL

14.39 234 mL

14.41 75.0% isopropyl alcohol by volume. Water (present as the smaller quantity) is the solute; isopropyl alcohol is the solvent.

14.43 300. mL alcohol; add enough water to 300. mL of the alcohol to give 500. mL of solution.

14.45 14.4 mL

14.47 a. 25 g glucose, the solute, and 475 g water
b. 25 g $NaHCO_3$, the solute, and 475 g water
c. 60. g sodium carbonate, the solute, and 1940 g water

14.49 350. g

14.51 1.5×10^{-5} parts per thousand (ppt)

14.53 840 ppb

14.55 0.002 mg/L, 2 ppb

14.57 4.6 g/L, 0.58% (v/v)

14.59 200. mL of antiseptic diluted with 150 mL H_2O

14.61 4.17 mL acid diluted with water to give a total volume of 500. mL

14.63 1.25 M

14.65 With 5 kg of each, we have a greater number of dissolved moles (and of molecules) of methanol, so the *methanol* sample is the more effective antifreeze.

14.67 a. 2 b. 1 c. 3 d. 1

14.69 Particles in solution have diameters of about 0.1 to 1.0 nm. Particles in suspension have diameters of 100 nm or more. Colloidal particles have intermediate diameters (about 1.0 to 100 nm).

14.71 a. aerosol b. sol c. solid foam d. foam

14.73 In a saturated solution, the solute concentration is at a maximum. In a dilute solution, the solute concentration is less than maximum.

14.75 The carbonated beverage goes ''flat'' more quickly at a higher temperature because gases are less soluble in liquids at higher temperatures.

14.77

	Solute	*Solvent*	*Solution State*
a.	Zn	Cu	solid
b.	NH_3	water	liquid
c.	acetic acid	water	liquid

14.79 a. soluble; Rb^+ is an alkali metal ion.
b. soluble; ammonium compounds are soluble.
c. insoluble; more than four carbon atoms per oxygen atom
d. soluble; Na^+ is an alkali metal ion.

14.81 760. g, 458 mL

14.83 20.0% (w/w)

14.85 0.004 ppm, 4 ppb

14.87 During *osmosis,* only the solvent molecules pass through a semipermeable membrane. During *dial-*

ysis, the semipermeable membrane allows passage of ions and small solute molecules along with the solvent but retains large molecules and colloidal particles.

14.89 Celery in a saltwater solution loses crispness because water flows out of the cell—across the cell membrane—toward the more concentrated solution. When the celery is placed in water alone, water flows back across the membrane into the cells, giving them greater rigidity and restoring crispness.

14.91 A phenomenon called the Tyndall effect occurs when the path of a beam of light is visible as it passes through a liquid or gas (such as air) because the light is scattered by colloidal particles. This phenomenon can often be observed by looking sideways at the beam of light from a movie projector or the beam of a strong spotlight or searchlight.

Chapter 15

Ex. 15.1 The three factors are (1) collision frequency—controlled by concentration and temperature, (2) orientation or geometry at the time of collision, and (3) collision energy must be greater than the activation energy.

Ex. 15.2 All three of these are everyday examples where lower temperatures cause the rates of chemical reactions to be slowed down. The hibernating animal's lower body temperature decreases the rate of metabolism and the need for food. In a refrigerator, the cold temperature slows down the reactions that occur when food spoils. Finally, the chemical reactions in a cold automobile battery may be so slow there is not enough energy released to start an engine.

Ex. 15.3 In the Haber ammonia synthesis, either removing ammonia (a product) or increasing the pressure would shift the equilibrium toward the right to give a greater equilibrium concentration of ammonia. However, increasing the temperature would shift the equilibrium toward the left and decrease the equilibrium concentration of ammonia.

15.1 *Chemical kinetics* is the term used for the study of reaction rates and factors that affect reaction rates.

15.3 *Collision frequency* is the rate at which particles collide

15.5 *Activation energy* is the minimum kinetic energy that colliding particles must have—or must overcome—in order for a reaction to occur.

15.7 The specific atom of a molecule that is to be transferred to a second atom or molecule must be at the site of the collision. The rate of reaction increases when colliding particles have the required collision geometry.

15.9 Decreasing the temperature lowers reaction rates because particles that move more slowly will collide less frequently. Also the slower moving particles may not have the minimum activation energy for effective collisions.

15.11 At low temperatures, chemical reactions occur more slowly, so at the low temperatures of hibernating animals, metabolism—the chemical reactions involved in the utilization of food—must occur more slowly. Because metabolism is slowed, smaller amounts of food are utilized.

15.13 Decreasing the concentration lowers the reaction rate because the frequency of particle collisions is lower when there are fewer particles to collide.

15.15 A *catalyst* is a substance that speeds up the rate of a chemical reaction without itself being consumed in the reaction. MnO_2.

15.17 In going from Browning to Kalispell there are two main routes. The first requires that we climb about 700 m to get over Logan Pass. The second route through Marias Pass requires a climb of only 300 m. This second route—with a lower energy barrier—is analogous to the use of a catalyst.

15.19 Sugar is metabolized in living cells by a series of small steps, each of which requires a specific enzyme as a catalyst.

15.21 MnO_2(a catalyst) speeds up the decomposition of H_2O_2.

15.23 MnO_2 lowers the activation energy for the decomposition of H_2O_2.

15.25 a. Chlorine atoms react with ozone to give ClO molecules and O_2 gas.
b. ClO molecules react with oxygen atoms to give Cl atoms and O_2 gas.
c. The Cl atom that reacts with ozone in step 1 is released in step 2 and is free to repeat the cycle again and again by reacting with—and decomposing—more ozone molecules.

15.27 a. physical equilibrium; at the boiling point

b. chemical equilibrium; when the rate of the forward reaction equals the rate of the reverse reaction.
c. physical equilibrium; when a saturated solution is present with excess sugar

15.29 Some molecules are melting while an equal number are freezing.

15.31 Before equilibrium is established, the concentrations of reactants (A + B) decrease while the concentrations of products (C + D) increase.

15.33 When equilibrium is established—in this analogy—the number of uncombined nuts and bolts is stabilized (constant) and the number of combined nuts and bolts is also stabilized (constant).

15.35 At equilibrium in the nuts and bolts analogy, the rate at which nuts and bolts are assembled—the forward reaction—is equal to the rate at which nuts and bolts are taken apart—the reverse reaction.

15.37 Before equilibrium is established, the concentrations of reactants, CO and O_2, decrease as the concentration of CO_2 increases. When equilibrium is established, the concentrations of CO, O_2, and CO_2 remain constant.

15.39 a. Equilibrium shifts toward the left to favor (increase) the production of reactants.
b. Equilibrium shifts toward the right to favor (increase) the production of products.
c. Equilibrium shifts toward the right to favor (increase) the production of products.

15.41 a. Increasing total pressure will have no effect on the position of the equilibrium (because the number of moles of gas does not change).
b. Increasing total pressure will shift equilibrium to the left to favor an increase in concentrations of reactants.
c. Increasing total pressure will shift equilibrium to the right to favor an increase in concentrations of products.

15.43 a. The equilibrium shifts to the right to favor an increase in concentrations of products.
b. The equilibrium shifts to the right to favor an increase in concentrations of products.
c. The equilibrium shifts to the right to favor an increase in concentrations of products.
d. no effect

15.45 a. The equilibrium shifts to the right to favor an increase in the concentrations of products.
b. The equilibrium shifts to the left to favor an increase in concentrations of reactants.

15.47 A catalyst increases the rate of a reaction by lowering the activation energy. With a lower activation energy, a greater fraction of reactants have sufficient energy to react, so the rate of the reaction increases.

15.49 The temperature of the exhaust gases is not great enough for high percentages of the gases to react. A catalyst is needed so the reactions can proceed more efficiently at lower activation energies.

15.51 a. $K_{eq} = \frac{[HCl]^2}{[H_2][Cl_2]}$ b. $K_{eq} = \frac{[CO]^2[O_2]}{[CO_2]^2}$

c. $K_{eq} = \frac{[O_3]^2}{[O_2]^3}$

15.53 The equilibrium lies to the right; the concentrations of products are greater than concentrations of reactants.

15.55 a, c. to the right, favoring products
b. to the left, favoring reactants
c. to the left, favoring reactants (but this K_{eq} is too close to 1 to tell for sure)

15.57 $K_{eq} = \frac{[HI]^2}{[H_2][I_2]}$

$$[H_2] = \frac{[HI]^2}{K_{eq}[I_2]} = \frac{[1.50]^2}{50[0.200]} = 0.225$$

15.59 The equilibrium constant increases from 50 to 69 as temperature decreases, so the products are favored more at the lower temperature.

15.61 The activation energy for the oxidation of glucose is much less than the activation energy for the photosynthesis of glucose.

15.63 In the fall of the year, when temperatures are lower, insects move more slowly because the surrounding temperature controls the body temperature of cold-blooded insects. Their biochemical reactions are also slowed.

15.65 In the high temperature and pressures present in an automobile engine, some nitrogen gas from the air combines with oxygen from the air to give oxides of nitrogen. Certain catalysts are used in a catalytic converter to decompose these oxides of nitrogen to give N_2 and O_2.

15.67 a, b. The equilibrium shifts to the right, favoring the production of methanol.

c. The equilibrium shifts to the left, favoring an increase in concentrations of reactants.
d. no effect

15.69 a. $K_{eq} = \dfrac{[NO]^2}{[N_2][O_2]}$

b. Higher engine temperatures tend to increase NO(g) pollution because adding heat shifts the equilibrium to the right for this reaction.

Chapter 16

Ex. 16.1 The six strong acids are hydrochloric, HCl, hydrobromic, HBr, hydroiodic, HI, sulfuric, H_2SO_4, nitric, HNO_3, and perchloric, $HClO_4$.

Ex. 16.2 An oven cleaner containing NaOH (a strong base) would be more dangerous to use than one containing NH_3 (a weak base). A strong base like NaOH can cause severe burns. Ammonia is not as dangerous to use because it is a weak base, but precautions should be taken when using either strong or weak bases.

Ex. 16.3 Substitute the known $[H^+]$ in the equation $pH = -\log [H^+]$ to obtain $pH = -\log(1.0 \times 10^{-3})$. The coefficient is 1.0, so the pH is equal to the number, 3, in the exponent. The pH of the vinegar solution is 3. The pH is less than 7 so it is acidic.

Ex. 16.4 Substitute the known $[H^+]$ in the equation $pH = -\log [H^+]$ to obtain $pH = -\log(5.7 \times 10^{-4})$. From the exponent, -4, we can make an approximation; the pH will be between 3 and 4. Use a calculator with a LOG key. Press 5.7, press the EE (or EXP) key, press 3 (the exponent), press the +/− key (to change the sign of the exponent), press the LOG key, and press +/− to obtain the negative log. Read the pH, 3.24 (3 significant figures).

Ex. 16.5 For a pH of 4.83, the $pOH = 14 - 4.83 = 9.17$.

Ex. 16.6 Substitute the known pH in the equation $pH = -\log [H^+]$ to obtain $3.60 = -\log [H^+]$. Multiply both sides by -1 to give $\log [H^+] = -3.60$ and take the inverse log (antilog) of both sides to find $[H^+]$. For a calculator with INV and LOG keys, enter 3.60, press the +/− key (to change the sign), press the EE (or EXP) key (to access the exponential mode), press INV then LOG to obtain the inverse log. Read the $[H^+]$, 2.51×10^{-4} (3 significant figures).

Ex. 16.7 The solution is basic (rule 3). This is a salt of a strong base, NaOH, and a weak acid, acetic acid, CH_3COOH.

16.1 Acids (1) taste sour, (2) cause litmus to change from blue to red, (3) react with active metals to produce hydrogen gas, and (4) react with bases to form water and salts.

16.3 Arrhenius attributed properties of acids to the presence of hydrogen ions.

16.5 A hydronium ion, H_3O^+, is a hydrogen ion, H^+, bonded to a water molecule. The terms are used interchangeably because all hydrogen ions are actually hydrated (they are joined to water molecules) to form hydronium salts.

16.7 a, d. acid b. base c, e. salt
f. none of these

16.9 a. H_3PO_4 (triprotic) b. HNO_3 (monoprotic)
c. H_2CO_3(diprotic) d. $CH_3CHOHCOOH$ (monoprotic)

16.11 Six strong acids: HCl (hydrochloric acid), HBr (hydrobromic acid), HI (hydroiodic acid), H_2SO_4 (sulfuric acid), HNO_3 (nitric acid), and $HClO_4$ (perchloric acid). Three weak acids: CH_3COOH (acetic acid), H_2CO_3 (carbonic acid), and HCN (hydrocyanic acid). There are others.

16.13 Strong acids are those that ionize completely in water; weak acids do not.

16.15 A weak acid is one that ionizes only slightly in water. After water is mixed with a concentrated acid, it is called a dilute acid.

16.17 Aqueous ammonia is NH_3 dissolved in water. A relatively small amount of the NH_3 reacts with water to produce OH^- ions and NH_4^+ ions in solution.

16.19 a, b. weak acid c, d. strong acid
e, f. weak base

16.21 Yes; acids spattered in the eyes can cause permanent damage. Concentrated acids are quite corrosive and can cause severe burns. Concentrated bases are quite caustic (they destroy body tissues).

16.23 $H_2SO_4(aq) + 2\ KOH \rightarrow K_2SO_4(aq) + 2\ H_2O$
The net ionic equation: $H^+ + OH^- \rightarrow H_2O$

16.25 a. $2\ HCl(aq) + Mg \rightarrow MgCl_2 + H_2$
b. $2\ HCl(aq) + MgO \rightarrow MgCl_2 + H_2O$
c. $2\ HCl(aq) + MgCO_3 \rightarrow MgCl_2 + CO_2 + H_2O$

16.27 An amphoteric metal hydroxide is one that will react with either strong acids or strong bases. Examples: $Al(OH)_3$, $Zn(OH)_2$.

16.29 a. Already balanced. $Al(OH)_3$ will react with a strong base.
b. $Al(OH)_3(s) + 3\ HCl(aq) \rightarrow AlCl_3(aq) + 3\ H_2O$
$Al(OH)_3$ will react with a strong acid; it is an amphoteric hydroxide.

16.31 $HCl(aq) + NaHCO_3(aq) \rightarrow NaCl(aq) + CO_2(g) + H_2O$
The net ionic equation:

$$H^+ + HCO_3^- \rightarrow CO_2 + H_2O$$

16.33 An acid, according to Brønsted–Lowry, is a proton donor. According to the Lewis definition, an acid is an electron pair acceptor.

16.35 Water as a base:
$HCl(aq) + H_2O \rightarrow H_3O^+(aq) + Cl^-(aq)$

16.37 $HCl(g) + NH_3(g) \rightarrow NH_4Cl(s)$
This is an acid–base reaction by Brønsted–Lowry and Lewis definitions, but not according to Arrhenius.

16.39 a. C_5H_5N is a Brønsted base (it accepts a proton). $C_5H_5NH^+$ is its conjugate acid.
b. C_6H_5OH is a Brønsted acid (it donates a proton). $C_6H_5O^-$ is its conjugate base.
c. $CH_3CHOHCOOH$ is a Brønsted acid (it donates a proton from the —COOH group). $CH_3CHOHCOO^-$ is its conjugate base.

16.41 a. HBr and HF are the acids. HBr is the stronger acid of the two; it gives up a proton more readily so there are more Br^- ions than F^- ions in solution.
b. HCN and HF are the acids. HF is the stronger acid of the two; it gives up a proton more readily so there are more F^- ions than CN^- ions in solution.

16.43 a. F^- and Br^- are bases. F^- is the stronger base of the two; it accepts a proton more readily so there is more HF than HBr in solution.
b. F^- and CN^- are bases. CN^- is the stronger base of the two; it accepts a proton more readily so there is more HCN than HF in solution.

16.45 Phenol, C_6H_5OH, acts as an acid, releasing a proton to form $C_6H_5O^-$ ions.

16.47 a. $[OH^-] = 1 \times 10^{-11}$ M; the pH is 3.0; the sample is acidic.
b. $[OH^-] = 1 \times 10^{-5}$ M; the pH is 9.0; the sample is basic.
c. $[H^+] = 1 \times 10^{-9}$ M; the pH is 9.0; the sample is basic.
d. $[H^+] = 1 \times 10^{-4}$ M; the pH is 4.0; the sample is acidic.

16.49 a. pH = 6.00 b. pH = 9.3 c. pOH = 11.00
d. pOH = 4.6

16.51 a. $[H^+] = 1.00 \times 10^{-3}$ M; pH = 3.0
b. $[H^+] = 1.00 \times 10^{-9}$ M; pH = 9.0
c. $[H^+] = 1.00 \times 10^{-4}$ M; pH = 4.0
d. $[H^+] = 1.00 \times 10^{-11}$ M; pH = 11.0

16.53 a. pH = 4.5 b. pH = 11.1 c. pH = 7.9
d. pH = 8.8

16.55 a. $[H^+] = 4.5 \times 10^{-9}$ M b. $[H^+] = 1.9 \times 10^{-3}$ M
c. $[H^+] = 1.3 \times 10^{-5}$ M d. $[H^+] = 1.2 \times 10^{-8}$ M

16.57 a. neutral; this is a salt of a strong base, KOH, and a strong acid, H_2SO_4.
b. basic; this is a salt of a strong base, $Ca(OH)_2$, and a weak acid, CH_3COOH.
c. acidic; this is a salt of a weak base, NH_3, and a strong acid, HCl.
d. acidic; this is a salt of a weak base, NH_3, and a strong acid, HNO_3.
e. basic; this is a salt of a strong base, NaOH, and a weak acid, H_2CO_3.

16.59 acidic; this is the salt of a weak base, NH_3, and a strong acid, HNO_3.

16.61 A buffer is prepared by using a weak acid and a salt of that acid or a weak base and a salt of that base.

16.63 acetic acid; it is the weak acid that contains the acetate ion.

16.67 In an acid–base titration, the concentration of an acid can be determined by adding just enough base, dropwise, to neutralize the acid. At the end point, the concentration of the acid can be calculated when concentration of the base and volumes of both acid and base are known.

16.69 basic; the salt formed is that of a strong base and a weak acid.

16.71 0.5000 M HCl

16.73 0.06896 M NaOH

16.75 a. triprotic (three acidic H's)
b. monoprotic (one acidic H)
c, d. diprotic (two acidic H's)

16.77 Although the percent of acid is the same for both acids, the sulfuric acid is a strong acid that is totally ionized, whereas the vinegar contains a weak acid (acetic acid) that is only slightly ionized.

16.79 $CaCO_3 + H_2SO_4 \rightarrow CaSO_4 + CO_2 + H_2O$

16.81 $NaHCO_3 + CH_3COOH \rightarrow NaCH_3COO^- + CO_2 + H_2O$

16.83 $CaCO_3$, $Mg(OH)_2$, and $Al(OH)_3$

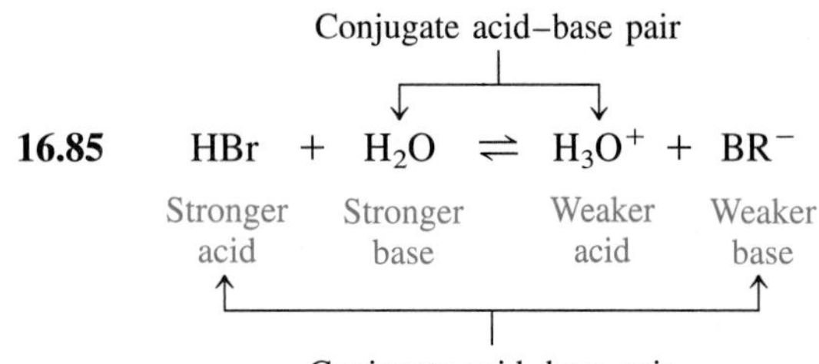

16.87 1225 tons H_2SO_4

16.89 This answer depends on the antacids you select.

16.91 a. Only the Lewis definition applies here. The $FeCl_3$ is the Lewis acid.
b. Brønsted–Lowry and Lewis definitions both apply. The H_2O is a Brønsted base (it accepts a proton) and a Lewis base (it donates an electron pair).

16.93 a. $[OH^-] = 1.00 \times 10^{-4}$ M; the pH is 4.0; the sample is acidic.
b. $[OH^-] = 1.14 \times 10^{-7}$ M; the pH is 7.1; the sample is very slightly basic (almost neutral).
c. $[H^+] = 1.00 \times 10^{-8}$ M; the pH is 8.0; the sample is basic
d. $[H^+] = 1.03 \times 10^{-6}$ M; the pH is 6.0; the sample is acidic.

Chapter 17

Ex. 17.1 The oxidation number of oxygen is −2. The overall charge is zero (because SO_2 is a compound), so the sum of charges is 0. If the oxidation number of sulfur is represented as x, the equation is $x + 2(-2) = 0$ so $x = +4$. The oxidation number of s is +4.

Ex. 17.2 The oxidation number of oxygen is −2 and there are three oxygen atoms. The sum of charges is −2 (the same as the ion charge). If the oxidation number of carbon is represented as x, the equation is $x + 3(-2) = -2$ so $x = +4$. The oxidation number of C is +4.

Ex. 17.3 Aluminum loses electrons and is oxidized (oxidation definition 3, Section 17.2), so it is the reducing agent, RA. Br_2 gains electrons and is reduced (reduction definition 4, Section 17.3), so it is the oxidizing agent, OA.

Ex. 17.4 The bracelet must be attached to the cathode to be plated. Reduction occurs when the gold (or any other metal) is deposited.

17.1 a. −2 b. +6 c. +3 d. +1

17.3 a. 0 b. +1 c. +5 d. +3

17.5 a. Oxidation occurs when an element or compound gains oxygen atoms.
b. Oxidation occurs when a compound loses hydrogen atoms.
c. Oxidation occurs when an atom or ion of an element loses electrons.
d. Oxidation occurs when there is an increase in oxidation number.

17.7 a–c. reduced d. oxidized

17.9 a, d. not redox b, c. redox

17.11 oxidized (oxidation number of carbon changed from $-\frac{1}{2}$ to 0)

17.13 a. $\underset{\text{RA}}{4\ Al} + \underset{\text{OA}}{3\ O_2} \rightarrow 2\ Al_2O_3$
b. $\underset{\text{OA}}{C_2H_2} + \underset{\text{RA}}{H_2} \rightarrow C_2H_4$
c. $\underset{\text{OA}}{2\ AgNO_3} + \underset{\text{RA}}{Cu} \rightarrow Cu(NO_3)_2 + 2\ Ag$
d. $\underset{\text{OA}}{CuO} + \underset{\text{RA}}{CO(g)} \rightarrow Cu + CO_2(g)$

17.15 a. S is oxidized. b. N is reduced.
c. Oxidizing agent is HNO_3.
d. Reducing agent is SO_2.

17.17 a. Fe is oxidized. b. Mn is reduced.
c. Oxidizing agent is MnO_4^-.
d. Reducing agent is Fe^{2+} ions.

17.19 a. H is oxidized. b. Mo is reduced.
c. Oxidizing agent is MoO_3.
d. Reducing agent is H_2.

17.21 a. Indoxyl is oxidized.
b. Oxidizing agent is O_2 gas.

17.23 a. oxidizing agent: Cl_2; reducing agent: bromide ions
b. oxidizing agent: $Cr_2O_7^{2-}$ ions; reducing agent: iron(II) ions

17.25 Oxidizing agents include oxygen gas, permanganate ions, dichromate ions, and hydrogen peroxide.

17.27 Sulfur is oxidized. Its oxidation number changes from +4 in SO_3^{2-} ions to +6 in SO_4^{2-} ions.

17.29 $Cu + 4\ H^+ + 2\ NO_3^- \rightarrow Cu^{2+} + 2\ NO_2 + 2\ H_2O$

17.31 $10\ Cl^- + 16\ H^+ + 2\ MnO_4^- \rightarrow 5\ Cl_2 + 2\ Mn^{2+}\ 8\ H_2O$

17.33 An electrolytic cell is an electrochemical cell that uses electricity to drive a chemical reaction that would otherwise not occur (Section 17.8).

17.35 a. reduction b. Pb c. at the cathode

17.37 a. oxidation b. Cd c. at the anode

17.39 a. oxidation; at the anode
b. reduction; at the cathode.
c. net reaction: $2\ H_2(g) + O_2(g) \rightarrow 2\ H_2O$

17.41 a. $Mg^{2+} + 2\ e^- \rightarrow Mg(s)$
b. $2\ Cl^- \rightarrow Cl_2(g) + 2\ e^-$

17.43 A nickel metal strip (one electrode) is inserted into a solution of $Ni(NO_3)_2$ that is connected by a salt bridge to a solution of $Cu(NO_3)_2$ that contains a copper metal strip (the second electrode). An electric current flows when the electric circuit is completed by connecting a conductor to the two electrodes. Anode (oxidation) half-reaction:

$$Ni(s) \rightarrow Ni^{2+} + 2\ e^-$$

Cathode (reduction) half-reaction:

$$Cu^{2+} + 2\ e^- \rightarrow Cu(s)$$

Net ionic equation:

$$Ni(s) + Cu^{2+} \rightarrow Ni^{2+} + Cu(s)$$

17.45 a. +1 b. +3 c. +5 d. 0

17.47 +4

17.49 +7

17.51 The hypochlorite ion, ClO^-, is acting as an oxidizing agent. The oxidizing agent is the one that gets reduced. Here the oxidation number of Cl is reduced from +1 in ClO^- to −1 in the chloride ion, Cl^-.

17.53 H_2SO_4 is the oxidizing agent; KI is the reducing agent.

17.55 $2\ I^- + H_2O_2 + 2\ H^+ \rightarrow I_2(s) + 2\ H_2O$

17.57 a. $Zn(s) \rightarrow Zn^{2+} + 2\ e^-$
b. oxidized c. anode

17.59 a. $Fe \rightarrow Fe^{2+} + 2\ e^-$
b. oxidation (electrons are lost)

Chapter 18

Ex. 18.1 Sulfur is formed. ${}^{32}_{15}P \rightarrow {}^{0}_{-1}e + {}^{32}_{16}S$

Ex. 18.2 The fraction remaining after 3 half-lives is

$$\frac{1}{2} \times \frac{1}{2} \times \frac{1}{2} = \frac{1}{2^3} = \frac{1}{8}$$

$$6.00\ g \times \frac{1}{8} = 0.75\ g\ \text{remain}$$

Ex. 18.3 Plutonium-238 is formed. ${}^{238}_{93}Np \rightarrow {}^{0}_{-1}e + {}^{238}_{94}Pu$

Ex. 18.4 Fermi and Segrè (physicists) bombarded uraium with neutrons and found four radioactive species, but incorrectly presumed they had produced element 93. (Actually, fission had occurred.) Hahn and Strassman (chemists) repeated the Fermi–Segrè experiment and identified the products as having masses about half that of uranium, and correctly concluded that the uranium atom had split. Meitner and Frisch calculated the energy released during fission of uranium and found it to be much greater than for any known nuclear reaction. They also concluded that the large number of neutrons released could split other uranium atoms—sustaining a chain reaction—and release enormous amounts of energy.

18.1 350; most elements have more than one naturally occurring isotope.

18.3 ${}^{4}_{2}He$; α

18.5 During alpha decay, the atomic number of the original nucleus decreases by 2 while the mass number decreases by 4.

18.7 Gamma rays—one type of high-energy radiation—have neither mass nor charge; they are emitted by unstable isotopes undergoing alpha or beta decay.

18.9 a. ${}^{60}_{27}Co$ b. ${}^{1}_{1}H$ c. ${}^{4}_{2}He$ d. ${}^{226}_{88}Ra$

18.11 a. ${}^{2}_{1}H$ b. ${}^{39}_{19}K$ c. ${}^{40}_{19}K$ d. ${}^{32}_{15}P$

18.13 **b** and **d** represent isotopes (atoms with the same atomic number).

18.15 ${}^{214}_{84}Po \rightarrow {}^{4}_{2}He + {}^{210}_{82}Pb$; lead

18.17 a. ${}^{238}_{92}U \rightarrow {}^{234}_{90}Th + {}^{4}_{2}He$
b. ${}^{234}_{90}Th \rightarrow {}^{234}_{91}Pa + {}^{0}_{-1}e$
c. ${}^{234}_{91}Pa \rightarrow {}^{234}_{92}U + {}^{0}_{-1}e$
d. ${}^{234}_{92}U \rightarrow {}^{230}_{90}Th + {}^{4}_{2}He$

18.19 1500 atoms remain; 750 atoms remain.

18.21 24.6 y

18.23 26 s

18.25 Ionizing radiation can cause vital molecules in living cells to be ionized or broken into fragments that cannot carry out normal functions.

18.27 Geiger counters and scintillation counters

18.29 80

18.31 Radon-219 and radon-220 decay very quickly (their half-lives are measured in seconds), but radon-222, with a longer half-life of 3.82 days, is of greater concern; its daughters—alpha-emitting polonium radioisotopes—can be trapped in the lungs where they can damage cells and cause cancer.

18.33 beta and gamma radiation

18.35 alpha particles

18.37 radon, cosmic rays, rocks and soil, and radiation from inside our bodies

18.39 ${}^{31}_{16}S \rightarrow {}^{0}_{+1}e + {}^{31}_{15}P$

18.41 a. $^{7}_{3}Li$ b. $^{1}_{1}H$ c. $^{1}_{1}H$

18.43 a. $^{240}_{95}Am$ b. $^{242}_{96}Cm$ c. $^{243}_{97}Bk$ d. $^{245}_{98}Cf$

18.45 The cancer cells of the thyroid are destroyed by the radiation.

18.47 Cobalt-60 and radium-226 are radioisotopes used in cancer therapy.

18.49 no

18.51 Plutonium-238 is used to power heart pacemakers.

18.53 The critical mass is the minimum amount of a fissionable material required to sustain a chain reaction.

18.55 neutrons

18.57 A moderator is used to slow down fast neutrons released by fissionable atoms; the slower moving neutrons are captured by nuclei that, in turn, undergo fission. Heavy water and graphite have been used as moderators.

18.59 uranium-235

18.61 When Fermi and Segrè bombarded uranium with neutrons, they thought they had produced a new element.

18.63 Differences between fission in nuclear bombs and fission in nuclear power plants: (1) Control rods are used in nuclear power plants, but not in bombs. (2) Nuclear bombs require a uranium-235 concentration of 85 to 90%, but nuclear power plants require uranium-235 enrichment to only 3%. (3) A moderator is used in nuclear power plants, but not in fission bombs.

18.65 During nuclear fusion, nuclei of small atoms must be brought together and united. Fusion takes place at the intense temperatures of the sun and in hydrogen bombs.

18.67 Technetium-99^{m} is used in diagnostic medical testing of the brain, liver, and lungs.

18.69 a. 30 protons, 32 neutrons, mass number 62, symbol Zn
b. 94 protons, 147 neutrons, mass number 241 symbol Pu
c. 15 protons, 17 neutrons, mass number 32, symbol P

18.71 a. $^{11}_{5}B$ b. $^{12}_{6}C$

18.73 3; the reaction shows that 1 neutron is captured and 3 neutrons are released so a chain reaction is possible.

18.75 This answer depends on your point of view. If you had relatives among the U.S. troops preparing for an invasion of Japan, you would probably agree with Truman's decision. If you lived in one of the cities to be bombed you would undoubtedly disagree with Truman's decision.

18.77 1

18.79 positron emission tomography; positrons are particles with a mass equal to that of an electron but with a positive charge, symbolized $^{0}_{+1}e$. When a positron collides with an electron in the human body, two gamma rays are produced.

Chapter 19

Ex. 19.1 methane, ethane, propane, butane

Ex. 19.2 Numbering from right to left, the double bond is at carbon number 2, methyl is at number 4, and the bromine is at number 5.

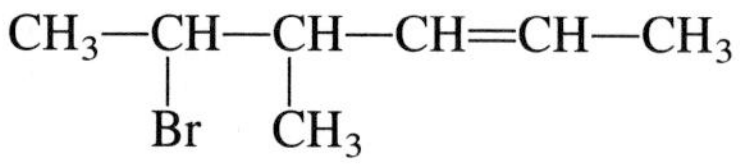

$$\begin{array}{l} CH_3-\underset{\displaystyle Br}{\underset{|}{CH}}-\underset{\displaystyle CH_3}{\underset{|}{CH}}-CH{=}CH-CH_3 \end{array}$$

Ex. 19.3

$$CH_3-\overset{\displaystyle CH_3}{\overset{|}{\underset{\displaystyle OH}{\underset{|}{C}}}}-CH_3 \qquad CH_3-CH_2-\overset{\displaystyle CH_3}{\overset{|}{\underset{\displaystyle OH}{\underset{|}{C}}}}-CH_3$$

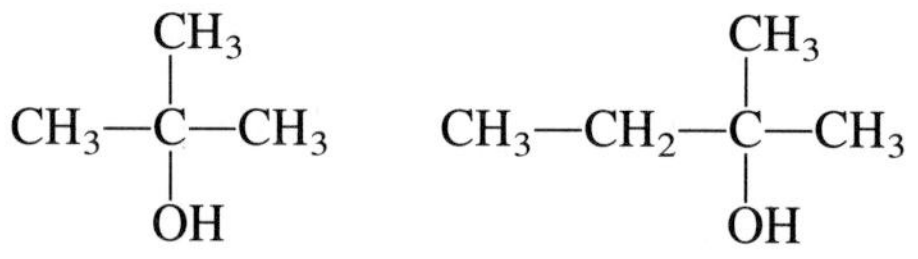

tert-Butyl alcohol 2-Methyl-2-butanol

Both alcohols are classified as tertiary alcohols.

Ex. 19.4 The common name is *sec*-butyl ethyl ketone because a *sec*-butyl group and an ethyl group are attached to the carbonyl group. These groups are named in alphabetical order.

19.1 With the development of thousands of synthetic organic compounds, the old definition of organic chemistry had to be expanded to include all carbon-containing compounds; originally the definition included only those compounds that are present in or produced by living organisms.

19.3 a. NaOH, an inorganic compound, would be expected to have the higher melting point.
b. KCl, an inorganic compound, would be expected to have the higher melting point.
c. $C_{20}H_{42}$, with the higher molar mass, would be expected to have a higher melting point than C_6H_{12}.

19.5 organic

19.7 a. 2; CH_3CH_3 b. 4; $CH_3CH_2CH_2CH_3$
c. 4; $CH_3\underset{\displaystyle CH_3}{\underset{|}{C}}HCH_3$ d. 5; $CH_2{=}CHCH_2CH_2CH_3$
e. 8; $CH_3CH_2CH_2CH_2\overset{\displaystyle CH_3}{\overset{|}{\underset{\displaystyle CH_3}{\underset{|}{C}}}}HCH_3$
f. 2; $HC{\equiv}CH$

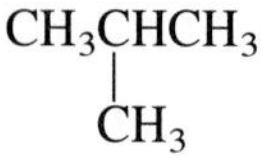

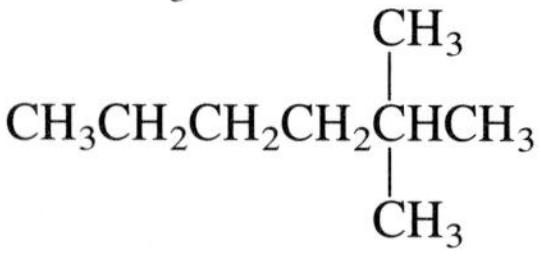

19.9 a. ethene b. hexane
c. bromoethane (ethyl bromide)
d. 2-chloropropane (isopropyl chloride)

19.11 a. same compound b, c. isomers

19.13 a–c. unsaturated d. saturated

19.15 a. $CH_3CH_2CH(CH_3)CH_2CH_3$, 3-methylpentane
b. $CH_3CH(CH_3)CH_2CH_2CH_3$, 2-methylpentane

19.17 $CH_3CH_2CH_2CH_2OH$, primary
$CH_3CH_2CH(OH)CH_3$, secondary
$CH_3C(CH_3)_2$—OH, tertiary
$CH_3CH(CH_3)CH_2$—OH, primary

19.19 hexane

19.21 a. CH_2=$CHCH(CH(CH_3)CH_3)CH_2CH_2CH_3$
b. CH_3—C(Cl)=C(Cl)—CH_3
c. CH_3—C(Cl)=C(Cl)—CH_3

19.23 a. aromatic b, c. aliphatic

19.25 [benzene resonance structure] ⟷ [benzene resonance structure]

Based on these structures, one would think benzene would behave like an alkene and undergo addition reactions, but this is not the case.

19.27 a. para b. ortho c. meta

19.29 a. Cl—C_6H_4—Cl b. NO_2—C_6H_4—NO_2

c. d.

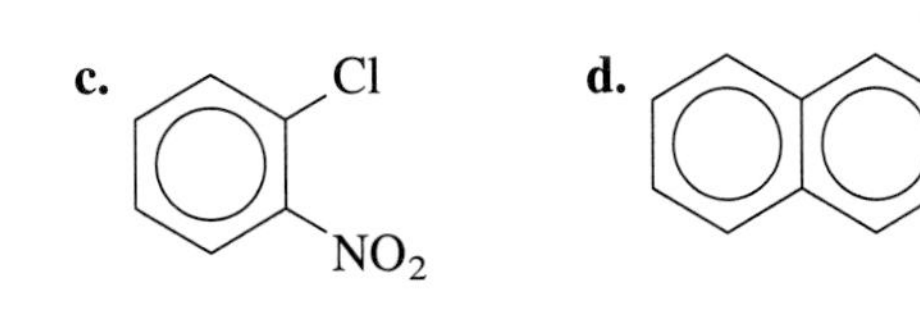

19.31 a. CH_3OH, primary
b. $CH_3C(CH_3)_2$—OH, tertiary
c. $CH_3CH_2CH(OH)CH_3$, secondary
d. $CH_3CH(CH_3)CH_2$—OH, primary

19.33 Destructive distillation involves heating a substance—such as wood—in the absence of oxygen. Methyl alcohol can be obtained from the destructive distillation of wood.

19.35 Denatured alcohol is ethanol that has been made unfit for use as a beverage by the addition of various poisonous and noxious substances.

19.37 86 proof (The proof is two times the percentage of ethanol.)

19.39 Commercial production of methanol involves the catalyzed reaction of hydrogen gas with carbon monoxide (obtained from partial oxidation of methane) at a high temperature and pressure.

$$CO + 2\,H_2 \xrightarrow{\text{catalyst}} CH_3OH$$

19.41 1,2-Ethanediol (called ethylene glycol) is the most common diol. It is the key ingredient in automobile antifreeze, but it is also used extensively in the manufacture of polyester fibers and films.

19.43 A phenolic compound is one that contains a benzene ring with an —OH group substituted on the ring. The general formula is Ar—OH, where Ar symbolizes a benzene ring or a structure that contains a benzene ring.

19.45 R—O—R

19.47 R—C(=O)—H All aldehydes contain the carbonyl group, —C(=O)—.

19.49 H—C(=O)—H Formaldehyde is used in embalming fluids, adhesives, and plastics. Formalin is a 40% solution of formaldehyde in water.

19.51 a. CH_3CH_2—C(=O)—CH_3 b. CH_3—C(=O)—H
c. CH_3—C(=O)—$CH(CH_3)CH_3$

19.53 A secondary alcohol can be oxidized to give a ketone.

19.55 a. $CH_3CH(OH)CH_2CH_2CH_3$
2-pentanol
b. CH_3CH_2OH
ethanol

19.57 $R-\overset{O}{\overset{\|}{C}}-OH$ is the general structural formula for a carboxylic acid.

19.59 $R-\overset{O}{\overset{\|}{C}}-OR$ is the general structural formula for an ester.

19.61 $CH_3CH_2CH_2\overset{O}{\overset{\|}{C}}-OH + HOCH_3 \longrightarrow$

Butyric acid Methyl alcohol

$CH_3CH_2CH_2\overset{O}{\overset{\|}{C}}-OCH_3 + H_2O$

Methyl butyrate Water

19.63 $-NH_2$ is the general structure of an amino group. An amine is a compound that can be formed by replacing one, two, or three hydrogen atoms of ammonia with an alkyl or aryl group.

19.65 a. $CH_3CH_2-NH_2$, primary

b. $CH_3-\overset{H}{\overset{|}{N}}-CH_3$, secondary

c. $CH_3-\overset{CH_3}{\overset{|}{N}}-CH_3$, tertiary

19.67 $CH_3\overset{O}{\overset{\|}{C}}-NHCH_3$ This is an amide.

19.69 candle wax

19.71 Methanol is metabolized to formaldehyde, and its long-term toxicity is quite high; furthermore, ingestion of as little as 30 mL of methanol can cause permanent blindness, so it cannot be used as a solvent for medicines. Ethanol can be metabolized—within limits—without these problems.

19.73 a. 2-butanol b. 2-pentanol
c. 3-methylbutanol d. ethoxyethane (diethyl ether)

19.75 a. butanone (ethyl methyl ketone)
b. propanol
c. 3-hexanone (ethyl propyl ketone)

19.77 a. $HO-\overset{O}{\overset{\|}{C}}CH_2CH_2CH_3$, butanoic acid

b. $HO-\overset{O}{\overset{\|}{C}}(CH_2)_{10}CH_3$, dodecanoic acid (It contains 12 carbon atoms.)

c. $CH_3CH_2CH_2\overset{O}{\overset{\|}{C}}-OCH_2CH_3$, ethyl butanoate

19.79 a. propylamine, primary
b. ethylmethylamine, secondary
c. trimethylamine, tertiary

Chapter 20

Ex. 20.1 Starch and cellulose are both natural polymers made up of glucose units. In starch, the glucose units are joined by alpha linkages, but in cellulose, the glucose units are joined by beta linkages. We humans have enzymes that break down starch, but we do not have enzymes that allow us to digest cellulose. Glucose and galactose are both aldohexoses. Their only structural difference is the right- and left-handed arrangement (the stereochemistry) of —OH and —H groups.

Ex. 20.2 The eight essential amino acids are phenylalanine, valine, leucine, isoleucine, methionine, threonine, tryptophan, and lysine.

Ex. 20.3 We trust that this course and this text have affected your life in numerous positive ways. The specific ways vary, but may include a better understanding of such things as pollution, solvents, acids and bases, chemicals in our bodies, among others. Chemistry affects everyone because we all depend on chemical changes. From metabolism to combustion, from food production to medicines, electrical lighting, sound equipment, automobiles, computers, photographs, paints, plastics, and so on, chemistry is everywhere.

20.1 dextrose or ''blood sugar''; it is a monosaccharide.

20.3 a. glucose b. maltose

20.5 Amylose, amylopectin, and glycogen are polysaccharides made up of glucose units joined by alpha linkages.

20.7 In starch, the glucose units are joined by alpha linkages, but in cellulose, the glucose units are joined by beta linkages.

20.9 The starch present in the soda cracker is broken up to give glucose units even while the cracker is in the mouth, and the glucose tastes sweet.

20.11 The configuration shown is L.

20.13 A saturated fatty acid is a long-chain carboxylic acid with its carbon atoms joined by single bonds. Examples: stearic and oleic acids.

20.15 A simple lipid is composed of a glycerol molecule joined to three fatty acid molecules by ester linkages.

20.17 a. monounsaturated; olive oil
b. saturated; palm oil
c. saturated; oil of nutmeg
d. polyunsaturated; liver

20.19 The iodine number is an overall measure of unsaturation of fats and oils.

20.21 The iodine number for corn oil, a vegetable oil, would be expected to be higher than the iodine number for beef tallow, an animal fat.

20.23 Proteins are present in every living cell.

20.25 An amino acid contains a carboxyl group, —COOH, and an amino group, $—NH_2$.

20.27 About 20 different amino acids are incorporated into proteins.

20.29 A polypeptide is made up of a string of about 10 or more amino acids. When the molar mass (molecular weight) of the polypeptide exceeds 10,000, it is called a protein.

20.31 the primary structure

20.33 Deoxyribonucleic acid (DNA) and ribonucleic acid (RNA) are the two types of nucleic acids. The DNA is found in the cell nucleus; the RNA is found in all parts of the cell.

20.35 Ribose and deoxyribose are the two sugars that can be in a nucleotide.

20.37 DNA; the two helixes are wound about one another and are held together by strong hydrogen bonding between the paired bases.

20.39 A vitamin is a specific organic compound that is required in the diet in small amounts for proper body functioning; the shortage of a vitamin results in a vitamin-deficiency disease.

20.41 Vitamins A, D, E, and K are the fat-soluble vitamins; they are nonpolar.

20.43 a. The hormone prolactin is a protein that maintains the production of estrogens and progesterone; it stimulates the formation of milk.
b. The hormone insulin is a protein that increases cell usage of glucose and increases glycogen storage.
c. The hormone aldosterone is a steroid that regulates salt metabolism and stimulates the kidneys to retain Na^+ and excrete K^+.
d. The hormone progesterone is a steroid that regulates the menstrual cycle and maintains pregnancy.

20.45 Glucose is the monosaccharide obtained from the hydrolysis of maltose.

20.47 a. lactose b. sucrose

20.49 Chemists and biochemists help in fighting genetic disease when they work to find which specific genes are linked to specific diseases. For example, a U.S.–Canadian research team identified the gene that causes cystic fibrosis, a common lethal genetic disease.

20.51 Ascorbic acid is vitamin C, calciferol is vitamin D_2, retinol is vitamin A, and tocopherol is vitamin E.

Credits for Photographs

page 6 Marc Romanelli/The Image Bank
Fig. 1.1 William Whitehurst/The Stock Market
Fig. 1.2 Charles West/The Stock Market
page 10 Dan McCoy/The Stock Market
page 14 John R. Wilson/U.S. Navy
Fig. 1.4 AP/Wide World

page 18 Kasz Maciag/The Stock Market
Fig. 2.1 Courtesy of NASA
Fig. 2.2 Kristen Brochmann/Fundamental Photographs
Fig. 2.3 (a, b) Runk/Schoenberger/Grant Heilman Photography, Inc.
(c) Breck P. Kent
Fig. 2.6 David W. Hamilton/The Image Bank
Fig. 2.7 (a) Charles Thatcher/TSW
(b) Stephen Frisch/Stock Boston
Fig. 2.8 Courtesy of Tropicana
Fig. 2.10 (a) Tom Tracy/TSW
(b) Courtesy of Reynolds Metals Co.
Fig. 2.11 Richard Megna/Fundamental Photographs
page 29 Richard Megna/Fundamental Photographs
page 32 Jacques Louis David (1748–1825), ''Antoine Laurent Lavoisier and His Wife, Marie'' (detail), oil on canvas. The Metropolitan Museum of Art, Purchase, Mr. and Mrs. Charles Wrightsman Gift, 1977 (1977.10)
page 33 H. W. Silvester/Rapho/Photo Researchers, Inc.
page 35 Ben Blankenburg/Stock Boston
Fig. 2.12 Prof. C. M. Lang, G. J. Schulfer, University of Wisconsin–Stevens Point

page 42 Chris Hackett/The Image Bank
Fig. 3.3 Richard Megna/Fundamental Photographs
Fig. 3.10 (a) Courtesy of Sartorius Instruments
(b, c) Courtesy of Mettler Instrument Corp.
Fig. 3.11 SIU/Photo Researchers, Inc.
page 64 Bachmann/Stock Boston

page 82 Stephen Frisch/Stock Boston
page 84 Courtesy of Aldrich Chemical Co.
Fig. 4.2 Prof. C. M. Lang, G. J. Schulfer, University of Wisconsin–Stevens Point
Fig. 4.6 Co: Russ Lappa/Photo Researchers, Inc.
Mg, Cu: Paul Silverman/Fundamental Photographs
Cl: Carey B. Van Loon
P: Richard Megna/Fundamental Photographs
S: Barry L. Runk/Grant Heilman Photography, Inc.
Fig. 4.8 (a) Jhaine Manske/The Stock Market
(b) Runk/Schoenberger/Grant Heilman Photography, Inc.
Fig. 4.9 Richard Megna/Fundamental Photographs
Fig. 4.10 Richard Megna/Fundamental Photographs
Fig. 4.11 Prof. C. M. Lang, G. J. Schulfer, University of Wisconsin–Stevens Point
Fig. 4.14 The Bettmann Archive
Fig. 4.18 Courtesy of M. Isaacson, Cornell University, and M. Ohtsuki, The University of Chicago
Fig. 4.19 Courtesy of IBM Thomas J. Watson Research Center, NY
Fig. 4.23 Roy Morsch/The Stock Market
Fig. 4.24 Stephen Frisch/Stock Boston

page 120 Craig Aurness/Westlight
Fig. 5.2 University of Cambridge, Cavendish Laboratory, #45
Fig. 5.4 (b) Carey B. Van Loon
Fig. 5.8 Courtesy of Burndy Library, Norwalk, CT
page 127 Culver Pictures
Fig. 5.12 Dick Canby/DRK Photo

Fig. 5.13 David Parker/Science Source/Photo Researchers, Inc.
page 131 Courtesy of NASA
page 132 Courtesy of The Perkin Elmer Corp.
Fig. 5.17 Courtesy of Wabash Instrument Co.
page 134 Richard Megna/Fundamental Photographs
page 135 Prof. C. M. Lang, G. J. Schulfer, University of Wisconsin–Stevens Point
page 138 Stephen Frisch/Stock Boston
Fig. 5.23 Courtesy of the Nobel Foundation, Stockholm
page 141 Tom Martin (left), H. P. Merten (right)/The Stock Market
Fig. 5.26 (a, b) Paul Silverman/Fundamental Photographs
page 146 Bancroft Library, University of California, Berkeley

page 160 Schneider Studio/The Stock Market
Fig. 6.2 (c) Chemical Education Publishing Company; reprinted with permission
Fig. 6.3 UPI/Bettmann
Fig. 6.15 (a, b) Courtesy of Mazda Information Bureau
Fig. 6.17 Owen Franken/Stock Boston
Fig. 6.20 (a, b) Courtesy of the Dial Corp.
Fig. 6.21 (a) Paul Silverman/Fundamental Photographs
(b) Barry L. Runk/Grant Heilman Photography, Inc.
Fig. 6.22 Courtesy of IBM Corp.
Fig. 6.24 Paul Silverman/Fundamental Photographs
page 179 (left) Ken Eward/Science Source/Photo Researchers, Inc.
(right) Michael Neveux/Westlight
Fig. 6.25 Courtesy of Xerox Corp.
Fig. 6.26 Paul Silverman/Fundamental Photographs
Fig. 6.28 Richard Megna/Fundamental Photographs
Fig. 6.31 Carey B. Van Loon
Fig. 6.33 Courtesy of Eveready
Fig. 6.35 Chris Baker/TSW
Fig. 6.37 Courtesy of BRK Electronics, Inc.

page 190 Jim Cummins/Allstock
Fig. 7.4 Dr. Jeremy Burgess/SPL/Photo Researchers, Inc.
Fig. 7.16 Courtesy of Farmland Industries
Fig. 7.17 Grant Heilman/Grant Heilman Photography, Inc.

page 224 Science Photo Library/Photo Researchers, Inc.
Fig. 8.2 Richard Megna/Fundamental Photographs
Fig. 8.3 Richard Megna/Fundamental Photographs
Fig. 8.4 Richard Megna/Fundamental Photographs
Fig. 8.5 Richard Megna/Fundamental Photographs
Fig. 8.6 Jack Plekan/Fundamental Photographs
page 242 Courtesy of Fisher Scientific
Fig. 8.7 Paul Silverman/Fundamental Photographs
page 247 Paul Silverman/Fundamental Photographs

page 254 Chuck Savage/The Stock Market
page 259 Richard Megna/Fundamental Photographs
Fig. 9.2 Barry L. Runk/Grant Heilman Photography, Inc.

page 282 Richard Megna/Fundamental Photographs
Fig. 10.1 (a) Gary Gladstone/The Image Bank
(b) Bob Daemmrich/Stock Boston
Fig. 10.2 Richard Megna/Fundamental Photographs
Fig. 10.10 John Mayer/Stock Boston
Fig. 10.11 Richard Megna/Fundamental Photographs
page 294 Scott Robinson/TSW
Fig. 10.12 Richard Megna/Fundamental Photographs
page 296 Courtesy of NASA
Fig. 10.14 Courtesy of Standard Fusee Corp.
Fig. 10.15 Richard Megna/Fundamental Photographs
Fig. 10.16 Peticolas/Megna/Fundamental Photographs
Fig. 10.17 Richard Megna/Fundamental Photographs
Fig. 10.18 Richard Megna/Fundamental Photographs
Fig. 10.19 (a, b) Richard Megna/Fundamental Photographs
page 309 Richard Megna/Fundamental Photographs

page 316 Charles Thatcher/TSW
Fig. 11.3 Richard Megna/Fundamental Photographs

page 342 John Eastcott/Yva Momatiuk/DRK Photo
Fig. 12.1 Courtesy of NASA
Fig. 12.5 (a) Ward's Scientific Company
Fig. 12.6 The Bettmann Archive
Fig. 12.12 (a, b) Richard Megna/Fundamental Photographs

page 378 Ira Kirschenbaum/Stock Boston
Fig. 13.7 Dwight Kuhn
Fig. 13.14 (a) Biophoto Associates/Photo Researchers, Inc.
(b) M. Wurtz/Biozentrum, University of Basel/ Science Source/Photo Researchers, Inc.
page 401 Courtesy of Dean Foods
Fig. 13.18 Ken Edward/Science Source/Photo Researchers, Inc.

page 410 Jim Corwin/Photo Researchers, Inc.
Fig. 14.1 Richard Megna/Fundamental Photographs
Fig. 14.6 (a–c) Richard Megna/Fundamental Photographs
Fig. 14.10 Washnik Studio/The Stock Market
page 427 Ken Lax/Photo Researchers, Inc.
Fig. 14.12 (a) Richard Megna/Fundamental Photographs
(b) Richard R. Hansen/Photo Researchers, Inc.
Fig. 14.13 Paul Silverman/Fundamental Photographs

page 440 Lawrence Manning/Westlight
page 447 Courtesy of Dr. J. Ernesto Molino, University of Minnesota Medical School
Fig. 15.6 (a, b) Richard Megna/Fundamental Photographs
page 459 Courtesy of Farmland Industries

page 470 Paul Silverman/Fundamental Photographs
Fig. 16.1 Richard Megna/Fundamental Photographs
Fig. 16.2 Richard Megna/Fundamental Photographs
Fig. 16.3 Paul Silverman/Fundamental Photographs
Fig. 16.4 Richard Megna/Fundamental Photographs
page 477 (left) Richard Megna/Fundamental Photographs
(right) Courtesy of Fisher Scientific
Fig. 16.5 Courtesy of S. C. Johnson Wax
Fig. 16.7 Richard Megna/Fundamental Photographs
page 481 Diane Schiumo/Fundamental Photographs
page 482 Richard Megna/Fundamental Photographs
page 484 Coco McCoy/Rainbow
Fig. 16.10 Paul Silverman/Fundamental Photographs
Fig. 16.11 Courtesy of Corning Inc.
page 501 Mike Moreland/Custom Medical Stock Photo

page 510 Richard Megna/Fundamental Photographs
Fig. 17.2 (a) Stephen Frisch/Stock Boston
(b) Porterfield/Chickering/Photo Researchers, Inc.
Fig. 17.3 (a) Bob Krist/TSW
(b) Pierre Koff/Westlight
(c) Aaron Haupt/Stock Boston
page 515 Courtesy of NASA
page 519 (top) The Bettmann Archive
(bottom) Courtesy of Goodyear Tire Company
page 523 Hank Morgan/Photo Researchers, Inc.
page 525 With permission of The Purdue Frederick Co., distributor of Betadine Microbiocides
page 526 Diane Schiumo/Fundamental Photographs
page 527 Tomas Sennett for World Bank

page 544 Jay Freis/The Image Bank
Fig. 18.3 Susan Leavines/Science Source/Photo Researchers, Inc.
Fig. 18.7 Hank Morgan/Photo Researchers, Inc.
Fig. 18.8 Courtesy of Tech/Ops Landauer, Inc.
page 555 (top) Richard Megna/Fundamental Photographs
(bottom) Courtesy of Landauer, Inc.
Fig. 18.11 (left) The Bettmann Archive
(right) Prof. C. M. Lang, G. J. Schulfer, University of Wisconsin–Stevens Point
page 561 Will and Demi McIntyre/Photo Researchers, Inc.
Fig. 18.13 Runk/Schoenberger/Grant Heilman Photography, Inc.
Fig. 18.14 Peticolas/Megna/Fundamental Photographs
Fig. 18.15 Yoav Levy/Phototake
Fig. 18.16 Courtesy of Du Pont Merck Pharmaceutical
page 567 Sygma
Fig. 18.17 Courtesy of Mrs. Laura Fermi
Fig. 18.21 Courtesy of National Air and Space Museum, The Smithsonian Institution
page 574 Sovfoto/Eastfoto
Fig. 18.23 Darrell Gulin/Allstock
page 579 Edward Bower/The Image Bank
Fig. 18.24 Princeton Plasma Physics Laboratory, Princeton University

page 586 Stephanie Stokes/The Stock Market
Fig. 19.3 Richard Megna/Fundamental Photographs
Fig. 19.4 Richard Megna/Fundamental Photographs
Fig. 19.5 Richard Megna/Fundamental Photographs
Fig. 19.6 (a) Richard Megna/Fundamental Photographs
(b) Courtesy of Bic Corp.
Fig. 19.7 Richard Megna/Fundamental Photographs
Fig. 19.8 Richard Megna/Fundamental Photographs
Fig. 19.9 Richard Megna/Fundamental Photographs
Fig. 19.11 Richard Megna/Fundamental Photographs
page 614 Guy Gillette/Photo Researchers, Inc.
Fig. 19.15 Diane Schiumo/Fundamental Photographs
Fig. 19.16 Richard Megna/Fundamental Photographs
Fig. 19.17 Richard Megna/Fundamental Photographs
Fig. 19.18 Richard Megna/Fundamental Photographs
Fig. 19.19 Richard Megna/Fundamental Photographs
Fig. 19.20 Paul Silverman/Fundamental Photographs

page 632 Douglas Struthers/TSW
Fig. 20.1 Robert Mathena/Fundamental Photographs
Fig. 20.3 Digital Art/Westlight

Fig. 20.7 Biophoto Associates/Science Source/Photo Researchers, Inc.
Fig. 20.8 Richard Megna/Fundamental Photographs
page 642 Dan McCoy/Rainbow
Fig. 20.9 Courtesy of Science Related Materials, Inc., Janesville, WI
Fig. 20.15 Courtesy of Harvard University Biological Laboratories
Fig. 20.16 Courtesy of Science Related Materials, Inc., Janesville, WI
page 653 Hank Morgan/Rainbow

Index

A

D

F

J

K

L

Q

R

S

Atomic Masses of the Elements

Based on carbon-12. A number in parentheses is the atomic mass of the most stable isotope of a radioactive element.

Name	*Symbol*	*Atomic Number*	*Atomic Weight*	*Name*	*Symbol*	*Atomic Number*	*Atomic Weight*
Actinium	Ac	89	227.0	Neon	Ne	10	20.18
Aluminum	Al	13	26.98	Neptunium	Np	93	237.0
Americium	Am	95	(243)	Nickel	Ni	28	58.69
Antimony	Sb	51	121.8	Niobium	Nb	41	92.91
Argon	Ar	18	39.95	Nitrogen	N	7	14.01
Arsenic	As	33	74.92	Nobelium	No	102	(259)
Astatine	At	85	(210)	Osmium	Os	76	190.2
Barium	Ba	56	137.3	Oxygen	O	8	16.00
Berkelium	Bk	97	(247)	Palladium	Pd	46	106.4
Beryllium	Be	4	9.012	Phosphorus	P	15	30.97
Bismuth	Bi	83	209.0	Platinum	Pt	78	195.1
Boron	B	5	10.81	Plutonium	Pu	94	(244)
Bromine	Br	35	79.90	Polonium	Po	84	(209)
Cadmium	Cd	48	112.4	Potassium	K	19	39.10
Calcium	Ca	20	40.08	Praseodymium	Pr	59	140.9
Californium	Cf	98	(251)	Promethium	Pm	61	(145)
Carbon	C	6	12.01	Protactinium	Pa	91	231.0
Cerium	Ce	58	140.1	Radium	Ra	88	226.0
Cesium	Cs	55	132.9	Radon	Rn	86	(222)
Chlorine	Cl	17	35.45	Rhenium	Re	75	186.2
Chromium	Cr	24	52.00	Rhodium	Rh	45	102.9
Cobalt	Co	27	58.93	Rubidium	Rb	37	85.47
Copper	Cu	29	63.55	Ruthenium	Ru	44	101.1
Curium	Cm	96	(247)	Samarium	Sm	62	150.4
Dysprosium	Dy	66	162.5	Scandium	Sc	21	44.96
Einsteinium	Es	99	(252)	Selenium	Se	34	78.96
Erbium	Er	68	167.3	Silicon	Si	14	28.09
Europium	Eu	63	152.0	Silver	Ag	47	107.9
Fermium	Fm	100	(257)	Sodium	Na	11	22.99
Fluorine	F	9	19.00	Strontium	Sr	38	87.62
Francium	Fr	87	(223)	Sulfur	S	16	32.07
Gadolinium	Gd	64	157.3	Tantalum	Ta	73	180.9
Gallium	Ga	31	69.72	Technetium	Tc	43	(98)
Germanium	Ge	32	72.59	Tellurium	Te	52	127.6
Gold	Au	79	197.0	Terbium	Tb	65	158.9
Hafnium	Hf	72	178.5	Thallium	Tl	81	204.4
Helium	He	2	4.003	Thorium	Th	90	232.0
Holmium	Ho	67	164.9	Thulium	Tm	69	168.9
Hydrogen	H	1	1.008	Tin	Sn	50	118.7
Indium	In	49	114.8	Titanium	Ti	22	47.88
Iodine	I	53	126.9	Tungsten	W	74	183.9
Iridium	Ir	77	192.2	Unnilennium	Une	109	(266)
Iron	Fe	26	55.85	Unnilhexium	Unh	106	(263)
Krypton	Kr	36	83.80	Unniloctium	Uno	108	(265)
Lanthanum	La	57	138.9	Unnilpentium	Unp	105	(262)
Lawrencium	Lr	103	(260)	Unnilquadium	Unq	104	(261)
Lead	Pb	82	207.2	Unnilseptium	Uns	107	(262)
Lithium	Li	3	6.941	Uranium	U	92	238.0
Lutetium	Lu	71	175.0	Vanadium	V	23	50.94
Magnesium	Mg	12	24.31	Xenon	Xe	54	131.3
Manganese	Mn	25	54.94	Ytterbium	Yb	70	173.0
Mendelevium	Md	101	(258)	Yttrium	Y	39	88.91
Mercury	Hg	80	200.6	Zinc	Zn	30	65.39
Molybdenum	Mo	42	95.94	Zirconium	Zr	40	91.22
Neodymium	Nd	60	144.2				